Industrial
MECHANICS

Fourth Edition

atp AMERICAN TECHNICAL PUBLISHERS
Orland Park, Illinois

Albert W. Kemp

Industrial Mechanics contains procedures commonly practiced in industry and the trade. Specific procedures vary with each task and must be performed by a qualified person. For maximum safety, always refer to specific manufacturer recommendations, insurance regulations, specific job site and plant procedures, applicable federal, state, and local regulations, and any authority having jurisdiction. The material contained herein is intended to be an educational resource for the user. American Technical Publishers assumes no responsibility or liability in connection with this material or its use by any individual or organization.

American Technical Publishers Editorial Staff

Editor in Chief:
    Jonathan F. Gosse
Vice President—Editorial:
    Peter A. Zurlis
Assistant Production Manager:
    Nicole D. Bigos
Digital Media Coordinator:
    Adam T. Shuldt
Art Supervisor:
    Sarah E. Kaducak
Supervising Copy Editor:
    Catherine A. Mini
Technical Editors:
    Kyle D. Wathen
    Julie M. Welch

Copy Editor:
    James R. Hein
Editorial Assistant:
    Sara M. Patek
Cover Design:
    Nicholas W. Basham
Illustration/Layout:
    Nicholas W. Basham
    Nicole S. Polak
    Bethany J. Fisher
    Thomas E. Zabinski
    Christopher S. Gaddie
Digital Resources:
    Robert E. Stickley
    Cory S. Butler

Boston Gear is a registered trademark of IMO Corporation. Griphoist is a registered trademark of Tractel, Inc. National Electrical Code® and NEC® are registered trademarks of the National Fire Protection Association, Inc. NFPA 70E is a registered trademark of the National Fire Protection Association, Inc. Quad-ring is a registered trademark of Minnesota Rubber and Gasket Co. Corporation. Super Glue is a registered trademark of Pacer Technology Corporation. Tapcon is a registered trademark of Illinois Tool Works Inc. Teflon is a registered trademark of The Chemours Company FC, LLC. Vise-Grip is a trademark of Irwin Industrial Tool Company. Quick Quiz, Quick Quizzes, and Master Math are trademarks of American Technical Publishers.

4 5 6 7 8 9 – 16 – 9 8 7 6 5

Printed in the United States of America

    ISBN 978-0-8269-3712-4
    eISBN 978-0-8269-9539-1

 This book is printed on recycled paper.

Acknowledgments

The author and publisher are grateful to the following companies and organizations for providing technical information and assistance.

Advanced Assembly Automation Inc.
Atlas Copco
Atlas Technologies, Inc.
ARO Fluid Products Div., Ingersoll-Rand
Autodesk, Inc.
Baldor Electric Co.
Bimba Manufacturing Company
Boeing Commercial Airplane Group
Boston Gear®
Briggs & Stratton Corporation
Carrier Corporation
Cincinnati Milacron
Clippard Instrument Laboratory, Inc.
Columbus McKinnon Corporation, Industrial Products Division
Cone Drive Operations Inc./Subsidiary of Textron Inc.
Cone Mounter Company
Continental Hydraulics
Cooper Industries, Inc.
Cooper Tools
CNH America
The Crosby Group, Inc.
Crowe Rope Industries LLC
Datastream Systems, Inc.
DeWALT Industrial Tool Co.
DoALL Company
Dow Corning Corporation
DPSI (DP Solutions, Inc.)
Eaton Corporation
Emerson Power Transmission
Engelhardt Gear Co.
Exxon Company
Fenner Drives
Flow Ezy Filters, Inc.
Fluke Corporation
Gast Manufacturing Company
The Gates Rubber Company
GE Motors & Industrial Systems
Greenlee Textron, Inc.
Harrington Hoists Inc.
Heidelberg Harris, Inc.
Honeywell's MICRO SWITCH Division
Humphrey Products Company
Hyster Company

Ingersoll-Rand Material Handling
Klein Tools, Inc.
Lab Safety Supply, Inc.
Lift-All Company, Inc.
Lift Tech International, Division of Columbus McKinnon
Lovejoy, Inc.
LPS Laboratories, Inc.
L.S. Starrett Company
Ludeca Inc.
Manufacturing & Maintenance Systems, Inc.
Martin Sprocket & Gear Inc.
Miller Fall Protection
Milwaukee Electric Tool Corp.
Mine Safety Appliances Co.
North American Industries, Inc.
NTN Bearing Corporation of America
Pacific Bearing Company
PLI LLC
Predict/DLI
Precision Brand Products, Inc.
Prince Manufacturing Corporation
Power Team, Division of SPX Corporation
Ratcliff Hoist Co.
Reed Manufacturing Co.
Salisbury
Saylor-Beall Manufacturing Company
SEW-Eurodrive, Inc.
Siemens
The Sinco Group, Inc.
Schneider Electric USA, Inc.
Snorkel
Sonny's Enterprises, Inc.
SPM Instrument, Inc.
Sprecher + Schuh
Staedtler, Inc.
The Stanley Works
Stork Technimet, Inc.
The Timken Company
Tractel Inc., Griphoist® Division
Vibration Monitoring Systems
Werner Ladder Co.
Wick Homes
W.W. Grainger, Inc.

Contents

Learner Resources

- Quick Quizzes™
- Illustrated Glossary
- Flash Cards
- Precision Measurement Techniques

- Chapter Reviews
- Media Library
- ATPeResources.com

Introduction

Industrial Mechanics is a comprehensive introduction to industrial mechanical principles, components, and circuits. The textbook provides the latest information on safety and system efficiency, and is designed as a resource for mechanics, technicians, maintenance personnel, and individuals in industrial training programs. This new edition has been reorganized for easier understanding of topics covered.

Industrial Mechanics covers workplace safety, tools, printreading, precision measurement, rigging and lifting, lubrication and bearings, flexible belt and mechanical drives, vibration and alignment, hydraulic and pneumatic principles and applications, and preventive maintenance programs. New topics include test tool safety and fastening methods. Expanded topics include electrical safety, electrical applications, hydraulic component testing, and hydraulic system troubleshooting.

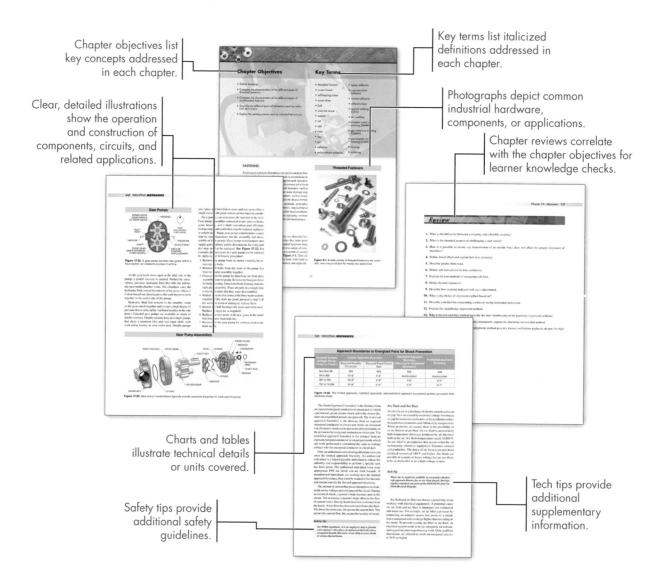

Chapter objectives list key concepts addressed in each chapter.

Key terms list italicized definitions addressed in each chapter.

Clear, detailed illustrations show the operation and construction of components, circuits, and related applications.

Photographs depict common industrial hardware, components, or applications.

Chapter reviews correlate with the chapter objectives for learner knowledge checks.

Charts and tables illustrate technical details or units covered.

Tech tips provide additional supplementary information.

Safety tips provide additional safety guidelines.

Industrial Mechanics Learner Resources include online resources that reinforce textbook content to promote learning and comprehension. These online resources can be accessed using either of the following methods:

- Key ATPeresources.com/quicklinks into a web browser and enter QuickLinks™ access code 462409.
- Use a Quick Response (QR) reader app to scan the QR code with a mobile device.

The learner resources include the following:

- **Quick Quizzes™** that provide interactive questions for each chapter, with embedded links to highlighted content within the textbook and to the Illustrated Glossary
- An **Illustrated Glossary** that provides a helpful reference to commonly used terms, with selected terms linked to textbook illustrations
- **Flash Cards** that provide a self-study/review tool of terms and definitions, DMM abbreviations and symbols, and fluid power symbols
- **Precision Measurement Techniques,** which are video-based tutorials that present common measurement techniques using precision measurement tools
- **Chapter Reviews** in interactive PDF format that provide learners with the opportunity to demonstrate comprehension of chapter objectives
- A **Media Library** that consists of videos and animations that reinforce textbook content
- **ATPeResources.com,** which links to online reference materials that support continued learning

To obtain information on related training products, visit the American Technical Publishers website at www.atplearning.com.

The Publisher

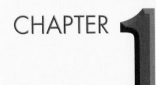

WORKPLACE SAFETY

Although workplace safety is the responsibility of all workers, technicians are often the first to be aware of safety problems. Much of the work performed by technicians involves preventing hazardous conditions for all facility workers and the general public. This responsibility requires them to understand and follow general safety rules established by regulatory agencies and other specific workplace rules. For technicians, following general safety rules requires understanding how to perform basic first aid, wearing proper personal protective equipment, safe use of tools, working safely in confined spaces, and proper use of fall-arrest systems.

SPM Instrument, Inc.

Chapter Objectives

- Explain workplace safety.
- Describe the basic first aid procedures used in a workplace.
- Explain the general safety rules that should be enforced in a workplace.
- List the different types of personal protective equipment (PPE).
- Define confined spaces.
- Explain the purpose of a fall-arrest system.

Key Terms

- code
- standard
- first aid
- earplug
- earmuff
- respirator
- explosive range
- lower explosive limit (lel)
- upper explosive limit (uel)
- lanyard
- deceleration distance
- free-fall distance
- safety net

WORKPLACE SAFETY

In the past, little emphasis was placed on safety in the workplace. More recently, organizations such as the Occupational Safety and Health Administration (OSHA), created in 1971, have been established to develop and enforce safety standards for industrial workers. In addition, organizations such as the National Institute for Occupational Safety and Health (NIOSH), the National Fire Protection Association (NFPA), and others have formed codes, standards, and guidelines for a safe and healthy work environment.

A *code* is a regulation or minimum requirement. A *standard* is an accepted reference or practice. OSHA industrial safety standards were put into effect in the Code of Federal Regulations, Title 29 and is identified as OSHA 29 CFR 1910—*Occupational Safety and Health Standards.* Today, a safe and healthy work environment has become law. Safety awareness and regulations have dramatically reduced the national industrial fatality occurrences to 6632 in 1994 and 4693 in 2011 according to the U.S. Bureau of Labor Statistics.

On-the-job safety begins with personal safety awareness and understanding of basic first aid, general safety rules, and personal protective equipment. Personal safety awareness is being aware of safety issues and potential hazards to individuals in the workplace. A personal awareness of on-the-job safety is vital in preventing workplace injuries. Technicians must have an understanding of certain health and safety risks and requirements and the steps needed to minimize those risks. The most common types of hazards found in industrial environments include electrical, chemical, mechanical, fall, and confined-space hazards. **See Figure 1-1.**

Personal safety awareness includes keeping a mindset towards safety while on the job. Risky activities can be performed during a task when a technician's stress level is high or there is an unbalance between work and personal life. Work injuries can occur because a person either takes a chance or takes their mind off their work. In addition, taking shortcuts when in a rush can make a person make poor decisions regarding their safety and the safety of others. For example, a technician in a hurry may feel that wearing safety glasses is not required for a task, leading to a temporary or permanent eye injury.

Safety Tip

Many industrial facilities have mandatory substance testing programs to determine if an employee that was injured on the job was impaired when the injury occurred.

Common Industrial Hazards

Fluke Corporation

ELECTRICAL   **CHEMICAL**

MECHANICAL   **FALL**   **CONFINED SPACE**

Figure 1-1. The most common types of hazards found in industrial environments include electrical, chemical, mechanical, fall, and confined-space hazards.

Basic First Aid

Accidents that cause injury to technicians can happen at any time and at any place. Immediate medical treatment is required for the victim, regardless of the extent of the injury. Often, first aid given immediately at the scene of an accident can improve the victim's chances of survival and recovery. *First aid* is help for a victim immediately after an injury and before professional medical help arrives. If someone is injured, steps taken to keep that person as safe as possible until professional help arrives include the following:

- Remain calm.
- Call 911 or the workplace emergency number immediately if a person is seriously injured.
- Never move an injured person unless a fire or explosives are involved. Moving an injured person may make the injury worse.
- Assess the injured person carefully and perform basic first aid procedures.
- Maintain first aid procedures until professional medical help arrives.
- Report all injuries to the supervisor.

A conscious accident victim must provide consent before care can be administered. To obtain consent, the victim must be asked if help can be provided. Once the victim gives consent, the appropriate care can be provided. If the victim does not give consent, care cannot be given. In such a case, 911 or the workplace emergency number should be called. Consent is implied in cases where an accident victim is unconscious, confused, or seriously injured. Implied consent allows an individual to provide care to an accident victim because the victim would agree to the care if he or she could.

Various states have enacted Good Samaritan laws to give legal protection to individuals who provide emergency care to accident victims. These laws are designed to encourage individuals to help others in emergencies. Good Samaritan laws vary from state to state. A legal professional or the local library should be consulted for information regarding the Good Samaritan laws in a particular state.

No attempt should be made to provide care without proper training. The American Red Cross may be consulted for training programs covering basic first aid and CPR. Basic first aid procedures can be used to help treat shock, bleeding, burns and scalds, choking, and poisoning.

Emergency eyewash stations and deluge shower units are used to flush chemicals and other contaminants from the eyes and body and provide first aid to burn victims.

Shock. Shock usually accompanies severe injury. Shock can threaten the life of an injured person if not treated quickly. Shock occurs when the body's vital functions are threatened by lack of blood or when the major organs and tissues do not receive enough oxygen. Symptoms of shock include cold, clammy, or pale skin, chills, confusion, nausea or vomiting, shallow breathing, and unusual thirst. To treat shock, the victim should lie down with his or her legs elevated if there is no sign of broken bones or spinal injury. The victim should also be covered to prevent chills or loss of body heat. Any obvious signs of bleeding should be controlled. A victim who is unconscious or bleeding from the mouth should lie on one side so breathing is easier. Professional medical help should be sought as soon as possible.

Bleeding. Bleeding is the most visible result of an injury. Most people can lose a small amount of blood without major problems. However, if a quart or more of blood is lost quickly, shock and/or death is possible. Bleeding is controlled by placing a clean, dry cloth on the wound and applying pressure with a finger or the palm of the hand, depending on the severity of the wound. The pressure should be kept on the wound until the bleeding stops.

If there are no signs of broken bones, the wound should be elevated above the victim's heart to slow the bleeding at the wound. Once the bleeding stops, the cloth against the wound should not be removed, as it could disturb the blood clotting and restart the bleeding. If possible, rubber or latex gloves should be worn before touching any blood, because touching blood involves health risks. If rubber or latex gloves are not available, a clean plastic bag may be used to cover the hands. If the injury is extensive, the victim may go into shock and should be treated for it. Professional medical help should be sought as soon as possible.

Burns and Scalds. Burns can be caused by heat, chemicals, or electricity. For burns caused by heat or chemicals, the burn should be immediately flushed with cool water for a minimum of 30 min. Ice should not be applied, because it may cause further damage to the burned area. This treatment should be maintained until the pain or burning stops. After flushing the burn with cool water, the burn should be covered with a clean cotton cloth. If a clean cotton cloth is not available, the burn should not be covered.

Clothing that is stuck to a burn should not be removed. The burn should not be scrubbed and soap, ointments, greases, and powders should not be applied. Blisters that appear should not be broken. The burn victim should not be offered anything to drink or eat. The victim should be kept covered with a blanket to maintain a normal body temperature. All burn victims should be treated for shock. Professional medical help should be sought as soon as possible.

A burn caused by electricity requires first ensuring that the victim is removed from the power source. Once the victim is clear of the power source, the victim is checked for any airway obstructions, and his or her breathing and circulation is checked. CPR should be administered if necessary and the individual administering the CPR is trained to do so. Once the victim is stable, the burn is flushed with cool water for a minimum of 30 min. The victim should not be moved, and the burn should not be scrubbed. Soap, ointments, greases, and powders should not be applied. After flushing the burn with cool water for 30 min, the burn should be covered with a clean cotton cloth. If a clean cotton cloth is not available, the burn should not be covered. The victim should be treated for shock and a normal body temperature should be maintained. Professional medical help should be sought as soon as possible.

Choking. Choking occurs when food or a foreign object obstructs the throat and interferes with normal breathing. Permission must be obtained from a conscious victim before providing care. The following steps are advised if a choking victim is unable to speak or cough:

1. Ask the victim if they are choking.
2. Shout for help if the victim cannot cough, speak, or breathe.
3. Call 911 or the workplace emergency number.
4. Perform abdominal thrusts. Abdominal thrusts are performed by wrapping the arms around the victim's waist, making a fist, and placing the thumb of the fist on the middle of the victim's abdomen just above the navel. The fist is grasped with the other hand and pressed into the abdomen with a quick upward thrust.

Poisoning. Poisons may be in solid, liquid, or gas form. If the poison has been ingested in a solid form, such as pills, any poison that is in the victim's mouth should be removed using a clean cloth wrapped around a finger. If the poison is a corrosive liquid on the skin, any clothing should be removed from the affected area and the skin should be flushed with water for 30 min. If the poison is in contact with the eyes, the victim's eyes should be flushed for a minimum of 15 min with clean water. When calling for medical help, the poisonous product container or label should be available. This enables the caller to answer questions about the poison.

When the poison is in the environment, the victim must be removed immediately from the poison source. Once the victim is removed from the poison source, the appropriate treatment should be administered based on the form of the poisoning. If the poison is a gas, a respirator may be required for protection when entering the area. The victim should be moved to fresh air and professional medical help should be sought as quickly as possible.

First Aid Kits. To administer effective first aid, adequate supplies must be maintained in a first aid kit. A first aid kit can be purchased stocked with the necessary supplies, or one can be made by assembling a kit containing adhesive bandages, butterfly closures, rolled gauze, nonstick sterile pads, and various first aid tapes. Additional items can be included as required, such as tweezers, aspirin, an additional analgesic, first aid cream, a thermometer, and an ice pack.

Safety Tip

When performing CPR, always remember "A-B-C," for "airway-breathing-cardio." First, check the victim's airway for blockage. Then, determine if victim is breathing. Lastly, check for the presence of a pulse. Once these three items are checked, proper CPR can be performed until first responders arrive on the scene.

General Safety Rules

Accidents are caused by the lack of attention to safety rules and regulations. Constant reminders are needed so individuals understand that safety is everyone's responsibility. Safety is learned by reading, observing, and practicing safe habits. Technicians should review safety materials related to the environment and equipment used on the job. Technicians should observe safe actions and practice safe work habits to avoid harmful and injurious activities.

Professional Behavior. Technicians should emulate professional behavior at all times. Pranks and horseplay in an industrial environment are serious industrial hazards that cause accidents through inattention, carelessness, and recklessness and may lead to injury or death. Nearly all workplaces have rules that make horseplay an offense punishable by discipline or discharge.

Stair Safety. Although ascending and descending stairs is a common function, many injuries occur from individuals falling down stairs. Ascending stairs is relatively safe because most people observe each step they take, and if a trip occurs, falling forward is relatively safe. However, injuries from descending stairs occur more frequently because if a trip occurs, an injury is more likely than when falling while ascending stairs. Caution must be used and attention given to each foot placement when descending stairs, especially when carrying objects. Holding a handrail when available improves safety.

Slip and Fall Prevention. Slips and falls are the second most common cause of accidents, accounting for 35% of nonfatal injuries, such as fractures and amputations, and 21% of over-three-day injuries, such as sprains, strains, and minor fractures. Slips and falls are generally due to poor housekeeping. Spills, tools or equipment, and electrical wires located in high traffic areas are tripping hazards. Poor or inappropriate footwear is also a cause of slips and falls. Slips and falls may be avoided by being aware of each step traveled, removing or cleaning obstacles so others may not be injured, and wearing the appropriate footwear.

Safe Lifting and Moving. Back injury is one of the most common injuries resulting in lost work time. Improper lifting and moving techniques can result in back injuries. Back injuries are extremely common but preventable through a combination of safe lifting and moving techniques.

Before lifting, the weight of the object should be determined and the lifting and moving method should be planned based on the object's size, shape, weight, and path of travel. Assistance should be sought when necessary. The entire pathway should be clear of obstacles, obstructions, and other hazards. The knees should be bent and the object grasped firmly. The object is then lifted while straightening the legs and keeping the back as straight as possible. Finally, movement can occur after the entire body is in the vertical position. The load should be kept as close to the body as possible. **See Figure 1-2.**

Carts or hand trucks should be used when objects are excessively heavy or oddly shaped. The load must be balanced to avoid tipping. When moving a load on a cart, the technician's body weight should be used while taking even, short steps.

Poor physical condition is a major contributor to all types of back problems. An individual's weight, strength, and flexibility all play a part in minimizing strain on the muscles and tendons that support the spinal column (core muscles) and contributing to or preventing back problems. When possible, forklifts should be used to move heavy loads and objects. **See Figure 1-3.**

Heat-Related Illnesses. Working in high-temperature environments can cause fatigue, fainting, muscle cramps, heat exhaustion, and heat stroke. In addition, working in high-temperature environments increases the risk of injuries due to accidents caused by slippery hands, sweat, fogged glasses, or dizziness.

Proper Lifting Technique

① BEND KNEES AND GRASP OBJECT FIRMLY

KEEP BACK STRAIGHT

② LIFT OBJECT BY STRAIGHTENING LEGS

③ MOVE OBJECT AFTER WHOLE BODY IS IN VERTICAL POSITION

Figure 1-2. Proper lifting requires that an individual uses his or her legs to lift an object.

Forklifts

W.W. Grainger, Inc.

Figure 1-3. When possible, forklifts should be used to move heavy loads and objects.

Muscle cramps usually occur during the first days of hot stressful work, especially if individuals are not accustomed to this type of work. To avoid muscle cramps, individuals should drink plenty of water to replace the fluids lost from sweating.

A loss of large amounts of fluid from the body can cause heat exhaustion. Individuals with heat exhaustion sweat, but their body cannot maintain the correct body temperature because of the amount of heat. In a person suffering from heat exhaustion, the body increases its heart rate and strengthens blood circulation. Individuals suffering from heat exhaustion may feel disoriented, feel dizzy, feel fatigued, have a headache, or have flu-like symptoms. The skin has a normal temperature but also has a damp and clammy feeling. Generally, the individual needs to rest, cool down, and drink plenty of liquids.

Heat stroke is the most serious heat-related health problem. Heat stroke occurs when the body stops adjusting to the hot temperature and sweat glands shut down. Symptoms of heat stroke include confusion, collapse, unconsciousness, dry mottled skin, and skin that is warm or hot to the touch. A heat stroke victim can die quickly. Immediate medical attention is required. An individual suffering from heat stroke should be moved to a cool area and cool water should be used to cool the individual.

Carts should be used to safely move objects that are heavy or oddly shaped.

Personal Protective Equipment

Personal protective equipment (PPE) is designed to protect technicians from serious workplace injuries or illnesses resulting from contact with chemical, radiological, physical, electrical, mechanical, or other workplace hazards. PPE includes safety glasses, hard hats, leather gloves, steel-toe work boots, earplugs, ear muffs, welding helmets and fire-resistant clothing (used for welding operations), and dust masks. **See Figure 1-4.** PPE regulations are listed in OSHA 29 CFR 1910 Subpart G—*Occupational Health and Environmental Control* and OSHA 29 CFR 1910 Subpart I—*Personal Protective Equipment,* which also references various standards for each type. For example, appropriate protective helmets are worn in areas with overhead hazards, and safety shoes with steel toes are worn to provide protection from falling objects.

Personal Protective Equipment (PPE)

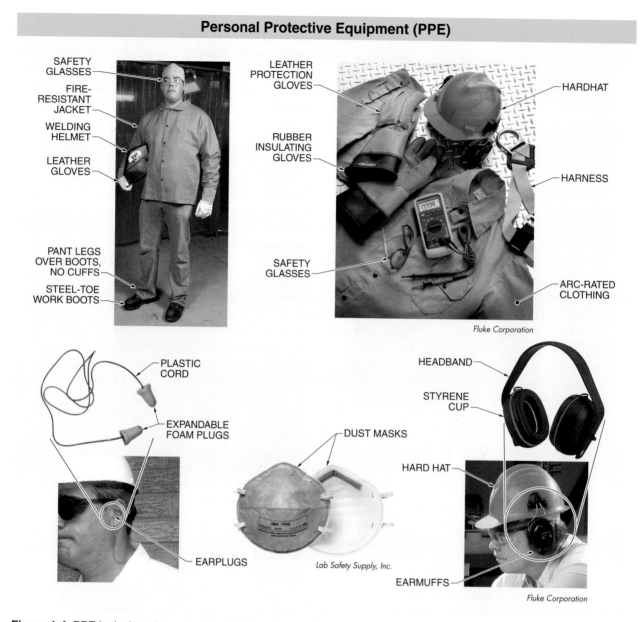

Figure 1-4. PPE is designed to protect technicians from serious workplace injuries or illnesses resulting from contact with chemical, radiological, physical, electrical, mechanical, or other workplace hazards.

Eye and Face Protection. PPE such as safety glasses with side shields, face shields, and goggles can protect technicians from the hazards of flying fragments, chips, hot sparks, optical radiation, splashes from molten metals, particles, sand, dirt, mists, dusts, and glare. OSHA 29 CFR 1910.133—*Eye and Face Protection* is designed to prevent or lessen the severity of injuries to technician's eyes and face. **See Figure 1-5.**

Eye protection must be maintained to provide protection and clear visibility. Pitted or scratched lenses should be replaced because they impair vision and are more likely to break. Eye protection should be cleaned and disinfected regularly. When not in use, eye protection should be kept in cases that keep them clean and prevent scratches.

If debris or chemicals get in the eyes or face, emergency eyewash and shower stations must be available to quickly flush the area with clean water. Requirements for these safety devices are listed in ANSI/ISEA Standard Z358.1, *Emergency Eyewash and Shower Equipment.*

Eye and Face Protection

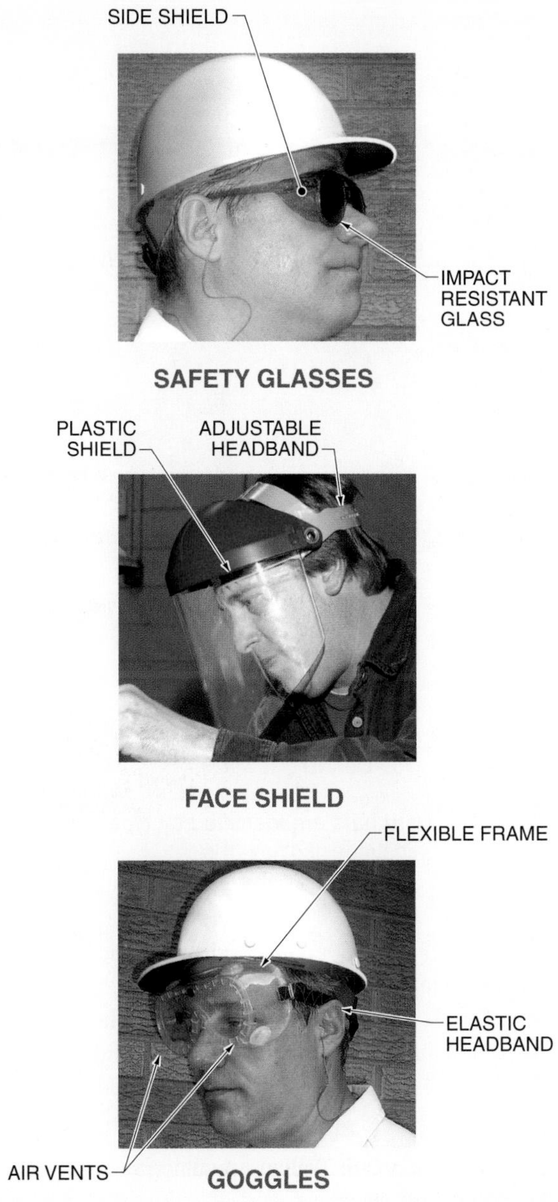

Figure 1-5. Eye and face protection must be worn to prevent injuries caused by flying particles, chemical splashes, dust and dirt, or optical radiation.

Dust may be present in the workplace during cutting, sanding, and buffing operations. Working in a dusty environment can cause eye injuries and present additional hazards to contact lens wearers. Either eyecup or cover-type safety goggles should be worn when dust is present. Safety goggles are the only effective type of eye protection from nuisance dust because they create a protective seal around the eyes.

The majority of impact injuries result from flying or falling objects or sparks striking the eye. Most of these objects are smaller than a pinhead and can cause serious injury such as punctures, abrasions, and contusions. While working in a hazardous area where the technician is exposed to flying objects, fragments, and particles, primary protective devices such as safety glasses with side shields or goggles must be worn. Secondary protective devices such as face shields are required in conjunction with primary protective devices during severe exposure to impact hazards.

Heat injuries may occur to the eye and face when technicians are exposed to high temperatures, splashes of molten metal, or hot sparks. Eye protection from heat is vital when workplace operations involve pouring, casting, hot dipping, furnace operations, and other similar activities. Burns to eye and face tissue are the main concern when working with heat hazards. Working with heat hazards requires eye protection such as goggles or safety glasses with special-purpose lenses and side shields. However, many heat hazard exposures require the use of a face shield in addition to safety glasses or goggles. When selecting PPE, consider the source and intensity of the heat and the type of splashes that may occur in the workplace.

A large percentage of eye injuries are caused by direct contact with chemicals. These injuries often result from an inappropriate choice of PPE that allows a chemical substance to enter from around or under protective eye equipment. Serious and irreversible damage can occur when chemicals contact the eyes in the form of splashes, mists, vapors, or fumes. When working with or around chemicals, the locations of emergency eyewash stations and how to access them with restricted vision must be known.

Welding operations create intense concentrations of heat, ultraviolet, infrared, and reflected light radiation. Eye and face protection are essential for technicians performing welding operations. Ultraviolet and infrared radiation produced by arc welding and cutting may cause flash burn, which is a burn to the eyes. Arc radiation can also cause burns to the face. The effects depend on the radiant energy wavelengths, intensity, and amount of exposure. Technicians performing welding operation must wear a suitable helmet with the proper filter plate shade number and safety glasses when arc welding or cutting. They must also take appropriate safety precautions when oxyfuel welding and cutting by wearing goggles or a face shield with the appropriate shade number.

Technicians performing welding operations should always be alert for welding arcs that can pose hazards to others near the welding operation. Technicians and visitors should use appropriate eye protection at all times when near welding operations. Other plant personnel can be protected by welding screens, curtains, or by remaining an adequate distance from the operation.

In arc welding and cutting, the filter plate shade number should be matched to the type of process and the welding current. In oxyfuel welding and cutting, the filter plate should be matched to the thickness of the material. When selecting filter lenses, a shade that is too dark to see the welding zone is selected. Lighter shades are then selected until one allows a sufficient view of the welding zone without going below the minimum protective shade. Standards for eye and face protection are specified in OSHA 29 CFR 1910.133(a)(5)—*Eye and Face Protection.*

Head Protection. Hard hats (protective helmets) provide protection from head impact, penetration injuries, and electrical injuries such as those caused by falling or flying objects, fixed objects, or contact with electrical conductors. Hard hats protect technicians by resisting penetration and absorbing the blow of an impact. Hard hat standards are identified by type and class for protection against specific hazardous conditions. **See Figure 1-6.** Standards for hard hats are specified in ANSI/ISEA Z89.1, *American National Standard for Industrial Head Protection.* Also, OSHA regulations require that technicians cover and protect long hair to prevent it from getting caught in machine parts such as belts and chains.

Foot and Leg Protection. Maintenance technicians perform many tasks that require handling objects that could injure a foot if dropped. Safety shoes with reinforced steel toes provide protection against injuries caused by compression and impact. Some safety shoes have protective metal insoles and metatarsal guards for protection from punctures from below, such as stepping on nails. Additional foot guards or toe guards may be used over existing work shoes.

Specialized safety shoes are available for certain situations, such as metal-free shoes for electrical work or shoes with soles that prevent slips on wet or oily surfaces. Rubber boots may be required in wet or chemically corrosive environments. Standards for protective footwear are specified in ASTM F2413, *Standard Specification for Performance Requirements for Foot Protection.*

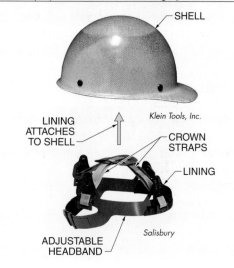

Hard Hats

Type	Impact Protection
I	Impacts from top
II	Impacts from top and side
Class	**Use**
G	General service, limited voltage protection
E	Utility service, high voltage protection
C	Special service, no voltage protection

SHELL

Klein Tools, Inc.

LINING ATTACHES TO SHELL

CROWN STRAPS

LINING

Salisbury

ADJUSTABLE HEADBAND

Figure 1-6. Hard hats are identified by type and class for protection against specific hazards.

In addition to foot guards and safety shoes, leggings made from leather, aluminized rayon, or other appropriate materials can help prevent injuries. These leggings protect technicians from hazards such as falling or rolling objects, sharp objects, wet and slippery surfaces, molten metals, hot surfaces, and electrical hazards.

Hearing Protection. When a technician is exposed to harmful sounds, such as sounds that are excessively loud or loud sounds over a long period of time, structures within the inner ear can be damaged. Inner ear damage is termed noise-induced hearing loss (NIHL). NIHL can be caused by a one-time exposure to a loud sound or by repeated exposure to sounds at various loudness levels over an extended period of time. The loudness of sound is measured in decibels. **See Figure 1-7.** An individual exposed to a sound level of 110 dB for longer than 30 min risks permanent hearing loss. No more than 2 hr of unprotected exposure to a sound level of 100 dB is recommended per OSHA 29 CFR 1926.52 (D)(1)—*Occupational Noise Exposure.*

Noise Level Exposure	
Decibel Level of Various Sounds	
130	Jet engine on ground
120	Reciprocating aircraft engine on ground
110	Punch press or pneumatic riveter
100	Maximum street noise
90	Loud shout
80	Diesel truck engine
74	Normal street noise
70	Barking dog or diesel locomotive
60	Average city office or conversational speech
50	Average city residence
40	Quiet office or average country residence
30	Turning page of newspaper
20	Rustle of leaves in breeze
10	Human heartbeat
0	Threshold of hearing

OSHA Noise Exposure Limitations	
Sound Level*	Duration Permitted†
90	8
92	6
95	4
97	3
100	2
110	0.5
115	0.25

* in dBA
† in hr

Figure 1-7. Hearing protection must be worn when noise levels exceed the time-weighted permissible exposures.

Wearing earplugs or earmuffs can help prevent hearing damage. An *earplug* is a compressible device inserted into the ear canal to reduce the level of noise reaching the eardrum. Exposure to high noise levels can cause irreversible hearing loss or impairment as well as physical and psychological stress. Earplugs made from foam, waxed cotton, or fiberglass wool are self-forming and usually fit into the ear well. Also available are individually molded or preformed earplugs. Earplugs must be cleaned regularly or replaced if they cannot be cleaned.

An *earmuff* is a device worn over the ears to reduce the level of noise reaching the eardrum. A tight seal around the earmuff is required for proper protection. To protect electrical technicians, some earmuffs have no metal parts. Electronic earmuffs can reduce certain sound frequencies while still allowing voices to be heard. Earmuffs must be washed and dried daily according to manufacturer's instructions.

Hand Protection. Hand injuries account for approximately one-third of all disabling job-related injuries each year. Over 80% of these injuries are caused by pinch points. Pinch point injuries vary from bruises, cuts, and fractures to amputations. Common sources of pinch points include rotating equipment that has meshing gear teeth, chains and sprockets, rollers, belts, and pulleys. **See Figure 1-8.** About 20% of pinch point injuries become infected.

Pinch points must be identified and properly guarded and technicians must be aware of their existence and potential danger. Any machinery being serviced should be properly locked out and tagged. Machines should never be repaired, cleaned, or adjusted while in motion or energized.

Approved hand protection, such as work gloves, can be useful in preventing cuts, bruises, and abrasions resulting from handling rough materials or sharp objects. The correct gloves must be worn for the application. Wearing the wrong type of glove for an application may cause the glove and hand to be caught in moving parts or machinery. Certain gloves can also be effective in minimizing hand infections such as dermatitis.

Technicians exposed to harmful substances through skin absorption, severe cuts or lacerations, severe abrasions, chemical burns, thermal burns, or harmful temperature extremes require hand protection. Rings and jewelry should not be worn while working on machinery.

Safety Tip

PPE specifically designed for use when working on or near energized electrical circuits and equipment is available. This equipment includes fire-resistant clothing, rubberized gloves with leather protectors, insulated mats for working/ standing on concrete surfaces, and work boots or shoes with insulated soles.

Body Protection. In some cases, technicians must shield most or all of their bodies against hazards in the workplace, such as exposure to heat, radiation, hot metals, scalding liquids, body fluids, hazardous materials or waste, and other hazards. In addition to fire-resistant wool and fire-resistant cotton, materials used in whole-body PPE include rubber, leather, synthetics, and plastic.

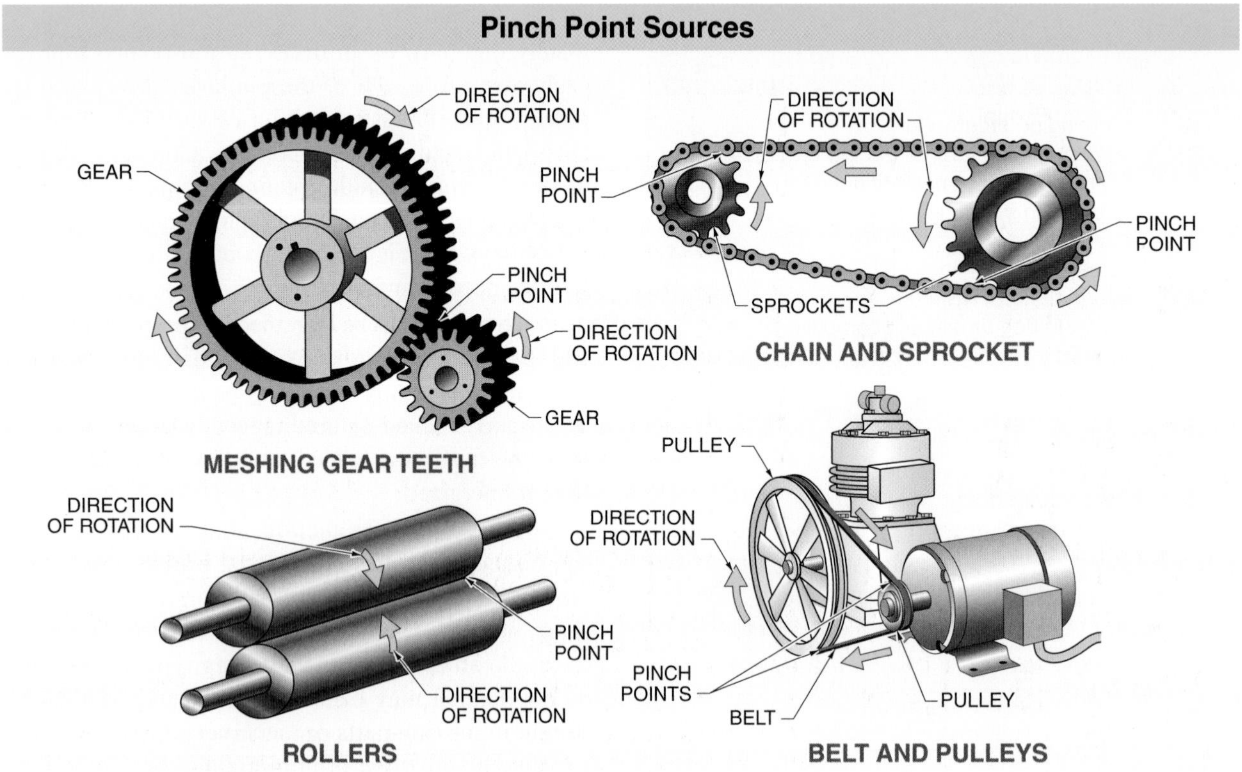

Pinch Point Sources

Figure 1-8. Common sources of pinch points include rotating equipment that has meshing gear teeth, chains and sprockets, rollers, belts, and pulleys.

Respiratory Protection. When engineering controls are not feasible, technicians must use appropriate respirators to protect against adverse health effects caused by breathing air contaminated with harmful dust, fog, fumes, mist, gas, smoke, spray, or vapor. A *respirator* is a device that protects the wearer from inhaling airborne contaminants. A respirator is selected based on the type of material or chemical hazard to which a technician is exposed. **See Figure 1-9.** Respirator selection should be made according to NIOSH 87-108, *Respirator Decision Logic* and ANSI/AIHA Z88.2, *Respiratory Protection.*

Respirators generally cover the nose and mouth or the entire face or head and help prevent respiratory illness and injury. A proper fit is essential for respirators to be effective. Respirators must be selected by qualified personnel. Improper use and/or selection of a respirator may result in serious injury or death. Required respirators must be NIOSH approved and medical evaluation and training must be provided before use.

PPE Selection. The PPE selected must ensure a level of protection greater than the minimum level required

to protect a technician from the hazard. Careful consideration must be given to comfort and fit. PPE must fit properly to be effective and must be comfortable to enable prolonged use. If the protective gear does not fit, it may not adequately protect the technician. When protective gear is uncomfortable it is hard to concentrate on the job and it may tempt technicians to remove it. Defective or damaged PPE shall not be used.

PPE alone should not be relied on to provide protection against hazards. PPE should be used in conjunction with guards, engineering controls, and sound manufacturing practices.

Confined Spaces

A *confined space* is a space large enough for an individual to physically enter and perform assigned work but has limited or restricted means for entry and exit and is not designed for continuous occupancy. Confined spaces include storage tanks, process vessels, boilers, ventilation or exhaust ducts, sewers, underground utility vaults, pipelines, and open top spaces more than 4′ in depth such as pits and ditches.

Respirators	
Air-Purifying	
Respirator	Description
 Disposable particulates mask	Low profile, lightweight, designed for limited use. Low-cost protection against dusts, mists, and fumes (not for mists containing gases, vapors, or nonabsorbed contaminants). Completely disposable. No cleaning or spare parts required.
 Reusable half-mask respirator	Lightweight, easy to maintain, very little restriction of movement or vision. Uses replaceable cartridges and filters. Limited number of parts. Protects against chemical hazards such as dust, fumes, mists, and vapors.
 Reusable full-face respirator	Offers greater eye and face protection than half-mask. Uses replaceable cartridges and filters. Easy to maintain (no intricate parts). Protects against chemical hazards such as dust, fumes, mists, and vapors.
 Powered air-purifying respirator (PAPR)	Cooler, less exhausting for worker. Provides easier breathing for higher productivity. Uses cartridges or filters. Face- or belt-mounted with a battery for power. Includes air blower that pulls air through the cartridges and filters into the face piece.
Supplied-Air	
 Airline respirator	Uses outside air source to keep worker cooler and offers greater protection than an air-purifying respirator. Available in two styles: constant flow and pressure demand. Uses Grade D air supply from ambient air pump, plant compressor, or bottled air. Not for use in IDLH (immediately dangerous to life or health) situations or where the oxygen content is less than 19.5%.
 Self-contained breathing apparatus (SCBA)	Provides greatest protection available. Pressurized bottle of air is carried on worker's back. For use in oxygen-deficient atmospheres, IDLH, and emergency situations. Available in two different types of cylinders: aluminum and composite. Provides good mobility with few restrictions because air source is carried on back.
 Emergency escape breathing apparatus (EEBA)	For use in escape situations only. IDLH and oxygen deficiency. Service life depends on a 5 min to 10 min bottle of air. Not designed for rescue use.

Color Code for Cartridges/Canisters	
Contaminant	Color Assigned
Acid gases only	White
Organic vapors only	Black
Ammonia gas	Green
Acid gases and organic vapors	Yellow
Radioactive materials (except tritium and noble gases)	Purple
Dust, fumes, and mists (other than radioactive materials)	Orange
Other gases and vapors (not listed above)	Olive

North Safety Products

Figure 1-9. Respirators are selected for the specific contaminants and concentrations present in a hazard area.

Confined-Space Hazards. Confined spaces are particularly susceptible to containing oxygen-deficient, toxic, or explosive atmospheres. Oxygen deficiency is caused by the displacement of oxygen by leaking gases or vapors, combustion or oxidation processes, oxygen absorbed by the vessel or stored products, and/or consumption by bacteria. Oxygen-deficient air can result in injury or death. **See Figure 1-10.** Common toxic gases include hydrogen sulfide, natural gas, and carbon monoxide. Toxic gases are released by cleaning solvents, chemical reactions, heated materials, and other sources.

Potential Effects of Oxygen-Deficient Atmospheres*	
Oxygen Content†	Effects and Symptoms‡
19.5	Minimum permissible oxygen level
16 – 19.5	Decreased ability to work strenuously. May impair condition and induce early symptoms in persons with coronary, pulmonary, or circulatory problems
12 – 14	Respiration exertion and pulse increase. Impaired coordination, perception, and judgment
10 – 11	Respiration further increases in rate and depth, poor judgement, lips turn blue
8 – 9	Mental failure, fainting, unconsciousness, ashen face, blue lips, nausea, and vomiting
6 – 7	Eight min, 100% fatal; 6 min, 50% fatal; 4 to 5 min, recovery with treatment
4 – 5	Coma in 40 seconds, convulsions, respiration ceases, death

* Values are approximate and vary with state of health and physical activities
† % by volume
‡ at atmospheric pressure

Figure 1-10. Oxygen-deficient atmospheres in confined spaces can be life threatening.

Combustible atmospheres are commonly caused by gases such as methane, carbon monoxide, and hydrogen sulfide. An increase in the oxygen level above the normal 21% increases the explosive potential of combustible gases. Finely ground materials, including carbon, grain, fibers, metals, and plastics, can also cause explosive atmospheres.

The *explosive range* is the difference between the lower explosive limit and the upper explosive limit of combustible gases. **See Figure 1-11.** The *lower explosive limit (LEL)* is the lowest concentration (air-fuel mixture) at which a gas can ignite. Concentrations below this limit are too lean to burn. The *upper explosive limit (UEL)* is the highest concentration (air-fuel mixture) at which a gas can ignite. Concentrations above this limit are too rich to burn.

Combustible Atmospheres

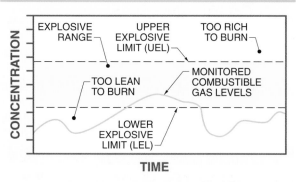

Figure 1-11. Atmospheres are potentially explosive when the concentration of a combustible gas is within a certain range.

Combustible gases at any concentration are a concern. Lean mixtures can collect in areas and reach combustible levels. Also, lean mixtures may still be toxic. Rich mixtures can be diluted with air and become combustible. Detection instruments are set to sense the presence of combustible gases at levels that forewarn technicians of potentially hazardous combustible atmospheres before the LEL is reached. **See Figure 1-12.**

Warning: Confined-space procedures vary for each facility. For maximum safety, always refer to specific facility procedures and applicable federal, state, and local regulations.

Confined-Space Permits. A *permit-required confined space* is a confined space that has specific health and safety hazards capable of causing death or serious physical harm. Permit-required confined spaces are spaces that have the potential to contain a hazardous atmosphere; contain a material that has the potential for engulfing an entrant; have an internal configuration such that an entrant could be trapped or asphyxiated by inwardly converging walls or a floor that slopes downward and tapers into a smaller cross-section; or contain any other recognized safety or health hazard.

OSHA 29 CFR 1910.146—*Permit-Required Confined Spaces,* contains the requirements for practices and procedures to protect individuals from the hazards of permit-required confined spaces. **See Figure 1-13.**

Confined-Space Atmosphere Testing

Mine Safety Appliances Co.

Figure 1-12. Confined spaces must be tested for oxygen level and air contaminants before entering.

Even if a confined space does not initially require a permit, individuals must be aware that the condition can change with tasks such as welding, painting, or solvent use within the confined space. Employers must evaluate the workplace to determine if spaces are permit-required confined spaces. If confined spaces exist in the workplace, the employer must inform individuals of their existence and location and of the danger they pose. This is accomplished by posting danger signs or by other equally effective means. In addition, the employer must develop a written permit-required confined-space program that specifies entry procedures, hazard identification, access restriction, hazard control, and monitoring of the space during entry.

Permit-Required Confined Spaces

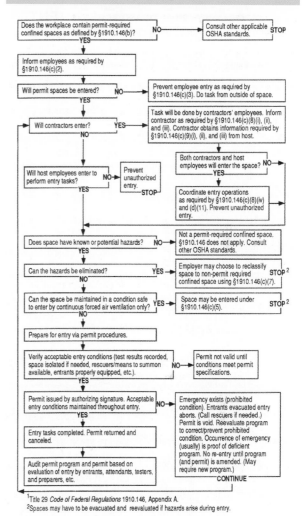

[1]Title 29 *Code of Federal Regulations* 1910.146, Appendix A.
[2]Spaces may have to be evacuated and reevaluated if hazards arise during entry.

Figure 1-13. Procedures for entering a confined space must follow established OSHA safety standards.

Entry-Permit Procedures. An entry permit must be posted at confined-space entrances or otherwise made available to entrants before entering a permit-required confined space. The permit is signed by the entry supervisor and verifies that preentry preparations have been completed and that the space is safe to enter. **See Figure 1-14.** A permit-required confined space must be isolated before entry. This prevents hazardous energy and materials from entering the space. Plant procedures for lockout/tagout of permit-required confined spaces must be followed. A plant may develop its own confined space permit, but OSHA requires that certain checklist items be included.

Confined-space Entry Permits

═══ **ENTRY PERMIT** ═══

✔ CONFINED SPACE   ✔ HAZARDOUS AREA

PERMIT VALID FOR 8 HOURS ONLY. ALL COPIES OF PERMIT WILL REMAIN AT JOB SITE UNTIL JOB IS COMPLETED

SITE LOCATION and DESCRIPTION   Bunker Fuel Oil Tank #2

PURPOSE OF ENTRY   Routine Maintenance/Inspection

SUPERVISOR(S) in charge of crews.   Type of Crew   Phone #
Michael Green   Maintenance   Shift II - X5924

* BOLD DENOTES MINIMUM REQUIREMENTS TO BE COMPLETED AND REVIEWED PRIOR TO ENTRY*

REQUIREMENTS COMPLETED	DATE	TIME	REQUIREMENTS COMPLETED	DATE	TIME
Lock Out/De-energize/Try-out	10/2	09:00	**Full Body Harness w/"D" ring**	10/4	08:00
Line(s) Broken-Capped-Blanked	10/2	11:00	**Emergency Escape Retrieval Equip**	10/4	08:00
Purge-Flush and Vent	10/3	09:00	**Lifelines**	10/4	08:00
Ventilation	10/3	10:00	Fire Extinguishers	10/4	08:00
Secure Area (Post and Flag)	10/2	08:00	Lighting (Explosion Proof)	10/4	08:00
Breathing Apparatus	10/4	08:00	Protective Clothing	10/4	08:00
Resuscitator - Inhalator	10/4	08:00	Respirator(s) (Air Purifying)	10/4	08:00
Standby Safety Personnel	10/4	08:00	Burning and Welding Permit	N/A	N/A

Note: Items that do not apply enter N/A in the blank.

** RECORD CONTINUOUS MONITORING RESULTS EVERY 2 HOURS

CONTINUOUS MONITORING** TEST(S) TO BE TAKEN	Permissible Entry Level	10/4				
PERCENT OF OXYGEN	19.5% to 23.5%	20.5	20.6	20.7	20.5	20.5
LOWER FLAMMABLE LIMIT	Under 10%	5	5	5	6	
CARBON MONOXIDE	+35 PPM	0	0	0	0	0
Aromatic Hydrocarbon	+ 1 PPM * 5PPM	2	1	2	1	1
Hydrogen Cyanide	(Skin) * 4PPM	N/A				
Hydrogen Sulfide	+10 PPM *15PPM	N/A				
Sulfur Dioxide	+ 2 PPM * 5PPM	3	2	2	2	2
Ammonia	* 35PPM	N/A				

* Short-term exposure limit:Employee can work in the area up to 15 minutes.

+ 8 hr. Time Weighted Avg.:Employee can work in area 8 hrs (longer with appropriate respiratory protection).

REMARKS:

GAS TESTER NAME & CHECK #	INSTRUMENT(S) USED	MODEL &/OR TYPE	SERIAL &/OR UNIT #
Marty James	Combination Gas Meter	Industrial Scientific	15A

SAFETY STANDBY PERSON IS REQUIRED FOR ALL CONFINED SPACE WORK

SAFETY STANDBY PERSON(S)	CHECK #	NAME OF SAFETY STANDBY PERSON(S)	CHECK #
Kate Washington	3312		
Tony Linder	3318		

SUPERVISOR AUTHORIZING ENTRY ALL ABOVE CONDITIONS SATISFIED   Michael Green

AMBULANCE 2800   Safety 4901

FIRE 2900   Gas Coordinator 4529/5387

Figure 1-14. A confined-space entry permit documents preparations, procedures, periodic monitoring, and required equipment.

Checklist items required by OSHA include the permit date, time, expiration, supervisor name, results of atmospheric testing both prior to entry and periodically during work, the applicable safety equipment, and communication and rescue procedures. Standby personnel, commonly referred to as a "hole watch," must remain just outside a confined space while technicians are inside. Hole watch personnel are responsible for periodically testing and recording the atmospheric conditions, monitoring the technicians inside the space, and assisting rescue personnel in the event of an emergency. The duration of entry permits must not exceed the time required to complete a task. The entry supervisor must terminate entry and cancel permits when a task has been completed or when new conditions exist. All canceled entry permits must be filed for at least one year.

Training is required for all personnel who are required to work in or around permit-required confined spaces. A certificate of training includes the technician's name, the signature or initials of the trainer, and the dates of training. The certificate must be available for inspection by authorized personnel.

Fall-Arrest Systems

A personal fall-arrest system is used to arrest (stop) a technician in a fall from a working level. Personal fall-arrest systems may be used, and are many times required for use, with equipment such as aerial lifts. Proper personal fall-arrest systems must be used when working at heights greater than 10′. Personal fall-arrest systems consist of an anchorage point, connectors, and a body harness and may include a lanyard, lifeline, deceleration device, or combinations of these devices. **See Figure 1-15.**

An anchorage point provides a secure point of attachment for lifelines, lanyards, or other deceleration devices. Anchorage points for personal fall-arrest systems must be independent of any anchorage point used to support or suspend platforms and must be capable of supporting 5000 lb minimum per person attached to the anchorage point or at least twice the anticipated load. Anchorage points must be designed and installed under the supervision of a qualified person. A qualified person can be a safety engineer or a supervising technician.

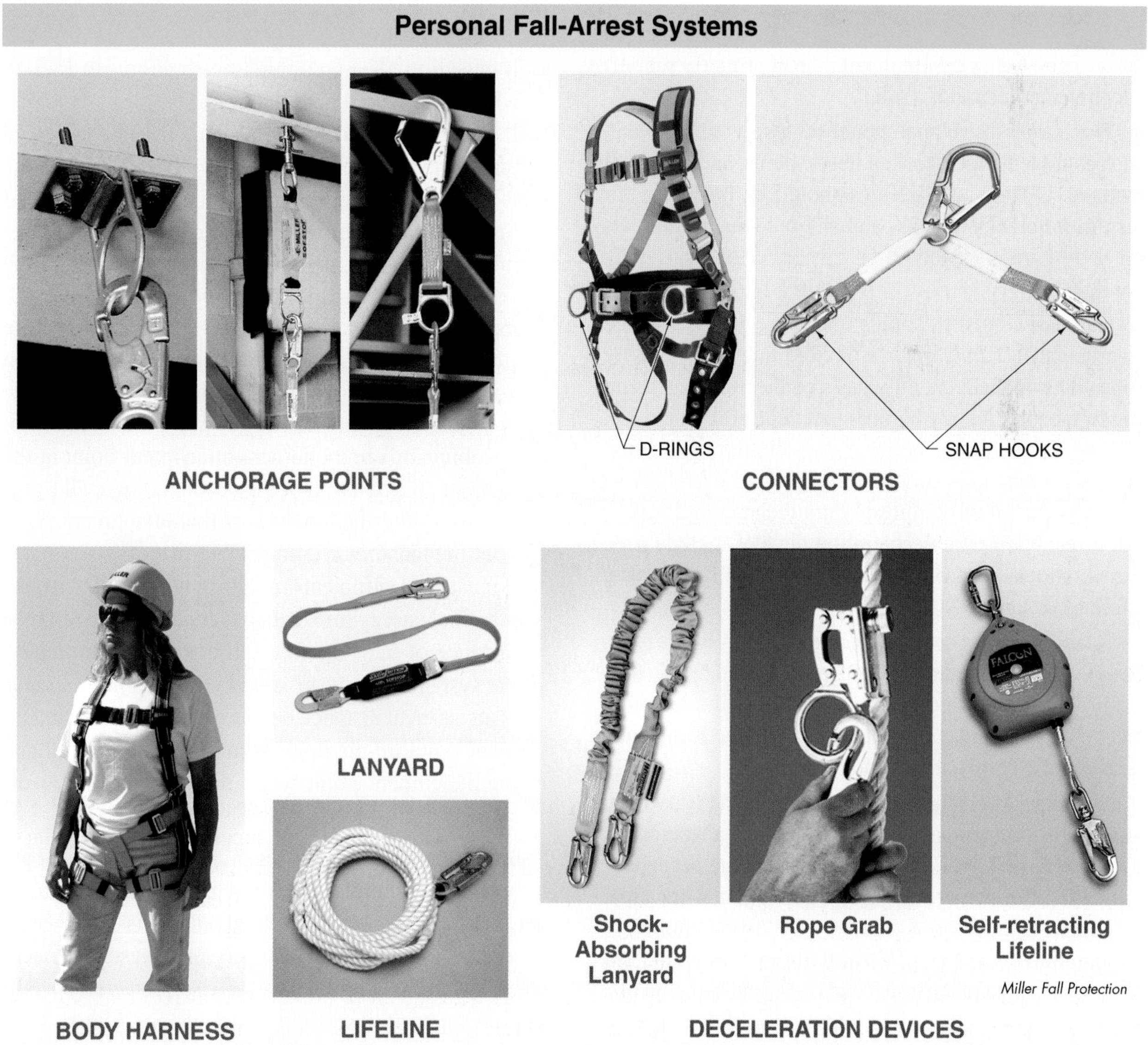

Personal Fall-Arrest Systems

ANCHORAGE POINTS

D-RINGS   SNAP HOOKS

CONNECTORS

LANYARD

Shock-Absorbing Lanyard   Rope Grab   Self-retracting Lifeline

Miller Fall Protection

BODY HARNESS   LIFELINE   DECELERATION DEVICES

Figure 1-15. Personal fall-arrest systems must be utilized when working at heights greater than 10′ when other means of fall protection are not provided.

Connectors such as D-rings and snap hooks must have a minimum tensile strength of 5000 lb and must be proof-tested to a minimum tensile load of 3600 lb. Connectors are normally attached to other personal fall-arrest system components such as anchorage points, body harnesses, lifelines, and lanyards. Only locking-type snap hooks are permitted to be used for personal fall-arrest systems. Snap hooks can be attached directly to webbing, rope or wire rope, a D-ring, each other, and horizontal lifelines.

When worn properly, body harnesses protect internal body organs, the spine, and other bones in a fall. The attachment point of a body harness should be located in the center of the back near shoulder level or above the technician's head. Body harnesses should be inspected before each use to ensure that they will properly support a technician in case of a fall.

The harness webbing should be inspected for wear, such as frayed edges, broken fibers, burns, and chemical damage. D-rings and buckles should be inspected for distortion, cracks, breaks, and sharp edges. Grommets on body harnesses should be inspected to ensure they are tight.

Harnesses must fit snugly and be securely attached to a lanyard. A *lanyard* is a flexible line of rope, wire rope, or strap that generally has a connector at each end for connecting a body harness to a deceleration device, lifeline, or anchorage point. A deceleration device dissipates a substantial amount of energy during a fall arrest or limits the energy imposed on a technician during a fall arrest. Common deceleration devices include rope grabs, shock-absorbing lanyards, and self-retracting lifelines or lanyards.

A *rope grab* is a device that clamps securely to a rope. A rope grab automatically engages a vertical or horizontal lifeline by friction to arrest the fall while allowing freedom of movement. A lanyard or lifeline is attached between the body harness and rope grab.

A shock-absorbing lanyard has a specially woven, shock-absorbing inner core that reduces fall arrest forces. The outer shell of the lanyard serves as the secondary lanyard. Lifelines are anchored above the work area, offering a free-fall path, and must be strong enough to support the force of a fall. Vertical lifelines are connected to a fixed anchor at the upper end that is independent of a work platform such as a scaffold. Vertical lifelines must never have more than one technician attached per line.

A self-retracting lifeline is a type of vertical lifeline. A self-retracting lifeline contains a line that can be slowly extracted from or retracted onto its drum under slight tension during normal technician movement. When a fall occurs, the drum automatically locks, arresting the fall. Horizontal lifelines are connected to fixed anchors at both ends. Technicians attach their lanyard to a D-ring on the lifeline, allowing them to freely move horizontally along the lifeline.

A lifeline must be properly terminated (anchored) to prevent the safety sleeve or ring from sliding off its end. The path of a fall must be visualized when anchoring a lifeline. Obstacles below and in the fall path can be deadly.

Personal Fall-Arrest System Requirements. When a personal fall-arrest system is used for fall protection, the system must be able to do the following:

- It must limit the maximum arresting force to 1800 lb when used with a body harness.
- It must be rigged so the technician can neither fall more than 6′ nor contact any lower level. **See Figure 1-16.**
- It must bring a technician to a complete stop and limit the maximum deceleration distance a technician falls to 3′-6″. *Deceleration distance* is the additional vertical distance a falling technician travels, excluding lifeline elongation and free-fall distance, before stopping, from the point at which the deceleration device begins to operate. The deceleration distance is measured as the distance between the location of a technician's body harness attachment point at the moment of activation of the deceleration device during a fall and the location of that attachment point after the technician comes to a full stop.
- It must have sufficient strength to withstand twice the potential energy impact of a technician free-falling a distance of 6′ or the free-fall distance permitted by the system, whichever is less. The *free-fall distance* is the vertical distance between the fall-arrest attachment point on the body harness before the fall and the attachment point when the personal fall-arrest system applies force to arrest the fall.

Safety Nets. A *safety net* is a net made of rope or webbing for catching and protecting a falling technician. A safety net must be used anywhere a technician is 25′ or more above the ground, water, machinery, or other solid surfaces when the technician is not otherwise protected by fall-arrest equipment or scaffold guardrails. **See Figure 1-17.** Safety nets must also be used when public traffic or other technicians are permitted underneath a work area that is not otherwise protected from falling objects.

Estimated Fall Distance

6' (LENGTH OF LANYARD OR LIFELINE)

(TOTAL 18'-6" FALL DISTANCE)

3'-6" (MAXIMUM DECELERATION DISTANCE)

6' (WORKER HEIGHT)

9'-6" (FALL-ARREST DISTANCE)

3' (SAFETY FACTOR)

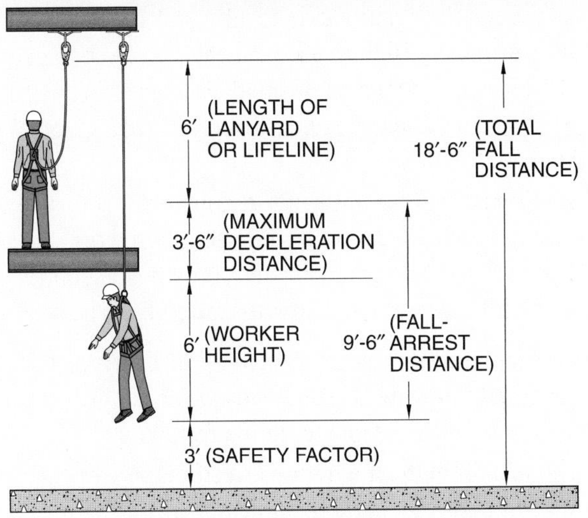

Figure 1-16. Always determine the estimated fall distance when selecting the proper personal fall-arrest system.

Safety Nets

The Sinco Group, Inc.

Figure 1-17. A safety net must be installed whenever a worker is 25′ or more above the ground, water, machinery, or other solid surfaces when technicians are not otherwise protected by fall-protection equipment or guardrails.

Review

1. What are the two main causes of injuries in the workplace?

2. What is first aid?

3. What is a Good Samaritan law?

4. What is the most visible result of an injury?

5. List the causes of burns.

6. List the steps to follow when assisting a choking victim.

7. What are the two main causes of slips and falls in the workplace?

8. What causes a large percentage of eye injuries?

9. What can be caused by repeated exposure to sounds at various loudness levels over an extended period of time?

10. What type of PPE should be worn to prevent hearing damage?

11. Explain how a respirator is selected.

12. Explain the difference between the lower explosive limit (LEL) and upper explosive limit (UEL).

13. Where must a confined-space entry-permit be placed?

14. What is the minimum working height that requires the use of a personal fall-arrest system?

15. Define safety net.

Digital Resources

ATPeResources.com/Quicklinks
Access Code 462409

CHAPTER 2

TOOLS AND TOOL SAFETY

When an industrial or mechanical technician performs work, the proper tools and equipment must be selected for each operation. The tools and equipment must be organized and readily available for use. Types of tools include hand tools, power tools, and test tools. To avoid injury and prevent damage to a tool, technicians should consult the operator's manual for proper operation and use before performing any work.

Milwaukee Electric Tool Corp.

Chapter Objectives

- List the different types of tools used for noncutting operations.

- List the different types of tools used for cutting operations.

- Explain the tasks that usually require the use of power tools.

- Describe the test tools used to measure and record quantities involving electrical current, motion, distance, speed, and temperature.

- Compare contact and noncontact temperature test tools.

- Explain safety precautions to follow when working with hand tools, power tools, and test tools.

- Describe the different types of ladders and their uses.

- Describe the different types of scaffolds and their uses.

- Explain the use of safety nets.

- Explain safety precautions to follow when working with ladders and scaffolds.

Key Terms

- hammer
- wrench
- screwdriver
- pliers
- punch
- vise
- screw extractor
- mechanical puller
- file
- tap
- chisel
- handsaw
- power drill
- circular saw
- reciprocating saw
- metal ladder
- fiberglass ladder

TOOLS

Various hand, power, and test tools are used by industrial and mechanical technicians for the maintenance, troubleshooting, and installation of different types of equipment. Different tools are designed for the efficient and safe completion of a specific job. Proper use of tools is required for safe and efficient work.

Specialized hammers are used for industrial operations such as removing slag from weld lines.

Hand Tools

Industrial and mechanical technicians use hand tools for twisting, turning, bending, cutting, stripping, attaching, pulling, and securing operations. Noncutting operations require the use of tools such as hammers, wrenches, screwdrivers, pliers, punches, vises, tape rules, screw extractors, and mechanical pullers for gears and bearings. Cutting operations require the use of tools such as files, taps and dies, chisels, wire stripper/crimper/cutters, handsaws, and hacksaws.

Hammers. A *hammer* is a striking or splitting tool with a hardened head fastened perpendicular to a handle. Common types of hammers used in industrial and mechanical applications include claw, ball peen, engineer's, and sledge. **See Figure 2-1.** A claw hammer is used to drive nails. A claw hammer can also be used to remove nails previously driven into a surface. Ball peen hammers of the proper size are designed for striking chisels and punches. Ball peen hammers can also be used for riveting, shaping, and straightening unhardened metal. Engineer's hammers are double-faced hammers used for driving large-size chisels, cold punches, rock drills, and hardened nails. Medium-size sledgehammers (5 lb to 8 lb) are used for driving stakes and other heavy-duty pounding.

Hammers

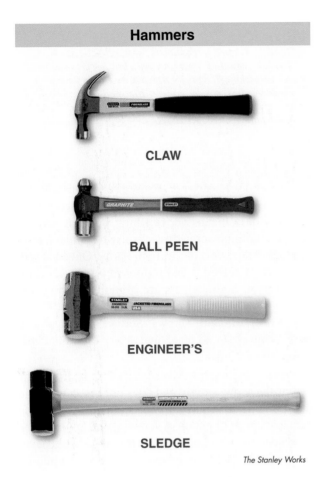

The Stanley Works

Figure 2-1. Common types of hammers used in industrial and mechanical applications include claw, ball peen, engineer's, and sledge.

Wrenches. A *wrench* is a hand tool with jaws at one or both ends that is designed to turn bolts, nuts, or pipes. Common wrenches used for industrial and mechanical applications include socket, adjustable, Allen (hex key), combination, and pipe. **See Figure 2-2.** Socket wrenches are used to tighten a variety of items, such as hex-head lag screws, bolts, and various electrical connectors. Adjustable wrenches are used to tighten items such as various-size hex-head lag screws, bolts, and large conduit couplings. Allen wrenches are used for tightening hex-head bolts. A combination wrench is a hand tool with an open-end wrench on one end and a closed-end box wrench on the other. Pipe wrenches can be straight, offset, strap, or chain. Pipe wrenches are used to tighten and loosen pipes and large conduit.

When using wrenches, a pipe extension or other form of "cheater" should never be used to increase the leverage of the wrench. A wrench is selected with an opening that corresponds to the size of the nut to be turned. A wrench with too large an opening for the workpiece can spread the jaws of an open-end wrench and batter the points of a box or socket wrench. Care should be taken in selecting inch wrenches for inch fasteners and metric wrenches for metric fasteners. It is a good practice to pull on a wrench handle before use. The safest wrench that can be used is a box or socket wrench because wrenches of these types cannot slip and injure a technician. A straight handle is better than an offset handle if conditions permit.

Wrenches

The Stanley Works

Figure 2-2. Wrenches have jaws at one or both ends and are designed to turn bolts, nuts, or pipes.

Tech Tip

To help reduce worker fatigue and provide uniform torque, many types of wrenches used in industrial applications are available with electric or pneumatic power.

Screwdrivers. A *screwdriver* is a hand tool with a tip designed to fit into a screw head for fastening operations. Industrial and mechanical technicians use screwdrivers in most installation, troubleshooting, and maintenance activities to secure and remove various threaded fasteners. Various types of screwdrivers are available, but the two main types of screwdrivers are the flat head and Phillips. Flat head and Phillips screwdrivers are available as standard and offset. Offset screwdrivers are also available as a combination with a Phillips head and a flat head. A screw-holding attachment can be used with standard screwdrivers. **See Figure 2-3.**

Tech Tip

> *Hand tools such as hammers, screwdrivers, scrapers, wrenches, chisels, pry bars, and wire brushes are available as nonsparking designs for use in areas where flammable vapors or dust may be present. Nonsparking hand tools are typically composed of brass, beryllium copper, bronze, bronze alloys, or copper alloys.*

Standard screwdrivers are used for the installation and removal of threaded fasteners. Offset screwdrivers provide a means for reaching difficult screws. A screw-holding screwdriver is used to hold screws in place when working in tight spots. Once started, the screw is released and tightened with a standard screwdriver. Screwdrivers are available with square shanks to which a wrench can be applied for the removal of stubborn screws. Screwdrivers can also have a thin shank to reach and drive screws in deep, counterbored holes.

When using a screwdriver, verify that the tip fits the slot of the screw snugly and does not project beyond the screw head. A screwdriver should never be used as a cold chisel or punch. A screwdriver should not be used near energized electrical wires and should never be exposed to excessive heat. A worn tip should be redressed with a file to regain a good, straight edge. A screwdriver that has a worn or broken handle should be discarded.

Pliers. *Pliers* are a hand tool with opposing jaws for gripping and/or cutting. Industrial and mechanical technicians use pliers for various gripping, turning, cutting, positioning, and bending operations. Common pliers include slip-joint, tongue-and-groove, long nose, locking, diagonal-cutting, lineman's, end-cutting, and self-adjusting pliers. **See Figure 2-4.**

Slip-joint pliers are used to tighten box connectors, lock nuts, and small-size conduit couplings. Tongue-and-groove pliers are used for a wide range of applications involving gripping, turning, and bending. Locking pliers (such as Vise-Grip™ pliers) are used to lock onto a workpiece. Locking pliers can be adjusted to lock at any size with any desired amount of pressure. The adjustable jaws of tongue-and-groove pliers enable adjustment to a wide range of sizes. Long nose pliers are used for bending and cutting wire and positioning small components. Diagonal-cutting pliers are used for cutting cables and wires that are too difficult to cut with side-cutting pliers.

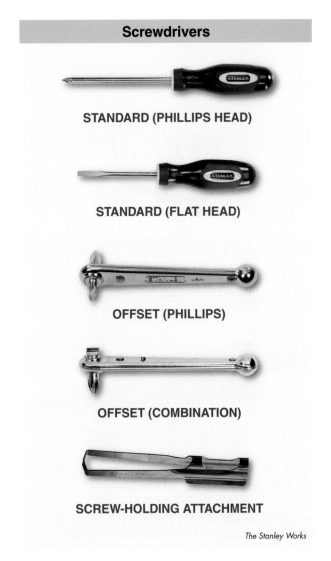

Screwdrivers

STANDARD (PHILLIPS HEAD)

STANDARD (FLAT HEAD)

OFFSET (PHILLIPS)

OFFSET (COMBINATION)

SCREW-HOLDING ATTACHMENT

The Stanley Works

Figure 2-3. Different types of screwdrivers are used for various turning operations.

Pliers

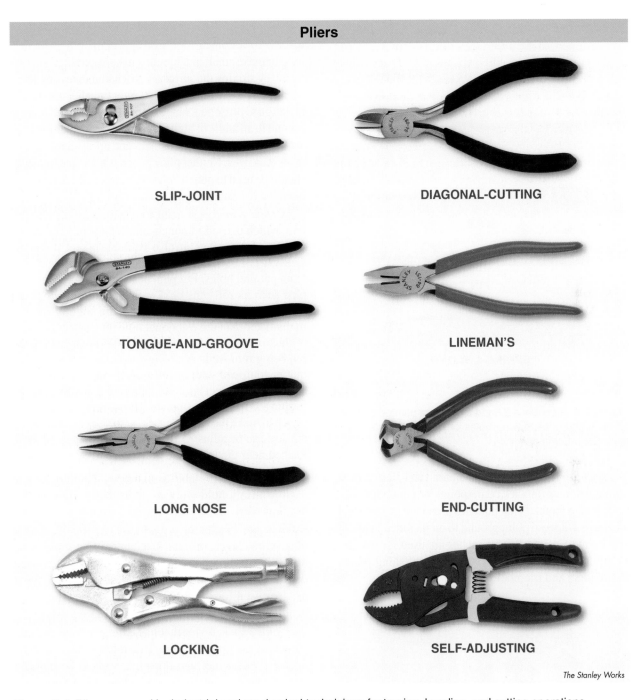

SLIP-JOINT

DIAGONAL-CUTTING

TONGUE-AND-GROOVE

LINEMAN'S

LONG NOSE

END-CUTTING

LOCKING

SELF-ADJUSTING

The Stanley Works

Figure 2-4. Pliers are used by industrial and mechanical technicians for turning, bending, and cutting operations.

Lineman's pliers are used for cutting cable, removing knockouts, twisting wire, and deburring metal pipe. End-cutting pliers are used for cutting wire, nails, rivets, etc., close to the workpiece. Self-adjusting pliers work in the same manner as slip-joint pliers but automatically adjust to the size of the workpiece and lock in place.

Punches. A *punch* is a hand tool with a pointed or blunt tip for marking or making holes or driving objects when struck by a hammer. Care should be used in selecting the proper punch for each operation. Punches include prick, center, solid, pin, and spring-loaded punches. **See Figure 2-5.**

PUNCHES

PRICK

CENTER

SOLID

PIN

SPRING-LOADED

The Stanley Works

Figure 2-5. A punch is a hand tool with a pointed or blunt tip for marking or making holes or driving objects when struck by a hammer.

A *prick punch* is a sharp, pointed steel shaft struck with a hammer to mark centerpoints or punch holes in light-gauge metal. Prick punches are made of tool steel and have a tapered point ground to an included angle of approximately 30°. Prick punches are used for making small dents or indentations and/or establishing points for dividers and trammel points.

Reed Manufacturing Co.

Copper vise-jaw caps can be used to protect vise jaws and surfaces of workpieces from scratching or marring.

A *center punch* is a steel hand tool with one end formed to a conical point of approximately 90°. It is used to make small indentations in hard surfaces, such as metal, by striking the surface with a hammer to establish a centerpoint for drilling a hole. Center punches are similar in design to prick punches, except that the tapered point is ground to an angle of approximately 90°. Center punches are manufactured in various sizes and are available in sets of various sizes. Neither prick punches nor center punches should be used to punch holes. Punches of these types are intended for establishing points only.

A *solid punch* is a hand tool with a blunt, circular end, which is used to punch small holes in light-gauge metal. Solid punches are also available in sets of various sizes. A *pin punch* is a punch used for removing pins from parallel holes. Pin punches are used for removing pins such as roll pins or dowel pins. Pin punches are not tapered like most punches, but are slightly smaller in diameter than the fractional hole they enter. The face of the pin punch is flat to ensure a solid seat against the pin being removed. A *spring-loaded punch* is a punch that is equipped with a spring. A spring-loaded punch can mark an object without the use of a hammer. When pushing down on the punch, the spring activates the punch to make the desired mark on the object. The spring tension can be adjusted by turning an adjustment knob located near the top of the punch.

Vises. A *vise* is a portable or stationary clamping device used to firmly hold work in place. Vises are used for holding work during inspection, assembly, forming, welding, and machining processes. Vises usually consist of a screw, lever, or cam mechanism that closes and holds two or more jaws around a workpiece. Although there are various types of vises, the most common types of vises used for mechanical work include pipe vises, bench vises, and machine vises. **See Figure 2-6.**

A *pipe vise* is a vise that has a hinge at one end and a hook at the opposite end. This allows the vise to be opened so that the pipe does not need to be inserted through the vise jaws to be worked on. A yoke pipe vise is bolted to a workbench. A clamp kit vise is a vise that contains a clamp for mounting. A clamp kit vise can be temporarily mounted for light-duty work without drilling holes. Yoke vises and chain pipe vises are available in portable workbench models.

A *bench vise* is a vise that can be attached to a bench top. Bench vises are the most common type of vise and can be used for most industrial and mechanical applications. A *machine vise* is vise made with high-grade body castings and ground-steel jaw plates. Machine vises are designed for heavier industrial processes such as milling, drilling, shaping, and grinding. Vises should adhere to the International Standards Organization (ISO) and The American National Standards Institute (ANSI) standards.

Vises

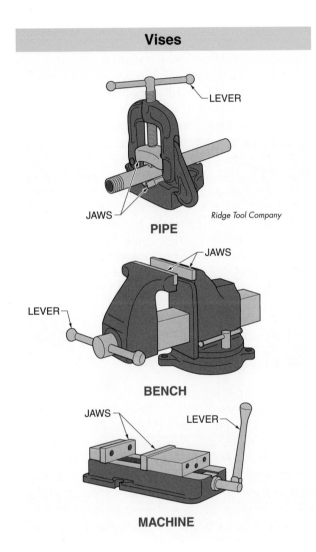

Figure 2-6. Common types of vises used in industrial mechanics include pipe vises, bench vises, and machine vises.

Screw Extractors. A *screw extractor* is a tool used to remove studs, bolts, or screws broken below or near the surface of a workpiece. Some screw extractors resemble reverse-threaded screws, while others resemble square tapered rods with chiseled edges. Screw extractors are available for most screw sizes. Screw extractors require the use of penetrating oil to assist in the removal of the broken screw. *Penetrating oil* is an industrial lubricant used to clean and loosen frozen parts. When penetrating oil is applied to a frozen or stuck part, it must be allowed

to soak into the work area for several minutes. The size of hole depends on the broken bolt or screw size and extractor size. For example, the screw extractor hole size for a ⅜″ screw requires a ¼″ hole for a reverse-threaded extractor and a ¹⁵⁄₆₄″ hole for a square extractor. Usually, the required hole size is stamped on the screw extractor. Removing a broken bolt or screw with a screw extractor is completed by applying the procedure:

1. Drill a small pilot hole in the center of the broken bolt or screw. **See Figure 2-7.**
2. Drill a second, larger extractor hole into the pilot hole.
3. Apply penetrating oil and use hammer to tap proper-size screw extractor into drilled hole.
4. Rotate the extractor counterclockwise for a broken bolt or screw with a right-hand thread (clockwise for a left-hand thread) and remove broken bolt or screw.

Note: Most screw threads used in industrial and mechanical equipment are right-handed.

Mechanical Pullers. A *mechanical puller* is a tool used to remove fitted machine parts. Fitted machine parts include parts such as sprockets, gears, pulleys, and bearings. Prying and hammering to remove a fitted machine part can damage both the fitted machine part and the matching machine part. The three basic types of pullers used for most applications are external pullers, internal pullers, and press pullers. **See Figure 2-8.** For a straight pull, a mechanical puller must be aligned with the shaft of the object to be removed. Each jaw must be parallel with the forcing screw. An off-center pull can bind parts, causing damage to equipment and harm to the user. As the forcing screw is rotated, the jaws pull the object to be removed.

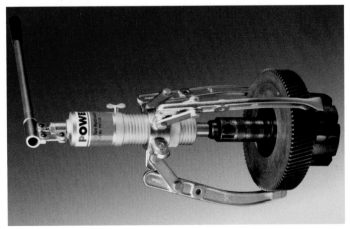

Power Team, Division of SPX Corporation

Mechanical pullers are available with hand pumps for removal of medium-size gears and bearings.

Screw Extractors – Broken Bolt Removal Procedure

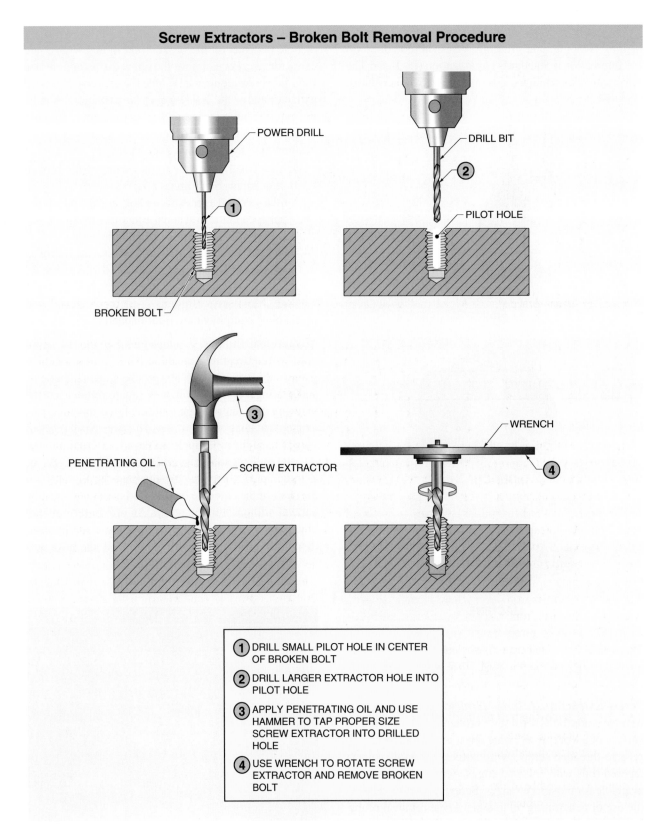

① DRILL SMALL PILOT HOLE IN CENTER OF BROKEN BOLT

② DRILL LARGER EXTRACTOR HOLE INTO PILOT HOLE

③ APPLY PENETRATING OIL AND USE HAMMER TO TAP PROPER SIZE SCREW EXTRACTOR INTO DRILLED HOLE

④ USE WRENCH TO ROTATE SCREW EXTRACTOR AND REMOVE BROKEN BOLT

Figure 2-7. Screw extractors are used to remove broken bolts and screws from equipment without causing damage to screw threads or equipment.

Mechanical Pullers

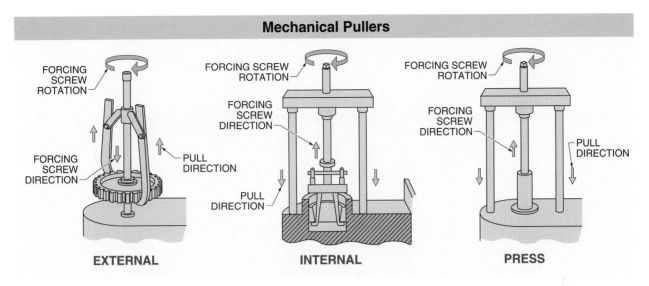

Figure 2-8. Mechanical pullers are used to remove fitted machine parts such as sprockets, gears, pulleys, and bearings.

Each type of puller has a forcing screw, puller arms (jaws), and a yoke (supporting arms). Pullers are available with either two or three puller arms. A three-arm puller offers smoother pulls, gives greater support, and is self-centering. An external puller is sometimes referred to as a universal gear puller, although it pulls more than gears. An external puller has a set of jaws that grip the back of the object from around the outside diameter.

Internal pullers are used to remove objects such as bearings or bushings from a bore. Internal pullers have a set of jaws that grip from the back by entering the inside of the object to be removed. Internal pullers have legs that mount on the stationary part while the forcing screw pulls the object to be removed. Most internal puller sets have interchangeable jaws for specific applications. For example, larger jaws can be attached to achieve greater coverage for removing bushings without breaking them.

Press pullers are used to remove shafts from within a bore. In most applications, a hole must be drilled and tapped in the center of the shaft to be removed to accommodate the assembly of the force screw. Press pullers have legs, which mount on the stationary part as the forcing screw is screwed into the end of the shaft. As the forcing screw is rotated, a pulling motion is applied to the shaft.

Files. A *file* is an abrasive tool with single or double rows of fine teeth cut into the surface. Files are used to remove burrs from sheets of metal, to square the ends of band iron, to straighten uneven edges, and for various other operations that require removing a small amount of metal. File parts include the point, edge, face, heel, and tang. File cutting faces can be single cut or double cut. **See Figure 2-9.** Single-cut files have a single set of teeth cut at an angle of 65° to 85°. Double-cut files have two sets of teeth crossing each other. Double-cut files are used for rough filing since they remove material faster than single-cut files. File handles are usually made of wood and are designed to fit the hollow of the hand. A metal ferrule on the end of the handle prevents it from splitting. The most common types of files used for industrial and mechanical projects include flat, mill, three square, round, square, half-round, and knife. **See Figure 2-10.**

Files are used to remove burrs and rough spots from metal workpieces.

File Cutting Faces

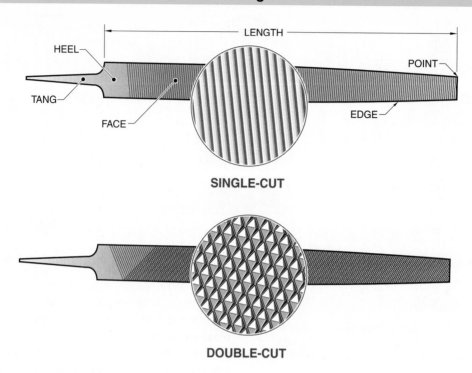

Figure 2-9. File cutting faces can be single cut or double cut.

A *flat file* is a file that is used to file flat surfaces as well as for other operations that require a fast-cutting file. A *mill file* is an all-purpose, single-cut file especially adapted for finish filing. A *knife file* is a file with a blade cross-section tapering from a square to a pointed edge. A knife file is used to smooth areas enclosed by an acute angle. It is suited for finishing sharp corners of grooves and slots where other files cannot fit. A *three-square file (three-cornered file)* is a file that has angles of 60° and is used for filing internal angles, clearing out corners, etc. A file card or brush with a scorer is used to remove particles that clog files.

Taps. A *tap* is a tool used to cut internal threads in a predrilled hole. Taps have a helical design, and are self-feeding once engaged into the material to be tapped (cut). Taps are used with a tap wrench to "pull" the tap into a workpiece. Taps are used to cut female threads into a rigid metal or plastic workpiece. The most common types of taps used for industrial or mechanical applications are the taper tap, plug tap, and bottom tap. **See Figure 2-11.**

A *taper tap* is a tap with a long, gradual taper that allows the tap to start easily. Taper taps are used for threading small-diameter holes and alloy steels because they cause less breakage. The taper also allows for a gradual cutting action, which keeps the tap square with a drilled hole.

A *plug tap* is a tap that is used after a taper tap to start a true and straight thread. Plug taps (intermediate taps) are used for through tapping and are used as the second tap in bottom tapping. Plug taps have good cutting capabilities due to their slight starting taper.

A *bottom tap* is a tap designed to cut threads at the bottom portion of a hole. Bottom tapshould not be used for cutting threads in an unthreaded hole. A lack of a taper at the cutting edge greatly lessens the starting capability of a bottom tap.

Tech Tip

When working with files, it is important that the object being filed is held firmly or in a vise so there is no movement. Movement of the object can cause the teeth of the file to cut down the object in an unsatisfactory way or even damage the file.

File Shapes

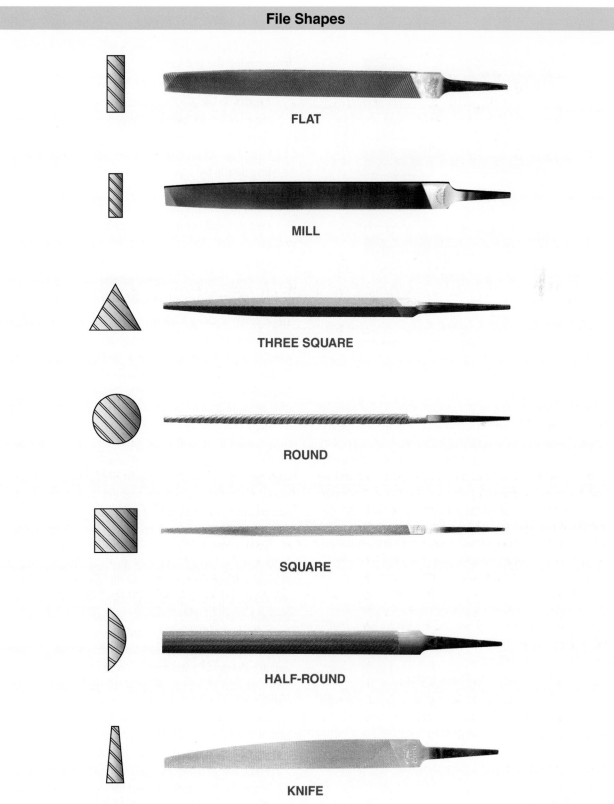

FLAT

MILL

THREE SQUARE

ROUND

SQUARE

HALF-ROUND

KNIFE

Cooper Industries, Inc.

Figure 2-10. Different files are used for various metalworking applications.

Taps

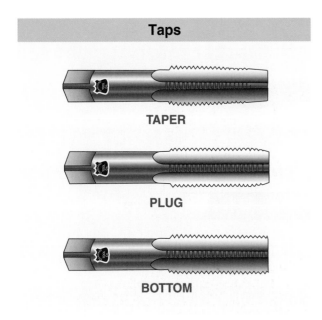

TAPER

PLUG

BOTTOM

Figure 2-11. Taps have a helical design, and are self-feeding once engaged into the material to be tapped (cut).

Cutting metal with taps requires the use of proper cutting fluids to reduce friction and to achieve cleanly formed threads. Friction due to improper use can cause cutting edges of taps to chip, break, or shatter. A properly tapped hole begins with a properly drilled hole and is completed by determining the number of threads per inch and diameter of the finished tapped hole. A tap-drill size chart can be used for this determination. **See Appendix.** When a tap-drill size chart is not available, tap-drill size is determined by applying the formula:

$$TD = MD - \frac{1}{N}$$

where
TD = tap-drill size
MD = major diameter
N = number of threads per inch

For example, what is the tap-drill size for a ¼″-20 threaded hole?

$$TD = MD - \frac{1}{N}$$
$$TD = 0.250 - \frac{1}{20}$$
$$TD = 0.250 - 0.050$$
$$TD = \mathbf{0.200''}$$

Checking the drill sizes on a drill index indicates that the tap drill for a ¼″-20 hole is a number 7 drill (0.201″). Tapping a hole with a taper tap is completed by applying the procedure:

1. Drill a hole perpendicular to the work surface. **See Figure 2-12.**
2. Place a taper tap in a tap wrench ("T" handle wrench).
3. Hold tap wrench square to the surface and start cutting action by applying cutting fluid and turning the tap clockwise.
4. After about two turns, back off the tap about a half-turn to break away metal turning chips.
5. Repeat steps 3 and 4 until hole is tapped to specifications.

Note: To prevent loading the hole with metal chips and cutting fluid, it may be necessary to periodically remove the tap to clean out the chips.

Dies. A *die* is a tool used to cut external threads on round rods. Dies are available in a variety of designs and are usually made from carbon tool steel or high-speed steel. Different types of dies are available for producing either English or metric threads. The most common types of dies are round or hexagonal, and are typically available as single piece design. **See Figure 2-13.** Dies are placed in a diestock (handle) for hand threading. The depth of the thread cuts can be adjusted with one or more adjustment screws in the die or die holder. Dies have a side with a 45° chamfer to prevent die-tooth breakage and to allow for a gentle cutting start. Dies have holes adjacent to the cutting threads to allow for the removal of metal turning chips.

When adjusting a die, the die can shatter if the adjusted opening is too large. If the adjusted opening is too small, the die can damage the workpiece and cause thread breakage. Dies without an adjustment screw are preset by the manufacturer and generally do not require adjustment. Hexagonal dies are used for tough cutting and are designed with thicker cross sections than round dies to permit cleaning and rethreading of threads (thread chasing). Hexagonal dies used for rethreading purposes are not to be used for cutting new threads on un-threaded material. Cutting external threads by hand with a single-piece die is completed by applying the procedure:

1. Place die in diestock.
2. Place the 45° starting chamfer side of the die on end of workpiece to be threaded.
3. Align die squarely with the starting point of the workpiece.

4. Apply cutting fluid, and rotate die clockwise (counterclockwise for left-hand threads). Cutting fluid is used to disperse heat and assist in the formation and removal of metal chips.

5. Apply equal pressure to both sides of the die to verify a straight cut.

6. Repeat steps 4 and 5 until workpiece is threaded to specifications.

Note: To prevent chipped die threads, avoid turning the die completely against the shoulder (bottom) of the workpiece.

Chisels. A *chisel* is a hand tool with a cutting edge on one end that is used to shape, dress, or work wood, metal, or stone. Chisels are commonly driven with a mallet or hammer. Chisel types used for industrial and mechanical applications include flat cold, cape, round nose, and diamond point. **See Figure 2-14.** A *flat cold chisel* is a chisel with a tempered cutting edge to maintain durability. A flat cold chisel is typically used for cutting sheet metal, rivets, and bolts, and in chipping operations. A *cape chisel* is a chisel with a thin, tapered face and a narrow cutting edge. Cape chisels are used for cutting slots, grooves, and keyways in deep corners of metal. A *round-nose chisel* is a chisel that has a round nose and is used for roughing out the concave surfaces of corners. Round-nose chisels are also used for cutting grooves. A *diamond-point chisel* is a chisel with a V-shaped blade that is less than 180°. Diamond-point chisels are used for cutting V-shaped grooves, for chipping corners, and sometimes for removing bolts where the heads have broken off.

Tech Tip

Powered tapping tools are available to help reduce worker fatigue and increase tapping accuracy for projects that require a high number of tapped products. Powered tapping machines include electric tap guns, power drill attachments, and hand tappers that can be attached to a workbench.

Hand Tapping

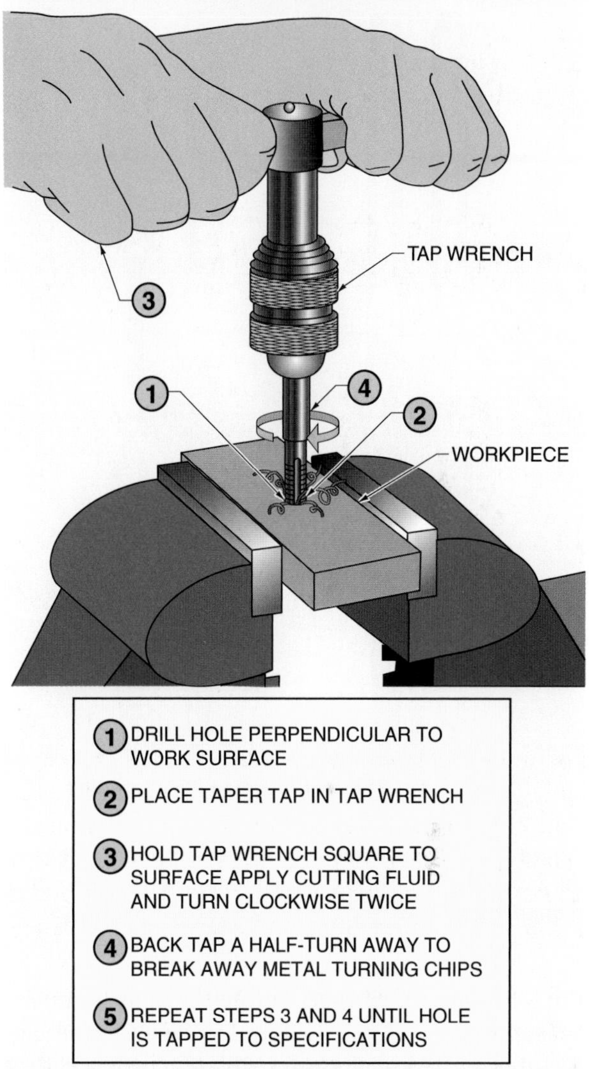

① DRILL HOLE PERPENDICULAR TO WORK SURFACE

② PLACE TAPER TAP IN TAP WRENCH

③ HOLD TAP WRENCH SQUARE TO SURFACE APPLY CUTTING FLUID AND TURN CLOCKWISE TWICE

④ BACK TAP A HALF-TURN AWAY TO BREAK AWAY METAL TURNING CHIPS

⑤ REPEAT STEPS 3 AND 4 UNTIL HOLE IS TAPPED TO SPECIFICATIONS

Figure 2-12. Hand tapping is performed by holding a tap square to a predrilled hole and turning in a clockwise motion.

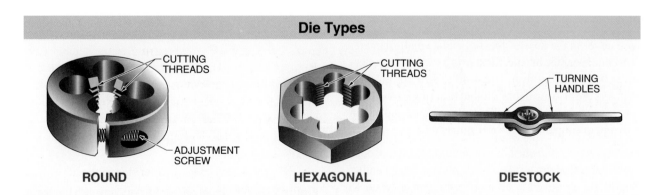

Die Types

Figure 2-13. Each type of die has features designed for specific threading applications.

Chisels

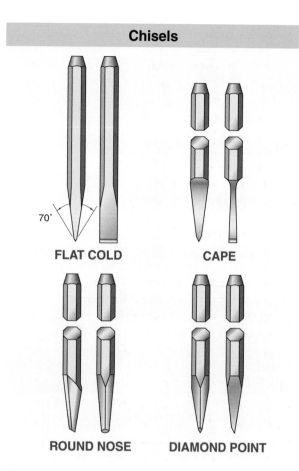

70°

FLAT COLD　　**CAPE**

ROUND NOSE　　**DIAMOND POINT**

Figure 2-14. Chisel types used for industrial and mechanical applications include flat cold, cape, round nose, and diamond point.

Wire Stripper/Crimper/Cutters. A *wire stripper/ crimper/cutter* is a tool used for the removal of insulation from small-diameter wire. Most wire strippers strip stranded wire from American Wire Gauge (AWG) size 22 to AWG size 10 and solid wire from AWG size 18 to AWG size 8. A wire stripper/crimper/cutter is also used to crimp wire connection terminals from AWG size 22 to AWG size 10. Most modern models are also designed with a wire cutter in the nose of the tool and a small-diameter bolt cutter near the handle. Most stripper/crimper/cutters shear bolts ranging from size 4-40 to size 10-32. Wire strippers are also available without the crimper and bolt-shear functions and are smaller and easier to handle than a stripper/crimper/cutter. **See Figure 2-15.**

Handsaws. A *handsaw* is a woodcutting hand tool consisting of a straight, toothed blade attached to a handle. The blade is moved back and forth against wood to produce cutting action. The main parts of a handsaw are the blade (including the toe and heel of the blade),

Wire Strippers

WIRE STRIPPER/CRIMPER/CUTTER

WIRE STRIPPER

Greenlee Textron, Inc.

Figure 2-15. Wire stripper/crimper/cutters are used for removing insulation from and cutting small-diameter wire.

teeth, back, and handle. **See Figure 2-16.** Although the basic construction of all types of handsaws is similar, there are differences in the length and shape of the blade and the number and shape of the teeth. Most handsaws have a straight back, but skewback saws (curved-back saws) are also available.

Handsaws

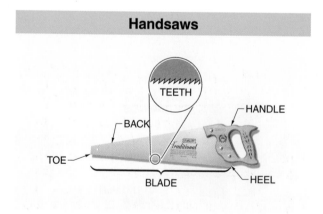

TEETH

HANDLE

BACK

TOE

BLADE

HEEL

Figure 2-16. The main parts of a handsaw are the blade (including the toe and heel of the blade), teeth, back, and handle.

The cut made by a handsaw is wider than the thickness of a saw blade to allow the blade to move freely through the material being cut. If the saw blade cannot move freely, the wood fibers pressing against the blade cause the saw to bind, making the cutting action difficult. Handsaw teeth are alternately bent from side to side, which provides a wider cut than the blade thickness (kerf).

High-quality handsaw blades are tapered from the blade top to the blade bottom. The top portion of the blade is thinner than the blade at the cutting edge, requiring less set in the teeth. A handsaw usually has a number printed on its blade indicating the number of teeth or points per inch. The lower the number, the larger the teeth, and the rougher the cut that is made. For example, an 8-point saw has larger teeth than an 11-point saw and produces a rougher cut than an 11-point saw. **See Figure 2-17.** When using a handsaw, position the workpiece face up to avoid splintering on the face. The back of the workpiece splinters due to the cutting action of the blade.

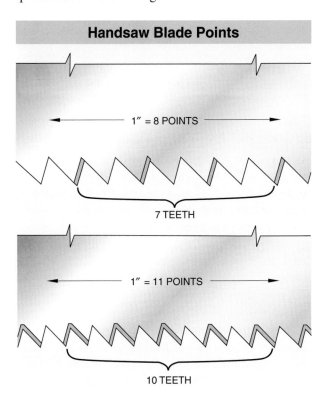

Handsaw Blade Points

1″ = 8 POINTS

7 TEETH

1″ = 11 POINTS

10 TEETH

Figure 2-17. An 8-point saw has larger teeth than an 11-point saw and produces a rougher cut than an 11-point saw.

The two types of cuts that can be made are cross cuts and rip cuts. A *cross cut* is a cut that is made against the direction of the wood grain and is made with full, even strokes at about a 45° angle. A *rip cut* is a cut that is made with the direction of the wood grain and is made with full, even strokes at about a 60° angle. Cutting material by hand with a handsaw is completed by applying the procedure:

1. Wear safety glasses or a face shield.
2. Choose a handsaw of the proper size and shape for the material to be cut.
3. Check the material to be cut for objects that can buckle or damage the saw, such as nails, screws, staples, and knots.
4. Hold the material to be cut firmly against a support such as a table, vise, or bench.
5. Hold the handsaw beside the desired cut location with the thumb upright on the saw handle while applying pressure against the blade.
6. Carefully and slowly start the cut by pulling the blade upward until the blade creates a starting cut.
7. With a partial cut made, set handsaw to proper cutting angle. Sawing is accomplished with pressure applied on the down stroke only.

Note: If the material being cut is long, a wedge can be used to spread the material apart to prevent blade bind and to assist in blade cleaning.

Hacksaws. A *hacksaw* is a metal-cutting hand tool with an adjustable steel frame for holding various lengths and types of blades. Blades are inserted with the teeth pointing away from the handle. Hacksaws can be straight handle or pistol-grip handle. **See Figure 2-18.** A straight-handle hacksaw is usually preferred for fine work. Either type of frame is adjustable for various blade lengths. Tension is applied to the blade to make it taut by means of a wing nut on the pistol-grip frame or by turning a threaded handle on the straight-handle hacksaw. Hacksaws are used to cut workpieces such as metal pipe, PVC pipe, conduit, small-diameter metal rods, rigid plastics, bolts, and nails.

Power Tools

While much of the work performed by industrial and mechanical technicians can be completed with hand tools, many tasks require the use of power tools. Tasks that usually require the use of power tools include drilling, cutting, and sawing. Most power tools can operate with either an AC power source or a DC power source (battery). Power tools that operate with a DC power source are also known as cordless tools. Cordless tools are used because of their mobility but typically do not have as much power as power tools that operate on AC power. Another disadvantage of cordless tools is that their batteries can become run down before the project is completed.

Projects that require a large amount of drilling, cutting, or sawing typically require the use of a tool powered by an AC power source. The most common power tools used by industrial and mechanical technicians include drills, circular saws, portable band saws, and reciprocating saws.

Hacksaws

STRAIGHT HANDLE

PISTOL-GRIP HANDLE

The Stanley Works

Figure 2-18. Common hacksaw designs are straight handle and pistol-grip handle.

Power Drills. A *power drill* is a power-driven rotary tool used with a bit with cutting edges for boring holes in materials such as wood, metal, or plastic. Some types of power drills, such as hammer drills, can drill at speeds up to 3000 rpm and simultaneously hammer into the material at up to 50,000 blows per minute. Hammer drills are used by industrial and mechanical technicians to drill holes in concrete walls, floors, and ceilings. Power drills can also be retrofitted with different types of attachments for operations such as screwing, grinding, and buffing. **See Figure 2-19.**

Circular Saws. A *circular saw* is a handheld or table-mounted power saw with teeth around the circumference of a circular blade that is rotated at high speed on a central axis or shaft. Circular saws are primarily used to cut wood but can also be used to cut certain types of plastic and metal. Circular saws are used to cut lumber or wood panels to length and width. A circular saw blade turns in a counterclockwise direction, cutting the material from the underside through the top. Circular saws are equally efficient for crosscutting and ripping lumber and panel products, and can be adjusted to cut

angles ranging from 45° to 90°. The size of a circular saw is based on the largest-diameter blade that can be properly installed. Circular saw blade diameters range from 4½″ to 12″, with 7¼″-, 7½″-, or 8¼″-diameter blades being the most commonly used. A variety of saw blades are available for different operations required for cutting wood and other materials.

Power Drills

Milwaukee Electric Tool Corp.

Figure 2-19. Power drills can be fitted with different types of attachments for operations such as screwing, grinding, and buffing.

Portable Band Saws. A *portable band saw* is a handheld power saw that has a flexible metal saw blade that forms a continuous loop around two parallel pulleys. Portable band saws use a continuous cutting action to slice through metals. These saws are used in the field to cut all-thread rod, strut channel, pipe, and other metal components. They come in both corded and cordless versions.

Most portable band saws have two speeds. The lower speed is used when cutting hard materials, and the higher speed is used when cutting soft materials. Portable band saws should be held firmly with both hands, and at least three teeth should be kept in the cut. When using a portable band saw, it is important to let the blade do the cutting and to not put too much pressure on the material with the blade. Excessive pressure can cause the blade to bind or break, leading to tool damage or worker injury.

Portable band saw blades are available in several pitches (teeth per inch). Using a blade with the correct pitch for the material being cut will extend the blade life.

The size, shape, and material being cut should be considered when selecting the proper blade. Soft materials require coarse pitch blades, and hard materials require fine pitch blades. Furthermore, thick materials require coarse pitch blades, and thin materials require fine pitch blades. **See Figure 2-20.**

Reciprocating Saws. A *reciprocating saw* is a multipurpose cutting tool in which the blade reciprocates (moves back and forth) to create the cutting action.

Reciprocating saw blades can be plunged directly into walls, floors, ceilings, and other resilient material. Reciprocating saws operate at 1700 to 2800 strokes per minute (at no load) and are used by industrial and mechanical technicians to make cuts into finished walls, ceilings, floors, and any other rigid materials. **See Figure 2-21.** Reciprocating saws can also be retrofitted with different types of saw blades for cutting small-diameter plastic or metal pipes, rods, and tubing.

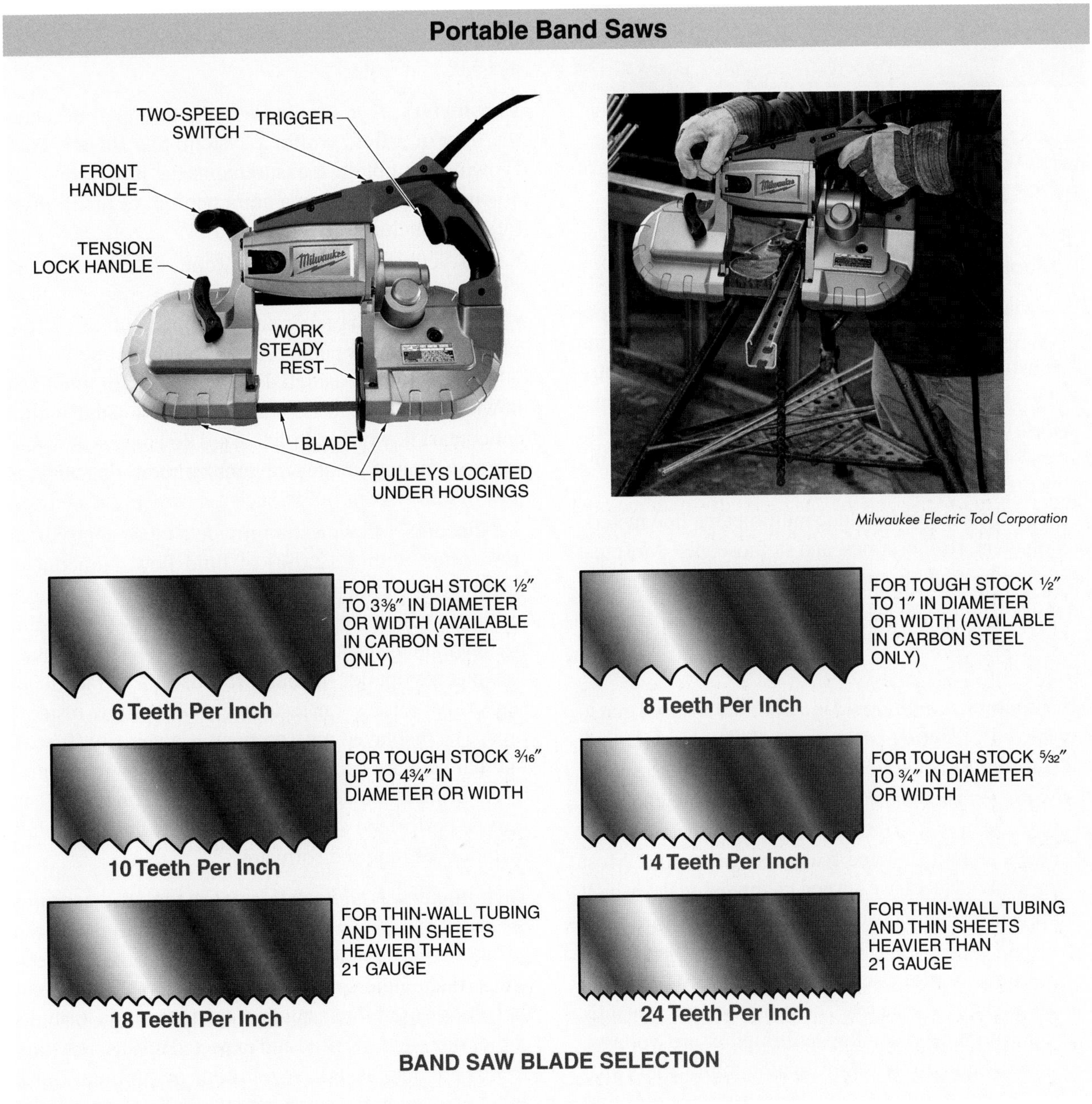

Portable Band Saws

TWO-SPEED SWITCH
TRIGGER
FRONT HANDLE
TENSION LOCK HANDLE
WORK STEADY REST
BLADE
PULLEYS LOCATED UNDER HOUSINGS

Milwaukee Electric Tool Corporation

6 Teeth Per Inch — FOR TOUGH STOCK ½″ TO 3⅜″ IN DIAMETER OR WIDTH (AVAILABLE IN CARBON STEEL ONLY)

8 Teeth Per Inch — FOR TOUGH STOCK ½″ TO 1″ IN DIAMETER OR WIDTH (AVAILABLE IN CARBON STEEL ONLY)

10 Teeth Per Inch — FOR TOUGH STOCK 3/16″ UP TO 4¾″ IN DIAMETER OR WIDTH

14 Teeth Per Inch — FOR TOUGH STOCK 5/32″ TO ¾″ IN DIAMETER OR WIDTH

18 Teeth Per Inch — FOR THIN-WALL TUBING AND THIN SHEETS HEAVIER THAN 21 GAUGE

24 Teeth Per Inch — FOR THIN-WALL TUBING AND THIN SHEETS HEAVIER THAN 21 GAUGE

BAND SAW BLADE SELECTION

Figure 2-20. Portable band saws use a continuous cutting action to slice through metals.

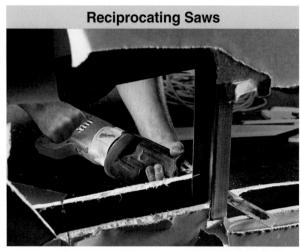

Reciprocating Saws

Milwaukee Electric Tool Corp.

Figure 2-21. Reciprocating saws are used by industrial and mechanical technicians to make cuts into finished walls, ceilings, and floors.

Test Tools

Industrial and mechanical technicians use various types of test tools when performing work. Many test tools are used to measure and record quantities involving electrical current, motion, distance, speed, and temperature. During tests, components can provide signals or responses that technicians can use while troubleshooting or recordkeeping.

Types of test tools include multimeters, flowmeters, tachometers, stroboscopes, and stethoscopes. Most test tools are designed to measure and display a particular quantity with the greatest amount of accuracy. The accuracy of a measurement is always within manufacturer specified limits.

Multimeters. A *multimeter* is a portable test tool that is capable of measuring two or more electrical quantities. Multimeters are either analog or digital.

An *analog multimeter* is a meter that can measure two or more electrical quantities and display the measured quantities along calibrated scales using a needle. Most analog multimeters have several calibrated scales which correspond to different selector switch settings (AC, DC, or R) and jack usage (mA jack or 10 A jack). **See Figure 2-22.**

It is important to use the correct scale when reading an analog multimeter. The most common quantities measured with analog multimeters are voltage, resistance, and current. Many analog multimeters also include scales and ranges for measuring decibels and checking batteries. The decibel scale is a scale that

indicates the comparison, or ratio, of two or more signal powers or voltages. The battery tester function allows the multimeter to display the charged condition (25%, 50%, 75%, or 100%) of a battery being tested.

A *digital multimeter (DMM)* is a meter that can measure two or more electrical quantities and display the measured quantities as numerical values. **See Figure 2-23.** Basic DMMs can be used to measure voltage, current, and resistance. Advanced DMMs include special functions, such as functions for measuring capacitance and temperature. The main advantages of a DMM over an analog multimeter are both the ability to record measurements as well as a digital display that is easier to read than a needle and scale.

Flowmeters. A *flowmeter* is a test tool that measures the flow of a fluid within a system. **See Figure 2-24.** Flowmeters obtain measurements as flow rates (the amount of fluid passing a point per unit of time). Flow rate units include the following:

- Gas—cubic feet per minute (cfm) or cubic feet per hour (cfh)
- Liquid—gallons per minute (gpm), cubic feet per second (cfs), or cubic feet per minute (cfm)

Flow can be measured directly with an in-line fixed flowmeter, or it can be closely approximated with a noncontact flowmeter. In-line fixed flowmeters are more accurate but may create problems by interfering with the flow of less viscous fluids, such as powders and slurries.

Noncontact flowmeters provide a close approximation (often within ±2.0%) of fluid flow. Noncontact flowmeters are ideal for testing or troubleshooting a system in which the indication of fluid flow is more important than the exact amount of fluid flow. Noncontact flowmeters operate through the use of a transducer sensor that allows a measurement of the flow inside a pipe. The displayed measurement is a velocity (ft/sec), but that velocity can be converted to a flow rate (gpm) by referring to manufacturer charts. The conversion can also be performed automatically by the flowmeter if it includes the required software.

Tachometers. A *tachometer* is a test tool that measures the speed of a moving object. Speed measurements can be taken using contact tachometers, photo tachometers, strobe tachometers, or laser tachometers. The type of tachometer used for an application depends on the units of measurement required and expected results. Rotating speeds are displayed in revolutions per minute (rpm), and linear speeds are displayed in feet per minute (ft/min, fpm) or meters per minute (m/min, mpm).

Analog Multimeters

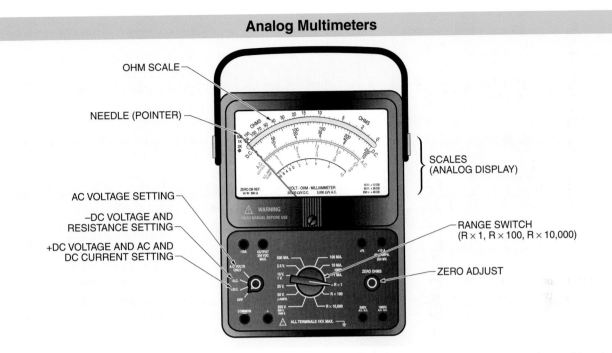

OHM SCALE

NEEDLE (POINTER)

SCALES
(ANALOG DISPLAY)

AC VOLTAGE SETTING

−DC VOLTAGE AND
RESISTANCE SETTING

+DC VOLTAGE AND AC AND
DC CURRENT SETTING

RANGE SWITCH
(R × 1, R × 100, R × 10,000)

ZERO ADJUST

Figure 2-22. Analog multimeters can measure two or more electrical quantities and display the measured quantities along calibrated scales using a needle.

Digital Multimeters (DMMs)

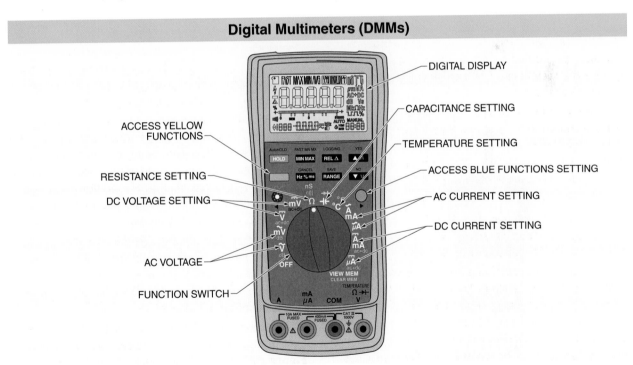

DIGITAL DISPLAY

CAPACITANCE SETTING

ACCESS YELLOW
FUNCTIONS

TEMPERATURE SETTING

ACCESS BLUE FUNCTIONS SETTING

RESISTANCE SETTING

DC VOLTAGE SETTING

AC CURRENT SETTING

DC CURRENT SETTING

AC VOLTAGE

FUNCTION SWITCH

Figure 2-23. DMMs can measure two or more electrical quantities and display the measured quantities as numerical values.

A *contact tachometer* is a test tool that measures the rotational or linear speed of an object through direct contact with the object. **See Figure 2-25.** Contact tachometers are used to measure the rotational speed of objects such as motor shafts, gears, belts, and pulleys, as well as linear speeds of moving conveyors and press webs. Contact tachometers measure speeds from 0.1 rpm to about 25,000 rpm.

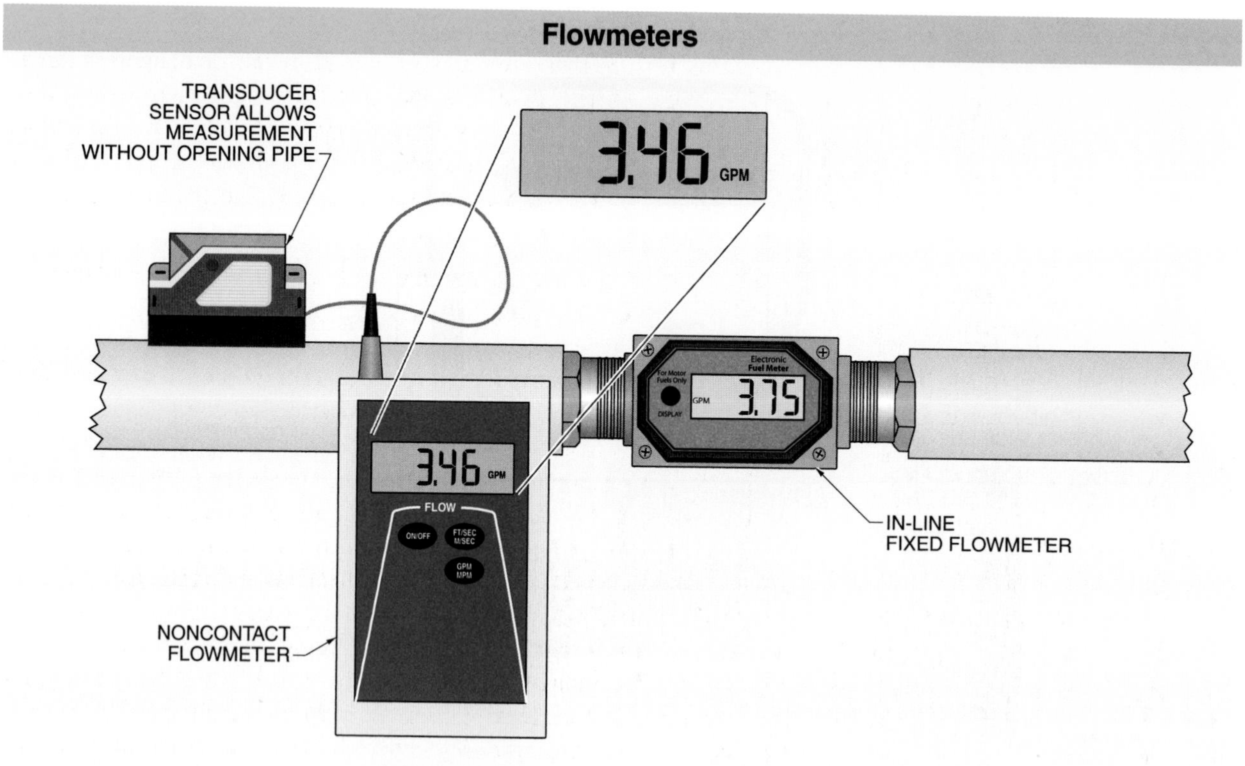

Flowmeters

TRANSDUCER
SENSOR ALLOWS
MEASUREMENT
WITHOUT OPENING PIPE

3.46 GPM

Electronic
Fuel Meter

For Motor
Fuels Only

GPM **3.75**

DISPLAY

3.46 GPM

FLOW

ON/OFF
FT/SEC
M/SEC

GPM
MPM

IN-LINE
FIXED FLOWMETER

NONCONTACT
FLOWMETER

Figure 2-24. Fluid flow can be measured directly with an in-line fixed flowmeter or can be closely approximated with a noncontact flowmeter.

Contact Tachometers

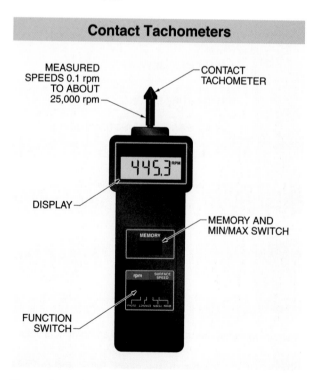

MEASURED
SPEEDS 0.1 rpm
TO ABOUT
25,000 rpm

CONTACT
TACHOMETER

445.3 RPM

DISPLAY

MEMORY

MEMORY AND
MIN/MAX SWITCH

rpm
SURFACE
SPEED

FUNCTION
SWITCH

PHOTO TURN CONTACT MIN/MAX

Figure 2-25. Contact tachometers measure the rotational or linear speed of an object, such as a motor shaft or gear, through direct contact with the object.

A *photo tachometer* is a test tool that uses light beams to measure the rotational speed of an object. **See Figure 2-26.** A photo tachometer measures speed by focusing a light beam on a reflective area or a piece of reflective tape on a rotating object and counting the number of reflections per minute. A photo tachometer is used in applications in which the rotating object cannot be reached or touched. It can measure speeds from 1 rpm to about 100,000 rpm, but it cannot measure linear speeds.

A *strobe tachometer*, similar to a photo tachometer, is a test tool that measures the rotational speed of an object by use of a flashing (strobe) light. Strobe tachometers are used to test shaft speeds for motors or pumps; abnormal vibrations; and alignment of components such as cams, shafts, couplings, and gears.

A *laser tachometer* is a test tool that uses a laser light to measure the rotational speed of an object. **See Figure 2-27.** A laser tachometer measures speed by focusing a laser light beam on a reflective area or a piece of reflective tape on a rotating object and counting the number of reflections per minute. Laser tachometers work well in areas of high ambient light because red laser light is easier to see than the white light emitted by photo tachometers.

Photo Tachometers

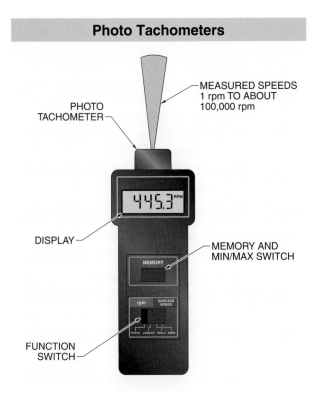

Figure 2-26. Photo tachometers use light beams to measure the speed of rotating objects.

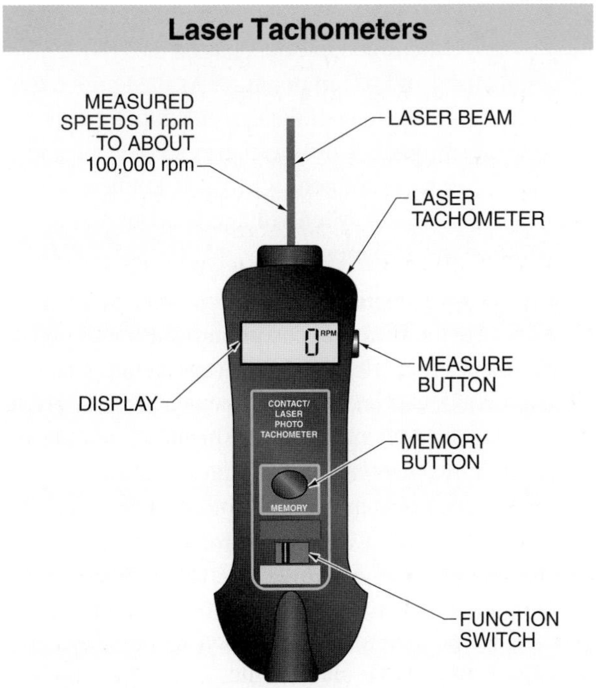

Laser Tachometers

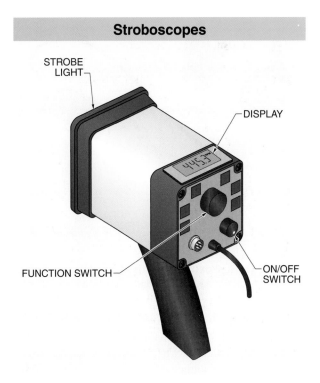

Figure 2-27. Laser tachometers can be used to measure the rotational speed of objects in hard-to-reach areas.

Laser tachometers can also be used to make speed measurements in hard-to-reach or dangerous areas. Because a laser light is a concentrated beam, laser tachometers are better than photo tachometers when making measurements in confined areas. Similar to photo tachometers, laser tachometers measure speeds from 1 rpm to about 100,000 rpm.

Stroboscopes. A *stroboscope* is a test tool that can capture a motionless image of a moving object for ease of inspection by use of a flashing light, or strobe light. **See Figure 2-28.** Stroboscopes operate by using an electronic pulse generator to control microsecond flashes of light to illuminate a moving object. To capture a motionless image, the frequency of the stroboscope flash has to match the object's frequency of motion. When the frequencies match, the flash creates an image of the object that appears still, or frozen in place, which allows for inspection of the component during operation. Stroboscopes can help technicians inspect slippage between shafts, conditions of belts and gears, and the alignment of couplings at all operating speeds. Stroboscopes, like strobe tachometers, can also measure rotational speed in revolutions per minute.

Stroboscopes

Figure 2-28. Stroboscopes capture motionless images of moving objects for ease of inspection by use of a flashing light, or strobe light.

CAUTION: Improper use of a stroboscope can cause serious harm due to the high frequency and brightness of the flashing light. The flashing light of a stroboscope can cause seizures or blackouts in personnel who are photosensitive or have an epileptic condition. Care should be taken so the strobe light does not enter the eye.

Stethoscopes. A *stethoscope* is a test tool used for detecting and locating abnormal noises within machines or equipment. Defective components, such as worn conveyor parts, degraded bearings, chipped gear teeth, and fluid leaks, often create abnormal noises. Stethoscopes can help locate defective components such as these by detecting the abnormal noises they generate. Abnormal noises made by moving components are significantly amplified with the use of a stethoscope. Environmental noise is reduced by using an insulated headset. **See Figure 2-29.**

Stethoscopes

Figure 2-29. Stethoscopes are used for detecting and locating abnormal noises within machines and components.

Industrial facilities generate various sounds that are not easily distinguishable or familiar. These sounds can even be problematic. By using an industrial stethoscope, unfamiliar sounds can be traced and located by an industrial technician. Once a sound is located, the technician can then start to further test and troubleshoot the component or machine.

Temperature Test Tools

Temperature test tools are used to measure the intensity of heat in certain types of equipment. Temperature measurements can be taken using contact or noncontact test tools. A *contact temperature probe* is a test tool that measures temperature at a single point through direct contact with the area being measured. Contact temperature probes use conduction to monitor temperature changes. A *noncontact temperature probe* is a test tool that measures temperature using convection or radiation.

There are many types of temperature test tools. The most common are thermocouples, change-of-state sensors, and infrared thermometers.

Thermocouples. A *thermocouple* is a device that produces electricity by heating two different metals that are joined together. Thermocouples are contact temperature probes. **See Figure 2-30.** Thermocouples have a wide temperature range from 328°F to over 1112°F.

Thermocouples are considered thermoelectric voltage devices and operate by producing voltage from the connection of two dissimilar metals (iron and constantan) as they are heated by the component being tested. The reaction at the junction of the two dissimilar metals produces a potential difference of a few millivolts (mV), normally about 1 mV to 50 mV. The amount of voltage produced by a thermocouple depends on the type of thermocouple used and its temperature. As the temperature rises, the output voltage of the thermocouple increases. The voltage produced by the thermocouple provides a temperature measurement when it is connected to a temperature meter or when voltage is converted into a temperature measurement.

Change-of-State Sensors. Change-of-state sensors, or phase-change temperature sensors, indicate the condition of equipment affected by a change in temperature. Change-of-state sensors are nonelectric devices available as labels, pellets, crayons, or liquid crystals. The most common are temperature-sensitive labels. **See Figure 2-31.**

Temperature-sensitive labels are labels that can be applied to certain devices to indicate temperature change. Temperature-sensitive labels are constructed of a temperature-sensitive material sealed behind a transparent heat-resistant material and in front of an adhesive backing. As the label is heated, the sensitive material dissolves and is absorbed by the adhesive backing, causing a color change. As a result, when a certain temperature is reached, the appearance of the label changes.

Thermocouples

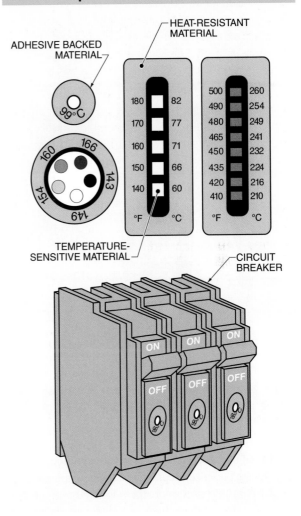

TEMPERATURE
MODULE

SET FOR
°F OR °C

THERMOCOUPLE

COMPONENT
MEASURED

MEASURING
(HOT) JUNCTION

IRON

CONSTANTAN

HEAT

REFERENCE
(COLD) JUNCTION

V_1

VOUT

J_1 J_2 V_2

**THERMOCOUPLE
CONSTRUCTION**

Figure 2-30. Thermocouples are used to measure temperatures at certain points of contact.

Temperature-Sensitive Labels

ADHESIVE BACKED
MATERIAL

HEAT-RESISTANT
MATERIAL

99°C

160   166
154   143
149

180	82
170	77
160	71
150	66
140	60
°F	°C

500	260
490	254
480	249
465	241
450	232
435	224
420	216
410	210
°F	°C

TEMPERATURE-
SENSITIVE MATERIAL

CIRCUIT
BREAKER

ON   ON   ON

OFF   OFF   OFF

Figure 2-31. Change-of-state sensors are nonelectric sensors that are available as temperature-sensitive labels.

Normally, these labels contain white or yellowish dots that change to black when temperatures are exceeded. Once the color has changed, it remains for the remainder of the label's use. Temperature-sensitive labels can be applied to components such as circuit breakers or temperature-sensitive equipment that is monitored often.

Infrared Thermometers. All materials or components emit infrared radiation in proportion to their surface temperatures. The higher an object's temperature, the greater the amount of infrared radiation emitted. An *infrared thermometer* is a handheld device that detects infrared emissions to measure temperature. Infrared (IR) thermometers are commonly referred to as noncontact temperature probes. **See Figure 2-32.**

Infrared Thermometers

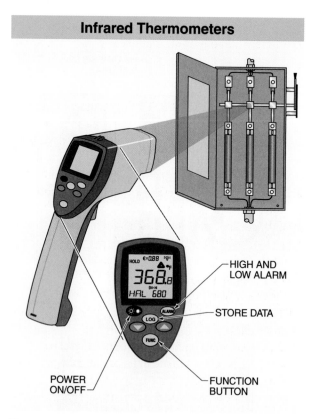

HIGH AND LOW ALARM

STORE DATA

POWER ON/OFF

FUNCTION BUTTON

Figure 2-32. An infrared thermometer is used as an inspection tool to check for the presence of a component malfunction.

Infrared thermometers are used as inspection tools to check for the presence of component malfunctions or to establish a baseline temperature for monitored components. Baseline temperature readings of circuit components can be scheduled and recorded. Any change in a baseline temperature can warrant further investigation. Infrared thermometers are commonly used to take measurements of electrical distribution systems, motors, bearings, switching circuits, and any other equipment where excessive heat can cause detrimental effects during operation.

With recent advancements in technology, infrared thermometers can provide advantages that other leading test tools cannot. For example, IR thermometers can measure temperatures far below freezing to well over 2000°F. IR thermometers also have a fast response time, allowing for correct temperature measurements in fractions of a second. The temperature of equipment that is too difficult to reach or access is easily measured with a clear line of sight, clean optics, and no dust or smoke surrounding the equipment. And since they are noncontact temperature probes, IR thermometers are less likely to have adverse effects on equipment.

The equipment is less likely to be damaged; there is less risk of contamination; and the temperature of the equipment is less likely to be changed by the thermometer.

TOOL SAFETY

Industrial and mechanical technicians must use hand tools, power tools, and test tools to successfully perform most tasks. To avoid personal injury and damage to the tools or equipment, proper safety precautions must be followed. A precaution that applies to hand, power, and test tools is the use of personal protective equipment. *Personal protective equipment (PPE)* is clothing and/or equipment worn by a technician to reduce the possibility of injury in the work area. The most common PPE used in industrial environments includes protective clothing, items worn for head protection, eye protection, hand protection, and foot protection. Additional PPE can include items worn for knee protection (knee pads), ear protection (ear plugs), and face protection (face shields). In addition to the use of PPE, separate precautions must be taken for hand tool, power tool, and test tool safety.

Hand Tool Safety

Hand tools are designed for specific purposes. Attempting to use a hand tool for a purpose other than what it is designed for or using it improperly can result in problems, such as incomplete projects, damaged or destroyed tools, damaged components, or injuries to workers.

According to the Bereau of Labor Statistics, the misuse of hand tools accounts for over 10% of all emergency room visits each year, many of which can develop into severe disabilities. For example, using a screwdriver on an unstable workpiece can result in tool slippage and possible injury. **See Figure 2-33.** Guidelines for safe use of hand tools include the following:
- Wear proper PPE.
- Dress for safety. When working with hand tools, wear long sleeves, work gloves, and heavy jeans.
- Pull tools rather than push them for greater control and balance.
- Never use a "cheater bar" or other device to increase leverage.
- Release stuck saws by pulling the saw toward the body.
- Fit screwdrivers snugly into the slot of a screw without extending over the edge.
- Work on a solid surface, such as a workbench.

- Never use a screwdriver for prying.
- Never use a screwdriver as a chisel.
- Keep tools clean, oiled, and free of rust.
- Keep drills sharp.
- Remove mushroomed heads on chisels.
- Repair or replace hammers with loose or splintered handles.
- Keep hand tools in a safe place. Each tool should have a place in a tool box or pouch.
- Never carry tools in a pocket unless the pocket is designed for that purpose.
- Never place pencils behind the ear or under a hat or cap.
- Carry sharp-edged cutting tools with the cutting point or edge in a downward direction away from the body.
- Use a safety checklist for hand tools and related equipment. **See Figure 2-34.**

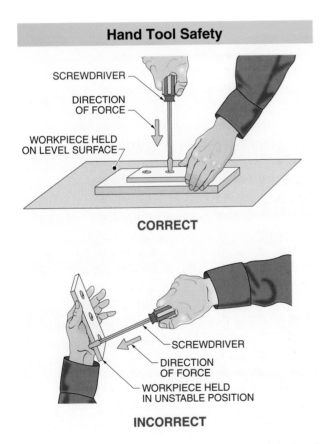

Hand Tool Safety

SCREWDRIVER

DIRECTION OF FORCE

WORKPIECE HELD ON LEVEL SURFACE

CORRECT

SCREWDRIVER

DIRECTION OF FORCE

WORKPIECE HELD IN UNSTABLE POSITION

INCORRECT

Figure 2-33. Improper hand tool use can result in hand injuries.

The most common reasons that technicians use incorrect tools or use tools improperly are inexperience and not following proper procedures (taking shortcuts).

Not following proper procedures is often the result of insufficient training or inexperience.

Power Tool Safety

Power tools use power from electric, pneumatic (air), and hydraulic (fluid) power sources. The advantage of using power tools is that certain tasks are quicker and easier to perform than by manual methods. Disadvantages of using power tools are that, because they can deliver a great deal of energy in a short period, they can cause injuries or errors in workmanship when used incorrectly. The basic safety guidelines that apply to hand tools also apply to power tools. The difference is that with power tools, errors and injuries can occur more quickly and be more severe if safety guidelines are not followed. Guidelines for safe use of power tools include the following:

- Read the manufacturer's manual before using any power tool.
- Have the proper tools available for the project.
- Use job-specific tools with the proper attachments.
- Wear proper PPE.
- Verify that all safety guards are in place and in working order.
- Verify that the tool switch is in the OFF position before connecting a tool to a power source.
- Properly position the tool before applying power.
- Keep the work area clear of personnel, debris, tools, and other items that can present a hazard.
- Do not wear loose clothing.
- Keep attention focused on the work.
- Have power tools inspected and serviced by a qualified service technician at regular intervals as specified by the tool manufacturer or by OSHA.
- Investigate any changes in the sound of the tool during operation.
- Verify that the outer protective jackets of power cords are free of nicks or damage.
- Verify that plugs are mechanically sound and not damaged.
- Verify that flexible power cords are securely fastened to the tool and the plug.
- Have tool maintenance performed at regular intervals and replace damaged or worn tools as required.
- Always work with another technician in hazardous or dangerous locations.
- Clean and lubricate all tools after use.
- Shut power OFF when work is completed.
- Use a safety checklist for power tools and related equipment. **See Figure 2-35.**

Checklist — Hand Tools and Related Equipment

☑ Tools and equipment (both company- and employee-owned) used by employees at their workplace are in good condition.

☑ Hand tools such as chisels and punches, which develop mushroomed heads during use, reconditioned or replaced as necessary.

☑ Broken or fractured handles on hammers, axes, and similar equipment replaced promptly.

☑ Worn or bent wrenches replaced regularly.

☑ Appropriate handles used on files and similar tools.

☑ Employees aware of the hazards caused by faulty or improperly used hand tools.

☑ Appropriate PPE (safety glasses, face shields, etc.) used while using hand tools or equipment which might produce flying materials or be subject to breakage.

☑ Jacks checked periodically to ensure they are in good operating condition.

☑ Tool handles wedged tightly in the head of all tools.

☑ Tool cutting edges kept sharp so the tool will move smoothly without binding or skipping.

☑ Tools stored in dry, secure locations where they won't be tampered with.

☑ Eye and face protection used when driving hardened or tempered studs or nails.

Figure 2-34. A safety checklist can be used for proper care and use of hand tools and related equipment.

Checklist — Power Tools and Related Equipment

☑ Grinders, saws, and similar equipment provided with appropriate safety guards.

☑ Power tools used with the correct shield, guard, or attachment, as recommended by the manufacturer.

☑ Portable circular saws equipped with guards above and below the base shoe.

☑ Circular saw guards checked to ensure they are not wedged up, thus leaving the lower portion of the blade unguarded.

☑ Rotating or moving parts of equipment guarded to prevent physical contact.

☑ All cord-connected, electrically operated tools and equipment are effectively grounded or of the approved double-insulated type.

☑ Protective guards in place over belts, pulleys, chains, sprockets, on equipment such as concrete mixers, and air compressors.

☑ Portable fans provided with full guards or screens having openings ½ inch or less.

☑ Hoisting equipment available and used for lifting heavy objects with hoist ratings and characteristics appropriate for the task.

☑ Ground-fault circuit interrupters provided on all temporary electrical 15 A and 20 A circuits used during periods of construction.

☑ Pneumatic and hydraulic hoses on power tools checked regularly for deterioration or damage.

Figure 2-35. A safety checklist can be used to ensure proper care and use of power tools and related equipment.

Grounded Power Tools. All power tools must be properly grounded unless they are approved, double-insulated tools. Power tools with a three-prong plug should be connected to a grounded receptacle. It is very dangerous to use an adapter to connect a three-prong plug to an ungrounded two-hole outlet unless a separate ground wire is connected to an approved grounded common return circuit. The ground ensures that any short will trip the circuit breaker or blow the fuse. **WARNING:** An ungrounded power tool has the potential to cause a fatal accident.

Double-insulated tools have two-prong plugs and are properly identified as such by the manufacturer on the nameplate. The electrical parts of the motor in a double-insulated power tool are provided with extra insulation to prevent electrical shock. Therefore, the tool does not have to be grounded. Both the interior and exterior should be kept clean and free from grease or dirt that might conduct electricity. Manufacturer recommendations should be followed regarding service and maintenance of all power tools.

Test Tool Safety

Test tools are designed for the purpose of testing and troubleshooting equipment in need of inspection or repair. In industrial facilities, the use of test tools in certain locations can raise concerns if the location becomes hazardous. Most test tools are considered "intrinsically safe" or "explosion proof" when operated in these locations.

Intrinsic safety (IS) is an electrical circuit design technique that provides explosion protection by eliminating arcing and heat that could ignite explosive atmospheres. The National Electric Code® (NEC®) defines an intrinsically safe circuit as a circuit in which any spark or thermal energy is incapable of igniting a mixture of flammable or combustible elements in air under prescribed test conditions. Intrinsic safety requirements for electrical equipment limit electrical and thermal energy to levels that are insufficient to ignite an explosive atmosphere.

Test tools that have self-contained power supplies could become an ignition source in hazardous locations. A location can become hazardous if the elements for an explosion are present. An *explosion* is a sudden reaction involving rapid physical or chemical decay accompanied by an increase in temperature, pressure, or both. Explosions occur when three elements meet: a flammable substance, an oxidizer, and an ignition source (energy). **See Figure 2-36.** All three elements must be present simultaneously in order for an explosion to occur. If any of the three are missing, an explosion cannot occur.

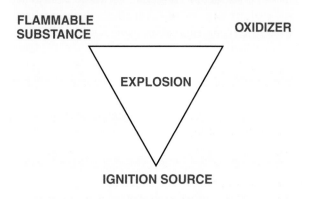

Explosion Requirements

FLAMMABLE SUBSTANCE

OXIDIZER

EXPLOSION

IGNITION SOURCE

Figure 2-36. The elements that must be present in order for an explosive reaction to occur are a flammable substance, an oxidizer, and an ignition source.

The improper use of any test tool, especially in hazardous locations, can be fatal. The manufacturer's manual for each test tool details safety precautions, warnings, and specifications and features for proper operating procedures. General safety precautions required when using test tools include the following:

• Identify hazards that could pose a risk.
• Exercise caution when workplace conditions are poor.
• Use proper PPE.

• Treat all equipment as energized until proven safe.
• Assure all isolating devices are in place and operate properly.
• Do not use unmaintained tools.
• Do not use tools inappropriately.
• Never assume a test tool is operating correctly. Check the test tool for proper operation.

Safety Tip

Precautions should be taken to ensure that a job site is safe and that adequate fire prevention strategies are put into place. Fire extinguishers should be on hand near all operations that pose a risk of fire. Fire extinguishers are classified as A, B, C, D, or K. These classifications indicate the type of combustible material and type of fire that an extinguisher is rated for. When using a fire extinguisher, it should be aimed at the base of a fire.

Hazardous Locations. A *hazardous location*, as defined by OSHA, is a location where flammable liquids, gases, vapors, or combustible dusts exist in sufficient quantities to pose a risk of an explosion or fire. Industries with potentially hazardous locations can include oil and gas refineries, petrochemical plants, sewage-treatment facilities, and spray-painting shops. The NFPA, the NEC®, and the Canadian Electrical Code (CEC) establish hazardous location classifications based on class, division, and groups.

Classes are designated as Class I, Class II, and Class III. They are based on the type of material present in the location. Class I locations contain flammable gases or vapors in sufficient quantities in the atmosphere that pose a risk of explosion or ignition. Class I locations can be found at petroleum processing plants due to the presence of gaseous hydrocarbons. Class II locations contain dust that is either electrically conductive or could be explosive when mixed with air. Class III locations are characterized by the presence of ignitable filings or flyings.

Hazardous locations are further classified into divisions that describe the conditions under which the hazardous material is present. Division I locations contain an explosive or ignitable material that is present under normal operating conditions. Division II locations contain hazardous substances that are handled or stored only under abnormal conditions, such as a location that may experience a containment failure that results in a spill or leak.

Hazardous Locations				
Class	Group	Hazaerdous Material	Autoignition Temperature	
			C°	F°
I	A	Acetylene	305	581
	B	Butadiene	420	788
		Ethylene Oxide	570	1058
		Hydrogen	500	932
	C	Acetaldehyde	175	347
		Cyclopropane	498	928
		Diethyl Ether	180	356
		Ethylene	450	842
		Isoprene	398	743
	D	Acetone	465	869
		Ammonia	651	1204
		Benzene	498	928
		Butane	287	550
		Ethane	472	882
		Ethanol	363	685
		Gasoline	246–280	475–536
		Methane	537	999
		Propane	450	842
		Styrene	490	914
II	E	Aluminum	650	1202
		Bronze	370	698
		Chromium	580	1076
		Magnesium	620	1148
		Titanium	330	626
		Zinc	630	1166
	F	Coal	610	1130
	G	Corn	400	752
		Nylon	500	932
		Polyethylene	450	842
		Sugar	350	662
		Wheat	480	896
		Wheat Flour	380	716

Figure 2-37. Hazardous materials are sorted by class and group.

Hazardous materials are sorted by class and group. Groups A, B, C, and D are used for Class I areas. Groups E, F, and G are used for Class II areas. **See Figure 2-37.** For example, Group A only includes acetylene, a gas that can create a highly dangerous explosion, whereas Group D contains propane, which is also hazardous but less violent than acetylene. The type of material in each group is designated with an autoignition temperature. An *autoignition temperature* is the minimum temperature at which there is sufficient energy for a material or gas to ignite spontaneously, without a spark, flame, or other source of ignition.

While test tools are not defined under the scope of hazardous locations, it is not recommended to use them in Class I locations unless the test tool is specifically recommended for the location. According to the International Society of Automation (ISA)—which provides guidance for the use of general-purpose, portable electronic products in certain hazardous (classified) locations—handheld portable electronic products can be used in Class I Division 2, Class III Division 1, and Class III Division 2 locations.

LADDERS

A *ladder* is a structure consisting of two siderails joined at intervals by steps or rungs for climbing up and down. Ladders are manufactured in lengths of 3′ to 50′. Ladders are constructed of metal or fiberglass. All ladders, regardless of the construction material, are manufactured to meet the same standards. Ladder purchase is generally decided by cost, use, and in some cases, portability. Industrial ladders include fixed, single, extension, and stepladders. **See Figure 2-38.**

All ladders must be used only for the purpose for which they are designed. Ladders must not be used as pry bars or horizontal platforms. All ladders must be equipped with nonslip safety feet such as butt spurs or foot pads. **See Figure 2-39.** A *butt spur* is a notched, pointed, or spiked end of a ladder which helps prevent the ladder butt from slipping. Butt spurs are generally attached to long ladders such as extension ladders. A *foot pad* is a metal swivel attachment with rubber or rubber-like tread which helps prevent the ladder butt from slipping. Foot pads are normally attached to short ladders such as stepladders.

Safety Tip

Temperature measurements can be used to detect a problem or the severity of a problem inside a machine or equipment in order to minimize the threat of fire or explosion.

Ladders

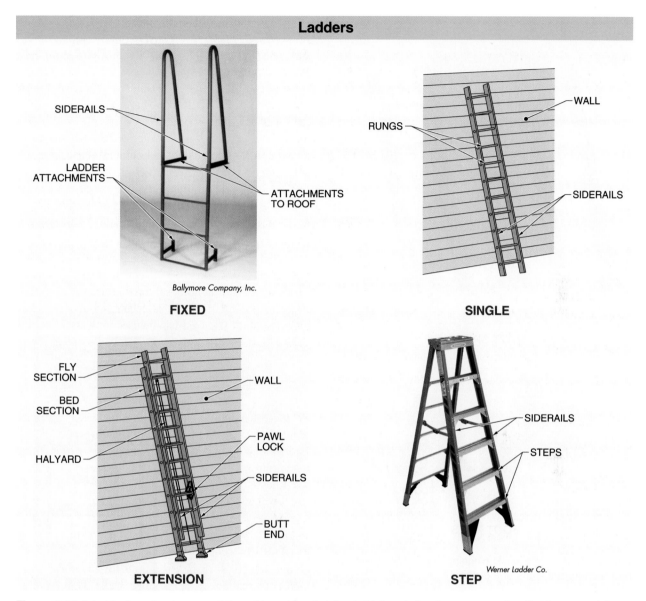

Figure 2-38. A ladder is a structure consisting of two siderails joined at intervals by steps or rungs for climbing up and down.

Metal Ladders

A *metal ladder* is a ladder constructed of metal (normally aluminum). The advantages of metal ladders include relatively light weight, extreme toughness, resistance to splintering or cracking when subjected to impact, and resistance to deterioration with age. These qualities reduce maintenance of metal ladders to inspection and lubrication, without the need for sanding and refinishing.

A disadvantage of metal ladders is their tendency to become very cold in winter and very hot in summer. In addition, metal ladders should not be used within 4′ of electrical circuits or equipment. Extreme caution is necessary if metal ladders are used near electrical power lines, service-entrance conductors, and electrical equipment because of their good electrical conductivity.

Fiberglass Ladders

A *fiberglass ladder* is a ladder constructed of fiberglass. Fiberglass ladders are rapidly becoming the most popular type of ladder, particularly in stepladders. Advantages of fiberglass ladders are that they do not conduct electricity when dry, can withstand considerable abuse, do not require surface finishing, and are more comfortable to use than metal ladders in cold or hot environments.

Ladder Safety Feet

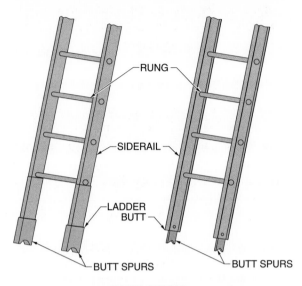

BUTT SPURS

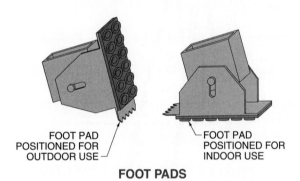

FOOT PADS

Figure 2-39. Ladders must be equipped with nonslip safety feet such as butt spurs or foot pads.

The disadvantages of fiberglass ladders are that they may crack and fail when overloaded, and may crack and chip when severely impacted.

Regulations and Standards

Regulations and standards for the use, design, and testing of ladders are published by various federal, state, and standards organizations. OSHA publishes federal industry standards for ladders in publications CFR Title 29, Part 1910.26, *Portable Metal Ladders* and Part 1910.27, *Fixed Ladders.*

ANSI publishes standards for ladders in ANSI ANSI A14.3-1992, *Ladders – Fixed – Safety Requirements* and ANSI A14.5-1992, *Ladders – Portable Reinforced Plastic – Safety Requirements.*

Duty Rating. *Ladder duty rating* is the weight (in lb) a ladder is designed to support under normal use. **See Figure 2-40.** The five ladder duty ratings are:

- Type IAA – Special duty, industrial, 375 lb capacity

- Type IA – Extra heavy-duty, industrial, 300 lb capacity

- Type I – Heavy-duty, industrial, 250 lb capacity

- Type II – Medium-duty, commercial, 225 lb capacity

- Type III – Light-duty, household, 200 lb capacity

Ladder Duty Ratings

Werner Ladder Co.

Figure 2-40. Ladder duty rating is the weight a ladder is designed to support under normal use.

Safety Tip

Over 11% of injuries that require time away from work are related to falling from ladders.

Climbing Techniques

Climbing may begin only after a ladder is properly secured. Climbing movements should be smooth and rhythmical to prevent ladder bounce and sway. Safe climbing employs the three-point contact method. In the three-point contact method, the body is kept erect, the arms straight, and the hands and feet make the three points of contact. Two feet and one hand, or two hands and one foot are in contact with the ladder rungs at all times.

Avoid reaching above shoulder level to grasp a rung to maintain balance and unobstructed knee movements. Each hand should grasp the rungs with the palms down and the thumb on the underside of the rung. Upward progress should be caused by the push of the leg muscles and not the pull of the arm muscles. When climbing, tools, parts, or equipment must be secured in a pouch or raised and lowered with a rope.

Fixed Ladders

A *fixed ladder* is a ladder that is permanently attached to a structure. Fixed ladders are commonly constructed of steel or aluminum. Fabrication of a fixed ladder, including design, materials, and welding, must be done under the supervision of a qualified licensed structural engineer. **See Figure 2-41.**

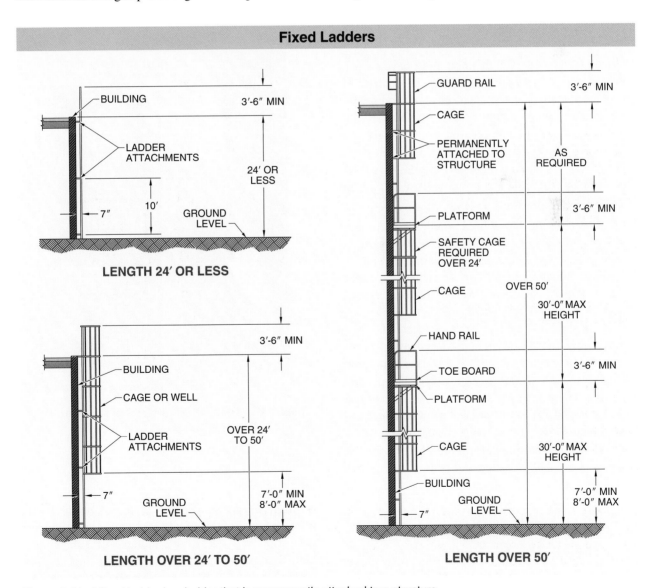

Fixed Ladders

Figure 2-41. A fixed ladder is a ladder that is permanently attached to a structure.

For fixed ladders, the width between siderails is normally 16″ and the spacing between rungs is 12″. Each rung cross-section must be a minimum of ¾″ in diameter. The rungs must be a minimum of 1″ in diameter if a steel fixed ladder is located in an unusually corrosive environment. Fixed ladder siderails for use in normal conditions must be made of material sized for gripping and are typically 2½″ wide and ⅜″ thick. Unusually corrosive environments require siderail material ½″ thick.

Fixed ladders are normally attached to a building or structure at spaces of 10′ or less with a minimum distance of 7″ from the center of the rung to the building. Siderails of a fixed ladder must extend a minimum of 3′-6″ above the landing for a walk-through ladder and 4′-0″ above the landing for a side access ladder. Each rung of a fixed ladder must support 250 lb.

Fixed ladders over 24′ in length must have a cage, well, or ladder safety system. A *cage* is a barrier or enclosure mounted on the siderails of a fixed ladder or fastened to the structure. Cages are also referred to as cage guards or basket guards. A *well* (shaft) is a walled enclosure around a fixed ladder. A well provides a climber the same protection as a cage. Cages must start no less than 7′-0″ and no more than 8′-0″ from the ground or platform.

A *ladder safety system* is an assembly of components whose function is to arrest the fall of a worker. Ladder safety systems consist of a carrier and its associated attachments. A *carrier* is the track of a ladder safety system consisting of a flexible cable or rigid rail secured to the ladder or structure. The carrier is attached to a safety sleeve. A *safety sleeve* is a moving element with a locking mechanism that is connected between a carrier and a worker's harness.

A cage, well, or ladder safety system must be provided where a single length of climb is greater than 24′ but less than 50′. The ladder must consist of multiple sections (50′ maximum each) if a cage or well is used. A landing platform must be provided at least every 50′ within the length of climb. A *platform* is a landing surface which provides access/egress or rest from a fixed ladder. The length of climb may be continuous if a ladder safety system is used. Rest platforms must be provided at maximum intervals of 150′ on fixed ladders using a ladder safety system.

Platforms should be a minimum of 24″ wide by 30″ long with railings and toe board. Railings around platforms are commonly 3′-6″ high. Adjacent ladder sections are offset at each landing by no more than 18″.

Fixed ladders are installed in a range between 60° and 90° from horizontal. The range between 60° and 75° is considered to be the fixed ladder substandard pitch range and is to be used only for special conditions. Fixed ladders are not to be installed over 90° from horizontal. **See Figure 2-42.**

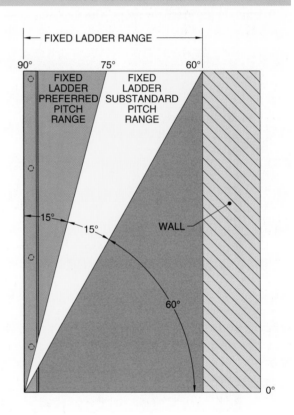

Fixed Ladder Installation

Figure 2-42. Fixed ladders are installed in a range between 60° and 90° from horizontal.

Single Ladders

A *single ladder* is a ladder of fixed length having only one section. Typical lengths of single ladders vary from 6′ to 24′. Single ladders offer the convenience of use by one person. However, they are limited in their versatility because a given length ladder may be safely used only within a small height range.

To ascend or descend a ladder safely, use the 3-point climbing method, in which one hand and two feet or two hands and one foot are in contact with the ladder rungs at all times.

Extension Ladders

An *extension ladder* is an adjustable-height ladder with a fixed bed section and sliding, lockable fly sections. The *bed section* is the lower section of an extension ladder. The *fly section* (first fly, second fly, etc.) is the upper section of an extension ladder. **See Figure 2-43.**

A *pawl lock* is a pivoting hook mechanism attached to the fly sections of an extension ladder. Pawl locks are used to hold the fly sections at the desired height. The three main parts of a pawl lock are the hook, finger, and spring. **See Figure 2-44.**

The rungs nest into the hook of each pawl, preventing downward movement of the fly section. The fly section is lowered by first raising the fly section just enough for a rung to pass below the finger of the pawl lock. The fly section is then lowered. When lowered, the rung forces the finger up, preventing the rungs from nesting into the hook. To hold the fly section in place, the pawl lock must be lowered slightly below a rung to allow the finger to drop. The fly is then raised slightly to allow the spring to force the pawl hook over the rungs.

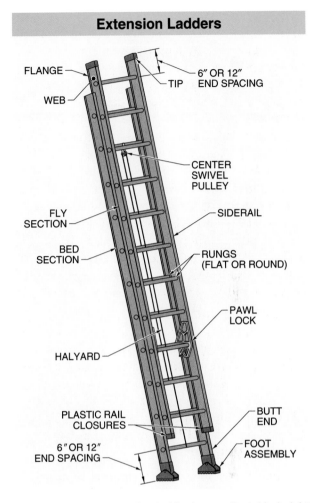

Figure 2-43. An extension ladder is an adjustable-height ladder with a fixed bed section and a sliding, lockable fly section(s).

Many fly sections are raised and lowered with the use of a halyard, halyard anchor, and pulley. A *halyard* is a rope used for hoisting or lowering objects. A halyard must be a minimum of ⅜″ in diameter with a minimum breaking strength of 825 lb. The halyard is threaded through the pulley attached to the top rung of the bed section. One end of the rope is attached to the bottom rung of the fly section and the other end is usually tied off at the bottom. **See Figure 2-45.**

Safety Tip

All hardware on ladders shall be made of aluminum, wrought iron, steel, malleable iron, or other material that is adequate in strength for the purpose intended and free from sharp edges and from sharp projections in excess of ¹⁄₆₄″.

Pawl Locks

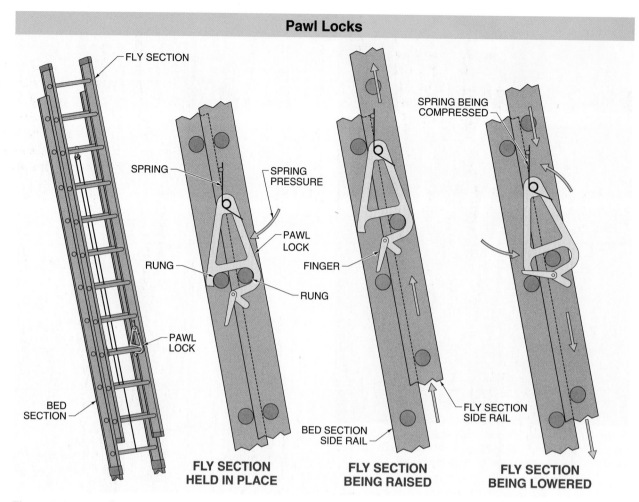

Figure 2-44. A pawl lock is a pivoting hook mechanism attached to the fly section(s) of an extension ladder.

Raising Ladders. Raising a ladder involves a smooth, proper, and safe operation. Care must be taken before beginning a raise to ensure that electrical conductors or equipment are not present.

Single or extension ladders may be raised with the ladder tip away from the building or with the ladder tip against the building. The method used is determined by whether the ladder is raised indoors or outdoors, the presence of overhead obstructions, the setup area, and the ladder size. **See Figure 2-46.**

To raise a ladder with the ladder tip away from the building, place the butt end of the ladder against the building with the fly section retracted and to the down side. Grasp the rung at the ladder tip with both hands. Raise the tip and walk under the ladder, grasping successively lower rungs while walking toward the building.

When the ladder is erect, hold the ladder against the wall by placing force with one hand at about eye level.

Place one foot approximately 10″ to 12″ in front of one of the ladder legs. Use the free hand to apply an upward pressure and an outward pull to slide the butt of the ladder to the foot. This procedure is continued until the ladder is in the approximately correct position and angle. Adjust the fly section for the proper height and readjust the ladder as necessary for the proper angle.

To raise a ladder with the ladder tip against the building, place the ladder tip against the building with the fly section fully retracted and to the upside. Standing with back to the wall, lift the ladder tip while pulling the butt end toward the wall. The ladder tip remains against the wall while repositioning for another lift and pull. This procedure is continued until the ladder is in the approximately correct position and angle. Adjust the fly section for the proper height and readjust the ladder as necessary for the proper angle.

Halyards

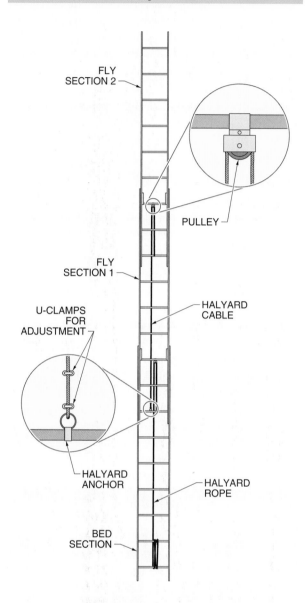

FLY SECTION 2

PULLEY

FLY SECTION 1

U-CLAMPS FOR ADJUSTMENT

HALYARD CABLE

HALYARD ANCHOR

HALYARD ROPE

BED SECTION

Figure 2-45. Extension ladder fly sections are raised and lowered with the use of halyards.

Never attempt to raise long extension ladders alone. At least two people are required to raise long extension ladders into position, with one person on each side of the ladder. Lifting and pulling are done together until the ladder reaches its final position. Three or more people may be required to raise extension ladders to overhead beams or pipes. Two people raise the ladder and one or two people firmly secure the butt end.

Raising Ladders

TIP AWAY FROM BUILDING

TIP AGAINST BUILDING

Figure 2-46. Single or extension ladders are raised with the tip away from or against the building.

Extension Overlap and Height. Extension ladders must have positive stops to prevent overextension of the fly section(s). The overlap of the fly section must be at least 3' for extension ladders up to 36', 4' for extension ladders over 36' and up to 48', and 5' for extension ladders over 48' and up to 60'.

Extension ladders are positioned on a 4:1 ratio (75° angle). For every 4′ of working height, 1′ of space is required at the base. *Working height* is the distance from the ground to the top support. The *top support* is the area of a ladder that makes contact with a structure. For example, a 12′ ladder should be placed at an angle that places the butt end 3′ from the wall. **See Figure 2-47.**

Selection of a properly-sized extension ladder requires knowledge of the vertical height of the top support. In applications where the top support is a roof eave, an additional working length is needed to obtain the required 3′ to 5′ extension beyond the top support.

The tip of a single or extension ladder should be secured at the top to prevent slipping and must be at least 3′ above the roof line or top support. Never stand on the top three rungs of a single or extension ladder. Ladders over 15′ should also be secured at the bottom.

Positioning Extension Ladders

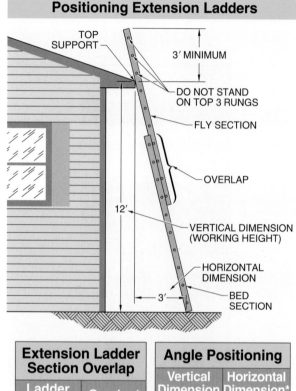

Extension Ladder Section Overlap	
Ladder Length*	**Overlap***
8 to 36	3
36 to 48	4
48 to 60	5

* in ft

Angle Positioning	
Vertical Dimension	**Horizontal Dimension***
8	2
10	2½
12	3
16	4
20	5
24	6
28	7
32	8
36	9
40	10
44	11

* in ft

Figure 2-47. Extension ladders are positioned on a 4:1 ratio.

Ladder Jacks. A *ladder jack* is a ladder accessory that supports a plank to be used for scaffolding. A *plank* is a board 2″ to 4″ thick and at least 8″ wide. A ladder jack is positioned on the inside or outside of the rungs. A ladder jack has hooks at the top and the bottom which attach to the ladder rungs. The hooks on the jacks are placed close to the siderails to allow suitable load force. A plank is placed on the horizontal projection, providing a work platform. **See Figure 2-48.** To prevent cracking and failing of a ladder jack assembly, use only IA, heavy-duty extension ladders with ladder jacks.

Ladder Jacks

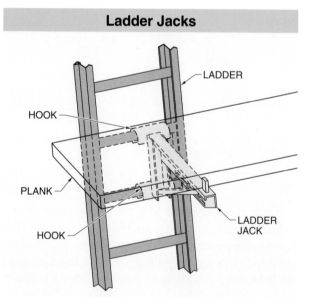

Figure 2-48. A ladder jack is a ladder accessory that supports a plank to be used for scaffolding.

Standoffs. A *standoff* is a ladder accessory that holds a single or an extension ladder a fixed distance from a wall. A standoff is attached to the ladder with adjustable U-bolts. Adjustable, non-slip tips on each end of the standoff help protect the wall surface from marring or scratching. Standoffs provide a comfortable working distance between a wall and the worker. Standoffs are particularly useful for painting around windows, etc. **See Figure 2-49.**

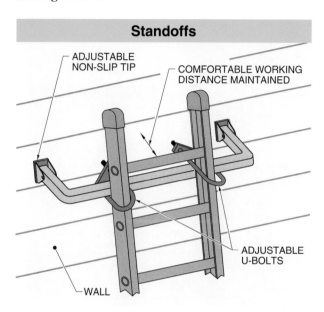

Standoffs

ADJUSTABLE NON-SLIP TIP

COMFORTABLE WORKING DISTANCE MAINTAINED

ADJUSTABLE U-BOLTS

WALL

Figure 2-49. A standoff is a ladder accessory that holds a single or an extension ladder a fixed distance from the wall.

Werner Ladder Co.

Ladder jacks are used on extension ladders to create a support for a platform.

Stepladders

A *stepladder* is a folding ladder that stands independently of support. Stepladders are commonly 2'-0" to 8'-0" in length. Because stepladders are easily portable and provide adequate working height for many applications, they are used more often than any other ladder. **See Figure 2-50.**

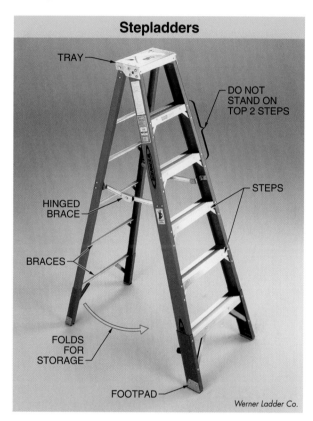

Stepladders

TRAY

DO NOT STAND ON TOP 2 STEPS

STEPS

HINGED BRACE

BRACES

FOLDS FOR STORAGE

FOOTPAD

Werner Ladder Co.

Figure 2-50. A stepladder is a folding ladder that stands independently of support.

Stepladders must be self-supporting, stable, and free from racking. *Racking* is the ability to be forced out of shape or form. Stepladders have steps instead of rungs. The steps must be horizontal when the ladder is open. Top-quality stepladders may have braces at the sides, folding shelves (trays) for utensils, platforms at the top with three-sided guard rails, or steps on both front and back of the ladder.

Never lean out or reach to one side of a stepladder. Reposition the stepladder so the work area can be easily and conveniently reached. Never stand on the top two steps of a stepladder. There is no support and it is easy to lose balance. Do not use metal stepladders near electrical conductors or equipment.

SCAFFOLDS

A *scaffold* is a temporary or movable platform and structure for workers to stand on when working at a height above the floor. Three basic types of scaffolds are pole, sectional metal-framed, and suspension scaffolds. **See Figure 2-51.**

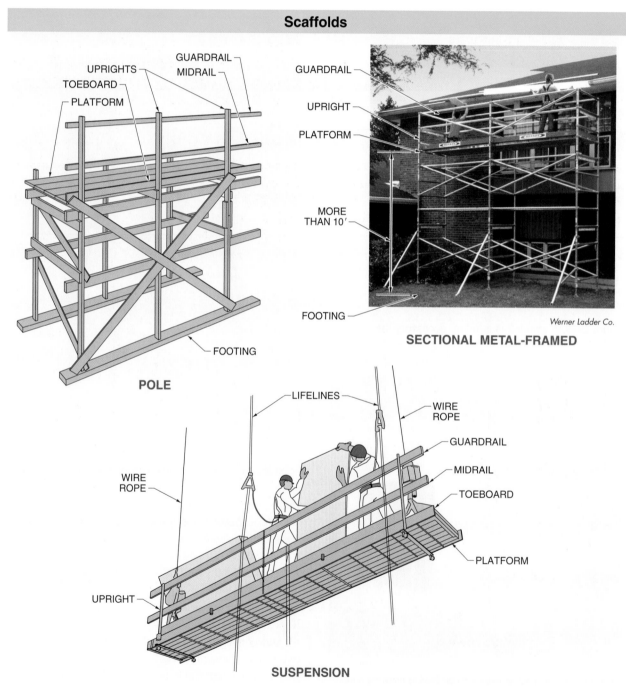

Figure 2-51. Three basic types of scaffolds are pole, sectional metal-framed, and suspension scaffolds.

A scaffold generally consists of wood planks or metal platforms to support workers and their materials. Scaffolds must not be supported by any unstable object, such as boxes, concrete blocks, or loose bricks. Scaffold footing must be sound and stable and must not settle or displace while carrying the maximum intended load. A *maximum intended load* is the total of all loads, including the working load, the weight of the scaffold, and any other loads that may be anticipated. Scaffolds and their components must be capable of supporting at least four times their maximum intended load.

All scaffolds 10′ or more above ground must have guardrails, midrails, and toeboards. A *guardrail* is a rail secured to uprights and erected along the exposed sides and ends of a platform. A *midrail* is a rail secured to uprights approximately midway between the guardrail and the platform. A *toeboard* is a barrier to guard against the falling of tools or other objects. Toeboards are secured along the sides and ends of a platform. Guardrails must be installed no less than 38″ or more than 45″ high, with a midrail. Guardrail and midrail support is to be at intervals of no more than 10′.

Regulations and Standards

OSHA regulations governing the use of scaffolds require that a scaffold may be erected, moved, or dismantled only under the supervision of a competent person. A *competent person* is a person capable of recognizing and evaluating employee exposure to hazardous substances or to other unsafe conditions and of specifying the necessary protection and precautions to be taken to ensure the safety of all employees. Refer to OSHA and ANSI for guidelines. OSHA industry standards are found in CFR Title 29 Part 1910.28 and 29, *Safety Requirements for Scaffolding* or Part 1926.451, *Scaffolding*. ANSI standards are found in ANSI A10.8-2001, *Safety Requirements for Scaffolding*.

Pole Scaffolds

A *pole scaffold* is a wood scaffold with one or two sides firmly resting on the floor or ground. **See Figure 2-52.** A *single-pole scaffold* is a wood scaffold with one side resting on the floor or ground and the other side structurally anchored to the building. A *double-pole scaffold* is

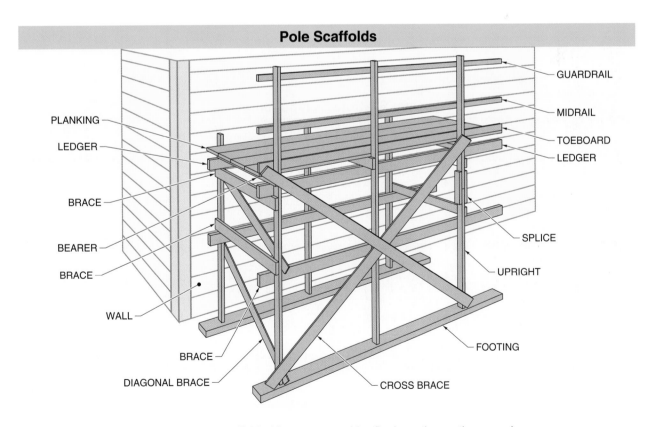

Pole Scaffolds

GUARDRAIL
MIDRAIL
TOEBOARD
LEDGER
SPLICE
UPRIGHT
FOOTING
CROSS BRACE
DIAGONAL BRACE
BRACE
WALL
BRACE
BEARER
BRACE
LEDGER
PLANKING

Figure 2-52. A pole scaffold is a wood scaffold with one or two sides firmly resting on the ground.

a wood scaffold with both sides resting on the floor or ground and is not structurally anchored to a building or other structure.

The uprights of pole scaffolds are assembled from wood or metal legs (poles). Uprights must be plumb and securely braced to prevent displacement or swaying. The poles are to be erected on suitable bases or footings, which must be strong enough and large enough to support the maximum scaffold load without settling or displacement. Unstable objects such as barrels, boxes, loose brick, or concrete blocks must not be used to support scaffolds. Steel plate supports are used under steel poles and a minimum of 2″ planking support is used under wood poles. Each base or footing must be of sufficient size and thickness to support at least four times the maximum intended load.

Pole scaffolds must be constructed with guardrails, midrails, toeboards, planking, bearers, ledgers, cross braces, diagonal braces, and footings. The complete assembly must be plumb, level, square, and rigid.

Toeboards prevent tools or materials from being knocked or kicked off of the platform. The bottom of toeboards should make contact with the platform and the top of the toeboard should be more than 3½″ from the platform. Bearers (putlogs) must be long enough to project at least 3″ over the ledgers. Ledgers must be long enough to extend over two pole spaces and must not be spliced between poles. Cross braces are assembled between the left and right uprights. Cross braces must be long enough to extend over two pole spaces. Both diagonal and cross braces are used to prevent buckling and lateral movement. Diagonal braces are included in the scaffold assembly between the inner and outer uprights (pole sets). All pole scaffolds must be constructed using minimum- and maximum-sized components according to their duty rating. **See Figure 2-53.**

Wood pole scaffolds are constructed from select clear lumber for maximum strength. Duplex-head nails are used to make dismantling easier. All nails must be driven in their full length and in directions where the pull is across their length, not with their length. Nails smaller than 8d common must not be used to construct scaffolds.

Scaffold platform planks consist of 2″ nominal structural planks. Maximum permissible planking spans vary according to wood thickness and width. For example, the maximum permissible span for a 2 × 10 plank on a light-duty scaffold is 10′. **See Figure 2-54.**

Pole Scaffold Components*						
Type	Poles	Bearers	Ledgers (Stringers)	Braces	Planking	Rails
Light-duty[†] single-pole	20′ or less – 2 × 4 60′ or less – 4 × 4	3′ width – 2 × 4 5′ width – 4 × 4	20′ or less – 2 × 4 60′ or less – 1¼ × 9	1 × 4	2 × 10	2 × 4
Medium-duty[‡] single-pole	60′ or less – 4 × 4	2 × 10	2 × 10	1 × 6	2 × 10	2 × 4
Heavy-duty[§] single-pole	60′ or less – 4 × 4	2 × 10	2 × 10	2 × 4	2 × 10	2 × 4
Light-duty* double-pole	20′ or less – 2 × 4 60′ or less – 4 × 4	3′ width – 2 × 4 5′ width – 4 × 4	20′ or less – 1¼ × 4 60′ or less – 1¼ × 9	1 × 4	2 × 10	2 × 4
Medium-duty[‡] double-pole	60′ or less – 4 × 4	2 × 10	2 × 10	1 × 6	2 × 10	2 × 4
Heavy-duty[§] double-pole	60′ or less – 4 × 4	2 × 10	2 × 10	2 × 4	2 × 10	2 × 4

* all members except planking are used on edge
[†] not to exceed 25 lb/sq ft
[‡] not to exceed 50 lb/sq ft
[§] not to exceed 75 lb/sq ft

Figure 2-53. Components of pole scaffolds are sized based on their duty rating.

Planking Spans		
Duty Rating	Working Load*	Permissible Span†
Light	25	10
Medium	50	8
Heavy	75	6

* in lb/sq ft
† in ft

Figure 2-54. Maximum permissible planking spans vary according to wood thickness and width.

Planking must extend between 6″ and 18″ over the end support. These dimensions ensure the prevention of tipping if a weight is placed on the end. The underside of the planks is cleated at both ends to the support to prevent the planks from sliding. A *cleat* is a narrow wood piece, nailed across another board or boards, to provide support or to prevent movement. Platform planks are laid side-by-side with a maximum opening between planks of 1″. If plank overlapping is necessary due to a continuous run, the planks must be secured from movement, overlapped a minimum of 12″, and supported in the center by a brace.

When platform planks are added as upward work progresses, the previous bearers and ledgers must remain for pole bracing. Bearers must be face nailed to the poles as well as toenailed to the ledgers. For heavy-duty scaffolds, the ledgers must be nailed to the poles and supported by cleats that are also attached to the poles.

In single-pole scaffolds, the bearer that butts against the wall must be supported by a 2 × 6 bearing block, a minimum of 12″ long, notched out the width of the bearer. The bearing block must be securely fastened to the wall. The bearer must be nailed to the bearing block and toenailed to the wall. **See Figure 2-55.**

Ladders must be provided and attached to the ends of a scaffold so their use will not subject the scaffold to tipping. Cross braces must not be used as a means of access. Scaffolds must not be used during storms or high winds or when covered with ice or snow. All tools and materials should be secured or removed from the platform before a scaffold is moved.

Tech Tip

All solid sawn scaffold planks must be of a "scaffold plank grade" and must be certified by a grading agency approved by the American Lumber Standards Committee.

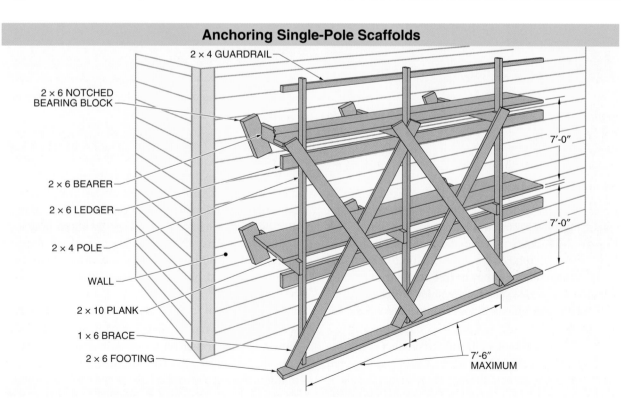

Anchoring Single-Pole Scaffolds

2 × 4 GUARDRAIL

2 × 6 NOTCHED BEARING BLOCK

2 × 6 BEARER

2 × 6 LEDGER

2 × 4 POLE

WALL

2 × 10 PLANK

1 × 6 BRACE

2 × 6 FOOTING

7′-0″

7′-0″

7′-6″ MAXIMUM

Figure 2-55. Single-pole scaffolds are anchored to the wall using notched bearing blocks.

Scaffolds over 25′ in height must be securely guyed or tied to the structure or building with No. 12 double-wrapped wire. A *guyline* is a rope, chain, rod, or wire attached to equipment as a brace or guide. **See Figure 2-56.** Guylines for scaffolds are wire ropes ¼″ in diameter or larger. They are commonly positioned at a 45° angle to the vertical. Guylines may be anchored temporarily in the ground by screw ground anchors or may be tied off securely to a structure. Where the height of the scaffold exceeds 25′, the scaffold must be secured at intervals no greater than 25′, vertically or horizontally.

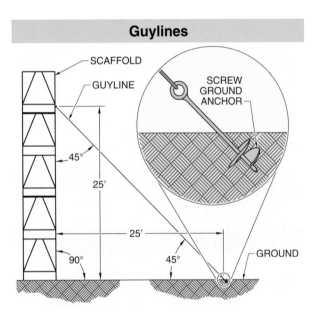

Figure 2-56. Scaffolds over 25′ in height must be secured with guylines.

Sectional Metal-Framed Scaffolds

A *sectional metal-framed scaffold* is a metal scaffold consisting of preformed tubes and components. **See Figure 2-57.** Sectional metal-framed scaffolds are also known as tube and coupler scaffolds. Sectional metal-framed scaffolds may be free-standing or mobile. They are easy to use, easily assembled with bolts, pins, or brackets, and may be fitted with locking casters for ease in moving.

Always check the manufacturer's product specifications sheet for scaffold load limits. Cross and diagonal braces of sectional metal frames allow several levels of planking installation. Planking may be wood, metal-reinforced wood, or metal. Each frame member is fastened together using nonslip clamps or pipe fittings. Prior to each use, the clamps or fittings must be checked and tightened if necessary.

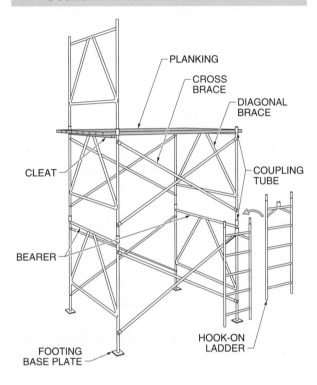

Figure 2-57. A sectional metal-framed scaffold is a metal scaffold consisting of preformed tubes and components.

When used as free-standing units, the height of a metal-framed scaffold must not exceed four times its minimum base dimension. For example, if the base of a scaffold measures 4′ × 8′, the maximum height is 16′ (4′ × 4 = 16′). Outriggers are sometimes used to increase the working height of a scaffold. Outrigger beams must rest on a sound foundation or on wood bearing blocks.

Mobile scaffolds are equipped with casters. A mobile scaffold may be moved with a worker on the platform. However, the worker on the mobile scaffold must be advised and aware of each movement in advance. The minimum dimension of the base, when ready for rolling, must be at least one-half of the height. Outriggers may be included as part of the base dimension. **See Figure 2-58.**

Mobile Scaffolds

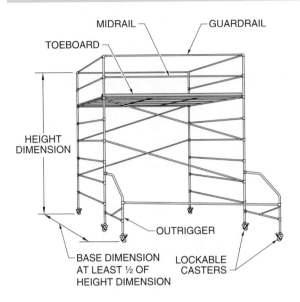

Figure 2-58. The minimum base dimension of a mobile scaffold must equal half the scaffold height before moving.

All tools and materials must be removed or secured before a mobile scaffold is moved. When mobile scaffolds are used on concrete flooring, the floor surface must be free from pits, holes, or obstructions that may create an unsafe condition. The surface must also be within 3° of level. After the scaffold has been moved, the casters must be locked to prevent movement while the scaffold is being used.

A hydraulic scissor lift is another type of mobile scaffold. A *hydraulic scissor lift* is a mobile hydraulically-operated platform controlled by remote switches attached at the platform. The platform is generally 4′ in its lowered position, with a maximum working height at about 20′. To offer a firm base, hydraulic scissor lifts may be equipped with screw-type outriggers. **See Figure 2-59.**

WARNING: Hydraulic scissor lifts should not be used outdoors in high-wind conditions. Extreme caution should be excercised when using scissor lifts outdoors.

Suspension Scaffolds

A *suspension scaffold* is a scaffold supported by overhead wire ropes. Suspension scaffolds are also referred to as swinging scaffolds. Suspension scaffolds use either the two-point or multiple-point suspension design.

A *two-point suspension scaffold* is a suspension scaffold supported by two overhead wire ropes. The overall width of two-point suspension scaffolds must be greater than 20″ but not more than 36″. **See Figure 2-60.**

A *multiple-point suspension scaffold* is a suspension scaffold supported by four or more ropes. Multiple-point suspension scaffolds must be capable of sustaining a working load of 50 lb/sq ft and are used mainly for repair and maintenance projects. They can be raised or lowered by permanently-installed, electrically-operated hoisting equipment. Multiple-point suspension scaffolds must not be overloaded.

Snorkel

Figure 2-59. A hydraulic scissor lift is a mobile hydraulically-operated platform controlled by remote switches attached at the platform.

Safety Tip

Only qualified and trained operators are to be authorized to use and operate mobile work platforms. Platforms must be elevated only on a firm, level surface.

Suspension Scaffolds

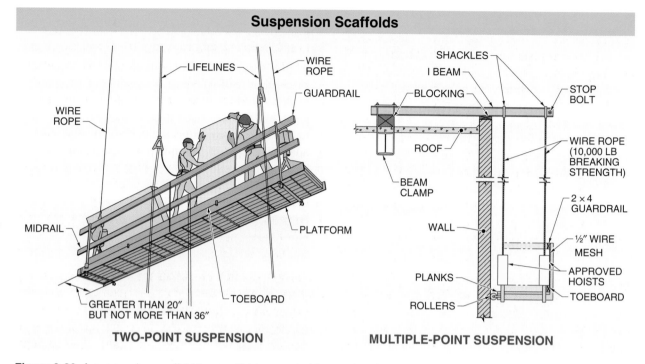

TWO-POINT SUSPENSION

MULTIPLE-POINT SUSPENSION

Figure 2-60. A suspension scaffold is a scaffold supported by overhead wire ropes.

Suspension scaffolds have a platform supported near the ends by overhead wire ropes. The ropes are attached to the platform by hangers or metal stirrups. Wire, fiber, or synthetic rope used for suspension scaffolds must be capable of supporting at least six times the maximum intended load. The hangers, U-bolts, brackets, and other hardware used for constructing a two-point or multiple-point suspension scaffold must be capable of sustaining four times the maximum intended load.

When power-driven hoisting equipment is used on suspension scaffolds, the power-driven equipment must have an emergency brake in addition to the normal operating brake. The emergency brake must operate automatically when the normal speed of descent is exceeded. The running end of the hoisting rope is attached to the hoist drum. At least four turns of rope must remain on the hoist drum at all times.

Suspension scaffold load testing is accomplished by raising the scaffold about 1′ off the ground and placing a load that is at least four times the normal workload on the platform. After approximately five minutes, check for cracked or splitting rope or sagging platforms. Do not use the scaffold if any of these conditions exist.

Lifelines and harnesses that can safely support a worker's weight must be provided for each worker. The lifeline is suspended from a substantial overhead structural member other than the scaffold, and should extend to the ground. Each worker's harness is tied to a lifeline by a lanyard, and to a fall prevention device that will limit the free fall to no more than 6′.

Safety Tip

Each person on a two-point suspension scaffold shall use an approved harness with a lanyard and shall be attached by means of a fall-arresting device to an independent lifeline.

Tractel Inc., Griphoist® Division
Suspended scaffolding systems contain powered traction hoists that are efficient and have a simple and constant traction principle that allows the hoist to be used in any orientation.

Safety Nets

A *safety net* is a net made of rope or webbing for catching and protecting a falling worker. A safety net must be used anywhere a person is working 25′ or more above ground, water, machinery, or any other solid surface when the worker is not otherwise protected by a lifeline, harness, or scaffolding. **See Figure 2-61.** Safety nets must also be used when public traffic or other workers are permitted underneath a work area that is not otherwise protected from falling objects.

The Sinco Group, Inc.

Figure 2-61. A safety net is a net made of rope or webbing for catching and protecting a falling worker.

Safety Net Requirements. Safety net size is generally 17′ × 24′, and may be coupled with another net to form a larger net. Netting mesh size for bodily fall protection is normally 6″ × 6″. Netting is constructed of ⅜″ No. 1 grade manila, ¼″ nylon, or ⁵⁄₁₆″ polypropylene rope. *Mesh* is the size of the openings between the rope or twine of a net.

In applications where workers or others are to be protected from falling tools or other objects, a lining of smaller mesh must be added to the fall protection net. The size and strength of the net lining mesh must restrict tools and materials capable of causing injury.

Net lining mesh must normally be less than 1″ and constructed of twine equal to or greater than No. 18. Installation of netting must have level border ropes and, when hung, no more than 3′ of sag should be allowed at the center of the net.

When two or more nets are secured together to form a larger net, lacing, drop-forged shackles, or safety hooks may be used, but must be less than 6″ apart. A drop-forged shackle or safety hook is to be used to attach nets to supporting structures, cables, or beams and must be spaced at intervals of no more than 4″. Border rope is to have a 5000 lb breaking strength when new. The minimum diameter for manila border rope is ¾″. The minimum diameter for synthetic border rope is ½″.

Safety Net Maintenance. Factors that affect net safety include environmental contaminants, sunlight, welding, mildew, abrasion, and impact loading. Contaminants from airborne chemicals create environmental conditions that affect net strength. Even though polypropylene and nylon are resistant to many acids and alkalis, moderate and unknown degradation can occur to rope used in these environmental conditions. Manila rope, being organic, degrades rapidly in a chemically-active environment.

Synthetic and natural fibers degrade in the presence of ultraviolet rays from sunlight or from arc welding. When safety nets are used regularly outdoors, an ultraviolet absorbing dye may be used for outer-layer protection. Welding slag or sparks may also harm safety nets because each is sufficient to burn the net.

Mildew and abrasion damage is caused by improper storage and rough handling. Storing safety nets in a warm, moist location causes mildew growth and a weakening of rope fibers. Dragging nets over rough or sharp surfaces abrades and degrades rope fibers. Also, impact loading is a form of damage created by the continuous shock of loads being dropped into the net. Even the impact from net testing may degrade the net's integrity.

Safety Net Testing. Nets must be impact load tested to assure that there is sufficient strength or that there has been no loss of strength. Impact load tests are first done on a sample by the manufacturer. Each safety net is certified by the manufacturer to withstand a 50′ drop of a 350 lb bag of sand, 24″ in diameter. On-the-job testing is also required by the user immediately following installation or after a major repair. Impact load testing must be done at six month intervals if the net is in regular use. Testing consists of dropping a 400 lb bag of sand, not more than 30″ (±2″) in diameter, from a height of 25′ above the net into the center of the net.

Personnel nets consist of strong, high tenacity nylon with a 3½″ diagonal net design made in accordance with ANSI A10.11-1989, Safety Nets Used During Construction, Repair, and Demolition Operations.

LADDER AND SCAFFOLD SAFETY

Safe use of ladders and scaffolds includes the use of fall-protection equipment. The three categories of fall-protection equipment include ladder climber fall protection, position protection, and scaffold worker fall protection.

Ladder Climber Fall Protection

Ladder climbers should use a carrier for fall protection. A *carrier* is the track of a ladder safety system consisting of a flexible cable or rigid rail secured to the ladder or structure. **See Figure 2-62.** The carrier is the track for the safety sleeve. A safety sleeve is a moving element with a locking mechanism that is connected between a carrier and the worker's harness. The connecting line between the carrier and the safety belt must be less than 9″. Fall-arrest devices utilize the worker's weight for activation. The closer a worker is connected to the fall-arrest device, the less distance traveled in a fall.

Carriers

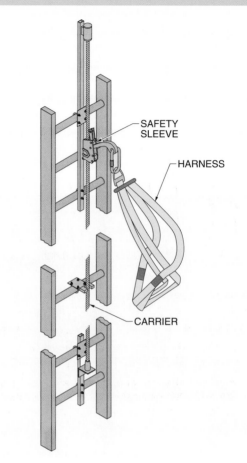

Figure 2-62. A carrier is the track of a ladder safety system consisting of a flexible cable or rigid rail secured to the ladder or structure.

Position Protection

Position protection supports a person in a working position by wrapping a body strap around a post, tree, or attachment to a structure. The weight and angle of the individual secures the worker. A position protection device holds an individual in position, but does not prevent falling. Position protection is commonly used by electrical lineworkers, telephone lineworkers, tree climbers, and window washers. **See Figure 2-63.**

Tech Tip

Suspension scaffolds shall be operated only by persons who have been instructed in the operation, use, and inspection of the particular suspended scaffold to be operated. Employers shall instruct and supervise their employees in the safe use of all equipment provided.

SIDE
D-RING
CONNECTS
TO SAFETY
ROPE

Salisbury

Figure 2-63. Position protection devices include harnesses containing side D-rings that are used to attach a safety rope around a post, tree, or attachment to a building.

Scaffold Worker Fall Protection

Scaffold worker fall protection may include items such as lifelines, harnesses, lanyards, and rope grabs. The proper fall-protection equipment should be worn when working at heights greater than the minimum safe distance from the ground. **See Figure 2-64.**

Lifelines are anchored above the work area, offering a free-fall path, and must be strong enough to support the force of a fall arrest. Vertical lifelines must never have more than one person attached per line and must be long enough to reach the ground or landing below the work area. The lifeline must then be terminated (tied up) to prevent the safety sleeve from sliding off of its end.

The path of a fall must be visualized when anchoring a lifeline. Use an anchored system without any obstructions to the fall. Obstacles below and in the fall path can be deadly.

Harnesses, when used properly, protect internal body organs, the spine, and other bones in a fall. Chest harnesses must not be worn for free-fall protection.

Harnesses must fit snugly and be attached to the lanyard at the person's back. A lanyard is used to attach a worker's harness to a lifeline. A *rope grab* is a device that clamps securely to a rope. Rope grabs contain a ring to which a lifeline can be attached. Rope grabs protect workers from falls while allowing freedom of movement.

Fall-Arrest Sequence

When a fall-arrest device is used, the breaking of a fall is preceded by a free fall and the taking up of slack between the harness and the safety device. This is followed by the distance of deceleration. Deceleration distance is generally 3½' to 4'. Always limit the total fall to 6' or less. For this reason, a lanyard must be kept high enough or short enough to limit the free fall.

A person may also be protected against a fall by tying off. *Tying off* is securely connecting a harness directly or indirectly to an overhead anchor point. Certain precautions must be made that lines and lanyards are not weakened by knotting or tying off to sharp or rough surfaces.

Ladder Safety

Proper maintenance of a ladder is critical due to its direct relationship to life safety. The following list of precautions should be observed for proper safety when using ladders.

- Use ladders only for the purpose for which they were designed.
- Inspect ladders carefully when new and before each use.
- Use leg muscles, not back muscles, for lifting and lowering ladders.
- Stand ladders on a firm, level surface.
- Face the ladder when ascending or descending.
- Exercise extreme caution when using ladders near electrical conductors or equipment. All ladders conduct electricity when wet.
- Ladders are intended for use by only one person unless specifically designated otherwise.
- Never use a ladder as a substitute for scaffold planks or for horizontal work.

Fall-Protection Equipment

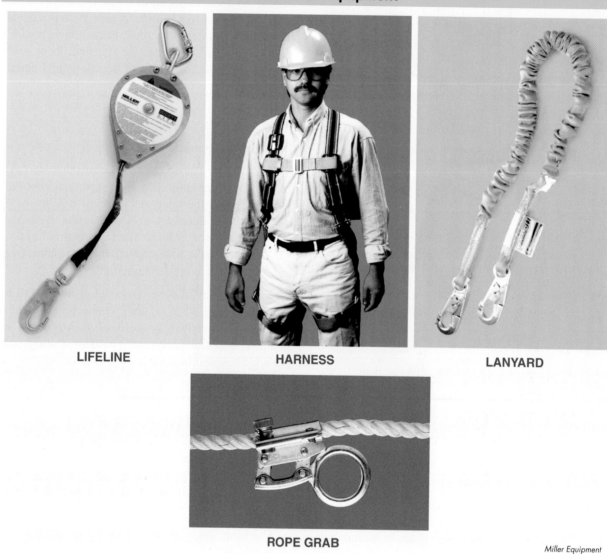

LIFELINE   HARNESS   LANYARD

ROPE GRAB

Miller Equipment

Figure 2-64. The appropriate fall-protection equipment should be worn when working at heights greater than the minimum safe distance from the ground.

- Always check for the proper angle of inclination before climbing a ladder.
- Verify that all pawl locks on extension ladders are securely hooked over rungs before climbing.
- Always check for proper overlap of extension ladder sections before climbing.
- Keep all nuts, bolts, and fasteners tight. Lubricate all moving metal parts as required.
- Ensure that stepladders are fully open with spreaders locked before climbing.
- Do not stand on the top two rails of a stepladder or on the top three rungs of an extension ladder.

- Use the three-point climbing method when ascending or descending a ladder.
- Never place a ladder in front of a door unless appropriate precautions have been taken.

Tech Tip

Only extra-heavy-duty (type IA) and heavy-duty (type I) ladders shall be used with ladder jacks. Medium-duty (type II) and light-duty (type III) ladders shall never be used with ladder jacks.

Scaffold Safety

OSHA regulations state that scaffolding may be erected, moved, altered, or dismantled only under the supervision of a competent person. The following precautions should be observed because the lives of workers depend on the construction of scaffolding.

- Use only 2″ nominal structural planking that is free of knots for scaffold platforms.

- Platform end extensions must be cleated with a minimum of 6″ extension and a maximum of 18″.

- Always observe working load limits. Scaffolds and their components must be capable of supporting four times the maximum intended load.

- Guardrails, midrails, and toeboards must be installed on all open sides and ends of platforms more than 10′ above the ground.

- Platform planks are to be laid with no openings more than 1″ between adjacent planks.

- Overhead protection must be provided for persons on a scaffold exposed to overhead hazards.

- Work must not be done on a scaffold during high winds or storms.

- Work must not be done on ice-covered or slippery scaffolds.

- Scaffolds with a height-to-base ratio of more than 4:1 must be restrained by the use of guylines.

- Mobile scaffolds must be locked in position when in use.

- All tools and materials must be secured or removed from the platform before a mobile scaffold is moved.

- All personnel in close proximity must be advised and aware of the movement of a mobile scaffold.

- Fall protection must be used on working heights of more than 10′.

- Safety nets for workers at any level over 25′ must be used when the workers are not otherwise protected.

- Safety nets restricting falling objects must be used when persons are permitted to be underneath a work area.

The Sinco Group, Inc.

Personnel nets are designed to withstand 350 lb dropped from a height of 50 ft. However, it is essential that nets be installed so a worker cannot fall more than 25′.

Review

1. What types of operations do industrial and mechanical technicians use hand tools for?

2. Why is a box or socket wrench the safest wrench that can be used?

3. What are the common types of pliers?

4. Describe the common punches used by mechanical technicians.

5. What are the most common types of vises used in mechanical work?

6. What is penetrating oil used for?

7. Why are double-cut files preferred over single-cut files when it comes to rough filing?

8. Describe the common taps used by mechanical technicians.

9. List two reasons why dies have a side with a 45° chamfer.

10. What tools are commonly used to drive chisels?

11. What are the two types of cuts that can be made with a handsaw?

12. Describe at least one advantage and disadvantage of using tools that operate on AC power.

13. What are the minimum and maximum diameters of circular blades? What diameters are most commonly used?

14. List at least five safety guidelines for using power tools.

15. What are the most common types of ladders used by industrial and mechanical technicians?

Digital Resources

ATPeResources.com/Quicklinks
Access Code 462409

CHAPTER 3

FASTENING METHODS

Many types of fastening methods are used in industrial facilities. These include mechanical fasteners, such as screws and rivets, along with industrial adhesives and welding. Fastening methods are available in a wide range of types, sizes, shapes, materials, strengths, procedures, and other variations, but the maintenance technician must select the best method for each application. Each has its own advantages and disadvantages.

Milwaukee Electric Tool Corp.

Chapter Objectives

- Define fastening.
- Compare the characteristics of the different types of threaded fasteners.
- Compare the characteristics of the different types of nonthreaded fasteners.
- Describe the different types of adhesives used by industrial technicians.
- Explain the welding process used by industrial technicians.

Key Terms

- threaded fastener
- screw thread
- self-tapping screw
- screw drive
- bolt
- machine screw
- washer
- nut
- nail
- rivet
- key
- pin
- adhesive
- polyurethane adhesive
- epoxy adhesive
- cyanoacrylate adhesive
- contact adhesive
- adhesive tape
- oxyfuel welding (OFW)
- arc welding
- shielded metal arc welding (SMAW)
- gas metal arc welding (GMAW)
- gas tungsten arc welding (GTAW)
- brazing
- soldering

FASTENING

Fastening is a process that affixes one part to another. Fastening methods are generally classified as nonpermanent, semipermanent, or permanent. Nonpermanent fasteners, such as a nut and bolt, can be readily disconnected without damaging the fastener. Semipermanent fasteners, such as cotter pins, may be disconnected, but some damage may occur to the fastener. Permanent fasteners, such as rivets, adhesives, or welds, are not intended to be disassembled.

There are many fastening types, methods, principles, and applications used in industrial facilities. Any technician selecting a fastening method must consider load conditions, assembly efficiency, accessibility, the operating environment, the material used, and overhaul and maintenance.

THREADED FASTENERS

The most common type of fasteners are threaded fasteners. A *threaded fastener* is a device that joins parts together with a screw thread. Threaded fasteners have several advantages. They are available in a variety of sizes, styles, strengths, and materials and are capable of joining similar or dissimilar materials. **See Figure 3-1.** They are easily installed in the shop or in the field with hand or power tools, and they are easily removed and replaced.

Threaded Fasteners

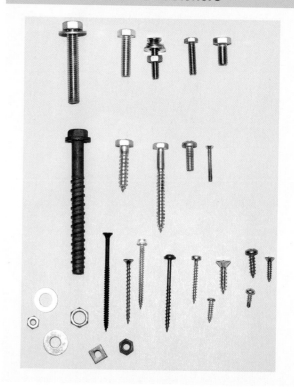

Figure 3-1. A wide variety of threaded fasteners are available, ensuring a solution for nearly any application.

A *screw thread* is a ridge in the form of a spiral on the internal or external surface of a cylinder or cone. **See Figure 3-2.** Most threads are right-handed, meaning that when viewed on end, the fastener moves away from the viewer when turned clockwise into an internal thread. Left-hand threaded fasteners are available for special applications.

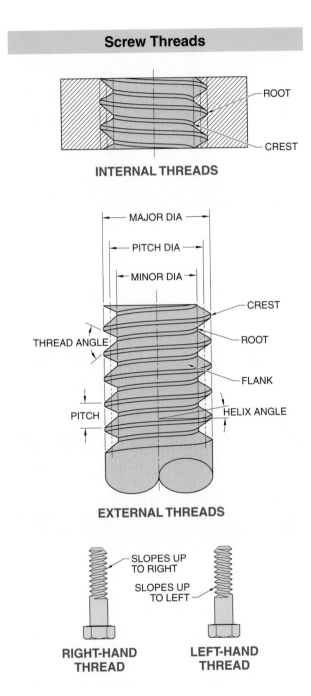

Screw Threads

INTERNAL THREADS

ROOT
CREST

MAJOR DIA
PITCH DIA
MINOR DIA

CREST
ROOT
THREAD ANGLE
FLANK
PITCH
HELIX ANGLE

EXTERNAL THREADS

SLOPES UP TO RIGHT
SLOPES UP TO LEFT

RIGHT-HAND THREAD
LEFT-HAND THREAD

Figure 3-2. A screw thread is a ridge in the form of a spiral on the internal or external surface of a cylinder or cone.

Threaded fasteners are broadly classified into two groups—self-tapping screws, and bolts and machine screws. However, the terminology for some threaded fasteners can be confusing as it is not entirely consistent or clearly defined. For example, the common use of the term "screw" may not clearly convey the type of fastener being referred to. Context may also be needed.

Self-Tapping Screws

A *self-tapping screw,* or *screw,* is a threaded fastener with a tapered point and screw threads that bite into a compressible or deformable material as it is driven. **See Figure 3-3.** *Tapping* is the forming of internal threads in a material. Depending on the material, sometimes a pilot hole must be drilled first. The hole must be an appropriate size for the screw. The hole must be smaller than the diameter of the threads so that the threads can cut into the material, but it must not be so small that the screw cannot displace enough material to advance. The screw is secured in place by the friction of its threads engaging the material.

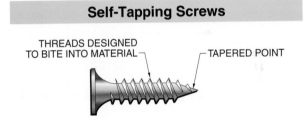

Self-Tapping Screws

THREADS DESIGNED TO BITE INTO MATERIAL
TAPERED POINT

Figure 3-3. A self-tapping screw has a tapered point and threads designed to bite into a material.

Examples of self-tapping screws are wood screws, sheet-metal screws, Tapcon® screws, and lag screws. Many are known simply as self-tapping screws, though there are many types that vary slightly depending on the type of material they are intended to screw into. For example, some have a notch on the side of the point to help cut into material, and some have double threads, one taller and one shorter, for engaging certain materials.

Self-tapping screws are available in a variety of materials, sizes, head shapes, and drives. The range of sizes varies for different types of screws, but most include diameter sizes from #2 to #12 (gauge sizes that are smaller than about ¼″) and then fractional inches. Lengths range from fractions of an inch to several inches. There are also many screw head styles and shapes. **See Figure 3-4.**

Screw Head Styles

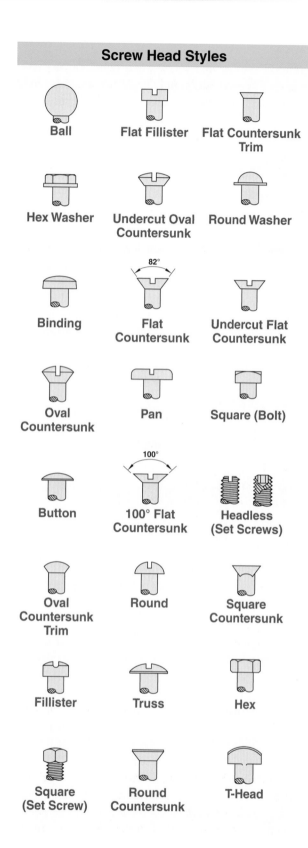

Figure 3-4. Screw head styles and shapes are designed to meet the requirements of different installations.

A *screw drive* is a standardized internal shape recessed in a screw head that allows the screw to be rotated with a matching tool. These tools include manual screwdrivers and electric impact drivers. There are many screw drives, each designed for certain characteristics, such as strength, self-centering, torque handling, ease of insertion, and tamper-resistance. **See Figure 3-5.**

Screw Drive Styles

Hex Socket

Phillips

Drilled Spanner

Slotted

Clutch

Pozidriv

One-Way

Fluted Socket

Frearson

Slotted Spanner

Figure 3-5. A screw is rotated into a material with a tool engaged into the screw's drive recess. Many different drive styles are available, though only a few are common.

Bolts and Machine Screws

In contrast to self-tapping screws, bolts and machine screws are intended to be used with preformed internal threads. The difference between a bolt and a machine screw (often referred to just as a "screw," though this term often leads to confusion with self-tapping screws) is not clearly defined. However, the industry, by general consensus, distinguishes them by size, thread length, drive style, and application. **See Figure 3-6.**

Bolts and Machine Screws

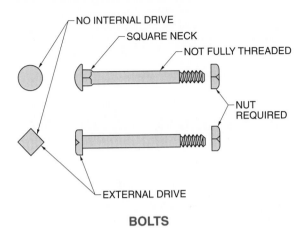

BOLTS

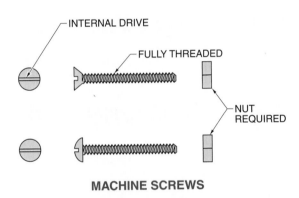

MACHINE SCREWS

Figure 3-6. The distinction between bolts and screws is not standardized or consistent. However, the industry, by general consensus, uses a few characteristics to designate threaded fasteners as one or the other.

A *bolt* is a large, partially threaded fastener with no internal drive recess that is intended to be used with a nut. A bolt is often threaded for only part of its length because the nut only engages the threads near the end. Bolts are installed in through holes, which are large enough for the threads to pass through without engaging. Therefore, they do not require internal drive recesses to advance them into the parts to be fastened. However, some bolts have features in the neck section that bite into compressible materials (such as wood) to resist rotation when the nut is tightened. These bolts have a flat or smooth, rounded head shape. Other bolts have a square or hex head, which can be used to drive the bolt

externally with a wrench, but it is usually used to resist rotation (held with a wrench) when the nut is tightened. These bolts are used when fastening incompressible materials, such as metal.

Common, general purpose bolts include hex head bolts and carriage bolts, along with their many variations. **See Figure 3-7.** A wide variety of specialty fasteners are also available for nearly any type of application.

A *machine screw*, or *screw*, is a fully threaded fastener with an internal drive recess that is intended to be engaged into an internally threaded material. However, fasteners commonly referred to as "screws" are still often secured with nuts, which does not strictly follow this definition and often leads to confusion. A machine screw is tightened from the head end with a driver tool that matches the screw's internal drive recess.

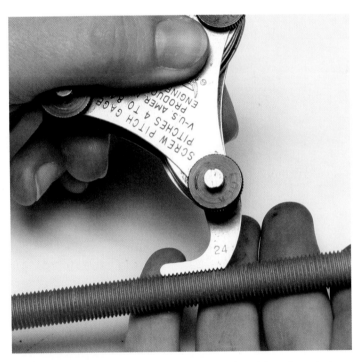

Thread pitch gauges are used to identify the number of threads per inch on a bolt or machine screw.

Bolt Grades. Bolt hardness and tensile strength are designated with grades, as specified by industry standards. *Tensile strength* is the maximum load in tension (pulling apart) that a material can withstand before breaking or fracturing. Bolt grades are identified by grade number, which is marked on the head with radial lines. **See Figure 3-8.**

Bolt Styles

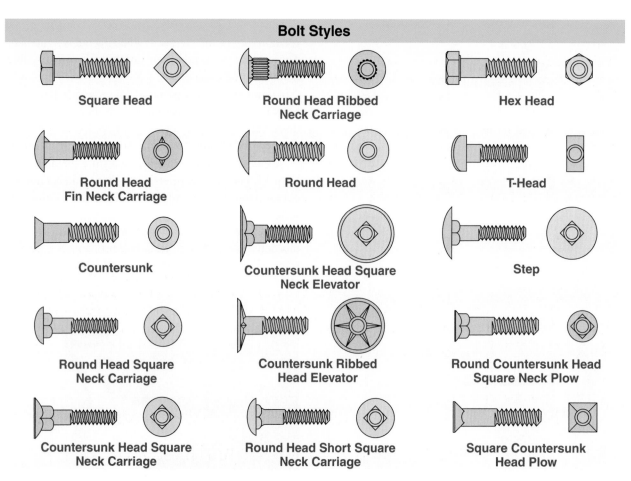

Square Head

Round Head Ribbed Neck Carriage

Hex Head

Round Head Fin Neck Carriage

Round Head

T-Head

Countersunk

Countersunk Head Square Neck Elevator

Step

Round Head Square Neck Carriage

Countersunk Ribbed Head Elevator

Round Countersunk Head Square Neck Plow

Countersunk Head Square Neck Carriage

Round Head Short Square Neck Carriage

Square Countersunk Head Plow

Figure 3-7. The most common bolt styles are similar, with only minor variations.

Bolt Grade Markings			
Bolt Head Marking	SAE/ASTM Definitions	Material	Minimum Tensile Strength*
NO MARKS	SAE/Grade 1 Grade 2 Indeterminate quality	Low-carbon steel	65,000
3 MARKS	SAE/Grade 5 ASTM A 449 Common commercial quality	Medium-carbon steel, quenched and tempered	120,000
5 MARKS	SAE/Grade 7	Medium-carbon alloy steel, quenched and tempered, roll threaded after heat treatment	133,000
6 MARKS	SAE/Grade 8 ASTM A 354 Best commerical quality	Medium-carbon alloy steel, quenched and tempered.	150,000

* in psi

Figure 3-8. Bolts are graded according to their minimum tensile strength. The grades are indicated on the head with markings.

Bolt and Machine Screw Sizes. Bolt and machine screw sizes include a nominal thread diameter and the number of threads per inch (TPI), separated by a hyphen. For example, ¼-20 screw has a nominal diameter of ¼″ and 20 TPI. Fasteners smaller than ¼″ in diameter use a gauge number designation from #0 to #12 for the size, similar to the self-tapping screw sizes. **See Figure 3-9.** Therefore, an #8-32 machine screw is 0.1640″ in diameter with 32 threads per inch. Metric fasteners use a similar designation system, except the diameters are in millimeters and the thread is indicated by pitch, or the distance between the threads. For example, an M20´2 screw has a diameter of 20 mm and a thread pitch of 2 mm per thread.

Bolts and machine screws are manufactured in both coarse thread and fine thread. For example, ¼″ screws are commonly available in 20 TPI (coarse) and 28 TPI (fine). Coarse threads are more durable and offer greater resistance to stripping or cross-threading. And due to the lower thread pitch, coarse thread allows for faster installation. Coarse-threaded fasteners are more common for most applications. However, fine-threaded fasteners are stronger than the corresponding coarse-threaded fasteners of the same hardness. Fine-threaded bolts are less likely to loosen under vibration and require less torque to tighten.

Washers

A *washer* is a small metallic disc with a hole in its center used under the head of a bolt or screw, and/or under a nut, to spread the load (tightening force) over a larger area. Some washers also prevent the fastener from unthreading unintentionally. The three basic types of washers are plain, spring lock, and tooth lock. **See Figure 3-10.**

Bolt and Machine Screw Sizes			
Diameter		Threads per Inch	
Gauge	Decimal Size*	Coarse	Fine
0	0.0600	---	80
1	0.0730	64	72
2	0.0860	56	64
3	0.0990	48	56
4	0.1120	40	48
5	0.1250	40	44
6	0.1380	32	40
8	0.1640	32	36
10	0.1900	24	32
12	0.2160	24	28
Size*			
¼″	0.2500	20	28
5⁄16″	0.3125	18	24
⅜″	0.3750	16	24
7⁄16″	0.4375	14	20
½″	0.5000	13	20
9⁄16″	0.5625	12	18
⅝″	0.6250	11	18
¾″	0.7500	10	16
⅞″	0.8750	9	14
1″	1.0000	8	12

* in in.

Figure 3-9. Bolt and machine screw diameters are designated with either gauge numbers or fractional inch sizes. For each size, there are standards for the number of threads per inch.

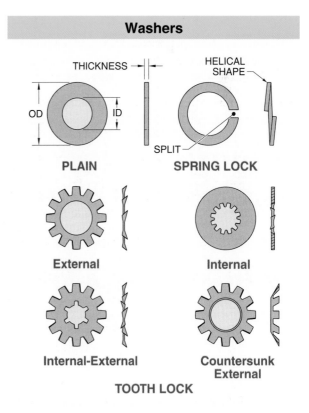

Washers

Figure 3-10. Washers are used under the head of a bolt or screw, and/or under a nut, to spread the load (tightening force) over a larger area. They may also prevent the fastener from unthreading unintentionally.

Plain Washers. Plain washers are round and flat. They are used under the head of a screw or bolt or under a nut to spread a load tightening force over a greater area. They are also used to prevent the marring of parts during assembly as a result of turning a screw, bolt, or nut. Plain washers are available in zinc, stainless steel, SAE zinc, fender, and SAE grade zinc.

Spring Lock Washers. Spring lock washers are split on one side and are helical in shape. They are made of hardened steel, bronze, or aluminum alloys. Spring lock washers have several functions. As springing devices, they maintain tension in the fastener and prevent its loosening from vibration or corrosion. Spring lock washers also act as hardened bearing surfaces and provide uniform load distribution.

Tooth Lock Washers. Tooth lock washers have hardened teeth along an edge that are bent and offset in opposite directions to bite or grip both the work surface and the bolt head or nut. This locks the fastener to the assembly or increases the friction between the fastener and the assembly. The teeth can be external, internal, internal-external, or countersunk external. Tooth lock washers also make good electrical contacts. Unlike spring lock washers, they do not provide spring action to counteract wear or stretch in the parts of an assembly.

An external tooth lock washer is the most commonly used tooth lock washer, but an internal tooth lock washer is used when necessary for the sake of appearance and to ensure engagement of the teeth with the bearing surface of the fastener. If additional locking ability is required or if there is need for a large bearing surface, such as over a clearance hole, then an internal-external tooth lock washer may be used. Countersunk external tooth lock washers are used with flat head and oval head machine screws.

Nuts

A *nut* is a small block of metal with an internally threaded hole that is threaded onto the end of a matching bolt or machine screw to assemble a stack of parts. When tightened, the nut pulls the bolt or machine screw in tension, which compresses the parts and increases friction between the internal and external threads. This helps keep the assembly together and the fastener from loosening.

There are many different types of nuts. **See Figure 3-11.** They vary greatly in materials and sizes, though these must be compatible with the bolt or machine screw they are used with. Most are square or hexagonal in shape so that they can be tightened with a wrench. Other features include nylon locking inserts, flanges (built-in washers), attached lock washers, or slots to allow for a cotter pin placement.

Common Nuts

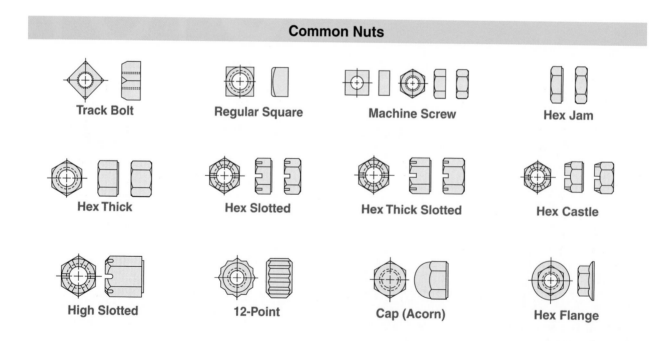

Track Bolt   Regular Square   Machine Screw   Hex Jam

Hex Thick   Hex Slotted   Hex Thick Slotted   Hex Castle

High Slotted   12-Point   Cap (Acorn)   Hex Flange

Figure 3-11. Nuts are used to secure a threaded fastener.

NONTHREADED FASTENERS

Various methods have been developed to fasten parts mechanically without threads. Common nonthreaded fasteners used are nails, rivets, keys, and pins.

Nails

A *nail* is a small metal, rod-shaped spike with a broadened circular head that is driven typically into wood with a hammer. Nails are usually made of steel, although they can also be made of aluminum, brass, and many other metals. They may also be coated or plated to improve corrosion resistance, to improve gripping strength, or for a decorative appearance. Without a coating, nails should not be used where they could rust from exposure to the environment.

The most common nail type is the round head nail. **See Figure 3-12.** Round head nails (sometimes referred to as framing nails) are used mostly for rough carpentry where appearance is not important but strength is essential. Round head nails are available as either common or sinker nails. Common nails are generally plain nails with smooth heads and no coating. Sinkers have a waffle pattern on the top of the head and come in a variety of coatings. The waffle pattern provides a nonskid surface for the hammer face, and the coating improves holding power.

Nails are driven into wood either by hand with hammers or mechanically with pneumatic- or combustion-powered nail guns. Nail guns typically require the use of special nails that are bound together in strips. **See Figure 3-13.** These tools are capable of causing serious injury or death and should be used only according to the manufacturer's instructions and by those wearing appropriate PPE. Nail guns include safety features to avoid unintentional release, but safe practices should always be followed to prevent accidents.

Nail Guns

Figure 3-13. Nails can be hammered in place by hand, but most are shot into wood with powered nail guns.

A nail derives its holding strength by friction with wood along its length. The nail also provides strength to two pieces of wood in shear (sliding force). Certain features or nailing techniques can maximize holding strength. **See Figure 3-14.** Grooves or serrations are added to some nails to improve their grip. An even more permanent grip is accomplished through the use of thermoplastic coatings. These coatings heat up through friction while a nail is driven, and then quickly cool to set and lock the nail in place. Driving adjacent nails at slight angles makes it difficult for the wood to pull away. Also, nails should be driven across the grain to increase friction with the wood.

Nails

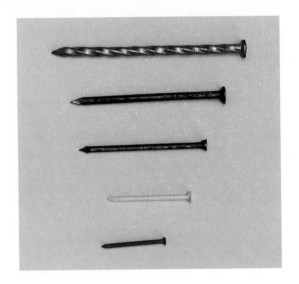

Figure 3-12. Nails are a basic fastener. They are available in many types, sizes, and materials.

Nailed Joint Strength

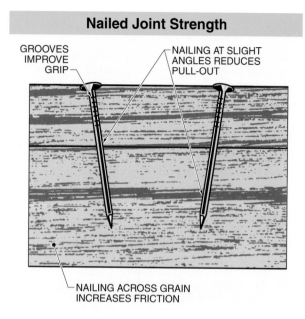

Figure 3-14. Some nail features and hammering techniques can be used to maximize a nail's holding strength in wood.

Nail Sizes

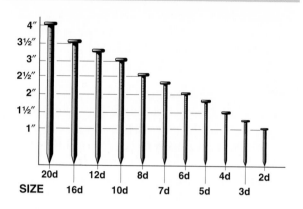

Figure 3-15. Nail lengths are designated by the penny unit (d). Nail diameter varies accordingly.

Nailing can often split wood, particularly if the wood is hard or the nail must be close to an edge. If wood is likely to split, a pilot hole slightly smaller than the diameter of the nail may need to be drilled. Alternatively, sometimes wood splitting can be avoided by blunting the point of a nail. This is done by tapping the point of the nail with a hammer. The blunt tip tends to cut the wood fibers as it enters, rather than wedge the fibers apart.

Similar nail types or categories are tacks, brads, or spikes. The terminology is related to length. Generally, nails under 1″ are called tacks or brads and are measured in fractions of an inch. Nails between 1″ and 4″ in length are referred to as nails and measured in inches or pennies (d). **See Figure 3-15.** For instance, a two-penny (2d) nail is 1″ long; a six-penny (6d) nail is 2″ long; and a 16-penny (16d) nail is 3-½″ long. The most commonly used sizes are 8d and 16d. Nails over 4″ are often called spikes.

Rivets

A *rivet* is a permanent mechanical fastener with a cylindrical shaft, a preformed head on one end, and a head on the other that is formed in place. Original rivet designs are installed by inserting the rivet through a hole and pressing or beating the inserted end to form a head. **See Figure 3-16.** This form of riveting is still popular. However, many other rivet types have been devised.

As with other threaded fasteners, there are many common head styles for rivets. **See Figure 3-17.** Low-profile domed heads are used for most applications. However, when softer materials are fastened, large flanged heads are recommended. Where flush surfaces are required, countersunk heads are available.

Solid Rivets. Solid rivets are used to assemble bridges, cranes, and building frames. They are also used with the structural parts of aircraft. Historically, rivets were heated and then hammered into shape by hand. Presently, solid rivets are pneumatically, hydraulically, or electromagnetically pressed into place, often without the need for preheating. When placed, the tail end is mushroomed to about 1.5 times the original shaft diameter. These fasteners require access to both sides of a structure.

Semi-Tubular Rivets. A semi-tubular rivet is similar to a solid rivet, but it has a partial hole at the tail end. **See Figure 3-18.** This end is rolled outward during installation and requires much less force than that required for forming solid rivet heads. Also, unlike the solid rivet, installation of a semi-tubular rivet does not expand the rivet shank to fill the assembly hole, which retains an allowance for pivoting.

Semi-tubular rivets are commonly used in lighter assemblies that cannot tolerate the force required to install solid rivets, such as circuit boards, brakes, ladders, binders, or lighting assemblies. They are available in various metals for mechanical or decorative reasons, such as brass, aluminum, stainless steel, or copper.

Riveting

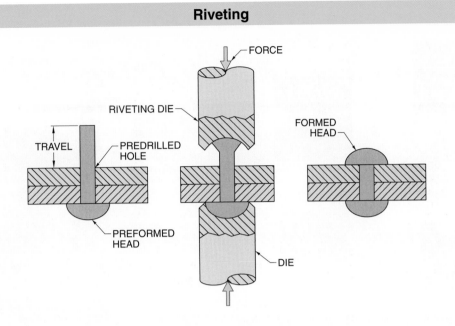

Figure 3-16. Riveting forms a permanent fastener in place by the use of force.

Rivet Head Styles

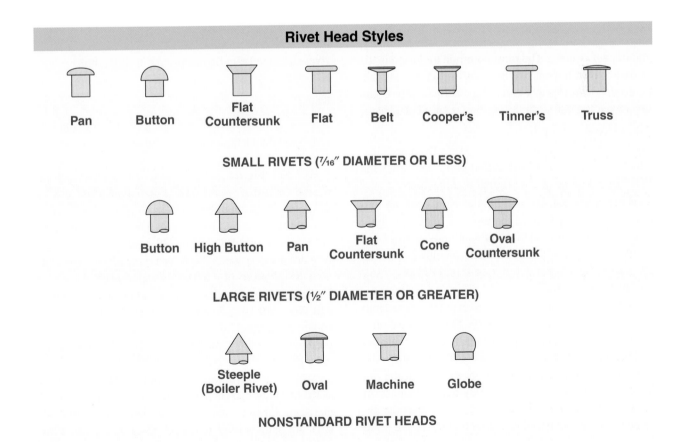

Pan   Button   Flat Countersunk   Flat   Belt   Cooper's   Tinner's   Truss

SMALL RIVETS (7/16″ DIAMETER OR LESS)

Button   High Button   Pan   Flat Countersunk   Cone   Oval Countersunk

LARGE RIVETS (½″ DIAMETER OR GREATER)

Steeple (Boiler Rivet)   Oval   Machine   Globe

NONSTANDARD RIVET HEADS

Figure 3-17. Rivet heads come in a variety of shapes and sizes.

Semi-Tubular Rivets

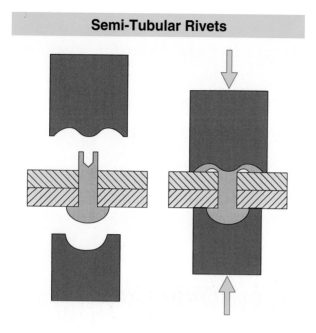

Figure 3-18. A semi-tubular rivet has a depression in the shaft on one end, which is used to form the end into a new rivet head.

Blind Rivet Installation

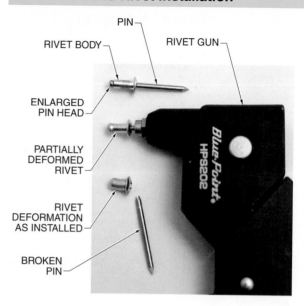

Figure 3-19. Blind rivets can be installed when only one side of an assembly is accessible.

Tools used to install a semi-tubular rivet include manual tools, tool sets, kick presses, robotic machinery, and impact riveters. The most common tool used is the impact riveter. Impact riveting uses impact force to fasten components together quickly and permanently. Some impact riveters can set both solid and semi-tubular rivets.

Blind Rivets. Blind rivets are light-duty rivets that are used to fasten thin materials to an object from one side. Blind rivets are commonly called pop rivets, after the POP® brand name of the original manufacturer.

Blind rivets consist of a tubular rivet body around a metal pin. They require the use of a blind rivet gun. The pin comes inserted in the tubular body, with a head larger than the tube diameter. There are notches on the pin's stem that allow the pin to be removed easily after the rivet is fastened.

To fasten a blind rivet, first the rivet pin is inserted into a blind rivet gun. **See Figure 3-19.** The rivet assembly is inserted fully into the predrilled hole, and the rivet gun handle is ratcheted several times. The gun grips the pin and pulls it through the rivet body, forcing the pin head to expand the softer rivet body material and form the second rivet head. As the pin head reaches the assembled components, resistance increases, and the force of continued ratcheting then breaks the pin. The fastening is then complete, and the broken pin can be discarded.

To remove a blind rivet, a drill is used to cut through the rivet head until it falls away. The remainder of the rivet can then be pulled or pried out of the hole.

Keys

A *key* is a small, removable piece of steel of standard shape and dimensions that is placed in a keyseat between a shaft and hub to provide a means for transmitting power. **See Figure 3-20.** A *keyseat* is a groove along the axis of a shaft or hub. The hub is typically part of a gear, pulley, or coupling. The key is placed partially in the shaft keyseat and partially in the hub keyseat. Keys are available in stock sizes in US customary and metric measurement systems. Sizes are typically in standardized designations of letters and/or numbers.

The three main types of keys include parallel, taper, and Woodruff keys. **See Figure 3-21.** Parallel keys are square or rectangular in cross section. Parallel keys are used for transmitting unidirectional torques in shafts and hubs that do not have heavy starting loads. Taper keys are used for transmitting heavy unidirectional torques in shafts and hubs that are reversed frequently and subject to vibration. Taper keys may be easily withdrawn. Woodruff keys are semicircular in shape. They may have a full radius or flat bottom. Woodruff keys are used for transmitting light torques or aligning parts on tapered shafts.

Keys

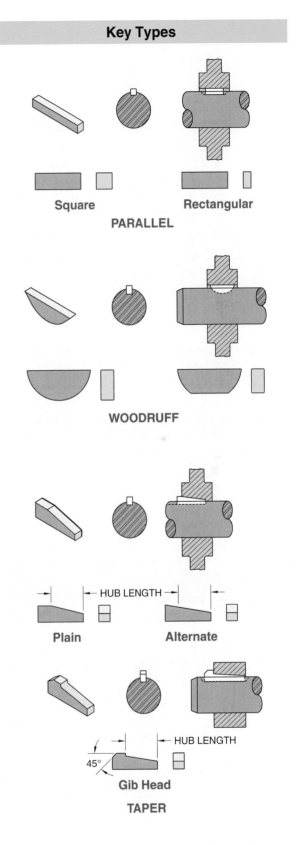

Figure 3-20. A key forms a mechanical connection between a shaft and hub, such as those on gears and pulleys.

Pins

A *pin* is a cylindrical fastener that is placed into a hole to secure the relative positions of two or more parts. **See Figure 3-22.** A wide variety of pin types, sizes, and materials are commercially available. **See Figure 3-23.** Standard pins include the following:

- Straight pins are usually fabricated from bar stock with square or chamfered ends. They are often used to transmit torque in round shafts.
- Taper pins have rounded ends and a uniform taper. Taper pins are used to transmit torque between shafts and hubs or to position parts.
- Clevis pins are used to attach clevises to rod ends and rigging. They are also used to serve as bearings. Clevis pins are held in place by cotter pins.
- Cotter pins are pins made from a looped piece of wire. After insertion through a hole, the ends are bent open to keep the cotter pin in place. Cotter pins are often used with clevis pins and slotted nuts to prevent disengagement.
- Spring pins rely on spring tension to hold themselves in place. Spring pins include slotted and coiled spring pins.
- Grooved pins are solid pins with three parallel, equally spaced grooves. The grooves provide a tight fit and a locking feature. There are several styles of grooves.

Tech Tip

Cotter pins are made from relatively soft metals, making them easy to install and remove. Some can be reused and some cannot.

Key Types

Figure 3-21. Key types are optimized for various applications.

Pins

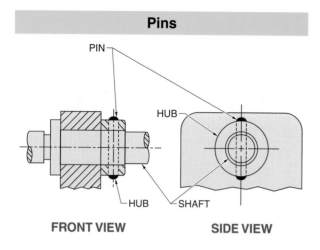

FRONT VIEW   **SIDE VIEW**

Figure 3-22. Pins are another way of connecting a hub to a shaft with aligned holes.

ADHESIVES

An *adhesive* is a substance that is used to bond materials together at the surface. Early adhesives were made of natural materials, but the rapid growth of manufacturing encouraged the development of synthetic adhesives with chemical (polymer) joining methods. These newer adhesives have a high degree of structural strength as well as a resistance to fatigue and severe environments.

The wide variety of off-the-shelf and custom-blended adhesives makes nearly any bonding task possible. **See Figure 3-24.** These high-strength and tough adhesives allow for lighter and more permanent joints than mechanical fasteners. However, adhesive joints cannot usually be disassembled without damage, making repairs or replacement of parts more costly or impossible. Also, bonded joints often fail instantaneously instead of progressively.

Pin Types

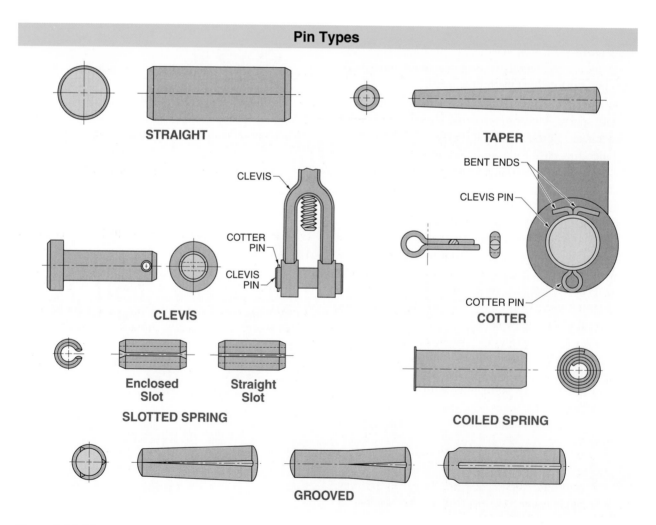

Figure 3-23. Pins are available in various types and with different features.

Adhesive Types										
Adhesive	**Components**	**Cure Time**	**Viscosity**	**Void-Filling**	**Flexibility**	**Heat Resistance**	**Cold Resistance**	**Thermal Resistance**	**Water Resistance**	**Metal Bonding**
Acrylic	one-part (UV or heat cure) two-part	medium to fast	medium	good	good	good	good	good	good	good
Anaerobic	one-part	medium	low	poor to fair	good	good	good	good	good	fair
Cyanoacrylate	one-part	fast	low	poor to fair	poor to fair	fair	fair	good	good	good
Epoxy	two-part	slow to medium	medium to high	excellent	fair	good	fair	good	good	good
Hot Melt	one-part	high	high	excellent	fair to good	poor to fair	fair	fair	fair	fair
Polyurethane	one-part two-part	medium	medium	good	good	fair	good	good	good	good
Polysulfide	one-part two-part	high	high	excellent	good	good	good	excellent	excellent	good
Silicone	one-part two-part	high	high	excellent	excellent	excellent	excellent	excellent	excellent	fair
Solvent-Base	one-part	low to medium	low to medium	poor to fair	good	good	good	good	good	good
Water-Base	one-part	low to medium	low to medium	poor to fair	poor to fair	fair	fair	poor	poor	poor to fair

Figure 3-24. A wide variety of adhesives are available, though only a few are common for industrial applications.

Polymer Adhesives

A *polymer* is a molecule made up of a chain of repeating units that are chemically bonded together. A *polymer adhesive* is a synthetic bonding substance that undergoes a chemical or physical reaction. In general, polymer adhesives are more flexible, have greater impact resistance, and are stronger than other types of adhesives. They can also support both alternating and static loads. Polymer adhesives are available in many varieties. The most commonly used are polyurethanes, epoxies, and cyanoacrylates.

Polyurethanes. A *polyurethane adhesive* is a durable synthetic resin polymer adhesive available in multiple variations, each optimized for different applications. Polyurethanes are very strong and flexible, and they bond well with many surface types. They are resistant to oils, fuels, acids, and chemicals, so they are popular for coating and sealing oil tanks, fuel tanks, and chemical processing equipment. Aluminum-filled polyurethane is used as a paint for mowers, bridges, trailers, and metal structures. Popular polyurethane adhesives include thread lock compounds, floor-tile adhesives, and paint primers.

Epoxies. An *epoxy adhesive* is a synthetic polymer resin adhesive that is chemically cured from two mixed liquids. One liquid is the resin and the other is a hardener. **See Figure 3-25.** When the two parts are mixed, curing (hardening) begins.

Epoxy formulations are available in many different liquid viscosities, setting times, final hardnesses, and other features. In general, epoxies are heat and chemical resistant and can bond plastics, wood, stone, fiberglass, and metal. Due to their electrical insulating qualities, they are a frequent choice for electronic and electrical applications. Because of their strong bonding quality, epoxy is the adhesive of choice for aircraft, marine, and auto manufacturers.

Epoxies

Figure 3-25. Epoxy adhesives come in two liquids, which must be mixed in the proper ratio to ensure proper curing.

Cyanoacrylates

Figure 3-26. Cyanoacrylate adhesives are very fast setting.

There is a special one-part epoxy that has high strength, offers excellent adhesion to metals, and is able to withstand harsh environmental and chemical conditions. However, this epoxy requires curing at temperatures between 250°F to 300°F.

Cyanoacrylates. A *cyanoacrylate adhesive* is an acrylic polymer resin adhesive that bonds very quickly. **See Figure 3-26.** This adhesive offers high mechanical strength with only small drops. The consumer version of this adhesive is known as Super Glue®.

Cyanoacrylates come in one part but are essentially two-part adhesives with water as the hardener. Curing begins by contact with the moisture found on the surface of the workpiece or in the environment. When placed on a perfectly dry surface or on certain plastics, this polymer cannot form a bond. Applying a thin layer of water or even breathing on a material may add enough moisture for a strong bond. The most common formulation is very thin and bonds only to nonporous surfaces with no gaps between the parts, though thick versions are also available.

However, cured cyanoacrylates have a low resistance to water and sunlight, so they are not suitable for joints that are exposed to the weather. Also, while strong under some stress, cyanoacrylates are more brittle than poly-urethanes and epoxies and may break under vibrations or high forces.

Cyanoacrylates can easily bond to skin, which usually contains enough moisture for curing. If this should happen, most cyanoacrylates dissolve with either acetone or hot water. Extreme care should be taken to prevent contact with the eyes.

Contact Adhesives

A *contact adhesive* is a flexible synthetic adhesive that is applied separately to two surfaces, which are then brought into contact. This adhesive is applied to both parts and is left to become dry to the touch. When the two surfaces are assembled, it provides an instant and permanent bond. Once the surfaces make contact, they cannot be separated without damaging the parts. There is no need for clamping, though pressure is usually applied to the surfaces to ensure contact. Contact adhesives are especially useful on nonporous surfaces such as plastic, glass, and metal.

Tech Tip

A type of cyanoacrylate adhesive is used in medicine as an alternative to sutures (stitches) for small wounds and incisions. The adhesive can be applied quickly, has lower infection rates, and reduces scarring. The small amounts used for this application are safe and gradually wear off after healing.

Many contact adhesives are extremely flammable. Proper PPE must be worn when using a contact adhesive, and the location where it is used must be properly ventilated. The manufacturer's recommendations and industry guidelines must always be followed.

Adhesive Tapes

Adhesive tape is a strip of material coated on one or both sides with an adhesive and used for the purpose of fastening. Adhesive tape backing is a continuous flexible strip of fabric, paper, metal foil, plastic film, or foam. A variety of adhesives can be applied to different types of backing, making the number of possible adhesive tape combinations vast. **See Figure 3-27.**

Adhesive Tapes

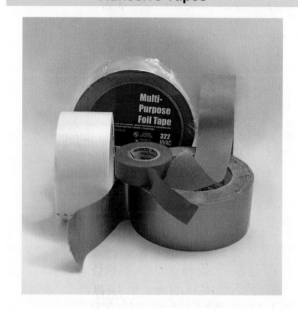

Figure 3-27. Adhesive tapes are available with different backing materials and adhesive types.

When applying adhesive tape, the mating surfaces should be cleaned of all oil, grease, water, rust, dust, and dirt. The tape is pressed firmly into position, eliminating all entrapped air. Whenever possible, tape should be applied in ambient temperatures above 40°F. Adhesion is generally better in higher temperatures. The two basic elements of adhesive tape is the tape backing and the adhesive.

Tape Backings. Tape backing is a highly engineered and essential component of adhesive tape. Tape backing is chosen for the application, the method of use, and its strength. For example, vinyl-backed electrical tape is electrically nonconductive and elastic. Fiberglass tapes are used in extreme high-temperature settings, and silicone tapes are used where waterproofing is necessary. Also, the basic strength of an adhesive tape is mostly based on the backing material. Common tape backings include the following:

- Vinyl or polyethylene plastic backing is elastic and nonconductive. It is used for piping, plumbing, and electrical work.
- Fiberglass, aluminum foil, or synthetic rubber backings are temperature resistant, which is useful with HVAC systems.
- Foam backing is gap-filling, which is useful for slightly irregular surfaces.
- Polyvinylchloride (PVC) is used for insulating tapes.
- Polypropylene backing offers good pulling strength, which makes it an ideal material for strapping, packaging, and bundling.
- Paper backing is easy to tear, making is useful for masking tape.

Other characteristics required of tape backing may include abrasion resistance, antistatic properties, permeability, sealing ability, or resistance to moisture, chemicals, and extreme temperatures. Adhesive tape backing can also be clear, color-coded, or reflective; thick or thin; and wide or narrow.

Tech Tip

In addition to the typical pressure-sensitive adhesive tape, other types of tape include water-activated and heat-sensitive tape.

Tape Adhesives. Tape adhesives are available in many varieties according to their fastening chemistries and types, including rubber, resins, and silicone. However, most adhesive tapes are applied with only pressure, which limits them to one-part chemistries with a sticky surface. One exception, though, would be water-activated tape adhesives. Common tape adhesives include the following:

- Rubber-based compounds, which provide waterproofing and flexibility
- Acrylic resins, which provide fast setting times with excellent environmental resistance
- Epoxy resins, which exhibit high strength and chemical resistance
- Silicone adhesives, which are flexible and have very high temperature resistances

Adhesive tapes are generally offered in rolls. A single-sided tape is typically manufactured with four layers: adhesive, primer, backing, and release coat. The release coat is what keeps the adhesive tape from sticking to itself when rolled. The primer allows the adhesive chemical to stick to the backing. Double-sided tapes are manufactured in layers as well, with a release liner that is removed upon use.

WELDING

Welding is a joining process that fuses materials by heating them to melting temperature. A filler metal is often used to reinforce a weld joint. Welding processes commonly used in maintenance work are oxyfuel welding (OFW) and shielded metal arc welding (SMAW). Other welding processes may be required for certain metals or specialized welding tasks.

Oxyfuel Welding

Oxyfuel welding (OFW) is a welding process that produces heat from the combustion of a mixture of oxygen and a fuel gas. The gases are mixed in a handheld torch and ignited, where they produce a very hot flame. The flame is then directed to the metal pieces to be melted and welded. Extra metal is added to the joint as needed from a rod held in the other hand.

Oxyfuel welding is commonly used for maintenance work because it does not require electricity. Tanks of the welding gases, the torch, and the accessories can be transported on a wheeled cart. Fuels commonly used with oxygen include acetylene, MAPP (methylacetylene-propadiene) gas, natural gas, and propane. *Oxyacetylene welding (OAW)* is an oxyfuel welding process that uses oxygen mixed with acetylene. **See Figure 3-28.**

Oxygen and acetylene can also be used for cutting operations. The metal to be cut is first preheated, and then pressurized oxygen is blown onto the heated metal, which rapidly oxidizes and burns it. The burned metal is also blown away by the oxygen, leaving a rough cut.

Arc Welding

Arc welding is a welding process that uses the electrical resistance of an arc bridging a gap to generate the heat necessary for melting a workpiece. All arc welding processes require a welding machine that connects to a facility's electrical supply and converts the power to the voltage, current, and waveform required for the welding

job. An electrode from the welding machine carries the electrical current to the workpiece, where it jumps the small gap and creates an arc. The current then returns to the machine through the grounding clamp, completing an electrical circuit. A shielding gas is needed to stabilize the arc and protect the molten metal from impurities in the air.

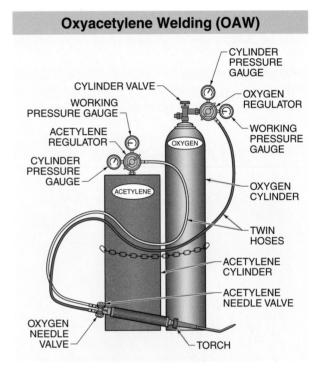

Oxyacetylene Welding (OAW)

Figure 3-28. Oxyacetylene welding (OAW) uses the combustion of acetylene and oxygen to generate the heat needed for welding. OAW is relatively portable, so it is often used for repairs.

Shielded Metal Arc Welding (SMAW). *Shielded metal arc welding (SMAW)* is an arc welding process in which the arc is shielded from impurities by the gases emitted from the decomposition of a consumable electrode covering. In SMAW, the electrode is a metal rod that melts to supply filler metal to the weld, and its coating decomposes into a shielding gas. **See Figure 3-29.** Generating the shielding gas from the coating is a convenient way to avoid the need for separate tanks of shielding gas. A variety of electrodes are available. They vary in metal and coating type for different weld requirements.

The SMAW process is commonly used in maintenance work because of its versatility, its low cost, and the minimal equipment required. It is also known as stick welding.

Shielded Metal Arc Welding (SMAW)

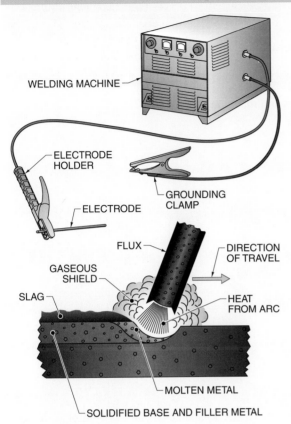

Figure 3-29. Shielded metal arc welding (SMAW) is also known as stick welding. This process uses a consumable electrode to generate an electric arc that is hot enough to weld metal.

Gas Metal Arc Welding (GMAW). *Gas metal arc welding (GMAW)* is an arc welding process that uses a continuous wire electrode. The wire electrode maintains the arc and is fed continuously by a wire feeder as it is consumed in the weld. The weld area is protected during the welding process by an inert gas shield. An *inert gas* is a gas that does not readily react with other elements. The gas is typically supplied by a tank. GMAW is also known as MIG (metal inert gas) welding.

A variation of GMAW is flux cored arc welding (FCAW). In this type of process, the wire electrode is actually a tiny tube containing a solid flux material that creates the shielding gas as it burns. A separate shielding gas is not needed, making this process relatively portable.

Gas Tungsten Arc Welding (GTAW). *Gas tungsten arc welding (GTAW)* is an arc welding process in which a shielding gas protects the arc between a tungsten electrode and the weld area. The electrode directs the arc and is not consumed during the welding process. Filler metal must be added separately. The GTAW process is also known as TIG (tungsten inert gas) welding.

Safety Tip

All welding work requires special PPE and safety procedures. Most importantly, dark filters on goggles and face shields must be used to protect the eyes from intense oxyfuel or arc light. Thick clothing and gloves protect the wearer from heat and debris.

Brazing and Soldering

Brazing and soldering are commonly used to join workpieces without melting the base metal. *Brazing* is a joining process that joins parts by heating a filler metal to temperatures greater than 840°F but less than the melting point of the base metal. *Soldering* is a joining process that joins parts by heating a filler metal to temperatures up to 840°F but less than the melting point of the base metal.

The filler metal is bonded to the weld parts by adhesion in brazing and soldering instead of fusion, as in welding. Therefore, brazing and soldering produce joints that are weaker than welded joints. However, these techniques are useful when the high heat required for welding may damage the parts. Both brazing and soldering require preparation of the joints for maximum surface contact with the filler metal. The preparation involves cutting, grinding, or filing.

Review

1. What are the general definitions of nonpermanent, semipermanent, and permanent fasteners?

2. Describe the primary difference between self-tapping screws, and bolts and machine screws.

3. Describe the generally accepted differences between bolts and machine screws.

4. How are bolt and machine screw sizes designated?

5. What is the purpose of a washer?

6. What features or techniques contribute to a nail's holding strength?

7. How are solid rivets installed?

8. How is a blind rivet installed?

9. How are the different types of keys used?

10. What are some common characteristics of polymer adhesives?

11. How is cyanoacrylate adhesive cured?

12. What is the general procedure for applying adhesive tape?

13. How is electricity used in arc welding processes?

14. What is the main difference between shielded metal arc welding (SMAW) and gas metal arc welding (GMAW)?

15. What is the difference between brazing and soldering?

Digital Resources

ATPeResources.com/Quicklinks
Access Code 462409

CHAPTER 4

PRINTREADING

Various drawings and prints are used in industry. Drawings are used to convey general information concerning devices and equipment. Prints are used to modify process control circuits and troubleshoot systems in commercial and industrial facilities. The drawing or print used depends on the work being performed. In addition to prints and drawings, specifications are used to define the materials and procedures required for a device or equipment.

Conventions and standards have been developed to ensure consistency among prints and to enhance communication between the industrial technicians repairing or assembling devices involved in a project. These conventions and standards create a framework for how information is displayed, organized, and interpreted on a print.

Chapter Objectives

- Describe the different types of drawings and their uses.
- Define print and the information included on prints.
- List the types of information displayed on print conventions.
- Describe the scales used to draw prints.
- Define specification.
- Describe the different types of sketching used on prints.
- Explain dimensioning.

Key Terms

- drawing
- pictorial drawing
- orthographic projection
- application drawing
- location drawing
- detail drawing
- assembly drawing
- instructional drawing
- sectional drawing
- plan
- print
- title block
- print convention
- drawing scale
- specification
- sketch
- picture plane
- orthographic sketch
- isometric sketch
- dimensioning
- dimension
- tolerance

DRAWINGS

A *drawing* is an assembly of lines, dimensions, and notes used to convey general or specific information as required by the application and use. Drawings such as pictorial drawings, application drawings, location drawings, and assembly drawings display enough information to produce a visual picture of what a device or component looks like, how it can be used or how it fits into a system, and where its major parts are located. Pictorial, application, and location drawings typically do not include dimensions but may include part numbers and some installation and/or mounting information. Drawings such as instructional drawings may include dimensions and callouts.

Drawings such as detail drawings and sectional drawings are used to provide as much detail as required for a clear understanding of the type, size, and dimensions of a device, component, machine, or system. Detail and sectional drawings are often used to show what surrounds a component being shown as well as to show the relationship between a component and the machine or system.

Pictorial, application, location, detail, assembly, instructional, and sectional drawings are drawn as viewed from a head-on view. A *head-on view* is a view when looking directly at an object from the same height as the object. **See Figure 4-1.**

Any drawing can be customized to show as much detail as required. For example, sectional drawings are drawn with a head-on view and as if the object were sliced open so the internal composition and construction of the object can be seen. Any drawing can include dimensions, part numbers, special notes, and/or any additional information as required to convey the information needed to do the specified work.

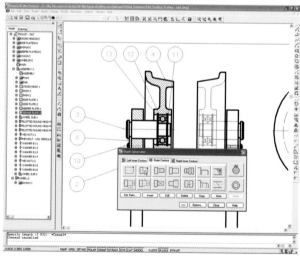

Autodesk, Inc.

Most mechanical drawings are generated on a computer through the use of specialized software programs.

Head-On Views

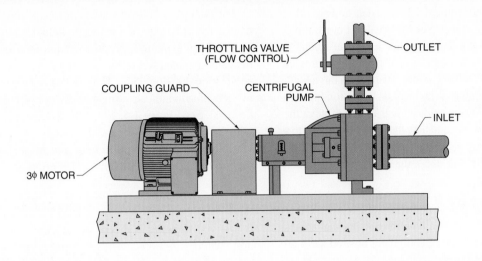

Figure 4-1. A head-on view is a view when looking directly at an object from the same height as the object.

Pictorial Drawings

A *pictorial drawing* is a three-dimensional drawing that resembles a picture. The picture can be drawn to highlight information and details required to locate connecting points, install or order a part, show the location of switches, fuses, batteries, or other parts, or provide a clear picture of how the device looks or fits into a larger system. A pictorial drawing is used to show the actual layout and position of all devices and components used in an application. In a pictorial drawing, devices and components are placed as near to their actual positions as possible. Objects in a pictorial drawing are typically drawn with great detail but can also be drawn in general outline form. **See Figure 4-2.**

Orthographic Drawings

When an object is drawn as a two-dimensional drawing, the object is drawn using orthographic projection. An *orthographic projection* is a type of drawing where the front, top, and side views of an object are projected onto flat planes. These views are generally at 90° (right) angles to one another.

Orthographic projection drawings include the dimensions and details required to convey the technical information of the object drawn. Typically, large orthographic drawings of large machines do not include

many dimensions. To avoid cluttering the drawing, only basic dimensions such as total length, width, and depth are included. Plans for large machines typically include multiple drawings that show all the dimensions.

Pictorial Drawings

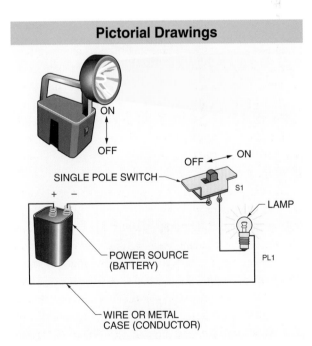

Figure 4-2. A pictorial drawing is used to show the actual layout and position of all devices and components used in an application.

Small objects drawn with orthographic views (typically front, right-side, and top views) include all the dimensions needed to convey the required technical data and provide a better picture of the object. **See Figure 4-3.** For example, industrial pushbuttons typically have a lockout means to prevent the pushbutton (start or run) from being pressed during downtime or repairs. A stop pushbutton has a locking attachment that holds it in the open condition (pushed in). Pictorial drawings provide a clear image of what the lockout device looks like and how the device is applied to a pushbutton. However, pictorial drawings do not provide information on the physical size of the device or component.

Application Drawings

An *application drawing* is a type of drawing that shows the use of a specific piece of equipment or product in an application. Application drawings show product use and are not intended to indicate component connections, wiring, dimensions, or actual size or shape. Application drawings present ideas on how to use a product in problem solving. They are used by manufacturers to promote products.

Application drawings are also used during troubleshooting to present ideas for the use of new or different components.

For example, a Hall effect sensor is a type of sensor that detects the proximity of a magnetic field. Hall effect sensors are used in many applications requiring a small magnetically operated sensor that can be used as a switch. An application drawing with a Hall effect sensor can be used in a machine to indicate if the machine is level or tilted (in degrees). **See Figure 4-4.** Typically, a magnet is installed with the Hall effect sensor and the switch is activated only when the machine is level. To fully activate the sensor, the magnet must be directly over the Hall effect sensor.

Location Drawings

A *location drawing* is a type of drawing used to position switches, buttons, terminal connections, and other features found on a device or component. For example, location drawings are often used in installation and operational manuals to show where to connect external wires and position indicating lamps, switches, and displays. **See Figure 4-5.**

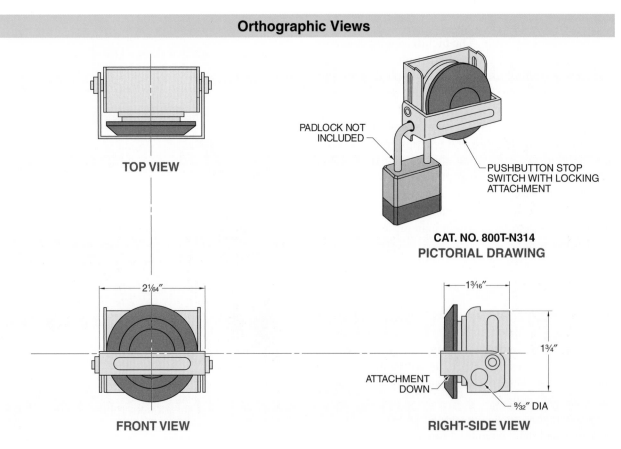

Orthographic Views

PADLOCK NOT INCLUDED

PUSHBUTTON STOP SWITCH WITH LOCKING ATTACHMENT

CAT. NO. 800T-N314
PICTORIAL DRAWING

TOP VIEW

FRONT VIEW
2¹⁄₆₄″

RIGHT-SIDE VIEW
1³⁄₁₆″
1¾″
⁹⁄₃₂″ DIA
ATTACHMENT DOWN

Figure 4-3. Smaller objects drawn with orthographic views include all the dimensions needed to convey the required technical data and provide a better picture of the object.

Application Drawings

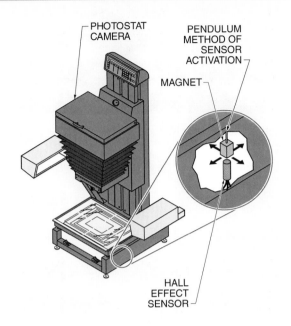

Figure 4-4. An application drawing with a Hall effect sensor (switch) can be used in a machine to indicate whether the machine is level or tilted (in degrees).

Typically, location drawings do not include dimensions. They include arrows and callouts showing the location of an object and a brief overview of what the object is used for. Location drawings can include simple callouts that state only what a switch (device) or component is, such as a "run lamp," or the callouts can provide as much information as space allows, such as "Caution: Run lamp is on when electric motor drive and motor are running."

Detail Drawings

A *detail drawing* is a type of drawing that shows as much information about a device or component as possible. A detail drawing shows how the individual parts of an object work together. Detail drawings are typically used during the construction and/or assembly of devices and components and are typically shown on service bulletins. **See Figure 4-6.**

Detail drawings can be drawn as pictorial drawings with many details and callouts. Dimensions and other information can be found on detail drawings. Detail drawings can also be drawn as orthographic projections to show one or more side views of an object.

Location Drawings

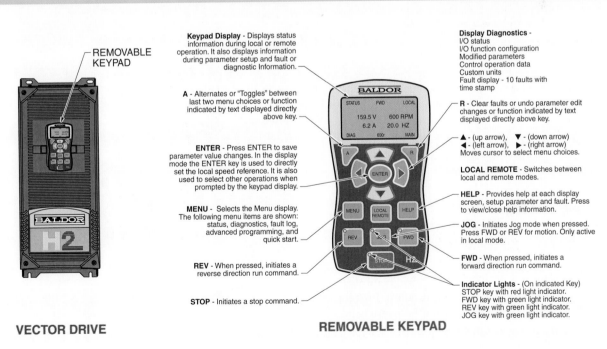

Figure 4-5. Location drawings include arrows and callouts showing the location of an object and a brief overview of what the object is used for.

Detail Drawings

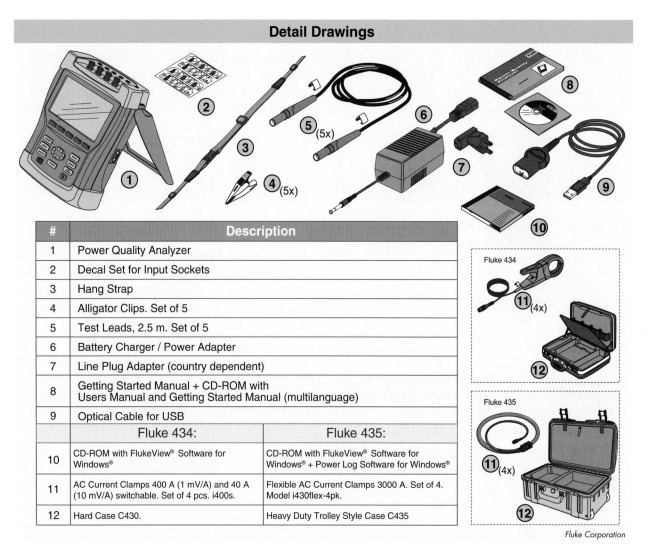

#	Description	
1	Power Quality Analyzer	
2	Decal Set for Input Sockets	
3	Hang Strap	
4	Alligator Clips. Set of 5	
5	Test Leads, 2.5 m. Set of 5	
6	Battery Charger / Power Adapter	
7	Line Plug Adapter (country dependent)	
8	Getting Started Manual + CD-ROM with Users Manual and Getting Started Manual (multilanguage)	
9	Optical Cable for USB	
	Fluke 434:	Fluke 435:
10	CD-ROM with FlukeView® Software for Windows®	CD-ROM with FlukeView® Software for Windows® + Power Log Software for Windows®
11	AC Current Clamps 400 A (1 mV/A) and 40 A (10 mV/A) switchable. Set of 4 pcs. i400s.	Flexible AC Current Clamps 3000 A. Set of 4. Model i430flex-4pk.
12	Hard Case C430.	Heavy Duty Trolley Style Case C435

Fluke Corporation

Figure 4-6. Detail drawings show as much information about a device or component as possible.

Assembly Drawings

An *assembly drawing* is a type of drawing that shows as closely as possible the way individual parts or components are placed together to produce a finished piece of equipment or result. Assembly drawings must show as much information as possible to allow for proper assembly of an object but must not be overwhelming in detail. When used, dimensions must be kept to a minimum. Dimensions are typically shown on separate detail drawings that easily relate back to the assembly drawing. **See Figure 4-7.**

Assembly Drawings

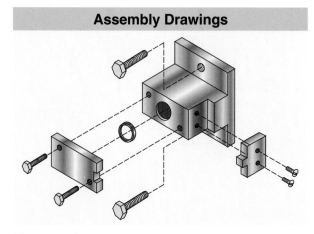

Figure 4-7. Assembly drawings show as closely as possible the way individual parts or components are placed together to produce a finished piece of equipment or result.

An assembly drawing may be an individual drawing or a series of drawings showing different steps or stages of an assembly process. The more complex the assembly, the greater the number of individual drawings required to convey information. Some assembly drawings are too complicated to follow or do not contain enough of the detail required to assemble the object.

Instructional Drawings

An *instructional drawing* is a type of drawing that is used to indicate how to do work using the simplest and/or safest method. Instructional drawings are typically pictorial drawings with arrows and callouts indicating the required work.

For example, two drive methods are used in facilities to move material along a conveyor. The two methods of conveyor operation are direct drive and roller drive. With the direct drive method, the material rides directly on a driven belt. The direct drive method is used for light loads. With the roller drive method, a belt underneath drives the rollers, and the material rides on top of them. The roller drive method is used for heavy loads.

When a conveyor belt is not tracking properly, it will drift to one side and become damaged by rubbing against the stationary parts of the conveyor. **See Figure 4-8.** To realign a conveyor belt, one side of the belt snub roller is adjusted forward or to the rear. A good instructional drawing will show how the conveyor belt is tracked using as few words as possible.

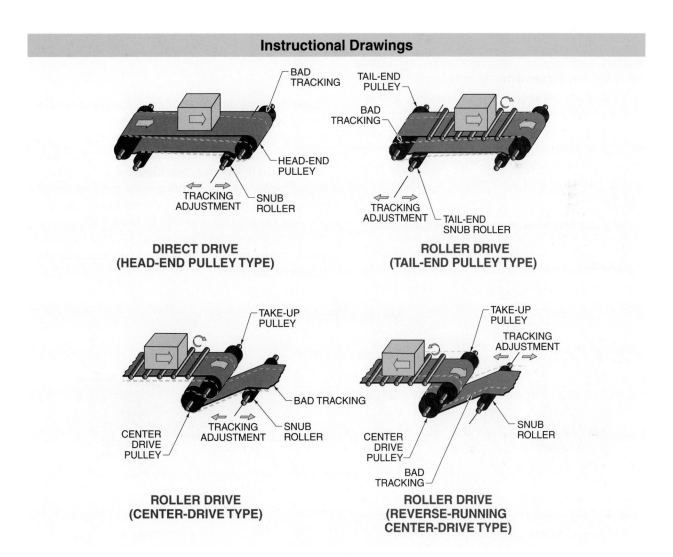

Instructional Drawings

DIRECT DRIVE (HEAD-END PULLEY TYPE)

ROLLER DRIVE (TAIL-END PULLEY TYPE)

ROLLER DRIVE (CENTER-DRIVE TYPE)

ROLLER DRIVE (REVERSE-RUNNING CENTER-DRIVE TYPE)

Figure 4-8. Instructional drawings are typically pictorial drawings with arrows and callouts indicating components and required work.

Sectional Drawings

A *sectional drawing* is a type of drawing that indicates the internal features of an object. An imaginary cutting plane is passed through the object perpendicular to the line of sight. The portion of the object between the cutting plane and the observer is removed, revealing the internal features of the object.

The views of sectional drawings can be pictorial or orthographic. **See Figure 4-9.** Section views contain hatch patterns and gradient fills that are drawn on all surfaces cut by the cutting plane to indicate the break. General-purpose hatch patterns are typically used and are drawn $\frac{1}{10}''$ (2.5 mm) apart at an incline of 45°. Hatch patterns are inclined in either direction unless the features of the object dictate that an angle other than 45° be used. The angle of the hatch patterns must not match the angle of any other lines on the drawing. Specific types of hatch patterns are used to identify specific types of material, such as steel, aluminum, wood, or thermal insulation. **See Figure 4-10.**

In a typical pictorial sectional drawing, general-purpose hatch patterns are used to indicate the type of material that the cutting plane cuts. As the cutting plane passes through various parts, hatch patterns are drawn at different angles to indicate the various parts.

Tech Tip

The current ANSI standard does not require section lining but does encourage the use of general-purpose section lining with adjacent or general notes explaining any material requirements.

Instructional drawings are used with industrial equipment such as scrap metal balers.

PRINTS

A *print* is a reproduction of original drawings created by an architect or engineer. A print is typically a part of a set of several prints. Original drawings are produced with traditional drafting instruments or with computer-aided design (CAD) software. **See Figure 4-11.** Traditional drafting involves the use of drawing tools and instruments such as pencils, triangles, compasses, scales, and T-squares to produce drawings by hand. CAD involves the use of a computer (typically a PC), CAD software, and a printer or plotter to produce original computer-generated prints.

Sectional Drawings

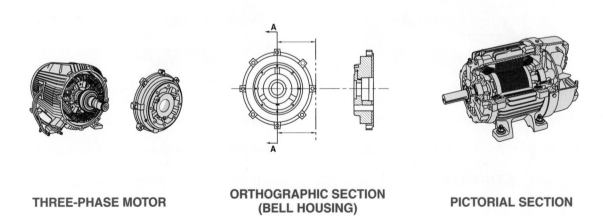

THREE-PHASE MOTOR

ORTHOGRAPHIC SECTION (BELL HOUSING)

PICTORIAL SECTION

Figure 4-9. Sectional drawings can have pictorial or orthographic views.

Hatch Patterns

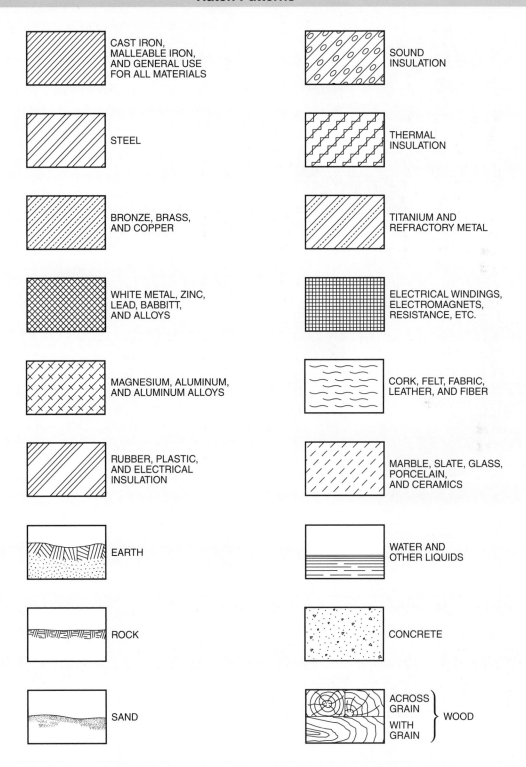

Figure 4-10. Specific types of hatch patterns are used to identify specific types of material, such as steel, aluminum, wood, or thermal insulation.

Original Drawings

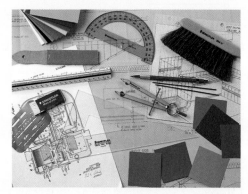

Staedtler, Inc.

**TRADITIONAL DRAFTING
INSTRUMENTS**

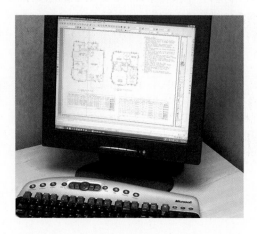

**COMPUTER-AIDED DESIGN (CAD)
SOFTWARE**

Figure 4-11. Original drawings are produced with traditional drafting instruments or computer-aided design (CAD) software.

Architects or engineers, through the use of CAD software, produce most original drawings. Drawing through CAD software has several advantages over traditional drawing, including the following:
- more accurate and consistent drawings
- ability to make changes to drawings
- ability to create four-color drawings
- ability to store and move drawings electronically
- ability to create drawings from other drawings through layering

Many types of drawings are used to create sets of prints. Sets of prints can include pictorial, orthographic, assembly, and sectional views as well as wiring diagrams, details, and schedules. Different types of prints are used to display the wide variety of information required to complete a project, such as a valve schedule being used to determine which type of actuator should be installed on a globe valve. Prints come in a variety of sizes. **See Figure 4-12.** Typically, a full-size set of prints (34″ × 44″) is used for construction purposes. Small sets of prints (8″ × 11″ or 11″ × 17″) are used for reference purposes. When reading prints, industrial technicians must be able to understand information presented in print title blocks, print line types, print abbreviations, print schedules, print divisions, and print revisions.

Print Sizes		
Flat Sheets		
Size	Width*	Length*
A	8.5	11
B	11	17
C	17	22
D	22	34
E	34	44
F	28	40

* in in.

Figure 4-12. Standard print sizes are designated by letters.

Print Title Blocks

A *title block* is an area on a working drawing or print used to provide information about the drawing or print. **See Figure 4-13.** The title block is located along the right side of a print or to the right and bottom. Information in the title block must be clearly understood before accessing information from a print. Title block information typically includes the following:
- company name and address
- print size
- print title
- print number
- revision version
- print sheet number
- print scale and weight
- initials of drafter, checker, and person issuing approval
- a Commercial and Government Entity (CAGE) code
- any revision or preperation information

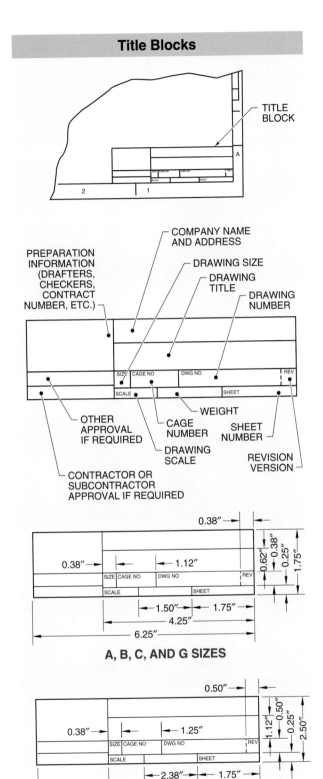

Figure 4-13. Information in the title block must be clearly understood before accessing information from a print.

A letter in the title block of a print identifies the size of the print as follows:

ANSI/ASME standard	Architectural Standard
• A = 8.5″ × 11″ sheet	Arch A = 9″ × 12″ sheet
• B = 11″ × 17″ sheet	Arch B = 12″ × 18″ sheet
• C = 17″ × 22″ sheet	Arch C = 18″ × 24″ sheet
• D = 22″ × 34″ sheet	Arch D = 24″ × 36″ sheet
• E = 34″ × 44″ sheet	Arch E = 36″ × 48″ sheet
• F = 28″ × 40″ sheet	Arch E1 = 30″ × 42″ sheet

Print Line Types

A *line* is a straight mark that begins at a starting point and stops at an end point. All prints are composed of lines to show the shape of the drawn object. A variety of line types are used to depict objects and items on prints. The type of line used depends on the type of object or item to be shown, the location of the object or item, and the type of dimensioning system being used. Basic line types are object, hidden, center, dimension, extension, leader, cutting plane, section, break, and phantom lines. Line types and their specific uses are defined by drafting standards. **See Figure 4-14.**

An *object line* is a line that indicates the visible shape of an object. Object lines are solid and are found without breaks. Object lines are typically the most common lines shown on a print. A *hidden line* is a line that represents the shape of an object that cannot be seen. Hidden lines are drawn as dashed lines and are used wherever there is a distinct change in the surface of an object.

A *centerline* is a line that locates the center of an object. Centerlines are thin dark lines broken into long and short dashes. Centerlines are used to locate the centers of windows, doors, and electrical enclosures and to indicate that an object is round or cylindrical in shape. Dimension lines, centerlines, and extension lines are used together to create a dimensioning system.

A *dimension line* is a line that is used with a written dimension to indicate size or location. Dimension lines are thinner than object lines. A dimension line typically has a gap for the placement of a dimension and has arrows at the ends. An *extension line* is a line that extends from the surface features of an object and is used to terminate dimension lines. Extension lines are drawn at 90° to the object and to the dimension lines. Extension lines do not touch object lines.

A *leader line* is a line that connects a written description such as a dimension, note, or specification with a specific feature of a drawn object. A leader line has an arrow at the end that contacts the edge of the object. Leader lines are drawn at any angle required to make the connection between the written description and the object.

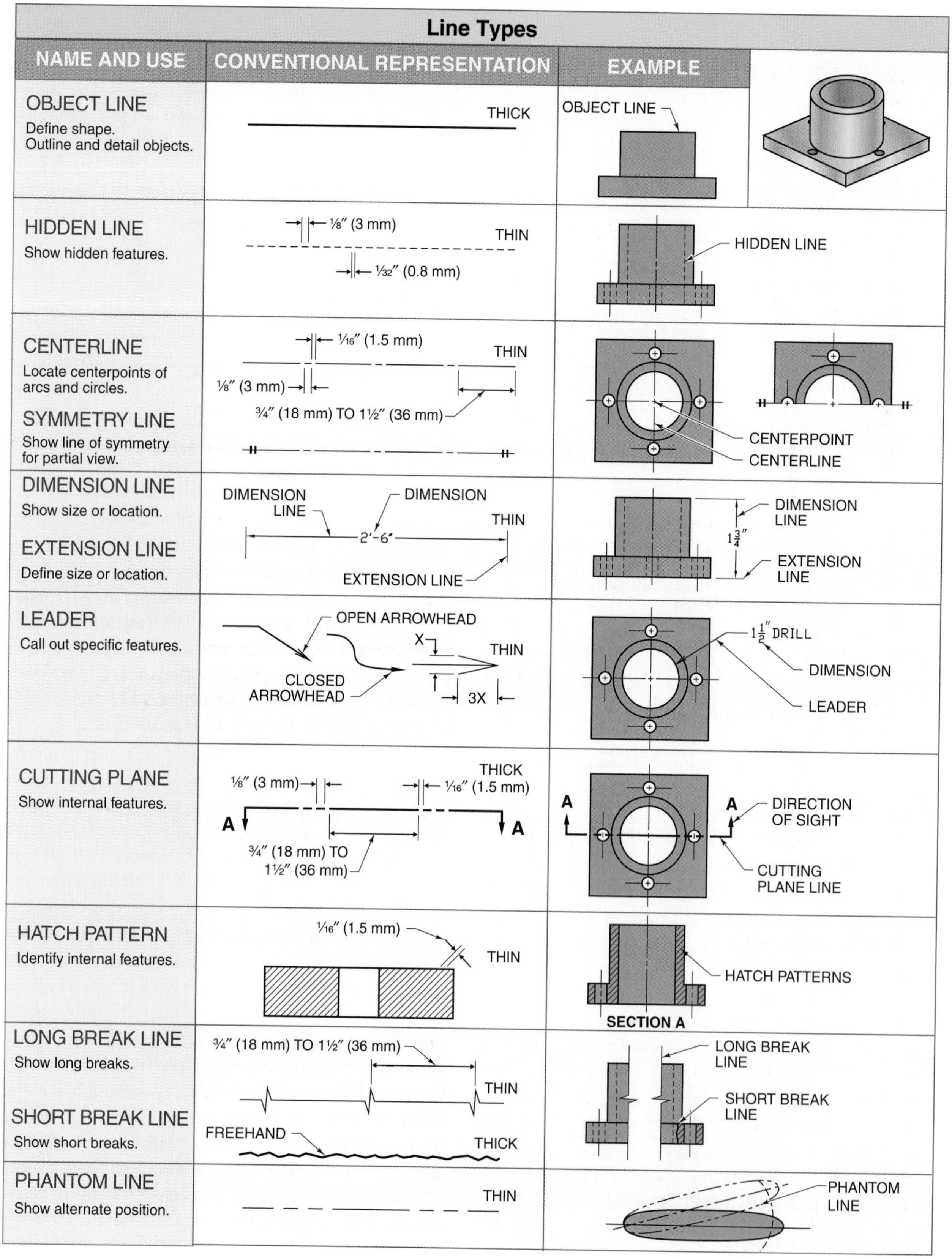

Line Types

NAME AND USE	CONVENTIONAL REPRESENTATION	EXAMPLE
OBJECT LINE Define shape. Outline and detail objects.	THICK	OBJECT LINE
HIDDEN LINE Show hidden features.	⅛″ (3 mm) THIN 1/32″ (0.8 mm)	HIDDEN LINE
CENTERLINE Locate centerpoints of arcs and circles. **SYMMETRY LINE** Show line of symmetry for partial view.	1/16″ (1.5 mm) THIN ⅛″ (3 mm) ¾″ (18 mm) TO 1½″ (36 mm)	CENTERPOINT CENTERLINE
DIMENSION LINE Show size or location. **EXTENSION LINE** Define size or location.	DIMENSION LINE DIMENSION 2′-6″ THIN EXTENSION LINE	DIMENSION LINE 1¾″ EXTENSION LINE
LEADER Call out specific features.	OPEN ARROWHEAD X THIN CLOSED ARROWHEAD 3X	1½″ DRILL DIMENSION LEADER
CUTTING PLANE Show internal features.	THICK ⅛″ (3 mm) 1/16″ (1.5 mm) A A ¾″ (18 mm) TO 1½″ (36 mm)	A A DIRECTION OF SIGHT CUTTING PLANE LINE
HATCH PATTERN Identify internal features.	1/16″ (1.5 mm) THIN	HATCH PATTERNS SECTION A
LONG BREAK LINE Show long breaks. **SHORT BREAK LINE** Show short breaks.	¾″ (18 mm) TO 1½″ (36 mm) THIN FREEHAND THICK	LONG BREAK LINE SHORT BREAK LINE
PHANTOM LINE Show alternate position.	THIN	PHANTOM LINE

Figure 4-14. Line types on prints have specific meanings.

A *cutting-plane line* is a line that indicates the path through which an object will be cut so that its internal features can be seen. A cutting-plane line is a thick line with arrows on each end at 90° that indicate the direction in which the resulting section will be viewed. A *section line* is a line that identifies the materials cut by a cutting-plane line in a section view. Section lines are typically drawn at an angle, which is different from all other lines on a print. Section lines can also consist of hatch patterns and gradient fills.

A *break line* is a line used to indicate internal features or to avoid showing continuous features of long or large objects. Break lines are drawn as a straight line with a zigzag in the middle or, if drawn freehand, as a jagged line. Break lines are used to eliminate a piece of an object where the whole length does not need to be shown. A *phantom line* is a line used to show a part's alternate positions or a repeated detail. Phantom lines are thin, broken lines alternating long dashes with pairs of short dashes.

Print Abbreviations and Symbols

A *print abbreviation* is a letter or group of letters that represents a term or phrase. Abbreviations allow information to be placed on a print without cluttering the print. Letters that are used for abbreviations are always capitalized. Abbreviations that form a word are followed by a period to avoid confusion. For example, "MET." is the abbreviation for metal. A *symbol* is a conventional representation of a quantity or unit. Symbols are not based on any specific language and can be easily recognized. For example, the symbol for diameter is Ø. **See Figure 4-15.**

Abbreviations are used in a wide variety of prints, such as electrical and electronic drawings; building elevations; and detail and assembly prints. Some abbreviations are universal, such as electrical/electronic abbreviations, and have the same meaning no matter what type of print the abbreviation is located on. Other abbreviations have meanings that are specific to a certain type of print.

In some cases, an abbreviation can have different meanings depending on which type of print the abbreviation is used on. For example, "ELEV" is the abbreviation for "elevation" on an elevation print and "elevator" on a mechanical or electrical print. Typically, the context of a drawing clarifies the meaning of the abbreviation. When two abbreviations appear for the same word, the first abbreviation and any information pertaining to it should be used. For example, because "FIN." and "FNSH" are both abbreviations for finish, "FIN." would be used. **See Appendix.**

Abbreviations and Symbols		
Abbreviation	**Symbol**	**Meaning**
CL	₵	CENTERLINE
DIA	Ø	DIAMETER
	×	REPETITIVE FEATURE
DEEP OR DP	⊥	DEPTH
CBORE OR SFACE	⊔	COUNTERBORE OR SPOTFACE
CSK OR CSINK	⌄	COUNTERSINK
R		RADIUS
SR		SPHERICAL RADIUS
SDIA	SØ	SPHERICAL DIAMETER

Figure 4-15. Abbreviations and symbols are used on prints to conserve space.

The use of abbreviations is standard among architects and engineers, and a list of these abbreviations is typically provided with a set of prints. When a large set of prints includes abbreviations, the abbreviations appear on a legend. The legend sheet is typically found at the beginning of a set of prints. In addition to abbreviations, the legend sheet can contain symbols, equipment identification information, and drawing conventions used throughout the set of prints. **See Figure 4-16.**

Tech Tip

Many symbols have been incorporated into the national and international standards for mechanical drawings. The use of symbols is preferred over the use of abbreviations when an option exists. For example, use the symbol for diameter, Ø, instead of the abbreviation DIA whenever possible. The use of symbols, rather than abbreviations, is required for math operations. For example, do use A = 12; do not use A EQL 12.

Print Schedules

Small amounts of information can be displayed (noted) on a print without causing the print to become cluttered. However, when a large amount of detailed information is required, a schedule is used. A *print schedule* is a chart used to conserve space and display information on a print in a concise and organized format. For example, because it is not practical to display in-depth light fixture information on a lighting floor plan, a lighting or fixture schedule is used to identify the various lighting fixtures used on a project. **See Figure 4-17.**

Legend Sheet Abbreviations

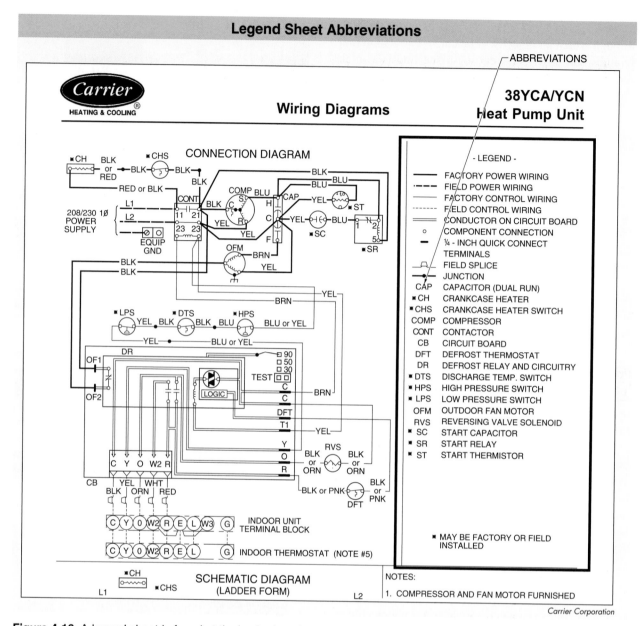

Figure 4-16. A legend sheet is found at the beginning of a set of prints and identifies the abbreviations used throughout the prints.

Print schedules resemble spreadsheets with information being displayed in columns and rows. A wide variety of schedules are found on architectural, electrical, and mechanical prints. Types of electrical schedules include information on fixtures, feeders, main switchboards, branch circuit panels, and transformers. Depending on the amount of information and room, multiple schedules, such as lighting fixtures and switch schedules, are grouped together on a single electrical print. The number and type of schedules increases with the complexity of the project.

Print Divisions

Prints are separated into different divisions to allow for quick and easy access to information. Print divisions are denoted by a capital letter. For example, architectural prints are denoted by a capital "A". In a similar manner, mechanical prints are denoted by an "M", structural prints by an "S", civil prints by a "C", and electrical prints by an "E". The number that comes after the capital letter denotes the sheet number. The print division and number are always found in the title block of a print. **See Figure 4-18.**

Schedules

	Lamps					Manufacturer
Type	**Qty.**	**Cat. No.**	**Description**	**Volt**	**Mounting**	**Cat. Number**
F1	3	F32T8 3500K	Recessed static troffer	120	Recess T-grid	METALUX 2GP-332A-120V-EB82
F2	3	F32T8 3500K	Recessed static troffer w/ self contained emer lighting	120	Recess T-grid	METALUX 2GP-332A-120V-EB82-EL4
F3			Not used			
F4	2	26W DTT	Compact fluorescent recessed downlight	120	Recess ceiling	HALO 8″ aperture H-7871-99870BA
F5	1 1	T5/8W T5/13W	Under cabinet task light	120	Under cabinet	METALUX 8F-CL-1-0813T5-EB-120
F6	1	100W A-21	Explosion proof incandescent	120	Surface ceiling	APPLETON AC1050
F7	1	100W A-21	Porc. keyless lampholder	120	Surface ceiling	PASS and SEYMOUR Series 44
F8	1	50W/Mogul MH	Surface wall pack	120	Wall 9′-6″	LUMARK MH-WL-50-120-FI-PE/MT
X1		LED	Polycarbonate exit; self powered; single face	120	Surface ceiling	SURELIGHT CCX-7-1-70-G-WH
X2		LED	Polycarbonate exit; self powered; double face	120	Surface ceiling	SURELIGHT CCX-7-2-70-G-WH

Figure 4-17. Schedules are used to conserve space and display information on a print in a concise and organized format.

Print Divisions

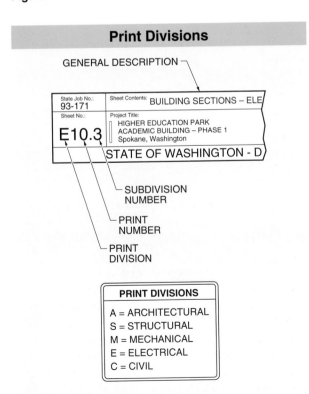

Figure 4-18. Print divisions are denoted in the title blocks by capital letters, such as "E" for electrical.

Print numbering begins with the number 1 within each division. For example, mechanical prints may run from page M1 to M65. Print divisions are subdivided when several pages apply to the same elements. For example, a set of electrical prints can begin with E1 but have pages subdivided as E1.1, E1.2, and E1.3.

Print Revisions

A *revision block* is a block that identifies the changes that have been marked on the drawing since its initial approval. A revision block is similar to a title block. Architects and engineers typically make several revisions to a set of prints during the course of a project. The revisions are labeled sequentially as 1, 2, and 3 or A, B, and C. The revision number or letter, a brief description of the revision, the date, and the initials of the person making the revision are all indicated in the revision block. **See Figure 4-19.**

A revision is identified in a drawing on a print by a revision symbol. The symbol is a number or letter inside a triangle, circle, or square. In some cases, a cloud is drawn around the revision symbol to indicate that a change was made. As new prints are issued, they must be checked for any new revisions. Typically, revision information is found next to the title block or in the upper right-hand corner of a print. Project delays, added costs, and penalties can result if the most recent set of prints is not used.

PRINT CONVENTIONS

A *print convention* is an agreed-upon method of displaying information on a print. Most print conventions were developed from years of practice. Conventions save space and prevent clutter on prints, provide for consistency between prints, and enhance communication between parties using the prints.

Print conventions typically govern the display of notes, detail symbols, section symbols, and column numbers and/or letters. A standard is an accepted reference or practice. Standards are written documents that provide a set of generally acceptable criteria to which work can be performed. Conventions often come from standards developed by a standards organization such as the NFPA or ANSI. Minor variations in print conventions exist and may depend on the personal preference of the architect or engineer.

Print Notes

A *note* is a sentence that provides drawing information that does not fit within the space of the drawing. The two types of notes are general (construction) notes and sheet notes. Notes are numbered sequentially by type.

Revision Blocks

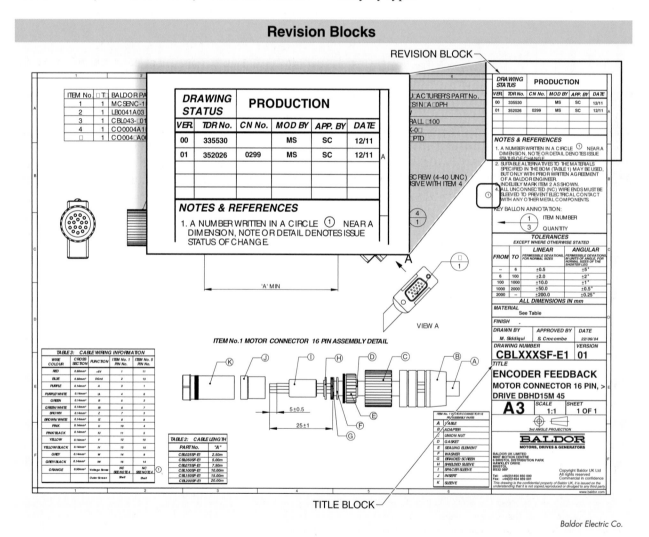

Baldor Electric Co.

Figure 4-19. Print revisions require a revision number or letter, a brief description of the revision, the date of the revision, and the initials of the person making the revision to be indicated in the revision block of a print.

A *general note* is a note that applies to the entire print on which the note appears. Although numbers are used to identify general notes, general notes are not represented by corresponding note symbols on the print. General notes cover broad or general topics, such as coordination between trades, contact information, and company- or contractor-specific procedures.

A *sheet note* is a note that applies to a specific item in the drawing on which the note appears. Sheet notes provide information about specific items, such as receptacle part numbers, motor overload numbers, and junction box information. Sheet notes are identified by a number and have a corresponding symbol on the print. The symbol (a sheet note number within a circle) identifies the item addressed by the sheet note. A leader line may be used to connect the symbol with the item, or the symbol may be located adjacent to the item. **See Figure 4-20.**

Tech Tip

The purpose of the National Fire Protection Agency (NFPA) is to reduce the worldwide burden of fire and other hazards on the quality of life by providing and advocating consensus codes and standards, research, training, and education. The NFPA's 300 codes and standards influence every building, process, service, design, and installation in the United States, as well as many of those used in other countries.

Print Section Views and Detail Drawing Symbols

Section views and detail drawings are enlarged drawings of an item or feature on a print, such as underground conduit installations, conduit support methods, or contact design. Section views and detail drawings provide additional information about an object. Section views and detail drawings are sometimes not included on the print from which they originate because of space considerations.

Section symbols and detail symbols on a print identify the item or feature that the section view or detail drawing is taken from. Section views and detail drawings have detail header symbols that correspond to the section view and detail drawing symbols found in the drawing space. The detail header symbols used for section views and detail drawings are similar. A section view or detail drawing header symbol may have the bottom of the circle divided into two parts. The left side denotes the sheet where the section view or detail drawing originated and the right side denotes where the view or drawing is located.

The symbol for a section view consists of a cutting-plane line drawn through the selected item and a circle divided in half by a horizontal line. An arrow on the cutting-plane line and/or arrow on the header circle indicates the direction in which the section is viewed. The upper half of the header circle contains the reference number or letter of the section view. The bottom half of the header circle contains the sheet the section view is found on.

The symbol for a detail drawing consists of a circle divided in half by a horizontal line. The upper half of the circle contains the reference number or letter of the detail drawing. The bottom half of the circle contains the number of the sheet on which the detail drawing is found. A leader line is drawn from the symbol to the item or feature that appears in the detail drawing.

The symbol for a detail header consists of a circle divided in half by a horizontal line that extends beyond the circle. **See Figure 4-21.** The upper half of the circle contains the reference number or letter for the section view or detail drawing. Typically, the bottom half of the circle contains the number of the sheet on which the section view or detail drawing is found. Alternately, the bottom half of the circle may contain the number of the sheet from which the section view or detail drawing originated. A description of the section view or detail drawing appears above the horizontal line that extends beyond the circle. The scale of the section or detail appears below the same horizontal line. Some section views and detail drawings are not drawn to a particular scale and will have the NTS (not to scale) abbreviation.

Section views and detail drawings are used to highlight details on a print such as conduit support methods.

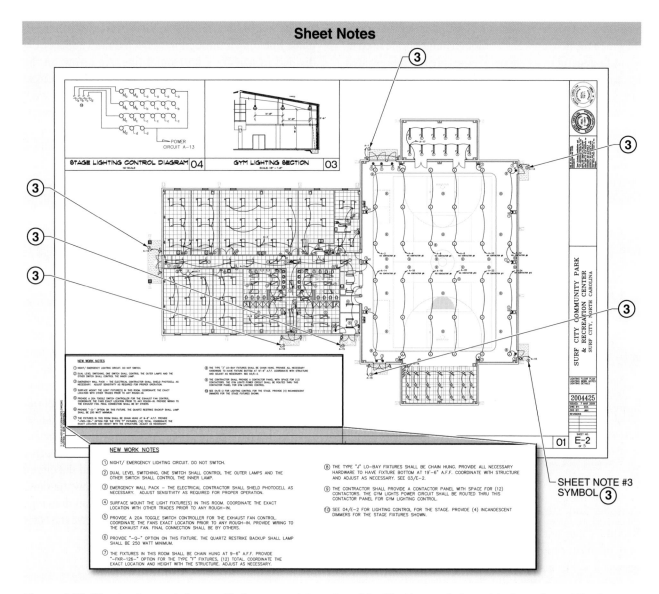

Figure 4-20. Sheet notes apply to specific items on prints and are identified by symbols next to the note and the item.

Print Building Column Numbers and Letters

Most commercial and industrial buildings are constructed using columns of reinforced concrete and structural steel. On a print, horizontal and vertical lines are drawn through the centers of the columns to form a center-to-center grid system. Typically, horizontal lines are identified by letters and vertical lines by numbers, but horizontal lines can be identified by numbers and vertical lines by letters.

A center-to-center grid system provides dimension information, identifies columns, and serves as a point of reference. A center-to-center grid system is used throughout a set of prints whether the prints are

architectural prints, electrical prints, or mechanical prints. **See Figure 4-22.**

Structural and architectural prints use the grid system to provide dimension information for placement of equipment, spacing between columns, and the distance between columns and major architectural features, such as walls. All dimensions are from the center of a column. Electrical and mechanical prints use the grid system as a point of reference. Column numbers and letters are frequently used in request-for-information (RFI) documents to identify a location. For example, an electrical room may be located at the intersection of line A and line 7 (A7).

Section View and Detail Drawing Print Symbols

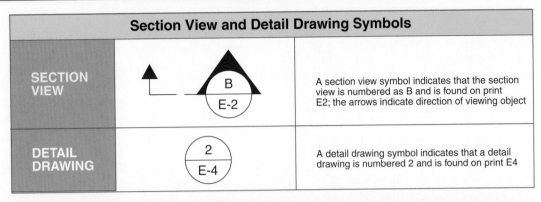

Section View and Detail Drawing Symbols

SECTION VIEW	B / E-2	A section view symbol indicates that the section view is numbered as B and is found on print E2; the arrows indicate direction of viewing object
DETAIL DRAWING	2 / E-4	A detail drawing symbol indicates that a detail drawing is numbered 2 and is found on print E4

Section View Detail Header and Detail Drawing Detail Header Symbols

SECTION VIEW DETAIL HEADER	B / E-2 DESCRIPTION SCALE: N.T.S.	A section view detail header symbol indicates the location of Section View B, what the object is in the section view, and the scale (NTS – not to scale) of the section view
DETAIL DRAWING DETAIL HEADER	2 / E-4 DESCRIPTION SCALE: ¼″ = 1″	A detail view detail header symbol indicates the location of Detail View 2, what the object is in the detail view, and the scale (four to one) of the detail view

Figure 4-21. Section views and detail drawings have detail header symbols that correspond to the section view symbols and detail drawing symbols found in the original drawing space.

Column Numbers and Letters

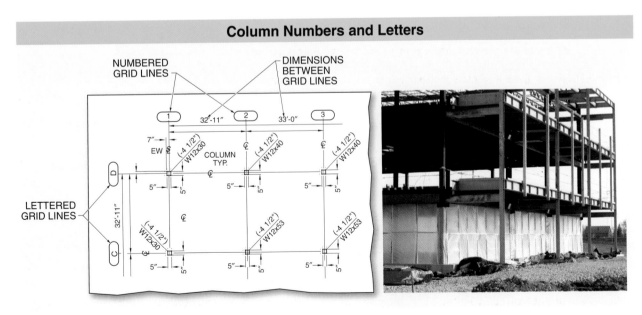

Figure 4-22. A center-to-center grid system made of numbers and letters is used in commercial buildings and industrial facilities to identify specific locations.

PRINT DRAWING SCALES

Prints are drawn at a reduced size in order to fit an entire project on standard-size sheets. Frequently, prints must be drawn to a specific scale. The actual dimensions of the objects on prints drawn to scale are reduced proportionally in order to maintain the correct relationship between the objects. Some prints are not drawn to scale. Prints that are not drawn to scale include dimensions where necessary. These prints display the notation "Not to Scale" or the abbreviation "NTS".

Different scales are used for various types of prints. The scale of a print is found in the title block. Architectural prints use scales ranging in size from ⅟₃₂″ = 1′ to 3″ = 1′. A common scale for commercial and industrial building plans is ⅛″ = 1′.

A *drawing scale* is a system of drawing representation in which drawing elements are proportional to actual elements. Dimensions of the drawing can be larger or smaller than the actual size of the object. For example, a typical scale as it appears on a drawing can be ¼″ = 1′, which indicates that one-quarter inch on the drawing represents one foot in the actual field. Both architect's scales and engineer's scales are used to create and read scaled drawings. Use of a scale helps to keep work in perspective and helps to ensure correct location of equipment during installation. Some types of drawings, such as electrical wiring diagrams and one-line diagrams, do not require use of a scale.

An *architect's scale* is a triangular scale used to draw objects to a particular size. Common architect's scales are triangular in cross section and have six edges. Five of the edges have two scales each, and one edge is divided into inches and fractions of an inch.

Architect's scales are read from left to right or right to left depending on the scale. A ¼″ = 1′-0″ scale is read from right to left beginning at the 0 on the right end of the scale. A ⅛″ = 1′-0″ scale is read from left to right beginning at the 0 on the left end of the scale. The same set of markings is used for both the ¼″ = 1′-0″ scale and the ⅛″ = 1′-0″ scale. The correct line in relation to the scale used must be read. For example, 18′-0″ on the ¼″ = 1′-0″ scale is on the line representing 57′-0″ on the ⅛″ = 1′-0″ scale.

An *engineer's scale* is a triangular scale used to draw large areas, such as property lines, on a building lot. Scales on an engineer's scale are 1″ = 10′, 1″ = 20′, 1″ = 30′, 1″ = 40′, 1″ = 50′, and 1″ = 60′. An engineer's scale is typically used for site plans. Since engineer's scales are used for drawing large areas, the inch divisions are omitted. **See Figure 4-23.**

Tech Tip

Rules and scales are available in variety of forms such as rigid or flexible rules, electronic linear scales, folding rules, drill point gauges, tape measures, and rule depth gauges. These rules should be used for their designed purpose. Rules provide an approximate measurement compared to precision gauges. Digital or electronic linear scales designed for machine tool applications provide a high degree of accuracy.

Scale Types

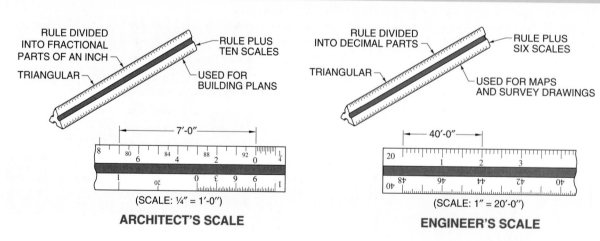

Figure 4-23. Both architect's scales and engineer's scales are used to create and read scaled drawings.

SPECIFICATIONS

A *specification* is additional information that is included with a set of prints. Specifications provide details that cannot be shown on a print. Specifications along with the various drawings describe the entire building, facility process, or project. Specifications contain information related to legal issues, building materials, equipment placement, installation, construction, and quality.

Architects and engineers develop specifications based on the requirements of the project owner, building, and/or facility process. Specifications list the codes, ordinances, and company policies that must be followed for the project. For example, "Perform all electrical work in full accordance with the National Electrical Code® (NEC®)" is part of a typical electrical specification.

Municipal building departments use a project's specifications and prints to verify that the proposed project complies with local building codes and zoning ordinances. Contractors use the specifications and prints to accurately bid on a project and then construct the project to meet the requirements of the owner.

Specifications are intended to supplement print drawings, and the specifications must agree with the set of prints. When a conflict exists between what the specifications state and what the prints indicate, the architect or engineer must be contacted for clarification.

A large set of specifications contains many sections. If a conflict exists between different sections of the specifications, the architect or engineer must be contacted. At times, a contractor may want to deviate from the specifications, such as by substituting one brand or model of electric motor drive for a different brand or model. Before deviating from the specifications, the contractor must obtain permission from the engineer. Failure to follow specifications can result in nonoperational systems, monetary penalties, and legal action.

The installation of electrical equipment is included in specifications such as the National Electrical Code® (NEC®).

Residential, commercial, and industrial projects require a set of specifications in addition to a set of prints. The size, format, and complexity of the specifications vary with the project. The specifications for a small residential project can consist of a page or two of requirements attached to the prints. **See Figure 4-24.** The specifications for a large commercial project may consist of a hundred pages or more of detailed requirements bound together as a book. Specifications for an industrial project typically have thousands of pages of detailed requirements.

TECHNICAL SKETCHING

A *sketch* is a two-dimensional visual representation of an object. Sketches give a fixed view of one object at a chosen angle and distance. Sketches are often drawn freehand and without the use of technical instruments so that they can be used to quickly convey information visually when words may not give a complete description.

Many technicians in industry must create a sketch to clearly convey information about a part that must be fabricated. Many manufacturing facilities have their own machine shops so that replacement parts can be fabricated in-house. Other facilities use local machine shops to fabricate replacement parts. Sketches must provide all the information required for a machinist to reproduce the required part. A sketch is generally used when a production machine is down and a replacement part must be fabricated quickly.

Information that is vital to a machinist includes the dimensions and tolerances of a part. This information gives a machinist an idea of how much time to spend on each step of the fabrication process and how long it will take to fabricate the part. General practice in the machining industry, unless otherwise noted, is that any fractional dimension is given a tolerance of $\pm\frac{1}{64}''$, and any decimal dimension is giving a tolerance of $\pm0.005''$. As the tolerance decreases, more time is required for the fabrication.

Sketching Principles

Information required for the fabrication and use of a part, such as dimensions, specifications, materials, tolerances, fits, or treatments, must be included on a sketch. A sketch is entirely different from a mechanically ruled or computer-generated drawing. A sketch may have a rough look, but should be clear, concise, and suited for its intended purpose. The following sketching rules are typically required:

• Choose a view that completely describes the shape of the part.

• Do not draw unnecessary views. Be brief and have as few lines as necessary.

• Dimensioning must not be obscure or crowded.

• Any special or added views or items that are used must be identified by explanatory notes.

A *picture plane* is a two-dimensional space on paper where the flattened image of a three-dimensional object is depicted. Because a sketch does not actually contain a third dimension, the picture plane is where the components of a sketch are arranged in order to create the illusion of space.

Tech Tip

A set of prints acts as a step-by-step guide to constructing a building. Some prints provide more detail than others, but all prints include certain basic information needed for the project. Tradesworkers must understand the details of individual drawings and how the drawings relate to other drawings in a set of prints.

Specifications

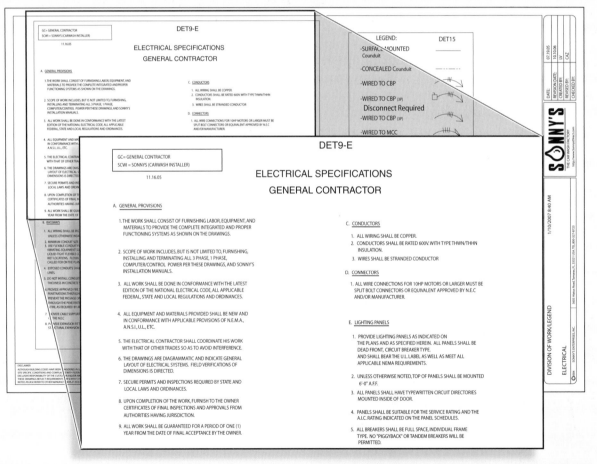

Sonny's Enterprises, Inc.

Figure 4-24. Specifications provide information about a drawing that cannot be shown on a print.

A sketch may contain contours, shapes, and forms that are combined to create a recognizable representation of an object. A *contour* is an identifying outline that separates all or part of an object from the background. A *shape* is the extent of all or part of an object that is often contained within a contour. Shapes are flat areas defined by an edge or perceived edge. A *form* is the shape and structure of all or part of an object and includes a sense of mass and volume. In a sketch, form contains the illusion of volume in an object. Volume is the three-dimensional space of all or part of an object. Volume is bound by planes and has the three dimensions of width, height, and depth. **See Figure 4-25.**

Wood pencils are often used in sketching. Wood pencils are designated as hard (H) or soft (B) with a number scale indicating the degree of hardness or softness. The hardness or softness of the lead is indicated with a stamp near one end of the pencil shaft. Hard leads create lighter, thinner marks on paper, whereas soft leads create darker, thicker marks on paper.

Soft leads are used primarily for sketching. Medium leads are used to draw general object lines. Hard leads are used to draw fine, precise lines. Grades of lead range from 6B (extremely soft) to 9H (exceptionally hard). Sketching is recommended with pencils that have a lead grade that is in the soft to middle range. Lead that is too soft will smudge easily, and lead that is too hard will make it difficult to put smooth marks on paper. The most commonly used lead grades for sketching include B, HB, F, and H. **See Figure 4-26.**

Sketches are drawn by adding components to the page in a methodical way. Addressing elements on the entire object or page at once creates a more accurate and balanced finished sketch. To begin a sketch, the layout and structure are blocked in using rough, light lines. *Blocking in* is a method of quickly marking the structure of an object on a sketch by breaking the subject into common shapes and lines. The placement, shape, and proportions of the major elements in the sketch should be blocked in accurately before more detail is added. Next, objects can be further depicted by adding more lines, tonal values, and textures. Finally, finishing details and dimensions should be added to the sketch. **See Figure 4-27.**

MULTIVIEW SKETCHES

Multiview sketches are two-dimensional representations of three-dimensional objects that are shown in true view. Orthographic sketches use multiple views to show the object from different orientations. Isometric sketches use one view in a rotated orientation to show multiple views of the object. The scale and proportions of the sketch are relative to the object being depicted and the level of detail necessary.

Tech Tip

Sketching is the ability to draw in order to communicate technical information with speed and accuracy. This ability involves fundamental drawing techniques to create a visual representation.

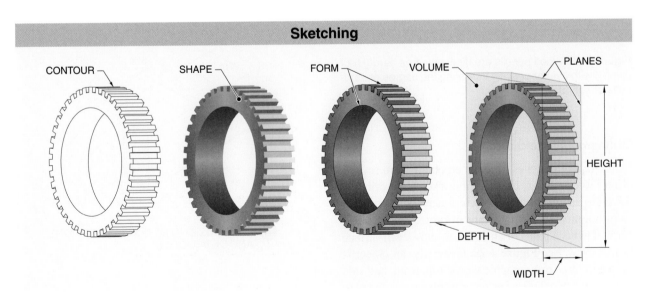

Sketching

Figure 4-25. Contour, shape, form, and volume are all used in a sketch to create a recognizable representation of an object.

Pencil Leads

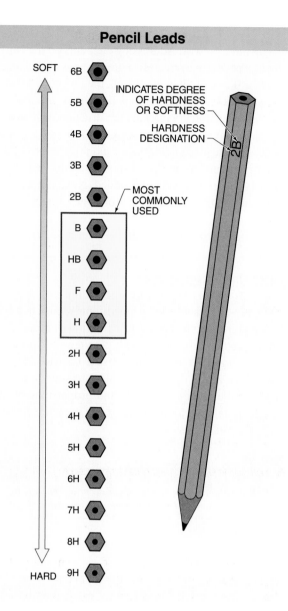

Figure 4-26. Wood pencils are designated as hard (H) or soft (B) with a number scale indicating the degree of hardness or softness.

Orthographic Sketches

Orthographic sketches describe an object from several different views. An orthographic sketch describes a three-dimensional object in two-dimensional space. In an orthographic sketch, a plane of projection is placed parallel to each plane of an object to show each surface in true view. **See Figure 4-28.** Generally, an object is shown in three orthographic views: top, front, and side views. Additional views or sections may be included to show more detail, when necessary.

Adding Components to a Sketch

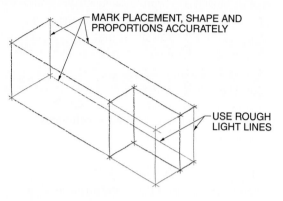

① BLOCK IN ROUGH COMPOSITION AND STRUCTURE OF OBJECT

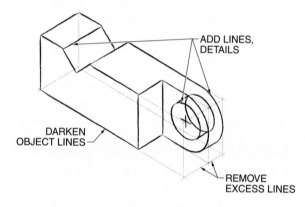

② FURTHER DESCRIBE ELEMENTS BY ADDING LINES AND DETAILS

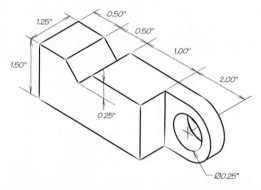

③ ADD FINISHING DETAILS, DIMENSIONS, AND DETAIL LINES

Figure 4-27. Sketches are drawn by adding components to the page in a methodical manner.

Orthographic Sketches

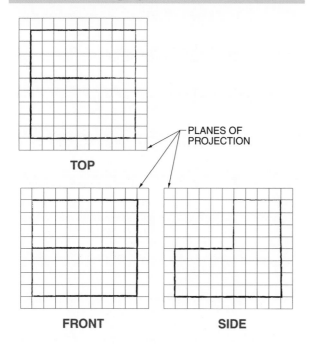

Figure 4-28. The minimum number of views in an orthographic sketch is top, front, and side.

Isometric Sketches

Isometric sketches are drawn around three axes from a horizontal line: vertical axis, a 30 degree axis to the right, and a 30 degree axis to the left. All three axes intersect at one point on the horizontal line. **See Figure 4-29.**

All horizontal lines in an isometric sketch are always drawn at 30 degrees and parallel to each other and are either to the left or to the right of the vertical. For this reason, all shapes in isometric sketches are not true shapes but distorted shapes. All vertical lines in an isometric sketch are always drawn vertically, and they are always parallel to each other.

Scales

Sketches are not drawn to any scale. Sketch size depends on the complexity of the object and paper size. Gridded or ruled paper is helpful when drawing parts to scale.

Proportions

A sketch should have its features in proportion to the actual object. If the sketch is not proportionate, the result can be machinist confusion and the object not being fabricated as required. Large areas can be compared with small areas by the use of a dowel or pencil. **See Figure 4-30.**

Isometric Sketches

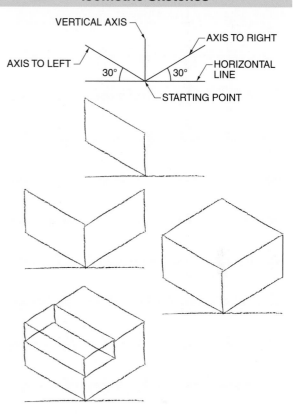

Figure 4-29. Axes of an isometric sketch begin at a point on the horizontal line.

DIMENSIONING

Dimensioning is a method of adding dimensions to a drawing to indicate the geometrical characteristics of an object. A *dimension* is a numerical value that gives the size, form, or location of objects on a drawing. A *size dimension* is a dimension that gives the overall size of an angle or feature. A *location dimension* is a dimension that uses the components of an object to locate an angle or feature on an object. Dimensions are generally shown in combination with lines, symbols, and notes to include all information to fully describe an object. Dimensions should be placed on a drawing so that they allow only one interpretation and include no repetition.

Tech Tip

Sketching requires a basic skill to visually communicate technical information with speed and accuracy. This skill combines sketching with fundamental drawing techniques to create a visual communication tool.

Estimating Proportions

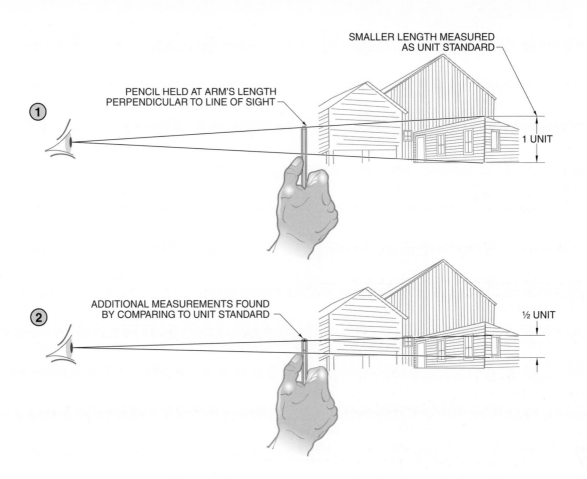

Figure 4-30. A pencil can be used as a unit of measure to aid in keeping sketches proportional.

Indicating Dimensions

Lines are used on a drawing to indicate the extent or area of an object being dimensioned. Lines used in dimensioning are thinner than object lines so they are visibly different from the lines in the object being drawn. Dimension lines, arrowheads, extension lines, and leaders are used with numbers to apply dimensions to an object. **See Figure 4-31.**

Dimension Lines. A *dimension line* is a line that is used with a written dimension to indicate size or location. Dimension lines are commonly broken for the placement of numerals giving the measurement. **See Figure 4-32.** If a horizontal dimension line is not broken, the numeral is placed above the dimension line.

Dimensions

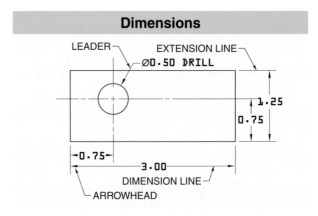

Figure 4-31. Dimension lines, arrowheads, extension lines, and leaders are used with numbers to apply dimensions to an object.

Dimension Lines

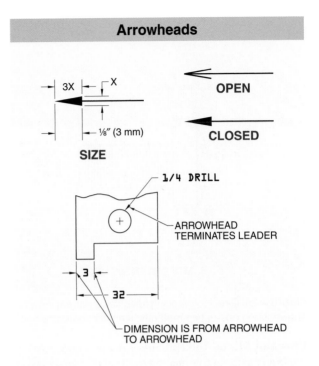

Figure 4-32. A dimension line is a line that is used with dimensions to show size or location and is commonly broken for the placement of numerals.

When placing dimension lines, the dimension line nearest the object being dimensioned should be at least ⅜″ (9 mm) from the object. Succeeding dimension lines should be no closer than ¼″ (6 mm) to the previous dimension line. Dimensions are aligned for appearance. When parallel dimension lines are used, the dimensions are staggered to avoid crowding. Dimension lines should not cross other dimension lines unless this is unavoidable.

Arrowheads. An *arrowhead* is a symbol that indicates the extent of a dimension. Arrowheads terminate dimension lines. Arrowheads should be drawn three times as long as they are wide with the width proportional to the thickness of the line. Arrowheads may be open or closed. When there is not adequate space, arrowheads may be placed outside the indicated dimension. **See Figure 4-33.**

Indicating Tolerancing

Tolerance is the permissible deviation from a given value or dimension. Tolerances should be included in a drawing because the variations in size and shape among manufactured objects can affect the interchangeability of parts. *Direct tolerancing* is the practice of specifying a dimension's permissible range directly within the dimensioning lines. The two common methods of direct tolerancing are limit dimensioning and plus and minus tolerancing.

Arrowheads

Figure 4-33. An arrowhead is a symbol that indicates the extent of a dimension and terminates dimension lines.

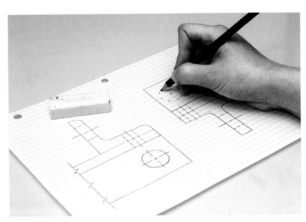

Using grid paper for sketching allows sketches to be consistent and easily understood. Preprinted grid paper is available in a variety of grid sizes with the ¼″ (quad-rule) most popular.

Limit Dimensioning. *Limit dimensioning* is a direct tolerancing method that includes only the maximum and minimum values of a dimension on a drawing. The two limits can be expressed in a single line containing the low limit and the high limit separated by a dash or in two lines showing the high limit above the low limit. **See Figure 4-34.** The high and low limits should be displayed with the same number of decimal places for uniformity. Zeroes can be added to limits for uniformity if necessary.

Direct Tolerancing

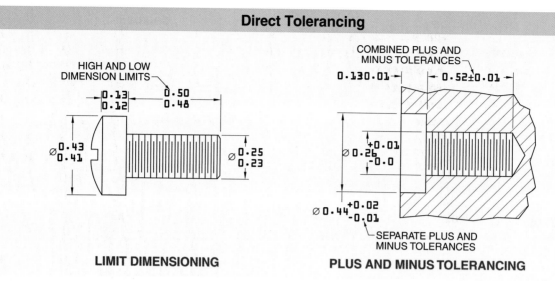

LIMIT DIMENSIONING   **PLUS AND MINUS TOLERANCING**

Figure 4-34. Direct tolerancing specifies a dimension's permissible range directly within the dimensioning lines and can be shown using one of two methods: limit dimensioning and plus and minus tolerancing.

Plus and Minus Tolerancing. *Plus and minus tolerancing* is a direct tolerancing method that provides an ideal dimension and includes the allowable deviations in the positive and negative directions. The tolerances are placed after the dimension with the positive tolerance always shown above the negative tolerance. The dimension and tolerances should be displayed with the same number of decimal places. If necessary, zeroes can be added to the dimensions and tolerances for uniformity. When the numerical tolerance is the same in both directions, the plus and minus values can be combined into a single expression using the ± symbol, such as 15.0 ±1.

If there are a large number of dimensions on a drawing that require the same degree of tolerance, a general note may be added to indicate the common tolerance for these dimensions. Only tolerances that differ from the general values need to be included on the drawing. **See Figure 4-35.**

Machinist Tolerances

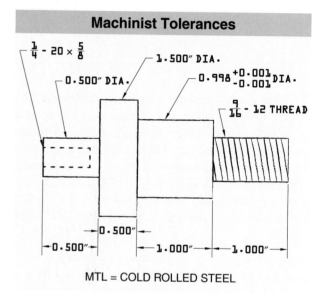

MTL = COLD ROLLED STEEL

Figure 4-35. Certain dimensioning offers a machinist the required tolerances and critical measurements.

Review

1. What are the common types of drawings used by industrial and mechanical technicians?

2. What information is commonly included in these drawings?

3. What are the different types of prints and how are they used?

4. What information is commonly found in a title block?

5. Define line and the basic line types.

6. Why are abbreviations and symbols used on prints?

7. What are the two types of notes used on a print?

8. What are the different scales commonly used and how are they selected for a print?

9. Why are specifications included with a set of prints?

10. What are typical rules to follow when sketching?

11. What types of lead pencils are used for sketching?

12. How is an isometric sketch drawn?

13. Why is it important to make sketches that are correctly proportioned?

14. Define dimensioning.

15. Why should tolerances be included on drawings?

Digital Resources

ATPeResources.com/Quicklinks
Access Code 462409

CHAPTER 5

PRECISION MEASUREMENT

Precision measurement of components and parts is required for many production, maintenance, and quality control operations. Technicians must be able to take measurements through the use of rules, protractors, calipers, and micrometers. Proper handling, calibration, and storage of precision measuring instruments help in consistently attaining the most accurate measurements, while prolonging the working life of the instruments.

Chapter Objectives

- Explain the two commonly used systems of measurement.
- Describe rules and their uses.
- List the different types of protractors used for measurement.
- List the different types of calipers used for measurement.
- Describe micrometers and their uses.
- Explain the importance of measuring instrument maintenance.

Key Terms

- precision
- accuracy
- english system
- decimal
- metric system
- rule
- protractor
- conventional protractor
- machinist's steel protractor
- reversible protractor
- caliper
- tolerance
- vernier scale
- dial caliper
- digital caliper
- micrometer
- outside micrometer
- inside micrometer
- depth micrometer

PRECISION MEASUREMENT

Manufacturing standards are set to ensure that products are manufactured with the highest degree of precision. *Precision* is the level of accuracy or mechanical exactness. *Accuracy* is the degree to which a measurement conforms to a specific standard. *Precision measurement* is a method of using measuring instruments to acquire accurate measurements. Precision measurement allows craftsman or professional tradesworkers to perform tasks that require small dimensional tolerances. The two most commonly used systems of measurement are the English system and the metric system.

English System

The English system is primarily used in the United States. This system uses the inch and foot as the basic units of linear measurement. A foot can be broken down into fractional units and is equal to 12 inches. An inch can be broken down into common fractional units of ½″, ¼″, ⅛″, ¹⁄₁₆″, ¹⁄₃₂″ and ¹⁄₆₄″. Each inch or fraction of an inch can be converted to a decimal for a greater degree of precision.

Decimals. A *decimal* is a number expressed in base 10. A *decimal fraction* is a fraction with a denominator of 10, 100, 1000, etc. The number 1 is the smallest whole number. Any number smaller than 1 is a decimal and can be divided into any number of parts. For example, an inch can be divided into 10 parts. A tenth of an inch is written 0.10″. Each tenth of an inch can also be divided into 10 parts indicating hundredths. A hundredth of an inch is written 0.01″. Continuing to divide by 10 will create extremely small decimal units.

When a decimal number is written, the places to the left of the decimal point indicate a whole number. The places to the right of the decimal point indicate parts of a whole number. For example, the first place to the right of the decimal point (0.X) represents tenths. The second place to the right of the decimal point (0.0X) represents hundredths. The third place to the right of the decimal point (0.00X) indicates thousandths. Precision measurement often requires numbers to be accurate to the fourth place to the right of the decimal point (0.000X), indicating ten-thousandths.

A number to the right of the decimal point has a definite value but may be read more than one way. For example, 0.100″ may be read as one tenth of an inch or one hundred thousandths of an inch. Also, 0.050″ may be read as five one-hundredths of an inch or fifty thousandths of an inch.

Precision measurement requires that decimal dimensions be read in thousandths. For example, 0.001″ is read one thousandth, 0.020″ is read twenty thousandths, and 0.300″ is read three hundred thousandths. A measurement of 1.321″ is read one inch, three hundred twenty-one thousandths.

Converting Fractions to Decimals. Many measurements and specifications are given in fractions or decimals. To change a fraction to a decimal, divide the numerator by the denominator. For example, to convert the fraction ⁷⁄₁₆ to a decimal, divide 7 (numerator) by 16 (denominator). When seven is divided by 16 the resulting decimal is 0.4375.

Metric System

The metric system is the most common measurement system used in the world. The metric system is based on divisions of 10. The standard unit of length in the metric system is the meter. A meter is equivalent to 39.37″. Prefixes are used in the metric system to represent multipliers. Metric prefixes simplify the notation of large and small numbers. For example, the prefix milli (m) is equivalent to 0.001, so 1 millimeter (1 mm) equals 0.001 of a meter. The most commonly used prefixes in precision measurement are milli (0.001), centi (0.01), and deci (0.1). A decimeter is ¹⁄₁₀ of a meter, a centimeter is ¹⁄₁₀ of a decimeter, and a millimeter is ¹⁄₁₀ of a centimeter.

To change a quantity to another prefix, the quantity is multiplied by the number of units that equals one of the original metric units (conversion factor). For example, to change 54 mm to centimeters (1 mm = ¹⁄₁₀ cm), multiply 54 by 0.1 to obtain 5.4 cm (54 mm × 0.1 = 5.4 cm).

To convert a unit without a prefix (base) to a unit with a prefix, the decimal point in the base is moved to the left or right and a prefix is added. The decimal point is moved to the left and a prefix is added to convert a large value to a simpler term. The decimal point is moved to the right and a prefix is added to convert a small value to a simpler term. The convenience of computing numbers based on 10 has made the metric system the most popular numbering system in the world. **See Figure 5-1.**

Conversion Table							
Initial Units	**Final Units**						
	kilo	hecto	deka	base	deci	centi	milli
kilo		1R	2R	3R	4R	5R	6R
hecto	1L		1R	2R	3R	4R	5R
deka	2L	1L		1R	2R	3R	4R
base	3L	2L	1L		1R	2R	3R
deci	4L	3L	2L	1L		1R	2R
centi	5L	4L	3L	2L	1L		1R
milli	6L	5L	4L	3L	2L	1L	

R = move the decimal point to the right
L = move the decimal point to the left

Figure 5-1. To convert a unit without a prefix (base) to a unit with a prefix, the decimal point in the base is moved to the left or right and a prefix is added.

English/Metric Conversion

English and metric measurements are converted from one system of measurement to the other by multiplying by the number of units of one system that equals one unit of measurement of the other system. The number of units is the conversion factor. Conversions are performed by applying the appropriate conversion factor. Equivalency tables are used to convert between measurement systems. **See Figure 5-2.**

To convert an English measurement to a metric measurement, multiply the measurement by the number of metric units that equal one of the English units (conversion factor). For example, to convert 0.500″ to millimeters, multiply 0.500 by 25.40 to obtain 12.70 mm (0.500″ × 25.40 = 12.70 mm).

To convert a metric measurement to an English measurement, multiply the measurement by the number of English units that equal one of the metric units (conversion factor). For example, to convert 650 mm to inches, multiply 650 by 0.03937 to obtain 25.5905″ (650 mm × 0.03937 = 25.5905″).

Precision measurement is typically required in metalworking processes and operations, manufacturing, maintenance, and quality control (QC). Precision measurement requires the use of measurement or inspection instruments such as rules, protractors, calipers, and micrometers.

Equivalent Measures of Length

Unit	Equivalent		Unit	Equivalent	
1 meter	39.37	inches	1 inch	0.0833	feet
	3.28	feet		0.02777	yards
	1.0936	yards		25.40	millimeters
	1000	millimeters		2.540	centimeters
	100	centimeters			
	10	decimeters			
1 centimeter	0.3937	inches	1 foot	12	inches
	0.0328	feet		1.333	yards
	10	millimeters		0.30480	meters
	0.01	meters		30.480	centimeters
1 millimeter	0.03937	inches	1 yard	36	inches
	0.001	meters		3	feet
				0.9144	meters

Figure 5-2. Equivalency tables are used to convert between measurement systems.

RULES

A *rule* is a measuring tool marked with even increment lines used for measuring length. Although not generally regarded as a precision instrument, a rule is used by many machinists to read measurements of 0.010″ or more. Most rules consist of a rigid strip of steel, plastic, or wood having a straight edge that is marked in increments of inches or centimeters. Rules are available in a variety of lengths up to 72″, with the most commonly used lengths being 6″, 12″, 18″, and 24″. **See Figure 5-3.**

Rule markings consist of incremental vertical lines of various lengths. Each incremental vertical line indicates a unit of measure. The longer the length of the incremental vertical line, the larger the unit of measurement. For example, on a rule with inch markings, the 1″ increment line is longer than the ½″ increment line, the ½″ increment line is longer than the ¼″ increment line, and the ¼″ increment line is longer than the ⅛″ increment line. **See Figure 5-4.**

Rules

Figure 5-3. Rules used for precision measuring are composed of aluminum or stainless steel, with the most commonly used lengths being 6″, 12″, 18″, and 24″.

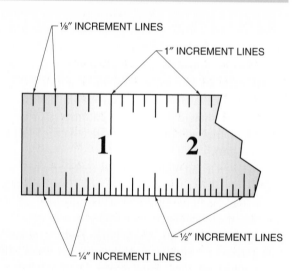

Rule Markings

Figure 5-4. An incremental vertical line indicates each unit of measure on a rule used for precision measurements.

Rules used for precision measurements are generally manufactured to offer the user a choice of scales. For example, one edge of a rule can be calibrated in increments of $\frac{1}{10}''$ (0.100″) and the opposite edge calibrated in increments of $\frac{1}{100}''$ (0.010″). The other side of the rule can have an edge calibrated in increments of $\frac{1}{32}''$ (0.032″) and the opposite edge calibrated in increments of $\frac{1}{64}''$ (0.015″). Some rules are available with English units on one edge and metric units on the opposite edge. Maintenance, production, and QC technicians must be able to identify and properly read all scales on a rule. **See Figure 5-5.**

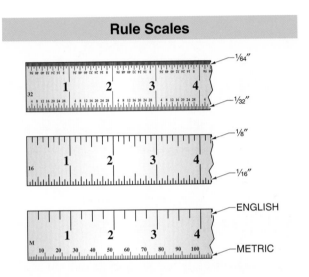

Figure 5-5. Rules used for precision measurements are generally manufactured to offer the user a choice of scales.

Tape Rules

A *tape rule*, or tape measure, is a measuring tool consisting of a long, continuous strip of fabric, plastic, or steel that is marked with evenly spaced increment lines. Tape rules are among the most commonly used measuring tools in the field. They come in various lengths, with 12′, 16′, 20′, 25′, 30′, and 35′ lengths being the most popular sizes. The strip is known as a blade. The blade of a tape rule can be ½″, ¾″, or 1″ wide.

The blade is coiled inside the case, and it can be extracted for making measurements. Once measurements are made, the blade can also be retracted manually or with a spring. Most tape rules also feature a blade lock and an end hook. A blade lock keeps the blade from retracting until the lock is released. The end hook is designed to keep the blade in place while making measurements.

To measure an inside dimension with a tape rule, the end hook is placed against one side of an object, and the back side of the case is placed against the other side. The width of the case is then added to the measurement. **See Figure 5-6.**

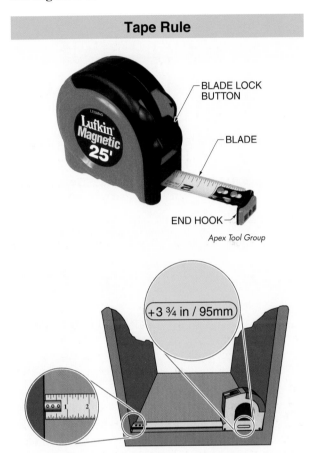

Figure 5-6. Tape rules are among the most commonly used measuring tools in the field, and they come in various lengths.

Reading Rules

Rules are read from left to right, generally from the left edge of the rule (the first inch line). The first inch line is also referred to as the zero mark. A measurement is read from zero (left edge) to the measure mark on the right of the workpiece to be measured. For example, if a $\frac{1}{10}''$ scale measurement is used, and the right edge of the workpiece to be measured ends on the fourth line after the 3″ mark, the measured reading is 3″ and $\frac{4}{10}''$ or 3.400″.

A more precise method of measuring with a rule is to place the 1″ index mark, rather than the end of the rule, at one edge of the workpiece to be measured. The edge of the measured workpiece should coincide with the center of the 1″ index mark. To complete the measurement, the measurement farthest to the right is read and 1″ is subtracted. **See Figure 5-7.**

Reading Rules

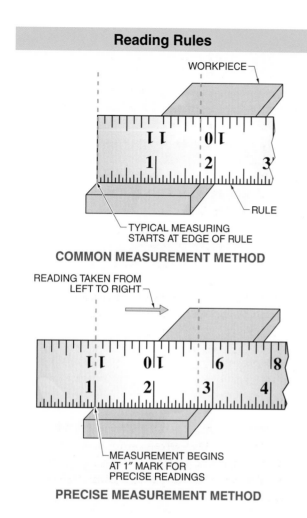

Figure 5-7. Although measurements made with a rule start from the rule's edge, more precise readings can be made if a measurement starts at the 1″ increment line and the inch is subtracted from the final reading.

PROTRACTORS

A *protractor* is a tool for measuring and laying out angles. Protractors can be circular or semicircular. Protractors measure angles of rays extending from a vertex. An *angle* is a geometric figure formed by two lines extending from the same point. A *ray* is a straight line intersecting a point (vertex) of an angle. A *vertex* is the point of intersection of the sides of an angle. **See Figure 5-8.**

Angles are measured in degrees (°). A full circle is 360°. One degree is equal to one 360th of a full circle. A small angle has fewer degrees than a large angle. The four angle types that are most common in industrial mechanics are acute, obtuse, reflex, and right angles.

An *acute angle* is an angle that contains less than 90°. An *obtuse angle* is an angle exceeding 90° but less than 180°. A *reflex angle* is an angle exceeding 180° but less than 360°. A *right angle* is an angle formed by two perpendicular lines and measures exactly 90°. The three common types of protractors used in the mechanical and metalworking trades are conventional, machinist's steel, and reversible. **See Figure 5-9.**

Protractors

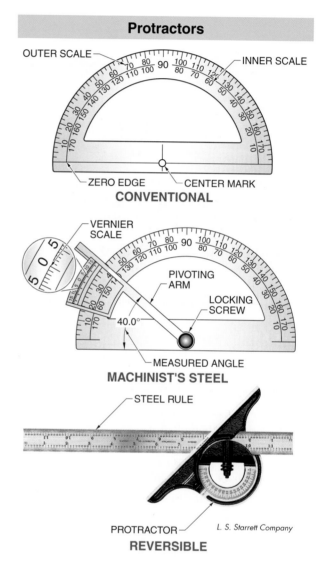

Figure 5-8. Protractors used for mechanical applications include conventional, machinist's steel, and reversible.

Conventional Protractors

A *conventional protractor* is a tool designed to measure printed angles. Conventional protractors can be composed of metal or plastic. The most common type of conventional protractor is made of transparent plastic. Conventional protractors are used for measuring angles on prints and drawings. They are semicircular in shape and have an outer scale, inner scale, zero edge, and center mark.

Outer scales are used to measure angles from left to right, and inner scales are used to measure angles from right to left. The zero edge is the measurement baseline and is aligned along one of the angle's rays. The bottom edge of the protractor is not used for measuring but usually has rule markings. The center mark is a small hole that is centered on the zero edge and is aligned directly over the vertex of an angle to be measured. To read a measurement with a conventional protractor, apply the following procedure:

1. Determine if the angle is acute or obtuse. If the angle is acute, the scale will read less than 90°. If the angle is obtuse, the scale will read greater than 90°.
2. Place the center mark of the protractor directly over the vertex of the angle.
3. Rotate the protractor, using the vertex as the pivot point, and align the zero edge with one of the angle's rays. The opposite ray must cross the protractor's scales.
4. Read the angle measurement of the ray that crosses the scale.

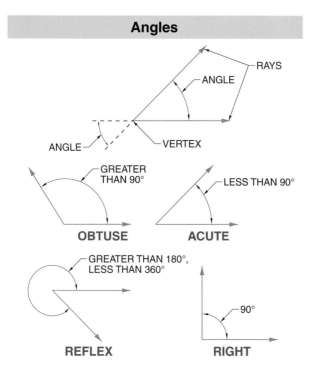

Figure 5-9. The four angle types that are most common in industrial mechanics are acute, obtuse, reflex, and right angles.

Machinist's Steel Protractors

A *machinist's steel protractor* is a tool used to measure or mark angle measurements on rigid workpieces. Machinist's steel protractors are typically composed of steel or stainless steel and may have a vernier scale. Machinist's steel protractors typically are semicircular with a pivoting arm and locking screws to lock the pivoting arm in place once a measured angle has been achieved. Locking the pivoting arm in place allows for the transfer of the same angle to other workpieces. To take a measurement with a machinist's steel protractor, apply the following procedure:

1. Place the flat side of the workpiece to be measured flat against the base of the protractor.
2. Adjust the adjustable arm until it comes to a stop against the workpiece to be measured.
3. Check that the workpiece to be measured is square against the protractor's base and adjustable arm.
4. Read the angle measurement on the scale.
5. Lock the adjustable arm in place once the desired angle has been achieved.

Reversible Protractors

A *reversible protractor* is a finely graduated tool for measuring angles on workpieces with small tolerances. A reversible protractor is usually one tool with a combination of different measuring devices attached to a steel rule and is sometimes referred to as a combination square set. A reversible protractor can be reversed to take measurements along either side of the steel rule. To take a measurement with a reversible protractor, apply the following procedure:

1. Place the base of the reversible protractor on one surface of the workpiece.
2. Verify that the protractor's locking screws are loose.
3. Place the workpiece against the rule or straight edge portion of the reversible protractor.
4. Rotate the rule to the angle of measurement while making square contact with the adjustment rule and the base.
5. Tighten protractor's locking screws.
6. Read angle measurement.

CALIPERS

A *caliper* is a hand tool with one fixed and one adjustable jaw. Markings for taking precision measurements are graduated on the body of the tool. Calipers are used to make precision inside diameter (ID), outside diameter (OD), and length measurements on components.

The parts of a caliper include the main (beam) scale, inside jaws, outside jaws, adjustment screw, calibration screw or switch, locking screw, and either a vernier, dial, or digital scale. The jaws are used to measure the workpiece. The adjustment screw moves the sliding scale along the length of the beam. The calibration screw or switch calibrates the caliper. A locking screw is used to lock the sliding scale in place once the measurement has been made. Some calipers are available with a depth measurement blade for taking depth measurements of hollow or tubular workpieces. Common calipers used by tradesworkers are vernier, dial, and digital calipers. **See Figure 5-10.**

For precision measurement purposes, the tolerance or accepted dimension of a workpiece normally determines the measuring tool required. Tolerance is the permissible range of variation in a dimension or given value. Any tool used for precision measurement purposes must be accurate to at least 10 times the tolerance required. For example, a rule may be a reliable measuring tool if the dimensions of an item are fractional in size. A caliper or micrometer may be a reliable measuring tool if the dimensions of an item must be within a tolerance of plus or minus 0.001″.

Because no two objects are manufactured precisely identical, features and dimensions are specified with an allowable range of variation, referred to as the tolerance range. The tolerance range, when met, enables an object to fit with another object without being the exact size specified and still allow the product to function as designed.

Tolerance ranges are generally shown with a plus (+) and minus (−) in written or drawn specifications. For example, the specification of a machined part may be 2.000″, ±0.010″. In this case, the desired dimension is 2.000″. However, parts with a dimension of 1.990″ or 2.010″ are within tolerance and acceptable. Tolerance indicated on a drawing may be of one range for fractional dimensions and another range for decimal dimensions. **See Figure 5-11.**

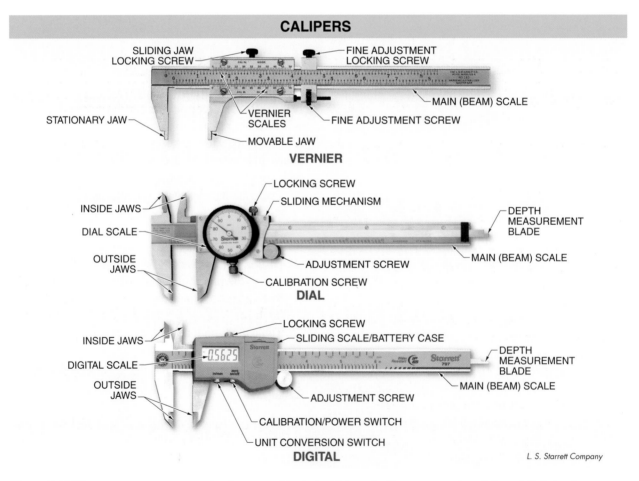

CALIPERS

SLIDING JAW LOCKING SCREW
FINE ADJUSTMENT LOCKING SCREW
STATIONARY JAW
VERNIER SCALES
FINE ADJUSTMENT SCREW
MAIN (BEAM) SCALE
MOVABLE JAW

VERNIER

INSIDE JAWS
LOCKING SCREW
SLIDING MECHANISM
DIAL SCALE
DEPTH MEASUREMENT BLADE
OUTSIDE JAWS
ADJUSTMENT SCREW
MAIN (BEAM) SCALE
CALIBRATION SCREW

DIAL

INSIDE JAWS
LOCKING SCREW
SLIDING SCALE/BATTERY CASE
DIGITAL SCALE
DEPTH MEASUREMENT BLADE
OUTSIDE JAWS
ADJUSTMENT SCREW
MAIN (BEAM) SCALE
CALIBRATION/POWER SWITCH
UNIT CONVERSION SWITCH

DIGITAL

L. S. Starrett Company

Figure 5-10. The most common types of calipers used for industrial applications are vernier, dial, and digital calipers.

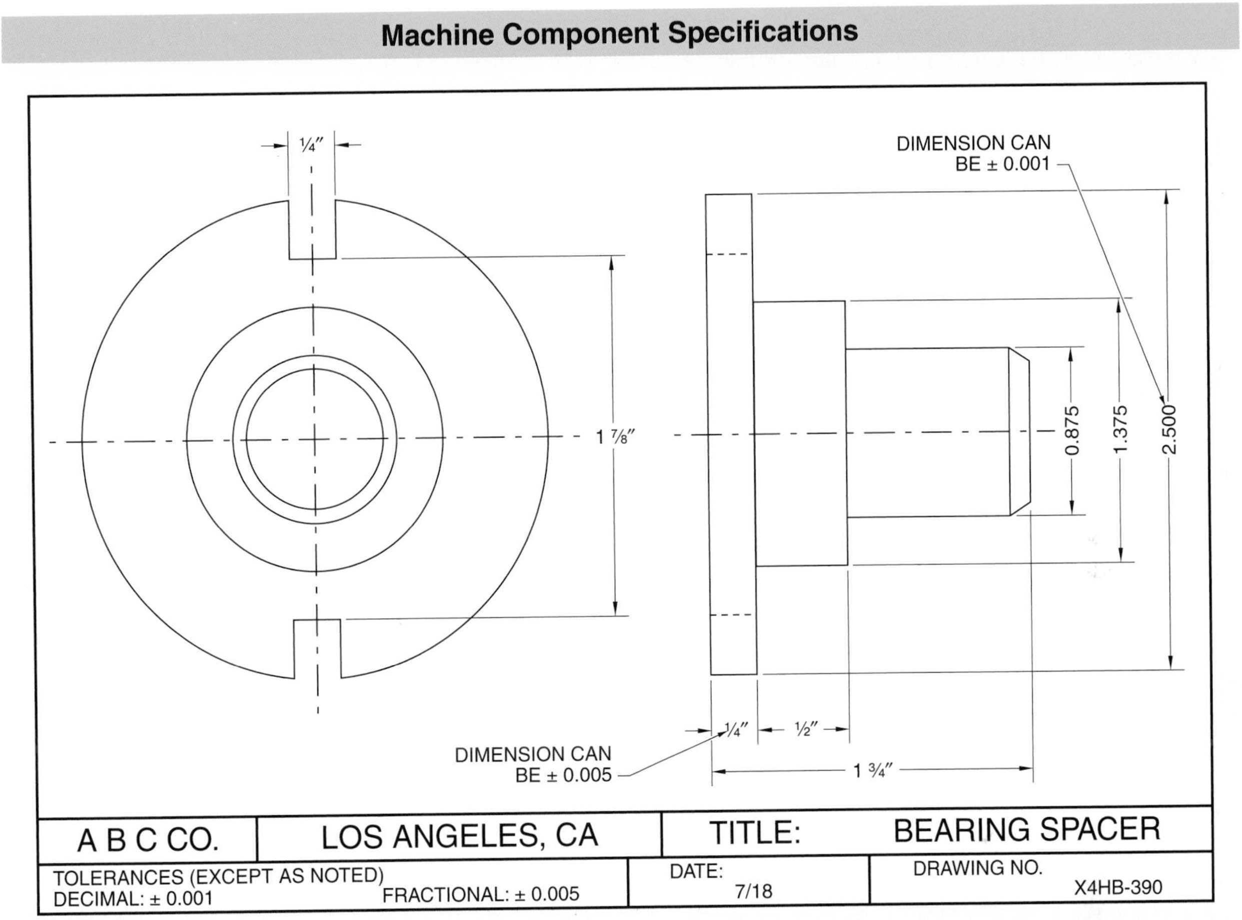

Machine Component Specifications

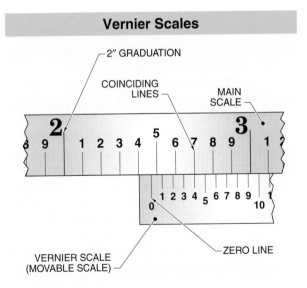

Figure 5-11. Machine component specifications provide tolerances that indicate the required precision of a dimension.

Vernier Calipers

The ability to measure to a greater accuracy than a rule allows began with the invention of the vernier scale. A *vernier scale* is a short auxiliary scale placed along the main scale of a measuring instrument to provide accurate fractional readings of the smallest division on the main scale.

A vernier caliper consists of a main scale and a parallel sliding vernier scale. **See Figure 5-12.** The main scale is fixed and has each unit normally divided into 10 subunits or decimal units. The vernier scale is also laid out with 10 units, accept they total one-tenth shorter in length than the main scale. Each vernier space is 1/100″ smaller than the main scale space.

English vernier caliper main scales have one inch increments with each inch divided into tenths (0.100″). The tenths are further divided into either 0.025″ section scales or 0.050″ section scales, which allows a caliper to dependably measure lengths as precise as 0.001″.

Vernier Scales

Figure 5-12. Accurate and precise measuring is possible with the use of a vernier scale.

Reading Vernier Calipers. Vernier calipers are operated by sliding the movable jaw containing the vernier scale along the main scale. The moveable jaw is slid along the beam by rotating a thumb wheel or pressing a thumb clamp. To obtain a reliable reading, the correct amount of pressure must be applied while taking a measurement. The correct amount of pressure is a firm pressure that is not too tight but not loose. More than one measurement should be taken to obtain a reliable reading.

After the jaws are closed on an object, the main scale is read to locate the position of the 0 on the left side of the vernier scale in relation to the main scale. If the 0 is positioned between the 1″ and 2″ lines on the main scale, the reading begins with 1.000″. **See Figure 5-13.** The measurement continues on to the tenth lines. If the 0 on the vernier scale is positioned between the second and third tenth lines, 0.200″ is added to 1.000″ for a reading of 1.200″.

The two sections between each tenth line are equal to 50 thousandths (0.050), for a total between tenth lines of 100 thousandths. The first section represents a measurement from 0 to 0.050 thousandths and the second section represents a measurement from 0.050″ to 0.100″. If a reading taken at this point is 1.200″ and the vernier 0 is observed between the 2 and 3 tenths lines, then the reading is 1.250″.

To increase the precision of a reading, the next reading is taken from the vernier scale. For example, if the vernier is observed between 0 and 50 the reading is determined by the line on the vernier scale that coincides with a line on the main scale. If the segment line 9 on the vernier scale lines up with a line on the main scale, then 0.013″ is added to the reading. In this example, if the reading from the main scale is 1.250″, the 0 on the vernier scale is in the first section between two tenth lines, and the segment line 13 lines up to a line on the main scale, the final reading is 1.263″.

It is possible for a vernier line to not match up exactly with a line on the main scale. When this occurs, the line closest to matching up is used. In this case, a reading cannot be off by more than 0.0005″.

Reading Vernier Calipers

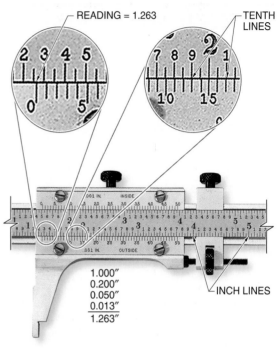

1.000″
0.200″
0.050″
0.013″
1.263″

L. S. Starrett Company

Figure 5-13. Readings from a vernier scale are added to the main scale reading to obtain a final reading.

Vernier calipers are chosen mainly due to their relatively low cost, reliability, and availability in frame ranges from 6″ up to 48″. These calipers are not normally affected by extreme temperatures, nor do they require a power source. Finally, many are purchased as reliable backup tools when other types of calipers fail.

Dial Calipers

A *dial caliper* is a caliper with a dial indicating gauge. Dial calipers are easier and quicker to read than vernier calipers and are commonly used by industrial technicians because of ease of use and durability. However, they must be calibrated prior to each use. Dial calipers show inches and tenths on a main scale (millimeters on the metric scale), similar to a rule. Dial calipers can take only one type of unit measurement. **See Figure 5-14.** For example, metric dial calipers can only take metric measurements. A separate English-unit dial caliper is used to take English measurements. The dial on an English-unit dial caliper indicates measurements in thousandths of an inch (0.001″).

Measuring with Dial Calipers

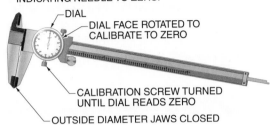

① CLOSE CALIPER JAWS AND CALIBRATE BY USING CALIBRATION SCREW TO ROTATE DIAL TO SET INDICATING NEEDLE TO ZERO.

DIAL

DIAL FACE ROTATED TO CALIBRATE TO ZERO

CALIBRATION SCREW TURNED UNTIL DIAL READS ZERO

OUTSIDE DIAMETER JAWS CLOSED

② CLOSE CALIPER JAWS GENTLY ON WORKPIECE TO BE MEASURED BY MOVING ADJUSTMENT SCREW. DO NOT OVERTIGHTEN CALIPER JAWS ON WORKPIECE.

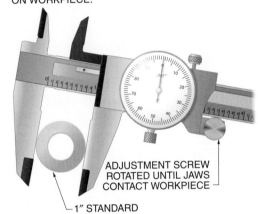

ADJUSTMENT SCREW ROTATED UNTIL JAWS CONTACT WORKPIECE

1″ STANDARD

③ READ MEASUREMENT BY READING VALUE ON MAIN SCALE AND ADDING IT TO VALUE ON DIAL SCALE. WHEN REQUIRED, TIGHTEN LOCKING SCREW TO PREVENT LOSING MEASUREMENT READING.

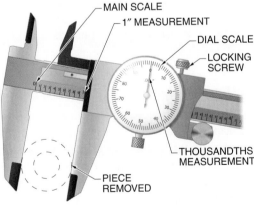

MAIN SCALE

1″ MEASUREMENT

DIAL SCALE

LOCKING SCREW

THOUSANDTHS MEASUREMENT

PIECE REMOVED

MAIN SCALE = 1.00″
+ DIAL SCALE = 0.00″
TOTAL DIMENSION = 1.00″

④ RECORD MEASUREMENT.

Figure 5-14. Dial calipers are commonly used by industrial technicians because of ease of use, durability, and their quickness to produce accurate measurements.

To take a measurement with a dial caliper, apply the following procedure:

1. Calibrate the caliper by using the calibration screw to set the indicating needle on the dial to 0.
2. Close the caliper jaws gently on the workpiece to be measured by moving the adjustment screw. Do not overtighten.
3. Read and record the measurement by reading the measurement on main scale and adding it to the measurement on dial scale.
4. If required, tighten the locking screw to avoid losing the measurement reading.

Digital Calipers

A *digital caliper* is a caliper that displays measurements with a digital electronic indicating gauge. Digital calipers are more precise and easier to read than vernier or dial calipers but require a DC power source (battery) for operation. Digital calipers are used for taking many precision measurements over a short period. Digital calipers have components similar to those of a dial caliper but with a digital display in place of a dial. Most digital calipers can take measurements in either English or metric units. Digital calipers indicate digital readings in thousandths of an inch (0.001″) or fiftieths of a millimeter (0.02 mm). To take a measurement with a digital caliper, apply the following procedure:

1. Turn the caliper ON.
2. Select a unit of measure by depressing the "in./ mm" switch until the desired unit of measure is displayed.
3. Calibrate the caliper by depressing the "ZERO," "ON/ZERO", or "CAL" switch, depending on the design of the caliper. The digital reading of a calibrated caliper displays 0.000.
4. Close the caliper jaws gently on the workpiece to be measured by moving the adjustment screw. Do not overtighten.
5. Read and record the measurement on the digital display.
6. If required, tighten the locking screw to avoid losing the measurement reading.

Note: When using calipers, several readings can be taken to verify accuracy of measurements. If the first and second readings are the same, the measurement is probably correct. If the first and second readings are not the same, retake the readings until two or more readings are the same.

MICROMETERS

A *micrometer* is a hand tool used to make high-accuracy measurements, often to the closest ten-thousandth of an inch (0.0001″). Micrometers are used for verification of component dimensions such as thickness, diameter, and depth. Micrometers are commonly used by production and maintenance technicians. Micrometers are precision measuring instruments designed to use either decimal divisions of an inch (English) or hundredths of a millimeter (metric). Some types of micrometers can take measurements in both English and metric units.

Micrometers are graduated to read to a thousandth of an inch (0.001″) and, with an additional vernier scale, can be read to one ten-thousandth of an inch (0.0001″). The accuracy of micrometer measurements is dependent on the quality of the tool, the care the tool receives, and the skill and knowledge of the technician using the tool. The main types of micrometers used by maintenance, production, and QC technicians are outside micrometers, inside micrometers, electronic micrometers, and depth micrometers. The frame determines if a micrometer is an outside, inside, or depth micrometer.

All micrometers are adaptations of a basic micrometer head. A basic micrometer head consists of an anvil, frame, spindle, spindle lock, reference line, barrel, barrel scale, thimble, thimble scale, and stop. **See Figure 5-15.** An *anvil* is the fixed measuring surface of a micrometer. A *spindle* is a precision-ground, moveable surface of a micrometer. A micrometer frame is attached between the anvil and spindle. Anvils and spindles are available with different faces to accommodate different types of workpieces. For example, a point micrometer has a pointed anvil and spindle for measurement of grooves or keyways. The stop can be a ratchet or friction mechanism that automatically stops spindle movement when pressure is applied to the spindle.

When measuring parts small enough to hold in the hand, a micrometer is held in a manner that allows for thumb and forefinger movement. **See Figure 5-16.** To hold a micrometer properly when measuring small parts, apply the following procedure:

1. Hold the micrometer by the frame with the fourth and fifth fingers.
2. Open the micrometer slightly greater than the thickness or diameter of the workpiece to be measured and insert the workpiece.
3. Use thumb and forefinger to hold the workpiece between the anvil and the spindle while using thumb and forefinger of the hand holding frame to turn the thimble and close the spindle against the workpiece.

4. Lock the micrometer in position.
5. Read the measurement.

Maintenance, production, and QC technicians must regularly perform micrometer scale readings. The most common types of micrometers used for regular maintenance, production, and QC tasks are outside micrometers, inside micrometers, electronic micrometers, and depth micrometers.

Micrometer Parts

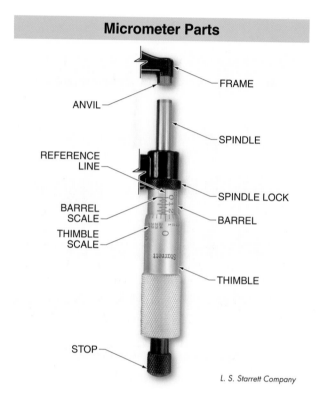

L. S. Starrett Company

Figure 5-15. All micrometers are adaptations of a basic micrometer.

Micrometer Scale Readings

Taking micrometer scale readings requires the lining up of vertical and horizontal lines, which are either engraved or stamped on the thimble and barrel. Each line is directly related to a decimal value of measurement. The thimble is rotated around the barrel so that the adjustable measuring surface (spindle) can extend to make contact with the surface to be measured. Light contact is used to make the final adjustment. A spindle is extended by a screw-driven mechanism that can be damaged if the thimble is rotated with excessive pressure against the workpiece.

Micrometers are based on the principle of a calibrated screw, where the pitch of the screw thread on the spindle of a micrometer is 40 threads per inch. If a micrometer screw is turned one revolution, it moves 0.025″. Also, one revolution of a micrometer thread is divided into 25 equal parts, which represents 1/25″ of 0.025″ (0.001″). Each type of micrometer is available in either English or metric versions.

Holding Micrometers

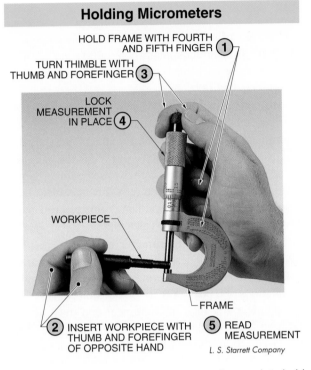

Figure 5-16. When measuring parts small enough to hold in the hand, a micrometer is held in a manner that allows for thumb and forefinger movement.

Micrometers that have fifty 0.001″ graduations to a barrel's rotation instead of twenty five 0.001″ graduations are available for a high level of accuracy. When the thimble is numbered 0 to 50, there are only two minor divisions on the barrel, each representing 0.050″. The distance between each mark on the thimble represents 0.001″. For one revolution, a thimble passes 50 marks, or 0.050″. A scale of 0.000″ to 0.050″ or 0.050″ to 0.090″ is the only difference between using a micrometer with a 0.050″ scale versus one with a 0.025″ scale. Once the inch and tenth scales are determined, the thimble reading is taken and added to the previous reading. The steps for determining and reading the inch and tenth scales are the same as the steps required for the 0.025″ scale.

Outside Micrometers

An *outside micrometer* is a micrometer used for measuring outside diameters and thicknesses of parts. The frame of an outside micrometer is U-shaped to allow the anvil and spindle to fit around a workpiece. The thimble rotation determines the movement of the spindle. The zero mark on an outside micrometer scale is at the spindle end, while the highest numbered division is at the thimble end of the micrometer. In addition to the barrel and thimble scales, many outside micrometers have a vernier scale located on top of the barrel. **See Figure 5-17.**

Outside Micrometers

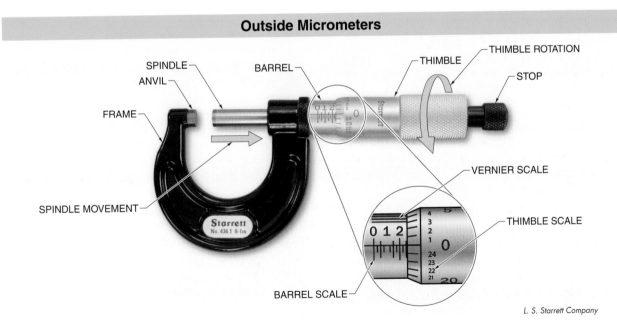

Figure 5-17. An outside micrometer is used for measuring outside diameters and thicknesses of parts.

On English vernier micrometers, the vernier scale can be used to measure as precisely as one ten-thousandth of an inch (0.0001″). On metric vernier micrometers, the vernier scale can be used to measure readings as precise as one one-thousandth of a millimeter (0.001 mm). To take a basic measurement of a workpiece using an outside micrometer with a vernier scale, apply the following procedure:

1. Read the uncovered division on the barrel closest to the thimble.
2. Read the number of graduations on the barrel after the uncovered division number.
3. Read the division on the thimble that is most closely aligned with the horizontal line on the barrel.
4. Read the vernier division on the barrel that is most closely aligned with the thimble division.
5. Add the readings from steps 1 through 4 for the final measurement.

For example, a thimble reading begins with 0.300″. There are also two minor division marks or lines visible, with each division representing 0.025″. Two divisions are read as 0.050″. The barrel scale reading is equal to 0.300″ + 0.050″ = 0.350″. The division on the thimble most closely aligned with the barrel is equal to 0.007″. Vernier division one, representing 0.0001″, aligns exactly with a graduation mark on the thimble scale. If the barrel scale reference line falls between two divisions on the thimble scale, and there is no vernier scale, the reading is taken from the closest thimble number to the barrel scale line. The reading, up to this point, is 0.300″ + 0.050″ + 0.007″ = 0.357″.

The tenth added to the last read thousandth mark is the number taken from one to nine on the vernier scale. This number is the only one that exactly coincides with a line on the thimble. The final reading is 0.300″ + 0.050″ + 0.007″ + 0.0001″ = 0.3571″. **See Figure 5-18.**

Note: When reading a vernier scale, verify that the vernier scale value is added to the final measurement rather than the thimble scale.

Inside Micrometers

An *inside micrometer* is a micrometer used to measure linear dimensions between two inside points or parallel surfaces. An inside micrometer is typically part of a manufacturer's kit that contains different length measuring rod extensions. Inside micrometers use measuring rod extensions attached to the micrometer head to allow for measurements between 2.000″ and 12.500″. The length of each measuring rod extension varies in increments of 1.000″.

Measurements are read when contact is made with the workpiece between the anvil and shoulder head of the micrometer. **See Figure 5-19.** By attaching measuring rod extensions, a single micrometer can take measurements ranging from a few inches to several feet. Some inside micrometer manufacturer's kits also include an attached handle that allows a micrometer head to be inserted within a bore, such as a pipe.

Scales on inside micrometer barrels are read in the same way as scales on outside micrometer barrels (from left to right) and operate using the same thimble graduations of 0.001″. A 0.500″ extension collar is used to complete any inch measurement and must always be used if the measurement is over 0.500″. **See Figure 5-20.** To use an inside micrometer to take a basic measurement of a workpiece ID, attach the appropriate measuring rod extension to the micrometer and apply the following procedure:

1. Place micrometer shoulder head against one surface of workpiece.
2. Adjust micrometer anvil face toward the opposite surface of the workpiece.
3. Use a gentle rocking movement to adjust the micrometer to best fit the workpiece dimensions. The fit is correct when the movement through the centerline of the workpiece has a slight drag between the workpiece surface and the micrometer anvil face.

Outside Micrometers With Vernier Scale

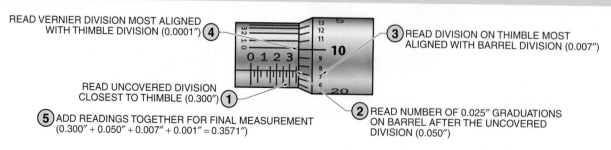

READ VERNIER DIVISION MOST ALIGNED WITH THIMBLE DIVISION (0.0001″) **4**

3 READ DIVISION ON THIMBLE MOST ALIGNED WITH BARREL DIVISION (0.007″)

READ UNCOVERED DIVISION CLOSEST TO THIMBLE (0.300″) **1**

5 ADD READINGS TOGETHER FOR FINAL MEASUREMENT (0.300″ + 0.050″ + 0.007″ + 0.001″ = 0.3571″)

2 READ NUMBER OF 0.025″ GRADUATIONS ON BARREL AFTER THE UNCOVERED DIVISION (0.050″)

Figure 5-18. A vernier scale can be used on a micrometer to measure as precisely as one ten-thousandth of an inch (0.0001″).

4. Turn thimble counterclockwise until firm contact is made with anvil, spindle, and workpiece.
5. Determine final measurement by adding the length of the rod, collar, and micrometer reading.
6. Take several measurements and determine an average reading.

Tech Tip

Tubular inside micrometers have measuring rod extensions attached to both ends of the thimble and barrel and are used for taking measurements on large parts.

Digital Micrometers

A *digital micrometer* is a micrometer with a digital electronic indicating gauge. Digital micrometers operate similarly to standard micrometers but are battery powered with a digital display from 0.000 to 0.999. The only markings on most digital micrometers are graduated lines (tenth scale) on the barrel. In addition to basic micrometer head components, a digital micrometer has separate switches that control power (ON/OFF), unit of measure (in./mm), a zero function (ZERO), and measurement hold function (HOLD). Digital micrometers are easier and quicker to read than standard micrometers and are available in the same formats. **See Figure 5-21.**

Inside Micrometers

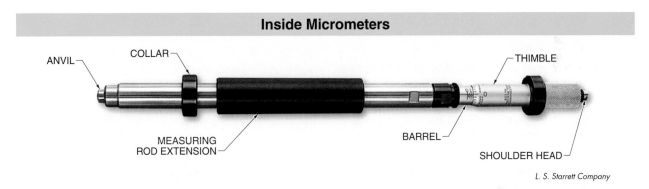

L. S. Starrett Company

Figure 5-19. An inside micrometer is used to measure linear dimensions between two inside points or parallel surfaces.

Inside Micrometer Measurement Procedures

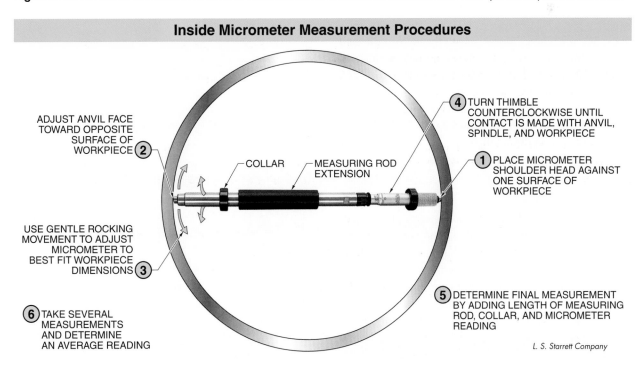

L. S. Starrett Company

Figure 5-20. Inside micrometer measurements are taken by adding the length of the measuring rod, collar, and micrometer reading.

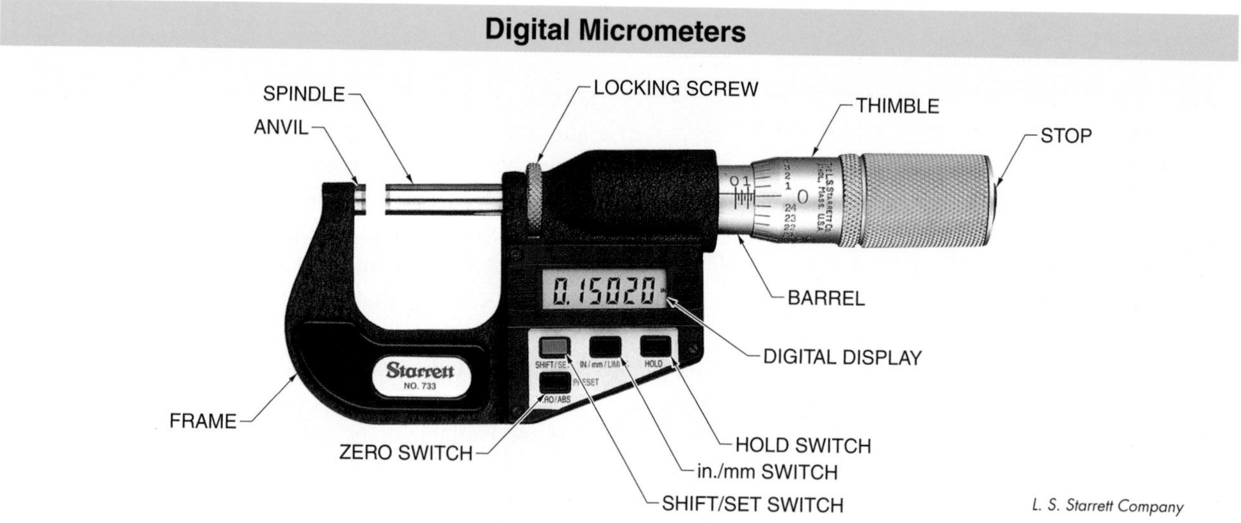

Digital Micrometers

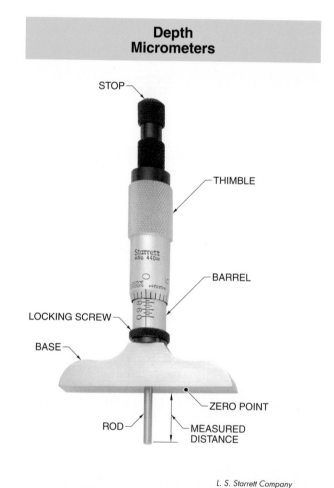

Figure 5-21. Digital micrometers operate similarly to basic micrometers but have a battery-powered digital display.

To take a basic measurement of a workpiece using a 0″ to 1″ outside digital micrometer, apply the following procedure:

1. Turn micrometer ON.
2. Select unit of measure by depressing in./mm switch until proper units are displayed on micrometer display screen.
3. Align thimble scale reference line with barrel scale reference line.
4. Calibrate micrometer by depressing the ZERO switch until display shows 0.000.
5. Place workpiece between anvil and spindle and turn thimble until firm contact is made with anvil, spindle, and workpiece. Tighten locking screw and depress HOLD switch as required.
6. Read uncovered division value on the barrel closest to the thimble.
7. Read value on digital display.
8. Add values together to obtain final reading.

Depth Micrometers

A *depth micrometer* is a precision measuring instrument that measures component depths and heights. Depth micrometers are designed with standard micrometer heads (thimble and barrel) attached to a flat "tee" base. The base design allows the micrometer to be positioned on the workpiece perpendicular to the surface from which the depth is measured (reference surface). **See Figure 5-22.** Depth micrometers are used to measure distances of workpieces that have critical inside dimensions, that have holes, slots, shoulders, and projections such as splines, filter housings, sleeve bearings, and bushings.

Depth Micrometers

L. S. Starrett Company

Figure 5-22. The base of a depth micrometer must be positioned perpendicular (square) to the reference surface of the workpiece.

Depth Micrometer Measuring Rod Extensions. Depth micrometers are generally offered as a set that includes interchangeable measuring rod extensions to accommodate different depth measurements. Depth micrometer measuring rod extensions typically vary by exactly 1″ and are calibrated by the set manufacturer. **See Figure 5-23.** To change the measuring rod extension of a depth micrometer, apply the following procedure:

1. Hold the thimble and unscrew the thimble cap.
2. Remove measuring rod extension from micrometer barrel.
3. Slide proper-size rod extension into barrel.
4. Position rod extension against the contact surface and gently rotate to ensure a positive-contact fit.
5. Screw thimble cap firmly onto the thimble. Thimble cap must not be overtightened.

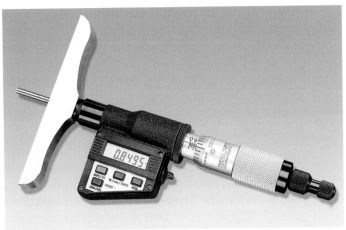

L. S. Starrett Company

Digital depth micrometers are used to take quick, accurate depth measurements.

Depth Micrometer Measuring Rod Extensions

STOP

THIMBLE

MEASURING ROD EXTENSIONS

Starrett
No. 440

BARREL

BASE

MEASURED DISTANCE

ZERO POINT

ROD

DEPTH MICROMETER

L. S. Starrett Company

Figure 5-23. Depth micrometers are generally offered as a set that includes interchangeable measuring rod extensions to accommodate different depth measurements.

Depth Micrometer Calibration. Basic calibration of a depth micrometer requires that the measurement or rod extension start at zero. If a depth micrometer does not have the thimble and barrel scales aligned at zero, false readings can be taken. To calibrate a depth micrometer, apply the following procedure:

1. Install 1″ measuring rod extension.
2. Set the thimble and barrel scales to 0 (0.000″).
3. Wipe all measuring surfaces clean of dust and grime and zero the thimble. Thimbles are cleaned before they are zeroed in order to place the rod surface on the exact same plane as the base surface to prevent trapping of any contaminants, which could cause a false reading.
4. After cleaning the micrometer, back the thimble off slightly (to less than zero) and place base against a cleaned, ground surface.
5. Apply firm pressure on base and slowly rotate the thimble toward zero. At zero, rotate the thimble stop (ratchet) until the stop slips or ratchets. If after several attempts a zero-zero reading is not obtained, major calibration adjustments must be made.

A major calibration adjustment is usually not required because of precision factory settings. However, when a calibration adjustment is necessary, the barrel is rotated using the wrench and written instructions supplied by the manufacturer.

Reading Depth Micrometers. Reading a depth micrometer is similar to reading an outside or inside micrometer, but the graduations on a depth micrometer barrel are numbered in the opposite direction as on a basic micrometer. Barrel graduations on a depth micrometer are numbered 0, 9, 8, 7, 6, 5, 4, 3, 2, 1, and 0. **See Figure 5-24.**

Reading Depth Micrometers

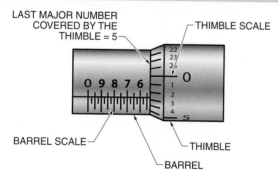

Figure 5-24. Sleeve graduations on a depth micrometer barrel are numbered in the opposite direction as a basic micrometer.

Reading a depth micrometer begins by determining which measuring extension rod to use. For example, a measurement between 2″ and 3″ requires installation of a 3″ rod. A measurement between 5″ and 6″ requires installation of a 6″ rod. Reading the depth micrometer then requires obtaining readings from the barrel and thimble scales. A depth micrometer base must be held securely in order to keep the rod from pushing the micrometer base away from the workpiece. **See Figure 5-25.** Each major division on the barrel scale represents 0.100″. The four minor divisions between them each indicate 0.025″. To read a depth micrometer after inserting a workpiece between the rod and the zero point, apply the following procedure:

1. Determine which numbered divisions on the barrel scale are covered by the thimble.
2. Determine how many unnumbered divisions (lines) are covered by the thimble. If one division is visible and three are not, 0.025″ is added to the reading. If two unnumbered divisions (lines) of the depth micrometer are covered, add 0.050″, if three are covered, add 0.075″.
3. Take a thimble scale reading by observing the graduation on the thimble that is closest to the reference line on the barrel scale.
4. Add thimble scale reading to barrel scale reading.

Note: Readings taken from a depth micrometer are reverse of those on an outside micrometer.

PRECISION MEASURING INSTRUMENT MAINTENANCE

Maintenance of precision measuring instruments, such as rulers, protractors, calipers, and micrometers, requires proper handling, calibration, and storage. Proper handling, calibration, and storage of precision measuring instruments ensure that the most accurate readings are consistently taken and that instrument life is indefinite.

Holding Depth Micrometers

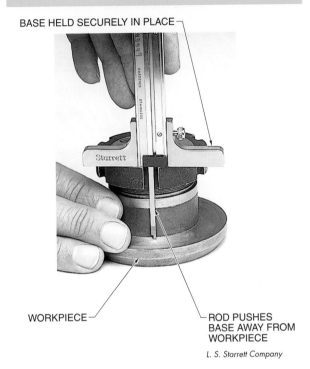

BASE HELD SECURELY IN PLACE

WORKPIECE

ROD PUSHES BASE AWAY FROM WORKPIECE

L. S. Starrett Company

Figure 5-25. A depth micrometer base must be held securely in order to keep the rod from pushing the micrometer base away from the workpiece.

Precision Measuring Instrument Handling

Precision measuring instrument handling requires proper cleaning and lubrication of the instrument before and after each use. Prior to precision measuring instrument use, the measuring surfaces should be cleaned with a lint-free cloth to remove dust and oil deposits. Depending on the accuracy of a desired measurement, dirt and other foreign debris particles can cause inaccurate measurements. After instrument use, a drop of light machine oil or all-purpose oil should be added to any screws or moving parts. For best results, a clean cloth containing light oil should be used to apply a light coat of oil to the instrument's surface. **See Figure 5-26.**

Take precautions to ensure that the instrument is not dropped on a hard surface. A fall of only several inches to a hard surface such as plate steel or concrete can be enough to cause misalignment between a precision measuring instrument's anvil, spindle face, or jaws, destroying the accuracy of the tool. When a measuring instrument is dropped it must be checked for proper calibration before use.

Precision Measuring Instrument Maintenance

CLOTH USED TO LIGHTLY COAT PRECISION MEASURING INSTRUMENT WITH OIL

OIL APPLIED TO CLOTH

Figure 5-26. Precision measuring instrument handling requires proper cleaning and lubrication of the instrument before and after each use.

Precision Measuring Instrument Calibration

Calibration is the verification of graduations and incremental values of a precision measuring instrument for accuracy and adjustments. Precision measuring instruments must be properly calibrated to ensure accurate readings and measurements. A *calibration standard* is a finely ground, precisely sized object that is used as a basis of dimensional comparison. Calibration standards are measured and used to adjust the measuring instrument to the proper calibration. Precision measuring instruments must be calibrated with calibration standards annually to ensure their accuracy. Calibration standards are typically used to calibrate micrometers. **See Figure 5-27.**

When an outside micrometer is properly calibrated, the zero mark on the thimble scale aligns with the zero mark on the barrel scale. Taking consistent, accurate measurements with an outside micrometer depends upon turning the ratchet stop correctly. Repeated practices of having the thimble zero mark align exactly with the zero mark on the barrel scale when closing a micrometer must be performed on a regular basis. If the ratchet is turned too hard or quickly, the thimble will drive the spindle too far. Getting the "feel" of a micrometer and determining when the ratchet stop turns to hit the exact zero point must be accomplished before taking a measurement. If a calibration adjustment is required, follow the manufacturer's instructions provided with the instrument.

Calibration Standards

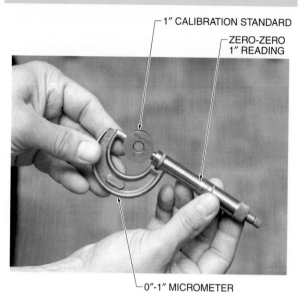

1″ CALIBRATION STANDARD

ZERO-ZERO
1″ READING

0″-1″ MICROMETER

Figure 5-27. Calibration standards are finely ground, precisely sized objects that are used as a basis of dimensional comparison.

The ambient temperature of the work area must be taken into consideration when using precision measuring instruments. The instrument must be placed in the work area for a period of 30 min to 60 min prior to use to allow it to adjust to the ambient temperature. For example, a micrometer that is stored at room temperature of 72°F cannot be used in a cold (50°F and lower) environment until it is placed in the cold environment and allowed to adjust to that temperature. Precision test instruments can give faulty readings because of expansion and contraction of the test instrument caused by temperature extremes. Allowing a test instrument to reach ambient temperature is typically specified when taking measurements using the one ten-thousandth of an inch (0.0001″) scale. Prior to using a precision measuring instrument that has been allowed to reach ambient temperature, it must be calibrated using calibration standards.

Precision Measuring Instrument Storage

Precision measuring instruments must be stored in a protected area away from contaminants such as dirt, dust, grease, moisture, and oil that could damage or bind mechanisms. Measuring instruments must be stored in a temperature-controlled environment and protected from incidental contact with other tools and equipment. Exposure of instruments to temperature extremes can cause premature oxidation on operating mechanisms and measurement surfaces. Incidental contact with other tools or equipment can cause permanent scratches and scuff marks that can affect measurement readings.

Most precision measuring instruments are provided with a rigid storage case with foam padding or a molded insert to fit the shape of the instrument to prevent incidental contact with other tools or objects. Proper care of precision measuring instruments can maintain their integrity and extend their life. For example, when storing micrometers and calipers, a gap must be maintained between the spindle face and anvil or caliper jaws to prevent contact against each other. Oxidation can form between these surfaces if they remain in contact for an extended period. **See Figure 5-28.**

Precision Measurement Instrument Storage

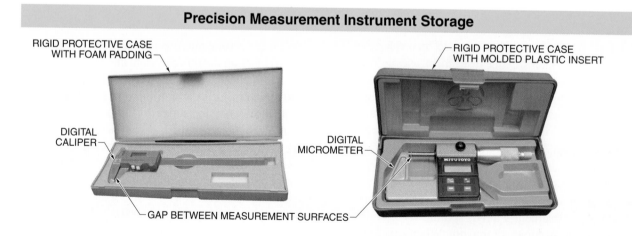

RIGID PROTECTIVE CASE
WITH FOAM PADDING

RIGID PROTECTIVE CASE
WITH MOLDED PLASTIC INSERT

DIGITAL
CALIPER

DIGITAL
MICROMETER

GAP BETWEEN MEASUREMENT SURFACES

Figure 5-28. Most precision measuring instruments are provided with a rigid, padded storage case to prevent incidental contact with other tools or objects.

Review

1. Define precision and accuracy and list the two most commonly used systems of measurement.

2. Explain how to convert a unit without a prefix (base) to a unit with a prefix.

3. Explain how to convert a metric measurement to an English measurement.

4. Define angle, ray, and vertex.

5. List three types of measurements calipers are used to take on components.

6. Define vernier scale.

7. List three factors that the accuracy of a micrometer measurement is dependent on.

8. Why is the frame of an outside micrometer U-shaped?

9. Describe the relationship between a thimble scale and barrel scale when an outside micrometer is properly calibrated.

10. Define digital micrometer and describe two ways in which it differs from a standard micrometer.

11. Describe the advantage of the base design on a depth micrometer.

12. What is the main difference between the graduations on a depth micrometer barrel scale and a basic micrometer barrel scale?

13. Define calibration standard.

14. What does proper handling, calibration, and storage of precision measuring instruments ensure?

15. Describe the proper way to adjust a precision measuring instrument to the ambient temperature before use.

Digital Resources

ATPeResources.com/Quicklinks
Access Code 462409

CHAPTER 6

RIGGING

Rigging is securing equipment in preparation for lifting by means of wire rope slings, chains, or web slings. Loads must be balanced and proper load weights calculated for safe lifting. Wire rope slings are used for lifting because of length and flexibility. Chain is used in situations in which other materials could be damaged by the load or operating environment. While repair and testing of rigging components are completed by the manufacturer, all rigging components should be inspected for damage and suitability before use.

Lift-All Company, Inc.

Chapter Objectives

- Explain the importance of load balancing during rigging.
- Explain how to calculate load weights.
- Define sling.
- Describe the basic construction of rope.
- Compare wire rope and fiber rope.
- Identify the basic elements used when working with rigging knots.
- List the common types of rigging knots.
- Describe the basic construction of webbing.
- Define round sling.
- Describe the basic construction of chain.
- Explain the inspection process of rigging components.

Key Terms

- rigging
- lifting
- center of gravity
- sling
- fiber rope
- rope lay
- static load
- wire rope
- cabling
- seizing
- socket
- fiber rope
- lay
- hitch
- bight
- splice
- knot
- webbing
- round sling
- chain
- shackle
- master link
- hook

RIGGING

The shape, weight, and location where a load is to be moved must be known prior to rigging and lifting. In many cases, the stresses and equipment needs for the lift may change as the load is moved and positioned. *Rigging* is securing equipment or machinery in preparation for lifting by means of wire rope slings, web slings, or chain. Rigging includes the planned and controlled movement of material or equipment resulting in no damage to equipment or personnel. All phases of a rigging job must be planned out prior to beginning a job. *Lifting* is hoisting equipment or machinery by mechanical means.

The weakest component determines the strength of the entire lifting system. For this reason, the limitation of every component used to move a load must be determined. Other considerations include the conditions (indoor or outdoor), travel path, equipment, and the skill level of the workers.

Load Balance

The shape of a load normally determines its center of gravity. The shape of a load may be symmetrical or asymmetrical. **See Figure 6-1.** A *symmetrical load* is a load in which one-half of the load is a mirror image of the other half. Symmetrical loads include straight pipe sections, motors, paper rolls, and sheet metal.

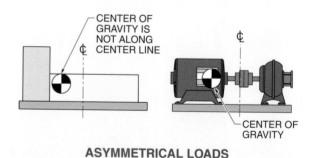

Center of Gravity

SYMMETRICAL LOADS

ASYMMETRICAL LOADS

Figure 6-1. The symmetry of a load is sometimes a visual clue to the location of the center of gravity.

An *asymmetrical load* is a load in which one-half of the load is not a mirror image of the other half. Asymmetrical loads include most machinery, motor and pump assemblies, pipe and valve assemblies, and engines.

Some loads are equipped by the manufacturer with lifting lugs for ease in lifting and transporting. A *lifting lug* is a thick metal loop (eyebolt) welded or screwed to a machine to allow balanced lifting. Lifting lugs eliminate the need for additional rigging.

Center of gravity is the balancing point of a load. The center of gravity of a load must be determined before lifting to prevent the loss of control of the load. Complex shapes and materials make it difficult to determine the center of gravity for most loads. Some manufacturers mark the center of gravity on their equipment, while others offer specification sheets that include center of gravity information. Calculations determining the center of gravity may be made from weight, shape, and material standard information. An educated guess may be made, placing the center of gravity in an approximate location.

A load may be lifted without chance of tipping or toppling once the center of gravity is determined (balanced load). **See Figure 6-2.** Load tipping occurs when a load is unsteady, unbalanced, or unstable.

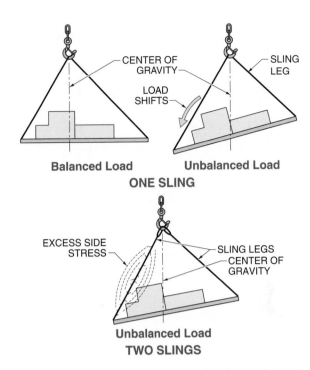

Figure 6-2. A load must be lifted from directly over the center of gravity in order to maintain balance.

Symmetrical loads are lifted using two slings by placing each sling at the sides of the load and lifting slightly to observe weight shifting. The slings are adjusted and the process repeated until the load is stable.

Load toppling occurs when a load is unbalanced because it is unstable or top heavy. Unbalanced loads always shift toward the center of gravity and create a dangerous condition when lifted. Most of the weight is placed on one sling if two slings are used to lift an unbalanced load.

Once the center of gravity has been determined, the lifting hook should be positioned directly over it. Lifting equipment is then connected to the load and the load lifted slightly to observe stability and weight shifting. The lifting equipment is readjusted towards the heavy (dipping) side of the load if an imbalance is observed. This procedure is continued until the load is balanced and stable. *Imbalance* is a lack of balance.

In addition to the vertical center of gravity, the horizontal center of gravity of a load must also be determined. This is a weight mass above a pivot point that causes a load to topple because it is top heavy. The lifting equipment must not be attached to the load at any point lower than the horizontal center of gravity. A load may be unstable and subject to toppling if lifting equipment is placed below the horizontal center of gravity of a load. **See Figure 6-3.**

The sling apex must be above the horizontal center of gravity if a load is to be lifted from the base of a machine or skid. The *sling apex* is the uppermost point where sling legs meet.

Vertical Placement of Lift Points

LIFT POINTS ABOVE
CENTER OF GRAVITY

LOAD MAY
TOPPLE

CENTER OF
GRAVITY

LIFT POINTS
BELOW CENTER
OF GRAVITY

LIFT
HOOK

LOAD MAY
TOPPLE

CENTER OF GRAVITY

SKID

**BALANCED
LOAD**

**UNBALANCED
LOAD**

Figure 6-3. The vertical location of the lift points in relation to the center of gravity also affects the stability of a load.

CALCULATING LOAD WEIGHTS

The weight of a load is calculated after considering the shape and size of the load. The weight of the load may be found on the data plate located on the equipment, on shipping documents, or on the manufacturer's product bulletin. Always ensure that the weight has not changed since the last printing or entry when data is obtained from printed documents. Weight changes often occur when equipment has been modified or when a product is left in a holding vessel.

Safety Tip

When calculating a total lift weight, the total weight must include the weight of the load and that of all rigging equipment including the crane block.

Load weights are normally determined by calculation if manufacturer's printed information is not available. Load weights may be calculated using stock material weight tables or the area, volume, and load material weight information. The full-load weight should include the weight of the rigging equipment. To obtain the full-load weight, add the total material weight to the weight of the rigging equipment.

Sprecher + Schuh

Overhead cranes driven by electric motors are used to move heavy metal beams.

Stock Material Weight Tables

Stock material weight tables are used when a load consists of basic stock materials such as steel or brass bar stock. Stock material weight tables are available listing the weight of materials by their linear, square, or cubic measurements in either the English or metric systems. **See Appendix.** For example, a 1″ diameter round steel bar weighs 2.67 lb/ft (from Weight of Steel and Brass Bar Stock table). **See Figure 6-4.**

Common stock materials include round and square bar, round and square tubing, I-beam, angle stock, tee stock, channel, and plate. Tables for these common shapes may be located through the American National Standards Institute (ANSI). The weight of stock material is found by referring to the proper stock material weight table and applying the following formula:

$$W = l \times w/ft$$

where

W = weight (in lb)

l = length (in ft)

w/ft = weight (in lb/ft)

Weight of Steel and Brass Bar Stock*

Diameter or Thickness†	Round Steel	Square Steel	Brass
¼	0.167	—	0.181
½	0.667	—	0.724
¾	1.50	—	1.63
1	2.67	3.4	2.89
1¼	4.17	—	4.52
1½	6.01	7.7	6.51
1¾	8.18	—	8.86
2	10.68	—	11.57
4	42.7	54.4	—
5	66.8	85.0	—
6	96.1	122.4	—
10	267.0	340.0	—
12	384.5	489.6	—

* in lb/ft
† in in.

Weight of Steel Plate*

Thickness†	Weight
¹⁄₁₆	2.55
⅛	5.1
³⁄₁₆	7.65
¼	10.2
⁵⁄₁₆	12.75
⅜	15.3
½	20.4
⅝	25.5
¾	30.6
1	40.8
1¼	51.0
1½	61.2
2	81.6

* in lb/ft
† in in.

Figure 6-4. Stock material weight tables list the weight of materials by feet, square feet, or cubic feet.

For example, what is the total material weight of a 10′ long, 2″ diameter brass bar?

$$W = l \times w/ft$$

$W = 10 \times 11.57$ (from Weight of Steel and Brass Bar Stock table)

$W = $ 115.7 lb

Load weights of multiple plates or sheets of steel, aluminum, or brass stock are calculated by finding the material weight using stock material weight tables, finding the square footage, and multiplying by the number of plates or sheets. The weight of multiple plates or sheets of stock is found by applying the procedure:

1. Find the weight of one sheet.
2. Find the area (in sq ft) of material. The area of the material is found by applying the following formula:

$$A = l \times w$$

where

A = area (in sq ft)

l = length (in ft)

w = width (in ft)

3. Find the total material weight. The total material weight is found by applying the following formula:

$$W = w/sq\,ft \times A \times q$$

where

W = weight (in lb)

$w/sq\,ft$ = weight per square foot (in lb)

A = area (in sq ft)

q = number of plates or sheets

For example, what is the total material weight of a load consisting of 35 4′ × 8′ sheets of steel that are ¹⁄₁₆″ thick?

1. Find the weight of one sheet.
   ¹⁄₁₆″ plate steel weighs 2.55 lb/sq ft (from Weight of Steel Plate table)

2. Find the area of the material.

$$A = l \times w$$
$$A = 4 \times 8$$
$$A = 32 \text{ sq ft}$$

3. Find the total material weight.

$$W = w/sq\,ft \times A \times q$$
$$W = 2.55 \times 32 \times 35$$
$W = $ 2856 lb

Safety Tip

Caution should be used whenever hoisting a load with an unknown center of gravity, since the load may tilt or shift. Never stand or place hands or feet under hoisted loads.

Material Weight Calculations

Material weight calculations are made using the volume of the object and the weight of the material if the load weight cannot be determined using a material weight table. The figures used in this method should be rounded off to allow rapid calculations. *Rounding off* is the process of increasing or decreasing a number to the nearest acceptable number. For material weight calculations, numbers are rounded to the nearest 50. Any numbers in question should be rounded up for added safety. For example, 490, 325, 782, 110, 231, and 506 can be added rapidly if the numbers are rounded off to 500, 350, 800, 100, 250, and 500.

Rounded-off calculations give dependable information for calculating load weights. With experience, rounded-off calculations become rapid and precise enough for safe rigging. The total material weight of loads with multiple regular shapes is estimated by applying the procedure:

1. Find the area of the vessel bottom. The area of the vessel bottom is found by applying the following formula:

$$A_b = \pi r^2$$

where

A_b = area of bottom

π = 3.1416

r^2 = radius squared

2. Find the area of the sides. The area of the vessel side is found by applying the following formula:

$$A_s = d_s \times \pi \times l_s$$

where

A_s = area of side

d_s = diameter of side

π = 3.1416

l_s = length of side

3. Find the total area of the vessel. The total area of the vessel is found by applying the following formula:

$$A_v = A_b + A_s$$

where

A_v = total area of the vessel

A_b = area of bottom

A_s = area of side

4. Find the total material weight. The total material weight is found by applying the following formula:

$$W = A_v \times w/sq\,ft$$

where

W = total material weight (in lb)

A_v = area of vessel

$w/sq\,ft$ = material weight per square foot

For example, what is the total material weight of an uncovered steel tank 4'-8" in diameter and 5'-11" deep with a wall thickness of ¼"? **See Figure 6-5.** *Note:* Dimensions are rounded off to 5' diameter and 6' depth. Pi is rounded to 3.

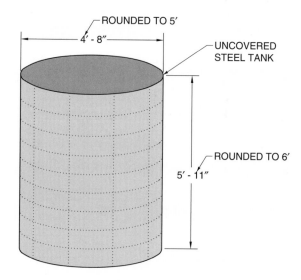

Figure 6-5. Numbers may be rounded off to allow rapid material weight calculations.

1. Find the area of the vessel bottom.

$$A_b = \pi r^2$$

$$A_b = 3 \times (2.5 \times 2.5)$$

$$A_b = 3 \times 6.25$$

$$A_b = 18.75 \text{ sq ft (round to 19 sq ft)}$$

2. Find the area of the sides.

$$A_s = d_s \times \pi \times l_s$$

$$A_s = 5 \times 3 \times 6$$

$$A_s = 90 \text{ sq ft}$$

3. Find the total area of the vessel.

$$A_v = A_b + A_s$$

$$A_v = 19 + 90$$

$$A_v = 109 \text{ sq ft (round to 110 sq ft)}$$

4. Find the total material weight.

$$W = A_v \times w/sq\,ft$$

$$W = 110 \times 10.2 \text{ (from Weight of Steel Plate table)}$$

$$W = \textbf{1122 lb}$$

SLINGS

A *sling* is a line consisting of a strap, chain, or wire rope used to lift, lower, or carry a load. Slings used to lift loads are made of various components. The main sling components lift the load. Main sling components include wire rope, fiber rope, chain, webbing, and round sling. **See Figure 6-6.** Other sling components include rigging hardware attachments such as clips, hooks, eyebolts, shackles, sockets, wedge sockets, web sling fittings, and master links. **See Figure 6-7.**

Safety Tip

The American Society of Mechanical Engineers (ASME) helps establish safe structural design of hoists and cranes.

Sling Hitches

A sling hitch consists of a section of the main component (rope, chain, etc.) with a loop at both ends. Basic sling hitches include vertical (single-leg), choker, U-basket, and bridle. Tables are used to determine the safe load capacities of specific rigging components such as rope, webbing, chain, etc., and may also rate their attachments. **See Figure 6-8.**

Lifting using a vertical sling results in a straight vertical pull. The straight connection between the hook and the load offers 100% load capacity of the single sling. A single-leg sling is used to lift loads such as pumps, motors, gear drives, or any device equipped with a single eyebolt or lifting lug that has not been modified.

Main Sling Components

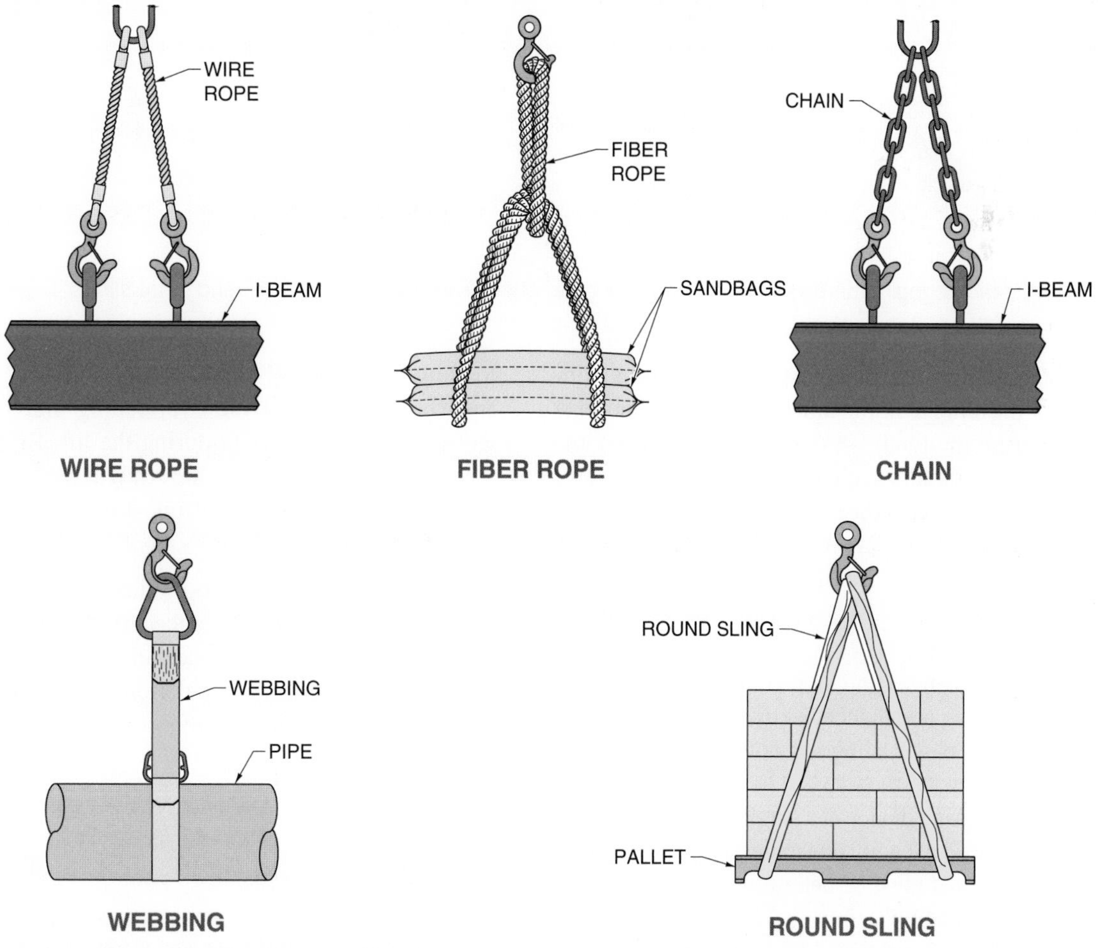

Figure 6-6. Main sling components include wire rope, fiber rope, chain, webbing, and round sling.

Rigging Hardware Attachments

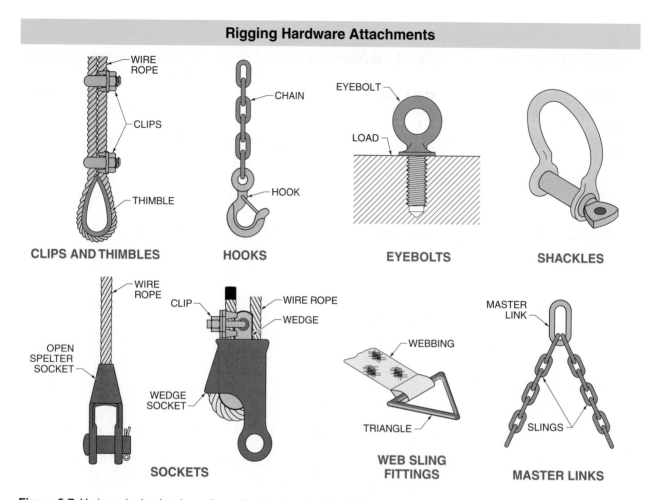

CLIPS AND THIMBLES   HOOKS   EYEBOLTS   SHACKLES

SOCKETS   WEB SLING FITTINGS   MASTER LINKS

Figure 6-7. Various rigging hardware is used with slings to attach slings to hoists, loads, and other slings.

A choker sling is created by slipping the loop from one end of the sling over the other end (choke junction) after wrapping the load. Choker sling loads are commonly used with loads such as pipes, bars, poles, etc. Choker sling load capacity is considerably less than that of a vertical sling. The reduced capacity is due to the angle of pull created at the choke junction. Each degree from the vertical 0° angle position increases tension on the sling.

A U-sling is a single line looped under the load. The ends of the U-sling are attached to different hooks of the lifting device. A basket sling uses one sling, similar to the U-sling. However, the two eye loops of the sling are attached to the lifting device at a single point. The basket sling has a reduced capacity due to the angles of the sling legs. The rated load must be determined by multiplying the vertical load rate by a sling angle loss factor.

A bridle hitch consists of two or more straight slings using identical sling constructions, length, and previous loading experience. Normal stretch must be the same for paired slings to avoid overloading individual legs and unbalancing the load during the lift. Bridle slings are used where more than one straight sling is required to make the lift, such as lifting drums, machinery, or lengths of material.

Load Weight Proportioning. When only two sling legs are used and the load is balanced, each sling can be assumed to carry half of the load weight. This information is then used to perform calculations and ensure that the sling is strong enough to support its portion.

Safety Tip

Welding rigging attachments can be hazardous. Knowledge of materials, heat treatment, and welding procedures are necessary for proper welding.

Basic Sling Combinations

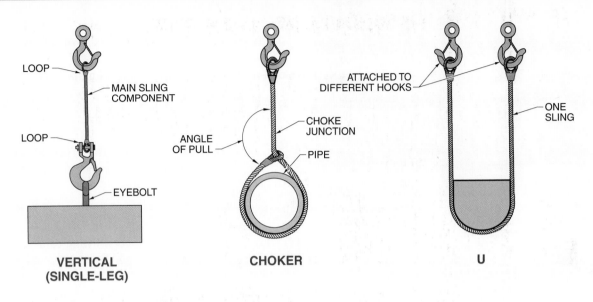

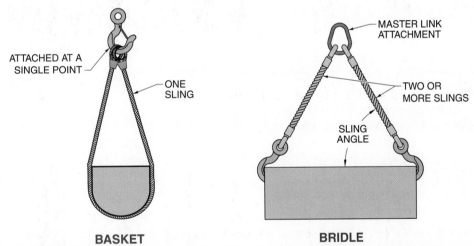

Sling Rope Load Capacity 6 × 36 Classification (2000 lb Ton)					
Rope Dia*	Choker	Vertical Load	2-Leg 30°	2-Leg 45°	2-Leg 60°
1¼	9.9	11.0	11.0	16.0	20.0
1⅜	12.0	14.0	14.0	19.0	24.0
1½	14.0	18.0	16.0	21.0	29.0
1¾	19.0	22.0	22.0	23.0	38.0
2	25.0	28.0	24.0	40.0	49.0

* in in.

Vertical Sling Component Load Capacity 6 × 19 Ips-fc Classification* (2000 lb Ton)				
Rope Dia†	Spelter/ Swaged	U-Bolt	Wedge	Mechanical Splice
¼	0.54	0.43	0.43	0.49
⅜	1.22	0.97	0.97	1.09
½	2.14	1.71	1.71	1.92
¾	4.76	3.80	3.80	4.28
1	8.36	6.68	6.68	7.52

* rates include safety factor of 5
† in in.

Figure 6-8. Slings are used in various arrangements depending on the size, shape, and weight of the load.

Lift-All Company, Inc.

When lifting a load with three slings, two of the slings must be able to each withstand half of the load.

Rigging can also be arranged with three or more sling legs. However, with more than two legs, an equal distribution of load weight across the slings becomes increasingly more difficult to ensure. Therefore, it is assumed that at least one of the legs carries no weight and provides stability only. **See Figure 6-9.** This adds a greater margin of safety that accounts for the likely unequal loading of the sling legs.

Each leg of a rigging arrangement with three legs is assumed to carry half of the load weight, and each leg of a rigging arrangement with four or more legs is assumed to carry one-third of the load weight. These load weight portions are then used to calculate actual forces on angular slings (if applicable) and ensure that the sling strength is adequate.

Angular Sling Load Capacity

Angular lifting tension increases and load capacity decreases as the sling angle decreases. The sling angle decreases as the lifted load widens. **See Figure 6-10.** The load capacity of choker, basket, or bridle slings is calculated by applying the following formula:

$$LC = vl \times l \times s$$

where

LC = load capacity (in t)

vl = vertical load rate (from Vertical Sling Component Load Capacity 6 × 19 IPS-FC Classification table)

l = number of sling legs (not more than two)

s = loss factor (from Sling Angle Loss Factors table)

For example, what is the total lifting capacity of a two-leg sling made of ⅜″, 6 × 19, IPS-FC wire rope with sling loops constructed of U-bolt clips and sling angles of 50°? *Note:* The vertical load rate for U-bolt clips is 0.97.

$$LC = vl \times l \times s$$
$$LC = 0.97 \times 2 \times 0.766$$
$$LC = \textbf{1.486 t}$$

ROPE

Rope is used for lifting because of its length and flexibility. Rope is flexible due to its construction of many wires or fibers and small strands. Rope is manufactured from wire, organic fibers, or synthetic fibers and is widely used for transporting loads.

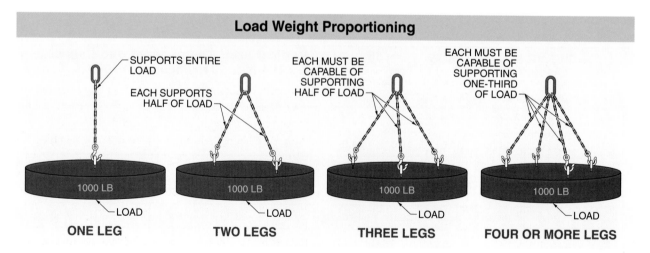

Figure 6-9. When calculating the distribution of load weight in sling arrangements with three or more legs, at least one of the legs is assumed to carry no load weight. As a result, the load is divided equally among the others legs.

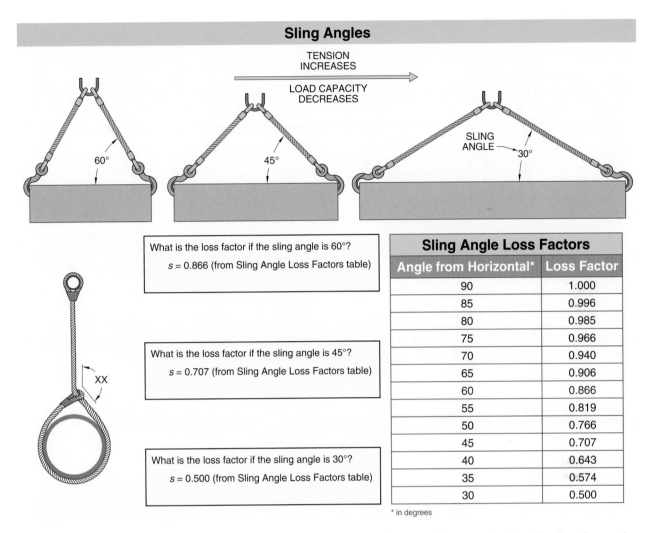

Sling Angles

TENSION INCREASES

LOAD CAPACITY DECREASES

60°

45°

SLING ANGLE 30°

What is the loss factor if the sling angle is 60°?

s = 0.866 (from Sling Angle Loss Factors table)

What is the loss factor if the sling angle is 45°?

s = 0.707 (from Sling Angle Loss Factors table)

XX

What is the loss factor if the sling angle is 30°?

s = 0.500 (from Sling Angle Loss Factors table)

Sling Angle Loss Factors	
Angle from Horizontal*	Loss Factor
90	1.000
85	0.996
80	0.985
75	0.966
70	0.940
65	0.906
60	0.866
55	0.819
50	0.766
45	0.707
40	0.643
35	0.574
30	0.500

* in degrees

Figure 6-10. When multiple slings are used, the sling angle affects the increased forces on the sling. Smaller sling angles result in greater sling forces.

Rope Construction

Fiber rope is constructed by twisting fibers into yarn, yarn into strands, and strands into rope. *Yarn* is a continuous strand of two or more fibers twisted together. A *strand* is several pieces of yarn helically laid about an axis. A *helix* is a spiral or screw shape form. Strands are twisted (laid) to form the rope. Wire rope is constructed by twisting wires into strands around a wire core. The strands are laid to form the wire rope. The strands are often laid around a fiber core. **See Figure 6-11.**

Lay and Strands. *Rope lay* is the length of rope in which a strand makes a complete helical wrap around the core. A rope lay also designates the direction of the helical path in which the strands are laid. The strands form a spiral to the right in a right lay rope. The strands form a spiral to the left in a left lay rope. The right and left lay rotation produces regular-lay or lang-lay rope. Regular-lay and lang-lay rope have different advantages.

A *regular-lay rope* is a rope in which the yarn or wires in the strands are laid in the opposite direction to the lay of the strands. A *right regular-lay rope* is a rope in which the strands are laid to the right and the yarn or wires are laid to the left. A *left regular-lay rope* is a rope in which the strands are laid to the left and yarn or wires are laid to the right. A *lang-lay rope* is a rope in which the yarn or wires and strands are laid in the same direction. A *right lang-lay rope* is a rope in which the yarn or wires are laid to the right and the strands are laid to the right. A *left lang-lay rope* is a rope in which the yarn or wires are laid to the left and the strands are laid to the left. Right regular-lay rope is generally used for rigging purposes due to its resistance to rotation. Lang-lay ropes are used where flexibility and fatigue resistance are required.

Rope Construction

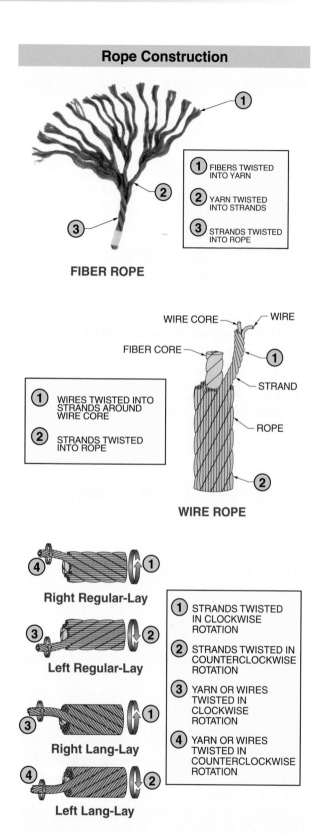

FIBER ROPE

① FIBERS TWISTED INTO YARN

② YARN TWISTED INTO STRANDS

③ STRANDS TWISTED INTO ROPE

WIRE CORE — WIRE

FIBER CORE

① STRAND

ROPE

① WIRES TWISTED INTO STRANDS AROUND WIRE CORE

② STRANDS TWISTED INTO ROPE

WIRE ROPE

Right Regular-Lay

Left Regular-Lay

Right Lang-Lay

Left Lang-Lay

① STRANDS TWISTED IN CLOCKWISE ROTATION

② STRANDS TWISTED IN COUNTERCLOCKWISE ROTATION

③ YARN OR WIRES TWISTED IN CLOCKWISE ROTATION

④ YARN OR WIRES TWISTED IN COUNTERCLOCKWISE ROTATION

Figure 6-11. Fiber rope is constructed by twisting fibers into yarn, yarn into strands, and strands into rope. Wire rope is constructed by twisting wires into strands around a wire core.

Rope Diameter. The diameter of wire rope is determined by the largest possible outside dimension. **See Figure 6-12.** The outside dimension (diameter) is the circle that fully encircles the rope. The rope is measured from the high spot on one side of the rope to the high spot on the opposite side using vernier calipers. New ropes are normally slightly larger in diameter than the specifications indicate.

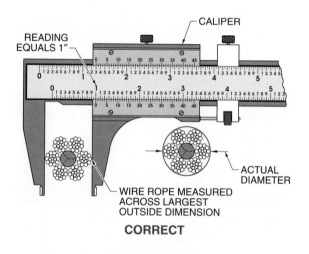

READING EQUALS 1″

CALIPER

WIRE ROPE MEASURED ACROSS LARGEST OUTSIDE DIMENSION

ACTUAL DIAMETER

CORRECT

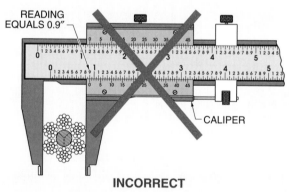

READING EQUALS 0.9″

CALIPER

INCORRECT

Figure 6-12. Wire rope diameter must be measured at its widest point.

Rope Strength

A rope loses strength during use due to moisture, temperature, chemical activity, and bending. The typical breaking strength of rope shown in charts or tables is based on new rope. Rope wear is indicated by abrasion marks, stretching or breaking of wire or fibers, or a reduction in rope diameter. **See Figure 6-13.**

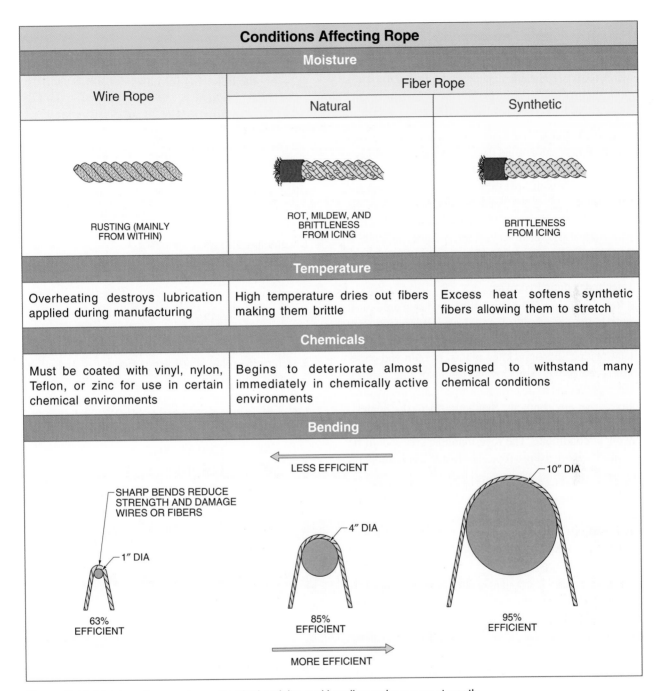

Figure 6-13. Moisture, temperature, chemical activity, and bending reduce rope strength.

Moisture. The effects of moisture vary between rope types. Wire ropes should be kept lubricated to prevent rusting. Moisture affects wire rope by causing it to rust from the inside out. Weakening from rust may not be indicated until the rope breaks.

Natural fiber rope may absorb moisture and decay or rot. Most rope manufacturers treat natural fiber rope with waterproofing. However, enough moisture may still be absorbed to significantly weaken a natural fiber rope when frozen. Natural fiber rope must be completely thawed before use.

Synthetic fiber rope is normally not affected by moisture because moisture is not absorbed by the fibers. However, the fibers may become brittle and weakened if synthetic rope is coated with ice. Ropes used for outdoor applications are more likely to experience moisture problems.

Temperature. Manufacturers supply data on the temperature limits of rope. Wire rope with a fiber core should not be used in temperatures over 180°F (82.2°C). Wire rope with a wire core is used in temperatures up to 400°F (204.4°C). The strength of fiber rope is rated for use in the temperature range of –20°F to 100°F (–28°C to 65°C).

Chemical Activity. Reaction to acids, alkalis, caustic solutions, or fumes cause rapid damage to rope. An *acid* is a sour substance with a pH value less than 7. An *alkali* is a bitter substance with a pH value greater than 7. Acids and alkalis turn certain materials to salts by means of corrosion. *Corrosion* is the action or process of eating or wearing away gradually by chemical action. A *caustic solution* is a liquid that causes corrosion. Common caustic solutions include potash (water and potassium hydroxide) and soda (water and sodium hydroxide).

Rope used in acidic or alkaline environments must be designed specifically for such use. Wire rope used in certain chemical environments such as battery shops, metal-plating shops, pickling plants, or pulp and paper mills is coated with vinyl, nylon, Teflon, or zinc. Natural fiber rope begins to deteriorate immediately in a chemically active environment. Synthetic fiber rope materials such as vinyl are manufactured to withstand many chemical conditions. The rope manufacturer should be consulted before using a rope in a chemical environment.

Bending. Bending subjects rope to stress. Small diameter bends can reduce strength efficiency by more than 50%. Rope efficiency depends on the bend ratio. The *bend ratio,* or (D/d) ratio, is the ratio between the diameter of a rope bend (D), such as over a pulley, and the nominal diameter of the rope (d).

The D/d ratio is compared to a rope bending efficiency chart to determine the rope's efficiency factor. **See Figure 6-14.** The information obtained for plotting the rope bending efficiency chart is established from static load tests applied to rope bent over stationary diameters.

A *static load* is a load that remains steady. A static load exerts a straight vertical pull on a rope. The rope's rated load is determined by applying the procedure:

1. Calculate D/d ratio. D/d ratio is calculated by applying the following procedure:

$$R_{bend} = \frac{D}{d}$$

where

R_{bend} = bend ratio

D = diameter of rope bend (in in.)

d = diameter of rope (in in.)

2. Determine the bending efficiency rating (from Rope Bending Efficiency chart).

3. Calculate the rope's rated load after bending by applying the following formula:

$$RL_{bend} = RL \times \eta_{bend}$$

where

RL_{bend} = rope bending load rating

RL = rope load rating

η_{bend} = relative efficiency rating

For example, what is the rated load after bending of a ⅜″ (0.375) rope which has a rated load of 1350 lb when traveling over a 6″ diameter pulley?

1. Calculate D/d ratio.

$$R_{bend} = \frac{D}{d}$$

$$R_{bend} = \frac{6}{0.375}$$

$$R_{bend} = \mathbf{16}$$

2. Determine relative efficiency rating.
   A D/d ratio of 16 = 90% (from Rope Bending Efficiency chart)

3. Calculate the rope's rated load after bending.

$$RL_{bend} = RL \times \eta_{bend}$$

$$RL_{bend} = 1350 \times 0.90$$

$$RL_{bend} = \mathbf{1215\ lb}$$

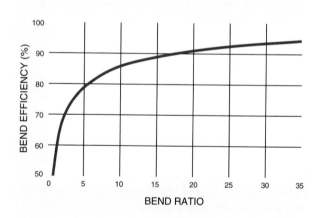

Rope Bending Efficiency

Figure 6-14. Rope bending efficiency increases with bend ratio.

Rope efficiency is increased when thimbles are used in rope ends or rope loops. A *thimble* is a curved piece of metal around which the rope is fitted to form a loop. The thimble increases the radius of the otherwise sharp bend and the rope is also protected from abrasion. Abrasion or cutting is also prevented if protection pads are placed over any rough or sharp corners of the load.

Working Load Limits. The strength rating of rope is its breaking strength. **See Figure 6-15.** The breaking strength of rope is obtained from tests where samples of rope are tensioned under increasing loads until they break. Many samples are tested and the results provide an average breaking strength for a particular type and size of rope.

Breaking Strengths of Selected Wire Ropes*			
Diameter†	**Improved Plow Steel**		**Extra-Improved Plow Steel**
	Fiber Core	**IWRC‡**	**IWRC‡**
¼	5340	5740	6640
⁵⁄₁₆	8300	8940	10,280
⅜	11,900	12,800	14,720
⁷⁄₁₆	16,120	17,340	19,900
½	20,800	22,400	26,000
⁹⁄₁₆	26,400	28,200	32,800
⅝	32,600	35,000	40,200
¾	46,400	50,000	57,400
⅞	62,800	67,400	77,600
1	81,600	87,600	100,800

* in lb, for uncoated, general purpose, rotation-resistant 6 × 19 (class 2) or 6 × 37 (class 3) wire rope
† in in.
‡ independent wire rope core

Figure 6-15. The breaking strengths of wire ropes vary by type, material, and size.

However, the breaking strength value cannot be used directly as a working load limit for lifting. The *working load limit (WLL)* is the maximum weight that a rigging component may be subjected to. Loading a rope up to its breaking strength offers no margin of safety for underestimated load weight, rope age, or other weakening conditions. Plus, slight manufacturing differences between the loaded rope and the actual tested samples mean that the working rope may break under slightly less load. Therefore, safe working load limits are established by dividing the breaking strength by a safety factor. **See Figure 6-16.** A *safety factor* is the ratio of a component's ultimate strength to its maximum allowable safe working load limit.

Safe Load Limits

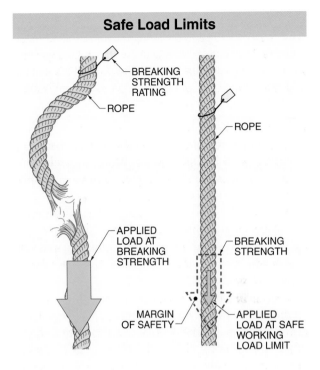

Figure 6-16. Rope must not be loaded beyond its safe working load limit, which is a fraction of its breaking strength.

Safety factors in rigging and lifting are normally between 5 and 8. Therefore, the breaking strength is 5 to 8 times greater than the allowable safe load limit. For example, if a safety factor of 5 is used for a rigging assembly with a breaking strength of 5000 lb, the working load limit of the rigging assembly is 1000 lb (5000 lb ÷ 5 = 1000 lb). A safety factor of 5 is used for steady or even loads. A safety factor of 8 is used for uneven loads or lifts that may shock the slings.

The nominal breaking strength of rope is often rated in tons. One ton equals 2000 lb. Rope strength or load weight may need to be converted between pounds and tons during calculations, depending on the available information. Either unit may be used, as long as it is the only weight unit used in the calculation. The rope breaking strength rating needed to safely lift a load is calculated with the following formula:

$$S_{break} = WLL \times SF$$

where

S_{break} = rope breaking strength (in lb)

WLL = working load limit (in lb)

SF = safety factor

For example, what is the rope breaking strength required to lift a 4000 lb milling machine with a single vertical sling? *Note:* A safety factor of 5 is used because the load is a steady lift without shock.

$$S_{break} = WLL \times SF$$
$$S_{break} = 4000 \times 5$$
$$S_{break} = \textbf{20,000 lb or 10 t}$$

According to reference tables of rope breaking strengths provided by manufacturers, an acceptable specification of wire rope for lifting the machine is ½″ improved plow steel, fiber-core rope.

Many factors must be considered when determining the necessary strength of rigging ropes. For example, the use of multiple slings divides the total load weight between the slings, which reduces the strength requirements. However, the use of other sling hitch arrangements, such as basket, bridle, or choker sling hitches, increases the forces on the sling due to sling angle. Also, bends and attachment hardware reduce a rope's efficiency, lowering its effective strength. After taking into account all these factors with additional calculations, the total force on the sling must not exceed the effective working load limit. **See Figure 6-17.**

Factors Affecting Working Load Limits

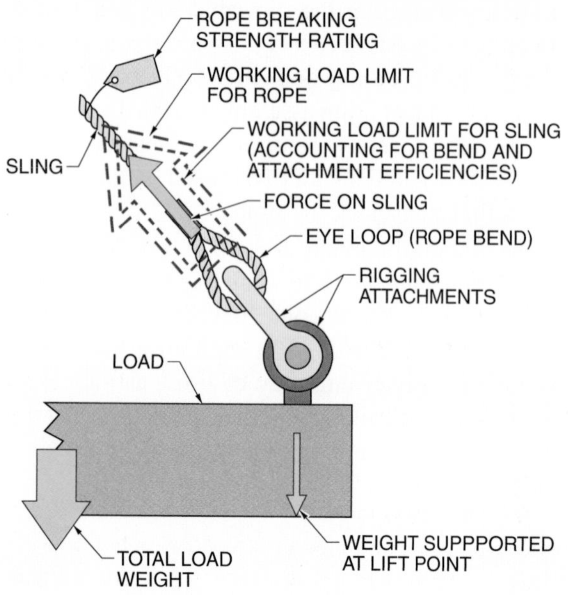

Figure 6-17. Many factors affect the actual force on a sling and the working load limit, which must not be exceeded.

Wire Rope

Wire rope is made of a specific number of strands wound helically (spirally) around a core. Each strand is made of a number of wires. Wire rope is classified according to the number of wires in a strand and the number of strands in the rope. For example, a 6 × 19 rope indicates a six-stranded rope with approximately 19 wires per strand. An 8 × 9 rope is an eight-stranded rope with approximately nine wires per strand. The second figure in the designation is nominal in that the number of wires in a strand may be slightly higher or lower.

A *nominal value* is a designated or theoretical value that may vary from the actual value. A 6 × 7 rope consists of six strands with 3 to 14 wires per strand. A 6 × 19 classification consists of six strands with 15 to 26 wires per strand. The 6 × 19 classification includes constructions such as 6 × 21 filler wire, 6 × 25 filler wire, and 6 × 26 Warrington-Seale. These constructions are classified as 6 × 19 rope, despite the fact that none of these constructions have 19 wires. Wire rope is selected based on the largest load weight, shock from acceleration or deceleration, speed, condition of the rope, attachments used, and temperature.

Wire Rope Construction. Wire rope is also classified by the design of the wire pattern. While many pattern variations exist, the most common wire ropes used for basic rigging are filler wire, Warrington, Seale, and Warrington-Seale. Each type uses wires of different sizes and offers more or less flexibility and wear than the others. **See Figure 6-18.**

Filler wire is wire rope that uses fine wires to fill the gaps between the major wires. Filler wire rope construction is the most flexible, but wears more than Warrington or Seale wire rope. *Warrington wire* is wire rope constructed of strands consisting of more than one size wire staggered in layers. Warrington wire rope is less flexible than filler wire rope, but wears better. *Seale wire* is wire rope that uses different size wire in different layers. Seale wire rope is less flexible than Warrington wire rope, but is the least susceptible to wear. A combination of Warrington and Seale wire provides the best wear and flexibility. The outer layer is Seale wire used for wear and the inner layers are Warrington wire used for flexibility.

Most rigging rope is constructed with a center (core) of various materials. Core materials include fiber, polyvinyl, thin wire rope, or a multiple-wire strand. Fiber cores (FCs) are made of sisal or manila fibers and are shaped to keep the strands in order and to act as a protecting cushion. Polyvinyl cores are used in areas of certain chemical or caustic solutions or fumes. Independent wire rope core (IWRC) is a small 6 × 7 wire rope with its own core of wire strands. IWRC resists crushing and offers consistent stretching. Multiple-wire strand core (WSC) is generally constructed of multiple wire strands similar to those used in wire rope construction. WSC is generally less flexible than IWRC. However, WSC and IWRC are rated 7½% stronger than other core types.

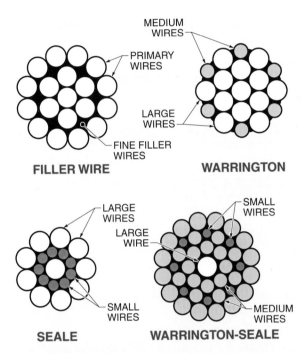

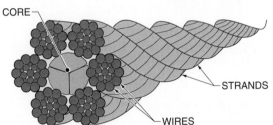

Figure 6-18. The wires in strands can be arranged in different patterns, which change the strength, flexibility, and wear characteristics.

Wire rope is made from several types of metals, including steel, iron, stainless steel, monel, and bronze. Wire rope manufacturers select the wire that is most appropriate for general requirements of the finished product. The most widely used material is steel with a high carbon content. This steel is available in a variety of grades, each having properties related to the basic curve for steel rope wire.

Grades of wire rope are traction steel (TS), mild plow steel (MPS), plow steel (PS), improved plow steel (IPS), extra improved plow steel (EIPS), and extra extra improved plow steel (EEIPS).

Galvanized (coated with zinc) wire rope may be required for harsh or corrosive environments. Typical rigging and lifting wire rope is uncoated (Bright). Galvanized wire rope is approximately 10% lower in strength than Bright rope.

Wire Rope Strength. The strength of rope used for safely lifting a load is determined by its breaking strength. **See Figure 6-19.** The breaking strength (ultimate strength) of rope is obtained from actual breakage tests. Safe load limits vary with different lifting applications but are generally established by dividing the breaking strength by a safety factor.

Once the proper wire rope size is determined, the load capacity of choker, basket, or bridle slings used to lift the load are determined. **See Figure 6-20.** The load capacity of a sling is found by applying the following formula:

$$LC = vl \times l \times s$$
where
LC = load capacity (in t)
vl = vertical load rate (from Sling Material Strength Capacities table)
l = number of sling legs (but not more than two)
s = loss factor (from Sling Angle Loss Factors table)

For example, what is the working load capacity of a wire rope sling using a ½″ 6 × 19 IPS/FC rope, a 5-leg sling, and a 70° load angle?

Note: The loss factor of a sling with a 70° load angle equals 0.940 (from Sling Angle Loss Factors table).

$$LC = vl \times l \times s$$
$$LC = 2.0 \times 2 \times 0.940$$
$$LC = \textbf{3.76 t}$$

Wire Rope Rotation. Due to the nature of its construction, rope must not be allowed to spin or rotate while being used. A special rotation-resistant rope is used in applications such as a single part line or situations where operating conditions require a rope to resist cabling. *Cabling* is a rope's attempt to rotate and untwist its strand lays while under stress.

Rotation-resistant ropes are available in single-layer and multi-layer strand classifications. A single-layer strand wire rope consists of a single layer of strands, without a core, where each strand supports one another. A multi-layer strand wire rope consists of two or more layers of strand laid in opposing directions.

Wire Rope Strength

Nominal Diameter*	Classification	Nominal Breaking Strength per 2000 lb Ton	
		IPS†	EIPS‡
¼	6 × 19 STANDARD HOISTING FIBER CORE	2.74	–
	6 × 19 STANDARD HOISTING IWRC	–	3.40
	8 × 19 SPECIAL FLEXIBLE HOISTING FIBER CORE	2.35	–
	18 × 7 NONROTATING	2.51	–
⅜	6 × 19 STANDARD HOISTING FIBER CORE	6.10	–
	6 × 19 STANDARD HOISTING IWRC	–	7.55
	8 × 19 SPECIAL FLEXIBLE HOISTING FIBER CORE	5.24	–
	18 × 7 NONROTATING	5.59	–
½	6 × 19 STANDARD HOISTING FIBER CORE	10.7	–
	6 × 19 STANDARD HOISTING	–	13.3
	8 × 19 SPECIAL FLEXIBLE HOISTING FIBER CORE	9.23	–
	18 × 7 NONROTATING	9.85	–

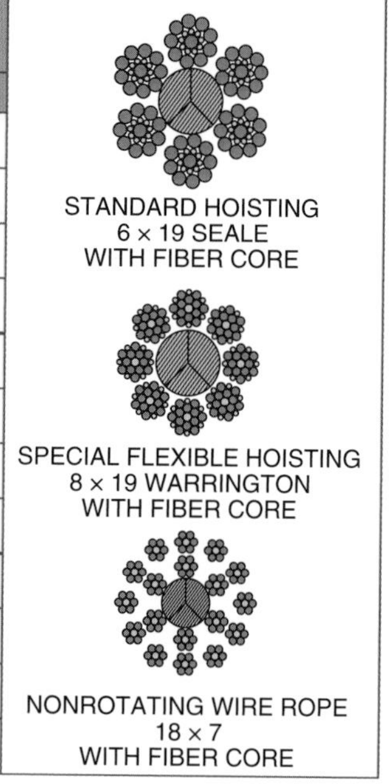

STANDARD HOISTING
6 × 19 SEALE
WITH FIBER CORE

SPECIAL FLEXIBLE HOISTING
8 × 19 WARRINGTON
WITH FIBER CORE

NONROTATING WIRE ROPE
18 × 7
WITH FIBER CORE

* in in.
† IPS - improved plow steel
‡ EIPS - extra improved plow steel

Figure 6-19. The strength of rope used for safely lifting a load is determined by its breaking strength.

Sling Material Strength Capacities*

	Rated Capacities (in Tons)‡		
6 × 19 ROPE DIA†	VERTICAL	CHOKER	BASKET
¼	0.51	0.38	1.0
5⁄16	0.79	0.60	1.6
⅜	1.1	0.85	2.2
7⁄16	1.5	1.1	3.0
½	2.0	1.5	4.0
9⁄16	2.5	1.9	5.0
⅝	3.1	2.3	6.2
¾	4.4	3.3	8.8
⅞	6.0	4.5	12.0
1	7.7	5.9	15.0

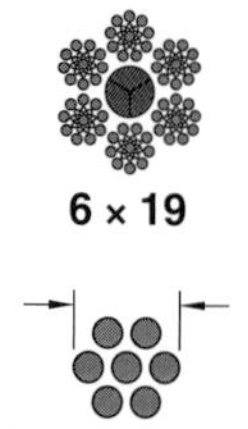

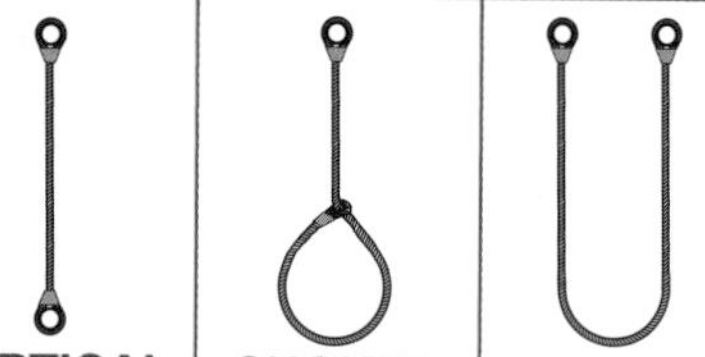

* improved plow steel/fc
† in in.
‡ rates include safety factor of 5

Figure 6-20. Rated strength capacities of 6 × 19 wire rope are based on the rope diameter and sling.

Cutting Wire Rope. The ends of wire rope must be secured by binding (seizing) to prevent raveling, unsafe loose wires, or strength reduction before cutting a wire rope. *Seizing* is the wrapping placed around all strands of a rope near the area where the rope is cut. **See Figure 6-21.** Seizing holds the strands firmly in place by the tight turning of seizing wire. This must be done twice to both ends before the cut.

Normally, one seizing on each side is sufficient for preformed wire rope. Common wire rope (those that are not preformed or are rotation-resistant) normally requires a minimum of two seizings on each side of the cut. *Preformed rope* is wire rope in which the strands are permanently formed into a helical shape during fabrication. Adequately seized ends prevent rope distortion, flattening, or strand loosening. Inadequate seizing shortens rope life by allowing uneven distribution of the strand load during lifting. Seizing is placed six rope diameters apart and the length of each seize should be equal to or greater than the rope diameter.

Tech Tip

Always maintain a safe working distance from a wire rope or ropes supporting a load, because a rope that snaps from being overloaded can act as a whip and cause serious injury to those within its range.

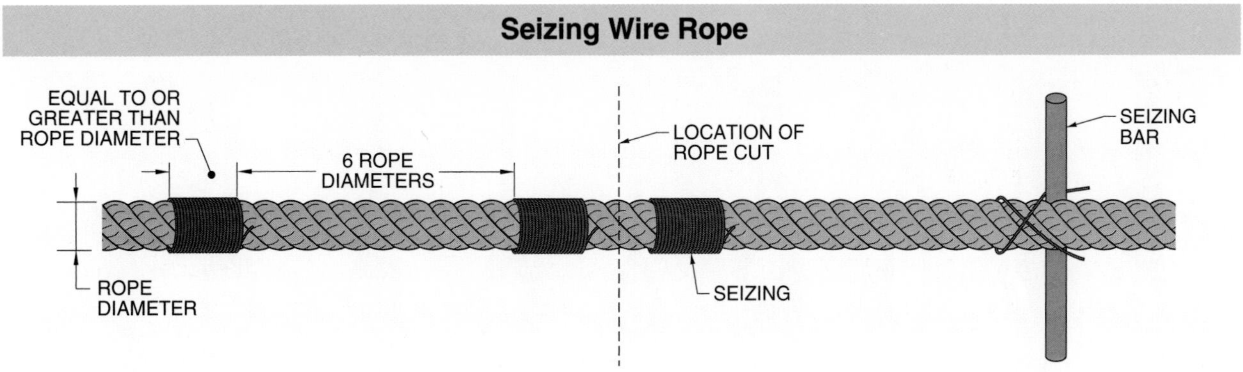

Seizing Wire Rope

Figure 6-21. Seizing is wire wrapping added to wire rope to prevent unraveling or loose wires.

The recommended method for seizing a wire rope is to lay one end of the seizing wire between two strands of the wire rope. The other end of the seizing wire is wrapped around the rope over the dead end of the seizing wire. In a looped wire or rope, the loose end is the dead end and the load-lifting portion or working end is the live end. A seizing bar is placed at a right angle to the rope. A *seizing bar* is a round bar ½″ to ⅝″ in diameter and about 18″ long used to seize rope. The live end of the seizing wire is brought around the back of the bar and the bar is twisted around the rope to wind the seizing wire around the rope. The wire is twisted in a tight helix, without overlapping, until the required seizing length is obtained. The seize is secured by twisting the ends of the seizing wire together.

Wire rope may be cut using a rope shear, an abrasive cutoff wheel, or an acetylene cutting torch. Shearing or abrasive cutting leaves a sharp edge that should be filed smooth. An acetylene cutting torch is preferred because the heat fuses the strands and strand wires together. Seizing specifications vary based on rope diameter. Always check manufacturer's specifications.

Wire Rope Terminations. Wire rope ends (terminations) are fastened to fittings or spliced into loops when wire rope is attached to a load. Common wire rope terminations include thimbles and sockets. Fittings, loops, or other attachments may not have the strength of the rope and may reduce the sling efficiency. **See Figure 6-22.**

A *thimble* is a curved piece of metal around which the rope is fitted to form a loop. A thimble protects the wire rope from sharp bends and abrasion. The length of wire rope that is looped back is determined by the loop style or manufacturer's specifications. The manufacturer's specifications should be consulted whenever possible.

In a looped rope, the loose end is the dead end, and the load-lifting portion is the live end. Loops are secured using a U-bolt clip (clamp). A U-bolt clip consists of a saddle, threaded U-section, and two nuts. The clip should be assembled with the threaded U-section contacting the dead-end section of the rope for maximum rope strength and to prevent damage to the live end of the thimble and clip assembly. Clip connections must be arranged, spaced, and assembled properly to maintain the strength of the rope. **See Figure 6-23.**

Assembling a thimble and the correct number of clips is determined by using manufacturer's specification charts. The first clip is placed approximately 4″ from the end of the rope. The other clips are spaced at a minimum of 6 rope diameters apart.

The U-section is assembled against the dead-end section. Rope damage occurs if the U-section is assembled against the live end. The first clip is installed 4″ from the dead end and the nuts are tightened. Next, the second clip is installed at the thimble and the nuts are finger-tightened. The remaining clips are assembled finger-tight. Finally, a strain is placed against the rope and the remaining nuts are alternately tightened.

A *socket* is a rope attachment through which a rope end is terminated. Wire rope sockets include swaged, speltered, and wedge designs. Swaged and speltered sockets are permanent wire rope attachments. Permanent wire rope attachments have the highest efficiency rating.

A *swaged socket* is a compressed socket assembled to the wire rope under high pressure. Swaged sockets are compressed in a hydraulic press. A *speltered socket* is a socket assembled by separating the wire rope ends after inserting the rope through the socket collar. Molten zinc or resin is poured into the collar, creating a solid assembly. Swaged and speltered sockets are 100% efficient due to the manufacturing and assembly process.

Wire Rope Terminations

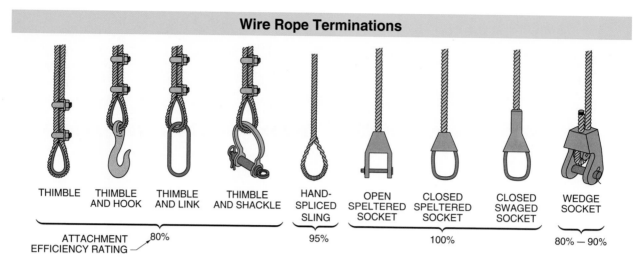

Figure 6-22. Common wire rope terminations include thimbles and sockets.

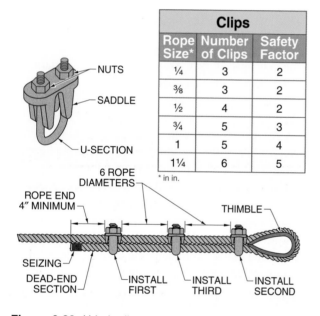

Clips		
Rope Size*	Number of Clips	Safety Factor
¼	3	2
⅜	3	2
½	4	2
¾	5	3
1	5	4
1¼	6	5

* in in.

Figure 6-23. U-bolt clip connections must be arranged, spaced, and assembled properly to maintain the strength of rope.

A *wedge socket* is a socket with the rope looped within the socket body and secured by a wedging action. Wedge sockets are popular because rapid position changes are possible and installation and dismantling processes are fairly easy. However, because of its design, a wedge socket can be incorrectly installed, creating a sharp bend on the live end of the rope. The live end must be in line with the socket. The exposed dead-end section must extend out of the wedge a minimum of eight rope diameters. **See Figure 6-24.**

Wedge Sockets

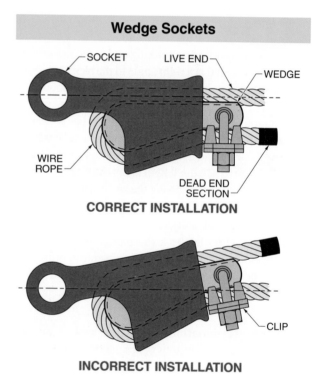

Figure 6-24. Wedge sockets hold tightly to a wire rope when it is tensioned.

Fiber Rope

Rope constructed of fibers is preferred for some applications because fiber is less likely to gouge or mar equipment surfaces than wire rope or chain. Fiber rope is classified by the materials used to construct the rope. Fiber rope can be made from either natural or synthetic fibers.

Natural fibers are obtained from plants. Plants used in the manufacture of fiber rope include cotton, hemp, manila, and sisal. Manila fiber is used predominantly for lifting and is derived from the abaca plant.

Natural fiber quality varies because the living plant quality varies. This affects the quality of the finished rope. The grade of rope is classified by the quality (grade) of fiber used. Common manila rope classifications include yacht rope, number 1, number 2, and hardware. Only yacht and number 1 class manila rope should be used for lifting. Number 2 and hardware classes of manila rope should not be used for lifting because, in some cases, the quality, type, and grade of fiber is unknown.

Synthetic materials used for rigging and lifting include nylon, polypropylene, and polyester. Synthetic ropes are used more commonly today because of the consistent quality of fiber from one rope to another. Also, the breaking strength of synthetic fibers is approximately twice that of manila fibers. Synthetic fibers are generally stronger than short natural fibers because the synthetic fiber is continuous throughout the length of the rope. **See Figure 6-25.**

Nominal Fiber Rope Strength*

Rope Dia†	Manila	Poly-propylene	Polyester	Nylon
¼	600	1200	1150	1500
⁵⁄₁₆	1000	2100	1750	2500
⅜	1350	3100	2450	3500
½	2600	4200	4400	6000
¾	5400	8000	9500	13,500
1	9000	13,500	16,000	23,500

* in lb
† in in.

NATURAL FIBER — SYNTHETIC FIBERS

Figure 6-25. Synthetic fibers are generally stronger than natural fibers.

Synthetic materials are used in ropes specifically for their special properties. Nylon is used primarily for strength, polyester for dimensional stability, fiberglass for electrical properties, vinyl for chemical properties, and polypropylene for flotation. Another advantage of synthetic fibers is that they do not mildew, rot, or decay as natural fibers do.

Fiber Rope Construction. Fiber rope is constructed by twisting fibers into yarn, yarn into strands, and strands

into rope. The first step in making fiber rope is to create a yarn by twisting the fibers in one direction. The yarn is twisted together in the opposite direction to create a strand. Strands are twisted (laid) in the reverse direction to create the rope. Reversing the twist of each step prevents the rope from unwinding. **See Figure 6-26.**

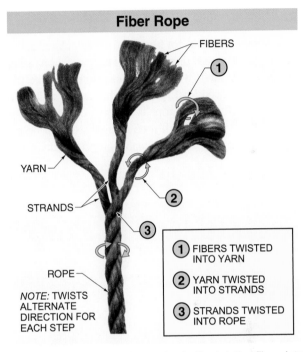

Fiber Rope

FIBERS

YARN

STRANDS

ROPE

NOTE: TWISTS ALTERNATE DIRECTION FOR EACH STEP

① FIBERS TWISTED INTO YARN
② YARN TWISTED INTO STRANDS
③ STRANDS TWISTED INTO ROPE

Figure 6-26. Fiber rope is constructed by twisting fibers into yarn, yarn into strands, and strands into rope.

Fiber rope is constructed of three or more strands (normally three) with or without a core. A *lay* is a complete helical wrap of the strands of a rope. The lay is used when inspecting a rope. Remove any rope from service if there is a measurable increase in lay length or reduction of rope diameter. For example, a rope classified as having three rope lays per foot must have a measurement of 4″ per lay.

The rope lay length measurement is used to determine if the rope has stretched any measurable amount due to age or load. The initial measurement is made when the rope is new. This gives a comparison figure for all future measurements. Measurements should be at a length of 1′ or 2′ and at the same area of rope at each measurement. The measurement should include whole lay lengths. For example, if four complete lays of a rope are recorded at 1′-4″, all future measurements would determine if the rope has or has not stretched beyond the 1′-4″ at four lays.

Fiber Rope Strength. Rope strength varies according to the degree of twist of each construction step. A high strength grade has a light degree of twist and is referred to as soft lay grade. Rope with a high resistance to abrasion is referred to as hard lay and has a high degree of twist. Ropes formed into cables use two or more three-strand ropes twisted together. Specific breaking strength of fiber rope varies greatly. Always consult the rope manufacturer's specifications when choosing rope for an application.

Safe rope strengths for fiber rope are calculated using the same method as for wire rope. The vertical breaking strength is divided by a safety factor to determine the maximum allowed load for a straight vertical pull. The safety factor for manila rope is 5, polypropylene 6, and polyester and nylon 9. All rope strengths should be increased where life, limb, or valuable property is involved.

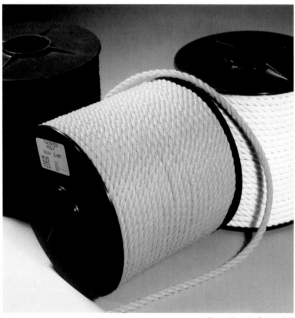

Crowe Rope Industries LLC

Polypropylene rope is a synthetic fiber rope that is available in a variety of colors, has a working temperature range of –20°F to 200°F, and excellent resistance against acids, alkalis, mildew, and rot.

Fiber Rope Applications. Binding (fastening) loads with a rope normally requires some form of hitch or knot. A *hitch* is the interlacing of rope to temporarily secure it without knotting the rope. *Knotting* is fastening a part of a rope to another part of the same rope by interlacing it and drawing it tight.

All knots involve changing direction of the rope axis and pinching against another part of the rope. For a rope, an *axis* is an imaginary straight line that runs lengthwise through the center of the rope. Each sharp change of direction weakens a rope. A knot in a rope can reduce rope efficiency by as much as 55%. For example, straight rope has an efficiency of 100%, an eye splice over a thimble has an efficiency of 90%, a short splice has an efficiency of 80%, and an overhand knot has an efficiency of 45%.

Most rope hitching and knotting terminology was derived from nautical (sailing) terms. Basic elements used when working with rigging knots include bight, loop, whipping, working end, working part, standing part, standing end, kinks, nips, and eye loops. **See Figure 6-27.**

A *bight* is a loose or slack part of a rope between two fixed ends. A *loop* is the folding or doubling of a line, leaving an opening through which another line may pass. The *working end* is the end of the working part of a rope. The *working part* is the portion of the rope where the knot is formed. The *standing end* is the end of the rope that is normally fixed to a permanent apparatus or drum or that is rolled into a coil. The *standing part* is the portion of the rope that is not active in the knot-making process.

The working end of a rope is protected from untwisting or raveling by whipping (seizing), splicing, or crowning. Synthetic fiber rope is finished off by sealing fibers together with a match or soldering iron.

Whipping (seizing) is tightly binding the end of a rope with twine before it is cut. **See Figure 6-28.** Rope is whipped by applying the following procedure:

1. Form a bight with the end of the twine and lay along the rope to be whipped.
2. Wrap twine tightly around the rope, gradually working toward the rope end. The turns are laid hard against each other without overlapping.
3. Tuck twine through the loop at the end of the rope.
4. Pull loop halfway through whipping by pulling other end of the twine.
5. Trim loose twine ends close to the turns.
6. Seal synthetic fibers with heat.

Tech Tip

Fiber rope must be protected from abrasion or cutting when it is used with equipment or loads that have rough surfaces or sharp corners. Protection pads should be placed over any part of the load that will contact the rope during lifting. To prolong rope life, verify that the load does not cause the rope to twist or bind.

Rope Terminology

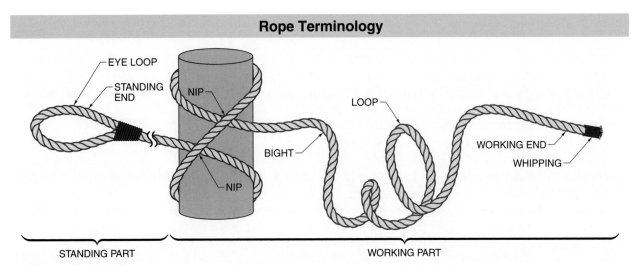

Figure 6-27. Most rope hitching and knotting terminology was derived from nautical (sailing) terms.

Whipping

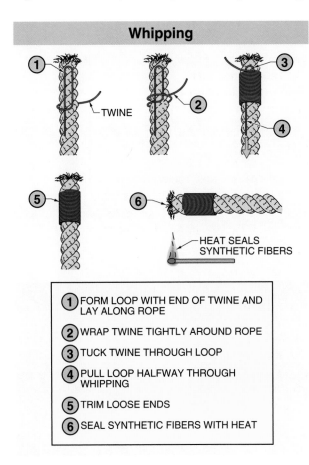

1. FORM LOOP WITH END OF TWINE AND LAY ALONG ROPE
2. WRAP TWINE TIGHTLY AROUND ROPE
3. TUCK TWINE THROUGH LOOP
4. PULL LOOP HALFWAY THROUGH WHIPPING
5. TRIM LOOSE ENDS
6. SEAL SYNTHETIC FIBERS WITH HEAT

Figure 6-28. Fiber rope whipping is completed before cutting in order to bind all the strands together.

A *splice* is the joining of two rope ends to form a permanent connection. **See Figure 6-29.** Splices are used to join the ends of two ropes of similar strength and thickness. Splices include the short splice and the long splice. A short splice uses an unlay of six to eight rope strands on each rope. An *unlay* is the untwisting of the strands in a rope. A short splice, when braided together, increases the rope diameter. This makes the short splice unsuitable for pulley use. A long splice uses an unlay of 15 turns and does not increase the rope diameter. A long splice is formed by applying the following procedure:

1. Unlay 15 turns and place a temporary whip on both rope standing parts.
2. Whip the strand ends.
3. Place the two rope ends (standing part terminations) together, alternating strands of one end with the strands of the other.
4. Remove the temporary whip from one rope and unlay one strand about 10 additional turns.
5. Fill the void in the grooves of the 10 turns with the matching strand of the other rope.
6. Remove the temporary whip from the other rope and unlay 10 additional turns.
7. Fill the void in the grooves of the 10 turns with the matching strand of the other rope.
8. Repeat unlay and void fill alternately in each direction.
9. Tie off each lay of strands using an overhand knot and begin tucking the strand from one rope through the strands of the other rope. A minimum of two tuck sets is required. A *tuck set* is the wedging of each strand of a rope into and between the other rope strands.
10. Clip the strand ends after rolling and pounding the splice.

Long Splices

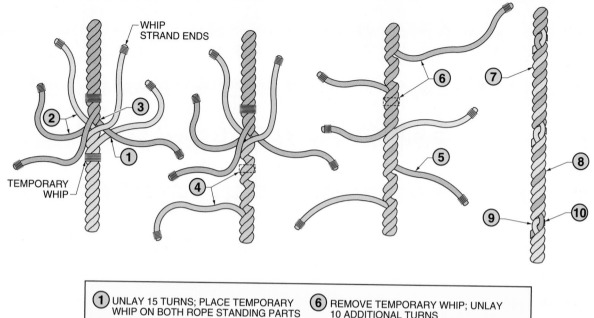

1. UNLAY 15 TURNS; PLACE TEMPORARY WHIP ON BOTH ROPE STANDING PARTS
2. WHIP STRAND ENDS
3. PLACE TWO ROPE ENDS TOGETHER
4. REMOVE TEMPORARY WHIP FROM ONE ROPE AND UNLAY 10 ADDITIONAL TURNS
5. FILL VOID WITH OTHER ROPE
6. REMOVE TEMPORARY WHIP; UNLAY 10 ADDITIONAL TURNS
7. FILL VOID WITH OTHER ROPE
8. REPEAT UNLAY AND VOID FILL ALTERNATELY IN EACH DIRECTION
9. TIE OFF EACH LAY
10. CLIP STRAND ENDS; TUCK LOOSE ENDS

Figure 6-29. A splice is the joining of two rope ends to form a permanent connection.

Splicing the working end of a rope into a crown or eye loop produces a workable, neat, and permanent rope termination. *Crowning* is a reverse strand splice that is used when an enlarged rope end is desired or not objectionable. **See Figure 6-30.** A rope crown termination is formed by applying the following procedure:

1. Unlay rope ends eight turns and whip the strand ends.
2. For a three-strand rope, loop strand 1 and lay strand 2 over strand 1 and down the side of the rope.
3. Lay strand 3 over strand 2 and through strand 1 loop.
4. Snug strands 1, 2, and 3.
5. Tuck strand 1 through strand 2 of the standing part of the rope.
6. Alternately tuck each strand. Trim ends. The crown of the rope becomes tighter with time and use.

Crowe Rope Industries LLC

Eye loops may be formed to attach hooks to the ends of a length of rope.

Crowning

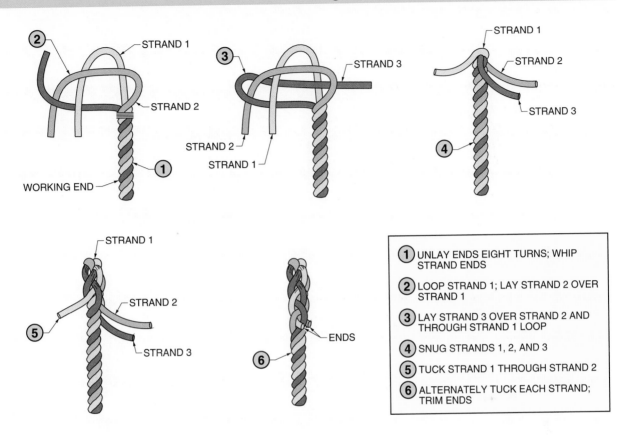

① UNLAY ENDS EIGHT TURNS; WHIP STRAND ENDS

② LOOP STRAND 1; LAY STRAND 2 OVER STRAND 1

③ LAY STRAND 3 OVER STRAND 2 AND THROUGH STRAND 1 LOOP

④ SNUG STRANDS 1, 2, AND 3

⑤ TUCK STRAND 1 THROUGH STRAND 2

⑥ ALTERNATELY TUCK EACH STRAND; TRIM ENDS

Figure 6-30. Crowning is a method of finishing a cut end of fiber rope without whipping.

An *eye loop* is a rope splice containing a thimble. **See Figure 6-31.** Eye loops are spliced for a permanent termination. A thimble is inserted for strength and wear and is held in place by whipping. Wire rope thimbles must not be used with fiber rope. An eye loop is formed by applying the following procedure:

1. Unlay four turns of strand. Place a temporary whip on the standing part and whip the strand ends.

2. Form the eye of thimble size.

3. Tuck strand 1 through the standing part at 90° to the lay of the rope.

4. Tuck strand 2 through the standing part in the same direction.

5. Turn the assembly over and tuck strand 3 through standing part.

6. Alternately tuck each strand through the standing part. Trim ends.

7. Remove the temporary whipping. Insert thimble and add whipping.

Knots and Hitches. A *knot* is the interlacing of rope to form a permanent connection. A hitch is the interlacing of rope to temporarily secure it without knotting the rope. Knots are designed to form a permanent connection that may be untied. Hitches are designed for quick release. Knots lose from 10% to 80% of the strength of a rope, depending on the knot used. A rope fails at the short bend in the knot if a rope fails under stress due to the presence of a knot.

Eye Loops

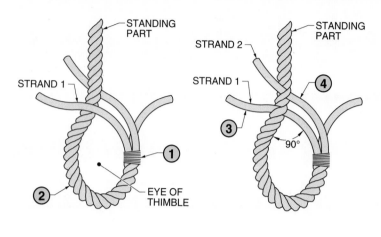

1. UNLAY FOUR TURNS; PLACE TEMPORARY WHIP ON STANDING PART AND WHIP STRAND ENDS

2. FORM EYE OF THIMBLE SIZE

3. TUCK STRAND 1 THROUGH STANDING PART AT 90°

4. TUCK STRAND 2 THROUGH STANDING PART

5. TURN ASSEMBLY OVER AND TUCK STRAND 3 THROUGH STANDING PART

6. ALTERNATELY TUCK EACH STRAND; TRIM ENDS

7. REMOVE TEMPORARY WHIPPING; INSERT THIMBLE AND ADD WHIPPING

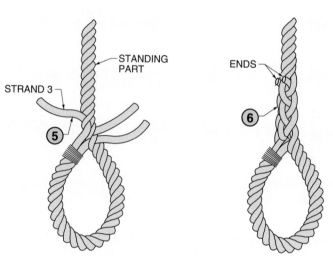
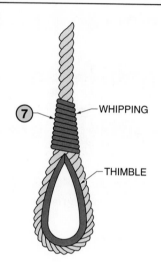

Figure 6-31. An eye loop is a rope splice containing a thimble.

Common rigging knots include the double hitch, slip, bowline, and wagoner's hitch knots. Many knots are variations using the basic half hitch knot. A *double hitch knot* is a knot with two half hitch knots. A *half hitch knot* is a binding knot where the working end is laid over the standing part and stuck through the turn from the opposite side. The double turn allows for two gripping nips. A *nip* is a pressure and friction point created when a rope crosses over itself after a turn around an object. A nip is an essential ingredient of any knot because of the pressure and friction. **See Figure 6-32.** A half hitch knot is formed by applying the following procedure:

1. Form a loop with the working end crossed over the standing part.
2. Tuck working end under and through loop.

A *slip knot* is a knot that slips along the rope from which it is made. A slip knot forms a noose which, when placed around an object, is progressively tightened by strain on the standing part. **See Figure 6-33.** A slip knot is formed by applying the following procedure:

1. Form a loop by placing the working end over the standing part.
2. Tuck working end under and through loop.
3. Pass standing part through loop.

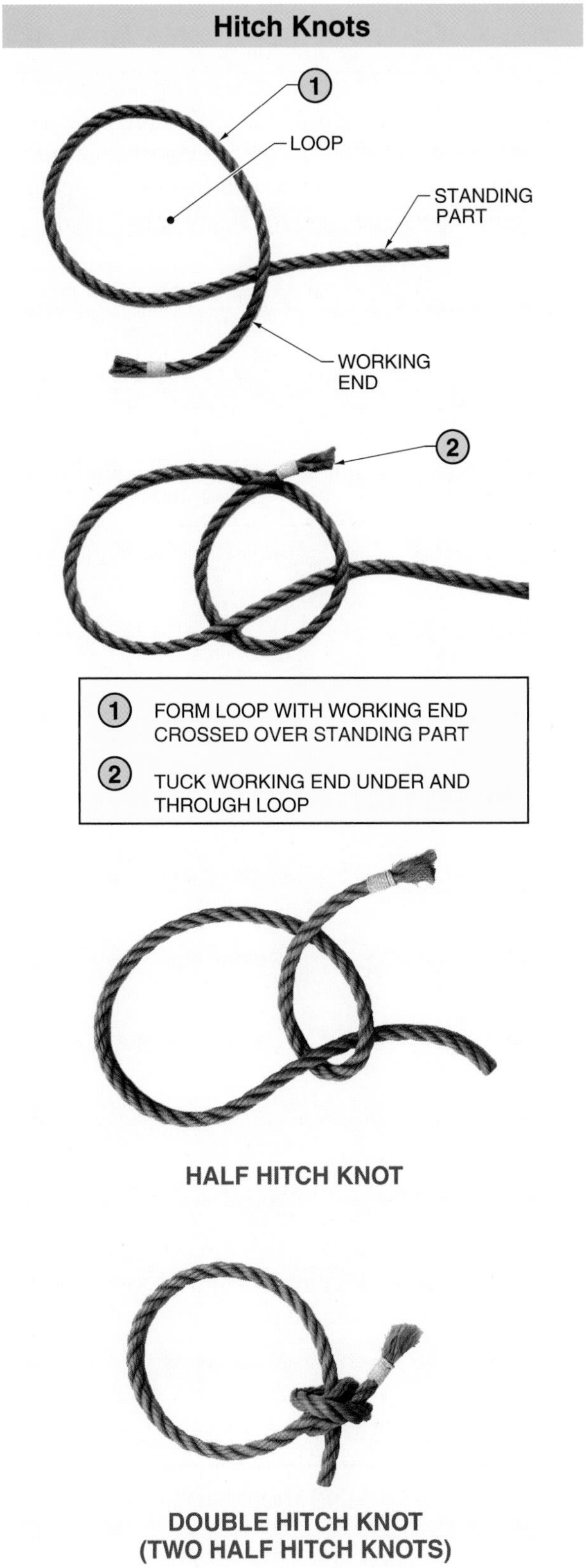

Hitch Knots

① — LOOP

STANDING PART

WORKING END

②

①	FORM LOOP WITH WORKING END CROSSED OVER STANDING PART
②	TUCK WORKING END UNDER AND THROUGH LOOP

HALF HITCH KNOT

DOUBLE HITCH KNOT (TWO HALF HITCH KNOTS)

Figure 6-32. Half hitch and double hitch knots are not secure knots but are the base formation of other knots.

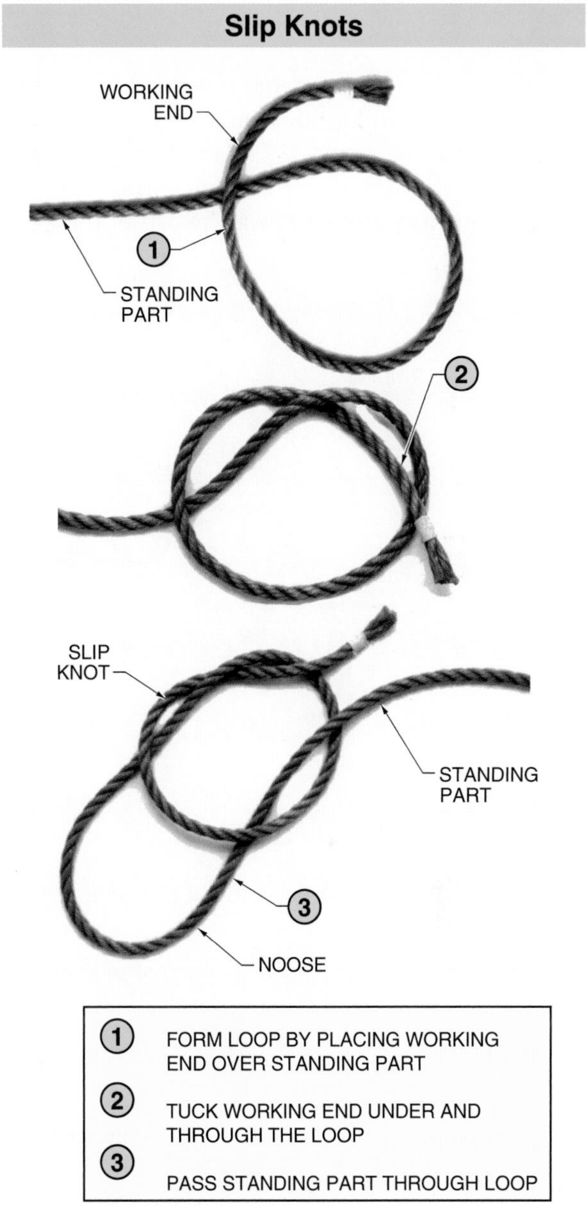

Slip Knots

WORKING END

STANDING PART

①

②

SLIP KNOT

STANDING PART

③

NOOSE

①	FORM LOOP BY PLACING WORKING END OVER STANDING PART
②	TUCK WORKING END UNDER AND THROUGH THE LOOP
③	PASS STANDING PART THROUGH LOOP

Figure 6-33. A slip knot is a knot that slips along the rope from which it is made.

A *bowline knot* is a knot that forms a loop that is absolutely secure. **See Figure 6-34.** The more strain placed on the rope, the stronger the knot becomes. The knot is easily released when needed. A bowline knot is formed by applying the following procedure:

1. Loop working part of rope over standing part. Allow enough rope to give the size loop required.
2. Thread the working end beneath and through the loop.

3. Pass the working end around the back side of the standing part.
4. Pass the working end back through the loop. Tighten by pulling the standing part and working end.

A *wagoner's hitch knot* is a knot that creates a load-securing loop from the standing part of the rope.

See Figure 6-35. A wagoner's hitch knot is formed by applying the following procedure:
1. Form a loop where the drawing loop is required by placing the working part on top of the standing part.
2. Form a second loop from the working part and insert into the first loop. Snug the knot assembly by pulling on the second loop and the standing part.

Bowline Knots

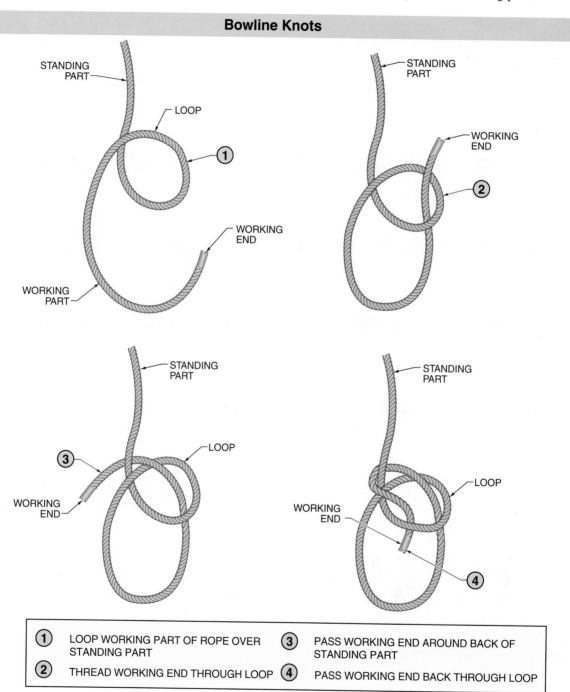

| ① | LOOP WORKING PART OF ROPE OVER STANDING PART | ③ | PASS WORKING END AROUND BACK OF STANDING PART |
| ② | THREAD WORKING END THROUGH LOOP | ④ | PASS WORKING END BACK THROUGH LOOP |

Figure 6-34. A bowline knot is a knot that forms a loop that is secure but easy to release.

3. Bring the working end through the second loop after passing the working end through load-securing hooks or loops.

4. Pull the working end tight and into a half knot to secure the load.

Hitches work by the pressure of rope being pressed together. The standing part of the rope is nipped (jammed) over the working part. Because of the friction created by nipping, the greater the pull on the rope, the more tightly the standing part nips the working part and prevents it from slipping through. Hitches should never be formed with slippery rope or wire. Hitches created from rope are the timber hitch, clove hitch, cat's-paw hitch, cow hitch, scaffold hitch, and blackwall hitch.

Wagoner's Hitch Knots

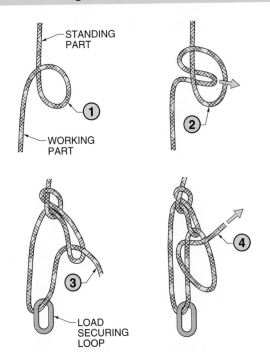

STANDING PART

WORKING PART

① FORM LOOP BY PLACING WORKING PART ON TOP OF STANDING PART

② FORM SECOND LOOP AND INSERT INTO FIRST LOOP; SNUG KNOT ASSEMBLY

③ BRING WORKING END THROUGH SECOND LOOP AFTER PASSING THROUGH LOAD HOOK OR LOOP

④ PULL WORKING END TIGHT

LOAD SECURING LOOP

Figure 6-35. A wagoner's hitch knot is a knot that creates a load-securing loop from the standing part of the rope.

A clove hitch is a quick, simple method of fastening a rope around a post, pole, or stake.

A *timber hitch* is a binding knot and hitch combination used to wrap and drag lengthy material. A timber hitch may be used to wrap and drag logs, pipes, beams, etc. **See Figure 6-36.** A timber hitch is formed by applying the following procedure:

1. Loop the working end around the standing part to form the binding knot.

2. Twist the working end in the direction of the lay of the rope three or four times.

3. Pull the working end tight to maintain form.

A *clove hitch* is a quick hitch used to secure a rope temporarily to an object. A clove hitch is used because it is attached quickly, holds firmly, and has a rapid release. **See Figure 6-37.** A clove hitch is formed by applying the following procedure:

1. Cross one hand over the other and grasp the rope with both hands.

2. Uncross the hands.

3. Bring the rope together to form two loops. Place over the object to be secured.

Timber Hitches

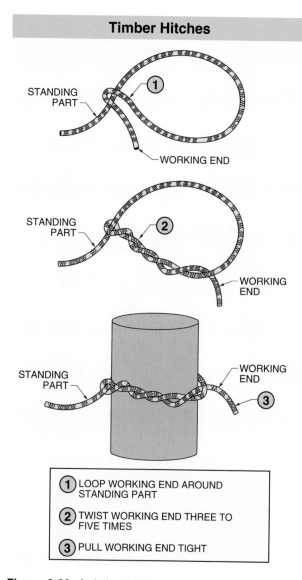

STANDING PART

WORKING END

STANDING PART

WORKING END

STANDING PART

WORKING END

1. LOOP WORKING END AROUND STANDING PART
2. TWIST WORKING END THREE TO FIVE TIMES
3. PULL WORKING END TIGHT

Figure 6-36. A timber hitch is a binding knot and hitch combination used to wrap and drag lengthy material.

A timber hitch, which does not jam and comes undone readily when the pull ceases, is used to tow or hoist cylindrical objects, such as logs, poles, etc.

Clove Hitches

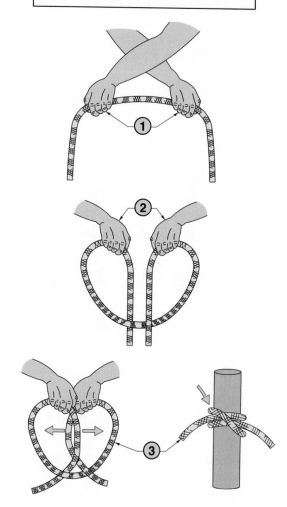

1. CROSS HANDS AND GRASP ROPE
2. UNCROSS HANDS
3. FORM TWO LOOPS; PLACE OVER OBJECT TO BE SECURED

Figure 6-37. A clove hitch is a quickly formed hitch used to secure a rope temporarily to an object.

A *cat's-paw hitch* is a hitch used as a light-duty, quickly formed eye for a hoisting hook. **See Figure 6-38.** A cat's-paw hitch is formed by applying the following procedure:

1. Grasp the rope with both hands, leaving plenty of bight.
2. Rotate both hands in the opposite direction and continue to rotate the two loops for two complete turns.
3. Place the eye over the end of a hook.

Cat's-Paw Hitches

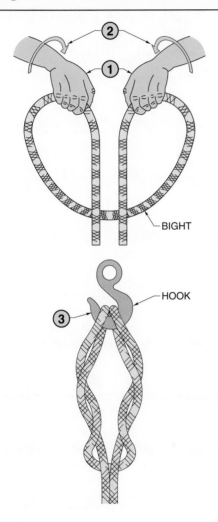

1. GRASP ROPE WITH BOTH HANDS
2. ROTATE WRISTS SEVERAL TIMES IN OPPOSITE DIRECTIONS
3. PLACE EYE OVER END OF HOOK

BIGHT

HOOK

Figure 6-38. A cat's-paw hitch is a quickly formed eye for light-duty lifting.

Cow Hitches

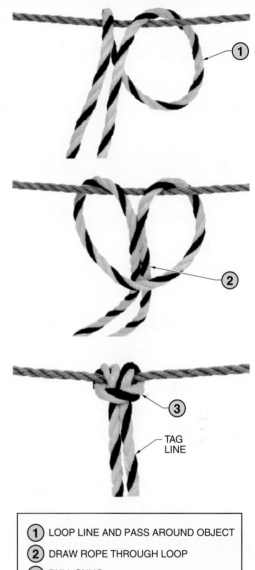

1. LOOP LINE AND PASS AROUND OBJECT
2. DRAW ROPE THROUGH LOOP
3. PULL SNUG

TAG LINE

Figure 6-39. A cow hitch is a hitch used to secure a tag line to a load.

A *cow hitch* is a hitch used to secure a tag line to a load. A cow hitch is made and released easily but is firm enough to steady loads. **See Figure 6-39.** A *tag line* is a rope, handled by an individual, to control rotational movement of a load. A cow hitch is formed by applying the following procedure:

1. Loop the line and pass the loop around the object.
2. Draw the rope through the loop.
3. Pull snug.

A *scaffold hitch* is a hitch used to hold or support planks or beams. **See Figure 6-40.** A scaffold hitch is made from a clove hitch and a bowline knot. A scaffold hitch is formed by applying the following procedure:

1. Attach a clove hitch to the object.
2. Tie the working end to the standing part using a bowline knot.

Scaffold Hitches

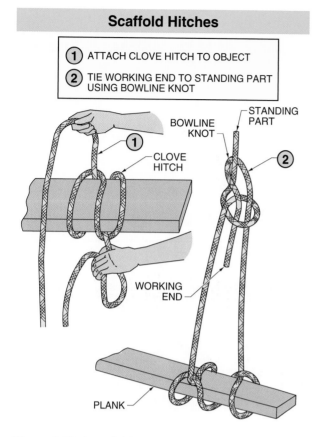

① ATTACH CLOVE HITCH TO OBJECT

② TIE WORKING END TO STANDING PART USING BOWLINE KNOT

Figure 6-40. A scaffold hitch is used to support planks or beams.

Blackwall Hitches

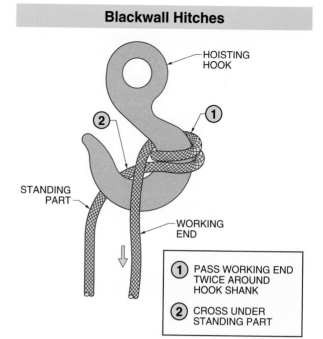

① PASS WORKING END TWICE AROUND HOOK SHANK

② CROSS UNDER STANDING PART

Figure 6-41. A blackwall hitch is a hitch made for securing a rigging rope to a hoisting hook.

A *blackwall hitch* is a hitch made for securing a rigging rope to a hoisting hook. **See Figure 6-41.** A blackwall hitch should be made from natural fiber ropes only because synthetic ropes may slip. A blackwall hitch is formed by applying the following procedure:

1. Pass the working end twice around the shank of a hook.
2. Cross it over the standing part in the mouth of the hook.

WEBBING

Webbing is a fabric of high-tenacity synthetic yarns woven into flat, narrow straps. **See Figure 6-42.** A *synthetic yarn* is yarn made of twisted, manufactured fibers such as nylon or polyester. Webbing is constructed in one to four plies, with protected edges and red warning cores to indicate wear or damage. Webbing for rigging purposes is normally used when maximum load damage protection is required. The softness of the webbing material, along with its wide and flat design, offers excellent protection for glass and polished or painted loads.

Webbing for rigging purposes is made of woven nylon or polyester with selvedges. *Selvedge* is a knitted or woven edge of a webbing formed to prevent unraveling. Most web sling damage starts on the edge and progresses across the web face.

The thickness (number of plies) determines the duty rating of webbing. A *ply* is a layer of a formed material. The number of plies is the number of thicknesses of load-bearing webbing used in the sling assembly. Slings are available in one- to four-ply construction, with widths ranging from 1″ to 12″. Generally, web slings are constructed of one or two plies with three- or four-ply slings reserved for special conditions.

In some cases, special colored yarns are woven into the webbing core to indicate excessive wear or cuts. The webbing must be removed from service when these warning cores of colored yarn are exposed.

Web sling construction consists of the length (reach), body, splice, and loop eye. The *web sling length* is the distance between the extreme points of a web sling, including any fittings. The *web sling body* is the part of the sling that is between the loop eyes or end fittings (if any). The splice of a web sling is the lapped and secured load-bearing part of a loop eye. The *loop eye* of a web sling is a length of webbing folded back and spliced to the sling body, forming an opening.

Webbing

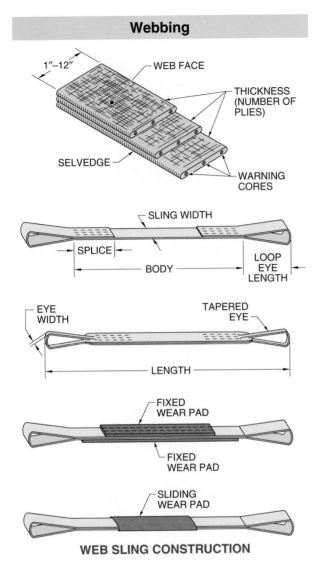

WEB SLING CONSTRUCTION

Figure 6-42. Webbing material consists of synthetic fibers woven into wide, flat straps. Multiple layers, or plies, can be sewn together to make stronger webbing.

Sling loop eyes may be too wide to properly fit into the bowl of a hoist hook. A tapered loop eye is formed by folding the webbing to a narrower width at its bearing point to accommodate the lifting device. Another common component of web slings is wear pads. A *wear pad* is a leather or webbed pad used to protect the web sling from damage. Wear pads are either sewn (fixed) or sliding for adjustable protection.

Basic web slings are assembled in various ways to be used as vertical, basket, or choker hitches. Web slings are fabricated in six configurations (Type I through Type VI). **See Figure 6-43.**

Lift-All Company, Inc.

Braided Tuflex roundslings from Lift-All are made from three (6-part) or four (8-part) individual Tuflex endless synthetic slings made from continuous loops of polyester yarn covered by a double wall tubular jacket.

Type I web slings are made with a triangle fitting on one end and a slotted triangle choker fitting on the other end. Type I web slings are used for vertical, basket, or choker hitches. Type I web slings are also used in cargo hold down situations where the webbing consists of a tightener and end fittings at each end. The strapping, being attached at two different hold down points, is then tightened over the cargo.

Type II web slings are made with a triangle fitting on both ends. Type II web slings are used for vertical or basket hitches.

Type III web slings are made with flat loops on each end with the loop eye openings in the same plane as the body sling. They are also known as eye-to-eye slings. Type III web slings are used for basket hitch applications or as a choker hitch by passing one eye around the load and through the opposite eye. The eye of a Type III web sling is generally flat but is available tapered to permit use on crane hooks.

Type IV web slings are made with both eyes twisted to form loop eyes that are at right angles to the plane of the body sling. They are also known as twisted eye slings. Type IV web slings are used for choker hitch applications.

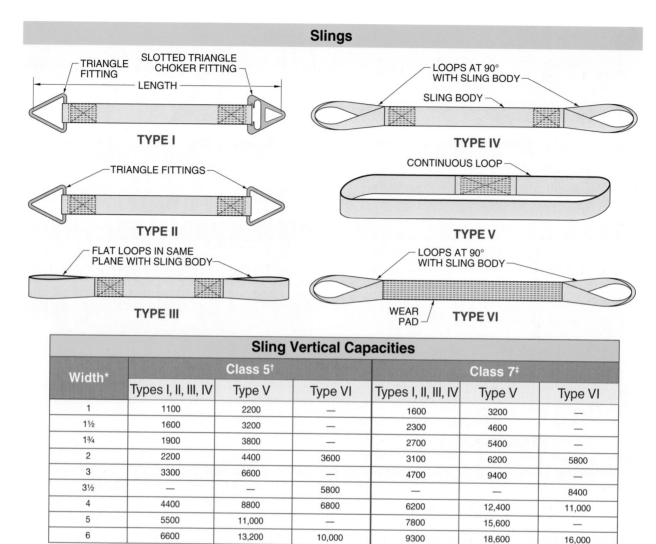

Figure 6-43. There are six primary web sling configurations that are available in various sizes and webbing types.

Type V web slings are endless web slings made by joining the ends with load-bearing splices. They are also known as grommet slings. Type V web slings are used for numerous applications and are the most widely used. Because they have an endless design, they may be used in a basket, vertical, or choker hitch application.

Type VI web slings are sometimes made from Type V slings by adding wear pads the length of the sling body. The wear pad of a type VI web sling may be on one side or both sides of the sling body and is only long enough to form eye loops at each end. The eye loops are at right angles to the plane of the web body. Type VI web slings are also known as reverse eye slings. Type VI web slings are used for rugged service, such as lifting irregularly shaped objects such as stone.

Webbing Strength

Sling vertical capacity tables give the minimum certified tensile strength for various classes, types, and number of plies in webbing. Webbing sling strength capacity is rated for one-ply or two-ply in Class 5 or Class 7. The specification of Class 5 or Class 7 is a specification for the manufacturer, with Class 7 being approximately 45% stronger than Class 5.

The following describes the figure content (sling diagrams and table):

TYPE I — TRIANGLE FITTING, SLOTTED TRIANGLE CHOKER FITTING, LENGTH
TYPE II — TRIANGLE FITTINGS
TYPE III — FLAT LOOPS IN SAME PLANE WITH SLING BODY
TYPE IV — LOOPS AT 90° WITH SLING BODY, SLING BODY
TYPE V — CONTINUOUS LOOP
TYPE VI — LOOPS AT 90° WITH SLING BODY, WEAR PAD

Sling Vertical Capacities

Width*	Class 5†			Class 7‡		
	Types I, II, III, IV	Type V	Type VI	Types I, II, III, IV	Type V	Type VI
1	1100	2200	—	1600	3200	—
1½	1600	3200	—	2300	4600	—
1¾	1900	3800	—	2700	5400	—
2	2200	4400	3600	3100	6200	5800
3	3300	6600	—	4700	9400	—
3½	—	—	5800	—	—	8400
4	4400	8800	6800	6200	12,400	11,000
5	5500	11,000	—	7800	15,600	—
6	6600	13,200	10,000	9300	18,600	16,000

* in in.
† minimum certified tensile strength of 6800 lb per in. of width
‡ minimum certified tensile strength of 9800 lb per in. of width

Webbing material strength is based on new webbing material. Factors that affect webbing strength include mishandling, environmental conditions, and ultraviolet light. Dragging webbing on the floor, tying it in knots, pulling it from under a load when the load is resting on the webbing, or dropping it with metal fittings are all types of mishandling. Bunching of material between the ears of a clevis, shackle, or hook also weakens webbing strength.

Chemicals create environmental conditions affecting webbing strength. These conditions cause varying degrees of degradation. The proper webbing material must be used in chemically active areas. Polyester is resistant to many acids, but is still subject to some degradation. Nylon is resistant to many alkalis but may be subject to moderate degradation. Webbing slings are not to be used if the end fittings are aluminum and alkalis or acid are present. Exposing synthetic webbing to ultraviolet light, such as sunlight or arc welding, affects the strength in varying degrees. The effects of ultraviolet degradation can occur without any visible indication.

The strength of webbing for vertical slings is used to determine strength capacities of choker, basket, and bridle sling hitches. Strength capacities are affected by the sling angle. Sling angle is measured from the horizontal to the point of attachment of multi-legged slings. The actual lifting capacity of a sling at a specific sling angle is found by applying the following procedure:

1. Determine sling vertical capacity. (from Sling Vertical Capacities table)
2. Determine sling angle loss factor. (from Sling Angle Loss Factors table)
3. Calculate lifting capacity. Lifting capacity is found by applying the formula:

   $LC = vl \times l \times s$

   where

   LC = lifting capacity (in lb)

   vl = sling vertical capacity (in lb)

   l = sling leg(s)

   s = sling angle loss factor

For example, what is the lifting capacity of a basket hitch using a 1″ wide, Class 7, Type V endless sling without fittings having a 45° sling angle? **See Figure 6-44.**

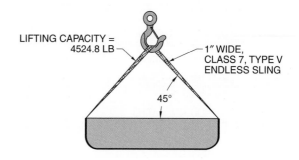

Figure 6-44. Sling load capacities vary based on the sling angles.

1. Determine sling vertical capacity.

   vl = 3200 lb (from Sling Vertical Capacities table)
2. Determine sling angle loss factor.

   s = 0.707 (from Sling Angle Loss Factors table)
3. Calculate lifting capacity.

   $LC = vl \times l \times s$

   $LC = 3200 \times 2 \times 0.707$

   $LC = \textbf{4524.8 lb}$

A choker hitch uses sling angle loss factors based on the angle measured from the vertical as the webbing passes through the choke eye. Choker hitch load capacity is based on sling vertical capacity, angle of choke, and the sling rated load factor. **See Figure 6-45.**

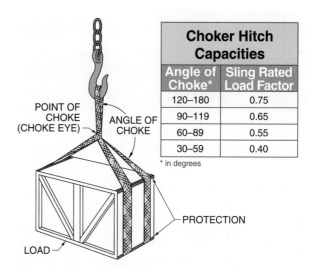

Choker Hitch Capacities	
Angle of Choke*	Sling Rated Load Factor
120–180	0.75
90–119	0.65
60–89	0.55
30–59	0.40

* in degrees

Figure 6-45. Choker hitch load capacity is based on the angle of choke as the sling body passes through the choke eye.

For example, what is the lifting capacity of a 2″, Class 5, Type V web sling choker hitch having a 110° angle at the point of choke?

1. Determine sling vertical capacity.

    vl = 4400 lb (from Sling Vertical Capacities table)

2. Determine sling angle loss factor.

    s = 0.65 (from Sling Angle Loss Factors table)

3. Calculate lifting capacity.

    $LC = vl \times l \times s$

    $LC = 4400 \times 1 \times 0.65$

    $LC = \mathbf{2860\ lb}$

Basic Rigging with Web Slings

Consideration must be given to the kind of load, its weight, and its center of gravity when selecting a web sling. The type of web sling selected and its use must be made with safety as the main consideration. For example, basket and choker hitches are used with web slings. **See Figure 6-46.**

Lift-All Company, Inc.
When using a basket hitch to lift a cylindrical load, the web slings must be attached to the ends of the load.

ROUND (TUBULAR) SLINGS

Round (tubular) slings make excellent choker slings because they are extremely flexible and conform to the shape of the load. Also, due to their construction, choker hitches do not bind or lock up, making sling release simple. The strength capacity of round slings is identified by color coding, which is an added safety feature.

Round Sling Construction

A *round sling* is a sling consisting of one or more continuous polyester fiber yarns wound together to make a core. Core yarns are wound uniformly to ensure even load-bearing distribution. A polyester round (tubular) sling is made of a core of continuous yarn, not woven, enclosed in a protective cover. **See Figure 6-47.**

The use of a round sling is similar to that of a Type I endless web sling when used as a choker or basket hitch. Round slings are also manufactured with fittings or coupling components. Additional sleeves are placed over the round sling to offer extra abrasion protection or to create a loop eye at each end (eye-and-eye design).

Bridle slings are assembled during manufacture by including more than one round sling leg to a master link. The tubular cover is color-coded to correspond with the rated strength capacity of the round sling. Each polyester round sling has rated capacities for use with vertical, choker, vertical basket, and 45° basket slings. The colors of round slings from smallest to largest are purple, green, yellow, tan, red, white, blue, and orange.

Round Sling Strength

A round sling is not to be used with a load greater than that marked on its identification tag. Identification tags generally carry the rated capacities for vertical, choker, and vertical basket slings. Any round sling that has a missing or unreadable identification tag must be removed from service. Capacities are rated similar to web slings.

For example, what is the lifting capacity of a round sling basket hitch with a green cover and 55° sling angle?

1. Determine sling vertical capacity.

    vl = 10,600 lb (from the Round Sling Color and Capacity Rating table)

2. Determine sling angle loss factor.

    s = 0.819 (from Sling Angle Loss Factors table)

3. Calculate lifting capacity.

    $LC = vl \times l \times s$

    $LC = 10,600 \times 2 \times 0.819$

    $LC = \mathbf{17,362.8\ lb}$

CHAIN

A *chain* is a series of connected metal links. Chain can be used for support, restraint, or transmission of mechanical power. Chain is used in situations in which other materials would be damaged by the load or environment, such as rough or raw castings or high-temperature processes.

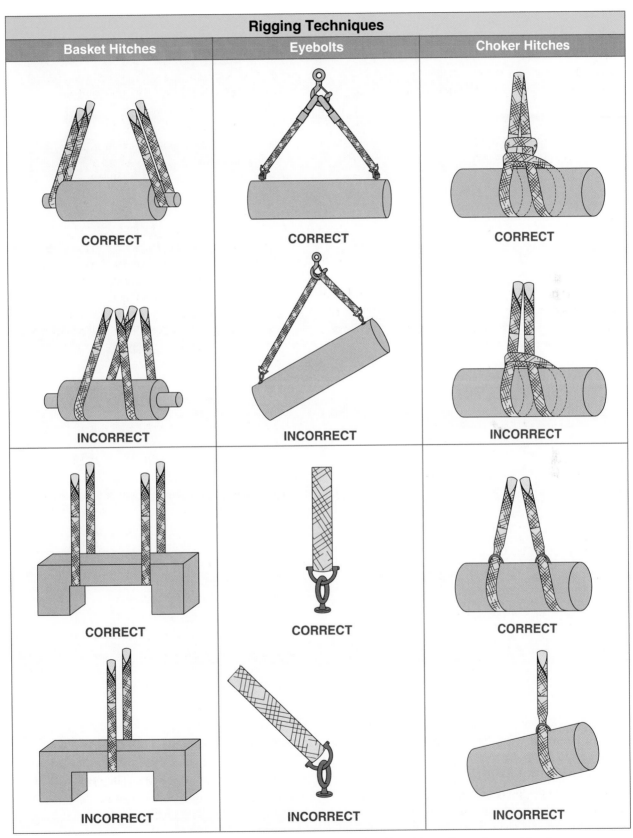

Figure 6-46. The type of web sling selected and its use must be made with safety as the main consideration.

Round Slings

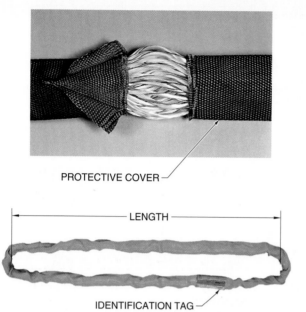

PROTECTIVE COVER

LENGTH

IDENTIFICATION TAG

Lift-All Company, Inc.

Round Sling Color and Capacity Rating*					
Round Sling Size No.	Color	Vertical	Choker	Vertical Basket	45° Basket
		Weight	Weight	Weight	Weight
1	Purple	2600	2100	5200	3700
2	Green	5300	4200	10,600	7500
3	Yellow	8400	6700	16,800	11,900
4	Tan	10,600	8500	21,200	15,000
5	Red	13,200	10,600	26,400	18,700
6	White	16,800	13,400	33,600	23,800
7	Blue	21,200	17,000	42,400	30,000
8	Orange	25,000	20,000	50,000	35,400
9	Orange	31,000	24,800	62,000	43,800
10	Orange	40,000	32,000	80,000	56,600
11	Orange	53,000	42,400	106,000	74,900
12	Orange	66,000	52,800	132,000	93,000

* in lb

Figure 6-47. Round slings are slings consisting of one or more continuous polyester fiber yarns wound together to make a core.

The use of chain for rigging is normally favored over wire rope because chain has approximately three times the impact-absorption capability of wire rope and is more flexible. Also, wire rope costs more than chain of similar strength and the life of wire rope is only 5% of chain life. The chain used for rigging or hoisting is considered a special chain because specific chain steel must be used in its manufacture.

Chain Construction

Chain and chain attachment strength depends on the composition of the steel from which they are made and the heat treatment process. The material and the manufacturing process determines the tensile, shear, and bending strength of chain.

Tensile strength is the maximum load in tension (pulling apart) that a material can withstand before breaking or fracturing. *Shear strength* is the ability of a material to withstand shear stress. *Bending strength* is a material's resistance to bending or deflection. The temperature used in the chain manufacturing process determines the metal's hardness. Steel becomes very hard and brittle when heated to a red color and quenched (dipped) in water. As the metal is heated, the grain (carbon structure) within the metal becomes unstructured (without a pattern). The structure is stilled (frozen) in an unstructured position when the metal is quenched.

Tempering allows the carbon structure to be held in more structured patterns. *Tempering* is the process in which metal is brought to a temperature below its critical temperature and allowed to cool slowly. Slow cooling methods include no quenching (air), quenching in oil, or quenching in salt for a very slow cool. Tempering of chain metal determines whether a chain gives or flexes under pressure or breaks with a snap of the chain.

Steel hardness is measured with a Brinell hardness tester and given a Brinell number ranging from approximately 150 for soft metal to 750 for hardened metal. A typical sling chain varies from 250 to 450 Brinell depending on the manufacturer. This is equivalent to a material tensile strength of 125,000 psi to 230,000 psi.

Steel used for rigging or hoisting chain is composed of premium quality, heat-treated, high-strength materials. This steel alloy contains 0.35% carbon maximum, 0.035% phosphorus maximum, and 0.040% sulfur maximum. A *steel alloy* is a metal formulated from the combining of iron with carbon and other elements.

Alloy material is chosen for sling chain to reduce the occurrence of fracturing of the metal. A *fracture* is a small crack in metal caused by the stress or fatigue of repeated pulling or bending forces. Through the combination of alloy metals and tempering, sling chain is capable of 15% to 30% elongation before breaking. This is a safety specification requirement for sling chain and is not to be used as a determination for replacement. Chain that has exceeded 1½% elongation from new should be removed from service. Any number of links may be used for measuring elongation, but the amount of used links should equal the number of new links. The chain should be taut during measurement. **See Figure 6-48.**

Tech Tip

Grade 80 and Grade 100 chain are the most common types of lifting chain. However, special situations may require different types. A chain's specifications must always be checked to determine whether the chain is appropriate for an application.

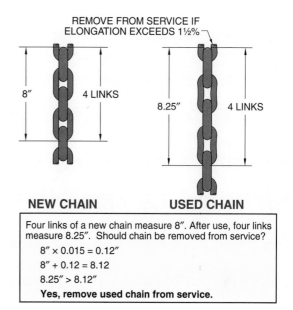

Four links of a new chain measure 8″. After use, four links measure 8.25″. Should chain be removed from service?

8″ × 0.015 = 0.12″
8″ + 0.12 = 8.12
8.25″ > 8.12″

Yes, remove used chain from service.

Figure 6-48. Chain should be removed from service if the measurement of used chain exceeds 1½% elongation from that of a new chain.

The National Association of Chain Manufacturers (NACM), in conjunction with the International Organization for Standardization (ISO), develops programs to standardize materials and processes for chain. The four main types of chain are binding chain, rigging chain, lifting chain, and hoist apparatus chain. Each chain, except for hoist apparatus chain, has a periodic embossing of a grade number or letter, indicating its capability. NACM specifies that identification must appear at least once every 36 links. **See Figure 6-49.** Each chain is classified as follows:

- Grade 43 – High-test steel chain having a carbon content of approximately 0.15% to 0.22%. A ½″ grade 43 chain is rated as having a working load limit (WLL) of 9200 lb. The *working load limit (WLL)* is the maximum weight that a rigging component may be subjected to. This chain is generally used for binding loads or tie downs and is embossed with an HT, $, $3, or M. Carbon steel chain is not to be used in overhead lifting.
- Grade 70 – High-strength transport binding chain having a carbon content of about 0.25% to 0.30%. A ½″ grade 70 chain is rated as having a working load limit of 11,300 lb. Grade 70 chain is generally used for critical load-securing applications such as binding and tie down on log and steel transport trucks and heavy equipment towing. This chain is embossed with a 7 or 70. Carbon steel chain is not to be used in overhead lifting.

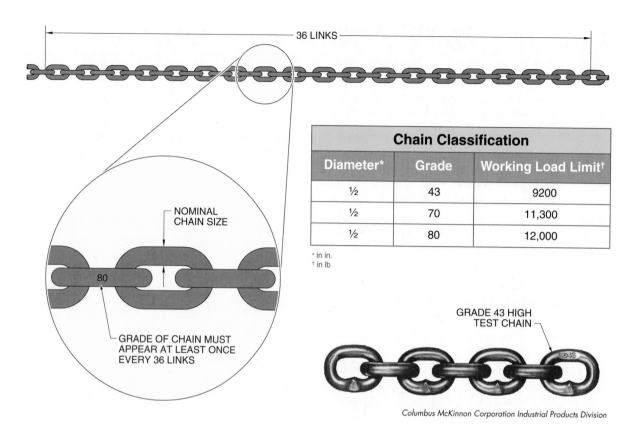

Chain Classification		
Diameter*	Grade	Working Load Limit†
½	43	9200
½	70	11,300
½	80	12,000

* in in.
† in lb

NOMINAL CHAIN SIZE

80

GRADE OF CHAIN MUST APPEAR AT LEAST ONCE EVERY 36 LINKS

GRADE 43 HIGH TEST CHAIN

Columbus McKinnon Corporation Industrial Products Division

Figure 6-49. Each chain, except for hoist apparatus chain, has a periodic embossing of a grade number or letter indicating its capability.

- Grade 80 – Alloy-steel chain is manufactured as a special-analysis alloy steel to provide superior strength, wear resistance, and durability. A ½″ grade 80 chain is rated as having a working load limit of 12,000 lb. **See Figure 6-50.** Grade 80 chain is used for all types of lifting slings, overhead lifting, or wherever maximum safety and strength is required.

Hoisting apparatus chain does not have a grade number because its physical measurements are more important than its strength. *Hoisting apparatus chain* is a precisely measured chain calibrated to function in pocket wheels used in manual or powered chain hoists. Hoisting apparatus chain is not to be used in sling or lifting applications. Hoisting apparatus chain is designated as hoist load chain and is not used in sling applications. Hoist load chain is a precisely measured chain designed to function in hand, air, or electric-powered hoist pockets.

In addition to the steel composition and heat treatment process, chain strength and working load limits are based on chain size. Nominal chain size is designated by the diameter of the material used to form the links.

Rigging Chain Strength

The working load capacity of chain is somewhat different from other slings because calculations are made using 1, 2, or 3 legs as a factor. A 4-leg factor is not used because a 4-leg (quad-branch chain sling), does not normally sustain the load evenly on each of its four legs. The maximum working load limits of a 4-leg chain are calculated as a 3-leg chain.

For example, what is the lifting capacity of a ¾″, 5-leg chain sling with a 30° sling angle?

1. Determine vertical capacity.

   $vl = 28{,}300$ lb (from Grade 80 Chain Load Limits table)

2. Determine sling angle loss factor.

   $s = 0.500$ (from Sling Angle Loss Factors table)

3. Calculate lifting capacity.

   $LC = vl \times l \times s$
   $LC = 28{,}300 \times 2 \times 0.500$
   $LC = \textbf{28,300 lb}$

Grade 80 Chain Load Limits*

Chain Size†	90° Vertical Load	60° Vertical Load	45° Vertical Load	30° Vertical Load	60° Quad Leg Load	45° Quad Leg Load	30° Quad Leg Load
7/32	2100	3600	3000	2100	5450	4450	3150
9/32	3500	6100	4900	3500	9100	7400	5200
3/8	7100	12,300	10,000	7100	18,400	15,100	10,600
1/2	12,000	20,800	17,000	12,000	31,200	25,500	18,000
5/8	18,100	31,300	25,600	18,100	47,000	38,400	27,100
3/4	28,300	49,000	40,000	28,300	73,500	60,000	42,400
7/8	34,200	59,200	48,400	34,200	88,900	72,500	51,300
1	47,700	82,600	67,400	47,700	123,900	101,200	71,500
1¼	72,300	125,200	102,200	72,300	187,800	153,400	108,400

* in lb
† in in.

Figure 6-50. Working load limits for slings using Grade 80 chain can be determined for a 90° vertical load or quad leg load up to 30° pull angle.

Rigging Chain Attachments

Typical connecting attachments between rigging and the hoisting device include shackles, master links, and hooks. A *shackle* is a U-shaped metal link with the ends drilled to receive a pin or bolt. The removal of the pin or bolt allows an opening for one or more loop eyes that can be attached to complete a sling. Shackles are made in a straight-U design (chain shackle) or a curved-U design (anchor shackle). A shackle may be used to make the connection between the rigging assembly and the hoisting hook. Shackle strength varies to conform to load weight. **See Figure 6-51.**

A chain shackle is used as a connector for a single lifting device such as a one loop eye for a vertical hitch. A chain shackle normally uses shackle washers as spacers to prevent side shifting. The anchor shackle uses a rounded eye to allow for more than one lifting device.

Like chain links, shackle strength is determined by its steel composition, heat treatment process, and rod diameter. Rigging shackles are made of alloy steel and are available in various strength capacities and rod sizes.

Cooper Tools

Care should be taken to select the type, grade, and size recommended by the manufacturer where attachments such as rings or hooks are designed for use with chain in sustaining loads.

Shackles

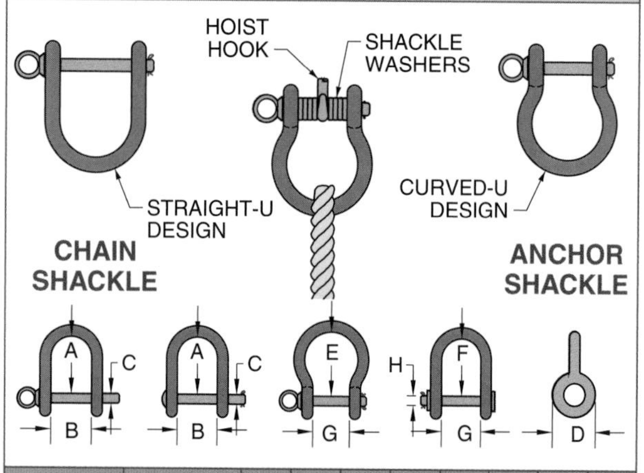

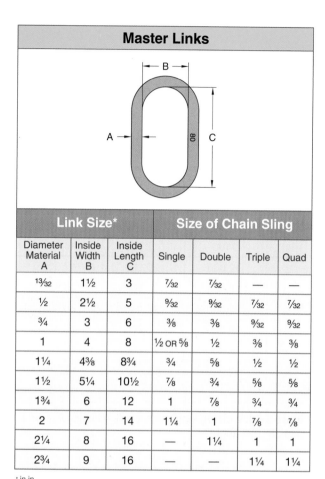

Working Load Limit*	Nominal Shackle Size†	A	B	C	D	E	F	G	H
½	¼	⅞	15/32	5/16	11/16	—	—	—	—
¾	5/16	1 1/32	17/32	⅜	13/16	—	—	—	—
1	⅜	1¼	21/32	7/16	31/32	—	—	—	—
1½	7/16	1 7/16	23/32	½	1 1/16	—	—	—	—
2	½	1⅝	13/16	⅝	1 3/16	1⅞	1⅝	13/16	⅝
3¼	⅝	2	1 1/16	¾	1 9/16	2⅜	2	1 1/16	¾
4¾	¾	2⅜	1¼	⅞	1⅞	2 13/16	2⅜	1¼	⅞
6½	⅞	2 13/16	1 7/16	1	2⅛	3 5/16	2 13/16	1 7/16	1
8½	1	3 3/16	1 11/16	1⅛	2⅜	3¾	3 3/16	1 11/16	1⅛
9½	1⅛	3 9/16	1 13/16	1¼	2⅝	4¼	3 9/16	1 13/16	1¼
12	1¼	3 15/16	2 1/32	1⅜	3	4 11/16	3 15/16	2 1/32	1⅜
13½	1⅜	4⅜	2¼	1½	3 5/16	5 3/16	4⅜	2¼	1½
17	1½	4 13/16	2⅜	1⅝	3⅝	5¾	4 13/16	2⅜	1⅝
25	1¾	5¾	2⅞	2	4⅛	7	5¾	2⅞	2
35	2	6¾	3¼	2¼	5	7¾	6¾	3¼	2¼

* in tons
† in in.

Figure 6-51. A shackle is a U-shaped metal link with the ends drilled to receive a pin or bolt.

In most cases, chain slings are assembled by the manufacturer. However, when a chain sling is required to be assembled, various attachments, such as a master link, connecting link, chain, and hook, may be purchased separately and then individually attached. A *master link* is a chain attachment with a ring considerably larger than that of the chain to allow for the insertion of a hook. **See Figure 6-52.** Master links are also large enough to incorporate more than one connecting link for creating extra-legged slings. Master links may be the connection between the hoisting hook and the rigged load. Master link capacity must be equal to or greater than that of any rigging components.

Master Links

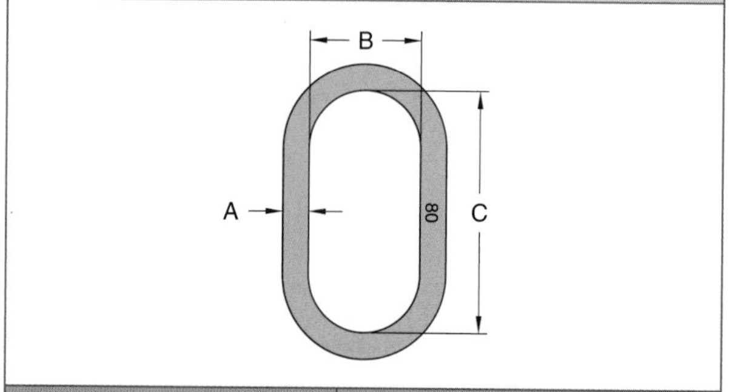

Link Size*			Size of Chain Sling			
Diameter Material A	Inside Width B	Inside Length C	Single	Double	Triple	Quad
13/32	1½	3	7/32	7/32	—	—
½	2½	5	9/32	9/32	7/32	7/32
¾	3	6	⅜	⅜	9/32	9/32
1	4	8	½ OR ⅝	½	⅜	⅜
1¼	4⅜	8¾	¾	⅝	½	½
1½	5¼	10½	⅞	¾	⅝	⅝
1¾	6	12	1	⅞	¾	¾
2	7	14	1¼	1	⅞	⅞
2¼	8	16	—	1¼	1	1
2¾	9	16	—	—	1¼	1¼

* in in.

Figure 6-52. A master link is a chain attachment with a ring considerably larger than that of the chain to allow for the intersection of a hook.

A *connecting link* is a three-part chain attachment used to assemble and connect the master link to the chain. In either case, sling attachments should have a working load limit equal to or greater than that of the alloy steel chain. For example, a ½″ Grade 80 alloy steel chain is rated as having a 12,000 lb WLL. However, a ½″ master link is rated as having a 4920 lb WLL and a ¾″ master link is rated at 10,320 lb WLL. Therefore, a 1″ master link with a 24,360 lb WLL must be used in a ½″ chain sling assembly if it is to be of equal or greater strength.

The WLL of the weakest component should be the working load limit of the entire sling assembly if a sling is required to be assembled using attachments with different working load limits. Makeshift hooks, links, or fasteners fashioned from bolts, rods, or other materials should not be used. The hook is a major link between the rigged load and the hoisting equipment. A *hook* is a curved or bent implement used for holding, pulling, or connecting rigging to loads or lifting equipment. **See Figure 6-53.**

Hooks

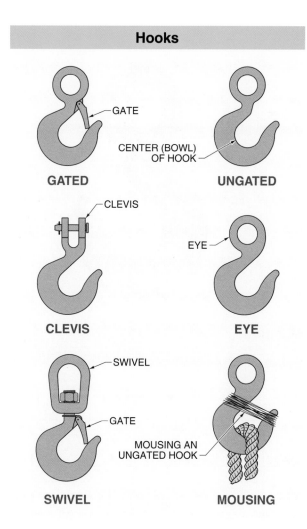

Figure 6-53. A hook is a curved or bent implement for holding, pulling, or connecting another implement.

Hoisting hooks used for rigging purposes include choker, grab, foundry, swivel, and sorting hooks. **See Figure 6-54.** A *hoisting hook* is a steel alloy hook used for overhead lifting and is connected directly to the piece being lifted. **See Figure 6-55.**

Hoisting Hooks

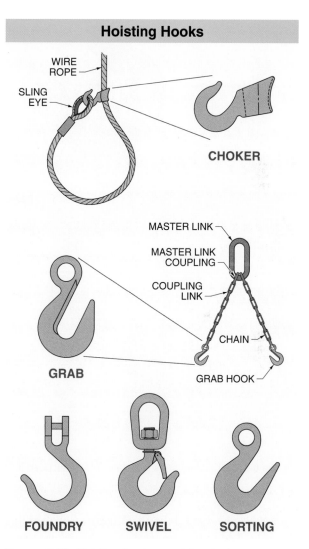

Figure 6-54. Hoisting hooks used for rigging purposes include choker, grab, foundry, swivel, and sorting hooks.

Hooks are made in various shapes, designs, and sizes and are normally forged of alloy steel. Connecting (load) hooks are quenched and tempered so they are the first to bend or give in an overload situation. The hook should be the weakest member of the hoisting equipment.

Deformation is the undesirable bending of a hook due to an applied force. Hook deformation becomes the measurable part of rigging overload. For this reason, hooks should never be heated above 800°F or become part of a welding situation. Heat applied to a hook can temper the metal, reducing its strength.

Hooks may be gated or ungated or contain a clevis, eye, or swivel. Some hooks are equipped with latches or gates to prevent the sudden release of shifted sling legs.

The hoisting hook attached to rigging may be designed to swivel. The swivel on a hoisting hook allows the load to turn without damaging or twisting the chain. A swivel hook is used only as a hook-positioning device and is not intended for load rotation. Special load rotation swivel hooks are available for such applications.

Hoisting Hook Capacity*

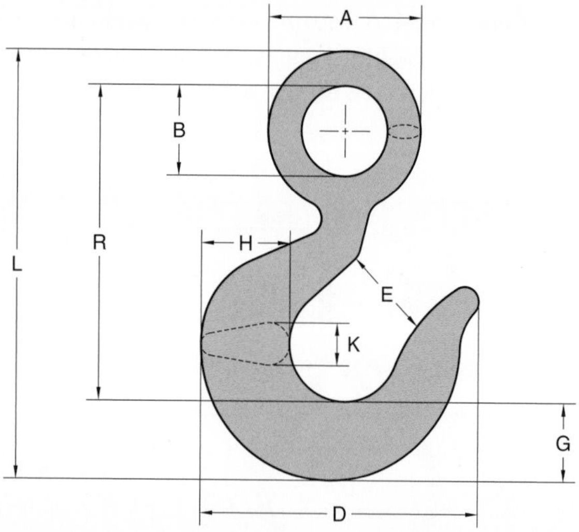

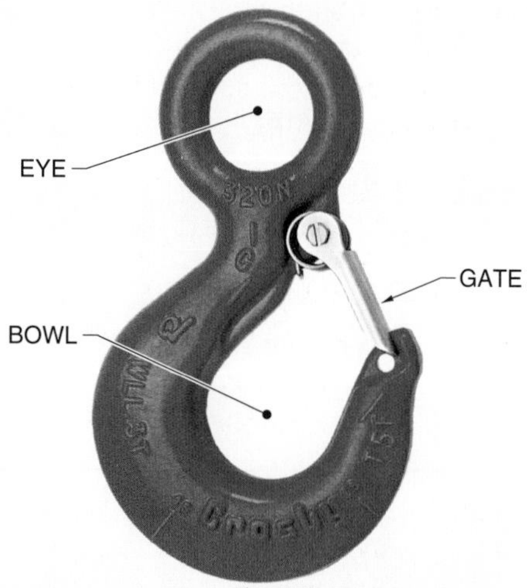

The Crosby Group, Inc.

Eye Hook Capacity

Capacity*	Dimensions								
	A	B	D	E	G	H	K	L	R
1.1	1.47	0.75	2.88	0.94	0.75	0.81	0.56	4.34	3.22
1.65	1.75	0.91	3.19	1.03	0.84	0.94	0.62	4.94	3.66
2.2	2.03	1.12	3.62	1.06	1.00	1.16	0.75	5.56	4.09
3.3	2.41	1.25	4.09	1.22	1.12	1.31	0.84	6.40	4.69
4.95	2.94	1.56	4.94	1.50	1.44	1.62	1.12	7.91	5.75
7.7	3.81	2.00	6.50	1.88	1.81	2.06	1.38	10.09	7.38
12.1	4.69	2.44	7.56	2.25	2.25	2.62	1.62	12.44	9.06
16.5	5.38	2.84	8.69	2.50	2.59	2.94	1.94	13.94	10.06
24.2	6.62	3.50	11.00	3.38	3.00	3.50	2.38	17.09	12.50
33	7.00	3.50	13.62	4.00	3.66	4.62	3.00	19.47	14.06
40.7	8.50	4.50	14.06	4.25	4.56	5.00	3.75	24.75	18.19

* in t

Figure 6-55. A hoisting hook is a steel alloy hook used for overhead lifting and is connected directly to the piece being lifted.

A *choker hook* is a sliding hook used in a choker sling and is hooked to the sling eye. A *grab hook* is a hook used to adjust or shorten a sling leg through the use of two chains. The grab hook is connected to a short length of chain that is attached to the master link. The longer (rigging) sling leg is shortened by engaging one of the links in the longer chain into the hook. A *foundry hook* is a hook with a wide, deep throat that fits the handles of molds or casting. A *sorting hook* is a hook with a tapered throat and a point designed to fit into holes.

The load supported by a hoisting hook should be supported from the center or bowl of the hook. Hook failure is likely to occur if the load shifts or is applied to the tip or from an area between the bowl and the tip. Tip loading of hoisting hooks greatly decreases their lifting capacity. Hooks should be chosen for their proper strength rating. Specific size dimensions allow for specific load capacity.

Safety Tip

Wire rope, shackles, rings, master links, and other rigging hardware must be capable of supporting, without failure, at least five times the maximum intended load. Where rotation resistant rope is used, the slings shall be capable of supporting, without failure, at least ten times the maximum intended load.

RIGGING COMPONENT INSPECTION

Inspection of rigging equipment covers inspection, recordkeeping, and storage. Maintenance of rigging components does not include making temporary repairs. Temporary repairs of rope, webbing, or chains should never be attempted. Damaged components should be removed from service and destroyed. Chain to be repaired by the manufacturer should be removed from service, tagged "Defective – Do Not Use," and separated from chains in use until they can be delivered to the manufacturer for repair.

An examination of all rigging equipment should be done initially, frequently, and periodically. Before any rigging component is placed in service, it shall be initially inspected to ensure that the specifications and condition requirements are correct. Frequent inspection is completed by the person using the rigging component each time it is used. Periodic inspections are conducted by designated, knowledgeable individuals.

The frequency of periodic inspection is determined by the service conditions and frequency of use. Experience, gained on the service life of the components, can also determine frequency of periodic inspection. However, periodic inspection of rigging components shall be conducted at least annually, with the exception of round slings, which shall be inspected at least monthly.

Wire Rope Inspection

Kinking, core protrusion, and bird caging may be encountered when inspecting wire rope. **See Figure 6-56.** *Kinking* is a sharp permanent bending. Kinking is normally caused by improper removal of wire rope from a spool or improper storage. Kinking weakens a wire rope and in many cases makes it useless.

Core protrusion is a damage condition of wire rope where compressive forces from within the rope force the strands apart. This happens when core material is squeezed out of the rope due to corrosion or degradation of the core. Core protrusion removes the support from the outer strands, reducing the efficiency of the rope.

Bird caging occurs from overloading, twisting, or squeezing when the rope is under load and is suddenly released. *Bird caging* is a damage condition of wire rope where the strands separate and open forming a shape similar to a bird cage.

Wire Rope Inspection

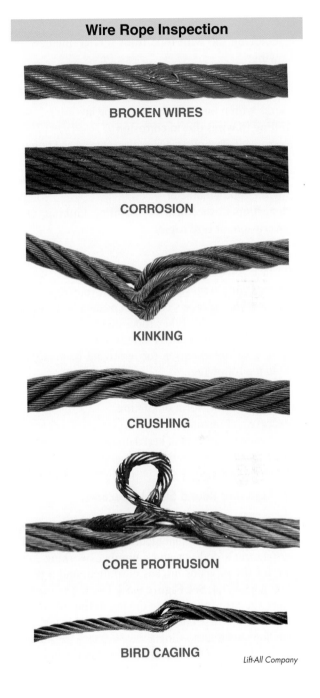

BROKEN WIRES

CORROSION

KINKING

CRUSHING

CORE PROTRUSION

BIRD CAGING

Lift-All Company

Figure 6-56. Those who examine wire rope and wire rope slings should look for evidence of excessive wear or other types of damage

Tech Tip

Per OSHA 1926.251, each wire rope used in hoisting or lowering or in pulling loads shall consist of one continuous piece without knots or splices except for eye splices in the ends of wires and for endless rope slings.

Conditions of wire rope that are considered reason for removing the wire rope from service include the following:

- one broken or cut strand
- ten broken wires in one rope lay length or five broken wires within one strand in one rope lay
- pitting in wires due to corrosion
- corrosive failure of one wire adjacent to an end fitting
- welding or weld splatter damage
- flat spots worn on the outer strands of the rope, reducing the rope's diameter
- distortion in the form of crushing, kinking, core protrusion, or bird caging

Fiber Rope Inspection

Fiber rope inspection is used to remove a rope from service before the rope's condition poses a hazard with continued operation. Fiber rope should be inspected monthly. **See Figure 6-57.** Conditions of fiber rope that are considered reason for removing from service include the following:

- cuts or broken fibers
- distortion in the form of kinking
- excessive abrasion or wear
- distortion in the form of heat damage, such as melting or charring on synthetic fiber rope

Web Sling and Round Sling Inspection

Web and round slings must include an identification tag that includes the manufacturer's name or mark, manufacturer's code or stock number, working load limits for the types of hitches permitted, and type of webbing material. **See Figure 6-58.** Use over time can obscure the printing or cause the tag to fall off, so the first part of an inspection is to check for this tag. If a tag is damaged but the sling appears otherwise acceptable, the sling may be returned to the manufacturer for testing and retagging.

Web slings include red warning yarns within the interior weave that are normally not visible. These are designed to show through if the webbing is damaged sufficiently to render it unfit for use. **See Figure 6-59.** The sling must be immediately removed from service if there is any sign of the warning yarns. However, lack of visible warning yarns does not indicate an acceptable sling condition. Some damage may not expose these yarns but can still weaken the sling. Round slings may not include any warning yarns, but the slings must be discarded if any of the interior load-bearing yarns are visible.

Fiber Rope Inspection

CUTS/BROKEN FIBERS

KINKING

EXCESSIVE WEAR

MELTING/CHARRING

Figure 6-57. Fiber rope inspection is used to remove a rope from service before the rope's condition poses a hazard with continued use.

Web Sling Identification Tags

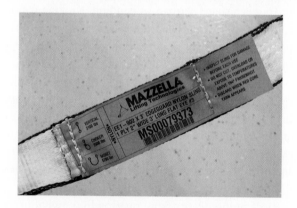

Figure 6-58. An identification tag includes the manufacturer's name or mark, manufacturer's code or stock number, working load limits for the types of hitches permitted, and type of webbing material.

Web Sling Warning Yarns

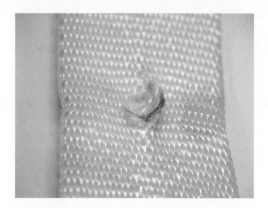

Figure 6-59. If the red warning yarns from webbing cores are visible, then the web sling has been overloaded or damaged and must be removed from service.

Further inspection of the sling should be done to check for physical, thermal, or chemical damage that can weaken the sling. **See Figure 6-60.** Excessive wear or abrasion can damage the stitching in the load-bearing splices or cause the surface fibers to fray and break. Significant cuts or tears are those that damage 50% of the longitudinal yarns in an area that extends ¼ of the webbing width or 100% of the yarns for ⅛ of the width. Edge cuts can be particularly weakening. Other types of physical damage include punctures, snags, or embedded particles.

Thermal damage, which can be caused by direct exposure to heat or from friction, is indicated by melting or charring of the webbing material. Exposure to certain chemicals can cause disintegration of the webbing fibers, depending on the fiber material.

Another serious problem with both web and round slings is knots. A knot alters the load-bearing characteristics of the yarns at that location, significantly weakening the sling. Knots are easily introduced when sorting long slings. Knots are often irreversibly tightened if not caught before the sling is loaded. A sling with any knot that has been tightened under load must be removed from service.

Tech Tip

Never attempt to repair a rope, sling, or chain that has been damaged. Damaged ropes, slings, and chains must always be discarded and replaced with new products.

Webbing Inspection

EXCESSIVE WEAR

SURFACE CUT

EDGE CUT

PUNCTURE

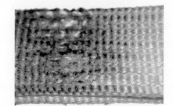

MELTING/CHARRING

ACID DAMAGE

Lift-All Company, Inc.

Figure 6-60. Web and round slings are weakened by many types of physical, thermal, and chemical damage.

Chain Inspection

Chain should be inspected annually. Repairs to rigging and hoisting chain should only be made and tested by the chain manufacturer. Never use mechanical coupling links or repair links to repair any sub-standard rigging or hoisting chain. **See Figure 6-61.** Conditions of chain that are considered reason for removing from service include the following:

- nicks, gouges, or wear having a depth in excess of the values given in maximum allowable link wear tables
- elongation
- hook throat opening in excess of 15%
- hook tip twisted more than 10% from the plane of the unbent hook
- hooks showing cracks or signs of abuse
- chain links bent or seated or flexing properly

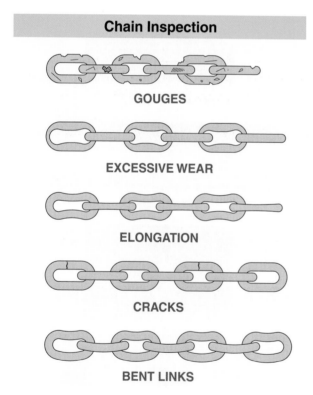

Chain Inspection

GOUGES

EXCESSIVE WEAR

ELONGATION

CRACKS

BENT LINKS

Figure 6-61. Signs of chain damage include gouges, wear, stretching, cracks, and distorted links.

RIGGING EQUIPMENT STORAGE

Proper care, use, and storage of rigging equipment prevents damage and ensures safety. Rope, webbing, and chain slings must be kept in an assigned area that is kept clean, neat, dry, and away from harmful fumes or heat. Synthetic webbing, round slings, and natural fiber rope should be stored out of sunlight and away from areas used for arc welding.

Kinking is prevented when slings are neatly hung from racks or lengthy rope is rolled onto spools. Slings or sling components must not be left lying where vehicles or forklifts may run over them or where heavy loads may be set on them. Slings should not be dragged over abrasive surfaces or sharp objects.

To prevent accidental use, damaged slings slated for manufacturer's repair must be stored in an area specifically posted, "Warning – Do Not Use." Damaged or worn sling or sling attachments that are not to be repaired should be immediately disposed of.

RIGGING COMPONENT RECORDKEEPING

Written inspection records and use of slings and sling components should be created for each new sling. Inspection records may cover the basic requirements of an annual inspection or be more comprehensive and frequent. The amount of rigging done by the user dictates the frequency and depth of recordkeeping.

A basic inspection document (logbook) begins by issuing a serial number to the sling. The basic form includes the date purchased, condition at purchase, type, rated strength, list of included attachments or components, date of periodic inspections, inspecting personnel, and comments for each inspection. **See Figure 6-62.**

Logbook inspection forms may also include dates and hours of each use, department where used, initial sizes and diameters, and sizes and diameters at each inspection, with part numbers given to each sling attachment. Although frequent inspections made before and after each use are visual, unusual or abnormal conditions should be reported and included in the logbook. All logbook entries, if comprehensive, offer enough historical information to chart sling degradation and life expectancy.

Tech Tip

Follow proper safety procedures when performing inspections on ropes, slings, and chains. Always wear leather gloves to protect hands from abrasions and safety glasses to protect eyes from loose strands or damaged chain links.

Rigging Inspection Records

Rigging Inspection

Date _____Jan. 14th_____ Department _____Shipping_____

Inspector _____J. Smith_____ Inspection Type _____Monthly_____

Rigging: Wire Rope / Fiber Rope / (Web Slings) / Round Slings / Chain / Other _____

Serial Number	Manufacturer	Size	Rated Capacity	Attachments	Condition	Removed from Service?
W-010	NB Slings	6"	9600	triangles	Good	N
W-029	NB Slings	4"	6400	none	½" edge tear	YES— destroyed
W-007	Lift-Tech	6"	16,800	choker/triangle	light wear	N
W-013	Lift-Tech	8"	19,200	none	unreadable tag	YES— returned to mfr.

Figure 6-62. The collection of inspection records for a particular piece of equipment offers enough historical information to chart equipment degradation and life expectancy.

Review

1. Define symmetrical load and asymmetrical load and list three examples of each.

2. List common stock materials.

3. How is wire rope classified?

4. How is fiber rope constructed?

5. Define seizing.

6. Why is fiber rope preferred over wire rope or chain in certain applications?

7. Define hitch.

8. When is webbing for rigging purposes normally used?

9. List three factors that affect webbing strength.

10. Define round sling.

11. Why is the use of chain for rigging normally favored over wire rope?

12. Describe what chain and chain attachment strength depend on.

13. Define master link.

14. List five conditions of wire rope that are considered reason for removing the wire rope from service.

15. Describe the contents of a proper basic inspection document (logbook).

Digital Resources

ATPeResources.com/Quicklinks
Access Code 462409

CHAPTER 7

LIFTING

Lifting is the hoisting of equipment using mechanical means. Hoists may be manually-operated or power-operated hoists. Overhead hoists and cranes are regulated by a large number of standards. Hoist safety requires inspection at frequent and periodic intervals.

Harrington Hoists Inc.

Chapter Objectives

- Explain the function of block and tackle.
- List the different types of hoists.
- Define drum wrap.
- Explain hoist safety.
- Describe the function of an eyebolt.
- Identify the different types of industrial cranes and their lift capacities.
- Explain crane operation.

Key Terms

- lifting
- block
- tackle
- reeving
- part
- mechanical advantage
- lead line
- hoist
- lift
- headroom
- reach
- drum wrap
- crossover
- hook drift
- eyebolt
- industrial crane

LIFTING DEVICES

Many lifting devices in use today use the centuries-old principle of the block and tackle (rope and pulley). Blocks and tackle were used to move sails, spars, and other components on sailing ships. Much of the block and tackle terminology used today is based on nautical applications. Today, blocks and tackle are primarily used for industrial lifting.

Lifting is the hoisting of equipment or machinery by mechanical means. Lifting is accomplished by using hand-operated or power-operated equipment. Industrial lifting equipment generally consists of the rigging assembly, hoist, and hoist support. Each component relies on the integrity of the other components. Lifting is attempted only after all components are determined to be safe.

Block and Tackle

Block and tackle is a combination of ropes and sheaves (pulleys) used to improve lifting efficiency. **See Figure 7-1.** A *block* is an assembly of one or more sheaves in a frame. *Tackle* is the combination of ropes and block assemblies arranged to gain mechanical advantage for lifting.

Block-and-tackle assemblies begin with the rope being reeved over a sheave. *Reeving* is the threading of a rope through a hole or opening or around a series of sheaves. The capacity, lifting speed, and lifting distance of a block and tackle is determined by the number of parts and the number of sheaves. A *part* is a rope length between the lower (hook) block and the upper block or drum. The greater the number of parts, the greater the lifting capacity. The dead end of the rope is usually attached to a becket. A *becket* is an attachment point, usually on a block, for the dead end of a hoisting rope. The remainder of the live rope end forms the lead line. A *lead line* is the part of a rope to which force is applied to hold or move a load. Lead lines are not counted as parts.

Mechanical Advantage. *Mechanical advantage* is the ratio of the output force to the input force of a device. In block-and-tackle applications, mechanical advantage allows a load to be lifted by applying a force less than the load weight. Forces can be either static (held) or dynamic (moving). Mechanical advantage of block and tackle is determined by the number of parts reeved. Block-and-tackle assemblies typically have one-part, two-part, or three-part reeving.

Block and Tackle

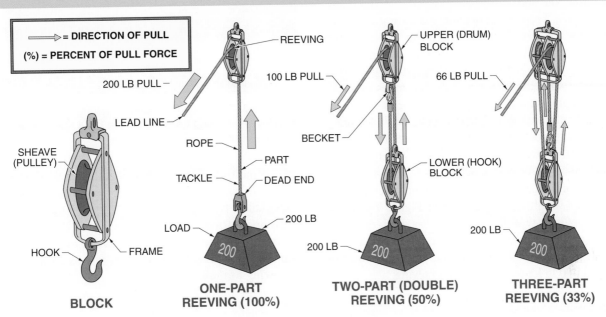

Figure 7-1. A block and tackle is a combination of ropes and sheaves (pulleys).

One-part reeving has one line between the load and a single, upper block. There is no mechanical advantage to one-part reeving because the force required to lift the load is equal to the weight of the load. Two-part reeving has two lines between the load and upper block. Two-part reeving reduces the lead-line force to 50% because both lines support one-half the weight of the load. Three-part reeving has three lines between the load and upper block. Three-part reeving reduces the lead-line force by 33% because each line supports one-third the weight of the load.

Under ideal conditions (no friction), the mechanical advantage of a block-and-tackle system equals the number of parts of rope that support the load. Therefore, a two-part block (reeving) system has an ideal mechanical advantage of 2:1.

A three-part block system (three-part reeving) consists of an upper block with two sheaves and a lower block with one sheave. Because three ropes support the load, a three-part block system has a mechanical advantage of 3:1

Static Forces. A force applied to the lead line is useful only when it is equal to or greater than the static (holding) force. A static force is great enough to hold a load stationary but not enough to lift the load. The amount of static force required to hold a load is calculated with the following formula:

$$F_S = \frac{W_{total}}{n}$$

where

F_S = static lead-line force (in lb)

W_{total} = total load weight, including rigging equipment (in lb)

n = number of parts

For example, what is the force required to hold a 500 lb load using a four-part reeving system? *Note:* The rope, block, and hook components total 30 lb.

$$F_S = \frac{W_{total}}{n}$$

$$F_S = \frac{530}{4}$$

$$F_S = \textbf{133 lb}$$

Lifting Forces. As a lead-line force exceeds the minimum static force, it overcomes friction in the sheaves or pulleys and begins to lift the load. The amount of additional force needed to begin lifting differs according to the number of pulleys, their bend ratio, and their bearing type. Each pulley adds friction to the system, which adds a practical limitation to the attempt of increasing the number of pulleys to a tackle arrangement for greater mechanical advantage.

Each pulley used has friction that must be overcome. Therefore, the number of pulleys affects the minimum lifting force. The number of pulleys is assumed to equal the number of parts. Bend ratio is a factor because a rope moves more easily over a larger pulley than a smaller one. The pulley bearing types have different friction characteristics. **See Figure 7-2.** The axles of plain bearing pulleys are pins held in the frame of the block. Alternatively, ball- or roller-bearing blocks hold pulley axles in reduced-friction bearings. Less additional force is required to overcome the friction of rolling bearing pulleys than plain bearing pulleys.

Friction in each pulley adds a certain percentage of the load's weight as resistance. **See Figure 7-3.** Plain bearing pulleys typically add 5% to 8%. Ball- or roller-bearing pulleys typically add 3% to 5%. For example, a pulley adding 6% in friction requires a 106 lb force to move a 100 lb load using one-part reeving.

Pulley Friction Contributions

Bend Ratio	Plain-Bearing	Ball- or Roller-Bearing
up to 15	8%	5%
15 to 20	7%	4%
greater than 20	6%	3%

Figure 7-3. The amount of friction resistance produced by a pulley when lifting depends on the bend ratio and type of pulley bearing.

For multiple-part reeving, the load is shared by each part, but the effect of friction compounds through each pulley. In order to simplify these calculations, an appropriate friction factor is determined from a table of friction percentages. **See Figure 7-4.** Then, the minimum lead-line lifting force is calculated with the following formula:

$$F_L = W_{total} \times f_{fr}$$

where

F_L = lifting lead-line force (in lb)

W_{total} = total load weight, including rigging equipment (in lb)

f_{fr} = friction factor

For example, what is the minimum force required to lift a 6000 lb load using an eight-part reeving system equipped with plain bearing pulleys? *Note:* The bend ratio is 17. Therefore, each pulley adds 7% of its load in friction and the associated friction factor is 0.21.

$$F_L = W_{total} \times f_{fr}$$
$$F_L = 6000 \times 0.21$$
$$\boldsymbol{F_L = 1260\ lb}$$

Due to the addition of friction factors, the true mechanical advantage of a block and tackle arrangement when lifting is less than the same arrangement when static. For example, to hold a 6000 lb load steady with an eight-part reeving system, the mechanical advantage is 8:1. The static lead-line force is only 750 lb. However, to lift the load, a minimum force of 1260 lb is required to overcome friction. The true mechanical advantage in this case is approximately 4.8:1 (6000 lb ÷ 1260 lb = 4.8).

Travel Distance. Block-and-tackle assemblies amplify force at the cost of distance. As more pulleys reduce the force required to lift a load, the distance the lead line must be pulled increases. The ratios are equal to the mechanical advantage. In other words, if a load must travel a certain vertical distance, then each of the rope parts above it must shorten or lengthen by the same

Pulley Bearings

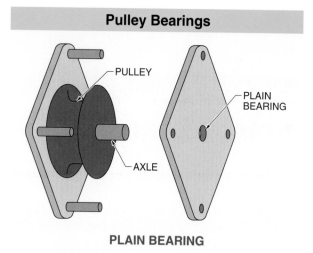

PLAIN BEARING

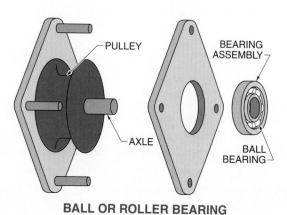

BALL OR ROLLER BEARING

Figure 7-2. The type of bearing holding the axle of a pulley affects the amount of friction in the pulley.

distance. For example, if a two-part reeve is used to lift a load 12″, the lead line must be pulled 24″, which is a 2:1 ratio. If the assembly is a three-part reeve, then the lead line must be pulled 36″, which is a 3:1 ratio.

Travel Speed. Travel speed is affected in the same manner as travel distance. If the lead line and the load are traveling different distances in the same amount of time, then the speeds must also be different. If the lead line travels 24″ in the same amount of time that the load travels 12″, then the load's speed is one-half the lead line's speed. Similarly, a three-part reeved load moves at one-third the speed of the lead-line.

For configurations of four parts or more, the speed of each pulley is different and gradually increases from slowest (the one nearest the dead end) to fastest (the one at the lead line). These differences may be a consideration when reeving between blocks. **See Figure 7-5.**

A load is raised 1′ if the lead line on a four-part reeve is pulled 4′. Also, the force required by the lead line is 20 lb (one-fourth of the load) if the load is 80 lb. The mechanical advantage of a four-part reeve under ideal conditions is a 4:1 ratio. **See Figure 7-6.** The distance the load moves is decreased by additional pulleys to the same extent that the ideal mechanical advantage is increased by additional pulleys.

Pulley Friction Factors						
Number of Parts	**Pulley Friction Contribution**					
	3%	**4%**	**5%**	**6%**	**7%**	**8%**
1	1.03	1.04	1.05	1.06	1.07	1.08
2	0.53	0.54	0.55	0.56	0.57	0.58
3	0.36	0.37	0.39	0.40	0.41	0.42
4	0.28	0.29	0.30	0.32	0.33	0.34
5	0.23	0.24	0.26	0.27	0.28	0.29
6	0.20	0.21	0.22	0.24	0.25	0.26
7	0.18	0.19	0.20	0.21	0.23	0.24
8	0.16	0.17	0.18	0.20	0.21	0.23
9	0.14	0.16	0.17	0.19	0.20	0.22
10	0.13	0.15	0.16	0.18	0.20	0.22
11	0.13	0.14	0.16	0.17	0.19	0.21
12	0.12	0.13	0.15	0.17	0.19	0.21

Figure 7-4. Pulley friction factors are used to calculate the lead-line force required to overcome pulley friction and lift a load.

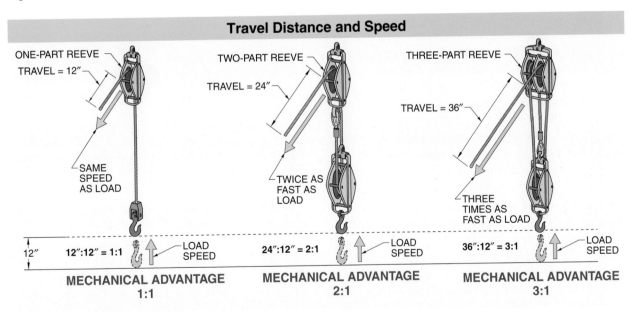

Figure 7-5. The ratios of travel distance and speed equate to the mechanical advantage of a block-and-tackle system.

Four-Part Reeving

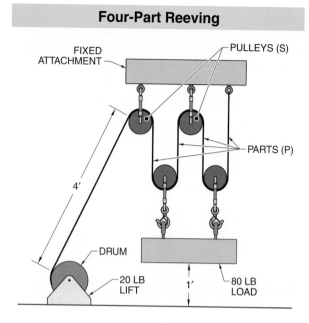

Figure 7-6. The mechanical advantage of a four-part reeve under ideal conditions is 4.

For example, a 20 lb lead line force lifts 40 lb with two-part reeving. The 20 lb force must be moved 3′ to move the 40 lb load 1½′. With three-part reeving, the 20 lb lead line force lifts 60 lb, but the 3′ lead line movement moves the load only 1′.

HOISTS

A *hoist* is a mechanical device used to provide the lifting force on lead lines. Hoists typically use gear drives to provide increased load capacity and safety over manual pulling. Another type of mechanical advantage gear drives provide is that a small input torque is amplified into a large output torque. *Torque* is the twisting (rotational) force of a shaft. The output torque winds a lead line onto a drum or across a sprocket to provide the linear pulling force that lifts the load. The mechanical advantage in torque is provided at the expense of speed.

Most hoists use worm gear or bevel gear drives. **See Figure 7-7.** A *worm gear drive* is a pair of gears consisting of a spiral-threaded worm (drive gear) and a worm wheel (driven gear) that are used extensively as a speed reducer. The worm gear must rotate many times in order to turn the worm wheel once, which provides a very high mechanical advantage. Also, the design prevents reverse rotation of the gears, so

when the input torque is removed the gears cannot slip in reverse.

A *bevel gear drive* is a pair of gears that mesh at an angle, usually 90°. The mechanical advantage of the drive depends on the relative sizes of the gears. A small gear driving a large gear provides a large advantage, while a pair of same-sized gears provides no advantage. Unlike the worm gear hoist, bevel gear hoists normally require a braking or locking mechanism to prevent reverse rotation.

Gear Drives

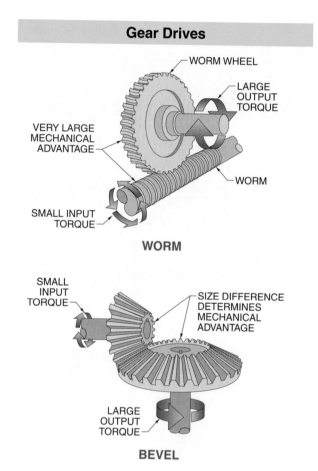

Figure 7-7. The two primary types of gear drives in hoists are worm gear drives and bevel gear drives.

Hoists are suspended overhead from a top hook or another type of attachment to a supporting structure. A support may be a fixed point on a building ceiling or structure or a movable point such as a trolley or crane. The attachment and the supporting structure must be strong enough to support the weight of the hoist, tackle, rigging, and the maximum load capable of being lifted with the equipment, given the appropriate safety factor.

Hoists are specified by their capacity rating and certain critical dimensions. **See Figure 7-8.** *Lift* is the distance between a hoist's upper and lower limits of travel. *Headroom* is the distance from the cup of the hoist's top hook to the cup of a hoist hook when the hoist hook is at its upper limit of travel. This dimension specifies how much space must be allowed under the support for the hoist. *Reach* is the distance between the cup of a top hook and the cup of a hoist hook when the hoist hook is at its lower limit of travel. Reach is the sum of the lift and the headroom.

with chain link pockets that are connected to the hoist mechanism. A *hoist chain* is the chain that raises the load. The rotation of the pocket wheel causes the hoist chain to raise or lower the load. The *hand-chain drop* is the distance between the lower portion of the hand chain to the upper limit of the hoist hook travel.

Hoist Dimensions

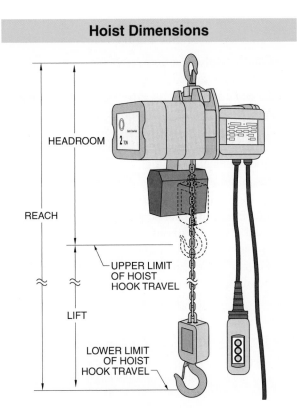

Figure 7-8. Hoists are selected based on their capacity rating and certain critical dimensions.

Hand-Chain Hoists

A *hand-chain hoist* is a manually operated chain hoist used for moving a load. Hand-chain hoists are suspended overhead from a top hook attached to a supporting structure. Supporting structures may be tripods, trolleys, cranes, or other fixed points. Hand-chain hoists are normally rated for ¼ t to 50 t.

A hand-chain hoist uses two chains: the hand chain and the hoist chain. **See Figure 7-9.** A *hand chain* is a continuous chain grasped by the operator to operate the pocket wheel. A *pocket wheel* is a pulley-like wheel

Hand-Chain Hoists

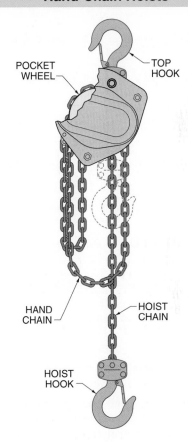

Figure 7-9. The input torque is provided to a hand-chain hoist by pulling on the continuous hand chain.

Lever-Operated Hoists

A *lever-operated hoist* is a lifting device that is operated manually by the movement of a lever. **See Figure 7-10.** Lever-operated hoists are used in confined and awkward areas and many are small enough to be kept in a toolbox. Lever-operated hoists are generally used to lift light loads. A light load is a load that weighs from 200 lb to 500 lb. An automobile engine is an example of a light load.

Lever-Operated Hoists

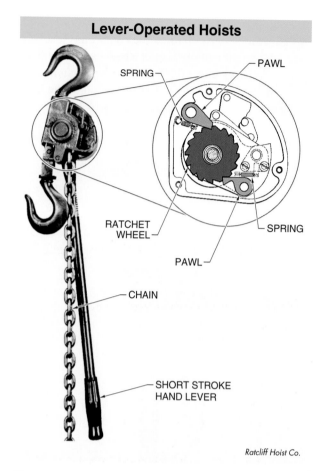

SPRING
PAWL

RATCHET
WHEEL

PAWL

SPRING

CHAIN

SHORT STROKE
HAND LEVER

Ratcliff Hoist Co.

Figure 7-10. Lever-operated hoists use a ratchet wheel to convert the back-and-forth motion of a lever to high output torque.

Two major types of lever-operated hoists are ratchet and slip-clutch models. Lever-operated hoists have ratchet mechanisms to prevent reversal. A *ratchet* is a mechanism that consists of a toothed wheel and a spring-loaded pawl. A *pawl* is a mechanism used to prevent the ratchet wheel from turning backwards. Some manufacturers design safety-slip clutches into lever-operated hoists to prevent unsafe overloading.

A *slip clutch* is a spring-loaded, friction-held fiber disc that is adjusted to slip at 125% to 150% of the hoist-rated load. Many slip clutches are preset by the manufacturer, while others are adjustable. Adjustable slip clutches must never be adjusted to hold over 175% of the hoist-rated load.

More than the hoist-rated load should never be lifted, except for clutch adjustment and testing. Before attempting to adjust and test a slip clutch, determine that the supporting structure is capable of safely supporting a load equal to 175% of the hoist-rated load plus the weight of the hoist.

Lever-operated hoists are subject to shock loads and severe conditions. Although the load may not be overweight, the shock can cause the friction brakes to slip or freeze. For this reason, some lever-operated hoist manufacturers believe a slip clutch may be unreliable or unsafe and do not manufacture slip-clutch models.

Power-Operated Hoists

A *power-operated hoist* is a hoist operated by pneumatic or electric power and uses either chain or wire rope as the lifting component. Chain or wire rope power-operated hoists are selected based on the heaviest load to be lifted. **See Figure 7-11.**

Pneumatic Hoists. A *pneumatic hoist* is a power-operated hoist operated by a geared reduction air motor. Pneumatic hoists have distinct advantages over electric hoists. An advantage is that there is no explosion hazard in areas such as paint spraying or petrochemical facilities because compressed air does not produce electric arcing.

A requirement for heavy-duty applications is the ability for unlimited start/stop situations. Air motors are inherently self-cooling and able to start, stop, and reverse without overheating and are ideal for high ambient temperature applications. *Ambient temperature* is the temperature of the air surrounding a piece of equipment.

Tech Tip

Standard cubic feet per minute (SCFM) is the flow rate of a gas at "standard" conditions of temperature, pressure, and relative humidity. However, when using pneumatic hoists, the conditions may vary between definitions of SCFM, so the specification of the pneumatic hoist and air compressor should always be checked.

Pneumatic controls have the advantage of variable speed to lift a load from slow to full speed while allowing smooth movements. A disadvantage of pneumatic operation is the chance that the operation will consume a greater amount of air than the available air supply. Because air compressor output is rated in horsepower, compressor horsepower must be converted to hoist standard cubic feet per minute (scfm) required by the hoist when selecting the required capacity compressor for a given hoist. **See Figure 7-12.**

Power-Operated Hoists

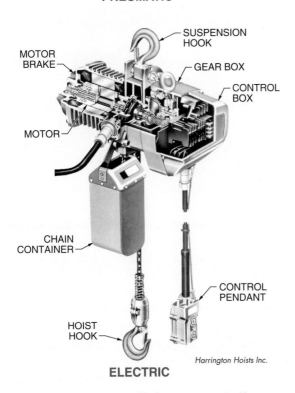

SELF-ADJUSTING DISC BRAKE

SUSPENSION HOOK

INTERNAL MUFFLER

INLET SWIVEL, STRAINER, AND FILTER

UPPER AND LOWER LIMIT STOPS

AIR MOTOR

CHAIN CONTAINER

HOIST HOOK

PILOT PENDANT CONTROL

Ingersoll-Rand Material Handling

PNEUMATIC

SUSPENSION HOOK

MOTOR BRAKE

GEAR BOX

CONTROL BOX

MOTOR

CHAIN CONTAINER

CONTROL PENDANT

HOIST HOOK

Harrington Hoists Inc.

ELECTRIC

Figure 7-11. Power-operated hoists are operated by pneumatic or electric power.

Airflow Requirements

Pipe Size*	SCFM	Min Comp HP	Lift Speed†	Capacity‡
⅜	48	12	50	½
¾	75	19	30	2
1	260	65	20	8
1¾	420	105	8	15

* in in.
† in fpm based on 90 psi line pressure with full speed at throttle
‡ 2000 lb ton

Figure 7-12. Hoist standard cubic feet per minute (scfm) must be determined to select the required capacity compressor for a given hoist.

The minimum compressor size for a hoist is found by applying the following formula:

$$HP = \frac{scfm}{4}$$

where

HP = horsepower

$scfm$ = standard cubic feet per minute

4 = constant

For example, what is the minimum compressor size required for a pneumatic hoist that requires 75 scfm?

$$HP = \frac{scfm}{4}$$

$$HP = \frac{75}{4}$$

$$HP = \textbf{18.75 HP}$$

When equipment such as a large hoist, pneumatic motor, or air drill is to be purchased to replace electrical units, the pneumatic units require a certain scfm for operation. To maintain a sufficient supply of compressed air, the scfm of new and existing equipment must not be higher than the present compressor system output. The scfm output of a compressor is found by applying the following formula:

$$scfm = 4 \times HP$$

For example, what is the scfm of a compressor rated at 15 HP?

$$scfm = 4 \times HP$$

$$scfm = 4 \times 15$$

$$scfm = \textbf{60 scfm}$$

Electric Hoists. Common industrial electric hoist capacities range from ¼ t to 20 t. Depending on the capacity of the hoists, the power supply range is 110 VAC, 208 VAC, 230 VAC, 460 VAC, or 575 VAC (volts alternating current). The operator's controls are low-voltage (normally 24 V) to provide electrical shock protection. Many electric hoists are equipped with mechanical and electrical safety overload protectors. Mechanical safety overload protectors (slip clutches) refuse to lift (by means of a slipping action) when a load is applied beyond their capacity. Electrical safety overload protectors are brake mechanisms activated when power is removed or lost. Hoist operating controls require pushbutton or lever action and an upper limit switch.

Lift-All Company, Inc.

Electric hoists can be used to move extremely heavy loads such as locomotives.

A pendant is used by the hoist operator to control load movement from the floor or other level beneath the crane. A *pendant* is a pushbutton or lever control suspended from a crane or hoisting apparatus.

A *limit switch* is a device that cuts off the power automatically at or near the upper limit of hoist travel. All pneumatic and electric hoists are equipped with an upper limit switch to prevent damage from overwrap. Limit switches are safety switches and are not to be used as stop switches. **See Figure 7-13.**

Lifting must be done under the control of an operator at all times. Sudden or jerky movements, rapid dropping, or high speeds in short distances can damage the load or lifting equipment. The movement of a load must always be in view of the hoist operator.

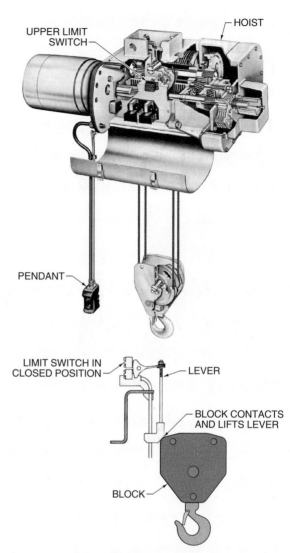

Lift Tech International, Division of Columbus McKinnon

Figure 7-13. Limit switches are devices that cut off the power automatically at or near the upper limit of hoist travel.

Drum Wrap. *Drum wrap* is the rope length required to make one complete turn around the drum of a hoist or crane. An overwrap condition occurs when the drum wraps enough rope or wire so that the load block comes in contact with the hoist or crane. If the load block comes in contact with the hoist or crane, ropes can be severely stretched or broken, causing the load to drop.

The proper direction for winding the first layer on a drum is determined by the lay of the rope. The rope lay determines if the rope is wound over or under on the spool. The rope should be overwound from left to right if the rope is anchored on the left and the rope is a right lay rope. **See Figure 7-14.**

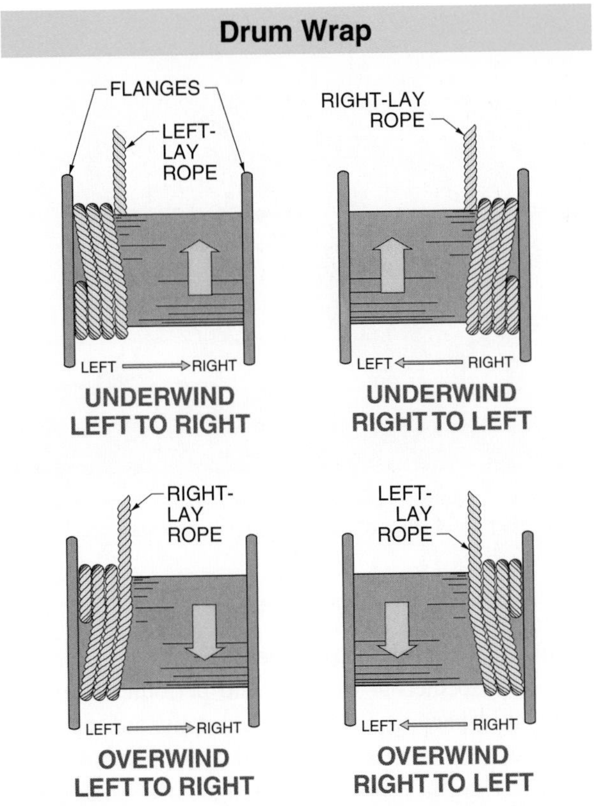

Drum Wrap

FLANGES
LEFT-LAY ROPE

RIGHT-LAY ROPE

LEFT ⟶ RIGHT
UNDERWIND LEFT TO RIGHT

LEFT ⟵ RIGHT
UNDERWIND RIGHT TO LEFT

RIGHT-LAY ROPE

LEFT-LAY ROPE

LEFT ⟶ RIGHT
OVERWIND LEFT TO RIGHT

LEFT ⟵ RIGHT
OVERWIND RIGHT TO LEFT

Figure 7-14. The pattern in which rope should be wound onto a drum depends on the rope lay and direction.

The drum must be observed to ensure a two-wrap minimum if the system is not equipped with a low-limit switch. The two-wrap minimum ensures the strength of a two-reeve pull against the rope and drum attachment. The attachment would not hold a load if all the rope is unwound from the drum. A drum that is rewound must be rewound in the same direction and the rope seated properly in the drum grooves if the drum is equipped with grooves. Incorrect winding damages the rope.

Care must be exercised when transferring rope from a reel to a hoist drum. A *reel* is a wooden assembly on which wire rope is wound for shipping and storage. During rope transfer, the unreeling process should be straight and under tension. A light squeezing pressure well away from the drum is used when guiding the wire rope onto the drum. Gloves should always be worn when handling wire rope. Wire rope should never be handled with bare hands.

While grooved drums offer few winding problems, smooth-face drums can be more difficult to wind and the proper procedure must be followed. The first layer

of rope should be wound with sufficient tension to ensure a close helix with each wrap being wound as close as possible to the preceding wrap. The first layer acts as a helical groove, which guides the successive layers. For this reason, the first layer should not be unwound on a smooth drum. As the rope is forced up to the second layer at the flange, a reverse helix is created, causing the rope to crossover. A *crossover* is one wrap winding on top of the preceding wrap. The crossover is the point at which the rope winds back over two rope grooves to advance. **See Figure 7-15.**

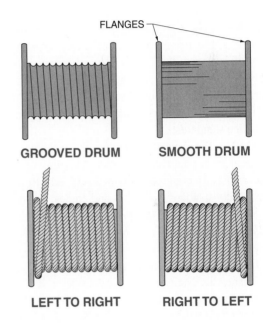

FLANGES

GROOVED DRUM   **SMOOTH DRUM**

LEFT TO RIGHT   **RIGHT TO LEFT**

Figure 7-15. A crossover is one wrap winding on top of the preceding wrap.

HOIST SAFETY

Overhead hoists and cranes are regulated by a large number of standards. A *standard* is a guideline adopted by regulating authorities. Regulating authorities that develop standards include OSHA, ANSI, ISO, CMAA, ASME, and NFPA. Standards organizations pertinent to lifting loads include the following:

- **OSHA** – The Occupational Safety and Health Administration specifies safety standards through the U.S. Department of Labor and the Occupational Safety and Health Act. OSHA is concerned with the development and enforcement of safety standards for industrial workers.

- **ANSI** – The American National Standards Institute is a standards-developing organization that adopts and copublishes standards that are written and approved by member organizations. ANSI branches out and connects its member organizations by unifying their adopted standards. ANSI manages United States participation in ISO standards activities. See ANSI A10.22-1990, *Rope-Guided and Nonguided Worker's Hoists – Safety Requirements.*
- **ISO** – The International Organization for Standardization is a nongovernmental international organization that is comprised of national standards institutions of over 90 countries (one per country). The ISO provides a worldwide forum for the standards developing process.
- **CMAA** – The Crane Manufacturers Association of America, Inc. is an organization of the leading crane manufacturers in the United States developed for the purpose of promoting standardization and providing a basis for proper equipment selection and use. CMAA is instrumental in establishing many crane-operating practice standards.
- **ASME** – The American Society of Mechanical Engineers helps establish safe structural design of hoists and cranes. In conjunction with ANSI, ASME develops safety standards for hoists and cranes. See ANSI/ASME B30.2-1990, *Overhead and Gantry Cranes (Top Running Bridge, Single or Multiple Girder, and Top Running Trolley Hoist).*
- **NFPA** – The National Fire Protection Association publishes the National Electrical Code® (NFPA 70), which contains standards for the practical safeguarding of persons and property from hazards arising from the use of electricity. Article 610 covers the installation of electrical equipment and wiring used in connection with cranes, monorail hoists, hoists, and all runways.

The two most prevalent standards in hoist safety and operating rules are that industrial hoists and cranes are not designed for and should not be used for lifting, supporting, or transporting humans, and loads or empty hook blocks should not be used over any individual, especially if the load is held magnetically or by vacuum.

Inspection Programs and Procedures

Verification that hoisting equipment meets current code requirements must be made if an inspection program is being developed and records of initial installation or subsequent modifications are not available. Verification includes a step-by-step procedure similar to that of inspecting and checking a new hoist.

Cables on locomotive cranes must be frequently inspected to ensure that there are no distorted, broken, or damaged cable strands.

Regularly Scheduled Inspections. Existing equipment must be inspected at frequent and periodic intervals. Frequent inspections are conducted daily and monthly. Daily inspections are accomplished with the assistance of a checklist and may not require the signature of the person making the inspection. **See Figure 7-16.** These checks are mostly visual and include identifying unusual sounds or temperatures that may indicate problems.

The exception to this procedure is the brake mechanism. The brake mechanism should be checked frequently by operating the hoist with and without a load and by testing its holding power at various levels. Any repairs or major adjustments performed as a result of the inspection must be recorded on a written report. The report should identify the hoist serviced and indicate work performed, the date, reason, individual performing the inspection, and the parts replaced.

Monthly checks are generally concerned with load-bearing components, such as the condition of hooks, wire rope, chain, and nut and bolt tightness. Periodic or semiannual inspections are conducted by specifically-designated personnel. Written reports of monthly and periodic inspections must be signed, dated, and placed in the equipment identification file. In many cases, the manufacturer of the equipment provides a checklist. This checklist can be used or it can serve as a guide in preparing a checklist better suited to a particular situation. In addition to inspection of hooks, wire rope, and chain, basic periodic or semiannual inspection checklists must include braking systems and limit switches.

Electric Hoist Checklist				
Item	Daily	Monthly	Semi-annually	Deficiencies
All functional operating mechanisms	✓	✓	✓	Maladjustment interfering with proper operation, excessive component wear
Controls	✓		✓	Improper operation
Safety Devices	✓		✓	Malfunction
Hooks	✓	✓	✓	Deformation, chemical damage, 15% in excess of normal throat opening, 10% twist from plane of unbent hook, cracks
Load-bearing components (except rope or chain)	✓	✓	✓	Damage (especially if hook is twisted or pulling open)
Load-bearing rope	✓	✓	✓	Wear, twist, distortion, improper dead-ending, deposits of foreign material
Load-bearing chain	✓	✓	✓	Wear, twist, distortion, improper dead-ending, deposits of foreign material
Fasteners	✓	✓	✓	Not tight
Drums, pulleys, sprockets			✓	Cracks, excessive wear
Pins, bearings, shafts, gears, rollers, locking and clamping devices			✓	Cracks, excessive wear, distortion, corrosion
Brakes	✓		✓	Excessive wear, drift
Electrical			✓	Pitting, loose wires
Contactors, limit switches, pushbutton stations			✓	Deterioration, contact wear, loose wires
Hook retaining members (collars, nuts) and pins, welds, or rivets securing them			✓	Not tight or secure
Supporting structure or trolley			✓	Continued ability to support imposed loads
Warning label	✓		✓	Removed or illegible
Pushbutton markings	✓		✓	Removed or illegible
Capacity marking	✓		✓	Removed or illegible

Figure 7-16. Lifting equipment must be inspected at frequent and periodic intervals.

Hook, wire rope, and chain inspections for hoists are identical to inspection for rigging components. Hook inspection is concerned with the throat opening, load-bearing thickness, tip twisting, chemical damage, and latch damage. Wire rope inspection covers reduction in rope diameter, broken wires, worn outside wires, distortion damage, weld spatter, and corroded rope. Examination for elongation, gouges, bending, and worn chain is included in chain inspection. Chain assemblies should not be stored where they may be subject to damage or where exposed to corrosive action.

Hoist Motor Brake Inspection

Braking systems must be inspected before each shift change or prior to use after periods of nonuse. Daily hoist brake inspections are concerned with braking integrity (hook drift). *Hook drift* is the slippage of a hook caused by insufficient braking.

To inspect for hook drift, the hoist is operated in the lifting and lowering direction without load on the hook. The hook is stopped to check operation of the hoist braking system. The drift of the hook should not exceed 1″ in either direction. The motor brake normally requires adjustment or lining replacement if hook drift exceeds 1″.

Tech Tip

For maximum safety for personnel, hoists and cranes must never be used to lift, support, or transport personnel unless using crane attachments specifically designed for this purpose. In addition, loads must never be suspended above the ground for any time period longer than required to transport the load from one point to another.

Weston Brake

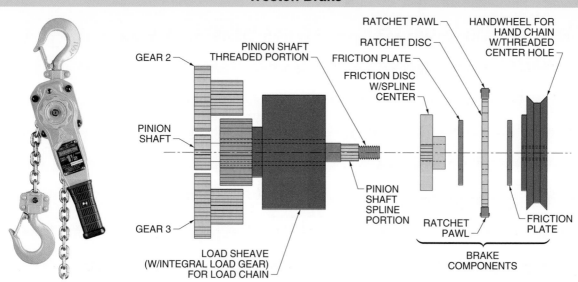

The mechanical brake employed in Harrington hoists is a Weston Brake. The Weston Brake was invented by Mr. Weston in the beginning of the 1900s and was introduced into the hoist industry around 1910. In one form or another, it is used in practically all hand hoists worldwide.

The Weston Brake is comprised of brake components, such as the friction disc with splined center hole, friction plates, ratchet disc, ratchet pawls, and handwheel with threaded center hole.

When the hand chain is used to rotate the handwheel in the forward (hoisting) direction, the threaded center portion of the handwheel screws the handwheel tighter onto the threaded portion of the pinion shaft. This squeezes the handwheel, the two friction plates, and the ratchet disc up against the friction disc. The friction disc, which is mated to the splined portion of the pinion shaft, cannot move along the pinion shaft because of the shoulder at the end of the splined portion of the pinion shaft. Therefore, the squeezing cinches the brake components together and the rotational motion imparted by the hand chain is transmitted to the pinion shaft.

The pinion shaft, which runs through a hole in the center of the load sheave, engages gear 2 and gear 3. When the pinion shaft rotates, it transmits its rotation to these two gears, which in turn are engaged to the geared portion of the load sheave. The rotation of gear 2 and gear 3 is transmitted to the load sheave and the load is lifted.

When the pulling on the hand chain ceases, as would happen between pulling strokes or when the load has reached its intended position, gravity acting on the load tends to cause the load sheave to rotate in the backward (lowering) direction. This rotational torque is transmitted through the gears to the pinion shaft and keeps the brake components cinched tightly together. With the pinion shaft and brake components cinched tightly together as a single body, the entire assembly attempts to rotate in the backward (lowering) direction. The ratchet pawls engage the ratchet disk and prevent this. Thus, the brake stops the load from lowering.

When the hand chain is used to rotate the handwheel in the backward (lowering) direction, the threaded center portion of the handwheel begins to back the handwheel off the threaded portion of the pinion shaft, which decompresses the brake components. This allows the pinion shaft to rotate in the backward (lowering) direction, which it begins to do by virtue of the load itself. As the load begins to fall and causes the pinion shaft to rotate in the backward direction, the threaded portion of the pinion shaft causes the handwheel and the other brake components to cinch up tight again, and the lowering of the load stops. The lowering of the load ceases until the handwheel is again rotated in the lowering direction. In this way, the lowering of the load is actually accomplished by a series of very small controlled falls that are perceived as a smooth motion.

Harrington Hoists Inc.

A mechanical hoist brake can contain a Weston load brake using two friction plates with four braking surfaces for positive brake action.

Typically, motor brakes are direct-acting (ON or OFF) disc-type mechanisms controlled by a rectified DC current. The air gap is set at a specific spacing (typically 0.03″). The air gap increases as the brake lining wears. Significant brake lining wear causes hook drift. In many cases, a limit switch is added to the assembly at the air gap location. The limit switch does not close and the hoist motor does not start when the gap reaches a worn limit (0.090″ in many cases). **See Figure 7-17.**

Power should be disconnected and locked out prior to adjusting the air gap. A brush or compressed air is used to remove accumulated brake lining dust. Specific attention should be directed to removing dust from the air gap between the coil and plates. Solenoid burning may occur if the air gap is not equal at all three adjusting points. Proper eye and breathing equipment must be used when using compressed air to clean the air gap. Hook drift occurs when adjustment springs are weak or broken or when overheating has warped the brake lining or compression plates.

Hoist Motor Brake Wiring Diagram

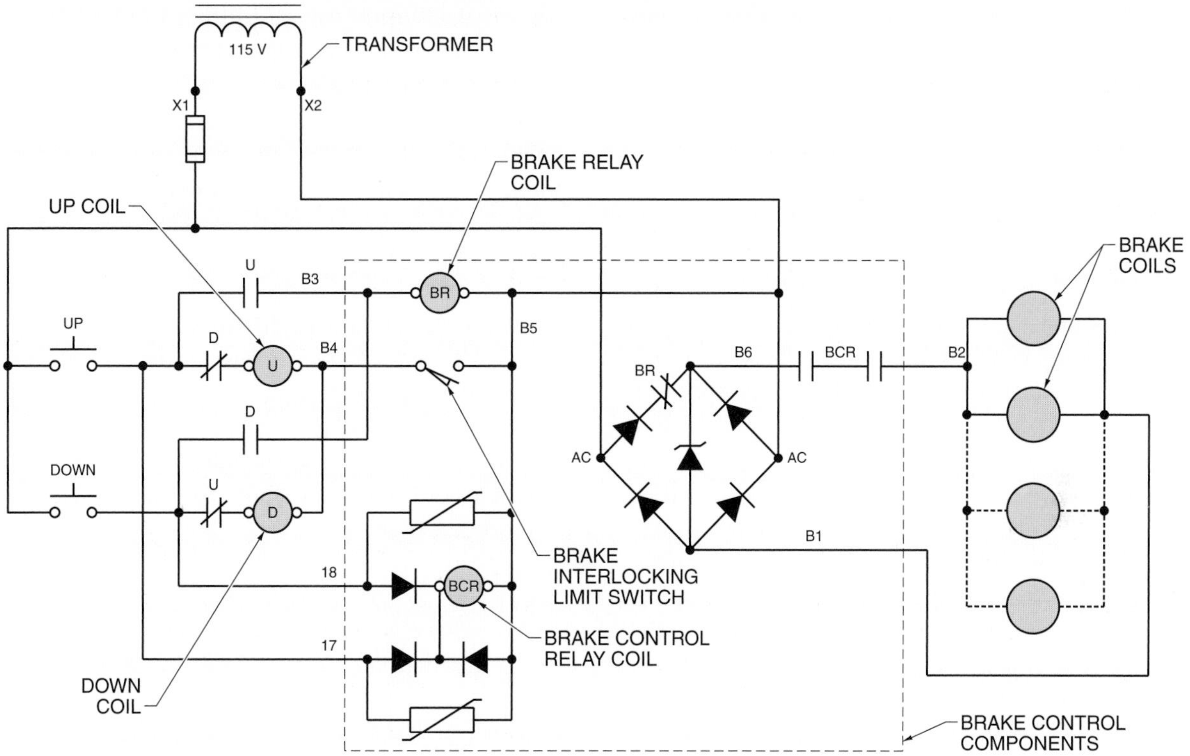

Figure 7-17. Most of the hoist motor control circuit is involved in the control of the brake mechanism. A limit switch prevents the motor from operating if activated, which occurs if the brake lining gap is excessive or misaligned.

Brake linings are inspected for warpage by laying them on a clean, flat, level surface. A straightedge is laid across the center and gaps are looked for as the straightedge is rotated. Typically, gaps of 1/32″ or more require that the brake linings be replaced. **See Figure 7-18.**

Conducting Load Tests

Warning signs and barriers must be used on the floor beneath a hoist, crane, or lifting system. A mechanical load brake test checks the hoist braking system for proper operation. All personnel should be alerted that a free-fall condition could exist during a mechanical load brake test. A rated capacity load is attached to the hoist hook. Before lifting the load, any slack is slowly removed from the line. The load is raised a few inches and stopped. If the load stops and the brakes hold, the load is raised and lowered several feet, while the hoist is stopped several times in each direction to check the brakes. Next, the hoist is checked with a load equal to 125% of its rated load capacity.

This check tests any load-limiting devices. Load-limiting devices for power-operated hoists are normally rated at 110% of the hoist's rated load capacity.

Brake Lining Inspection

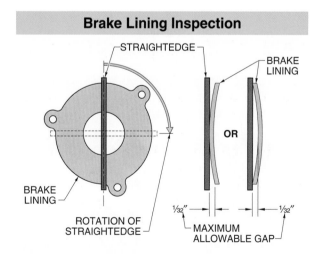

Figure 7-18. A straightedge is placed across the brake linings to check for warpage.

Hoist limit switches and other safety devices should be checked daily or at the start of each shift when the hoist is in operation. Hoists should never be operated without the protection of a properly functioning limit switch. Limit switches should be checked without a load on the hook and in the low speed on multiple-speed hoists. Hoists with one-speed operation should be inched into the limit switch. *Inching* is slow movement in small degrees.

Hoist limit switches are checked by activating the hoist to run in the lifting direction. The hoist is inched up near its upper limit of travel until the limit switch arm or weight is lifted. This stops the load block. A continuity tester can be used to indicate open or closed circuits without the need for electrical power if the limit switch appears to be faulty.

After the switch has been checked or corrected, all guards and safety devices are installed and reactivated. Finally, warning or out-of-order signs and maintenance equipment are removed. Following each inspection, a maintenance report form is used to report results of all approved tests and checks.

EYEBOLTS

An *eyebolt* is a bolt with a looped head. Eyebolts, like lifting lugs, are used primarily as lifting tools. The two basic types of eyebolts are formed steel and forged steel. Formed steel eyebolts are not strong enough for heavy weights and should not be used in lifting applications. Only forged steel eyebolts should be used for lifting applications. Three common types of forged eyebolts are machinery, regular nut, and shoulder nut eyebolts. **See Figure 7-19. See Appendix.**

The regular nut eyebolt and the shoulder nut eyebolt (where the shoulder is not used) should never be used for angular lifting. Angular lifting force should only be applied to an eyebolt that is firmly supported by a shoulder. **See Figure 7-20.**

Attaching eyebolts is accomplished by screwing the eyebolt into a threaded hole or inserting an eyebolt shank through a hole and securing it with a nut. The machinery eyebolt is the most commonly used eyebolt. Machinery eyebolts must be screwed in until the shoulder is tight against the load.

Always inspect an eyebolt before use and never use one that shows signs of wear, fatigue, or damage or one that is bent or elongated. Inspect for clean threads and sharp shoulder corners. Washers must be used to create a tight shoulder-to-load fit if the shoulder is not sharp and the load has not been countersunk. An eyebolt shank must never be undercut.

Tech Tip

Always use shoulder nut eyebolts for angular lifting, and never use an eyebolt that shows signs of wear or damage.

After a machinery eyebolt has been installed and tightened, the eye of the eyebolt must be perpendicular (at 90°) to the sling line for any degree of side pull. Adjusting the pivot of the eyebolt for proper alignment is accomplished by placing a shim under the shoulder. The shim thickness determines the angle of unthread rotation and varies according to the eyebolt size. *Unthread rotation* is counterclockwise rotation of an eyebolt having right-handed threads or the clockwise rotation of an eyebolt having left-handed threads. Unthread rotation can be anywhere from 5° to 90° to ensure a right angle to the sling line.

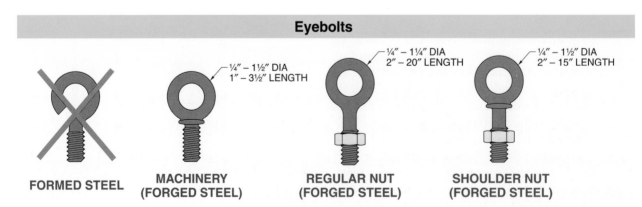

Eyebolts

¼" – 1½" DIA
1" – 3½" LENGTH

¼" – 1¼" DIA
2" – 20" LENGTH

¼" – 1½" DIA
2" – 15" LENGTH

FORMED STEEL

MACHINERY
(FORGED STEEL)

REGULAR NUT
(FORGED STEEL)

SHOULDER NUT
(FORGED STEEL)

Figure 7-19. An eyebolt is a bolt with a looped head.

Angular Lifting

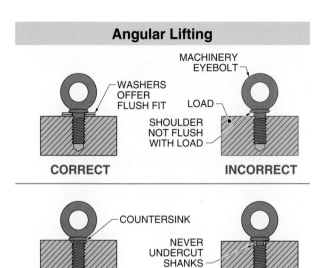

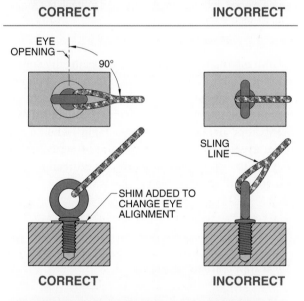

| Shim Thickness For 90° Rotation ||
Eyebolt Size*	Shim Thickness*
¼	0.0125
5⁄16	0.0139
3⁄8	0.0156
½	0.0198
5⁄8	0.0227
¾	0.0250
7⁄8	0.0278
1	0.0312
1¼	0.0357
1½	0.0417

* in in.

Figure 7-20. Angular lifting force should only be applied to an eyebolt that is firmly supported by a shoulder.

As a sling moves from a vertical to an angular position (decreased sling angle), the capacity of the eyebolt is reduced. Each degree of change requires a reevaluation of the eyebolt's safe limits. The safe limit is increased if the sling angle is increased. The safe limit is reduced if the sling angle is reduced. **See Figure 7-21.**

Eyebolt Load Capacities

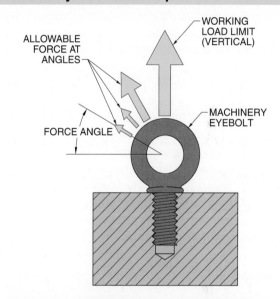

| Sling Eyebolt Capacity Loss ||
Force Angle*	Allowable Force as Percentage of Working Load Limit (Vertical)
90	100%
60 to 89	65%
45 to 59	30%
30 to 44	25%

* in degrees

Figure 7-21. Eyebolts have the greatest load capacity when used with vertical forces. The capacity at other angles is a percentage of the vertical working load limit.

For example, what is the percent change in eyebolt capacity of a sling that has been relocated from 55° to 65°?

A 55° angle equals a 30% reduction and a 65° angle equals a 70% reduction in eyebolt capacity (from Sling Eyebolt Capacity Loss table). The percent change is 40% (70% − 30% = 40%). The eyebolt capacity is reduced an additional 40% with a total capacity loss of 70%.

Nut eyebolts may be regular or shoulder. A regular nut eyebolt is normally screwed into the load and held firm using a secured hex nut. **See Figure 7-22.** A tapped hole in the load should be at a minimum depth of 2½ times the eyebolt diameter. The eyebolt should be screwed into the load at a minimum depth of 2 times the eyebolt diameter. Two nuts (one on either side of the load) must be used if the load thickness is less than one eyebolt diameter of threads. One nut is required when the load thickness is greater than one eyebolt diameter of threads. A regular nut eyebolt may also be slid through an unthreaded hole in a load and held in place by two nuts firmly secured against each other on the bottom of the load with a third nut secured on top of the load.

When a shoulder nut eyebolt is used, the threaded portion of the shank must protrude through the load sufficiently to allow full engagement of the nut. Washers must be used to take up the excess between the nut and the load if the unthreaded portion of the shoulder nut eyebolt protrudes so far that the nut cannot be tightened securely against the load. The washers must exceed the distance between the bottom of the load and the last thread of the eyebolt. The shoulder of the eyebolt must be tightly secured against the load surface. **See Figure 7-23.**

Eyebolt Loads

Eyebolt load ratings are affected by the angle of pull on the eyebolt. A straight vertical pull of a shoulder nut eyebolt offers 100% of the eyebolt load rating. An angular pull of 45° reduces the eyebolt load rating by as much as 65%.

Shoulder Nut Eyebolts

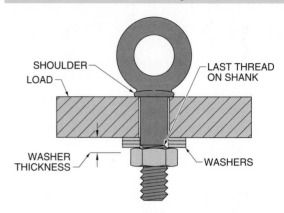

Figure 7-23. Shoulder nut eyebolts have a shoulder that can be placed tight up against the load surface but require a nut to secure the opposite end.

Until international standards are specifically defined and adhered to by all manufacturers, actual eyebolt load ratings will vary significantly between manufacturers. Specific load rating information must be obtained from distributors' or manufacturers' catalogs when a particular eyebolt is to be used. General lifting capacities of eyebolts may be used as a primary estimation when designing lifting assemblies.

Tech Tip

Always inspect eyebolts before use and never use an eyebolt if its eye is bent or elongated.

Regular Nut Eyebolts

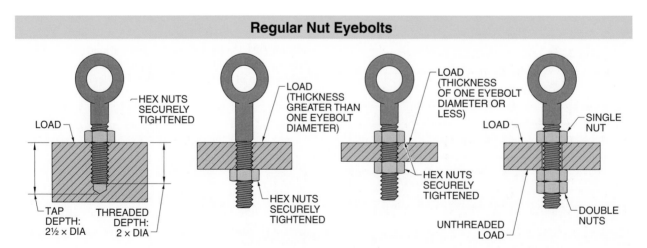

Figure 7-22. A regular nut eyebolt is normally screwed into the load and held firm using a secured hex nut.

Typical eyebolt angular lift capacity is calculated using a constant of 0.21 for sling angles of less than 45° and 0.25 for sling angles greater than 45°. The constant is multiplied by the eyebolt manufacturer's working load limit for vertical lifts. **See Figure 7-24.** This figure is divided by the sling angle loss factor to determine the working load capacity. **See Appendix.** Working load capacity is found by applying the following formula:

$$L = \frac{c \times wl}{s}$$

where

L = working load capacity (in lb)

c = constant (0.21 for sling angles less than 45°; 0.25 for sling angles greater than 45°)

wl = eyebolt working load limit (in lb)

s = sling angle loss factor (from Sling Angle Loss Factors Table)

For example, what is the working load capacity of a 50° bridle sling using a ¾″ shoulder nut eyebolt?

$$L = \frac{c \times wl}{s}$$

$$L = \frac{0.25 \times 5200}{0.766}$$

$$L = \frac{1300}{0.766}$$

$$L = \textbf{1697.128 lb}$$

A simplified method of adjusting the working load limit for eyebolts used with an angle sling is to make a 30% or 25% adjustment to loads over or under 45°. Working loads with a sling angle over 45° are adjusted by 25%. Working loads with a sling angle under 45° are adjusted by 30%. For example, a ¾″ eyebolt having a sling angle of 50° is multiplied by 0.25 (25%), giving the adjusted working load limit of 1300 lb (5200 × 0.25 = 1300 lb).

Force is applied slowly after slings have been attached to the eyebolts. The load is watched carefully and the operator must be prepared to stop if the load shows signs of shifting or buckling. Buckling occurs if a load is not stiff enough to resist the force of angular loading. Ensure a bridle sling combination is used when using two-leg slings. Never reeve slings from one eyebolt to another. Reeving slings alters the load strength of sling materials. **See Figure 7-25.**

Shoulder Nut Eyebolt Working Load Limits—Vertical Lifts	
Size*	Working Load Limit†
¼	500
⁵⁄₁₆	800
⅜	1200
½	2200
⅝	3500
¾	5200
⅞	7200
1	10,000
1¼	15,200
1½	21,400

* in in.
† in lb

Figure 7-24. Eyebolt working load limits are given by the manufacturer and are based on the size of the eyebolt.

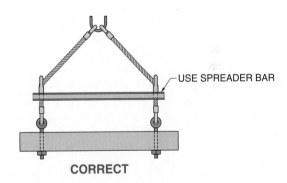

USE SPREADER BAR

CORRECT

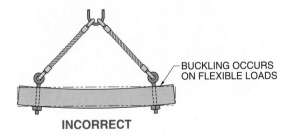

BUCKLING OCCURS ON FLEXIBLE LOADS

INCORRECT

Figure 7-25. Buckling occurs if a load is not stiff enough to resist the force of angular loading.

INDUSTRIAL CRANES

An *industrial crane* is a crane with structural beam supports for lifting equipment. Cranes are instrumental in moving parts or equipment from one location to another. Industrial cranes are classified depending on their service. Industrial cranes are classified as gantry cranes, jib cranes, overhead cranes, telescopic-boom cranes, and lattice-boom cranes.

Gantry Cranes

A *gantry crane* is a crane with bridge beams supported on legs. **See Figure 7-26.** The legs are supported by end trucks that normally travel on floor rails. Floor rails are small-gauge railroad rails, which are recessed into the floor or set directly on top of the floor surface.

An *end truck* is a roller assembly consisting of a frame, wheels, and bearings generally installed or removed as complete units. Some gantry crane legs are supported by wheels for movement across flat floor surfaces.

Gantry cranes may be single-leg or double-leg gantry cranes. A single-leg gantry crane is normally a supplemental crane for a large capacity overhead crane. In this application, one side of the bridge beam is attached to a leg which rolls on a floor rail and the other side is attached to an overhead crane rail.

Double-leg gantry cranes are used on two parallel floor rails. In some cases, the double legs are on wheels, which allows the gantry crane to operate across any flat floor surface.

Jib Cranes

A *jib crane* is a crane that is mounted on a single structural leg. The three basic types of jib cranes are wall-mounted, base-mounted, and mast. **See Figure 7-27.** Wall-mounted jib cranes are top-braced or cantilevered and may have a stationary or a partial rotating boom. A *cantilever* is a projecting beam (boom) or member supported at only one end.

Base-mounted jib cranes are free-standing cranes on a heavily supported base mounting. The boom of a base-mounted jib crane may be stationary or be capable of 360° rotation.

Mast jib cranes have one structural leg (mast) mounted to the floor and/or ceiling. Mast jib cranes are cantilevered, underbraced, or top-braced. Mast jib cranes may be stationary or installed with upper and lower bearings for partial or 360° rotation.

The booms on all three types of cranes serve as tracks for hoisting mechanisms. Prior to lifting a jib crane load, slack must be taken up slowly. Removing slack minimizes shock to the crane boom. Rotating the boom on a jib crane should be done slowly and easily to prevent damage to the load, surroundings, or individuals.

Gantry Cranes

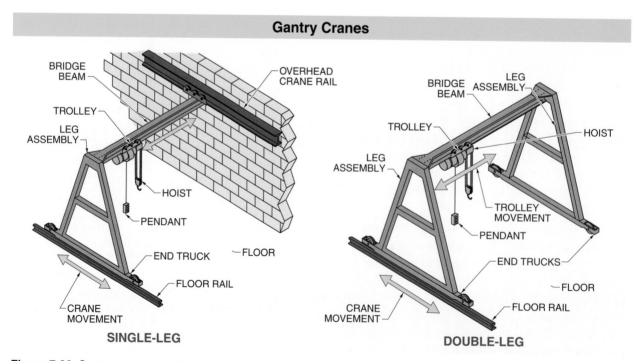

SINGLE-LEG

DOUBLE-LEG

Figure 7-26. Gantry cranes consist of a hoist on a horizontal beam, which is supported by legs that roll along the floor.

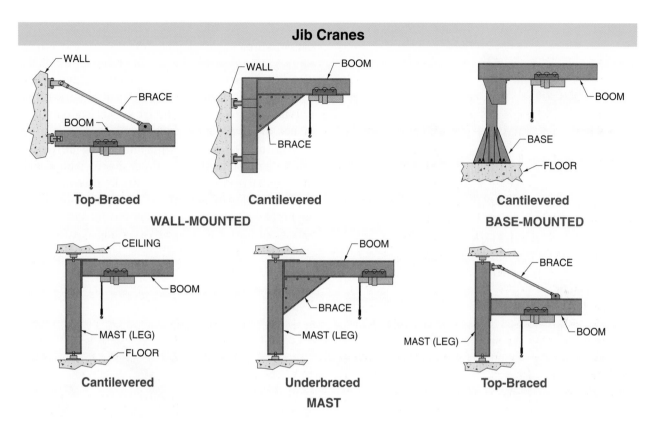

Figure 7-27. Jib-crane booms are supported at only one end but can be configured in several different designs.

Baldor Electric Company

Gantry cranes are used for lifting large loads.

Overhead Cranes

An *overhead crane* is a crane that is mounted between overhead runways. Three common overhead crane configurations are the top-running crane with top-running hoist, the top-running crane with underhung hoist, and the underhung crane with underhung hoist. **See Figure 7-28.** Overhead cranes are normally operated from the cab or pendant pushbutton station. The *cab* is a compartment or platform attached to a crane in which an operator may ride.

Top-running cranes with top-running hoists are the most common overhead crane configuration. A *bridge girder* is the principal horizontal beam that supports the hoist trolley and is supported by the end trucks. A *hoist trolley* is the unit carrying the hoisting mechanism that travels on the bridge girder. End trucks are units consisting of truck frame, wheels, bearings, axles, etc., which support the bridge girder(s).

Tech Tip

While many overhead cranes are operated by an operator from a remote location, in large industrial facilities, such as steel mills, the crane operator controls the crane's lifting action and horizontal movement from an operator's cab that is attached to the crane's bridge girder.

Overhead Cranes

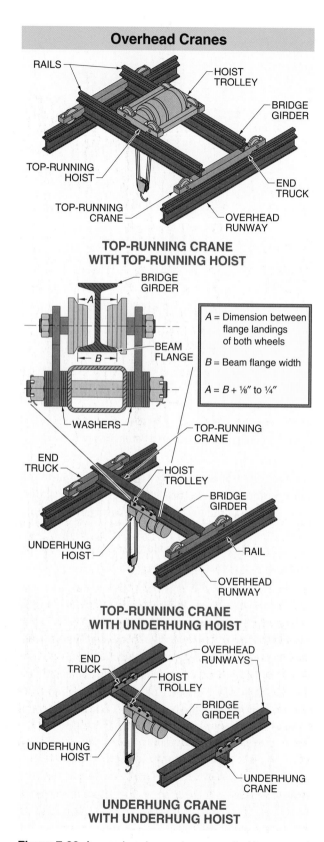

**TOP-RUNNING CRANE
WITH TOP-RUNNING HOIST**

A = Dimension between flange landings of both wheels

B = Beam flange width

A = B + ⅛″ to ¼″

**TOP-RUNNING CRANE
WITH UNDERHUNG HOIST**

**UNDERHUNG CRANE
WITH UNDERHUNG HOIST**

Figure 7-28. An overhead crane is a crane that is mounted between overhead runways.

End trucks on top-running cranes travel on small gauge railroad rails mounted on top of the overhead runways. A *runway* is the rail and beam on which the crane operates. Either single- or double-bridge girders support the hoist trolley. The hoist trolley travels on rails mounted on top of the bridge girders in top-running cranes with top-running hoists.

Top-running cranes with underhung hoists are top-running single-bridge girder cranes with underhung hoist trolleys. The end trucks travel on rails mounted on top of the runways. A single-bridge girder supports the hoist trolley, which travels on the upper surface of the lower flange of the bridge girder.

Some types of mobile cranes have puncture-proof wheels to allow ease of movement over areas that could cause damage to ordinary wheels.

Underhung cranes with underhung hoists are another common crane configuration. This design has a single- or double-bridge girder supporting an underhung hoist trolley. Single- or double-bridge girders may support the hoist trolley, but the hoist trolley only travels on one bridge girder. The hoist trolley travels on the upper surface of the lower flange of the bridge girder.

Hoist trolleys should be inspected frequently and periodically. The inspection should include observation of the condition of the wheel bearings and checking for flat spots, cracks, or breaks on the trolley wheels and the trolley bumpers. Periodic inspections should include checking the dimension between wheel flanges.

On most hoist trolleys, the dimension between the flange landings of both wheels (A) must be at least ⅛″ greater than the beam flange width (B) and not more than ¼″. The addition or removal of washers may be necessary to obtain the proper dimension. Washers should be distributed equally between flanges to ensure that the hoist is centered under the bridge girder. Some hoist trolleys are not adjustable. The manufacturer's manual should be checked for proper use.

Telescopic-Boom Cranes

A *telescopic-boom crane* is a crane with an extendable boom composed of nested sections. The boom sections are extendable and retractable, allowing a wide range of boom lengths. **See Figure 7-29.** When fully retracted, the crane is easily transported on a street-legal vehicle. The boom movement and extension is typically powered by integrated hydraulic systems.

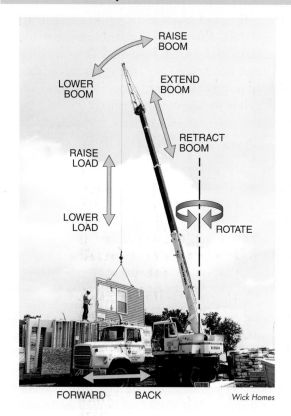

Telescopic-Boom Cranes

Wick Homes

Figure 7-29. Telescopic-boom cranes have the ability to extend or retract the boom, providing significant versatility to their reach.

Lattice-Boom Cranes

A *lattice-boom crane* is a crane with a boom constructed from a gridwork of steel reinforcing members. The lattice structure provides a very strong boom that is light for its size. **See Figure 7-30.** The boom may be composed of one or multiple sections that must be assembled onto the crane body when on site. This makes a lattice-boom crane less transportable between sites, though its vehicle platform allows it to move around while on site. For this reason, lattice-boom cranes are often deployed for long-term use at large construction sites.

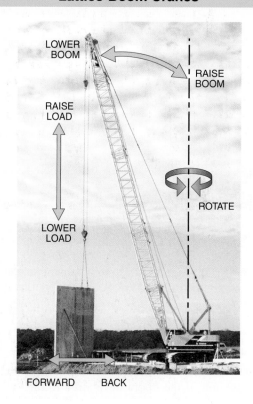

Lattice-Boom Cranes

Figure 7-30. The boom of a lattice-boom crane is constructed of a lightweight, open structure that allows these cranes to have particularly long booms.

Crane Lift Capacities

A mobile crane may be specified by its maximum lift capacity, such as a "90-ton crane." However, because of the necessary horizontal reach for most lifts, the actual lifting capacity is likely much less. The capacity of a crane decreases the farther it must reach from its base because the moment (tipping) force increases. **See Figure 7-31.** This distance is determined by the length and angle of the boom. The capacity at each point in the lift must be considered. For example, a certain load located 20′ from the crane must be moved to a new location 40′ away. Since the farther distance allows less lifting capacity, it determines the maximum allowable load weight for the entire lift.

Mobile Crane Lift Capacity

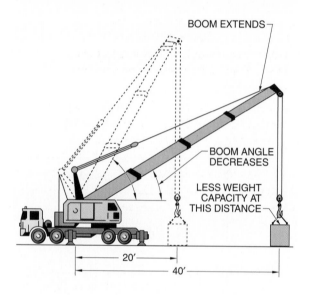

Figure 7-31. The lift capacity of a mobile crane is determined primarily by the horizontal distance of the load from the crane.

Other factors that affect the lift capacity are the amount of counterweight, the use of outriggers, and the direction of the boom in relation to the vehicle base (over the side or over the rear). Counterweights balance the moment (tipping) force caused by the load, so more counterweight increases the lifting capacity (to a point). Outriggers increase the effective footprint of the crane, improving stability. The direction of the boom also affects stability because cranes are typically not as wide as they are long. Charts of lift capacities that account for all of the applicable factors are available with the crane manufacturer's specifications and on cards located inside the operator's cab. **See Figure 7-32.**

Lift Capacity Charts

Boom Length (ft)	Radius (ft)	Boom Angle (°)	Maximum Allowable Loads (lb)		
			With Outriggers	Without Outriggers	
				Over Side	Over Rear
40	12	73	220,000	100,400	139,700
	15	68	185,000	71,800	97,400
	20	60	149,000	48,100	54,600
	25	51	118,000	35,800	47,900
	30	41	86,000	28,300	37,800
	35	29	67,000	23,200	31,100
60	15	76	180,000	71,000	97,000
	20	71	147,000	47,300	63,900
	25	65	118,500	35,000	47,200
	30	60	86,200	27,500	37,100
	35	54	67,400	22,400	30,300
			54,700		25,500

Figure 7-32. Lift capacity charts are used to determine the maximum allowable load weight for a particular crane with certain lift distances.

Many cranes include an alarm that sounds if the attempted lift weight exceeds the lift capacity for the current boom length and orientation. This safety feature must never be ignored. Overloading a crane can damage the boom or cause the entire crane to tip over, which can result in injuries and property damage.

Crane Operation

Each crane operator is held directly responsible for the safe operation of a crane. Lifting may be performed after ensuring that the condition of all rigging, hoisting, and crane components are within specifications.

Once the hoist is brought directly over the load's center of gravity, the lines or chains are checked to ensure they are not twisted, overwrapped, or unseated. The rigging is connected to the block hook, ensuring that the hook latch is closed and the lifting devices are fully seated in the hook bowl. The hook should be started upward slowly until all slack has been removed from the slings when slings are used to lift a load. The load is lifted slowly until it is clearly off the floor and properly balanced. At this time, the operator may increase the lifting speed. A crane should not be used for a side pull. **See Figure 7-33.** The crane is relocated if the load does not appear to be directly under the crane after the slack is removed. A swinging load is dangerous.

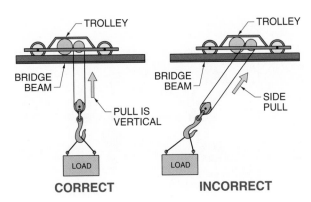

Figure 7-33. Loads must be lifted only from directly above. Side pulling is not permitted.

Cranes should be moved smoothly and gradually to avoid abrupt, jerky movements. When loads are lowered, the lowering speed should be decreased gradually. A load that is beyond the load capacity of any of the rigging, lifting, or crane components should not be lifted. Hoist brakes should be tested with the load a few inches off the floor. This is especially necessary if the load is near or at capacity. The load is checked for drift in the raise and lower positions. Drift is the slippage of a load caused by insufficient braking. The load is returned to the floor and the supervisor is notified if a drift of 1″ or more is noticed. A suspended load must not be left unattended.

All crane controls are placed in the OFF position if power is interrupted during operation. After power is restored, all controls are checked for satisfactory operation and correct direction before use.

To be safe and efficient, crane operation requires skill, the exercise of extreme care and good judgment, alertness, and concentration. Crane operators must adhere to proven safety rules and practices as outlined in applicable and current ANSI and OSHA standards.

Hand Signals. Many cab-operated cranes require the assistance of an additional person. The assistant, working in conjunction with the cab operator, gives crane and hoist communication in the form of standard hand signals. **See Figure 7-34.**

Loads should not be moved unless standard crane signals are clearly given, seen, and understood. The operator must pay particular attention to the required moves signaled by the assistant. The operator takes signals only from the assistant. The only exception to this rule is that the operator must obey a stop signal, at all times, no matter who gives it. A stop signal may be given by anyone.

General Practice Conditions. Generally, no person should be permitted to operate a crane who cannot speak the appropriate language, read and understand the printed instructions, or who is not of legal age to operate the equipment. Anyone who is hearing or eyesight impaired or may be suffering from heart or other ailments which might interfere with safe performance should not operate a crane. The operator must have carefully read and studied the operation manual and have been properly instructed.

With the mainline switch open (power OFF), the crane operator should operate each master switch or pushbutton in both directions to get the "feel" of each device and to determine that they do not bind or stick in any position. The operator should report the condition to the proper supervisor immediately if any switch or pushbutton binds or sticks in any position.

North American Industries, Inc.

Overhead bridge cranes are available with variable-speed control for precise increases and decreases in travel speed to prevent swinging loads caused by ON/OFF controls.

Hand Signals for Crane Operators

STOP	EMERGENCY STOP	HOIST	RAISE BOOM

With arm extended horizontally to the side, palm down, arm is swung back and forth.	With both arms extended horizontally to the side, palms down, arms are swung back and forth.	With upper arm extended to the side, forearm and index finger pointing straight up, hand and finger make small circles.	With arm extended horizontally to the side, thumb points up with other fingers closed.

SWING	RETRACT TELESCOPING BOOM	RAISE THE BOOM AND LOWER THE LOAD	DOG EVERYTHING	LOWER
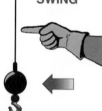				
With arm extended horizontally, index finger points in direction that boom is to swing.	With hands to the front at waist level, thumbs point at each other with other fingers closed.	With arm extended horizontally to the side and thumb pointing up, fingers open and close while load movement is desired.	Hands held together at waist level.	With arm and index finger pointing down, hand and finger make small circles.

LOWER BOOM	EXTEND TELESCOPING BOOM	TRAVEL/TOWER TRAVEL	LOWER THE BOOM AND RAISE THE LOAD	MOVE SLOWLY

With arm extended horizontally to the side, thumb points down with other fingers closed.	With hands to the front at waist level, thumbs point outward with other fingers closed.	With all fingers pointing up, arm is extended horizontally out and back to make a pushing motion in the direction of travel.	With arm extended horizontally to the side and thumb pointing down, fingers open and close while load movement is desired.	A hand is placed in front of the hand that is giving the action signal.

USE AUXILIARY HOIST	CRAWLER CRANE TRAVEL, BOTH TRACKS	USE MAIN HOIST	CRAWLER CRANE TRAVEL, ONE TRACK	TROLLEY TRAVEL
				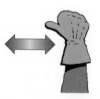
(whipline) - With arm bent at elbow and forearm vertical, elbow is tapped with other hand. Then regular signal is used to indicate desired action.	Rotate fists around each other in front of body; direction of rotation away from body indicates travel forward; rotation towards body indicates travel backward.	A hand taps on top of the head. Then regular signal is given to indicate desired action.	Indicate track to be locked by raising fist on that side. Rotate other fist in front of body in direction that other track is to travel.	With palm up, fingers closed and thumb pointing in direction of motion, hand is jerked horizontally in direction trolley is to travel.

Figure 7-34. If a crane operator lacks a good view of the load at all times, an assistant nearby can provide crane directions through hand signals.

Review

1. Define mechanical advantage.

2. What is the mechanical advantage of a one-, two-, and three-part block and tackle assembly?

3. Explain travel distance and travel speed, and how both compare to mechanical advantage.

4. Define bevel gear.

5. What type of areas are the best for the use of lever-operated hoists?

6. Define power-operated hoist.

7. List one advantage a pneumatic hoist has over an electric hoist.

8. Define drum wrap.

9. What are the two most prevalent standards in hoist safety and operating rules?

10. Describe how to inspect for hook drift.

11. Prior to use, what should a technician be inspecting for on an eyebolt?

12. How is specific load rating information obtained when a particular eyebolt is to be used?

13. List and describe the five classifications of industrial cranes.

14. When is the only time a crane operator can obey a signal from someone other than the assistant?

15. What should a crane operator do if any switch or pushbutton binds or sticks?

Digital Resources

ATPeResources.com/Quicklinks
Access Code 462409

CHAPTER 8

LUBRICATION

Lubrication maintains a fluid film between solid surfaces to prevent their physical contact. Lubricants reduce friction, prevent wear, act as a coolant for moving parts, act as a barrier under load pressure, prevent adhesion or galling of materials, and prevent corrosion. Lubricants are classified as a gas, liquid, semisolid, or solid. Lubricants must be distributed within a mechanical apparatus so all parts requiring lubrication receive the proper amount. Lubrication programs should be established within an organization to ensure that the criteria needed for dependable operation are met.

LPS Laboratories, Inc.

Chapter Objectives

- Describe lubrication.
- Explain the coefficient of friction.
- Describe gas lubricants.
- Describe liquid lubricants and their properties.
- Describe semisolid lubricants and their properties.
- Describe solid lubricants and their properties.
- Identify common types of lubricant application and contamination.
- Explain why lubrication programs should be established within an organization.

Key Terms

- lubrication
- lubricant
- coefficient of friction
- boundary lubrication
- chemisorption
- gas lubricant
- liquid lubricant
- viscosity
- shear strength
- viscosity index
- synthetic fluid
- semisolid lubricant
- solid lubricant
- submersion system
- wick system
- drip system
- centralized system
- oil analysis
- wear particle analysis

LUBRICATION

Lubrication is the process of maintaining a fluid film between solid surfaces to prevent their physical contact. A *lubricant* is a substance placed between two solid surfaces to reduce their friction. Friction occurs when an object in contact with another object tries to move. For example, walking requires friction between the feet and the floor in order to move. Stopping requires even more friction. Walking and stopping are more difficult if the friction is reduced by placing wheels, such as rollerblades, under the feet.

In addition to reducing friction, a lubricant is used to prevent wear, act as a coolant for moving parts, act as a barrier under load pressure, prevent adhesion or galling of materials, and prevent corrosion. Machines and tools depend on lubrication to ensure smooth and safe operation. Lubricants are classified as gas, liquid, semisolid, or solid. In many cases, a lubricant is a mixture of the different classes. Fluid lubricants, which include gas, liquid, and semisolid lubricants, must create a film between material surfaces to prevent contact with each other. **See Figure 8-1.**

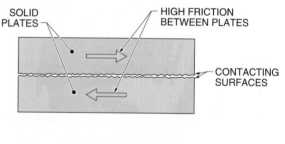

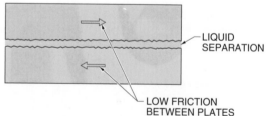

Figure 8-1. Fluid lubricants must create a film thick enough to completely separate moving surfaces.

Safety Tip

High-pressure injection of grease through the skin, such as that from a grease gun, can cause serious delayed damage to soft tissue. A physician should be contacted immediately regardless of the size or appearance of the wound.

Dow Corning Corporation

Chain and open gear lube is a synthetic grease used for lubricating high-speed gears and chains and slow-moving bearings.

Coefficient of Friction

Lubrication generally involves coating surfaces with a material (grease, oil, etc.) that has a lower coefficient of friction than the original surfaces. The *coefficient of friction* is the measure of the frictional force between two surfaces in contact. It is the relationship between the weight of an object and the force required to move it. Coefficient of friction is determined by applying the formula:

$$f = \frac{F}{N}$$

where

f = coefficient of friction

F = force at which sliding occurs (in lb)

N = object weight (in lb)

For example, what is the coefficient of friction of a 25 lb object resting on a horizontal surface that requires 10 lb to move?

$$f = \frac{F}{N}$$

$$f = \frac{10}{25}$$

$$f = \mathbf{0.40}$$

Greater force is required to move a body from rest (static condition) than the force required to keep it in motion (kinetic condition). However, once a lubricated body is in motion, less force is required to keep it in motion than if it were unlubricated. Generally, a static condition between two solid objects means neither object is moving. However, a static condition relating to coefficient of friction refers to the forces required to start a solid object in motion. **See Figure 8-2.**

For example, two unlubricated solid steel objects have a static coefficient of 0.8. This drops to a kinetic coefficient of 0.4 once sliding has started. However, when the two steel objects are lubricated, the static coefficient is 0.16 and the kinetic coefficient drops to 0.02.

Coefficients of Friction				
Material	**Unlubricated**		**Lubricated***	
	Static	**Kinetic**	**Static**	**Kinetic**
Aluminum-to-Aluminum	1.3	—	0.3	—
Copper-to-Copper	1.0	0.3	0.08	0.02
Graphite-to-Graphite	0.1	0.06	—	—
Nylon-to-Nylon	0.25	0.1	—	—
Steel-to-Steel	0.8	0.4	0.16	0.02
Teflon®-to-Teflon®	0.04	0.03	—	—

* values are approximations and vary according to lubricant type

Figure 8-2. The coefficient of friction is the measure of the frictional force between two surfaces in contact.

Boundary Lubrication

Boundary lubrication is the condition of lubrication in which the friction between two surfaces in motion is determined by the properties of the surfaces and the properties of the lubricant other than viscosity. *Viscosity* is the measure of a fluid's resistance to flow. These properties determine the behavior of the sliding system and its kinetic coefficient of friction. Effects of boundary lubrication occur mostly during stopping, starting, and periods of severe operation, where most sliding system failure is caused by inadequate lubrication during these moments.

Boundary lubrication occurs when molecules are adsorbed on metal surfaces through an exchange of electrons. This process occurs with metals that are either lubricated or unlubricated. Unlubricated metals are metals without the addition of a lubricant. However, even metals that have been cleaned of all foreign material contain a degree of lubrication. This lubrication is in the form of water vapor, adsorbed gases, or contaminants from handling. These lubricants affect the material's coefficient of friction. **See Figure 8-3.**

Unlubricated Metal Characteristics	
Copper-to-Copper	
Condition of Contact	Coefficient of Friction
Inert gas (no oxygen)	10.5
Oxygen	2.0
Moist air	1.2
Water	1.0
Steel-to-Steel	
Condition of Contact	Coefficient of Friction
Inert gas (no oxygen)	Seizure
Oxygen	0.8
Moist air	0.6
Water	0.5

Figure 8-3. Cleaned, unlubricated metals contain a degree of lubrication from elements in the atmosphere, such as oxygen and moisture.

Chemisorption. In most cases, chemicals are added to a lubricant to enhance its qualities or to blend with certain chemicals in the lubricant to create chemisorption. *Chemisorption* is a chemical adsorption process in which weak chemical bonds are formed between liquid or gas molecules and solid surfaces. Certain additives applied to a lubricant increase its chemisorption rate. This produces a thicker boundary layer and forms a solid, low shear-strength film. For example, when stearic acid is mixed with iron oxides in the presence of water, a film of iron stearate is formed on a specific metal surface by chemisorption. Another example is when the chemisorption of fatty acids is enhanced to form a film of soap on metal surfaces when water and various oxides

are combined in the presence of oxygen. This film is the second lubrication barrier that must be broken before permanent metal destruction begins. The first barrier is the viscosity of the liquid lubricant. **See Figure 8-4.**

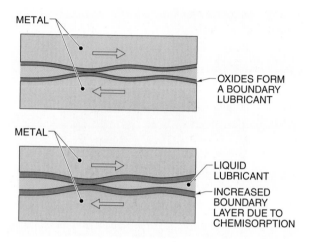

Figure 8-4. Oxides that form on sliding metals are considered boundary lubricants and are thickened and strengthened by the addition of certain chemicals.

Gas Lubricants

A *gas lubricant* is a lubricant that uses pressurized air to separate two surfaces. Other gases such as nitrogen or helium are used where inert properties are required. An *inert gas* is a gas, such as argon or helium, that does not readily combine with other elements. For example, nitrogen is an inert gas.

Gas lubricants are commonly used in low-friction, high-speed, high-technology applications such as grinding spindles, air-turbine dentist's drills, and computers. In addition to the low-friction and high-speed advantages, gas lubricants offer less wear and are able to operate in temperatures from –400°F to over 3000°F. Disadvantages of gas lubricants include the high-precision, high-cost relationship, and the requirement of a very clean gas supply.

Tech Tip

Proper lubrication minimizes machine breakdown and the resulting down-time while increasing savings in labor, production time, and costly repairs. Savings from proper lubrication can amount to 10 to 100 times the cost of lubricants purchased per year.

Liquid Lubricants

A *liquid lubricant* is a lubricant that uses a liquid, such as oil, to separate two surfaces. Liquid lubricants are the preferred lubricants because of their reliability, versatility, and flexibility. Besides reducing friction, liquid lubricants are used for heat removal (in combustion engines) or as a sealer (in hydraulic cylinders and pumps). Liquid lubricants include animal/vegetable oils, petroleum fluids, and synthetic fluids.

Animal/Vegetable Oils. Until 1860, animal and vegetable oils were the primary substances used for lubrication. Fish oils, animal tallow, and sperm oil were, and in some cases still are, used for lubrication because their fatty oils provide superior slipperiness (oiliness) and smoothness. Animal and vegetable oils are used mostly in the food industry. Food grade lubricants are lubricants approved for use on food machinery. Food grade lubricants can contact food being processed without being detrimental to human health.

Animal and vegetable fatty oils are applied to other lubricants to increase their load-carrying capabilities because of their great slipperiness. These lubricant compounds are used in automatic transmission fluids, industrial gear oils, and marine engine oils. However, due to their organic origin, animal and vegetable oils can support bacteria and require sterilization. Also, germicides and antiseptic agents must be added to reduce rusting and bad odors when used in water-soluble lubricants such as cutting oils.

Animal and vegetable oils contain fatty acids which tend to form more fatty acids and gums as they oxidize. Care must be taken to prevent an increase in the fluid's oxygen content. Also, these lubricants must be replenished regularly because they break down.

Petroleum Fluids. A *petroleum fluid* is a fluid consisting of hydrocarbons. A *hydrocarbon* is any substance that is composed mostly of hydrogen and carbon. Petroleum is composed of 12% hydrogen and 85% carbon, with a small amount of other elements such as oxygen, sulfur, nitrogen, etc. Petroleum fluids make up approximately 90% of the total lubricants used. Petroleum is not as susceptible to oxidation, bacteria, and the formation of acid and gummy residues as animal or vegetable oils.

Petroleum is formed by an evolutionary process that takes many millions of years. This process begins with oil vapor given off into the atmosphere by plants. The oil vapor, sometimes seen as a blue haze over heavily vegetated areas, settles or is washed to the ground by rain or snow. The oil works its way deep into rock voids where it is concentrated and processed under high temperatures and pressures. **See Figure 8-5.**

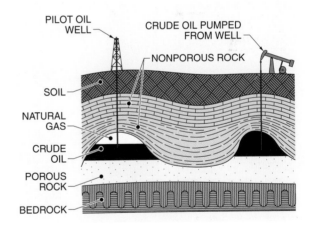

Figure 8-5. All animals and plants release hydrocarbons that, when absorbed into the earth's surface, combine to form pools of gas and crude oil.

Exxon Company

Synthetic lubricants are used in refinery hot liquid pumps to reduce pump failures.

Safety Tip

Never pressurize, cut, weld, drill, grind, or expose empty lubricant containers to heat. Empty lubricant containers retain residue and may explode and cause injury or death. Do not attempt to clean the residue because even a trace of remaining material constitutes an explosive hazard.

The concentrated fluid body (crude oil) works its way back to the Earth's surface through natural oil and gas seeps or through drilling. The earth's plants release approximately 175 million tons of hydrocarbons into the air each year. The evolution of the hydrocarbons into petroleum takes from approximately 50 million to 500 million years.

Crude hydrocarbons are found in various physical forms ranging from a light gas such as methane to a heavy tar. To maintain a consistent, stable, uniform, and reliable lubricating liquid, the crude oil is processed in steps of heating, distilling, and filtering. The final step of processing a petroleum lubricant is the application of certain additives for individual and special applications. New compounds are being developed daily to handle average consumer needs. Recent submarine, subterranean, and outer space equipment applications have added to the research and development process. **See Figure 8-6.**

Lubricant Additives. Additives are used to intensify and improve certain characteristics of a base oil for specific applications. Additives protect a machine from harm, maintain the integrity of the lubricant, and improve the physical properties of the lubricant, such as odor or color control. Additives are included by the manufacturer after performing considerable testing.

Additives include oxidation inhibitors (to provide long bearing or gear life), rust inhibitors (to prevent rust), fatty materials (to improve film strength), powdered lead or graphite (to prevent galling), viscosity index improvers (to ease machine movement in cold weather), and demulsifiers (to separate out water). A technician must understand the various choices and characteristics of the different lubricants before specifying a certain lubricant and must also understand the damage that can occur to a machine if its lubricant is arbitrarily changed.

Crude Oil Processing

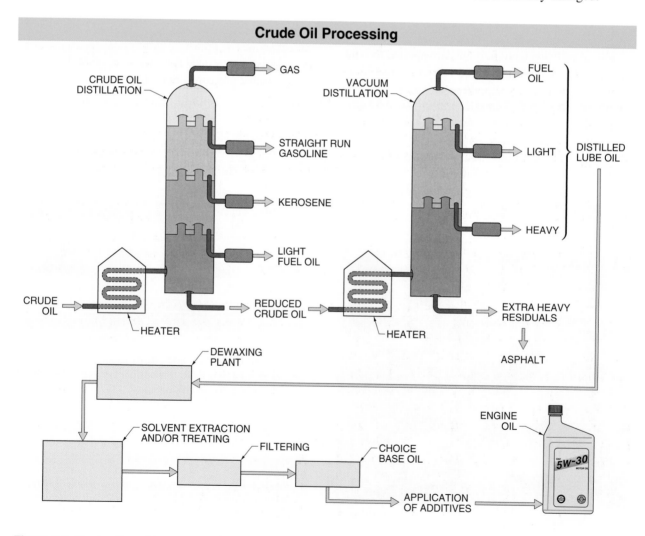

Figure 8-6. Crude oil must be processed to become a uniform and dependable product.

Exxon Company

Special lubricating oils are widely used in aluminum cold rolling applications because they are designed to minimize aluminum staining during the annealing process.

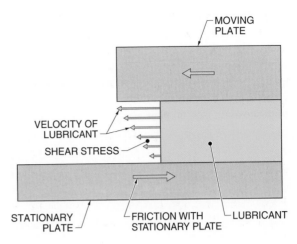

Figure 8-7. Shear strength, a liquid's ability to remain as a separator between solids in motion, relies greatly on the liquid's viscosity.

Viscosity. *Viscosity* is the measure of a fluid's resistance to flow. The flow rate is the most important property of a lubricant. Flow rate is directly proportional to the viscosity of a fluid. In general, the viscosity of a fluid is its thickness. For example, syrup is more viscous (thicker) than water. Viscosity affects fluid flow rate and frictional and thermal properties. Higher viscosity fluids offer greater frictional resistance, thereby increasing thermal activity.

Gear oil viscosity should be high enough to protect tooth surfaces and low enough to offer a good heat transfer. Generally, high-viscosity lubricants result in greater film thickness. Low-viscosity lubricants offer thin film thickness. Maintaining any separation between moving parts ensures equipment protection. However, as loads increase or viscosities decrease due to a rise in temperature, the protective layer can be ruptured as a result of decreased shear strength.

Shear strength is the ability of a material to withstand shear stress. *Shear stress* is stress in which a material is subjected to parallel, opposing, and offset forces. For example, a lubricant is contained between a moving plate and a stationary plate. The lubricant in contact with the moving plate moves with, and has the same speed as, the plate in motion. The lubricant in contact with the stationary plate is slowed by friction with the stationary plate. **See Figure 8-7.** The influence of the moving plate to move the lubricant is the same as the influence of the stationary plate to stop the lubricant movement.

Shear strength relies greatly on the liquid's viscosity. As the viscosity of a lubricant decreases or as forces increase, the shear strength of a lubricant is weakened or broken and metal to metal contact is made. A liquid lubricant's shear strength may be weakened when the liquid's viscosity decreases because of increasing temperatures. The viscosity/temperature characteristic for each liquid lubricant is cataloged and given a viscosity index number. *Viscosity index* is a scale used to show the magnitude of viscosity changes in lubrication oils with changes in temperature. Under basic conditions, as the temperature of oils increases, their viscosity decreases.

Additives must be applied to a lubricant to stabilize the viscosity/temperature characteristic because one of the main functions of a lubricant is cooling. For example, automobile engine oils must maintain even viscosity/temperature characteristics. Low-viscosity lubricants are better for low-temperature engine starting. The low viscosity is also beneficial while the engine is warming up. However, once the engine and oil heat up, engine damage may occur if the oil viscosity is greatly reduced. For this reason, oils are prepared by blending improvers with special base materials to create thermal stability.

Commercial lubricating oil is categorized by six different groups with each group designed for a specific application. **See Figure 8-8.** Lubricating oil is given an SAE viscosity rating based on its ability to flow at a specific temperature. The SAE viscosity rating is a number assigned based on the volume of a base oil that flows through a specific orifice at a specified temperature, atmospheric pressure, and time period. A high viscosity rating results from a small volume of oil flowing through the orifice caused by high resistance to flow. A low viscosity rating results from a large volume of oil flowing through the orifice caused by low resistance to flow. The higher the viscosity rating number, the thicker the oil. For example, a 40 weight oil is thicker than a 10 weight oil. The viscosity rating number assigned to an oil does not change, but oil viscosity can change with temperature and ambient pressure.

Oil Groups/Application
Group A: Automotive
SAE 10W SAE 20W SAE 30 SAE 40 SAE 50
Group B: Gear Trains and Transmissions
General Purpose Oils
Group C: Machine Tools
SUS 75 SUS 80 SUS 90 SUS 140 SUS 250
Group D: Marine Propulsions and Stationary Power Turbines
Turbine Oils
Group E: Turbojet Engines
Aviation Oils
Group F: Reciprocating Engines
Aviation Oils

Figure 8-8. Commercial lubricating oil is categorized by six different groups with each group designed for a specific application.

During startup, oil is cool and does not flow easily. As the machine and oil warms, it flows more easily. For this reason, most machine wear occurs during startups when the cool oil provides less lubrication. Some oils may require a longer machine operating time before proper oil flow to bearing surfaces occurs. This situation can cause premature wear during startup when lubrication may be deficient.

Oil film thickness decreases with an increase in oil temperature and can be completely depleted in high operating temperatures. Oil specified for an application should flow at low temperatures but still protect the engine at high ambient and/or operating temperatures. Oil manufacturer's recommendations use standards provided by the Society of Automotive Engineers (SAE), the American Society for Testing and Materials (ASTM), and the American Petroleum Institute (API). These organizations provide standards in both viscosity and additive packages for the majority of lubricants manufactured worldwide.

Always follow the recommendations of the machine manufacturer or the recommendations of the oil manufacturer when selecting an oil for an application. In general, operating temperature, frequency of stopping and starting, shock producing actions, and any unusual conditions must be considered when selecting a lubricant.

Synthetic Fluids. A *synthetic fluid* is a lubricant, often with a petroleum base, that has improved heat and chemical resistance compared to straight petroleum products. Synthetic lubricants are higher priced than petroleum lubricants but have certain advantages over petroleum lubricants. One major advantage is that synthetic lubricant viscosity changes less with temperature changes. Another advantage is that the oxidation rate of a synthetic lubricant is more stable at higher temperatures.

Synthetic lubricants evolved out of petroleum product shortcomings, such as a relatively low temperature stability, high oxidation rate, and short shelf life. Many of the additives designed for petroleum lubricant enhancement are now used in synthetic lubricants.

The various synthetic lubricant compositions behave differently and have improved qualities over petroleum lubricants. However, some are not compatible when mixed. A mixture of certain synthetic lubricants with petroleum lubricants, or one synthetic composition with another, may cause rapid deterioration of seals or an increase in oxidation. During oil changes, old lubricant should be replaced with the same type of lubricant.

In some cases, a recommended change may have to be made. Great care must be taken when switching lubricants. During replacement of a petroleum product with a synthetic product, any residual amounts of petroleum product left in a machine may be enough to cause a chemical reaction and deteriorate seal material. In most cases, a flushing fluid is required. Consult both lubricant manufacturers for the correct replacement method.

Synthetic lubricants, due to their increased efficiency, high performance, and long life, will be used in an increasing number of applications despite the initial cost savings of petroleum lubricants. In addition, if properly maintained, machine life can be extended by using synthetic lubricants while petroleum supplies are decreasing.

Certain types of diesel locomotives use up to 360 gallons of lubricating oil for proper operation.

Semisolid Lubricants

A *semisolid lubricant* is a lubricant that combines low-viscosity oils with thickeners, such as soap or other finely dispersed solids. A *dispersed solid* is a solid that is finely ground in order to be spread. Dispersed solids such as soap, clay, lead, etc. must be able to adsorb or trap lubricating oils. A *grease* is a semisolid lubricant created by combining low-viscosity oils with thickeners, such as soap or other finely dispersed solids. The oil and thickeners in grease act as a whole with only slight bleeding of the oil.

Semisolid lubricant base oils may consist of mineral oils, silicones, diesters, esters, or fluorocarbons. Thickeners may consist of soaps from aluminum, sodium, lithium, or calcium. Complex soaps may be used with other solids such as clay, graphite, Teflon, or lead. When thickeners are mixed, one thickener is predominant, such as soap, and the other thickeners are included as an additive. Each grease base has its own specific characteristics. For example, aluminum soap offers clarity, calcium soap is water-resistant, lithium soap allows high-temperature use, clay is used for extreme temperatures, and fiber is added to resist being thrown off.

Grease is classified by thickener grade. The National Lubricating Grease Institute (NLGI) has established a series of nine consistency grades. The higher the NLGI number, the stiffer the grease and the less penetration it has. **See Figure 8-9.** Each group is designed for a specific temperature range and purpose. For example, grades NLGI 00 and 0 can be used at temperature as low as –30°F. Grades NLGI 1 and 2 are recommended for applications operating in the temperature range of 0°F to 350°F. Grades NLGI 00, 0, and 1 are recommended for centralized lubrication systems because of their relatively high flow characteristics. The increasing NLGI grade corresponds to an increase in the percentage of thickener. Grades 0, 1, and 2 are the most widely used in industry. The more fluid grades, such as 000 and 00 are used where thickened oil is desired, such as in gearboxes, where leakage may occur when using oil lubricants.

NLGI Grease Grades		
NLGI Grade	**Penetration***	**Stiffness**
000	1.75 – 1.87	VERY SOFT
00	1.57 – 1.69	
0	1.32 – 2.30	
1	1.22 – 1.33	
2	1.04 – 1.16	
3	0.86 – 0.98	
4	0.68 – 0.80	
5	0.51 – 0.62	VERY HARD
6	0.33 – 0.45	

* in in.

Figure 8-9. The National Lubricating Grease Institute (NLGI) has established a series of nine consistency grades for grease.

Effects of Temperature

Grease consistency varies with a change in temperature. Grease typically softens as temperatures increase. As temperatures increase, greases become soft enough to separate the oil from the thickener. This is known as the grease dropping point. The *grease dropping point* is the maximum temperature a grease withstands before it softens enough to flow through a laboratory testing orifice. In practical applications, as machine speeds or loads increase the temperature of a grease, the oil within the grease is generally lightened enough to run off or bleed away from the grease's thickener, leaving behind a hardened, dark-colored substance. The dropping point of greases vary according to type. Some special greases do not exhibit a dropping point. **See Figure 8-10.**

Grease Characteristics

Grease Thickener	Oil Viscosity*	Dropping Point†	Maximum Usable Temperature†
Soap Base			
Aluminum	275	195	175
Calcium	300	180	175
Lithium	300	340	300
Nonsoap Base			
Colloidal silica	400	‡	‡
Bentonite	400	‡	‡

* SSU @ 100°F
† in °F
‡ these products bleed oil but do not exhibit a melting point

Figure 8-10. Grease base thickeners are chosen according to the temperature and viscosity demands placed on the lubricating product.

The viscosity of grease also changes with a change in its shear rate (applied pressure). As shear rates increase, the viscosity of the thickener decreases to the point of the base oil. This is common in many high-speed machines. A grease with an extreme pressure additive is used in applications where high speeds are combined with high loads.

PLI LLC

A self-contained electromechanical positive displacement lubricating system contains a motor, gearbox, piston pump, microprocessor, and battery pack. The system ejects lubricant at over 200 psi and is used in applications where frequent, reliable lubrication is required.

Solid Lubricants

A *solid lubricant* is a material such as graphite, molybdenum disulfide, or polytetrafluoroethylene (PTFE) that shears easily between sliding surfaces. Most solid lubricants are used to provide a dry film in a high-load, low-speed application. Solid lubricants may be used with semisolid lubricants (greases), which fuse or bond the lubricant to the substrate, or as a solid block.

Graphite, produced from coal, is milled to obtain grayish black crystalline flakes. Graphite has low shearing forces, especially between each graphite flake layer. The graphite flake layers run parallel to the bearing surface. Although graphite slides easily over itself, it also adheres well to bearing surfaces. Graphite requires moisture, such as atmospheric humidity, to be effective. This makes graphite a poor lubricator where humidity is absent, such as aerospace applications. Graphite has a low coefficient of friction and an ability for service in temperatures up to 500°C. Graphite may be used dry, mixed with oil or grease, or as a solid block.

Molybdenum disulfide is a common substance that is similar in appearance to graphite. Molybdenum disulfide is found as an ore in molybdenite. Molybdenum disulfide does not require the presence of moisture. This allows it to be used in dry atmospheres, high vacuums, and in high temperatures. This makes molybdenum disulfide perfect for aerospace applications. Molybdenum disulfide becomes a bonded film when used with the binder combination of ceramic and resin. Its film thickness is generally 0.001″ or 0.002″ and this thickness is typically its limitation for life.

Polytetrafluoroethylene (PTFE) is a long-chain polymer produced from ethylene, which is a coal by-product. A *polymer* is a molecule made up of a chain of repeating units that are chemically bonded together. PTFE is reinforced by being combined with other substances, such as glass fibers, rayon, or other synthetics. This offers outstanding strength capabilities with low wear and a low coefficient of friction. PTFE is used in pharmaceutical and food plants as well as chemical industries because it is nontoxic and has excellent chemical stability. PTFE is commonly used for nonstick surfaces of household cooking utensils.

LPS Laboratories, Inc.

Low temperature bearing grease is effective to –58° F and is used in refrigeration equipment, mechanical bridges, automatic safety gates, and motor starters.

LUBRICANT APPLICATION

Lubricants must be distributed within a mechanical apparatus so all parts requiring lubrication receive the proper amount. Except in sealed units where there is a void of oxygen, lubricants require replenishment or replacement through occasional topping or by means of a feed system.

Oil Application

Oil must be replenished because oil, being a liquid, does not cling to and remain on the sides of moving parts. In addition, splashing or sloshing vaporizes the oil, thus reducing the amount of oil present for an application. Oil lubrication may be applied to bearing elements by submersion, wick, drip, or centralized systems. **See Figure 8-11.**

Submersion Systems. A *submersion system* is a lubrication system in which the bearings are submerged below oil for lubrication. Oil submersion systems allow oil to be carried throughout load bearing surfaces. The level of oil in most systems is critical to prevent churning or drag when the level is too high or inadequate lubrication accompanied by high temperatures when the level is too low. Proper oil levels require that ball or roller bearings be immersed halfway up the lowest ball or roller, and gears be immersed to twice the tooth height. Where gears are designed in a vertical train, the oil level should be just below the shaft of the lowest gear. When oil transfer is inadequate using these methods, ring, chain, or splash devices may be added to assist oil movement.

Wick Systems. A *wick system* is a lubrication system that uses capillary action to convey oil to a bearing surface. *Capillary action* is the action by which the surface of a liquid is elevated on a material due to its relative molecular attraction. This is seen in the raising of a fuel through the wick in an oil lamp. In a wick system, wicks or felt pads consisting of pores or spaces allow oil to penetrate and spread from an oil source to bearing surfaces. Oil feed rate is controlled by the thickness of felt pads, number of wick strands, or the length of material immersed in the oil.

Safety Tip

In applications where repeated skin contact with lubricants occurs, use protective skin creams, such as silicon-base creams, applied to clean hands prior to contact.

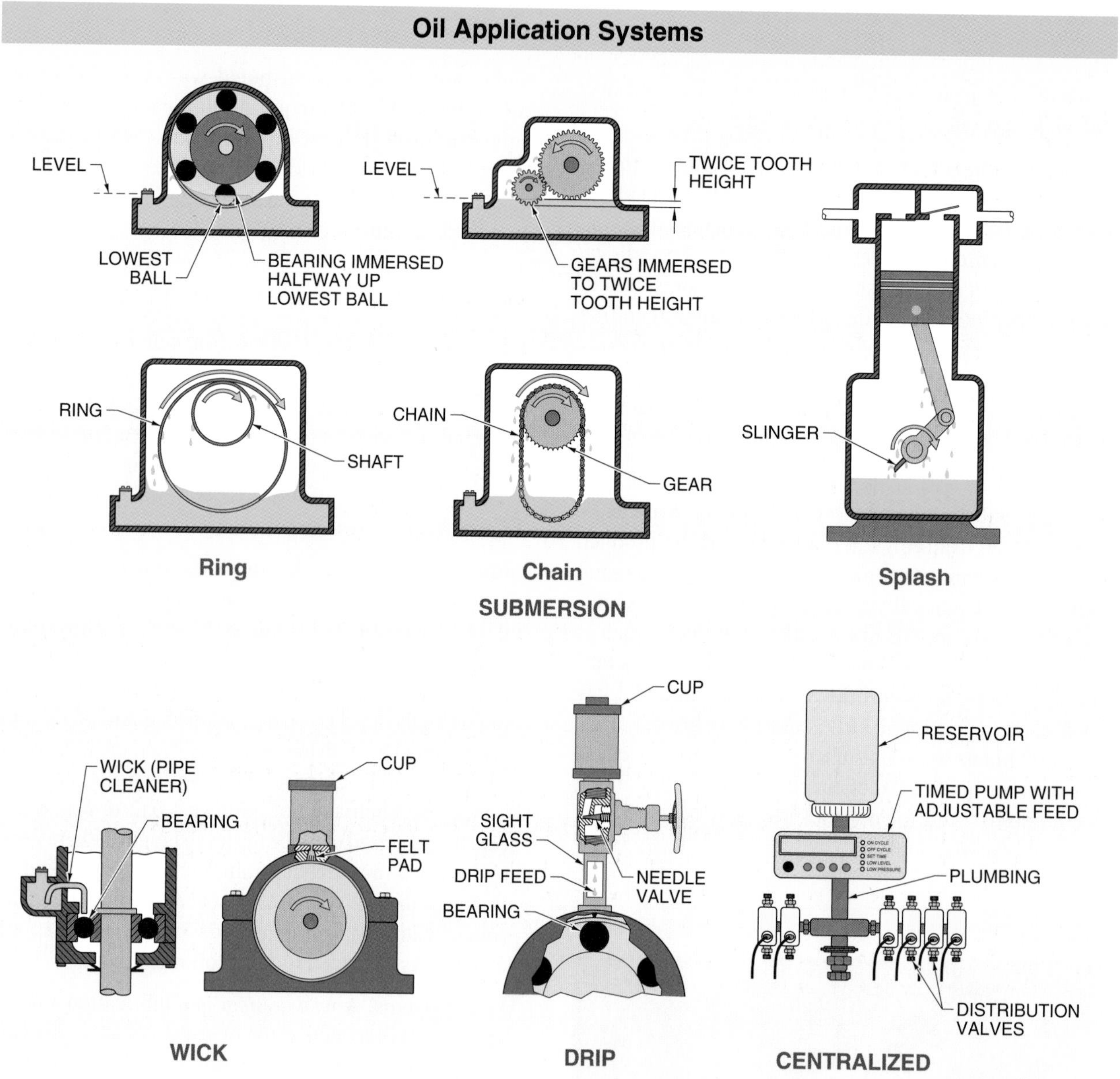

Oil Application Systems

Figure 8-11. Lubricant delivery methods vary according to the design, speed, and accessibility of the machinery.

Drip Systems. A *drip system* is a gravity-flow lubrication system that provides drop-by-drop lubrication from a manifold or manually filled cup through a needle valve. The needle valve is adjusted for flow regulation. A drip system offers flow regulation in addition to a higher rate of liquid flow than that of wick systems. Most drip systems are equipped with a sight glass for liquid level and dripping motion observation.

Centralized Systems. A *centralized system* is a lubrication system that contains permanently installed plumbing, distribution valves, reservoir, and pump to provide lubrication. These systems, although initially high in cost, simplify lubrication processes. The initial cost is offset by longer running, dependable operating equipment with more reliable lubrication periods and less possibility of contamination.

Safety Tip

Empty lubricant drums should be completely drained, properly bunged, and promptly returned to a drum reconditioner. All other empty lubricant containers should be disposed of in an environmentally safe manner and in accordance with local, state, and federal regulations for petroleum distillates.

Grease Application

For proper operation, rolling elements must be thoroughly coated with grease. However, they must also have the correct quantity of grease. Overgreasing leads to overheating, aerating, and churning of the grease, resulting in early bearing failure. The total space available in a rolling element bearing should contain no more than 50% grease lubricant.

The grease of all open, unsealed bearings must be replenished because the oil from all unsealed greases eventually bleeds off from the thickeners. This leaves the thickeners, having no value, to solidify and burn. Also, all greases eventually oxidize and corrode, which is evident by their dark color and burnt oil smell. In sealed bearings, relubrication is not practical. For this reason, bearing failure will occur over time, especially when used in high-temperature applications. Greases are applied by grease guns, grease cups, or centralized systems. **See Figure 8-12.**

A *grease gun* is a small hand-operated device that pumps grease under pressure into bearings. A *grease cup* is a receptacle used to apply grease to bearings. The receptacle is packed with grease. The cap is rotated to force the grease into bearings. A centralized system contains permanently installed plumbing, reservoir, and air supply to provide the lubrication. The air supply provides pressure above a diaphragm in the reservoir. The pressure forces the grease below the diaphragm into a pipe which is routed to the devices requiring lubrication. The grease used in a centralized system should be one grade softer than is otherwise required. For example, a No. 0 grease is used instead of the required No. 1 grease when used in a centralized system.

Motor Regreasing. Motors equipped with grease fittings and drain plugs must be regreased using a low-pressure grease gun. **See Figure 8-13.** Motors are regreased by applying the following procedure:

1. Wipe the grease fitting and drain plug on the motor. Also wipe the grease gun nozzle and expel a small amount of grease from the gun to ensure that grease is uncontaminated.
2. Remove the drain plug and clean any hardened material from the port.
3. Add grease to the motor until new grease is expelled from the drain plug port.
4. Run the motor without the drain plug for approximately 10 minutes to expel excess grease.
5. Clean and replace the drain plug.

Lubricant Contamination

Lubricant contamination is the main cause of mechanical system failure. Dirt and other abrasive materials that contaminate lubricant wear moving components and bearing surfaces as the equipment operates. Mechanical system lubricants must remain free of damaging contaminants to ensure the useful life of the drive system. A maintenance program should be used to provide a routine schedule for lubricant changes and/or filtering. Some mechanical systems use oil purifiers to clean and recycle lubricating oil while the equipment is operating. Oil purifiers can reduce maintenance, lubrication, and disposal costs.

Grease Application Methods

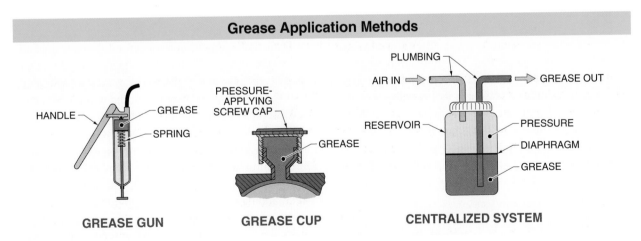

Figure 8-12. Applying grease may be accomplished through hand-operated pumping methods or by centralized systems serving whole facilities.

Motor Regreasing

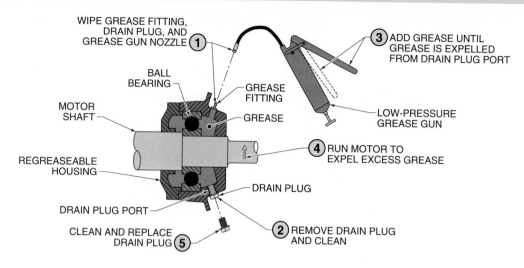

Figure 8-13. Regreasing a motor may include purging the old grease from between the shaft and the housing.

Contamination of the lubricant can also reduce the effectiveness of the lubricant. Common contaminants include dirt and water. Lubricant contaminated with dirt subjects moving components to a constant flow of abrasives. Water that mixes with the lubricant also reduces the effectiveness of the lubricant. The water causes bearing components to rust, increasing friction and eventually causing bearing failure. Sources of water can include condensation and environments with high humidity. Periodic oil changes are necessary to remove water from the lubricant. Oil that is contaminated with water has a milky appearance.

LUBRICATION PROGRAMS

A lubrication program should be established within an organization to ensure that the criteria needed for dependable operation are met. The lubrication program should establish the parameters of each lubricant used and include training of personnel in the methods of application of each lubricant. The program should establish the scheduling and frequency requirements needed by each machine. Finally, recorded results of equipment failure should be reviewed frequently to determine possible faulty lubrication or lubrication practices.

Safety Tip

Minimize exposure to petroleum and synthetic-base hydrocarbons because these liquids and vapors pose potential human health risks that vary from person to person.

Oil Analysis

Oil analysis is a predictive maintenance technique that detects and analyzes the presence of acids, dirt, fuel, and wear particles in lubricating oil to predict equipment failure. Lubricating oil analysis is performed on a scheduled basis. An oil sample is taken from a machine to determine the condition of the lubricant and moving parts. Oil samples are commonly sent to a company specializing in lubricating oil analysis. **See Figure 8-14.**

Equipment commonly used for oil analysis is a spectrometer. A *spectrometer* is a device that vaporizes elements in the oil sample into light. The light is separated into a spectrum and then converted into electrical signals, which are processed and displayed by a computer.

Data management software for lubricant analysis allows the user to be connected (via modem) directly to an analysis laboratory where oil samples are tested and results are sent to the user providing a quick, comprehensive analysis and trending. The analysis commonly includes oil viscosity, particle count, wear particle concentration analysis, and wear particle analysis. Analysis information allows the user to determine if wear is occurring, what component is affected, what is causing the wear, and how far damage has progressed.

The lubricant condition rating is identified as normal, marginal, or critical based on the results of the sample, comparison with previous data, and the machine condition analyst's experience with the particular type of equipment. A normal condition rating indicates the lubricant is within expected levels and require no corrective action. A marginal condition rating indicates that critical physical properties and/or trace elements are outside expected levels and require minor maintenance action such as increased sampling frequency. A critical condition indicates that the majority of physical properties are outside the expected levels. A lubricant and/or wear condition problem exists that requires definitive maintenance action.

Wear Particle Analysis. *Wear particle analysis* is the study of wear particles present in the lubricating oil. While lubricating oil analysis focuses on the condition of the lubricating oil, wear particle analysis concentrates on the size, quantity, shape, and composition of the particles produced from worn parts. The equipment condition is assessed by monitoring wear particles in the lubricating oil. Normal wear occurs as equipment parts are routinely in contact with each other. An increase in the frequency and size of wear particles in the lubricating oil indicates a worn part or predicts possible failure.

For example, lubricating oil samples having consistent wear particle readings over a period of time provide a baseline measurement. An increase in wear particles may indicate premature wearing of parts. Large, sharp wear particles indicate parts sheared in the equipment. Fractured wear particles indicate broken parts in the equipment.

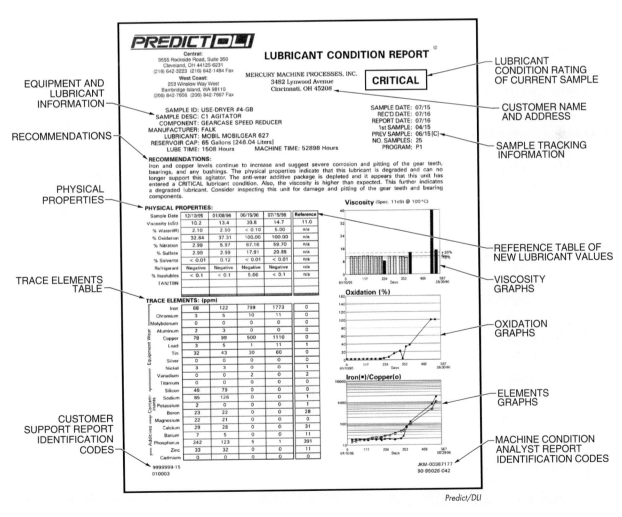

Figure 8-14. Lubricating oil analysis is used to predict potential equipment malfunction or failure.

Review

1. Aside from reducing friction, list three additional ways lubricant helps ensure smooth and safe operation of tools and machines.

2. When do the effects of boundary lubrication mostly occur?

3. Define chemisorption.

4. What is a liquid lubricant used for and why is it the preferred lubricant?

5. List three examples of lubricant additives and explain how each one protects a machine from harm.

6. What does the viscosity index scale show?

7. What are SAE viscosity ratings based on?

8. List two advantages of using synthetic fluid as opposed to petroleum-based lubricant.

9. Define semisolid lubricant.

10. Explain how an increase in temperature affects grease consistency.

11. Determine why graphite is useful as a solid lubricant and explain why it is not useful when humidity is absent.

12. Explain why oil must be replenished in order to ensure all parts within a mechanical apparatus receive the proper amount of lubrication.

13. List the four systems associated with oil application.

14. Define oil analysis.

15. What does the practice of wear particle analysis focus on?

Digital Resources

ATPeResources.com/Quicklinks
Access Code 462409

CHAPTER 9

BEARINGS

Bearings guide and position moving parts to reduce friction, vibration, and temperature. Machine efficiency and accuracy depends on proper bearing selection, installation and handling, and maintenance procedures. Bearings are classified as rolling-contact or plain bearings. Rolling contact bearings include ball, roller, and needle bearings. Successful bearing installation requires cleanliness, correct bearing selection, mounting methods, tool use, and tolerance specifications.

NTN Bearing Corporation of America

Chapter Objectives

- Explain the functions of a bearing.
- List the common types of rolling-contact bearings and describe their components.
- Describe plain bearings.
- Explain how to properly remove bearings.
- Identify common causes of bearing failure.
- Explain how to properly prepare bearing parts.
- List the factors to consider when installing bearings.

Key Terms

- bearing
- radial load
- axial load
- rolling-contact bearing
- ball bearing
- roller bearing
- needle bearing
- race
- plain bearing
- spalling
- false Brinell damage
- fretting corrosion
- galling
- fluting
- bearing mounting
- preloading
- threaded cup follower
- tapered bore bearing

BEARINGS

A *bearing* is a machine part that supports another part, such as a shaft, which rotates or slides in or on it. A bearing guides and positions moving parts to reduce friction, vibration, and temperature. The length of time a machine retains proper operating efficiency and accuracy depends on proper bearing selection, installation and handling, and maintenance procedures. Bearings are available with many special features, but all incorporate the same basic functioning parts.

Bearings are designed to support radial, axial, and radial and axial loads. A *radial load* is a load in which the applied force is perpendicular to the axis of rotation. For example, a rotating shaft resting horizontally on, or being supported by, a bearing surface at each end has a radial load due to the weight of the shaft itself. **See Figure 9-1.** In a radial load, the shaft should have negligible end-to-end movement. An *axial load* is a load in which the applied force is parallel to the axis of rotation. For example, a rotating vertical shaft has an axial load due to the weight of the shaft itself.

Radial and axial loads occur when a combination of the two loads are present. For example, the shaft of a fan blade is supported horizontally (radial load) and is pulled or pushed (axial load) by the fan blade. Bearings are classified as rolling-contact (anti-friction) or plain bearings.

Rolling-Contact Bearings

A *rolling-contact (anti-friction) bearing* is a bearing composed of rolling elements between an outer and inner ring. Rolling-contact bearings are referred to as anti-friction bearings because they are designed to roll on a film of lubricant, which separates the metal components. All rolling-contact bearings are given a life expectancy (fatigue life). *Fatigue life* is the maximum useful life of a bearing. The fatigue life of a bearing is determined by expected speed, temperature, lubrication, and load rate standards. These standards are recommended by the Anti-Friction Bearings Manufacturers Association (AFBMA).

Standard bearings never wear out or reach their fatigue life. Average bearings normally exceed the life of the machine where they are installed. Bearing failure is generally the result of a deviation from bearing fatigue life standards. Properly maintained bearings do not wear out. They are meant to eventually, if run long enough, fail due to fatigue. Close examination of a failed bearing often provides evidence to the cause of failure. For example, dark discolored metals indicate high temperatures, rusting surfaces indicate high moisture and/or improper lubrication, and split or fractured rings indicate an improper fit or assembly.

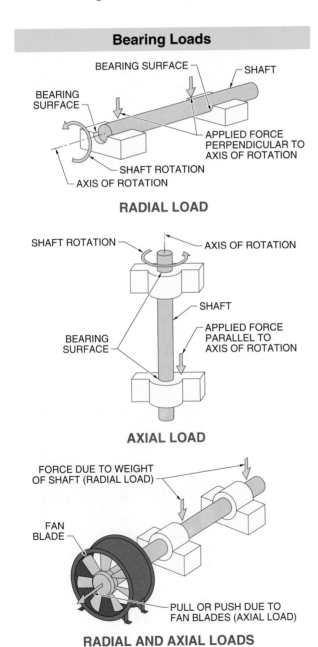

Bearing Loads

RADIAL LOAD

AXIAL LOAD

RADIAL AND AXIAL LOADS

Figure 9-1. Bearings are designed to support radial loads, axial loads, or radial and axial loads.

Early bearing failure due to a minor deviation from bearing fatigue life standards is regarded as bearing service life. *Service life* is the length of service received from a bearing. Service life is generally shorter than fatigue life. This is due to less-than-optimal operating conditions. For example, doubling the speed of a bearing reduces its service life to one half its fatigue life. Doubling the load on a bearing reduces its service life by 6 to 8 times.

Rolling-contact bearings include ball, roller, and needle bearings. A *ball bearing* is an anti-friction bearing that permits free motion between a moving part and a fixed part by means of balls confined between inner and outer rings. A *roller bearing* is an anti-friction bearing that has parallel or tapered steel rollers confined between inner and outer rings. A *needle bearing* is an anti-friction roller-type bearing with long rollers of small diameter. **See Figure 9-2.**

The rolling-contact bearing categories may be further divided into more specific designs or configurations. For example, ball bearing races may be made deeper for supporting radial and axial loads or additional rolling elements (balls or rollers) may be installed to support heavier loads.

Rolling-Contact Bearing Construction. Ball and roller bearings are constructed of an outer ring (cup), balls or rollers, and an inner ring (cone). The outer ring is generally slid or pressed easily into a housing. The inner ring is generally pressed on a shaft with a tighter fit than the outer ring. Ball bearing rings are designed with a groove known as a race. A *race* is the track on which the balls of a bearing move.

Needle bearings contain an outer ring (cup) and rollers. The rollers are retained in a cage and bear directly on the rotating shaft. Bearing precision and cost is determined by the smoothness of the ground surfaces (grade of finish) and the quality of tolerances. Better finishes produce less friction, lower temperatures, smoother movements, and longer bearing life, but usually cost more.

Additional bearing components include cages, bearings, seals, and snap rings. Cages are devices used to hold the balls or rollers in place. Bearings are designed as open for lubrication injection, or as sealed which hold lubricant for the life of the bearing. Seals are used to retain lubrication as well as prevent contamination from dust, dirt, or other solids. Snap rings allow the bearing to be inserted into a housing and held at a certain depth.

Rolling-contact (Anti-friction) Bearings

BALL

ROLLER

NEEDLE

Figure 9-2. Rolling-contact (anti-friction) bearings include ball, roller, and needle bearings.

Ball Bearings. Ball bearings are anti-friction bearings that permit free motion between a moving part and a fixed part by means of balls confined between inner and outer rings. Ball bearings are selected based on the application of the bearing. Ball bearings may be designed for light or heavy loads, radial or axial loads (or combination of each), or harsh or clean environments.

General-use ball bearings are designed as single-row radial, single-row angular-contact (axial), or double-row radial or axial based on the direction of applied force. **See Figure 9-3.**

A *radial bearing* is a rolling-contact bearing in which the load is transmitted perpendicular to the axis of shaft rotation. Single-row radial bearings have a single row of balls and may be designed with or without loading slots. A *loading slot* is a groove or notch on the inside wall of each bearing ring to allow insertion of balls. Bearings with loading slots are referred to as maximum capacity bearings due to the ability to add the maximum number of balls. Applying axial (thrust) loads to maximum capacity bearings causes rapid damage. A *Conrad bearing* is a single-row ball bearing without a loading slot that has deeper-than-normal races. Conrad bearings allow for axial and radial loads. Installing Conrad bearings in the wrong direction results in immediate damage.

An *angular-contact bearing* is a rolling-contact bearing designed to carry both heavy axial (thrust) loads and radial loads. The ability to provide axial (thrust) support is due to the high shoulder on one side of the outer race. This high-shouldered race, acting as a seat for the balls, provides high thrust load capacities in one direction only.

Bearings that are designed for thrust loads must be installed in only one direction to prevent the load from separating the bearing components. These bearings have a face and back side for ease in identifying the thrust direction. The back side receives the thrust and is marked with the bearing number, tolerance, manufacturer, and, in some cases, the word "thrust".

Emerson Power Transmission

Self aligning tapered roller bearings handle a combination of radial and thrust loads and are used in a wide variety of applications such as chemical processing, sawmills, and pulp and paper mills.

Ball Bearings

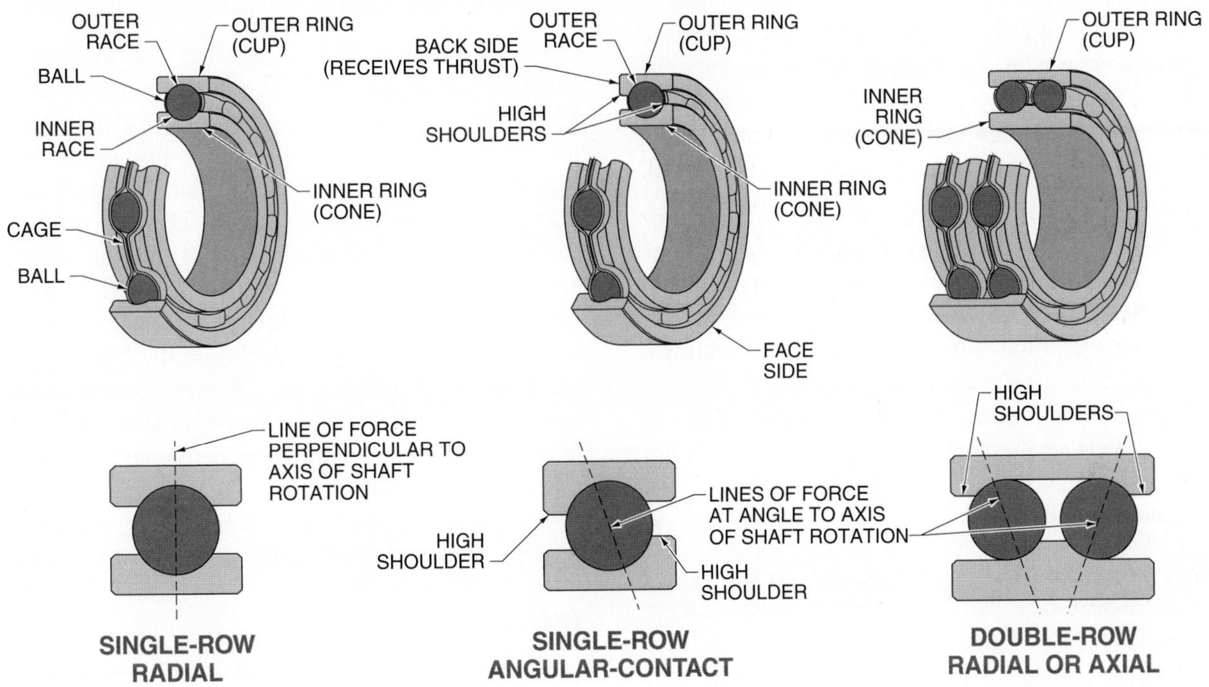

Figure 9-3. General-use ball bearings are designed as single-row radial, single-row angular-contact (axial), or double-row radial or axial.

Double-row bearings, also known as duplex bearings, are matched pairs of angular-contact bearings. They are capable of heavy radial and thrust loads in both directions. Double-row bearings are designed as matched sets and are identified based on their configuration, such as back-to-back, face-to-face, etc. Never use two single-row bearings when replacement requires the use of double-row bearings. Pairing unmatched single-row bearings adversely affects shaft rotation.

Angular-contact bearings are always used in pairs, sometimes at opposite ends of a spindle shaft. Combined sets are installed back-to-back, face-to-face, or back-to-face. **See Figure 9-4.** Bearing numbers normally indicate that each set has been ground to mount as a pair. Mixing different manufacturer's bearings is not recommended because each angular-contact ball bearing has a ground finish on the back and face.

Ball bearings are installed with one ring being a press fit and the other a push fit. A press fit requires the force of an arbor or hydraulic press to install the ring. A push fit allows the ring to be slid into place by hand. Generally, the press-fit ring is pressed onto or into the rotating part, and the push-fit ring is installed onto or into the stationary component.

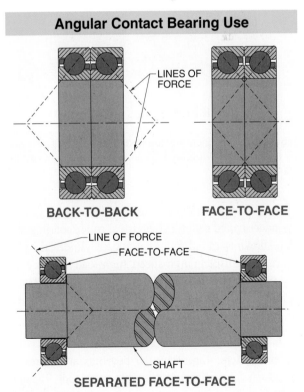

Angular Contact Bearing Use

Figure 9-4. Angular-contact bearings are used in sets.

Both inner and outer rings may be press fit where large bearings encounter high speeds or high loads. Under normal load conditions, ball bearings generally have 0.00025" interference per inch of shaft when the inner race is press fit. *Interference fit* is fit in which the internal member is larger than the external member so that there is always an actual interference of metal. Less interference is required when the outer race is press fit.

Roller Bearings. Roller bearings are anti-friction bearings that have cylinder-shaped or tapered steel rollers confined between an outer ring (cup) and an inner ring (cone). **See Figure 9-5.** Roller bearings are designed for loads and applications similar to those of ball bearings. Roller bearings are designed for heavy radial and axial loads. Cylindrical rollers are used for radial loads and tapered rollers are used for radial and axial loads. Roller bearings are precision devices and must be kept clean and handled with care.

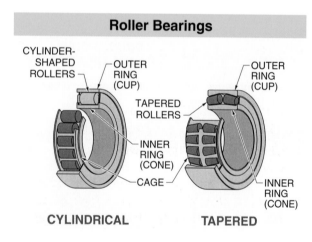

Roller Bearings

CYLINDER-SHAPED ROLLERS — OUTER RING (CUP)

TAPERED ROLLERS

INNER RING (CONE)

CAGE

OUTER RING (CUP)

INNER RING (CONE)

CYLINDRICAL   **TAPERED**

Figure 9-5. Cylindrical and tapered roller bearings are used for heavy loads. Tapered roller bearings are also used for axial loads.

A *cylindrical roller bearing* is a roller bearing having cylinder-shaped rollers. Cylindrical roller bearings are also known as radial roller bearings. Cylindrical roller bearings are used in high-speed, high-load applications and may contain as many as four rows of rollers. These bearings have a high radial load capacity but are not designed for axial loads or misalignment.

A *tapered roller bearing* is a roller bearing having tapered rollers. Tapered roller bearings are normally used for heavy radial and adjustable axial loads. Tapered roller bearings are available in more than 20 configurations from many different manufacturers. The different

tapered roller bearings are generally interchangeable and may be cross-referenced because most manufacturers subscribe to the International Standards Organization (ISO).

Needle Bearings. Needle bearings are anti-friction roller-type bearings with long rollers of small diameter. Needle bearings, similar to cylindrical roller bearings, are chosen for relatively high radial loads. Needle bearings are characterized by their rollers being of small diameter compared to their length. The ratio of length to diameter may be as much as 10 : 1. Needle bearings are often used in applications with limited space. Needle bearings normally have tightly packed rollers without separators or inner rings. Special needle bearings that are used for oscillating motion in aircraft elements may have separators with inner rings. Needle bearing cases may have machined surfaces or be drawn and formed for roller retention. Needle bearings are generally press fit. A firm, even, and square press during installation prevents damage to the bearing case.

Tech Tip

Depending on the material used, plain bearing load capacity can greatly exceed the load capacity of rolling-contact bearings because of the increased area of surface contact with a shaft.

Plain Bearings

A *plain bearing* is a bearing in which the shaft turns and is lubricated by a sleeve. Plain bearings are used in areas of heavy loads where space is limited. Plain bearings are quieter, less costly, and if kept lubricated, have little metal fatigue compared to other bearings. **See Figure 9-6.**

Plain bearings may support radial and axial (thrust) loads. In addition, plain bearings can conform to the part in contact with the bearing because of the sliding rather than rolling action. This allows the plain bearing material to yield to any abnormal operating condition rather than distort or damage the shaft or journal. A *journal* is the part of a shaft, such as an axle or spindle, that moves in a plain bearing. The sliding motion of a shaft or journal, whether rotating or reciprocating, is generally against a softer, lower friction bearing material.

The service life of a plain bearing depends on the surface condition of the shaft. A shaft with nicks, gouges, scratches, or rough machine marks wears a plain bearing rapidly. In addition, a shaft that is ground too fine does not allow lubricant retention and wears once the lubricant is squeezed out.

Plain Bearings

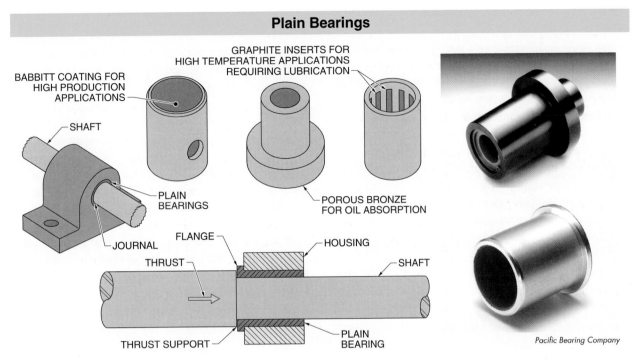

Figure 9-6. Plain bearings provide sliding contact between mating surfaces.

Plain Bearing Materials. Special materials, or combinations of materials, must be selected for plain bearings because of the momentary metal-to-metal contact that occurs during shaft stopping and starting. Plain bearing material must be corrosion- and fatigue-resistant, able to handle running loads and thermal activity, and compatible with other materials used.

Tin-base and lead-base babbitt metals are the best metals for plain bearing loads. *Babbitt metals* are alloys of soft metals such as copper, tin, and lead and a hard material such as antimony. Copper-leads, bronze, and aluminum base metals are used for plain bearings requiring increased load-carrying capacities. Babbitt metals are used in a thin layer over a steel support for heavy commercial applications, such as armature bearings used in hand drills.

The various metals used for plain bearings are chosen because of their ability to perform under specific conditions. Copper-lead bearings are designed for their ability to withstand high temperatures and high loads, such as engine connecting rod bearings. Porous bronze bearings become self-lubricating when impregnated with oil. These bearings are capable of absorbing oil equal to 30% of their total volume. Porous bronze bearings must be relubricated even though they are used in confined locations or where supplying lubricant is difficult.

Nylon and Teflon® bearings are designed for light loads because these materials readily deform under heavy loads. However, they are ideally suited for chemical or high-temperature applications and require no lubrication. Carbon-graphite bearings are designed to operate in temperatures exceeding 700°F and withstand 300 psi of load force without a lubricant. Carbon-graphite bearings are ideally suited for oven applications. Metal bearings with machined grooves containing graphite inserts may be used with high temperature applications requiring lubrication.

Bearing hardness must also be considered in addition to operating conditions. Normal wear and scoring must take place on the less costly bearing surface, not on the surface of the shaft or journal. For this to occur, plain bearings must be at least 100 Brinell points softer than the shaft or journal. A Brinell hardness test measures the hardness of a metal or alloy by hydraulically pressing a hardened steel ball into the metal to be tested and then measuring the area of indentation. The Brinell hardness number is found by measuring the diameter of the indentation and finding the corresponding hardness number on a calibrated chart.

BEARING REMOVAL

Proper tools and maintenance procedures are required when working with precision elements such as bearings. A clean working environment prevents dust, dirt, or other solids from contaminating the bearing, shaft, or housing.

Many bearing failures are due to contaminants that have worked their way into or around a bearing before it has been placed in operation. Internal abrasive particles permanently indent balls, rollers, and raceways. This alters the shape of the surface and begins bearing erosion. Bearing tolerances are such that a solid particle of a few thousands of an inch (0.003″) lodged between the housing and the outer ring can distort raceways enough to reduce critical clearances.

Work benches, tools, clothing, wiping cloths, and hands must be clean and free from dust, dirt, and other contaminants. For example, the rolling elements and races of high precision aircraft bearings can be contaminated the moment they are touched by bare hands. The acid from the skin surface is enough to begin corrosion of the metallic surfaces.

Bearing removal is more difficult than bearing installation. A firm, solid contact must be made for bearing removal. Bearings should always be removed from a shaft with even pressure against the ring. The removing pressure should be applied to the inner ring when the bearing is press fit on a shaft. The removing pressure should be applied to the outer ring when the bearing is press fit in a housing. Bearings are removed from shafts or housings using bearing pullers, gear pullers, arbor presses, or manual impact. These methods enable easy bearing removal and reduce the damage to the bearing. **See Figure 9-7.**

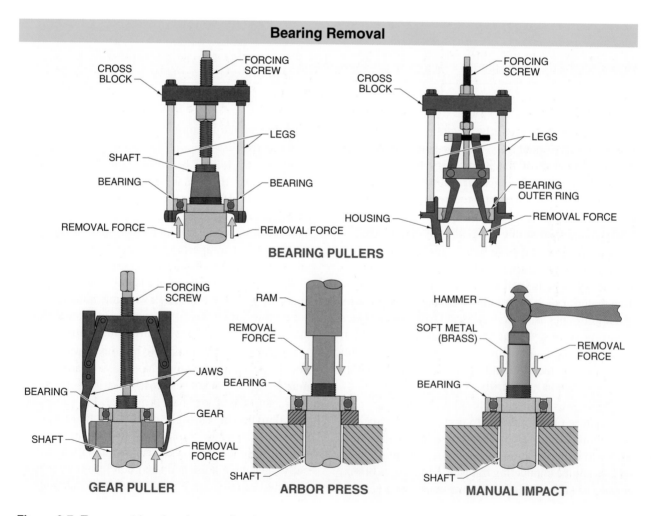

Bearing Removal

BEARING PULLERS

GEAR PULLER   **ARBOR PRESS**   **MANUAL IMPACT**

Figure 9-7. To prevent bearing damage, bearing removal forces should be applied to the back side of the ring that is pressed in place.

Power Team, Division of SPX Corporation

Hydraulically powered mechanical bearing pullers are available with pumps that develop pressures up to 10,000 psi for effortless bearing removal.

Extreme caution must be taken to prevent damage to any bearing part. Most damage during removal goes unnoticed. Also, mark each part as to its location when disassembling a bearing housing assembly so each part can be reassembled the way it was originally positioned.

The use of a hammer and chisel to pry a bearing off of its shaft usually results in damage and contamination. Discard any bearing that was difficult to remove. Bearing damage is possible if the wrong puller or removal method is used. Check for designs intended to hold a bearing in place, such as snap rings, set screws, or pins before applying force for bearing removal. Wear eye protection when using any device that applies force, such as pullers, presses, vises, or hammers. Never strike a bearing directly with a hammer because these are both hardened metals that can shatter when struck together.

Bearing Failure Investigation

Bearing service life, which is normally shorter than fatigue life, may be shortened for many reasons. For example, the load may be too heavy, alignment may be poor, installation procedures may have been improper, or the environment around the machinery may be excessively dirty. Whatever the reason for a bearing failure, each bearing provides a indication as to the cause of its damage. In many cases, it is possible to examine a damaged bearing to determine the cause of failure. This information is useful in taking corrective action to prevent recurring failure. Examining a damaged bearing is most reliable when damage or wear is at an early stage.

Analyzing a bearing to determine the cause of failure is sometimes difficult due to the stages of problems and symptoms. For example, a bearing may have a dark discolored appearance of failure due to high temperatures. However, this bearing may have gone through a series of other failures first, such as spalling or fretting corrosion.

Conditions providing clues when analyzing bearing failure include spalling, false Brinell damage, excessive temperature damage, fretting corrosion, misalignment wear, thrust damage, or electrical pitting and fluting.

Spalling. *Spalling* is the flaking away of metal pieces due to metal fatigue. *Metal fatigue* is the fracturing of worked metal due to normal operating conditions or overload situations. A ball or roller that is under a load and rolls over a bearing race momentarily distorts the metal of the ball or roller and race. This distortion or flexing occurs 4 million times in a 40 hour week if the bearing is rotating at 1700 rpm. Eventually, millions of microscopic fractures form and the bearing begins to spall. **See Figure 9-8.**

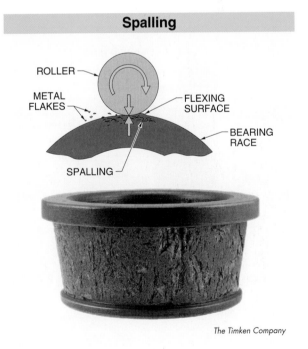

The Timken Company

Figure 9-8. Spalling is the flaking away of metal pieces due to metal fatigue.

Even though there are conditions of advanced destruction that produce little evidence to the initial problem, there are also conditions where clues may be used to correct an ongoing problem. For example, in a bearing failure due to spalling, the flaked metal created excess friction, overheated the lubricant, broke down the lubricant to an acidic condition, and burned up the bearing. Little evidence is given to determine the cause of the failure because the bearing was totally destroyed. Total destruction of the bearing and machine downtime could have been prevented if the bearing was analyzed at the spalling stage. A determination would still be needed to establish whether the flaking was due to overload or if the bearing had reached its life expectancy.

False Brinell Damage. *False Brinell damage* is bearing damage caused by forces passing from one ring to the other through the balls or rollers. False Brinell damage occurs on poorly installed bearings and bearings such as motor bearings or wheel bearings that sit on shelves that vibrate or are roughly transported over distances without rotation. False Brinell damage is also caused by pressure applied to the ring that has a loose fit during bearing removal. The vibration and hammering on a non-rotating bearing causes marks or indentations on the race that are spaced exactly the same distance apart as the balls or rollers. **See Figure 9-9.**

Excessive-Temperature Damage. As the temperature of steel increases, it discolors, turning from silver to blue to black. In addition, the hardness of steel decreases with an increase in temperature. Bearings at excessive temperatures deform more than normal which creates greater resistance and friction. **See Figure 9-10.**

Excessive-Temperature Damage

MELTED AND DEFORMED METAL

BLUE/BLACK COLORED STEEL

The Timken Company

Figure 9-10. Excessive temperatures result in melted and deformed bearing metal.

False Brinell Damage

FORCE

INNER RING

FALSE BRINELL DAMAGE

OUTER RING

VISE

SHAFT

INDENTATIONS

The Timken Company

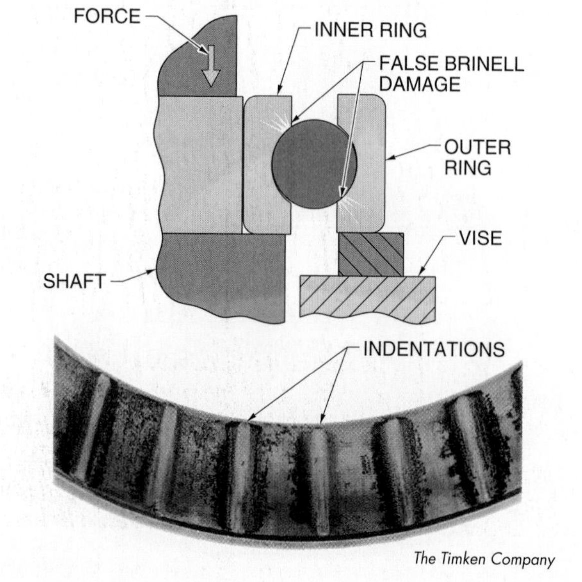

Figure 9-9. False Brinell damage is the indentations on the inner and outer ring races caused by the balls or rollers during rough handling or improper removal.

Another sign that bearings have overheated is the presence of solid or caked lubricant. The darkened, brittle grease has gone through the stages of being heated to its dropping point, allowing the thickener to be baked and burned. The *dropping point of grease* is the temperature at which the oil in the grease separates from the thickener and runs out, leaving just the thickener. This condition is created due to poor alignment, contaminated lubricant, overloading, or high speeds.

Fretting Corrosion. *Fretting corrosion* is the rusty appearance that results when two metals in contact are vibrated, rubbing loose minute metal particles that become oxidized. In many cases, fretting is a normal condition that appears as the discoloration on the outer surface of the outer ring between the outer ring and the housing. This happens as moisture from the air settles between the two contacting and unprotected metal surfaces.

Fretting corrosion becomes harmful when the oxidation breaks down supporting wall surfaces, creating looseness. Fretting corrosion is also harmful as its oxidation particles (oxides) mix with and break down the bearing lubricant. **See Figure 9-11.**

Fretting Corrosion

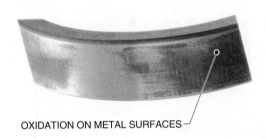

OXIDATION ON METAL SURFACES

NTN Bearing Corporation of America

Figure 9-11. Fretting corrosion is the rusty appearance that results when two metals in contact are vibrated, rubbing loose minute metal particles that become oxidized.

Misalignment Wear. Bearing surfaces that are misaligned appear as worn surfaces on one side or opposing sides of a bearing. Rollers in roller bearings can leave wear marks on one side of the bearing inner race. The roller may also show high and low trails on the inside of the outer race. Misalignment wear may also appear as lack of fretting on two sides of the outer surface of the outer ring. **See Figure 9-12.**

Misalignment Wear

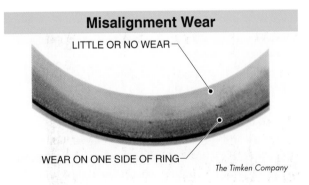

LITTLE OR NO WEAR

WEAR ON ONE SIDE OF RING

The Timken Company

Figure 9-12. Bearing surfaces that are misaligned appear as worn surfaces on one side or opposing sides of a bearing.

Safety Tip

Contact local lubricant suppliers for lubricant recommendations. Failure to maintain proper lubrication of a bearing can result in equipment failure, creating a risk of serious bodily harm.

Thrust Damage. *Thrust damage* is bearing damage due to axial force. Thrust damage on ball bearings appears as marks on the shoulder or upper portion of the inner and outer races and will be anywhere from a slight discoloration to heavy galling. *Galling* (adhesive wear) is a bonding, shearing, and tearing away of material from two contacting, sliding metals. The amount of galling is proportional to the applied load forces. Thrust damage on plain bearings appears as heavy wear at the bearing ends. **See Figure 9-13.**

Thrust Damage

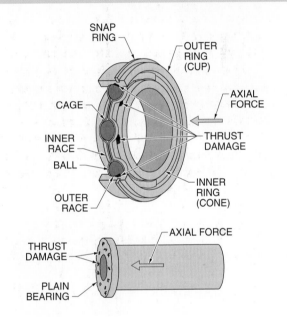

SNAP RING

OUTER RING (CUP)

CAGE

AXIAL FORCE

INNER RACE

THRUST DAMAGE

BALL

INNER RING (CONE)

OUTER RACE

AXIAL FORCE

THRUST DAMAGE

PLAIN BEARING

Figure 9-13. Thrust damage in ball bearings appears as marks on the shoulder or upper portion of the inner and outer races or as heavy wear at the bearing ends of plain bearings.

Electrical Pitting and Fluting. Electrical current passing through a machine can be harmful to humans, other electrical components, and machine bearings. Current may pass through machine parts without harm to humans. This happens as current passes from its introduction, such as static electricity created by the manufacturing process, electrical system feedback, or welding currents, to a grounded connection. When it passes through bearings, the current can etch or pit bearing surfaces. Mild electrical currents may not etch the metal, but can create high enough temperatures as current transfers through the bearing to burn and break down lubricants.

Welding current damage is observed as short pitted lines on balls or rollers that were stationary when the current was present. **See Figure 9-14.** The race has corresponding damage, but this is not normally observed unless the bearing is destroyed. Electrical feedback created by certain forces throughout plant electrical usage, faulty wiring, and static electricity can be prevented from flowing through a machine if extra grounding is provided. Extra grounding of a machine can be as simple as running a wire from the machine to a pneumatic or water line.

Electrical Pitting And Fluting

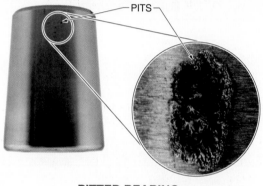

PITS

PITTED BEARING

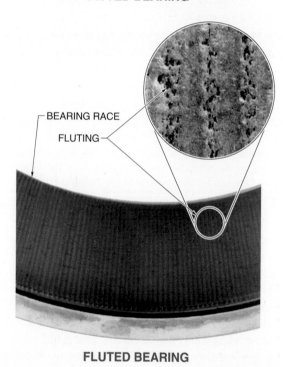

BEARING RACE

FLUTING

FLUTED BEARING

The Timken Company

Figure 9-14. Welding on a machine that is not grounded properly can permanently damage the machine bearings.

Warning: Never ground a machine by connecting a wire from the machine to a gas or oil pipe.

Fluting is observed in roller bearings that were rotating while welding currents passed through them. *Fluting* is the elongated and rounded grooves or tracks left by the etching of each roller on the rings of an improperly grounded roller bearing during welding. In roller bearings, fluting is caused by electrical arcing and pitting the length of each roller in the bearing. Damage from welding can be prevented by attaching the welding ground clamp in a location where no bearings are between the ground and the weld.

The Timken Company

Bearings are inspected after they are cleaned and dried by removing one roller to enable the inner race, cage, and rollers to be examined for damage.

PARTS PREPARATION

After all bearing parts have been dismantled, they should be cleaned and spread out on a clean surface for inspection. Cleaning is accomplished by dipping or washing the housing, shaft, bearing, spacers, and other parts in a clean, nonflammable cleaning solvent. Remove all traces of dirt, grease, oil, rust, or any other foreign matter. Dry all parts with a lint-free cloth.

Caution should be taken when using part cleaning solutions. Seals, O-rings, and other soft materials may deteriorate due to incompatibility. Check the cleaning solution label or contact the supplier before use. If in doubt, a lightweight, warm mineral oil is a good cleaning and flushing fluid. Care should be taken to clean housing and shaft bearing seats, corners, and keyways.

Bearings may be blown dry using clean, dry compressed air. Do not allow the air pressure to spin the bearings because this scratches the surfaces and the bearings may fly apart. Wipe all clean and dry parts with a lightweight oil and wrap or cover them to protect from dust and dirt. Inspect all parts carefully for nicks, burrs, or corrosion on shaft seats, shoulders, or faces. Closely inspect bearing components for indication of abnormality or obvious defects. Check both inner and outer races for cracks and the balls or rollers for wear or breaks.

Remove nicks, corrosion, rust, and scuffs on shaft or housing surfaces that may make assembly unsatisfactory. Check that corners and bearing seats are square and that all diameters are round, in tolerance, and without runout. Replace any worn spacers, shafts, bearings, or housings. Check the housing shoulder to ensure that it is square enough to clear the bearing corner.

Once lightly oiled, inspection of a bearing is accomplished by holding the inner ring while rotating the outer ring. This inspection may show uneven tolerances, particle contamination, or chipped elements. Replace any bearing when there is doubt about the condition of the bearing.

SPM Instrument, Inc.

Bearings must be handled and installed with care to prevent damage because a greater number of bearings are damaged during installation than during use.

BEARING INSTALLATION

Successful bearing installation requires cleanliness, correct bearing selection, mounting methods, tool use, and tolerance specifications. Proper bearing assembly is required for proper bearing performance, durability, and reliability.

Caution: More bearings are destroyed, damaged, and abused during the installation stage than from malfunction during their life expectancy. Often, mechanics handle bearings as though it is impossible to damage them by force, dirt, or misalignment. Within a few short seconds, carelessness can destroy the protective measures of any bearing.

Proper Bearing Selection

Bearings are often selected without the use of manufacturer's specifications. Certain factors other than dimensions must be observed when a replacement bearing is chosen by comparison of a removed bearing instead of from an equipment manual or parts book. Factors to be considered include the exact replacement part number, the type and position of any seal, the direction of force and positioning of a required high shoulder, and whether a retaining ring is required. Early failure is possible if a replaced bearing, even though it is dimensionally correct, lacks any other requirement. The factors in bearing selection include the load type and orientation of the bearing being replaced. Close attention must also be paid any special features of a removed bearing, such as special seal configurations, spherical outside diameters, noncircular bores, etc.

All types of bearings, especially rolling-contact bearings, are available in many design variations and vary greatly in internal design. Always ensure that a bearing is a direct replacement for the one required by the manufacturer when replacing the bearing. Check the equipment manual for the proper bearing number. A previous bearing replacement may have been incorrect. Bearing numbers on the box should be checked before opening the box and bearings in the box should be checked to ensure the number on the bearing corresponds to the number on the box. Duplex mounted bearings must not be mismatched and must be from the same manufacturer.

Know and identify angular contact bearings and their position within an assembly. For example, replacing a Conrad bearing with a general-purpose bearing may produce rapid deterioration of the bearing. This is because the Conrad bearing may be used for axial support and general-purpose bearings cannot provide axial support.

Bearing Mounting

Bearing mounting procedures have a great effect on performance, durability, and reliability. Precautions should be taken to allow a bearing to perform without excessive temperature rise, noise from misalignment or vibration, and shaft movements. During installation, force must be applied uniformly on the face or ring that is to be press fit. Any method that presses the bearing on squarely without damage may be used. Press fits may be accomplished by using a piece of tubing, steel plate, and hammer; an arbor press; or a hydraulic ram. Wood should not be used if there is any possibility of contaminating a bearing with wood splinters or fibers.

When reinstalling bearings, insert the lightly oiled shaft in the bearing and line up the marks made at disassembly. Start the new or used bearing assembly by hand. If pipe is used as an installation tool, press the bearing onto the shaft by placing the pipe only on the press-fit ring. This pipe must be the proper diameter, clean, and have both ends cut square. **See Figure 9-15.** Bearings that can be mounted in either direction should be mounted with the part number facing out for ease of future identification.

Bearings must be mounted so their inner and outer race is not distorted, their rolling elements do not become bound, and each machine part remains in its proper relationship. This is accomplished by proper mounting alignment (trueness) of machined parts, using proper machine tolerances (internal clearances), and allowing for additional tolerances due to thermal movement.

Tech Tip

Damage can easily occur internally to a bearing if removal or installation force is passed from one bearing ring to the other through the ball or roller set.

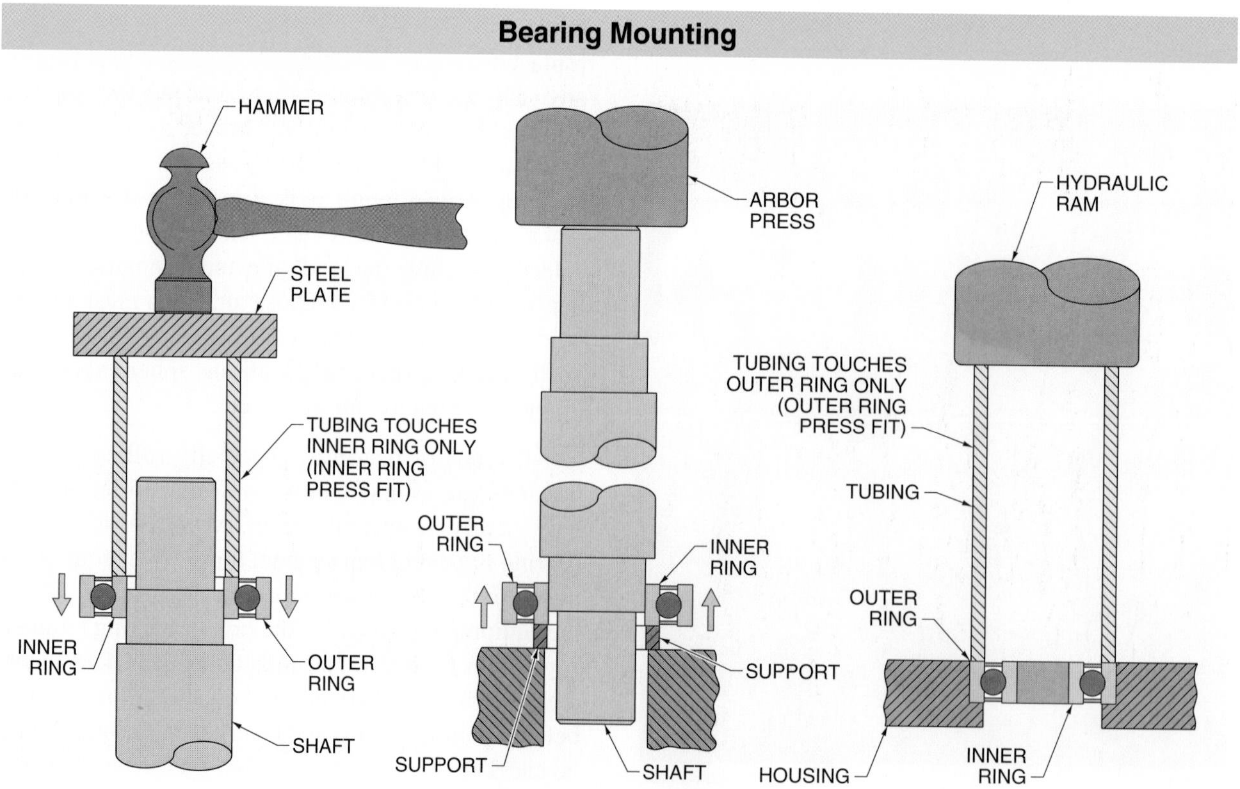

Bearing Mounting

Figure 9-15. Pressure applied when mounting must be applied squarely to the ring being pressed.

NTN Bearing Corporation of America

Uneven wear between a bearing inner ring, outer ring, and rollers is an indication of poor bearing mounting.

Bearings are mounted with the rotating ring pressed on a shaft or in a housing. The nonrotating ring is pushed or slid on a shaft or in a housing and may be held in place by a snap ring. Generally, the rotating element of a machine is the shaft with the inner ring pressed on to the shaft. However, if the outer part of the machine is the rotating element, the outer ring is press fit. The exception to this rule is with the use of heavy-duty cylindrical roller bearings where the extra loads require that both rings be press fit. The press fit ring is then clamped to prevent axial movement. The inner ring of a bearing mounted on a shaft is generally held in place by a locknut, a clamp plate, or a castle-nut and cotter pin. **See Figure 9-16.**

Rolling-Contact (Anti-Friction) Bearing Mounting

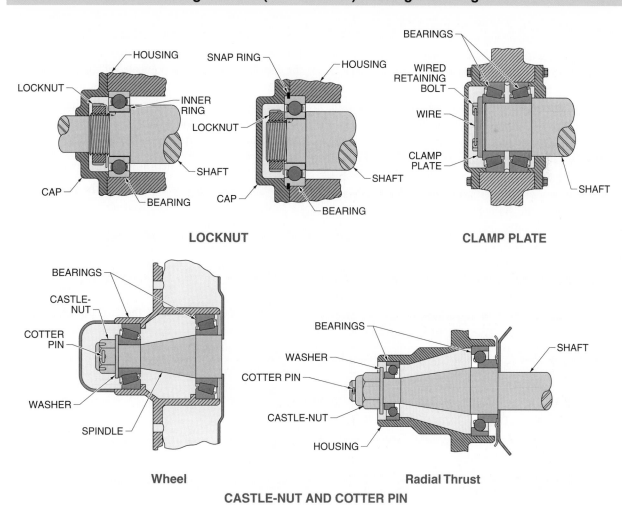

Figure 9-16. Bearing mounting requires the rotating ring to be mounted securely using locknuts, clamp plates, or castle-nuts and cotter pins.

Precision Class Bearings. Precision class bearings require quality mounting and control of shaft, housing, and bearing deflection. Deviations of more than 0.0005″ in precision class assemblies can cause significant vibration. For example, deviations of guided missile bearing applications must be limited to less than 0.00002″. Precision class bearings are generally marked with their high points of runout. Outer rings may be marked with a "V" on each ring and each ring should be aligned with the other. Check for the bearing manufacturer's runout marking. Inner race runout may be identified by a dot, a copper dot, or some other mark at the high point. After identifying the high point on the shaft, place the bearing runout mark 180° from the shaft high point.

Bearing Mounting Using Temperature. During bearing installation, it may be necessary to increase the inner ring diameter by heating the bearing. Regardless of bearing size, increasing the diameter with heat is the simplest way to mount a bearing that must be press fit on a shaft. Prelubricated bearings must not be heated for installation. Heating bearings may be accomplished using a light bulb, an oven, clean hot oil that has a high flash point, a hot plate, or induction heat. The light bulb, oven, and induction heating methods are most reliable because their temperature is easy to control. The hot oil method is less reliable because the oil temperature is harder to control and the oil is difficult to keep clean. The light bulb heating method is the best and most economical. This method uses a light bulb placed in the bore of a bearing to provide heat to increase the diameter of the bearing. The temperature is controlled by the length of time the bulb is placed in the bearing. This method works well because the inner ring is the only component to be brought to high heat, allowing handling and assembly using the outer ring. **See Figure 9-17.**

Temperatures must be even throughout the inner bearing ring and controlled to between 175°F and 200°F. Torches heat unevenly, distort diameters, and must not be used. Temperatures over 250°F may reduce the hardness of bearing metals, resulting in early failure. Bearing temperatures may be determined using thermal crayons, which melt at specific temperatures. Induction heaters, operating electrically and rapidly, leave metals magnetized. All small metal particles are drawn to the assembly if bearings are not demagnetized.

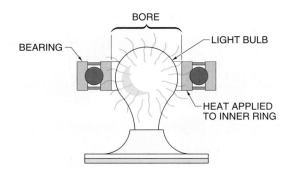

Figure 9-17. The light bulb heating method enlarges the diameter of a bearing inner ring through an increase in the ring temperature.

Freezing may be required to reduce the outside diameter of a bearing to allow installation into a housing when heating methods are not possible. Freezing causes shaft sizes to be reduced to allow installation of prelubricated bearings. A mixture of dry ice and alcohol can be used to lower the temperature of a bearing to –20°F. Condensation forms on bearings in areas where ambient conditions are humid. Corrosion is prevented if the condensate is wiped or blown off after assembly, followed by thorough lubrication. Corrosion should not be a problem if components were lightly oiled before freezing. Precautions that must be taken when mounting bearings include:

- Know the bearing function in a machine.
- Keep all bearings wrapped or in their sealed container until ready to use. Reusable bearings should be treated as new.
- Maintain clean tools, hands, and work surface, and work in a clean environment.
- Use clean, lint-free cloths when wiping bearings.
- Never attempt to remove the rust preventive compound used by the manufacturer unless specifically recommended.
- Use the best bearings available within reason. The life and reliability of a bearing is generally related to its cost.
- Always follow the instructions of the heating equipment manufacturer when bearings are to be heated for assembly.
- Use rings, sleeves, or adaptors that provide uniform, square, and even movements.
- Prevent cocking during bearing installation by starting races evenly on the shaft without pressure devices.

- Never strike the bearing with a wooden mallet or wooden block.

- Never apply pressure on the outer ring if the inner ring is press fit and never apply pressure on the inner ring if the outer ring is press fit.

- Be careful not to abuse, strike, force, press on, scratch, or nick bearing seals or shields.

Understanding Bearing Function. No universal bearing exists that can do all of the functions and applications required in industry. In many cases, a review of the machine function and its bearing requirements may indicate if proper bearings are being used.

Provisions made for thermal expansion within a machine are generally published by the machine manufacturer and are listed as space tolerances between housing, bearing components, and shaft. Greater space tolerances are allowed for plain bearings than for rolling-contact bearings because plain bearings are more susceptible to damage from higher temperatures.

Rolling-contact bearings must be firmly mounted so end play and shaft expansion and contraction due to thermal activity is minimized. *End play* is the total amount of axial movement of a shaft. To work properly, all bearings require axial and radial operating clearance, but are not intended to move or flex under load. Tapered roller bearings are adjusted to a specified end play or end lateral movement with the use of a dial indicator. Bearing movement is erratic, noisy, and damaging if end play is excessive. **See Figure 9-18.**

In some cases, zero clearance or preloading of tapered roller bearings is necessary. *Preloading* is an initial pressure placed on a bearing when axial load forces are expected to be great enough to overcome preload force, thereby resulting in proper clearances. Preloading is accomplished by the use of shims or adjustable nut settings. If preloading is too tight, lubrication is squeezed out, metal-to-metal contact occurs, and the increase in temperature damages the bearing. If preloading is too loose, bearing movement is sloppy and damaging. In some cases, preloading is measured with a torque wrench to determine the force required to rotate the bearing assembly. Proper bearing adjustment specifications are generally provided by equipment manufacturers.

Cone Mounter Company

Proper bearing installation is required for maximum bearing life. Heating a bearing before installation expands the inner race, allowing easy bearing mounting on a shaft.

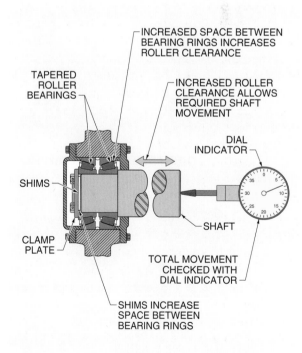

Figure 9-18. Tapered roller bearings are adjusted to a specified end play with the use of a dial indicator.

Tapered Roller Bearing Adjustment. Tapered roller bearings are generally arranged in pairs for opposed mounting. This ensures that thrust force in either direction is taken by one of the bearings. Careful control of the clearance between bearing faces is required to prevent excessive shaft movement. Adjustments are accomplished using methods such as a threaded cup follower, ring spacers (shims), nut and lock washer, double-nuts, or castle-nut and cotter pin. **See Figure 9-19.**

A *threaded cup follower* is a tapered bearing gap adjusting device that is used to adjust shaft endplay by controlling the amount of clearance between the bearings. A threaded cup follower is a dish-shaped cap that has a threaded outside diameter and a machined face. This adjusting device is screwed into a housing until its machined face makes contact with the bearing cup (outer ring). The threaded cup follower is then screwed in until the proper shaft end play is achieved. Once proper adjustment has been met, a locking screw is secured to prevent cup follower movement.

Special attention must be given to the arrangement of axial/radial bearing pairs. Assembly design may require back-to-back, face-to-face, or tandem arrangement. Ring spacer (shim) assemblies are generally used where a nonadjustable (NA) application exists and the ring spacer thickness is determined by the equipment manufacturer. Ring spacers are placed beneath end-cap assemblies to provide adjustment between the tapered roller bearing and the machined face of the end cap. In most cases, clearance is increased as ring spacers are added, and clearance is decreased as ring spacers are removed.

Nut and lock washer adjustment is made using a keyed washer, keyed shaft, and a nut with a group of notches on the outside diameter. The more notches designed into the nut, the finer the assembly may be adjusted. Upon final adjustment, a tab on the washer is bent over the corresponding notch on the nut. This assembly is held firm by a tab on the washer engaged in the shaft keyway.

The double-nut method of adjusting the running clearance of a tapered bearing assembly is accomplished using a keyed shaft, two notched nuts, and a keyed washer. With the bearing assembled onto the

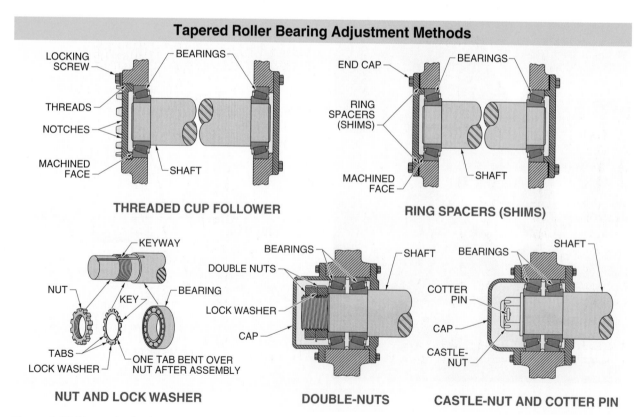

Tapered Roller Bearing Adjustment Methods

THREADED CUP FOLLOWER

RING SPACERS (SHIMS)

NUT AND LOCK WASHER

DOUBLE-NUTS

CASTLE-NUT AND COTTER PIN

Figure 9-19. Tapered roller bearings are adjusted using a threaded cup follower, ring spacers (shims), nut and lock washer, double-nuts and cotter pin.

shaft and into the housing, the first nut is screwed in until proper bearing clearance has been established. The keyed washer is then slid onto the shaft and the second nut tightened against the washer and the first nut. Care must be taken when tightening the second nut that the first nut does not turn, losing the clearance. After the second nut has been tightened, a tab is bent over both nuts, locking them in place.

Similar to the nut and lock washer assembly, a castle-nut and cotter pin uses a notched (slotted) nut for adjustment. Upon final adjustment, a castle-nut slot is aligned with a hole through the shaft, and a cotter pin is inserted to hold the assembly firm.

Tapered Bore Bearings. A *tapered bore bearing* is a bearing that uniformly increases or decreases from one face to the opposite face. Tapered bore bearings are used directly on tapered shafts or on straight shafts using tapered sleeves. Tapered sleeves (split adapter sleeves) allow firm mounting of tapered bore ball and roller bearings. The internal clearance between the shaft and the bearing is reduced as a locknut is tightened, forcing the bearing onto the tapered sleeve. A tabbed washer is bent over a flat on a locknut to hold the assembly firm. Tapered bore bearings using a removable sleeve must always be firm against the shaft shoulder. **See Figure 9-20.**

MACHINE RUN-IN

A machine run-in check should be made after bearing assembly is complete. A run-in check starts with a hand check of the torque of the shaft.

WARNING: Ensure power is locked out when manually rotating a shaft. Unusually high torques normally indicate a problem with a tight fit, misalignment, or improper assembly of machine parts. Listen carefully for clicks, squeals, or thumps and investigate. Restore machine power and listen for unusual noises. High noise levels may indicate excessive loading or cocked or damaged bearings. Correct the problem before continuing.

Final checks are accomplished by measuring machine temperatures. High initial temperatures are common due to bearings being packed with grease, which produces

excess friction. Generally, turning a machine OFF to cool down before restarting resolves the high temperature. Run-in temperatures should decrease to within recommended ranges. Any machine with temperatures that continue to run high should be corrected before proceeding. Continued high temperatures are generally a sign of tight fit, misalignment, or improper assembly. After a machine is placed back in operation, record motor amperage readings and bearing temperature for proactive and predictive maintenance procedures.

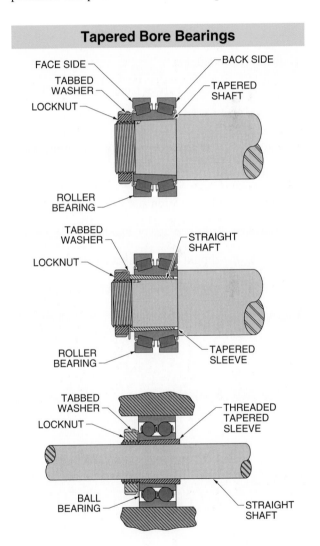

Figure 9-20. Tapered bore bearings are used directly on the tapered shafts or straight shafts using tapered sleeves.

Review

1. Define radial load, axial load, and radial and axial loads and give an example of each.

2. Why are rolling-contact bearings referred to as anti-friction bearings?

3. What do dark discolored metals, rusting surfaces, and split or fractured rings indicate about failed or failing bearings?

4. Why are bearings designed for thrust loads installed in only one direction?

5. Why is mixing different manufacturer's bearings typically not recommended?

6. Define plain bearing.

7. What two types of babbitt metals work best for plain bearing loads?

8. Explain how the Brinell hardness test works.

9. Why is it important to never strike a bearing directly with a hammer during bearing removal?

10. Define spalling.

11. Define fretting corrosion and list two examples of when it becomes harmful.

12. How does thrust damage on ball bearings appear?

13. Which way should bearings that can be mounted in either direction face and why?

14. List at least eight different precautions that should be taken when mounting bearings.

15. Define tapered bore bearing.

Digital Resources

ATPeResources.com/Quicklinks
Access Code 462409

CHAPTER 10

FLEXIBLE BELT DRIVES

Flexible belt drives are systems in which resilient flexible belts are used to drive one or more shafts. Belts used for power transmission in flexible belt drives include V-belts, double V-belts, and timing belts. V-belts that are not aligned properly are destroyed prematurely due to excessive side wear, broken or stretched tension members, or rolling over in the pulley. The main concerns when working with flexible belt drives are the safety of the technician working on the machine, co-workers working in the area of the machine, and the machine itself.

The Gates Rubber Company

Chapter Objectives

- Define flexible belt drives.
- Describe the characteristics of V-belts.
- Describe the characteristics of V-belt pulleys.
- Describe the characteristics of double V-belts.
- Describe the characteristics of timing belts.
- Define variable-speed belt drives.
- Explain how the speed change of belts and pulleys is accomplished.
- Identify common belt drive safety guidelines.

Key Terms

- flexible belt drive
- v-belt
- nominal value
- molded notch belt
- belt pitch line
- v-belt pulley
- offset misalignment
- nonparallel misalignment
- angular misalignment
- belt deflection method
- double v-belt
- timing belt
- pitch length
- circular pitch
- variable-speed belt drive
- lockout
- tagout
- blockout

FLEXIBLE BELT DRIVES

A *flexible belt drive* is a system in which a resilient flexible belt is used to drive one or more shafts. Flexible belt drives offer convenient power transmission between two or more shafts on a machine. A *machine* is an assembly of devices that transfers the force, motion, or energy input from one device to a force, motion, or energy output at another. Belts used for power transmission in flexible belt drives include V-belts, double V-belts, and timing belts.

V-Belts

A *V-belt* is an endless power transmission belt with a trapezoidal cross section. V-belts are made of molded fabric and rubber for body and bending action. V-belts contain fiber or steel cord reinforcement (tension members) as their major pulling strength material. The *tension member* is the load-carrying element of a belt which prevents stretching. V-belts become thinner and weaker as tension members break.

V-belts are resilient, quiet, and able to absorb many shocks because they are made of cloth and rubber.

V-belts are generally classified as standard or high-capacity. Standard V-belts are designated as A, B, C, D, or E. High-capacity V-belts are designated as 3V, 5V, or 8V. The letter or number designation also indicates the cross-sectional dimension and thickness of the belt. **See Figure 10-1.**

V-belts run in a pulley (sheave) with a V-shaped groove. V-belts transmit power through the wedging action of the tapered sides of the belt in the pulley groove. The wedging action results in an increased coefficient of friction. **See Figure 10-2.** V-belts do not normally contact the bottom of the pulley. A pulley or belt that has worn enough so that the belt touches the bottom of the pulley should be replaced. The belt becomes shiny and slips and burns if allowed to bottom out. More than one belt may be used if additional power transmission is required. However, each belt must be of the same type and size.

V-Belt Sizes. V-belt replacement, whether for preventive maintenance or equipment breakdown, starts with proper identification and sizing of the belt being replaced. The technician can prevent many premature belt failures by selecting the proper belt, belt size, and installation procedure.

V-Belt Classifications

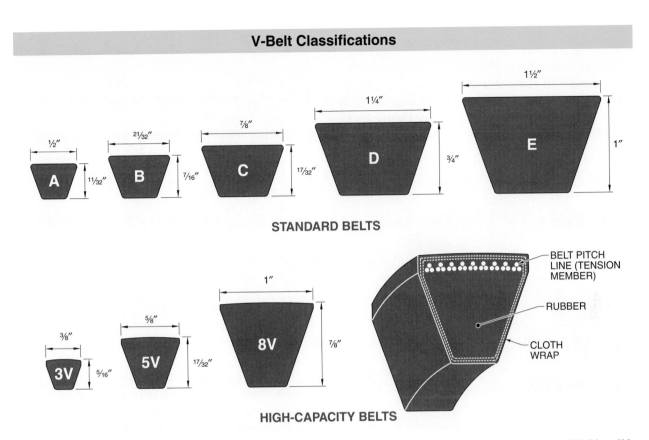

Figure 10-1. Standard V-belts are designated as A, B, C, D, or E and high-capacity V-belts are designated as 3V, 5V, or 8V.

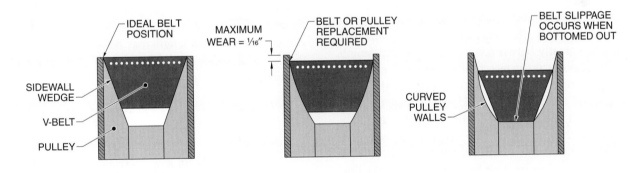

Figure 10-2. V-belts transmit power through the wedging action of the tapered sides of the belt in a pulley groove.

Standard V-belt sizes are identified by letter/number combinations. Letters identify the cross-sectional dimensions and numbers identify the nominal length. A *nominal value* is a designated or theoretical value that may vary from the actual value. For example, a flexible belt with a nominal length of 24″ may be significantly different when compared to another flexible belt with a nominal length of 24″. An X following the letter indicates that the belt is a molded notch belt. A *molded notch belt* is a belt that has notches molded into its cross-section along the full length of the belt. These notches provide for cooler operation along with greater flexibility.

Manufacturers measure each belt under tension and mark the belt with an additional code number because belt fibers and rubber allow some stretching. Typically, the code numbers following the manufacturer's name indicate the effective (working) belt length in tenths of an inch.

The additional code numbers begin with 50, indicating that the belt is within nominal size tolerance range. A number is added or subtracted from 50 for each $\frac{1}{10}''$ over or under the nominal length. For example, a BX60 55 belt is a $\frac{21}{32}''$ notched belt, 60" long, manufactured $\frac{5}{10}''$ longer than the nominal length. **See Figure 10-3.**

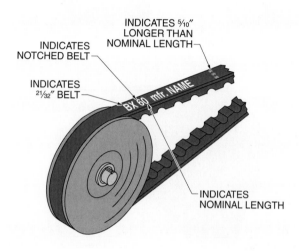

Figure 10-3. Standard V-belt sizes are identified by letter/number combinations.

Even though a belt may be classified as being the same nominal length, code numbers indicate that belts can vary greatly between actual and nominal length. Code numbers must match or be within a matching number range when more than one belt is used as a set. In general, sets should be made of belts with the same code range numbers. **See Figure 10-4.** For example, if a set of three belts are 300" long and one belt reads D300 50, the other two belts must be the same or within a range of $\frac{4}{10}''$ of the first belt. These belts may read anywhere from D300 48 to D300 52.

V-belts are selected based on the pulley shape, pulley size, and the outside circumference around the pulleys. The easiest method of determining a belt size is to locate the belt letter/number code stamped on the top surface of the belt. A belt and sheave groove gauge may be used to determine the proper belt cross section if the number is not legible. A *belt and sheave groove gauge* is a gauge that has a male form to determine the size of a pulley and a female form to determine the size of a belt. A set of gauges, generally on a chain or ring, consists of various belt and pulley sizes.

Recommended Belt Set Length Variations

Code Range†	Nominal Lengths*				
	A	B	C	D	E
2	26 – 180	35 – 180	51 – 180	—	—
3	—	195 – 315	195 – 255	120 – 255	144 – 240
4	—	—	270 – 360	270 – 360	270 – 360
6	—	—	390 – 420	390 – 660	390 – 660

* in in.
† for matching numbers

Figure 10-4. Variations in length for belt sets allow only a few tenths of an inch over or under the belt's nominal length.

V-belt length may be determined by measuring the outside circumference using flexible tape while the belt is still on the pulleys. Generally, the correct belt length is the next standard belt shorter than the measurement. For example, an A belt having a 27" outside circumference is replaced with an A26 belt. The reduced length is closer to the actual belt pitch line. *Belt pitch line* is a line located on the same plane as the belt tension member. Belt pitch line is the effective length of a belt. A more accurate belt length may be determined using the pulley diameters. **See Figure 10-5.** Belt length is found by applying the formula:

$$L = 2 \times C + 1.57 \times (D + d) + \frac{(D - d)^2}{4 \times C}$$

where
L = belt length (in in.)
2 = constant
C = distance between pulley centers (in in.)
1.57 = constant
D = large pulley diameter (in in.)
d = small pulley diameter (in in.)
4 = constant

Safety Tip

Belts should be replaced if there are signs of cracking, fraying, unusual wear, or loss of teeth in a timing belt.

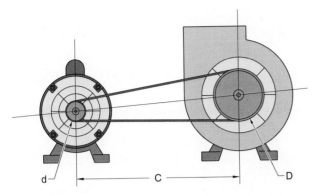

Figure 10-5. Belt length is calculated using the distance between pulley centers and the pulley diameters.

For example, what is the belt length for two pulleys 5″ and 10″ in diameter that are 30″ apart at their centers?

$$L = 2 \times C + 1.57 \times (D + d) + \frac{(D - d)^2}{4 \times C}$$

$$L = 2 \times 30 + 1.57 \times (10 + 5) + \frac{(10 - 5)^2}{4 \times 30}$$

$$L = 60 + 1.57 \times 15 + \frac{5^2}{120}$$

$$L = 60 + 23.55 + \frac{25}{120}$$

$$L = 83.5 + 0.208$$

$$L = \mathbf{83.758''}$$

V-belt Forces. V-belt forces are constantly changing as a belt bends around a pulley. Tension forces develop at the top of the belt and compressive forces build at the bottom of the belt. The amount of tension depends on the belt construction and the pulley diameter. Pulley diameters that are too small greatly reduce the life of the belt. Recommended minimum pulley diameters are used to reduce bearing loads and provide the longest possible belt life. For example, the minimum diameter for a pulley used on a system containing a 5 HP drive motor operating at 1750 rpm is 3.0″. **See Figure 10-6.**

The Gates Rubber Company

Abnormal wear on the bottom surface of a belt is caused by a belt too small for the pulleys or by worn or dirty pulleys.

Recommended Minimum Pulley Diameters*				
Motor HP	Motor Speed†			
	870	1160	1750	3500
½	2.2	—	—	—
¾	2.4	2.2	—	—
1	2.4	2.4	2.2	—
1½	2.4	2.4	2.4	2.2
2	3.0	2.4	2.4	2.4
3	3.0	3.0	2.4	2.4
5	3.8	3.0	3.0	2.4
7½	4.4	3.8	3.0	3.0
10	4.4	4.4	3.8	3.0
15	5.2	4.4	4.4	3.8
30	6.8	6.8	5.2	—
75	10.0	10.0	8.6	—
100	12.0	10.0	8.6	—

* in in.
† in rpm

Figure 10-6. Recommended minimum pulley diameters are used to reduce bearing loads and provide the longest possible belt life.

V-Belt Pulleys

A *V-belt pulley* is a pulley with a V-shaped groove. V-belt pulleys are manufactured to exact dimensions. All pulley dimensions, regardless of the manufacturer, should be machined to the same standard sizes. This allows various manufacturer's parts to be interchanged. **See Figure 10-7.**

V-belt pulleys may be fixed bore or tapered bore pulleys. A *fixed bore pulley* is a machine-bored one-piece pulley. Fixed bore pulleys have an integral hub cast into the pulley. Fixed bore pulleys are chosen by specific bore, outside diameter (OD), and groove dimensions. A *tapered bore pulley* is a two-piece pulley that consists of a tapered pulley bolted to a tapered hub (bushing). The assembly becomes as sound as a press fit when the two pieces are bolted together. The close fit is due to the extreme pressure created from the angled force of the tapered mating surfaces. The angled force prevents fretting corrosion and freezing of parts. The tapered mating surfaces also allow easy removal. For this reason, no lubricant is required. Also, using a lubricant for assembly may cause pulley breakage due to hydraulic (grease or oil) forces created when tightening pulley bolts. **See Figure 10-8.**

Standard V-Belt Pulley Groove Dimensions

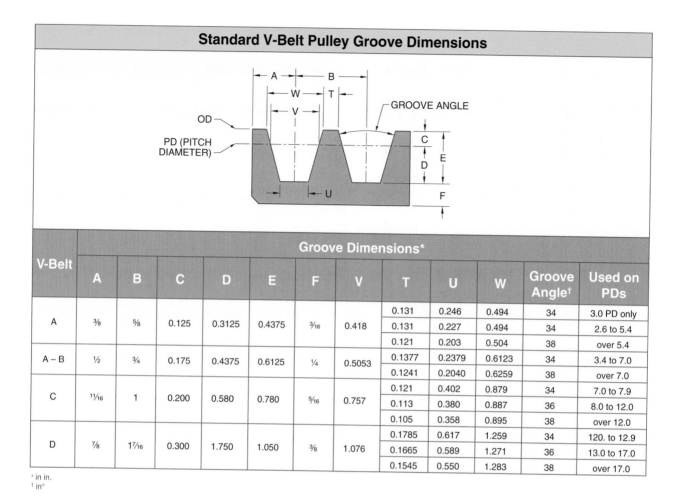

V-Belt	A	B	C	D	E	F	V	T	U	W	Groove Angle†	Used on PDs
A	⅜	⅝	0.125	0.3125	0.4375	3⁄16	0.418	0.131	0.246	0.494	34	3.0 PD only
								0.131	0.227	0.494	34	2.6 to 5.4
								0.121	0.203	0.504	38	over 5.4
A – B	½	¾	0.175	0.4375	0.6125	¼	0.5053	0.1377	0.2379	0.6123	34	3.4 to 7.0
								0.1241	0.2040	0.6259	38	over 7.0
C	11⁄16	1	0.200	0.580	0.780	5⁄16	0.757	0.121	0.402	0.879	34	7.0 to 7.9
								0.113	0.380	0.887	36	8.0 to 12.0
								0.105	0.358	0.895	38	over 12.0
D	⅞	17⁄16	0.300	1.750	1.050	⅜	1.076	0.1785	0.617	1.259	34	120. to 12.9
								0.1665	0.589	1.271	36	13.0 to 17.0
								0.1545	0.550	1.283	38	over 17.0

* in in.
† in°

Figure 10-7. Standard dimensions used by pulley manufacturers allow parts to be interchanged.

Tapered bore pulleys are installed by sliding the flange end of the hub on a shaft and inserting the key. The hub is positioned on the shaft and the set screw is lightly snugged over the key. The large end of the tapered bore is slid onto the hub and the unthreaded bolt holes on the pulley are lined up with the threaded bolt holes in the hub. The cap screws are screwed into the hub and tightened alternately and evenly. Never tighten enough to close the gap between the hub flange and the pulley. To ensure proper alignment and square-ness to prevent wobble, pull up screws evenly and progressively to approximately one-half the required torque values and check alignment before completing torque settings. Finally, apply the proper torque to the screws and tighten the set screw on the key. Torque values are given for safely assembling the pulley to each hub size. **See Figure 10-9.** Different size tapered bore hubs require particular size keyseats and fit on a particular size bore.

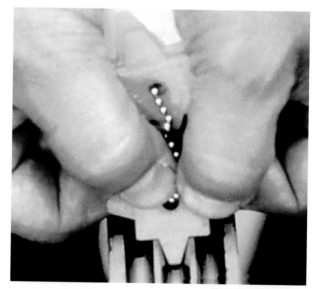

The Gates Rubber Company

A pulley gauge is used to check a pulley for wear. A pulley with 1⁄16″ wear on either side of the gauge should be replaced.

V-Belt Pulleys

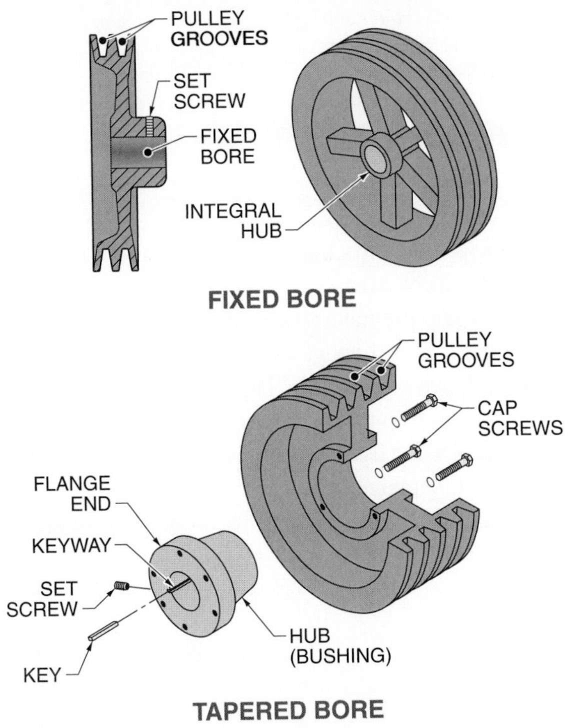

PULLEY GROOVES

SET SCREW

FIXED BORE

INTEGRAL HUB

FIXED BORE

PULLEY GROOVES

CAP SCREWS

FLANGE END

KEYWAY

SET SCREW

KEY

HUB (BUSHING)

TAPERED BORE

Figure 10-8. V-belt pulleys may be fixed bore or tapered bore pulleys.

Recommended Tapered Bore Hub Torque Values		
Hub Size	Cap Screw Size and Thread	Torque*
JA	No. 10 – 24	3
SH – SDS – SD	¼ – 20	6
SK	⁵⁄₁₆ – 18	10
SF	⅜ – 16	20
E	½ – 13	40
F	⁹⁄₁₆ – 12	50
J	⅝ – 11	90
M	¾ – 10	150
N	⅞ – 9	200
P	1 – 9	300
W	1⅛ – 7	400
S	1¼ – 7	500

* in lb/ft

Figure 10-9. Overtightening cap screws can result in pulley or bushing breakage.

Some applications require the flange to be on the outside of the pulley assembly. In these applications, always leave room between the pulley and the motor or gear drive for tool movement. Pulleys should be placed as close as possible to the shaft bearing to prevent overhung loads. An *overhung load* is a force exerted radially on a shaft that may cause bending of the shaft or early bearing and belt failure. **See Figure 10-10.**

Overhung Loads

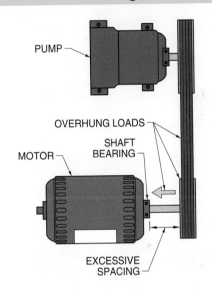

PUMP

OVERHUNG LOADS

SHAFT BEARING

MOTOR

EXCESSIVE SPACING

Figure 10-10. Bearings, pulleys, and belts are damaged from excess vibrations when the distance between the shaft bearing and the pulley is excessive.

Pulley Alignment. V-belts that are not aligned properly are destroyed prematurely due to excessive side wear, broken or stretched tension members, or rolling over in the pulley. Pulley misalignment may be offset, nonparallel, or angular. **See Figure 10-11.**

Offset misalignment is a condition where two shafts are parallel but the pulleys are not on the same axis. Offset misalignment may be corrected using a straightedge along the pulley faces. The straightedge may be solid or a string. Offset misalignment must be within ¹⁄₁₀″ per foot of drive center distance. *Nonparallel misalignment* is misalignment where two pulleys or shafts are not parallel. Nonparallel misalignment is also corrected using a string or straightedge. The device connected to the pulley that touches the straightedge at one point is rotated to bring it parallel with the other pulley so that the two pulleys touch the straightedge at four points.

Angular misalignment is a condition where two shafts are parallel but at different angles with the horizontal plane. Angular misalignment is corrected using a level placed on top of the pulley parallel with the pulley shaft. Angular misalignment must not exceed ½″.

Pulley Misalignment

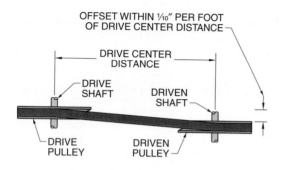

OFFSET WITHIN ⅟₁₀″ PER FOOT OF DRIVE CENTER DISTANCE

DRIVE CENTER DISTANCE

DRIVE SHAFT

DRIVEN SHAFT

DRIVE PULLEY

DRIVEN PULLEY

OFFSET

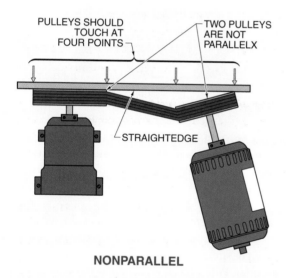

PULLEYS SHOULD TOUCH AT FOUR POINTS

TWO PULLEYS ARE NOT PARALLELX

STRAIGHTEDGE

NONPARALLEL

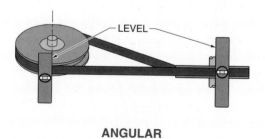

LEVEL

ANGULAR

Figure 10-11. Pulley misalignment is corrected by placing a straightedge across the pulleys and adjusting the position of the equipment so the pulleys touch the straightedge at four points.

V-Belt Tensioning. Proper alignment and tension of V-belts produces long and trouble-free belt operation. Excessive tension produces excessive strain on belts, bearings, and shafts, causing premature wear. Too little tension causes belt slippage. Belt slippage causes excessive heat and premature belt and pulley wear. The best tension for a V-belt is the lowest tension at which the belt does not slip under peak loads. Do not use belt dressing as a remedy for pulley wear or belt slippage. Two methods used for tensioning a belt include the visual adjustment and belt deflection methods. **See Figure 10-12.**

V-Belt Tensioning Methods

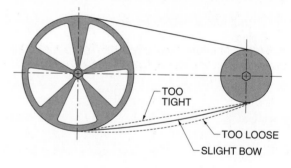

TOO TIGHT

TOO LOOSE

SLIGHT BOW

VISUAL ADJUSTMENT

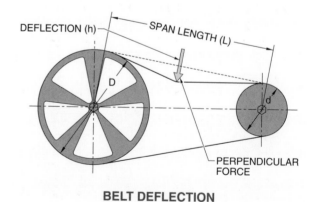

DEFLECTION (h)

SPAN LENGTH (L)

D

d

PERPENDICULAR FORCE

BELT DEFLECTION

Figure 10-12. Belt tensioning methods include the visual adjustment and belt deflection methods.

The Gates Rubber Company

Alignment is checked visually after pulleys are installed by placing a straightedge along the outside face of both pulleys.

The *visual adjustment method* is a belt tension method in which the tension is adjusted by observing the slight sag at the slack side of the belt. The visual adjustment method is used to roughly adjust belt tension. The belt is placed on the pulleys without forcing the belt over the pulley flange. Never force or pry a belt over the pulley flange. Apply tension to the belt by increasing the drive center distance until the belt is snug. Run the machine for approximately 5 min to seat the belt. Apply load to the belt and observe the slack side of the belt. A slight sag is normal. A squeal or slip indicates that the belt is too loose.

The *belt deflection method* is a belt tension method in which the tension is adjusted by measuring the deflection of the belt. The belt is installed and tension is applied by increasing the drive center distance. The span length (L) between the pulley centers and the deflection height (h) is measured. Proper deflection height is $\frac{1}{64}''$ per inch of span length. A perpendicular force at the midpoint of the span length is applied to measure for proper tension. Proper deflection height is found by applying the formula:

$$h = L \times \tfrac{1}{64}''$$

where

h = deflection height (in in.)

L = span length (in in.)

$\frac{1}{64}''$ = constant (0.0156″)

For example, what is the proper belt deflection of an assembly using a 10″ pulley and a 5″ pulley having a span length of 36″?

$$h = L \times \tfrac{1}{64}''$$
$$h = 36 \times 0.0156$$
$$h = \mathbf{0.562''}$$

New V-belts seat rapidly during the first few hours. Check and retension new belts following the first 24 hr and 72 hr of operation. During retensioning, determine the deflection height from the normal position by using a straightedge or stretched string across the pulley tops. Belt tension tools are also available. Belt tension tools allow for faster, simpler, and more accurate tension checks.

Double V-Belts

A *double V-belt* is a belt designed to transmit power from the top and bottom of the belt. Power must be able to be transmitted in the usual and reverse bend position. Double V-belts are used where the pulleys in a system are required to rotate in opposite directions. An example of a double V-belt system is that of the serpentine belt drive used in the automotive industry. **See Figure 10-13.**

Double V-belts, like standard V-belts, are identified by a letter-number combination, where the letter designates width and the number identifies nominal length. Like standard V-belts, double V-belts are rated as A, B, C, or D belts, but due to their double sides are designated as AA, BB, CC, or DD belts. For example, a double V-belt identified as DD180 has a 1¼″ width and a 180″ length.

Double V-belt tensioning is determined by the belt's width, speed in feet per minute (fpm), and the diameter of the pulley used. The tension is measured on the tight side of the drive system because every belt drive has a slack or loose side and a tight side. Basic recommended tensioning for double V-belts may be taken from double V-belt tension tables. **See Appendix.**

Timing Belts

A *timing (synchronous) belt* is a belt designed for positive transmission and synchronization between the drive shaft and the driven shaft. Timing belts consist of tension members, neoprene backing and teeth, and nylon facing. **See Figure 10-14.**

Double V-Belts

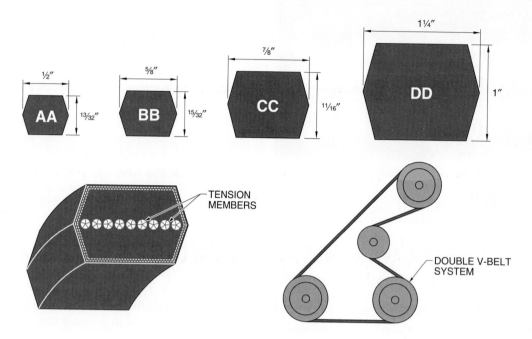

Figure 10-13. Double V-belts are designed with tension members in the center, which allow the belt to flex in both directions.

Timing Belts

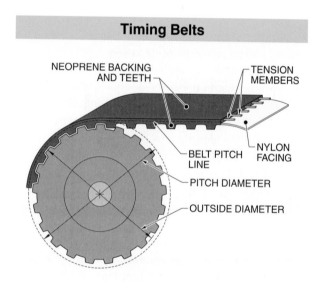

Figure 10-14. Timing belts are constructed of neoprene rubber for flexibility, tension members for strength, and nylon facing to offer low coefficient of friction.

Timing belts and their matching pulleys (sprockets) are toothed for force driving similar to chain drives. Many equipment manufacturers are replacing timing belts in power transmission systems where chains and sprockets were used previously. This is because timing belts are quieter than chain drives, require no lubrication, are able to operate efficiently in drive ranges up to 600 HP, and run at speeds over 10,000 fpm. Similar to gear drive operation, timing belts enter and leave pulleys in a smooth rolling motion with very low friction. Timing belts are identified by tooth profile, pitch length, circular pitch, and nominal width.

The four timing belt tooth profiles are the trapezoidal, double trapezoidal, curvilinear, and modified curvilinear. **See Figure 10-15.** A *trapezoidal belt* is a timing belt containing trapezoidal-shaped teeth. Trapezoidal belts are the most common timing belt used in industrial applications. A *double trapezoidal belt* is a timing belt containing two trapezoidal-shaped sets of teeth. Double trapezoidal belts are designed for serpentine drive applications where shaft rotations must run in opposite directions. A *curvilinear belt* is a timing belt containing circular-shaped teeth. Curvilinear belts were designed to provide increased capacity over a trapezoidal belt. A *modified curvilinear belt* is a timing belt containing modified circular-shaped teeth. Modified curvilinear belts were designed to maximize load life capacities while optimizing materials and tooth shape.

Timing Belt Tooth Profiles

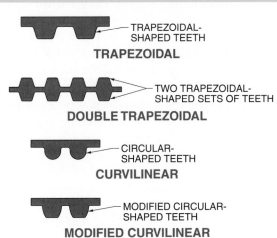

TRAPEZOIDAL-
SHAPED TEETH
TRAPEZOIDAL

TWO TRAPEZOIDAL-
SHAPED SETS OF TEETH
DOUBLE TRAPEZOIDAL

CIRCULAR-
SHAPED TEETH
CURVILINEAR

MODIFIED CIRCULAR-
SHAPED TEETH
MODIFIED CURVILINEAR

Figure 10-15. The four timing belt tooth profiles are the trapezoidal, double trapezoidal, curvilinear, and modified curvilinear.

Belt pitch length is the total length of the timing belt measured at the belt pitch line. *Circular pitch* is the distance from the center of one tooth on a timing belt to the center of the next tooth measured along the belt pitch line. Belt circular pitch is directly related to the cross section of the belt. The six basic cross sections of timing belts are mini-extra light (MXL), extra light (XL), light (L), heavy (H), extra heavy (XH), and double extra heavy (XXH). **See Figure 10-16.**

Fenner Drives

Product transfer lines use timing belts to synchronize rollers for the efficient transfer of product.

Timing Belt Cross Sections

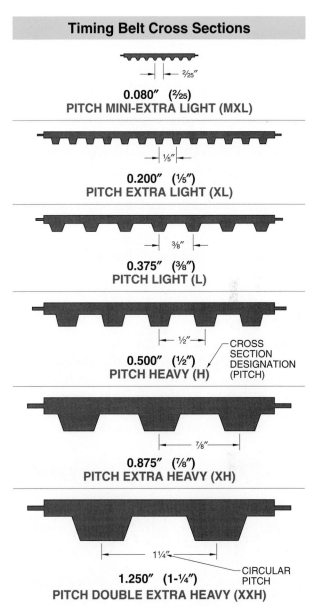

2/25″
0.080″ (²⁄₂₅)
PITCH MINI-EXTRA LIGHT (MXL)

1/5″
0.200″ (¹⁄₅″)
PITCH EXTRA LIGHT (XL)

3/8″
0.375″ (³⁄₈″)
PITCH LIGHT (L)

1/2″
0.500″ (¹⁄₂″)
PITCH HEAVY (H)

CROSS
SECTION
DESIGNATION
(PITCH)

7/8″
0.875″ (⁷⁄₈″)
PITCH EXTRA HEAVY (XH)

1¼″
1.250″ (1-¼″)
CIRCULAR
PITCH
PITCH DOUBLE EXTRA HEAVY (XXH)

Figure 10-16. Trapezoidal timing belts are classified by their standard length, cross section designation, and circular pitch.

Timing belt pitch lengths and widths are identified by standard numbers. Identifying a timing belt includes pitch length, section, and width designations (nominal width times 100). The nominal pulley width is the maximum standard belt width the pulley accommodates. For example, an XL section belt with a pitch length of 8.000 and a standard width of 0.38 is specified as an 80 XL038 synchronous belt. Belt pitch lengths, number of teeth, length tolerances, and size designations may be taken from American National Standards Institute (ANSI) tables. **See Appendix.**

VARIABLE-SPEED BELT DRIVES

A *variable-speed belt drive* is a mechanism that transmits motion from one shaft to another and allows the speed of the shafts to be varied. Variable-speed belt drives use wide V-belts and spring-loaded adjustable cone-faced pulleys. Variable-speed belt drives operate by adjusting the pulley width, which changes the pulley pitch diameter. This increases or decreases the shaft speed without changing the drive mechanism (motor) speed. **See Figure 10-17.**

Variable-speed belt drives are used where speed variations cannot be obtained using conventional V-belts. V-belts used for variable-speed drives are generally thin, wide, rigid, and are capable of operating in adjustable pulley assemblies.

Changing Speeds

Changing the speed of a belt and pulley is accomplished using an adjustable (sliding) motor base. The motor base is adjusted to increase the pulley center-to-center distance, which increases the belt tension. The increased belt tension forces the flanges apart, which forces the belt to ride lower in the pulley groove. The motor base may also be

adjusted to decrease the pulley center-to-center distance, which decreases the belt tension. The decreased belt tension allows the spring to force the flanges together, which forces the belt to ride higher in the pulley groove. Speed control is accomplished by the belt riding high or low in the pulley groove. Most variable-speed belt drives require the drive to be running during adjustment of the pulley width.

Variable-speed belt drive horsepower ratings range from fractional to over 100 HP and are available in speed ratios of 1.15:1 to over 9:1. Variable-speed drives are categorized as stationary-control or motion-control drives. Stationary-control drives require the drive to be stopped in order to manually adjust the pulley width. This adjustment usually requires drive component disassembly. Motion-control drives are distinguished by one or two movable flanges. The stationary flange is fixed to a central sleeve in drives where only one flange moves during adjustment. The movable flange maintains pressure on the belt by a spring. **See Figure 10-18.**

Safety Tip

The correct belt tension is the least amount of tension that enables the belt to run without slipping when a full load is applied.

Variable-Speed Belt Drives

LOW SHAFT SPEED

HIGH SHAFT SPEED

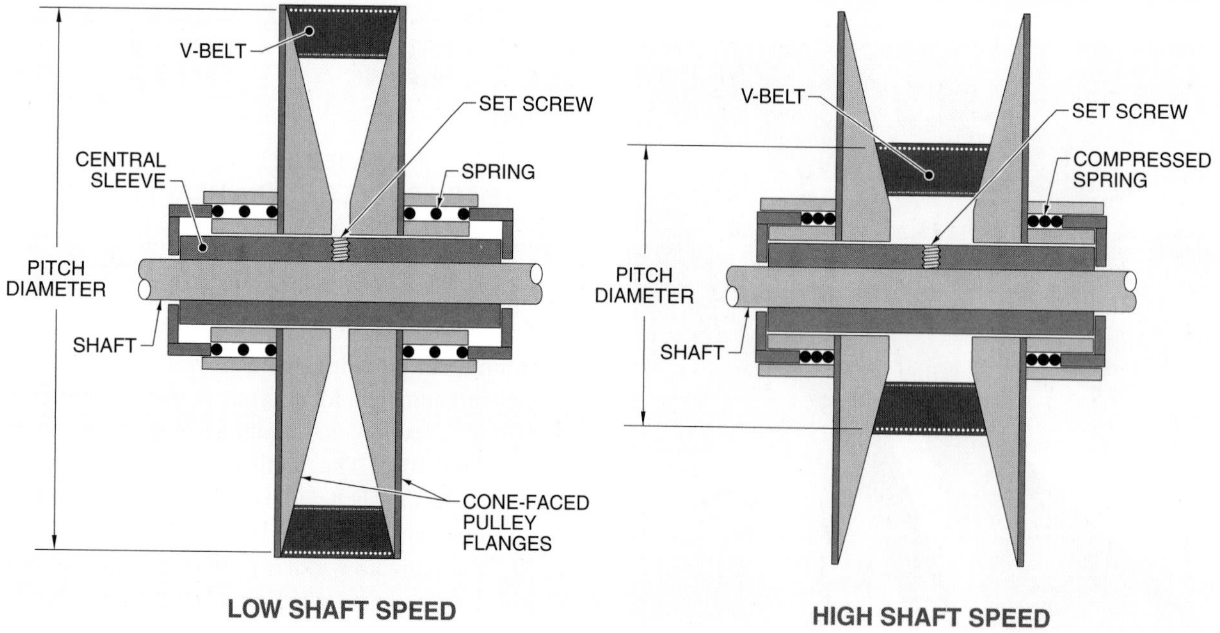

Figure 10-17. In a variable-speed belt drive, if the width of the pulley is increased, the pitch diameter is decreased, thus increasing the speed of the shaft.

Motion-Control Drives

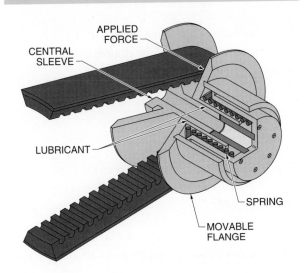

Figure 10-18. In a variable-speed drive, a layer of lubricant must be maintained for separation of parts and prevention of fretting corrosion due to the high forces applied on the central sleeve by the movable flange.

In motion-control drives where two flanges move, both flanges ride on a central sleeve. Each flange is backed up by a spring that exerts equal force on the belt. Extreme caution is required when attempting to disassemble any variable-speed drive assembly due to spring compression of the pulleys.

The Gates Rubber Company

Change all belts on a multiple belt drive even when only one belt requires replacement. New belts that are run with old belts carry more of the load, which shortens belt life.

Calculating Pulley Speed and Size

A change in pulley size may be required for an increase or decrease in pulley output speed. In addition, the diameter of destroyed or missing pulleys must be determined for proper replacement. Pulley size is determined using calculations based on the speed and size of the driven or drive pulley.

The speed or diameter of a driven or drive pulley may be determined by using the appropriate formula. The fourth value (speed or size) can be determined by using the correct formula for calculating speed or diameter when the other three factors (speed and size) are known. Any one value may be determined by rearranging the formula when the other three values are known. Driven pulley speed is based on the diameter and speed of the drive pulley and the diameter of the driven pulley. **See Figure 10-19.**

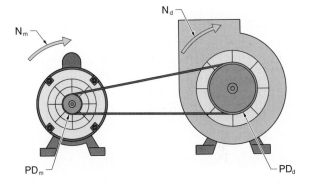

Figure 10-19. The speed or diameter of a driven or drive pulley may be determined by solving for any one value when the other three are known.

Drive pulley speed and diameter is denoted by a subscript m ($_m$). Driven pulley speed and diameter is denoted by a subscript d ($_d$). Driven pulley speed is found by applying the formula:

$$N_d = \frac{PD_m \times N_m}{PD_d}$$

where
N_d = driven pulley speed (in rpm)
PD_m = drive pulley diameter (in in.)
N_m = drive pulley speed (in rpm)
PD_d = driven pulley diameter (in in.)

For example, what is the speed of an 8″ driven pulley if the diameter of the drive pulley is 12″ and its speed is 150 rpm?

$$N_d = \frac{PD_m \times N_m}{PD_d}$$

$$N_d = \frac{12 \times 150}{8}$$

$$N_d = \frac{1800}{8}$$

$$N_d = \textbf{225 rpm}$$

Calculating driven pulley diameter is based on knowing the speed of the driven pulley and the diameter and speed of the drive pulley. Driven pulley diameter is found by applying the formula:

$$PD_d = \frac{PD_m \times N_m}{N_d}$$

For example, what is the diameter of a driven pulley rotating at 225 rpm if the diameter of the drive pulley is 12″ and its speed is 150 rpm?

$$PD_d = \frac{PD_m \times N_m}{N_d}$$

$$PD_d = \frac{12 \times 150}{225}$$

$$PD_d = \frac{1800}{225}$$

$$PD_d = \textbf{8″}$$

Calculating drive pulley speed is based on knowing the diameters of the drive and driven pulleys and the driven pulley speed. Drive pulley speed is found by applying the formula:

$$N_m = \frac{PD_d \times N_d}{PD_m}$$

For example, what is the speed of a 12″ drive pulley if the diameter of the driven pulley is 8″ and its speed is 225 rpm?

$$N_m = \frac{PD_d \times N_d}{PD_m}$$

$$N_m = \frac{8 \times 225}{12}$$

$$N_m = \frac{1800}{12}$$

$$N_m = \textbf{150 rpm}$$

Calculating drive pulley diameter is based on knowing the speed of the drive pulley and the speed and diameter of the driven pulley. Drive pulley diameter is found by applying the formula:

$$PD_m = \frac{PD_d \times N_d}{N_m}$$

For example, what is the diameter of a drive pulley rotating at 150 rpm if the diameter of the driven pulley is 8″ and its speed is 225 rpm?

$$PD_m = \frac{PD_d \times N_d}{N_m}$$

$$PD_m = \frac{8 \times 225}{150}$$

$$PD_m = \frac{1800}{150}$$

$$PD_m = \textbf{12″}$$

FLEXIBLE BELT DRIVE SAFETY

The main concerns when working with flexible belt drives, or any machine, are the safety of the technician working on the machine, co-workers working in the area of the machine, and the machine itself. Unsafe acts may jeopardize the health and welfare of those around the machine but can also damage or destroy the equipment that supports the technicians' employment. Safe work habits related to working on any mechanism in motion requires wearing proper clothing, maintaining a clean environment, and removing and locking out any energy supply.

Proper Clothing

Loose or bulky clothes, loose sleeves, and untucked shirts can cause serious injury while working on flexible belt drives and rotating machinery. Lab coats, neckties, loose belts, and loose long hair should never be worn when working around drive systems.

A technician is properly dressed for safety when wearing safety glasses for eye protection, safety toe shoes for protecting the toes against dropped heavy objects, and gloves to avoid being cut by sharp or nicked pulley edges. A technician must recognize any condition that may do harm and dress accordingly.

Tech Tip

A rapid drop in belt tension normally occurs during the run-in period. Tension new drives with a ½ greater deflection force than the maximum recommended force. Check tension frequently during the first day of operation.

The Gates Rubber Company

Never pry belts off their drives. Belt tension is adjusted by moving the motor and tightening the motor mount to the correct torque. Personal injury or machine damage may result if the pry bar slips.

Clean Environment

Access to and around machinery must be safe. Floors must be free of oil, clutter, or obstructions. Good footing and balance is necessary while working on any machinery. Also, clean environments make it possible for technicians to devote full attention to the assigned task while not being obstructed by irrelevant objects.

Removing and Locking Out Energy Supplies

Any person involved in the service or repair of drive systems shall be responsible for placing the equipment inoperable. Assurance must be made that any energy source that has been placed inoperable cannot be inadvertently made operable by any other person.

On January 2, 1990, OSHA 29 CFR 1910.147, *Control of Hazardous Energy Sources (Lockout/Tagout)* became effective. This standard provides rules designed to protect industrial workers who operate, service, and repair power equipment and machinery. Controlling energy sources include lockout, tagout, and blockout procedures. Failure to place equipment energy sources inoperable before working on them is a major cause of serious injury and death. **See Figure 10-20.**

Placing energy sources in a position of total safety may require either lockout, tagout, or blockout of equipment or possibly all three. *Lockout* is the process of removing a source of power and installing a lock that prevents the power from being turned ON.

Tagout is the process of placing a tag on a locked-out power source that indicates that the power may not be restored until the tag is removed. *Blockout* is the process of placing a solid object in the path of a power source to prevent accidental energy flow.

Mechanical lockout is required of any machine that is operated by any energy source other than electricity. Machines that are activated by compressed air or steam have valves that control movement. These valves need to be locked out and the system may also need to be bled to release any back pressure.

Additionally, coiled springs, spring-loaded devices, or suspended loads may need to be released so that their stored energy will not result in inadvertent movement. Chain or clamp-off large fan blades or turbines while changing belts or aligning. Any gust of wind or air flow may turn the blades enough to injure a worker.

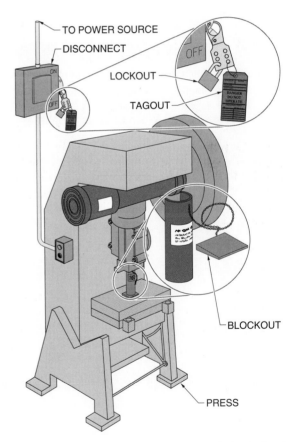

Figure 10-20. Controlling energy sources includes lockout, tagout, and blockout procedures.

Review

1. Define V-belt and identify its tension members.

2. List one sign that typically indicates a V-belt pulley should be replaced.

3. Describe a molded notch belt and list the services it provides.

4. What is the easiest method for determining a belt size?

5. Describe how tapered bore pulleys are installed.

6. Define overhung load.

7. Describe three ways improperly aligned V-belts are destroyed prematurely.

8. Describe the best tension for a V-belt.

9. Explain the procedures used for adjusting tension with the visual adjustment method and the belt deflection method.

10. Why is the tension of a double V-belt measured on the tight side of the drive system?

11. Explain what a timing (synchronous) belt is designed for and list two reasons why many equipment manufacturers are replacing sprockets and chains in favor of them.

12. Describe a typical V-belt used for variable-speed drives.

13. What is driven pulley speed based on?

14. List the safe work habits related to working on any mechanism in motion.

15. Define lockout, tagout, and blockout.

Digital Resources

ATPeResources.com/Quicklinks
Access Code 462409

CHAPTER 11

MECHANICAL DRIVES

Mechanical drives transmit power from one point to another. Mechanical drives use gears for the conversion of energy or the transmission of power. Gears may be spur, helical, rack, herringbone, bevel, miter, worm, or hypoid gears. Gear drives are designed for high speed, low speed, high thrust or radial loads, changing the angle of power, or compactness. Gear drives lose some of their efficiency to friction, but typically remain as high as 95% efficient.

SEW-Eurodrive, Inc.

Chapter Objectives

- Describe the function of a mechanical drive.
- Calculate forms of energy produced by mechanical drives.
- Calculate gear speed for gears found in a mechanical drive.
- List and describe common tooth forms and gear terminology.
- Demonstrate how to measure backlash.
- Identify eight common gear types.
- Describe common types of gear wear.

Key Terms

- mechanical drive
- torque
- horsepower
- gear train
- drive gear
- driven gear
- idler gear
- compound gear train
- backlash
- spur gear
- helical gear
- rack gear
- herringbone gear
- bevel gear
- miter gear
- worm gear
- hypoid gear
- hunting tooth
- abrasive wear
- corrosive wear
- electrical pitting
- rolling
- scuffing
- fatigue wear

MECHANICAL DRIVES

The law of conservation of energy states that energy can be neither created nor destroyed, but it can be converted from one form to another by appropriate mechanical means. *Mechanical* is pertaining to or concerned with machinery or tools. A *mechanical drive* is a system by which power is transmitted from one point to another. A *gear* is a toothed machine element used to transmit motion between rotating shafts. Gears are used in mechanical drives for the conversion of energy or the transmission of power. In addition to transmitting power, gears convert energy through changing shaft directions, reducing or increasing speed, and changing output torque.

Gear-driven mechanical drives are also used to transmit positive mechanical energy. Gear-driven mechanical drives are often chosen over belt drives because the efficiency of a gear drive is greater than that of a belt drive or friction disc. A *friction disc* is a device that transmits power through contact between two discs or plates. One disc or plate consists of a high frictional material, while the other disc or plate is fabricated from a soft metal. For example, one friction disc may be fabricated of fiberglass strands embedded in an epoxy base and the other friction disc may be fabricated from mild steel or brass. Belt drives or friction discs lose most of their efficiency through slippage. Gear drives lose some of their efficiency to friction but typically remain as high as 95% efficient. **See Figure 11-1.** Energy and efficiency is the principle of work where work input equals useful work output plus work done against friction.

Transformation of Energy

A gear or gear drive is the rotational mechanism within a machine. A machine is an assembly of devices that transfers the force, motion, or energy input from one device to a force, motion, or energy output at another. The force applied by a rotational mechanism is measured as its torque.

Torque. *Torque* is the twisting (rotational) force of a shaft. Torque is the amount of force at a distance from the center of a shaft required to achieve work. Torque is produced when turning something, such as removing a bottle cap or turning a doorknob. Torque is equal to force times the distance from the point of rotation to the point the force is applied. Torque is found by applying the formula:

$$T = F \times D$$

where

T = torque (in lb-ft)

F = force (in lb)

D = distance (in in. or ft)

For example, what is the torque developed if a 60 lb force is applied at the end of a 2′ lever arm?

$$T = F \times D$$

$$T = 60 \times 2$$

$$T = \textbf{120 lb-ft}$$

Machine Element Efficiency	
Mechanical Component	**Efficiency***
Common bearings (single)	97
Ball bearings	99
Roller bearings	98
Belting	96 – 98
Bevel gears, including bearings	
• Cast teeth	92
• Cut teeth	95
Hydraulic couplings	98
Hydraulic jacks	80 – 90
Overhead cranes	30 – 50
Transmission chains (high grade)	98

* in %

Figure 11-1. Friction, slippage, or a combination of both significantly reduces the efficiency of a mechanical component.

In many cases, torque is expressed as a measurement of pound-inches (lb-in.). For example, a 50 HP hydraulic motor may have 20,000 lb-in. of torque. The inch value may be converted to a foot value by dividing the inch value by 12. For example, 20,000 lb-in. of torque equals 1666.667 lb-ft (20,000 ÷ 12 = 1666.667 lb-ft).

This formula can be used to determine the amount of torque required to rotate an object such as a doorknob or screw. However, the torque developed or given by a rotating machine such as a motor is determined using the machine's energy output rating (horsepower) and its speed in revolutions per minute (rpm). The torque (in lb-ft) of a rotating machine is found by applying the formula:

$$T = \frac{5252 \times HP}{rpm}$$

where

T = torque (in lb-ft)

HP = horsepower

rpm = revolutions per minute

5252 = constant (33,000 lb-ft ÷ π × 2)

For example, what is the available torque supplied by a 1 HP, 1750 rpm motor?

$$T = \frac{5252 \times HP}{rpm}$$

$$T = \frac{5252 \times 1}{1750}$$

$$T = \frac{5252}{1750}$$

$$T = \textbf{3 lb-ft}$$

Torque can be measured at the different components of a machine, but it is not a measure of work performance. Torque is the measurement of overcoming a resistance. This is because torque is not dependent on time. For example, it is possible to have a specific torque valve that offers high speed, low speed, or no movement at all. Torque with an added time element, such as revolutions per minute, becomes horsepower. *Horsepower* is a unit of power equal to 746 W or 33,000 lb-ft per minute (550 lb-ft per second). **See Figure 11-2.**

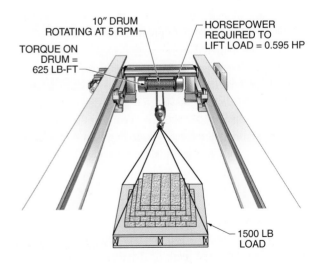

Figure 11-2. Horsepower required to lift a load is determined by speed and torque.

Horsepower is used to measure the energy produced by an electric motor while doing work. Horsepower required to overcome a load is found by applying the formula:

$$HP = \frac{T \times rpm}{5252}$$

For example, what is the horsepower required to turn a winch containing a 10″ drum at 5 rpm if a 1500 lb load is placed on the winch? *Note:* The torque at the drum equals 625 lb-ft (5 × 1500 ÷ 12 = 625 lb-ft).

$$HP = \frac{T \times rpm}{5252}$$

$$HP = \frac{625 \times 5}{5252}$$

$$HP = \frac{3125}{5252}$$

$$HP = \mathbf{0.595\,HP}$$

Gear Speed. A *gear train* is a combination of two or more gears in mesh used to transmit motion between two rotating shafts. A *drive gear* is any gear that turns or drives another gear. A *driven gear* is any gear that is driven by another gear. For example, on a two-gear machine, one gear is the drive gear and the other gear is the driven gear. On a three-gear machine, two gears are drive gears (first and second) and two gears are driven gears (second and last). The second gear is the driven gear when referenced to the first gear and is the drive gear when referenced to the third gear. The speeds of the drive and driven gears are inversely proportional to the number of teeth on each gear. For example, a gear with a large number of teeth rotates at a slow speed and a gear with a small number of teeth rotates at a fast speed.

Although some gear drives are designed to deliver an increase in speed (rpm) over the input (drive) gear, the majority of gear drives are designed to produce a speed reduction. In gear transmission, the speed of the driven gear depends on the speed of the drive gear and the number of teeth of the drive and driven gears. **See Figure 11-3.** The speed of a driven gear is found by applying the formula:

$$N_2 = \frac{T_1 \times N_1}{T_2}$$

where

N_2 = speed of driven gear (in rpm)
T_1 = number of teeth on drive gear
N_1 = speed of drive gear (in rpm)
T_2 = number of teeth on driven gear

For example, what is the speed of a 50 tooth driven gear if the drive gear has 18 teeth and rotates at 100 rpm?

$$N_2 = \frac{T_1 \times N_1}{T_2}$$

$$N_2 = \frac{18 \times 100}{50}$$

$$N_2 = \frac{1800}{50}$$

$$N_2 = \mathbf{36\ rpm}$$

This formula may be rearranged to find any one value when the other three are known. For example, the number of teeth required on a driven gear may be found by multiplying the number of teeth on the drive gear by the speed of the drive gear and dividing by the required speed of the driven gear. The number of teeth required on a driven gear to produce a required output speed is found by applying the formula:

$$T_2 = \frac{N_1 \times T_1}{N_2}$$

where

T_2 = number of teeth on driven gear
T_1 = number of teeth on drive gear
N_1 = speed of drive gear (in rpm)
N_2 = speed of driven gear (in rpm)

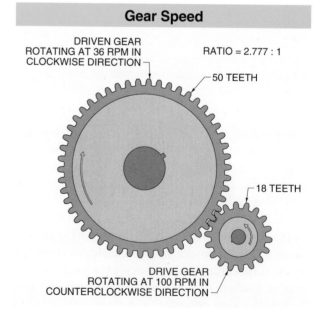

Gear Speed

DRIVEN GEAR ROTATING AT 36 RPM IN CLOCKWISE DIRECTION

RATIO = 2.777 : 1

50 TEETH

18 TEETH

DRIVE GEAR ROTATING AT 100 RPM IN COUNTERCLOCKWISE DIRECTION

Figure 11-3. The speed of the driven gear depends on the speed of the drive gear and the number of teeth of the drive and driven gears.

For example, what is the number of teeth on a driven gear required to produce 25 rpm if the drive gear having 40 teeth rotates at 75 rpm?

$$T_2 = \frac{N_1 \times T_1}{N_2}$$

$$T_2 = \frac{75 \times 40}{25}$$

$$T_2 = \frac{3000}{25}$$

$$T_2 = \textbf{120 teeth}$$

The speed of the gears in a gear train is inversely proportional to the ratio of the number of teeth of the gears. A *ratio* is the relationship between two quantities or terms. The ratio of one quantity to another is the first divided by the second. The colon (:) is the symbol used to indicate a relation between terms. For example, the ratio of 10:4 is $^{10}/_4$, which equals 2.5 (10 ÷ 4 = 2.5).

An *inverse ratio* is the reciprocal of a given ratio. The inverse ratio of 10:4 is 4:10, or $^4/_{10}$, which equals 0.4 (4 ÷ 10 = 0.4). A *proportion* is an expression of equality between two ratios. Proportions are either direct proportions or inverse proportions. A *direct proportion* is a statement of equality between two ratios in which the first of four terms divided by the second equals the third divided by the fourth. For example, in the proportion 8:4 = 12:6, both ratios equal 2 (8 ÷ 4 = 2 and 12 ÷ 6 = 2). An increase in one term results in a proportional increase in the other related term.

An *inverse proportion* is a proportion in which an increase in one quantity results in a proportional decrease in the other related quantity. For example, two gears with 50 and 18 teeth respectively are meshed. The 18 tooth gear rotates at 100 rpm and the 50 tooth gear rotates at 36 rpm. The ratio of the number of teeth is 50:18, or $^{50}/_{18}$ and the inverse ratio of speeds is 100:36, or $^{100}/_{36}$. Using the inverse ratio, the two ratios are equal at 2.777:1.

Idler Gears. An *idler gear* is a gear that transfers motion and direction in a gear train but does not change speeds. Two external gears in mesh operate in opposite directions unless an idler gear is used. Idler gears are drive and driven gears. Idler gears are generally used when the shafts of the driven gear and the drive gear are too far apart and the use of large gears is impractical. **See Figure 11-4.**

Idler Gears

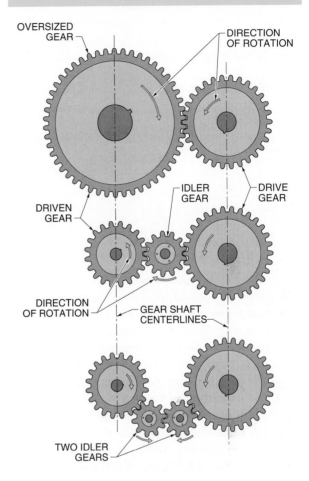

Figure 11-4. The use of idler gears in a gear train allows proper rotation of the output gear and offers compactness between gear shafts within the gear housing.

Gear-driven mechanical drives are often chosen over belt drives for use in industrial machines because their efficiency ranges between 92% and 95%.

Two idler gears may be placed between the drive and driven gears when opposite rotation between the drive and driven gears is required. This configuration of the gear train, barring extra losses from friction, provides opposite rotations without affecting speed.

Tech Tip

Gear drives that are shut down for longer than one week should be run at least 10 min each week to keep the gears coated with oil and help prevent rusting due to moisture condensation.

Compound Gear Trains

A *compound gear train* is two or more sets of gears where two gears are keyed and rotate on one common shaft. Compound gear trains produce higher speeds in less space than gear trains using simple gearing. **See Figure 11-5.**

In a four-gear compound gear train, gear 1 is the first drive gear. Gear 1 drives gear 2, which is the first driven gear. Gear 2 is keyed to the same shaft as gear 3, which is the second drive gear. Gear 2 and gear 3 are different sizes and contain a different number of teeth, but rotate at the same speed. Gear 3 drives gear 4, which is the second driven and output gear. Drive and driven gear determination is required when calculating compound gear train output speed. The formula for calculating the output speed of a compound gear train is based on the number of teeth of the drive gears multiplied by the speed of the first drive gear divided by the number of teeth of the driven gears. The output speed of a compound gear train is found by applying the formula:

$$N_4 = \frac{T_1 \times T_3 \times N_1}{T_2 \times T_4}$$

where

N_4 = speed of output gear (in rpm)
T_1 = number of teeth on first drive gear
T_3 = number of teeth on second drive gear
N_1 = speed of first drive gear (in rpm)
T_2 = number of teeth on first driven gear
T_4 = number of teeth on output gear

For example, what is the output speed of a 20 tooth output gear in a compound gear train in which gear 1 (first drive gear) contains 50 teeth and rotates at 25 rpm, gear 2 (first driven gear) contains 25 teeth and rotates at 50 rpm, and gear 3 (second drive gear) contains 75 teeth and rotates at 50 rpm?

$$N_4 = \frac{T_1 \times T_3 \times N_1}{T_2 \times T_4}$$

$$N_4 = \frac{50 \times 75 \times 25}{25 \times 20}$$

$$N_4 = \frac{93,750}{500}$$

$$N_4 = \textbf{187.5 rpm}$$

Compound Gear Train—Output Speed Determination

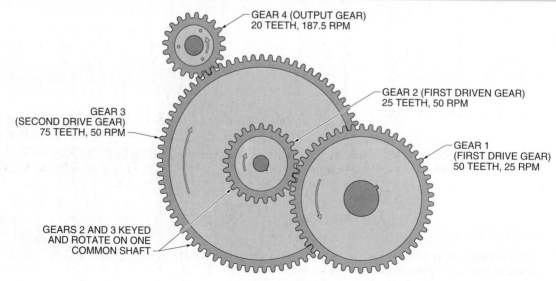

GEAR 4 (OUTPUT GEAR)
20 TEETH, 187.5 RPM

GEAR 2 (FIRST DRIVEN GEAR)
25 TEETH, 50 RPM

GEAR 3
(SECOND DRIVE GEAR)
75 TEETH, 50 RPM

GEAR 1
(FIRST DRIVE GEAR)
50 TEETH, 25 RPM

GEARS 2 AND 3 KEYED
AND ROTATE ON ONE
COMMON SHAFT

Figure 11-5. Compound gear trains produce higher speeds in less space than gear trains using simple gearing.

In compound gear trains, the number of teeth on the output gear is based on the number of teeth on the first and second drive gears and the speed of the first drive gear divided by the number of teeth of the first driven gear and the speed of the output gear. **See Figure 11-6.** The number of teeth on the output gear in a compound gear train is found by applying the formula:

$$T_5 = \frac{T_1 \times T_4 \times N_1}{T_3 \times N_2}$$

where

T_5 = number of teeth on output gear

T_1 = number of teeth on first drive gear

T_4 = number of teeth on second drive gear

N_1 = speed of first drive gear (in rpm)

T_3 = number of teeth on first driven gear

N_2 = speed of output gear (in rpm)

For example, how many teeth does an output gear rotating at 100 rpm have if connected to 5 gears in a compound gear train where gear 1 (first drive gear) contains 30 teeth and rotates at 300 rpm, gear 2 (idler gear) has 20 teeth, gear 3 (first driven gear) and gear 4 (second drive gear) have the same shaft and contain 50 and 35 teeth respectively?

$$T_5 = \frac{T_1 \times T_4 \times N_1}{T_3 \times N_2}$$

$$T_5 = \frac{30 \times 35 \times 300}{50 \times 100}$$

$$T_5 = \frac{315,000}{5000}$$

$$T_5 = \textbf{63 teeth}$$

GEAR FORM AND TERMINOLOGY

To transmit power smoothly from one gear to another, a special tooth form is used to allow sliding without damage or jerky motion. A *tooth form* is the shape or geometric form of a tooth in a gear when seen as its side profile. For example, the tooth form of a rack gear consists of three flat surfaces. The tooth form of a spur gear has a flat surface and two curved (involute) surfaces. *Rack teeth* are gear teeth used to produce linear motion. Smooth power transmission is accomplished using an involute tooth form. An *involute form* is a tooth form that is curled or curved. This tooth form is used for gear teeth to provide a uniform motion and straight line of action. **See Figure 11-7.**

Compound Gear Train—Output Teeth Determination

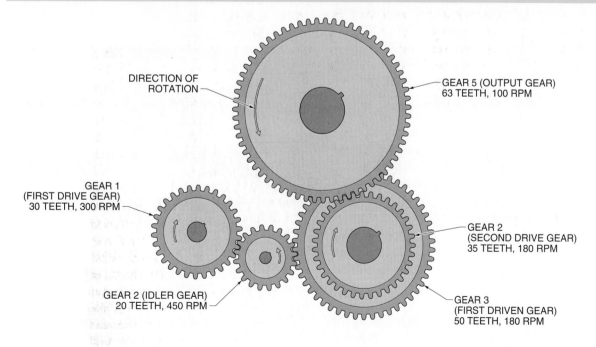

DIRECTION OF ROTATION

GEAR 5 (OUTPUT GEAR)
63 TEETH, 100 RPM

GEAR 1
(FIRST DRIVE GEAR)
30 TEETH, 300 RPM

GEAR 2
(SECOND DRIVE GEAR)
35 TEETH, 180 RPM

GEAR 2 (IDLER GEAR)
20 TEETH, 450 RPM

GEAR 3
(FIRST DRIVEN GEAR)
50 TEETH, 180 RPM

Figure 11-6. The idler gear is disregarded when determining the number of teeth or speed calculation in a compound gear train.

Involute Tooth Form

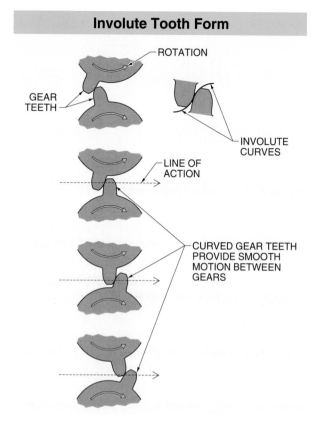

Figure 11-7. The involute tooth form on a gear allows a smooth and uniform motion between gears.

The involute form is a basic profile for gear teeth and is used on most gears. The involute form and gear tooth terminology are defined and standardized by the American National Standards Institute (ANSI) and the American Gear Manufacturers Association (AGMA). **See Figure 11-8.**

Certain terms are used when fabricating, specifying, or defining a gear's function. Gear terminology includes pitch circle, operational pitch point, pinion, diametral pitch, circular pitch, base diameter, clearance, working depth, and face width. *Pitch circle* is the circle that contains the operational pitch point. An *operational pitch point* is the tangent point of two pitch circles at which gears operate. A *pinion* is the smaller gear of a pair of gears, especially when engaging rack teeth. *Diametral pitch* is the ratio of the number of teeth in a gear to the diameter of the gear's pitch circle. *Circular pitch* is the distance from a point on a gear tooth to the corresponding point on the next gear tooth, measured along the pitch circle. The diametral pitch of a gear may be found if the pitch diameter and number of teeth are known. Diametral pitch is found by applying the formula:

$$DP = \frac{T}{D}$$

where

DP = diametral pitch

T = number of teeth

D = pitch diameter (in in.)

This formula may be rearranged to find any one value when the other two are known.

For example, what is the diametral pitch of a gear that has 30 teeth with a pitch diameter of 6″?

$$DP = \frac{T}{D}$$

$$DP = \frac{30}{6}$$

$$DP = \mathbf{5}$$

Gear Terminology

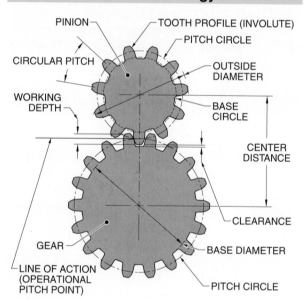

Figure 11-8. Gear terminology is standardized by ANSI and AGMA.

Base diameter is the diameter from which the involute portion of a tooth profile is generated. *Clearance* is the radial distance between the top of a tooth and the bottom of the mating tooth space when fully mated. Clearance is directly related to a gear's working depth. *Working depth* is the depth of engagement of two gears. Clearance is required because gears bind without clearance. Clearance is the gap between the base diameter of one gear and the top of the mating tooth. Proper clearance is generally accomplished when backlash measurements are met.

The *face width* is the length of the teeth in an axial plane. This is generally the width of any spur gear. However, the length or face width of angled gears, such as helix or spiral angled gears, are somewhat longer because the measurement is made along the cone distance of these gears.

Cone Drive Operations Inc./Subsidiary of Textron Inc.

Worm gear drives are manufactured with a double-enveloping design in which the worm and gear wrap around each other. This provides high shock resistance, low backlash, and increased load carrying capacity as more tooth area are in contact and more teeth are in mesh than other worm gear designs.

BACKLASH

Backlash is the play between mating gear teeth. Backlash allows for any errors in tooth profile, gear mounting, or shaft or gear runout. Backlash prevents gears from making contact on both sides of the teeth, causing seizing and failure, and allows for lubrication.

The amount of backlash does not affect the involute action of the gear, but too little backlash can cause overheating or overloading and too much backlash can cause damage from tooth slamming, especially when drives reverse frequently. Backlash measurement is made at the pitch diameter and is checked after rotating the gears in mesh to the point of closest engagement. *Pitch diameter* is the diameter of the pitch circle. Determining the proper

amount of backlash is based on the diametral pitch. A chart may be used to determine backlash tolerances when the diametral pitch is known. **See Figure 11-9.**

| Recommended Gear Backlash ||
Diametral Pitch	Backlash*
3	0.009 – 0.014
4	0.007 – 0.011
5	0.006 – 0.009
6	0.005 – 0.008
8	0.004 – 0.006
10	0.003 – 0.005
12	0.003 – 0.005
16	0.002 – 0.004
20	0.002 – 0.004
24	0.002 – 0.004

* in in.

Figure 11-9. Recommended backlash for mating gears is based on the gear diametral pitch.

Measuring Backlash

Backlash is generally measured by holding one gear stationary while rocking the engaged gear back and forth. The movement is measured with the stem of a dial indicator in the plane of rotation on the driven gear and at the pitch diameter. Other methods of measuring backlash include using thickness gauges or a lead wire. Thickness (feeler) gauges are slid into the spacing between meshing teeth without using force or rotation. The gauge thickness is read for backlash tolerance. Lead wire may be placed between meshing teeth and the gears rotated once. The lead is squeezed and sized to the spacing between the meshed teeth. A micrometer is then used to measure the formed (flattened) lead to determine backlash tolerance. **See Figure 11-10.**

Tech Tip

Gear-driven mechanical drives should be checked daily for oil leaks and unusual noises. If oil leaks are present, the drive should be shut down, the cause of the leakage corrected, and the oil level checked. If any unusual noises occur, the unit should be shut down until the cause of the noise has been determined and corrected.

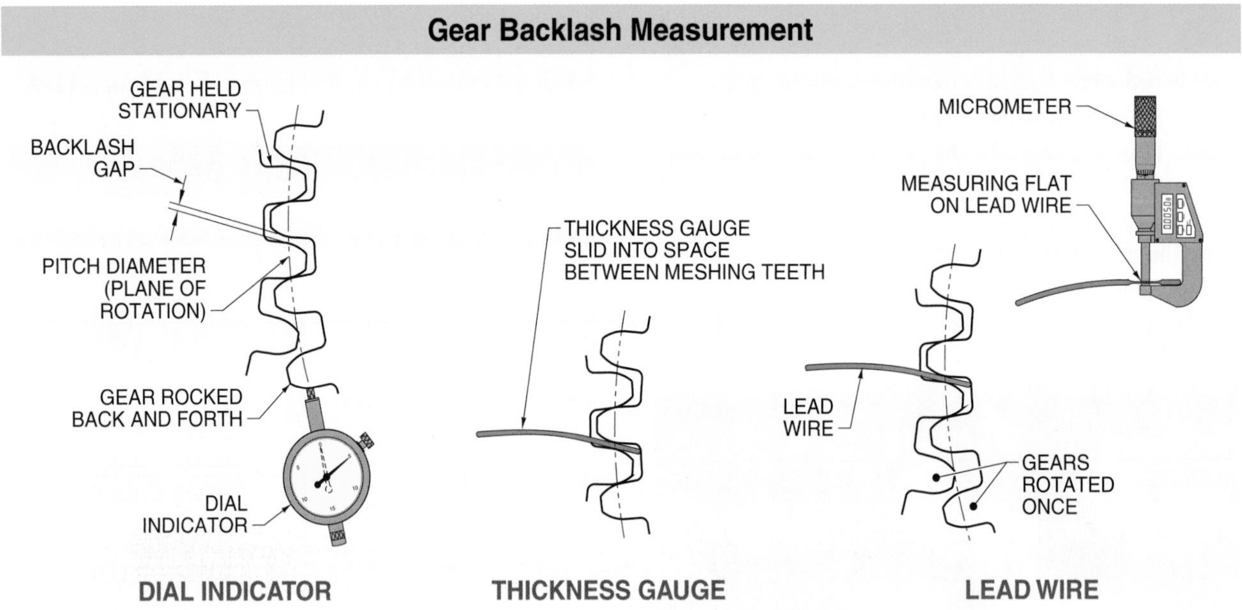

Gear Backlash Measurement

DIAL INDICATOR

THICKNESS GAUGE

LEAD WIRE

Figure 11-10. Backlash between two gears in mesh is measured using a dial indicator, thickness gauge, or lead wire.

GEARS

Gear drives are designed for high speed, low speed, high thrust or radial loads, changing the angle of power, or compactness. The transmission of power, torque, or angle is accomplished by the use of spur, helical, rack, herringbone, bevel, miter, worm, or hypoid gears. **See Figure 11-11.**

Spur

A *spur gear* is a gear that has straight teeth that are parallel to the shaft axis. Spur gears are the most commonly used gear and were originally designed for internal clock works. Spur gears are included in most mechanical drives and large industrial machinery. Spur gears are used to transmit power from one parallel shaft to another shaft where there is no end thrust or axial displacement.

Spur gears are excellent transmission gears due to their ability to slide and mesh from one gear size into another to change speeds. Spur gears of different diameters and numbers of teeth are interchangeable as long as they are of the same pitch. Spur gears are rougher running and noisier than other gears and usually run at slower speeds to reduce vibration and noise.

Helical

A *helical gear* is a gear with teeth that are cut at an angle to its axis of rotation. The steeper the angle, the quieter the gear. Side movement (thrust) is developed when a helical gear is used. Thrust energy caused by the tooth load must be supported by thrust bearings. The direction and amount of thrust depends on gear rotation and the direction of the helix. Helical gear drive angles may be anywhere from 0° to 90°.

Helical gears are quieter and smoother running than spur gears because pressure is transferred gradually and uniformly as successive teeth are meshed. Also, power is more widely distributed because it is placed on several teeth in mesh. The increased area of distribution allows for finer tooth sizes and equalized tooth wear.

Rack

A *rack gear* is a gear with teeth spaced along a straight line. Rack gears are used with pinion gears (spur gears) to convert rotary motion to linear motion or linear motion to rotary motion depending on which gear is the drive gear and which gear is the driven gear. Unlike the pinion gear, rack gear teeth do not have to be involute because the rack gear does not rotate. Rack and pinion gears may be either spur or helical in tooth form.

Gears

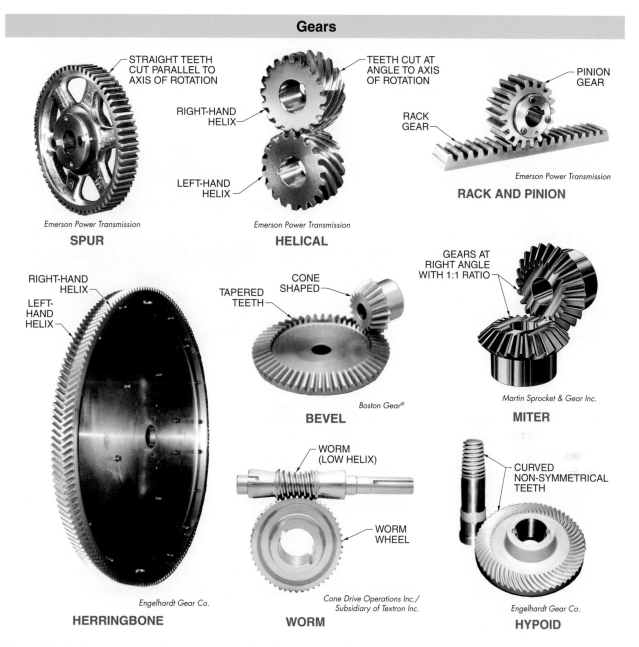

Figure 11-11. The gear used in a mechanical drive is determined by the required gear speed, load placed on the gear, angular requirements, and space constraints.

Herringbone

A *herringbone gear* is a double helical gear that contains a right- and left-hand helix. End thrust is avoided by the teeth being at opposite angles. Herringbone gears are used in parallel shaft transmission, especially where smooth, high speeds are required. Smooth transmission is possible due to the gradual overlapping mesh of the teeth and is made even smoother when the right- and left-hand helix is offset by one half pitch. Herringbone gears are often used for connecting steam turbines to electrical generators, pumps, propeller shafts, etc. because they are able to run at high speeds.

Tech Tip

The operating temperature of a gear drive is the temperature of the oil inside the gear case. Under normal conditions, the maximum operating temperature should not exceed 180°F.

Bevel

A *bevel gear* is a gear that connects shafts at an angle in the same plane. The involute tooth form is used on bevel gears when a drive requires large ratios using high torque, high speed, and non-parallel shafts. Bevel gears are cone-shaped with teeth that are tapered. The tapered teeth allow uniform clearance along the length of the teeth even though the gears are meshed on an angle. Mounting of bevel gears should be rigid enough to allow only a maximum separation of 0.006″. Proper meshing and backlash of bevel gears is accomplished using shims to position each gear axially.

Extra power transmission and noise reduction is possible on bevel gears by using spiral tooth bevel gears. The tooth spiral allows the teeth to engage with one another gradually. The continuous pitch line contact of spiral bevel gears makes it possible to obtain superior performance with high speeds and silent operations.

Miter

A *miter gear* is a gear used at right angles to transmit horsepower between two intersecting shafts at a 1:1 ratio. Only miter gears with the same pitch, pressure angle, and number of teeth can be operated together. However, more than two miter gears may be used in sets, such as in automotive differentials. Similar to bevel gears, miter gears should also have thrust bearings to absorb axial thrust.

Worm

A *worm* is a shank having at least one complete tooth around the pitch surface. A *worm gear* is a set of gears consisting of a worm (drive gear) and a wheel (driven gear) that are used extensively as a speed reducer. The worm is of such a low helix angle that it cannot be reversed. The driven gear cannot drive the worm because the gearing automatically locks itself against backward motion. Also, because of the low helix angle, a proportionate increase in torque is offered between the drive and driven gears.

Worm gears may be coarse-pitch or fine-pitch. Coarse-pitch worm gears are the most commonly used worm gears. Coarse-pitch worm gears are primarily used in industrial applications because they can transmit power efficiently, provide considerable mechanical advantage, and transmit power at a significant reduction

in velocity. Fine-pitch gears are used to transmit motion rather than power. For this reason, tooth strength is seldom an important factor with fine-pitch worm gears.

Emerson Power Transmission

Bevel or miter gears are used in right-angle gearboxes to change the angle of power for use in applications where space is limited.

Multiple threads are added to the worm when additional power is required from a coarse-pitch worm gear. More threads added to the worm produce greater tooth contact, thereby allowing increased power. Adding more teeth reduces the input/output ratio. The number of helical threads on a worm can be as high as seven. The ratio of a worm gear is found by applying the formula:

$$R = \frac{T}{T_w}$$

where

R = input/output ratio

T = number of teeth on worm wheel

T_w = number of threads on worm

For example, what is the input/output ratio of a 50 tooth speed reducer with a 3 thread worm?

$$R = \frac{T}{T_w}$$

$$R = \frac{50}{3}$$

$$R = \mathbf{16.667}$$

Worm gears are unique in that wear does not destroy the tooth form. The worm continues to keep a proper form even through wear. Unlike other gears, worm gears tend to wear in rather than wear out. Due to the sliding motion on the worm's helix, high heat is generated and must be kept to a minimum through proper lubrication.

Hypoid

A *hypoid gear* is a spiral bevel gear with curved, nonsymmetrical teeth that are used to connect shafts at right angles. The hypoid gear has the tooth angle of a helical gear, the base angle of a bevel gear, and the straight tooth of a rack gear. The hypoid gear is related to the worm gear due to the extreme pressure angle of the pinion. Hypoid gears, used widely in automotive differentials, provide greater sliding angles than spiral bevel gears for smoother and quieter operation.

GEAR WEAR

The sliding and meshing of gear teeth under load causes gear failure due to wear. Gear manufacturers design certain parts of a gear train to wear out or break sooner than others. This is done because some gears are physically easier to replace than others or one gear may be cheaper than another in a set. For example, worm gears are made of hardened steel while their matching worm wheels are made of bronze. At times, cast iron or nonmetallic gears are placed in a train of steel gears to wear out first.

Hunting Teeth

A *hunting tooth* is a tooth added to mesh with every tooth on the mating gear to produce even tooth wear. Hunting tooth choice is based on gear train ratio and the teeth per gear. For example, a pair of gears in a 3:1 ratio set have 60 and 20 teeth. In this case, wear is uneven because every 3 rotations of the pinion (20 teeth) produces one rotation of the driven gear (60 teeth). However, gears with 61 and 20 teeth (3.05:1 ratio) produce evenly distributed wear. Hunting teeth ratios should be used when gears in a pair or train are subject to uneven or cyclic loads such as in indexing or crankshaft use.

Gear Wear Identification

Proper identification of gear wear and its causes can prevent many hours of equipment downtime.

Identification is also a useful indicator for upgrading a preventive maintenance program. For example, lubrication may need to be changed if a gear is scuffing. In addition, lubricant may require more frequent changing if there are signs of abrasive wear. Early recognition and correction of gear wear may also prevent extensive equipment damage.

The AGMA has compiled gear wear identification standards as a guide to provide a common language on gear wear and to provide a means to document gear appearance as gears wear or fail. Gear wear may be abrasive wear, corrosive wear, electrical pitting, rolling and scuffing, or fatigue wear.

Abrasive Wear. *Abrasion* is the removal or displacement of material due to the pressure of hard particles. *Abrasive wear* is wear caused by small, hard particles. Abrasive wear is caused by particle-contaminated oil, where particles of metal, sand, scale, or other abrading material grind and scratch gear teeth as they make contact. The scratches appear as parallel furrows oriented in the direction of sliding. **See Figure 11-12.**

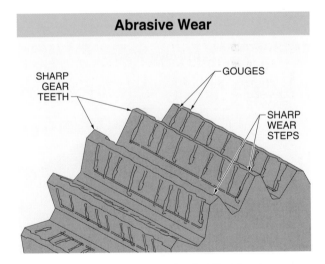

Abrasive Wear

SHARP GEAR TEETH

GOUGES

SHARP WEAR STEPS

Figure 11-12. Abrasive wear is identified by sharp wear steps from the original surfaces, gouges in the direction of sliding, and tooth tips with sharp edges.

Abrasive material may enter a gear housing due to a harsh environment or may have been left in the gear housing as residual casting scale when the component was manufactured. Corrective action taken to prevent abrasive wear includes:

- Drain and flush residual oil.

- Scrape, flush, and wipe the internal surfaces of the gear housing.

- Clean out and flush any oil passages.

- Refill the housing with a light flushing grade oil and run without load for approximately 10 min.

- Clean breathers and replace seals and filters if suspected contamination was from the environment.

- Drain the flushing oil and refill with correct oil.

Corrosive Wear. *Corrosion* is the action or process of eating or wearing away gradually by chemical action. *Corrosive wear* is wear resulting from metal being attacked by acid. Acid is usually formed when the oil temperature becomes high enough to boil and separate acid-forming resins from the oil. High temperatures may be the result of an overloaded system, the wrong lubricant, or old lubricant that has broken down.

Corrosive wear may be initially identified by a stained or rusty appearance. More advanced corrosive wear is identified by rust-colored deposits along with extensive acid-etched pits. **See Figure 11-13.**

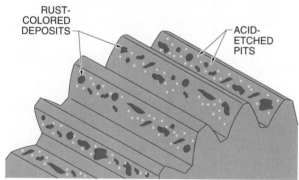

Corrosive Wear

Figure 11-13. Corrosive wear is identified by rust-colored deposits along with acid-etched pits.

Corrosion attacks the entire gear, but wear is greatest on working surfaces because the build-up of corrosion itself becomes a partial insulator to the non-working surfaces. Corrosive wear corrective action includes:

- Reduce the load if the system is overloaded.

- Upgrade the system if the system is overloaded and the load cannot be reduced.

- Use an extreme-pressure lubricant if a system is overloaded and the load cannot be reduced nor the system upgraded.

- Check to see if the wrong grade of lubricant is being used. Contact the machine manufacturer or an oil company representative for proper lubricant specifications.

- Check the frequency of oil changes. It may be necessary to increase the oil change frequency.

Electrical Pitting. *Electrical pitting* is the cratering and burn damage caused by electric arc discharge between mating metal components. The temperatures produced are high enough to locally melt gear tooth surfaces. Damage from electrical pitting may be caused by improperly grounded electrical connections, high static charges, or improper welding connections. Electrical pitting is identified by many small craters surrounded by burned or fused metal. **See Figure 11-14.**

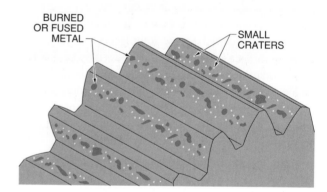

Electrical Pitting

Figure 11-14. Electrical pitting appears as small craters surrounded by burned or fused metal.

Electric current on gears may break down lubricant if the current is not high enough to etch metal but is high enough to locally burn and break down lubricants. Electrical pitting corrective action includes:

- Place a ground clamp on the same side of a gear box when welding.

- Run grounding straps from a machine to rigid electrical or pneumatic piping to reduce static electricity created by manufacturing processes.

- Check the electrical system for proper installation and grounding.

Exxon Company

Gear wear and machine downtime is minimized by ensuring the machine is not overloaded, the proper lubricant is used, and the lubricant is changed at the correct frequency.

Rolling and Scuffing. *Rolling* is the deforming of metal on the active portion of gear teeth caused by high contact stresses. Rolling is a displacement of surface materials that forms grooves along the pitch line and burrs on the tips of drive gear teeth. *Scuffing* is the severe adhesion that causes the transfer of metal from one tooth surface to another due to welding and tearing. Scuffing generally occurs in localized patches due to the surface area of meshed teeth being mismatched or misaligned. Rolling and scuffing are created when gear teeth do not mesh properly and are progressive, meaning that wear continues and worsens until total damage has occurred. **See Figure 11-15.**

Rolling and Scuffing

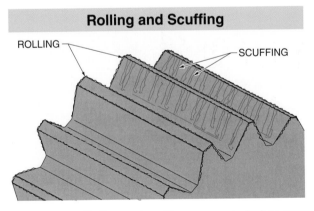

Figure 11-15. Rolling and scuffing are created when gear teeth do not mesh properly and continue until total damage has occurred.

Improper adjustments include radial/axial misalignment, improper end play, out of tolerance backlash, and manufacturer's defect. In gear teeth that do not mesh correctly due to misalignment, the gear wears at high points and removes metal until a mating profile is established. In cases such as manufacturer's defect, once a mating profile has been established, wear lightens or ceases. However, all other misalignments, if not corrected, continue to wear to gear destruction.

Fatigue Wear. *Fatigue wear* is gear wear created by repeated stresses below the tensile strength of the material. Fatigue may be identified as cracks or fractures. A *fatigue crack* is a crack in a gear that occurs due to bending, mechanical stress, thermal stress, or material flaws. A *fatigue fracture* is a breaking or tearing of gear teeth. Fatigue cracks usually culminate in a fracture when the fatigue crack grows to a point where the remaining tooth section can no longer support the load. Fatigue wear begins at the first moment a gear is used. Fatigue wear is repeated minute deformations under normal stress (normally unseen and immeasurable) that eventually produce cracks or fractures. **See Figure 11-16.**

Fatigue Wear

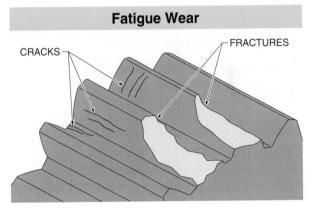

Figure 11-16. Fatigue wear is repeated minute deformations under normal stress that eventually produce cracks or fractures.

Sufficient alternating stresses (vibrations) produce rapid fatigue fracture in industrial gear trains. Other destructive noises, vibrations, overloading, and grinding of gears and gear surfaces must be identified early to be effectively corrected. Proper installation procedures, effective lubrication techniques, and regular periodic inspections with service can produce a successful, profit-oriented mechanical installation.

Review

1. In addition to transmitting power, how else do gears convert energy?

2. Why are gear-driven mechanical drives often chosen over belt drives to transmit positive mechanical energy?

3. How is torque produced?

4. What is torque used to measure and when does it become horsepower?

5. How does the number of teeth on the drive and driven gear relate to the speed of a machine?

6. What is the difference between a direct proportion and an inverse proportion?

7. When are idler gears generally used?

8. Explain the function of each gear in a four-gear compound gear train.

9. Define tooth form.

10. List and describe several commonly used gear terms associated with fabricating, specifying, or defining a gear's function.

11. Describe two different methods for measuring backlash.

12. Why are helical gears quieter and run smoother than spur gears?

13. Define worm gear and explain why a driven gear cannot drive it.

14. How is corrosive wear caused?

15. Define rolling and scuffing and explain how they can cause total gear damage.

Digital Resources

ATPeResources.com/Quicklinks
Access Code 462409

CHAPTER 12

VIBRATION

More than 80% of all rotary equipment failures are related to vibration. Vibration characteristics include cycle, displacement, frequency, phase, velocity, and acceleration. Vibration characteristics are used to determine machine misalignment, unbalance, mechanical looseness, resonance, bearing wear, or gear defect. A successful vibration monitoring program compares initial vibration readings with present vibration reading to see if an increase in readings is developing.

SPM Instrument, Inc.

Chapter Objectives

- Explain how vibration affects machines.
- Describe the vibration characteristics that define the dynamic properties of unwanted motions in a machine.
- Explain how the mechanical condition of a machine is determined.
- Explain the vibration analysis process.
- Explain how to establish a successful vibration monitoring program.

Key Terms

- vibration
- alignment
- misalignment
- vibration cycle
- displacement
- frequency
- phase
- vibration velocity

- vibration acceleration
- velocity transducer
- accelerometer transducers
- displacement transducers
- vibration signature
- time domain
- frequency domain
- frequency spectrum

VIBRATION

Vibration is a continuous periodic change in displacement with respect to a fixed reference. All objects on earth are constantly experiencing vibration. A vibration is a force that starts from a neutral position, travels a displaced distance to a positive upper limit (peak), reverses its direction to return to neutral, travels a displaced distance to a positive lower limit (peak), and returns to neutral. **See Figure 12-1.**

More than 80% of all rotary equipment failures are related to vibration. Vibration can break down the resiliency of seals and increase bearing and equipment temperatures. Vibration in one location may react with and add to vibration from another location. This vibration may be magnified, resulting in equipment damage.

Vibration also significantly reduces the expected life of bearing and rotating shaft seals. With recent motor designs, the effects of vibration become more critical because bearing and bearing fit tolerances have decreased, motor speeds have increased, and motor support frames have become lighter.

An estimate of vibration problems shows that 50% to 60% of damaging machinery vibrations are the result of shaft misalignment, 30% to 40% of damaging vibrations are the result of equipment unbalance, and 20% are the result of resonance. *Resonance* is the magnification of vibration and its noise by 20% or more. Resonance is often coupled with vibration from other sources. Less common vibration sources include bent shafts, loose parts, oil whirl, and defective bearings.

Industrial motors are rugged and able to handle heavy or continuous loads. For this reason, motors are manufactured with larger rotor shafts, which require larger bearings even though the rotors remain lightweight. The combination of lightweight rotors and large bearings should lengthen bearing service life considerably. In reality, previously allowable misalignment causes vibration that prematurely wears shaft seals, contaminates bearings, and shortens bearing life to less than 10 years.

Tech Tip

Vibration limits can be determined by comparison to standards developed by engineering standards organizations, manufacturer associations, or governmental bodies such as the American Petroleum Institute, American Gear Manufacturers Association, or American National Standards Institute.

Vibration

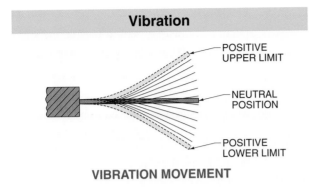

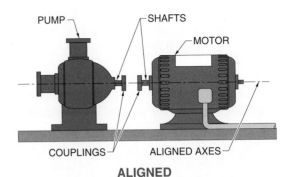

VIBRATION MOVEMENT

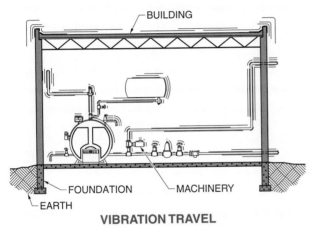

VIBRATION TRAVEL

Figure 12-1. Vibration is a continuous periodic change in a displacement with respect to a fixed reference.

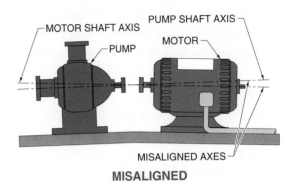

MISALIGNED

Figure 12-2. Properly aligned rotating shafts reduce vibration and add many years of service to pump/motor seals and bearings.

Alignment is the location (within tolerance) of one axis of a coupled machine shaft relative to another. *Misalignment* is the condition where the axes of two machine shafts are not aligned within tolerances. Properly aligned rotating shafts reduce vibration and add many years of service to pump/motor seals and bearings. **See Figure 12-2.**

Vibration Effects

Understanding vibration effects and establishing a testing, data collection, and analysis program goes beyond preventive maintenance and is a major tool in predictive maintenance programs. A complete vibration analysis program is one of the more complex and expensive predictive techniques. The cost of sophisticated electronic instruments able to collect, analyze, and store data is added to the cost of training personnel to interpret the data. However, the investment of total company commitment, power, and resources can provide a considerable payback through reduced equipment costs and less frequent machine downtime.

Many vibrations are unbearable for human comfort. These include loud noises and the shaking of buildings or vehicles. The magnitude of vibrations felt by humans is extremely small. The most damaging effect of vibration, especially in the case of machinery, occurs in ranges outside of human perception. These vibrations can produce fatigue failure in machine and structural elements, increase general wear of parts, and develop vibration through building foundations that are great enough to cause destruction or annoyance in another location.

Machinery Vibration

Machines vibrate even when in the best operating condition. This vibration and noise is generally due to minor defects or matching parts that are out of tolerance, such as clearances in bearings. Machinery vibration is a complex combination of signals created by a variety of internal vibration sources. Monitoring these signals is only the detection part of a maintenance program.

A complete maintenance vibration program requires technicians to have a basic knowledge of how machines work, their common problems and how to repair them, the ability to recognize and pinpoint mechanical problems early and accurately, and the ability to understand and use applied technology diagnostics in determining a specific problem, its severity, and the machine part being affected.

The combined vibration and noise from every machine is different. The component producing the vibration can be identified when certain vibration signals are separated from the others. The measurements are monitored, recorded, and compared to previous measurements. The sign of developing mechanical problems may be determined when the vibration or noise readings continue to rise.

VIBRATION CHARACTERISTICS

Characteristics offered in vibration analysis become the clues toward describing and detecting unwanted motions in a machine. They are the symptoms used in determining any significant variation and reflect the true mechanical condition of a machine. Vibration characteristics such as cycle, displacement, frequency, phase, velocity, and acceleration define the dynamic properties of machine misalignment, unbalance, mechanical looseness, resonance, bearing wear, or gear defect.

Vibration characteristics become valuable in determining machine condition because an unwanted vibration is caused by either a change in direction or amount. The resulting characteristic is determined by the manner in which its forces are generated, with each cause of vibration having its own peculiar characteristic. For example, misalignment vibration is generally characterized by a 2× (two times) running speed vibration frequency with high axial levels. Unbalance is usually characterized by a sinusoidal frequency of 1× (one time) running speed with an increase in amplitude with an increase in speed. Gear or bearing problems may be characterized as having a vibration frequency equal to the number of teeth on a gear or balls in a bearing multiplied by their rotational frequency.

Most vibration characteristics do not stand alone. When there is a change in one, there is generally a corresponding change in another. Changing the rotational frequency of a machine also changes the displacement, phase, and velocity characteristics of the machine.

For this reason, all running parameters such as load, pressure, speed, etc. must be, within reason, the same each time a condition-monitoring measurement is taken.

Vibration Cycle

Vibration, when measured, is referred to by its cycle or amplitude. A *vibration cycle* is the complete movement from beginning to end of a vibration. *Vibration amplitude* is the extent of vibration movement measured from a starting point to an extreme point. Amplitude may be measured as peak or peak-to-peak. *Peak* is the absolute value from a zero point (neutral) to the maximum travel on a waveform. *Peak-to-peak* is the absolute value from the maximum positive travel to the maximum negative travel on a waveform. A *waveform* is a graphic presentation of an amplitude as a function of time. The waveform shows a spectrum of a vibration. A *spectrum* is a representative combination of the amplitude (total movement) and frequency (time span) of a waveform. Vibration cycles continue as long as the object is disturbed. **See Figure 12-3.**

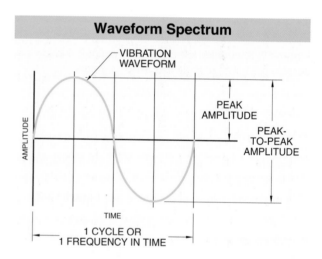

Figure 12-3. A waveform shows the frequency and amplitude of a vibration.

Displacement

Displacement is the measurement of the distance (amplitude) an object is vibrating. *Peak-to-peak displacement* is the distance from the upper limit to the lower limit of a vibration. Peak-to-peak displacement is measured to determine the severity of a vibration. Peak-to-peak displacement is expressed in mils, where 1 mil equals one-thousandth of an inch (.001″). **See Figure 12-4.** Displacement damage is similar to bending or flexing a twig. Increasing the amount of flex (displacement) increases its likeliness to snap.

Displacement

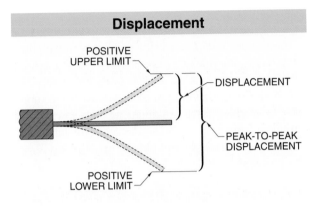

POSITIVE
UPPER LIMIT

DISPLACEMENT

PEAK-TO-PEAK
DISPLACEMENT

POSITIVE
LOWER LIMIT

Figure 12-4. Displacement is the measurement of the distance (amplitude) an object is vibrating.

Frequency

Frequency is the number of cycles per minute (cpm), cycles per second (cps), or multiples of rotational speed (orders). An *order* is a multiple of a running speed (rpm) frequency. Orders are commonly referred to as the order of rotation, such as 1× (one time) for running speed, 2× (two times) for twice the running speed, etc.

Orders are used for vibration analysis because most vibration problems are related to the running speed of a machine. For example, a vibration reading of 32 × 1800 is identified as a problem gear having 32 teeth and rotating at 1800 rpm. This reading has a frequency of 57,600 (32 × 1800 = 57,600) or an order of 32× running speed. Frequency damage is similar to bending a wire back and forth until it breaks. Increasing the frequency (rate of bending) reduces the time it takes the wire to break.

Machine-related frequencies correlate with the rotating speeds of the machine order and are expressed as either cycles per minute (cpm) or Hertz. *Hertz (Hz)* is a measurement of frequency equal to 1 cps. For example, 1 Hz is equal to 1 cps and 50 Hz is equal to 50 cps. Frequencies read as cycles per minute are divided by 60 sec to be converted to Hertz. For example, a motor operating at 1740 rpm is rotating at 29 revolutions per second (1740 rpm ÷ 60 = 29 Hz). Vibration frequencies within the range of human hearing fall between 15 Hz and 20,000 Hz. Machinery frequencies generally fall below 15,000 Hz.

Vibration frequencies within a machine are significant when analyzing the condition of the machine. Knowing and identifying the various frequencies within a machine enables a technician to pinpoint the part that may be at fault and determine the problem. Different machine problems create different vibration frequencies. Vibration due to gear problems is easily determined because the vibration generally occurs at a frequency equal to the meshing of gear teeth.

For example, a motor/gear unit indicates a vibration frequency of 16,000 cpm. The motor rotates at 800 rpm and has 4 gears with 15, 20, 40, and 28 teeth respectively. The problem component is found by dividing the frequency reading by the number of teeth in each gear. As a general rule, this frequency reading is indicating tooth wear on the 20 toothed gear (16,000 ÷ 20 = 800). **See Figure 12-5.**

Rolling-contact bearings also have high frequency readings. The frequency readings are often equal to the number of balls or rollers times the shaft speed. A bearing produces high frequency vibration even when only one of the rollers in a bearing is defective.

Worn or loose plain bearings produce a vibration frequency equal to twice the rotation of a shaft (order of 2× speed). The frequency increases as the bearing or shaft wears. The frequency may increase to an order as high as 10× speed, but does not increase consistently.

Heidelberg Harris, Inc.

Complex automatic machinery must be kept running to justify its cost. A vibration monitoring program requires reliable, detailed, and readily available information to enable maintenance personnel to replace failing components with the least downtime to the production process.

Machine Gear Vibration

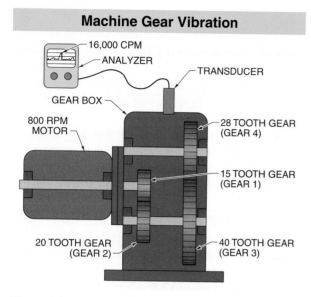

Figure 12-5. Vibration due to gear problems generally occurs at a frequency equal to the meshing of gear teeth.

Phase

Phase is the position of a vibrating part at a given moment with reference to another vibrating part at a fixed reference point. Phase readings are expressed in degrees from 0° to 360°, with one complete vibration cycle equaling 360°. Phase readings are a convenient method of comparing one vibration to another on a machine. *Note:* The end opposite the shaft is the front of the motor. When viewed from the front, forward rotation is clockwise and reverse rotation is counterclockwise. **See Figure 12-6.**

Phase readings are most commonly used in determining rotor unbalance where a rotor may be out of balance in one direction at shaft end and out of balance in the other direction at the opposite shaft end. Unbalance vibration has an order of 1× speed and may be equal rotor unbalance, opposing forces rotor unbalance, or coupling unbalance.

Equal rotor unbalance is the unbalance of weighted force across one side of a rotor or armature. Equal rotor unbalance produces a measurable vibration in only one direction. *Opposing forces rotor unbalance* is the unbalance of weighted forces on opposing ends and sides of a rotor or armature. Opposing forces rotor unbalance produces a measurable vibration in two directions.

Coupling unbalance is an unequal radial weight distribution where the mass and coupling geometric lines do not coincide. Coupling unbalance is an unbalance that occurs in different radial planes at opposite ends of a machine, similar to opposing forces rotor unbalance. **See Figure 12-7.** The problem is normally due to unbalance when the frequency of vibration is equal to the rotation of a shaft or an order of 1× speed. Unbalance always gives a radial reading of 1× speed.

Tech Tip

Use permanently installed transducers and remote measuring terminals for applications in which the bearings cannot be reached directly.

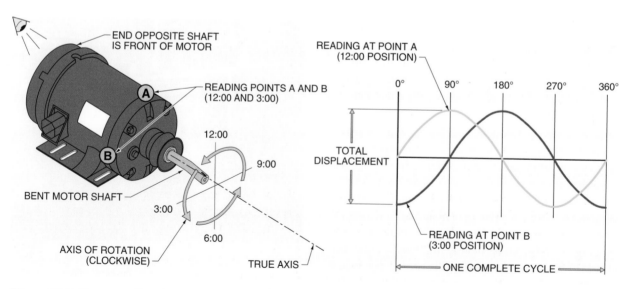

Figure 12-6. Phase readings between two signals of identical frequencies are helpful in determining unbalance or bent motor shafts.

Unbalanced Vibrations

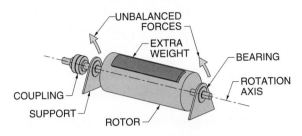

EQUAL ROTOR UNBALANCE

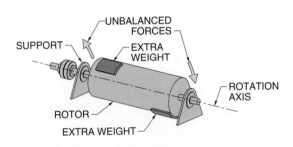

OPPOSING FORCES ROTOR UNBALANCE

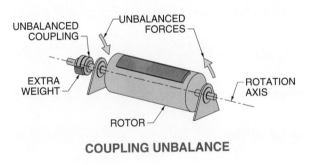

COUPLING UNBALANCE

Figure 12-7. Unbalance forces have an order of 1× speed and vary according to phase.

Vibration Velocity

Vibration velocity is the rate of change of displacement of a vibrating object. The amplitude of a vibration is measured as the maximum value of its distance moved in relation to the time of movement. Vibration velocity is measured in inches per second peak (in./sec). The value recorded is its maximum value (peak velocity) when traveling through the neutral position.

Velocity is an excellent indicator of damage because it is proportional to the extent of component damage and not the speed of the machine. The reason velocity does not rely on the machine speed is because it is proportional

to the energy content of the vibration. Velocity damage occurs from the repeated forceful cycles of flexing or fatigue.

Vibration Acceleration

Vibration acceleration is the increasing of vibration movement speed. It is the time rate of change of velocity. **See Figure 12-8.** The rate of acceleration reaches its maximum value as an object goes beyond its maximum limits of displacement. The peak value of acceleration is measured in units of g peak, where 1 g is equal to 386 inches per second squared (1 g = 386 ips^2). One g is also the international standard of acceleration produced by the force of gravity at the earth's surface and is used to indicate the force an object is subjected to when accelerated. Acceleration parameters are useful, especially with high frequencies, because accelerated forces at high frequency are extreme forces that ultimately cause a bearing to fail or lubricants to break down.

Vibration Acceleration

Figure 12-8. Vibration acceleration is the increasing of vibration movement speed.

Oil Whirl

Oil whirl is the buildup and resistance of a lubricant in a rolling-contact bearing that is rotating at excessive speeds. Oil whirl has a frequency of less than one-half the speed (rpm). Oil whirl occurs when a shaft is turning so fast that it attempts to roll over the lubricant rather than squeeze it out of the way. Oil whirl vibration may have a frequency rate of less than 4500 cpm while the shaft is rotating at 8600 rpm. **See Figure 12-9.**

Oil Whirl

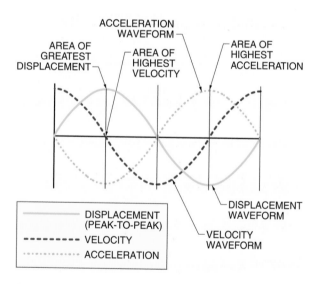

Figure 12-9. A shaft vibrates at less than 1× speed when it is turning so fast that it tries to roll over its lubricant.

Vibration Severity

Displacement, velocity, and acceleration are all direct measures of the severity of machine vibration. In most cases, taking only one of the three measurements is required to sufficiently describe machine vibration condition. Although displacement, velocity, and acceleration are directly related, the measurement taken is generally determined by the frequency (cpm) of the vibration. **See Figure 12-10.**

Figure 12-10. The phases of displacement, velocity, and acceleration are always 90° apart from each other.

Displacement is difficult to determine at high frequencies, but is a good choice for determining amplitudes of low-speed equipment. Displacement is best suited for frequencies between 1 cpm and 60,000 cpm.

Velocity amplitudes are mostly used because the velocity range of 600 cpm to 60,000 cpm is typical of most industrial facility equipment. Acceleration parameters are useful in the higher frequency ranges between 18,000 cpm and 600,000 cpm.

VIBRATION MEASUREMENT METHODS

The internal mechanical condition of a machine is determined by the vibration measurement method. Vibration measurement methods are developed by choosing the vibration transducer (pickup), the proper placement of the transducer for the measurement to be taken, and analysis of the measurement readings.

SPM Instrument, Inc.

Hand-held test instruments are used for comprehensive machine condition monitoring.

Vibration Transducers

Vibration measurement takes the variety of internal vibrations and their complex signals and converts them into readings through the use of a vibration transducer and an analyzer. A *transducer* is a device that converts a physical quantity into another quantity, such as an electrical signal or a graphic display. A *vibration analyzer* is a meter that pinpoints a specific machine problem by identifying its unique vibration or noise characteristics. Various analyzers are available for gathering the required data. Each can transform complex signals into an understandable display useful for diagnosis. Analyzers can range from simple walk-around vibration meters to the more sophisticated tracking analyzers having a permanent link between transducer and analyzer which provides 24 hr inspection, detection, and protection.

Transducers are similar to a microphone in that a physical movement, such as sound, is converted into an electrical signal. The choice of transducer is generally determined by the frequency of the vibration. The correct transducer must be selected to obtain a signal that represents a vibration accurately. Vibration transducers include velocity, accelerometer, and displacement transducers. **See Figure 12-11.** Vibration severity charts are used to determine if the readings refer to a smooth or rough condition. **See Appendix.**

Velocity Transducers. A *velocity transducer* is an electromechanical device that is constructed of a coil of wire supported by light springs. Velocity transducers are the most common transducer used because they operate in the frequency range of most industrial applications. The coil surrounds a permanent magnet that moves following the motion of vibration. The cut of magnetic flux between the coil and the magnet creates a voltage in the coil. This voltage, expressed in millivolts per inch per second, is the output to the analyzer. **See Figure 12-12.** Voltage output is proportional to relative movement (vibration). The faster the movement, the higher the voltage output. Internal moving parts of a velocity transducer require

recalibration because they wear out or change over a period of time. Velocity transducers give reliable results in a hand-held device at frequency rates of 600 cpm to 60,000 cpm (10 Hz to 1000 Hz).

Accelerometer Transducers. An *accelerometer transducer* is a device constructed of quartz crystal material that produces electric current when compressed. Quartz crystals are strongly piezoelectric, becoming polarized with a negative charge on one end and a positive charge on the other when subjected to pressure. *Piezoelectric* is the production of electricity by applying pressure to a crystal. Forces on piezoelectric crystals produce a voltage output proportional to vibration accelerations producing a display in gravitational (g) forces. **See Figure 12-13.**

Accelerometer transducers are becoming more popular than velocity transducers because they can operate at frequencies between 120 cpm and 600,000 cpm (2 Hz to 10,000 Hz), have no moving parts, and are more rugged than velocity transducers. They are not greatly affected by stray magnetic fields, which makes them useful around AC motors.

Tech Tip

Modern electronic accelerometer transducers include protection circuits which prevent electronic damage due to use beyond specified range or accidental dropping.

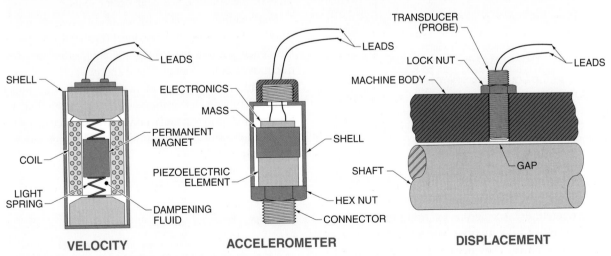

Vibration Transducers

Figure 12-11. Transducers are similar to a microphone in that physical movement, such as sound, is converted into an electrical signal.

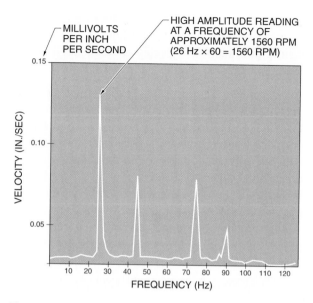

Figure 12-12. Velocity transducers relay a voltage output signal in millivolts, relative to vibration movement.

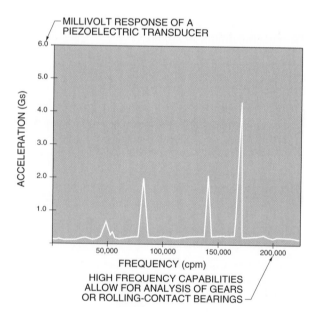

Figure 12-13. Accelerometer transducers can produce currents indicating up to 70 g of acceleration and measure frequencies up to 600,000 cpm.

Accelerometer transducers are also popular because they are small, lightweight, and are able to withstand high temperatures. Accelerometer transducers cannot be hand-held like velocity transducers because they can pick up interference. They must be firmly connected to the machine through the use of a mounting fixture that is glued, screwed, or clamped to the machine.

Baldor Electric Company

Minimizing machine vibration reduces the damaging vibration transmitted to other nearby machinery.

Displacement Transducers. A *displacement transducer* is a mechanical sensor whose gap-to-voltage output is proportional to the distance between it and the measured object (usually a shaft). Displacement transducers are used mostly as permanent installations to continuously monitor machine displacement vibration. Also known as proximity pickups, displacement transducers detect displacement of a vibrating shaft as the shaft opens and closes a gap between the probe tip and shaft as it rotates. Gap settings vary according to transducer size and shaft material, but are typically 0.020″, 0.030″, 0.050″, 0.060″, or 0.100″. Gap readings may be taken whether the shaft is rotating or not. The ability to measure shaft vibration relative to the bearing housing offers an inside view of the condition of bearings or seals.

Gap readings are taken by means of a magnetic field (flux) being set up at the transducer (probe) tip. Magnetic field energy changes as the gap changes during vibration movements. Magnetic field energy is converted to a signal amplitude (displacement signal) at the transducer. A displacement measuring system senses the distance between the probe tip and a conductive surface using the transducer, an oscillator/amplifier, and a meter/scope. **See Figure 12-14.**

Tech Tip

Displacement transducers are designed to measure the actual movement of a machine shaft relative to the transducer and therefore, must be rigidly mounted to a stationary structure to gain accurate, repeatable data.

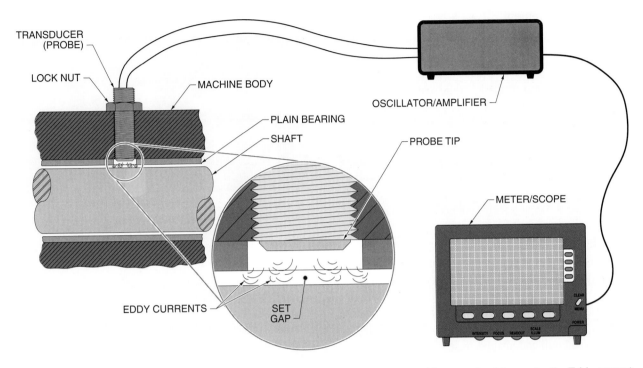

Figure 12-14. Displacement transducers are sent an electrical current to set up eddy currents at the probe tip. Eddy current voltages sent back to the meter as an output voltage vary according to the gap distance.

An *oscillator* is a device that generates a radio frequency (RF) field that, when sent to the transducer tip, creates eddy currents. An *eddy current* is an electric current that is generated and dissipated in a conductive material in the presence of an electromagnetic field. As the shaft vibrates relative to the transducer, energy changes modulate the oscillator voltage. An amplifier receives the feedback signal and amplifies the output signal to a meter, analyzer, or monitor. **See Figure 12-15.**

Transducer Selection

Selecting the proper transducer (velocity, accelerometer, or displacement) is determined by the design of the machine being monitored and the severity of the vibration. Velocity transducers detect vibration from defective components, loose parts, and rolling-contact bearings in the low-frequency range. Accelerometer transducers are helpful in detecting high-frequency defects in bearings, gears, or fan or turbine blades. These measurements are also helpful in detecting structural movements. Displacement transducers are used where dynamic stresses are found or where clearance within motors, gear boxes, or any mechanism using fluid film bearings can be measured. Displacement transducers may also be used to measure machine unbalance.

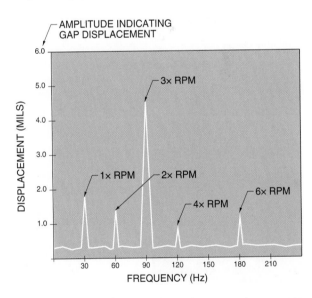

Figure 12-15. Displacement transducers produce an amplitude that represents the total peak-to-peak displacement of a vibrating shaft.

Tech Tip

When mounting transducers with cables, secure the cables to the vibrating structure to minimize cable or connector fatigue failures and loss of data.

Transducer Placement

Vibration may occur in any direction. Transducers must be placed in a position to directly receive a vibration. Transducers must also be placed at the exact same location on a machine to ensure accurate readings. The three major directions vibration travels within a machine are horizontally (radial), vertically (radial), or axially.

Horizontal measurements are taken with transducers placed in the horizontal plane (X axis) with the axis of rotation. Vertical measurements are taken with transducers placed in the vertical plane (Y axis) with the axis of rotation. Axial measurements are taken with transducers placed at the centerline (A axis) with the axis of rotation. **See Figure 12-16.**

Transducer placement depends on the direction of the vibration to be detected. Radial in the vertical plane (Y axis) is generally chosen over radial in the horizontal plane (X axis) or axial if only one measurement is chosen. This is because most vibration is heightened by gravitational pull and also because it is normally easier to probe from the top. A complete measurement uses all three positions. Transducers used to measure radial vibration must be attached within 3″ of the bearing.

Tech Tip

Transducers should be placed on or as near as possible to bearings because vibration forces are transmitted through the bearings.

SPM Instrument, Inc.

The early detection of bearing damage reduces the risk of breakdowns and enables maintenance personnel to plan the replacement and reduce the required downtime.

BASIC VIBRATION ANALYSIS

To analyze individual component condition, a vibration signature is read from an analyzer. A *vibration signature* is a set of vibration readings resulting from tolerances and movement within a new machine. Machine wear may then be plotted through the use of its vibration signature. Periodic analysis begins after a signature is established. The vibration signature of a machine in good operating condition provides a baseline measurement against which future measurements may be compared. A change in the vibration signature of a machine indicates the beginning of a defect.

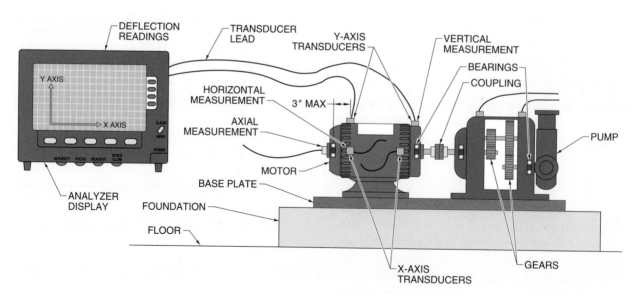

Figure 12-16. Transducers used to measure radial vibration must be placed within 3″ of bearings for transducer measurements and analyzer displays to be representative of machine condition.

Analysis of vibration signals allows the specific nature of problems to be found and an assessment for repair to be made while the machine is operating. A steady, continuous vibration signature change enables a technician to project machine condition and allow scheduled machine repairs in advance of machine failure.

Limits must be set to provide a basis for determining the condition of a machine. These limits must be close enough to normal values to allow corrective action before operating conditions begin to cause damage. Original limits are normally recommended limits supplied by charts or the equipment manufacturer. As data is developed or when necessary limits have been reached often, the limits should be adjusted and refined to more realistic values.

Limits must also be set according to the load. Radial loads wear gradually and continually until failure. This gradual wear is detected as a displacement change. Early corrective action is possible even though a radial position change due to bearing wear may not produce an increase in vibration amplitudes.

Thrust loads generally fail without much notice, where the only detectable movement prior to total failure is in the loss of lubrication space and some metal compression (about 2 mils). For this reason, limits must be set closer to the original limits for axial forces than for radial forces.

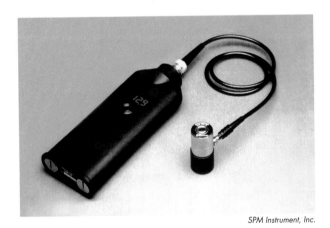

SPM Instrument, Inc.

A vibrameter is a hand-held test instrument used for periodic measurements to detect unbalance, misalignment, and other mechanical faults in rotating machines.

A basic vibration analysis system consists of a signal pickup (transducer), signal recording device, signal analyzer, analysis software, and a computer for data storage. Relatively inexpensive units are available consisting of a programmable data collector and its analysis software. Any analysis system requires the use of a computer.

Signal analyzers include Fast Fourier Transform (FFT) or dynamic signal analyzers (DSAs). A *Fast Fourier Transform (FFT)* is a calculation method for converting a time waveform into a series of frequency vs. amplitude components. A *Fast Fourier Transform (FFT) analyzer* is a microprocessor capable of displaying the FFT of an input signal. A *dynamic signal analyzer* is an analyzer that uses digital signal processing and the FFT to display a dynamic vibration signal as a series of frequency components. The DSA is an analyzer that uses both amplitude and frequency for display and has the capability to display both low frequencies and high frequencies on the same screen using a logarithmic scale. A *logarithmic scale* is an amplitude or frequency displayed in powers of ten.

Reading Amplitudes

A three-step vibration detection and analysis process is required before the characteristics presented by an analyzer showing unbalance, defective bearings or gears, etc. are placed into a maintenance record. **See Figure 12-17.** The three steps are the following: conversion of vibrations to electrical signals, reduction of electrical signals into component form, and identification of individual defect frequencies from component signals.

Converting vibration into an electrical signal is performed by the transducer. However, to achieve an accurate account of vibration condition, the correct transducer must be selected, located, and installed. An important consideration for transducer selection is the amplitude of the vibration parameters and the machine speed. Transducer selection is made regarding displacement, velocity, or acceleration.

Reducing the electrical signals into component form is best accomplished with a display using a conversion between the linear amplitude spectra and the logarithmic amplitude spectra. *Linear amplitude spectra* are amplitude signals displayed in equal increments. *Logarithmic amplitude spectra* are amplitude signals displayed in powers of 10. Signals viewed in linear spectra are summed together, making a waveform next to impossible to read. Reducing the many electrical signals into component form is accomplished by switching the linear spectra to the logarithmic spectra. **See Figure 12-18.**

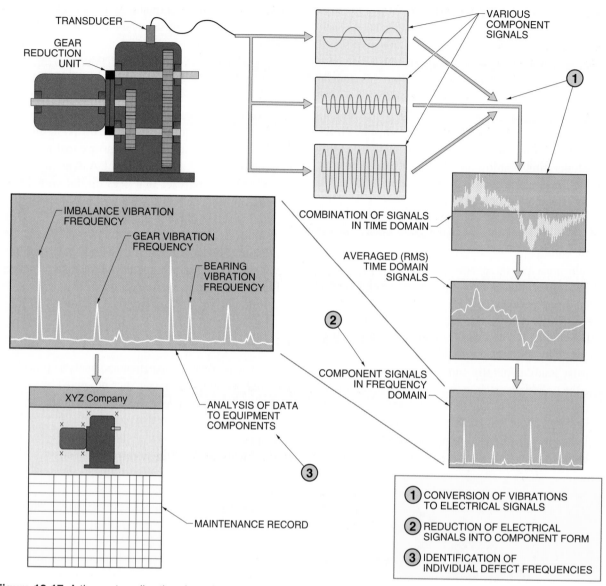

Figure 12-17. A three-step vibration detection and analysis process is required before the characteristics presented by an analyzer are placed into a maintenance record.

Linear spectra, although easier to read, may not show the value of a high scale and give an indication to the value of small signals. Logarithmic spectra allow all frequencies to be visible by compressing the large signal amplitudes and expanding the small ones. Most analyzers have the ability to shift between the linear and logarithmic amplitude scales. This offers the choice of viewing the amplitude of a single component on linear spectra or viewing logarithmic spectra for a full range of vibration data.

Identifying individual defect frequencies from component signals is the key to a good analysis. Each machine offers a unique set of component signals because the vibration measured is a response to a defect force, not the force itself. Amplitudes may be read as peak-to-peak, zero-to-peak, or root mean square (rms). **See Figure 12-19.**

Root mean square (rms) is the square root of the sum of a set of squared instantaneous values. Rms averages and smooths a signal containing high peaks, making the output more representative of unbalance or misalignment problems. An rms vibration signal produces a time-averaged amplitude proportional to the area within a time domain waveform. The extent of vibration may be read in either the time domain or the frequency domain. Rms amplitude of a true peak value may be determined by multiplying 0.707 by the peak value. For example, the rms value of a true peak of 3 is equal to 2.121 (3 × 0.707 = 2.121). When a true peak value is required, as in the case of measuring shaft vibration, a total rms peak-to-peak sine wave is multiplied by 1.414. For example, when a shaft vibration using an rms peak-to-peak signal shows a 4 mil waveform, the true peak-to-peak shaft displacement is equal to 5.656 mil (4 mil × 1.414 = 5.656 mil). The true peak value is one half of 5.656 mil or 2.828 mil (5.656 mil ÷ 2 = 2.828 mil). Dynamic signal analyzers perform rms averaging digitally on successive vibration spectra.

based on displacement characteristics. Peak is used with acceleration characteristics offering the greatest sensitivity required for defect detection of bearings or gears. Rms velocity characteristics are used when component levels vary significantly, such as with unbalance or misalignment. Vibration (noise) levels that vary greatly, such as in misalignment, are not statistically accurate with one measurement. More than one measurement is required to obtain accurate results.

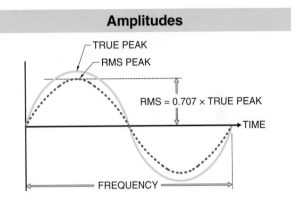

Figure 12-19. Smoothing a mix of many amplitudes and frequencies is accomplished by rms averaging.

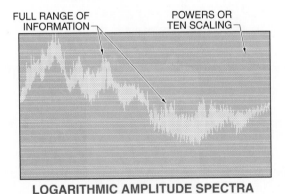

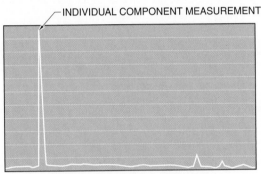

Figure 12-18. Logarithmic spectra allow the viewing of large and small signal amplitudes that are not normally visible on the same display.

Each amplitude value is used in a comprehensive vibration analysis and monitoring program because assessment of vibration severity is based on frequencies. Peak-to-peak is used as a shaft vibration analysis

Time Domain. *Time domain* is the amplitude as a function of time. Time domain signals are signals that appear as cycles where each cycle or wave occurs in a certain time period (generally in milliseconds). Time domain is displayed as amplitude versus time, with amplitude on the vertical axis and time on the horizontal axis.

The time domain is useful when a single source vibration, such as rotor unbalance or component looseness, is displayed. Although the time domain is used when a single vibration source is suspected, most machine vibration is a complex mix of many vibration frequencies, where each vibration requires an individual identification. The overall peak-to-peak vibration (displacement) of a machine is the sum (total) of the various individual vibrations of the machine.

For example, a gear transmission may have 2 mils of vibration occurring at 2× speed because of looseness, 2 mils of vibration occurring at 1× speed because of unbalance, and 1 mil of vibration at a high frequency due to bearing displacement. The total vibrating energy amplitude is approximately 5 mils peak-to-peak. **See Figure 12-20.**

Time Domain

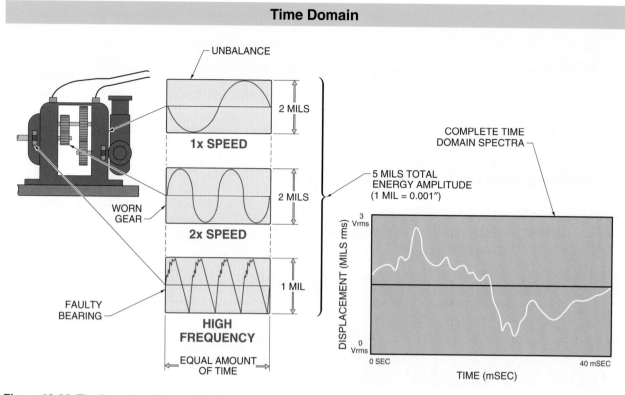

Figure 12-20. The time domain is useful when a single source vibration, such as rotor unbalance or component looseness, is displayed.

More than one vibration displayed in the time domain appears as a complex sum of amplitudes and is difficult to read. A particular frequency range that is not affected by other frequencies may be read by filtering out the other frequencies. A *filter* is a device that limits vibration signals so only a single frequency or group of frequencies can pass.

Frequency Domain. *Frequency domain* is the amplitude versus frequency spectrum observed on an FFT analyzer. Frequency domain is best understood by viewing a time domain cycle from a three-dimensional perspective. The frequency domain takes a combined multiple source vibration viewed in the time domain and separates the individual vibrations based on their frequencies. Each amplitude is read from left to right as an increase in frequency. **See Figure 12-21.**

A *frequency spectrum* is a representation of the frequency and content of a dynamic signal. The frequency spectrum is displayed with the vertical axis consisting of the amplitude (calibrated in convenient units) and the horizontal axis graduated in multiples of running frequencies. Abnormal characteristics related to balance, bearings, gears, and alignments are generally read

within their separate frequencies. For example, basic unbalance displayed on a frequency domain analyzer is seen as an amplitude at the frequency of 1× speed.

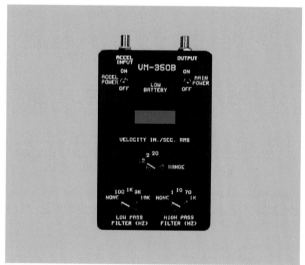

A vibration meter measures vibration velocity and contains internal high pass and low pass frequency filters that allow the operator to pinpoint the vibration source.

Frequency Domain

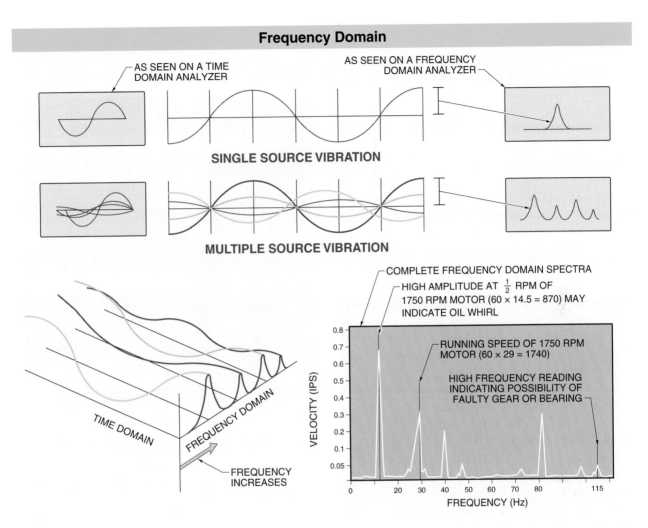

AS SEEN ON A TIME DOMAIN ANALYZER

AS SEEN ON A FREQUENCY DOMAIN ANALYZER

SINGLE SOURCE VIBRATION

MULTIPLE SOURCE VIBRATION

TIME DOMAIN

FREQUENCY DOMAIN

FREQUENCY INCREASES

COMPLETE FREQUENCY DOMAIN SPECTRA

HIGH AMPLITUDE AT $\frac{1}{2}$ RPM OF 1750 RPM MOTOR (60 × 14.5 = 870) MAY INDICATE OIL WHIRL

RUNNING SPEED OF 1750 RPM MOTOR (60 × 29 = 1740)

HIGH FREQUENCY READING INDICATING POSSIBILITY OF FAULTY GEAR OR BEARING

VELOCITY (IPS)

FREQUENCY (Hz)

Figure 12-21. The frequency domain takes a combined multiple source vibration in the time domain and separates the individual vibrations based on their frequencies.

Current vibration analysis equipment is able to scan through a range of frequencies using digital filtering and display amplitudes at various frequencies. Complex vibration signals can be broken down using filters into simpler vibration components, resulting in clarification of the vibration.

Dynamic signal analyzers are able to amplify and display signals that are so small they would not show up on a time domain spectra. A DSA can simultaneously display a vibration that is 1000 times greater than another vibration. *Dynamic range* is the ratio between the smallest and largest signals that can be analyzed simultaneously.

GE Motors & Industrial Systems

Machine alignment is critical to long equipment life because small amplitude vibration caused by motor shaft deflection may drive out lubrication from contacting parts resulting in accelerated wear.

Tech Tip

When taking vibration readings, ensure that the transducer is pointed accurately because transducers are most sensitive along their central axis.

VIBRATION MONITORING PROGRAMS

The key to establishing a successful vibration monitoring program is to understand that the only objective for taking and recording readings is to compare previous readings with present readings and to see if an increase in readings is developing. Mechanical trouble is generally the reason why machine vibration increases.

To maintain consistency, every effort should be made to take readings under the same conditions. For example, consecutive readings should be taken as a machine is under a load if the first readings were taken while the machine was under a load (during production).

Vibration monitoring programs may be short-term or long-term. A short-term program includes permanently installing sensors on critical equipment such as turbines or high-speed rollers to give an alarm at the first sign of trouble. A long-term monitoring program includes manually monitoring critical areas of plant equipment to provide information on gradual or impending problems. Regardless of the program used, there must be recognized trends and alarms to issue appropriate warnings.

Technicians should be able to accurately identify specific gears within a failing gearbox, interpret resonance problems creating damage to bearings, shafts, and couplings, and identify improper bearing installation and shaft alignment. Skilled personnel are able to diagnose unbalance, angular misalignment, eccentric gears, or defects to inner or outer bearing races. Vibration characteristics charts are beneficial towards pinpointing a specific component defect. **See Appendix.**

Diagnosing machinery problems can be complex, requiring extensive training and experience. However, smaller scale programs may be instituted and should be used only as a progressive development towards a complete vibration monitoring program. Developmental stages consist of personnel training, facility layout, machinery layout, machinery component layout, and establishing files and data retrieval systems. Forms are also required for generating machine component specifications such as motor sizes, types, dimensions, speed, and coupled devices such as pumps, gearboxes, belt drives, etc. This includes specifications such as the type of gear, number of teeth, etc. Other forms include those for scheduling vibration checks and recording machinery maintenance history and vibration checks.

Exxon Company

Vibration monitoring programs help reduce wear and maintenance costs on vital machines and provide increased operating life.

Recordkeeping

Historical data is recorded and kept on a vibration check sheet. A vibration check sheet contains the basic machine configuration to be checked and its checkpoints, equipment type and location, check interval (weekly, monthly, etc.), check made (displacement, velocity, or acceleration), individual making the checks, date checked, initial horizontal, vertical, and axial readings, test equipment used, and a place to record the periodic checks. **See Figure 12-22.**

An interpretation may be made from a vibration check sheet as recordings stay the same or change. Recordings may be graphed to develop a trend. *Trending* is a graphic display used for interpretation of machine characteristics. Picture or graphic plotting provides a visual display of recordings. The visual display, when updated at each recording, provides an indication as to whether the component check is showing a gradual life expectancy decline or a rapid deterioration.

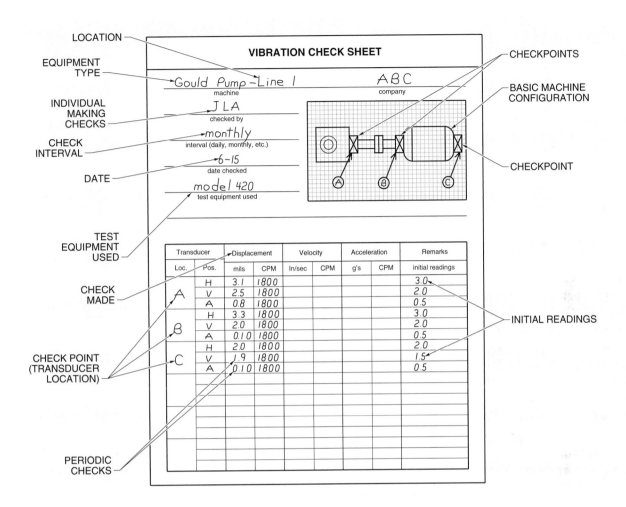

LOCATION

EQUIPMENT TYPE

INDIVIDUAL MAKING CHECKS

CHECK INTERVAL

DATE

TEST EQUIPMENT USED

CHECK MADE

CHECK POINT (TRANSDUCER LOCATION)

PERIODIC CHECKS

CHECKPOINTS

BASIC MACHINE CONFIGURATION

CHECKPOINT

INITIAL READINGS

VIBRATION CHECK SHEET

Gould Pump – Line 1 — machine

ABC — company

JLA — checked by

monthly — interval (daily, monthly, etc.)

6-15 — date checked

model 420 — test equipment used

Transducer		Displacement		Velocity		Acceleration		Remarks
Loc.	Pos.	mils	CPM	In/sec	CPM	g's	CPM	initial readings
A	H	3.1	1800					3.0
	V	2.5	1800					2.0
	A	0.8	1800					0.5
B	H	3.3	1800					3.0
	V	2.0	1800					2.0
	A	0.10	1800					0.5
C	H	2.0	1800					2.0
	V	1.9	1800					1.5
	A	0.10	1800					0.5

Figure 12-22. The numerical data entered on a chart indicates if machine conditions are staying the same or are worsening.

Review

1. Describe some of the negative effects vibration can have on equipment.

2. Why is it important that the axes of two rotating machine shafts have proper alignment?

3. List the requirements needed in a complete maintenance vibration program.

4. Explain why characteristics offered in vibration analysis are valuable in determining unwanted motions in a machine.

5. Define vibration cycle and identify how long it continues.

6. Explain how frequency damage is similar to bending a wire back and forth until it breaks.

7. Define phase.

8. Describe why velocity is an excellent indicator of damage.

9. Explain why acceleration parameters are useful, especially with high frequencies.

10. Why are velocity transducers the most commonly used transducer?

11. List several reasons why accelerometer transducers are becoming more popular.

12. When are displacement transducers typically used?

13. Explain why identifying defect frequencies from component signals is the key to a good analysis.

14. Explain what time domain is and how it is displayed.

15. Describe the contents of a vibration check sheet.

Digital Resources

ATPeResources.com/Quicklinks
Access Code 462409

CHAPTER **13**

ALIGNMENT

More than 50% of vibration problems are caused by misaligned machinery. Vibration from misaligned shafts has a direct impact on the operating costs of a facility. Misaligned shafts require more power, create premature seal damage, and cause excessive forces on bearings. This leads to early bearing, seal, or coupling failure. Alignment specifications must be much more accurate now than in the past because operating speeds have increased, material weights have been reduced, and bearing tolerances have increased. Various methods, such as dial indicator fixtures and electronic and laser measuring devices, have been developed to achieve running condition accuracy.

Ludeca Inc.

Chapter Objectives

- Compare the types of misalignment.

- Describe how to properly maintain and prepare equipment alignment.

- Explain how the elements related to anchoring machinery affect the alignment of equipment.

- Explain how dial indictors function.

- Describe the methods available for aligning machinery.

Key Terms

- alignment
- misalignment
- soft foot
- coupling
- foundation
- base plate
- bolt bound
- dowel effect
- jack screw

- runout
- dial indicator
- straightedge alignment method
- rim-and-face alignment method
- reverse dial method
- electronic reverse dial method
- laser rim-and-face alignment method

MISALIGNMENT

Alignment is the location (within tolerance) of one axis of a coupled machine shaft relative to that of another. *Misalignment* is the condition where the centerlines of two machine shafts are not aligned within tolerances. Properly aligned rotating shafts reduce vibration and add many years of service to equipment seals and bearings. Misalignment of a coupling by 0.004″ can shorten its life by 50%. **See Figure 13-1.**

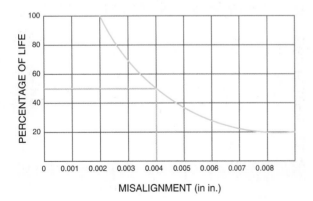

Figure 13-1. Misalignment of a coupling by 0.004″ can shorten its life by 50%.

Poor condition of equipment, such as worn bearings, bent shafts, stripped mounting bolts, bad gear teeth, or insufficient foundations or base plates, can create enough vibration to render any alignment effort useless. Once vibration starts, rapid wear of other components begins.

Misalignment exists when two shafts are not aligned within specific tolerances. Misalignment may be offset or angular. *Offset misalignment* is a condition where two shafts are parallel but are not on the same axis. *Angular misalignment* is a condition where one shaft is at an angle to the other shaft. Shaft misalignment is usually a combination of offset and angular misalignment. **See Figure 13-2.**

Offset and angular misalignment may be in the vertical or horizontal planes or both. Misalignment may be in the vertical offset, horizontal offset, vertical angularity, or horizontal angularity. Most misalignments are a combination of each. Many offsets in alignment are caused by improper machine foundations, weak supports, forces received from machine piping, soft foot, or thermal expansion. *Soft foot* is a condition that occurs when one or more feet of a machine do not make complete contact with its base. *Thermal expansion* is the dimensional change in a substance due to a change in temperature.

Misalignment

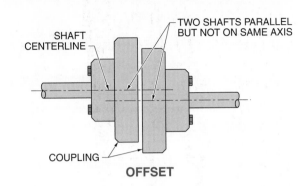

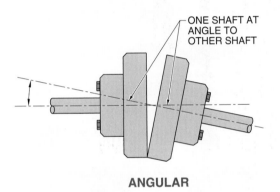

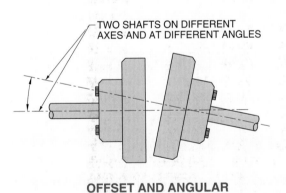

Figure 13-2. Misalignment may be offset or angular, but is generally a combination of the two.

Rotating machines are generally connected by couplings. A *coupling* is a device that connects the ends of rotating shafts. A *flexible coupling* is a coupling with a resilient center, such as rubber or oil, that flexes under temporary torque or misalignment due to thermal expansion. Flexible couplings can allow enough vibration to cause excessive wear to seals and bearings. Where flexible couplings are used, shaft alignment should be as accurate as it would be if solid couplings are used.

ALIGNMENT SEQUENCE

Vibration is now recognized as detrimental and industry is making many moves toward correction. Fundamental steps to reducing vibration include purchasing quality equipment, setting up good preventive maintenance procedures, and maintaining proper equipment alignment techniques and tolerances.

The objective of proper alignment is to perfectly couple two shafts under operating conditions so all forces that cause damaging vibration between the two shafts and their bearings are removed. The objective is to align the shafts, not the couplings. Aligning the couplings may result in misaligned shafts with aligned but irregularly shaped couplings.

Each component that directly or indirectly affects the proper alignment of machinery must be identified and considered before actual alignment begins. Considerations for good alignment include the following:

- Proper preparation of foundations, foundation base plates, and machinery
- Proper machinery anchoring
- Proper machinery movement during alignment
- Soft foot
- Proper use of alignment methods or procedures

EQUIPMENT PREPARATION

Machinery to be aligned that is connected electrically must be locked out first. Before working on the equipment, challenge the electrical functions by testing the start switch. *Challenging* is the process of pressing the start switch of a machine to determine if the machine starts when it is not supposed to start. On completion of the lockout challenge, place all switches in OFF position. Also, any product pumps to be aligned must be blocked out to prevent product flow in the piping. **See Figure 13-3.**

Machine Foundation and Base Plates

Aligning any equipment begins with the foundation and base plate to which the equipment is anchored. A *foundation* is an underlying base or support. A *base plate* is a rigid steel support for firmly coupling and aligning two or more rotating devices. Foundations must be level and strong enough to provide support without movement. Base plates must be rigid enough to firmly support the equipment without stress and be securely anchored to the foundation.

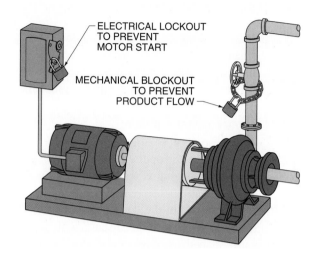

Figure 13-3. Lockout all electrical energy and blockout all mechanical energy before beginning alignment.

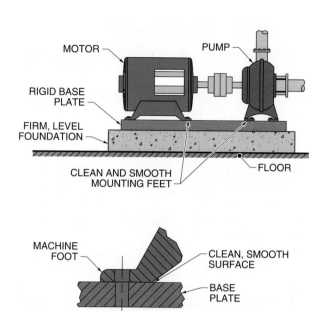

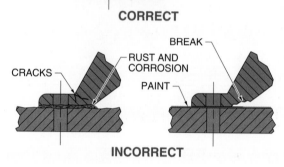

Figure 13-4. A clean, firm, and level base plate and foundation is required for proper alignment to ensure minimal flexing between machines.

Originally, equipment base plates were made of thick cast iron that was strong enough to support the equipment in all operating conditions. The mounting surfaces were machined level. Currently, many equipment base plates are only sheet metal or plate metal welded or bolted to angle iron or I-beams. Flexing base plates must not be used. Good alignment is wasted when a flexing base plate is used. **See Figure 13-4.**

The feet on machines such as motors, gearboxes, pumps, etc., must be checked for cracks, breaks, rust, corrosion, or paint. This equipment should be bolted to a base plate, not anchored to concrete. The contacting surfaces between the motor, pump, gearbox, etc., and the base plate must be smooth, flat, and free of paint, rust, or foreign materials. Inspect all areas for burrs, rust, cracks, breaks, or any other damage. Finally, inspect the couplings, shafts, and bearings for damage, contamination, or inaccurate sizes.

ANCHORING MACHINERY

Each element related to anchoring machinery has a direct effect on the alignment forces of the equipment. Machine anchoring must consider piping and plumbing and the condition of anchoring components such as bolts and washers.

Machinery such as industrial grinders must be properly anchored to prevent damage to workpieces and equipment from machinery vibrations.

Flexible plumbing connections are used on rotating machine piping to reduce stress caused by thermal expansion due to hot fluid processes.

Piping Strain

Pipe and conduit connections, if improperly installed, can produce enough force to affect machine alignment. Thermal expansion created by the temperature of liquids and reaction forces from piped products can produce enough force to affect machine alignment. To ensure that any transmission of an outside force does not affect the proper alignment of machines, machines should initially be aligned unattached from any piping if possible. Therefore, all plumbing must be properly aligned and have its own permanent support even when unattached. In some cases, flexible plumbing connections are necessary to separate stresses and vibrations between pump/motor and product lines. **See Figure 13-5.**

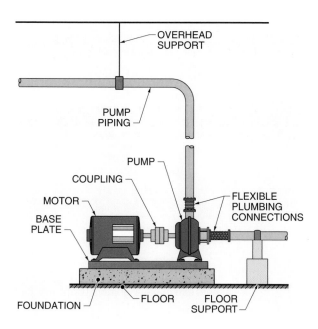

Figure 13-5. Pump piping must be independently supported to prevent angular forces from working against bearing and alignment tolerances.

Anchoring

Anchoring is any means of fastening a mechanism securely to a base or foundation. Firm but adjustable anchoring of mechanisms on a base plate is accomplished using the proper mechanical fasteners (bolts, screws, nuts, etc.). Adverse anchoring includes bolt-bound machines, bolts or bolt holes not of the proper size to allow for sufficient movement, and improper washers that create a dowel effect. Improper use, type, or fit of the anchoring bolts can make alignment of any machine impossible. **See Figure 13-6.**

Bolt Bound. *Bolt bound* is the prevention of the horizontal movement of a machine due to the contacting of the machine anchor bolts to the sides of the machine anchor holes. Bolt bound bolts prevent horizontal movement of a machine in any needed horizontal direction. Work is not wasted or duplicated if this condition is checked before any alignment checks begin. In many cases, bolt bound conditions may be checked using a straightedge placed along the machine shafts to determine if enough horizontal movement is available for proper alignment.

Loosening all anchoring bolts and shifting both machines to align the base plate mounting holes usually works if it is detected that sufficient movement is not available at the machine to be shimmed. The anchoring bolts must be turned down or the machine mounting holes enlarged if repositioning both machines does not work.

Anchoring bolts may be turned down without decreasing the tensile strength as long as the cut does not go beyond the root diameter (bottom portion) of the threads. Also, mounting holes may be enlarged, but this usually leads to the dowel effect. In some cases, a combination of undercutting the bolt and enlarging the holes may be necessary.

Dowel Effect. *Dowel effect* is a condition that exists when the bolt hole of a machine is so large that the bolt head forces the washer into the hole opening on an angle. Angled washers force the bolt to the center of the hole, making any horizontal movement impossible. Dowel effect is corrected by using machined washers 2 to 3 times thicker than the original washer. This prevents any deformation of the washer by bolt forces.

Always use the correct torque on an anchoring bolt to prevent excessive bolt stretch and reduce the possibility of distorting the base plate or machine frame.

Anchoring Characteristics

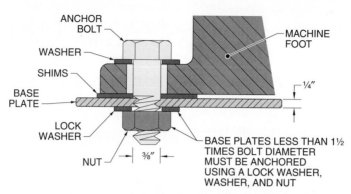

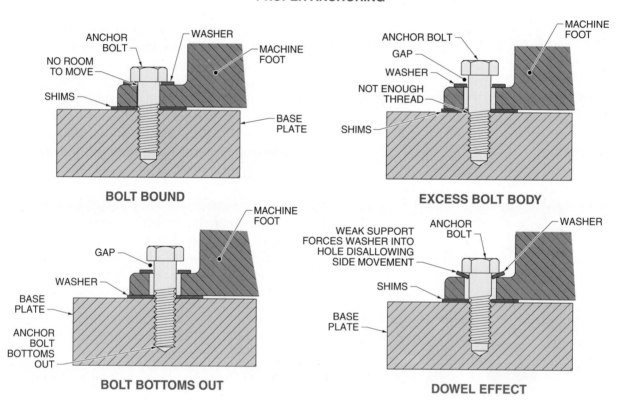

PROPER ANCHORING

BOLT BOUND

EXCESS BOLT BODY

BOLT BOTTOMS OUT

DOWEL EFFECT

Figure 13-6. Improper bolt diameter, excessive bolt body or length, and weak washers can make alignment of any machine impossible.

Proper Bolt Installation. A base plate that is drilled and tapped to anchor a machine must be a minimum thickness of 1½ times the root diameter of the anchoring bolts. The threaded depth must be a minimum of 1½ times the root diameter of the bolt when a base plate is thicker than 1½ times the root diameter of the bolt. Mounting bolts require the use of nuts and lock washers if the base plate is less than 1½ times the bolt diameter.

Always select anchor bolts that have the correct length of the unthreaded portion of the bolts. The bolt may run out of thread and leave an incomplete and loose anchor if the unthreaded portion is too long. Also, the bolt may bottom out in the base plate hole leaving a loose anchor if the threaded portion of the bolt is too long. Always use the correct grade and size anchor bolts to properly secure the machine frame to the base plate.

Controlled Machine Movements

The choice of which machine to move during alignment, the proper tools to use to prevent equipment damage, and the components used for precise movements must be known for an accurate, fast, and damage-free alignment. This knowledge includes the proper use and installation of jack screws and the proper use and selection of spacers and shim stock.

Machine Movement Specification. Generally, the heaviest machine or the machine attached to plumbing is the anchored machine and is referred to as the stationary machine (SM). The motor is generally the machine moved and is referred to as the machine to be shimmed (MTBS). Either machine may be the moved and shimmed machine. Regardless of which machine is moved, the SM must initially be higher than the MTBS to allow for proper vertical alignment. A good practice is to initially install the SM using 0.125″ shims under each foot. This practice requires raising the MTBS, but prevents any vertical movement requirements of the SM. **See Figure 13-7.**

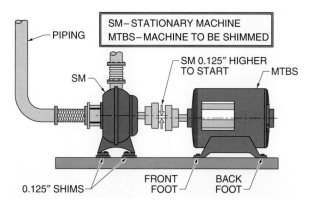

Figure 13-7. A machine is chosen as the MTBS because it is easier to move than the other machine, which may be larger or connected to piping.

Jack Screws. A *jack screw* is a screw inserted through a block that is attached to a machine base plate allowing for ease in machine movement. Jack screws should be installed on the base plate to easily and accurately control the horizontal movement of a machine. Typical jack screw blocks are 1½″ × 2½″ rectangular blocks made from ½″ thick steel (larger or thicker depending on the size of the machine) and are bolted or welded to the sides of the base plate with the jack screws directly in-line with each mounting hole. Each jack screw block is drilled and tapped to allow for a ½″ bolt assembly. **See Figure 13-8.**

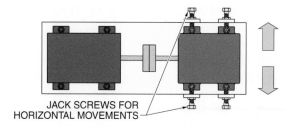

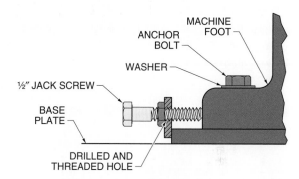

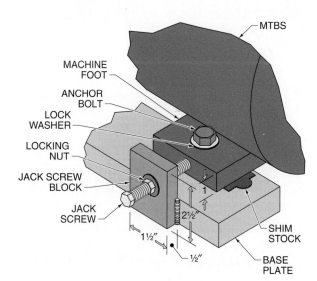

Figure 13-8. A jack screw is a screw attached to a block that is bolted or welded to a machine base plate to allow for ease in machine movement.

Jack screws are used for machine movement only. To prevent additional forces from being applied, jack screw pressure must be backed off after each tightening of the anchor bolts. Jack screws are invaluable when a machine is to be moved just a few thousandths of an inch.

A screwdriver or crowbar can be used for machine movement if it is not possible to install jack screws or if they are not available. An easy, steady prying force is safer and less damaging than a blow from a hammer. Use only a soft-blow hammer if a hammer is necessary.

Checking for Shaft Runout. *Runout* is a radial variation from a true circle. Any shaft that runs eccentric to the true centerline of a machine by more than 0.002″ makes achieving tolerance impossible and should be corrected. *Eccentric* is out-of-round or that which deviates from a circular path. An eccentric shaft produces high vibrations similar to those caused by a bent shaft. An eccentric shaft may be determined by the use of a dial indicator. A *dial indicator* is a device that measures the deviation from a true circular path. Correction may be accomplished by changing machines or by having the shaft recut. **See Figure 13-9.**

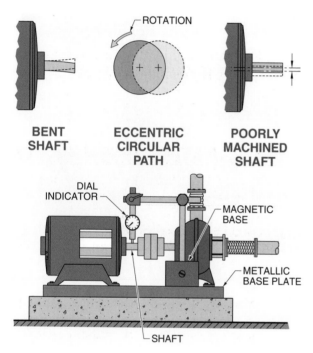

BENT SHAFT

ECCENTRIC CIRCULAR PATH

POORLY MACHINED SHAFT

DIAL INDICATOR

MAGNETIC BASE

METALLIC BASE PLATE

SHAFT

Figure 13-9. An eccentric shaft produces high vibrations similar to those caused by a bent shaft. An eccentric shaft may be determined by the use of a dial indicator.

Shim Stock. It is rare for any machine to have total contact of all of its feet with the base plate and also be within tolerance. Shims and spacers are used to adjust the height of a machine. *Shim stock* is steel material manufactured in various thicknesses, ranging from 0.0005″ to 0.250″. Shim stock can be purchased as a sheet, roll, or in precut shapes. A *spacer* is steel material used for filling spaces over ¼″. The feet of a machine must be firmly anchored to the base plate without creating excessive forces or movements between mating shafts. To prevent stacking inaccuracies, limit the amount of shims or spacers on each foot to five or less. Always choose the combination that uses the least amount of shims or spacers when different shim or spacer combinations can be chosen. Any spacer over 0.250″ (¼″) should equal 1 piece and may be mild steel or stainless steel. Any spacer 0.250″ (¼″) or under is considered a shim and must be stainless steel.

For example, a spacer and shim combination is required to raise a motor 0.683″. To keep the spacer to 1 piece and the shim stack to 5 or less, the spacer selected should be ⅝″ (0.625″) with the remaining 0.058″ taken up by shims. Shim stacks used are sometimes based on the sizes available in a shim pack set. Variations for the 0.058″ shim stack include one 0.050″ and two 0.004″ or two 0.025″, one 0.005″, one 0.002″, and one 0.001″. Extra spring is added each time a shim is added to a stack. The best combination is the one that uses the fewest shims.

Precut stainless steel shims are recommended for alignment purposes. Cutting shim stock for use can create rough edges and burrs, making a successful alignment improbable. Also, any material other than stainless steel is smashed under load and vibration or rusts and corrodes.

Good shim packs are laser cut with each size printed (not stamped) on the shim. Stack 4 or 5 shims with the printed size reversed on every other shim when checking for thickness accuracy. This arrangement, when checked with a micrometer, produces a true size and condition of the stack. **See Figure 13-10.**

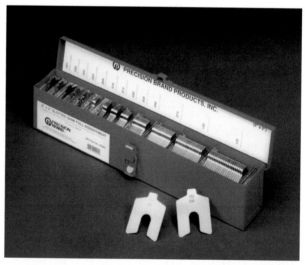

Precision Brand Products, Inc.

Slotted shim assortment kits are economical, safe, and accurate, and reduce costs by eliminating hand cutting, material waste, and shim preparation.

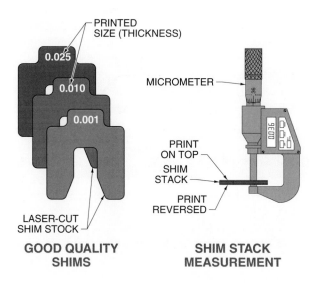

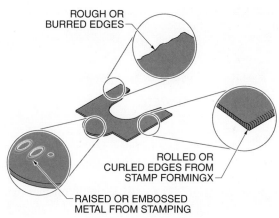

GOOD QUALITY SHIMS

SHIM STACK MEASUREMENT

POOR QUALITY SHIMS

Figure 13-10. Good shim quality ensures proper and firm machine-to-base contact.

Soft Foot

Soft foot is a condition that occurs when one or more machine feet do not make complete contact with the base plate. Distorted frames creating internal misalignment due to soft foot is a major reason for bearing failure. This internal misalignment and distortion loads the bearings and deflects the shaft. **See Figure 13-11.**

Tech Tip

All four corners of a machine foot should be checked with feeler gauges when a dial indicator check indicates soft foot. Angular soft foot is present if the total gap measured for any corner is different from the other corners. Any angularity greater than 0.001″ per inch of foot must be corrected before alignment is started.

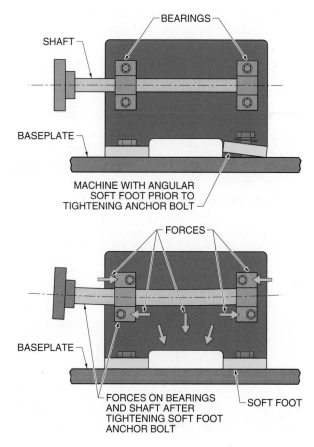

Figure 13-11. Uncorrected soft foot twists and distorts a machine enough to cause destructive forces on rotor and shaft bearings.

An angled or raised foot twists and distorts a machine frame when anchored. Distorted frames cause internal misalignment to bearing housings, shaft deflection, and distorted bearings, resulting in premature bearing and coupling failure. This condition also creates great difficulty in shaft alignment. Before aligning, any machine that has soft foot must be shimmed for equal and parallel support on all feet.

As a soft foot bolt is tightened, the shaft of the machine is deflected, which loads the bearings. This deflection can easily cause enough back-and-forth vibrations and pressures to damage the bearings, seals, and shaft. A shaft that rotates at 1800 rpm in a machine with a soft foot condition deflects 30 times each second or 108,000 times per hour. Soft foot should be checked on the MTBS and SM because soft foot can occur on any machine. Soft foot tolerance must be within 0.002″ of shaft movement. Soft foot may be parallel, angular, springing, or induced. Each is independent of the others. All four conditions may exist on the same machine. **See Figure 13-12.**

Soft Foot

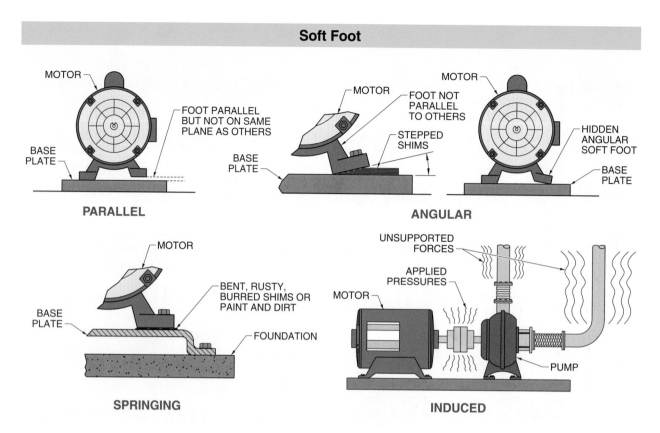

Figure 13-12. Soft foot may be parallel, angular, springing, or induced.

Parallel. *Parallel soft foot* is a condition that exists when one or two machine feet are higher than the others and parallel to the base plate. This condition occurs when a machine leg is short or when spacers of different thicknesses are used.

Correcting parallel soft foot is accomplished by first rocking the machine from side to side and determining the gap under the high foot using feeler gauges. A *feeler gauge* (thickness gauge) is a steel leaf at a specific thickness. Feeler gauges determine the air gap between two solids within thousandths of an inch.

Shim stock equal to the thickness of the soft foot gap is placed under the high foot and all four machine feet are rechecked. The shims should be moved from under the first foot to the other rocking foot if soft foot is noticed under a foot that was not shimmed. Check again for soft foot. The shim stock should be divided between the feet that were rocking if soft foot is noticed under two opposing feet. Checking more than once for proper parallel spacing is done as an attempt to find true, or close to true, parallel spacing without creating angular soft foot.

Angular. *Angular soft foot* is a condition that exists when one machine foot is bent and not on the same plane as the other feet. Generally, one corner of the angular foot is touching the base plate. Angular soft foot is usually the result of the machine being roughly handled or dropped or having uneven mounting pads due to poor machining or welding.

The machine may be sitting high on one side and low on the other if all four feet appear to have angular soft foot in the same direction. Angular soft foot may be determined when a 0.002″ feeler gauge can be placed under one side of a foot but not under the other side of the same foot.

Correction of angular soft foot is accomplished by machining the foot to be on the same plane as the other feet or by step shimming the foot to fill the gap. Step shimming begins by determining the direction and amount of slope and filling the sloping void with a series of steps (usually a maximum of 5 or 6). This is done by measuring the largest portion of the gap and dividing by 5 or 6, giving the thickness of each step. Finally, place each shim by hand, in steps, to fill the gap without lifting the machine. Check shaft movement using a dial indicator while tightening the machine bolts and correct as indicated from dial movement.

Springing. *Springing soft foot* is a condition that occurs when a dial indicator at the shaft shows soft foot, but feeler gauges show no gaps. This condition occurs from shims that are burred or bent, corroded bases or feet, dirt, grease, or rust between the feet, shims, and base plate, or too many shims. The machine acts as if it were mounted on springs due to each imperfection.

Prevention of springing soft foot is accomplished by using solid bases that are cleaned to the metal by removing all paint, grime, rust, and corrosion. Layers of grease can even act as a spring. The top and bottom of the machine feet must also be free of rust, paint, and grime. Shims used must be flat, clean, and without stamping imperfections. After cleaning, check for out-of-tolerance movement using a dial indicator at the shaft as the bolts are tightened.

Induced. *Induced soft foot* is soft foot that is created by external forces such as coupling misalignment, piping strain, tight jack screws, or improper structural bracing. Coupling forces from vertical or horizontal misalignment are noticed when couplings are difficult to bolt up or a spring or snap is noticed as couplings are disconnected.

Any external force in any direction to coupled pipe flanges on a pump strains the machine. This condition may be seen when checking for soft foot before, during, and after piping connections or when checking for structural bracing strain.

Lovejoy, Inc.

Shaft couplings allow drive and driven equipment connection and provide protection against misalignment, vibration, and shock.

Measuring Soft Foot

The two methods of measuring soft foot are the at-each-foot method and the shaft deflection method. Both methods use a dial indicator with a magnetic base to determine movement. More than one set of readings must be made, in the same direction of movement, to ensure that readings are constant. Two identical readings are assumed correct if three sets of readings are taken and one set is different from the other two. All other feet must remain securely tightened and the coupling uncoupled when checking each foot for soft foot. Check and correct angular soft foot before either method is used. The shaft deflection method is easier because the indicator does not interfere with loosening the anchoring bolts. **See Figure 13-13.**

Measuring Soft Foot

SHAFT DEFLECTION METHOD

MOTOR

DIAL INDICATOR

MOTOR SHAFT

BASE PLATE

MACHINE FOOT

MAGNETIC BASE

AT-EACH-FOOT METHOD

Figure 13-13. Soft foot may be measured at the shaft or each foot of a machine.

At-Each-Foot Method. In the at-each-foot method, soft foot is checked at each foot of the machine. A dial indicator with a magnetic base is secured to the base plate at the foot to be tested. The dial indicator is adjusted so the stem is above and perpendicular to the top of the foot. Ensure that all four feet are anchored firmly. With the dial adjusted to zero, watch the dial movement as the bolt for that foot is slowly but steadily loosened. The required shim thickness can be determined from the indicator movement if the foot rises and the dial moves more than 0.002″. Place shims beneath the foot equal to the amount of the indicator movement and re-tighten the bolt. Repeat this process until the movement is less than 0.002″. Relocate the dial indicator to another foot and repeat the above procedure until all four feet have been checked and corrected.

Shaft Deflection Method. In the shaft deflection method, the shaft is checked for deflection when anchoring bolts are tightened or loosened. Critical distortion has not occurred if there is no measured deflection even though a foot has movement. Correction is necessary if there is movement. This method is quicker and more accurate than the at-each-foot method.

A magnetic base dial indicator is secured to the base plate and adjusted so the stem is above and perpendicular to the top of the shaft or coupling, whichever is the farthest from the MTBS. The farther the dial indicator is from the first set of feet, the greater the dial movement, which increases accuracy. Ensure that all four feet are anchored firmly. Zero the dial on the indicator and slowly loosen the bolt on the first foot. A dial that indicates 0.003″ rise requires the placement of shim stock beneath that foot totaling 0.003″. Tighten the bolt and check the foot again. Continue checking and shimming each foot individually until shaft movement is within the 0.002″ tolerance.

Check all feet for hidden angular soft foot when the fourth or final foot has been corrected and one foot rises when double-checking. Once soft foot conditions have been corrected, shaft alignment may begin. Shaft alignment is difficult or impossible if soft foot conditions have not been corrected to within tolerance.

Thermal Expansion

For proper alignment, two coupled shafts must be on the same horizontal and vertical plane under operating conditions. However, there could be a significant change in physical dimensions when there is a change in operating condition temperature and thermal expansion results. *Thermal expansion* is the dimensional change in a substance due to a change in temperature. A temperature change between startup and running conditions can influence machine alignment because metal expands when heated and contracts when chilled. **See Figure 13-14.**

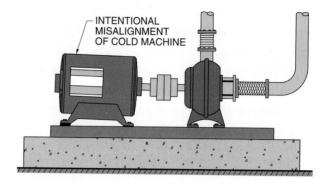

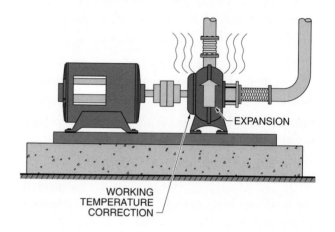

Figure 13-14. Temperature differences from cold startup to working temperature influence the position of one machine relative to another.

Temperature conditions that can change enough to affect critical alignment measurements may be caused by the temperature of fluid being pumped, excessive room or ambient temperature, or loaded motor temperatures. Most materials expand when heated and contract when chilled. For example, a piece of steel exactly 12″ long and 2″ in diameter in a room temperature of 72°F grows to 12.008″ long and 2.001″ in diameter when heated to 172°F. This change is affected by the material and how much the temperature has changed.

Change in material length is calculated to determine specific tolerances to absorb thermal expansion and accommodate different materials. A thermal expansion constant is given based on the material used. Constants include 0.0000063 for cast iron, 0.000009 for stainless steel, and 0.0001 for plastic. Thermal expansion is found by applying the formula:

$$\Delta L = L \times \Delta T \times C$$

where

ΔL = change in length (in in.)

L = original length (in in.)

ΔT = change in temperature (in °F)

C = material constant

For example, what is the change in length of a pump and cast iron frame motor combination when the operating temperature of the motor increases from 75°F to 140°F and the motor measures 15″ from its base to the shaft center? *Note:* The temperature change equals 65°F (140°F – 75°F = 65°F).

$$\Delta L = L \times \Delta T \times C$$
$$\Delta L = 15 \times 65 \times 0.0000063$$
$$\Delta L = \mathbf{0.006''}$$

The vertical plane of the motor should be reduced by 0.006″ because the motor shaft rises as it gets warmer (operating temperature). At times, the SM, which may be a pump that pumps hot liquid, is the machine that rises. Compensating shims may be added under the MTBS when this is the case.

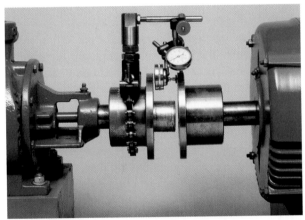

Manufacturing & Maintenance Systems, Inc.

Dial indicator probes must be mounted perpendicular to the contact surface when used for machine alignment.

DIAL INDICATORS AND ALIGNMENT

A *dial indicator* is a precise, jeweled movement instrument that measures the amount of misalignment between two machine shafts. A dial indicator is similar to a watch and must be treated as such if it is to be an accurate, useful tool. Dial indicators are required to be mounted perpendicular to their contacted surface. A dial indicator

10° out of perpendicular results in an immediate error of 2%. Blows from mallets or hammers on a machine into an indicator damage the indicator or throw off readings because indicators have jeweled movements. Always use slow, forceful movements when adjusting into an indicator.

Tech Tip

To prevent incorrect readings, position a dial indicator on the circumference of a shaft so that the centerline of the indicator probe runs through the centerline of the shaft.

Dial Indicator Use

Dial indicators are read to obtain the total dial movement (TDM) of a beginning and ending indicator reading between two machine shafts. Indicator readings do not have to begin at zero to determine a TDM. A TDM is found by subtracting one indicator reading from the other and finding the total difference between the two dial values. It does not matter which reading is subtracted from the other. Positive and negative signs are accounted for when determining the TDM. However, the total difference will always be a positive value.

For example, the TDM of an indicator that has a high reading of +0.022″ and a low reading of +0.006″ is 0.016″ (0.022″ – 0.006″ = 0.016″). The same holds true for readings that are negative. An indicator reading of –0.006″ is subtracted from a reading of –0.022″ to give a TDM of 0.016″ (–0.022″ – [– 0.006″] = –0.016″). The total difference is 0.016″. The same method is used when a reading is positive and the other reading is negative. For example, an indicator reading of +0.022″ is subtracted from a reading of –0.006″ to give a TDM of –0.028″ ([–0.006″] – +0.022″ = –0.028″). The total difference is 0.028″. **See Figure 13-15.**

Verifying Dial Readings. Indicators must be run through their movement as many as three times to ensure a proper reading. Dial movement must be in the same direction. False readings are given if shaft rotation is reversed. This is due to the play or tolerances of indicator moving parts.

Aligning shafts that rotate must be accomplished using an indicator to check both vertical (up and down) and horizontal (back and forth) positioning. Straight movements (not angular) are known as offset movements. Dial indicator offset measurements are twice their actual offset. For example, a total indicator reading of –0.020″ is indicating a shaft centerline offset of –0.010″.

Dial Indicator Readings

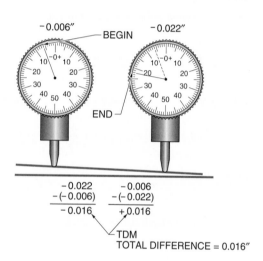

+0.022″ — BEGIN   +0.006″ — END

PROBE —

```
 + 0.006          + 0.022
-(+ 0.022)       -(+ 0.006)
 - 0.016          +,0.016
```
TDM
TOTAL DIFFERENCE = 0.016″

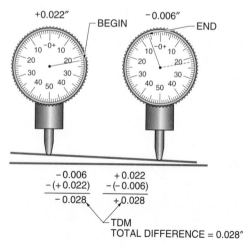

−0.006″ — BEGIN   −0.022″

END

```
 - 0.022          - 0.006
-(- 0.006)       -(- 0.022)
 - 0.016          +,0.016
```
TDM
TOTAL DIFFERENCE = 0.016″

+0.022″ — BEGIN   −0.006″ — END

```
 - 0.006          + 0.022
-(+ 0.022)       -(- 0.006)
 - 0.028          +,0.028
```
TDM
TOTAL DIFFERENCE = 0.028″

Figure 13-15. Dial indicators are read using the total movement of the dial needle and do not have to begin at zero.

An offset is checked by placing the indicator tip at the top of the shaft or coupling and zeroing the indicator. The top position is known as the 12:00 position. The indicator and shaft that it is attached to are rotated to the farthest position, known as the 3:00 position. Both shafts must be rotated with the coupling unrestricted.

Both coupling halves must be detached from each other and must be turned together. Turning the shafts together gives a true condition of the shaft centerlines and is not affected by coupling faces that may not be square with the plane of the shaft or coupling rims that may not be machined round or concentric with their inside dimension. Any alignment using improper methods results in having good readings on poorly aligned machinery.

An easy way to verify whether 12:00, 3:00, 6:00, and 9:00 readings are correct is to add the 3:00 and 9:00 readings and compare the sum to the 6:00 and 12:00 readings. The sum of 3:00 and 9:00 readings should equal the 6:00 reading if the 12:00 reading started at zero. Readings may be taken at only three positions when complete rotation is not possible. **See Figure 13-16.**

The shaft being checked is above the centerline of the shaft on which the indicator is mounted if the indicator reading at 6:00 is negative. The shaft being checked is below the centerline of the shaft on which the indicator is mounted if the indicator reading at 6:00 is positive. The shaft being checked is below the centerline of the shaft on which the indicator is mounted if the angular reading at 6:00 is negative. The shaft being checked is above the centerline of the shaft on which the indicator is mounted if the angular reading at 6:00 is positive.

Manufacturing & Maintenance Systems, Inc.

Dial indicators are used during machine alignment procedures to indicate the amount of misalignment between two machine shafts.

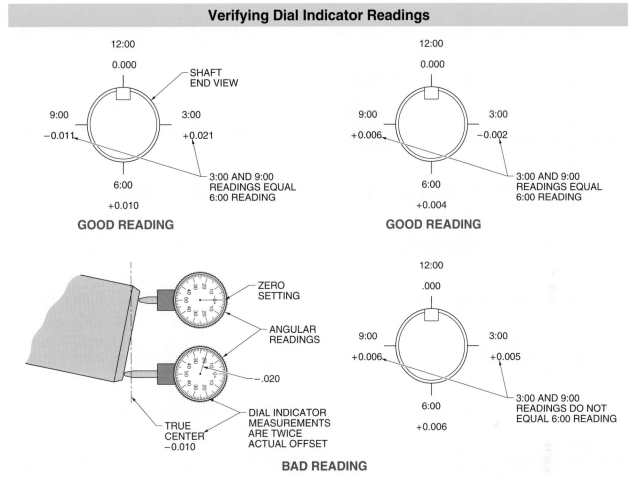

Figure 13-16. Good indicator readings are those where the sum of the 3:00 and 9:00 readings equals the 6:00 reading.

Indicator Rod Sag

To indicate the alignment of one shaft to another, the indicating device that indicates the alignment of one shaft must be clamped or strapped to the opposing shaft. The entire assembly generally consists of the clamp or strap, a riser rod, two 90° rod couplings, a spanning rod, and an indicating device. The indicating device may be a dial indicator or an electronic indicator. The greater the distance between the rod couplings, the more the weight of the indicating device creates a sag in the spanning rod. **See Figure 13-17.** This sag, if not accounted for, can measure readings incorrectly. A ⅜″ diameter spanning rod at a distance of 8″ between couplings can measure readings incorrectly by 0.010″.

Electronic indicators calculate rod sag when measurements are keyed in. Rod sag from a dial indicator must be determined by the technician. Accurate indicator rod sag is determined by first establishing the distance between rod couplings during the alignment measurement. All parts (rods, couplings, indicator) are assembled on a solid shaft or pipe using the established coupling distance. A solid shaft is helpful in determining rod sag because it is not misaligned. The dial indicator is adjusted to get a reading off the bar and then zeroed at the 12:00 position. The dial shows a negative reading at the 6:00 position equal to twice the actual amount of rod sag when the bar is rotated 180°. Record half of the total reading so it may be subtracted from the actual alignment readings. The actual sag is –0.005″ if the readings on the bar indicated a total sag of 0.010″.

For example, an actual alignment reading indicates a rotational misalignment of 0.024″. This is divided by two, giving a vertical reading of 0.012″. The indicator rod sag of 0.005″ is then subtracted from the vertical reading, giving an actual vertical offset reading of 0.007″ (0.012″ – 0.005″ = 0.007″).

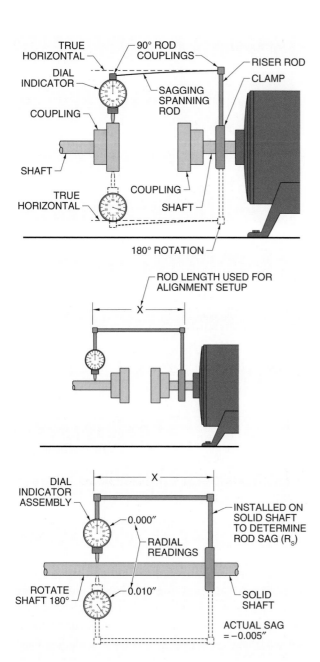

Figure 13-17. Gravity and the weight of the dial indicator produce a change in indicator readings from rod sag toward and away from the point reading.

ALIGNMENT METHODS

Five methods are available to align machinery, each having its own degree of accuracy. The five methods include straightedge, rim-and-face, reverse dial, electronic reverse dial, and laser rim-and-face methods. **See Figure 13-18.**

All alignment methods require that a specific order of adjustment be made. Any attempt to align a machine outside of the specific order is considered trial and error adjustment, which can only lead to frustration. The specific order of shaft alignment is: angular in the vertical plane (up and down angle), parallel in the vertical plane (up and down offset), angular in the horizontal plane (side to side angle), and parallel in the horizontal plane (side to side offset).

Once angular in the vertical plane and parallel in the vertical plane have been corrected, they generally are not lost when angular in the horizontal plane and parallel in the horizontal plane are in the process of being corrected. This step-by-step process is used regardless of the alignment method. Always double check each corrective move.

Accuracy Expectations

The choice of alignment method is based on cost, accuracy required, ease of use, and time required to perform the alignment. The accuracy of any alignment is based on the skill level of the individual doing the alignment and the alignment method used. For example, straight-edge measurements are usually made without the knowledge of coupling irregularities and require the feel of thickness gauge measurements. Therefore, the accuracy of straight-edge alignment generally is no better than 1/64″.

Dial indicators and electronic measuring devices (except laser) measure in the thousandths of an inch, which allows for an accuracy of alignment within 0.001″. Laser alignment methods are generally exact and quick with a possible accuracy of 0.0002″.

Alignment Tolerance. Alignment tolerance requirements of two or more shafts are based on the speed (rpm) of the motor or drive unit. At times, a manufacturer may indicate the alignment tolerance for its machine. A shaft alignment tolerances chart may be used if manufacturer tolerances are not available. **See Figure 13-19.** A shaft alignment tolerances chart indicates suggested tolerances by speed in thousandths per inch. Some shaft alignment tolerances charts show the tolerance for angularity in degrees, minutes, and seconds rather than mils or inches. This means that before and after each adjustment, an indicator reading must be taken and converted into angles of degrees, minutes, and seconds.

Alignment Methods

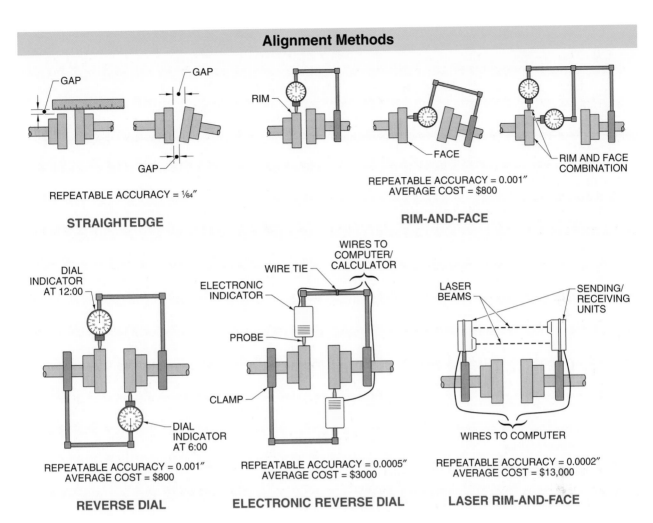

Figure 13-18. The five methods available to align machinery include straightedge, rim-and-face, reverse dial, electronic reverse dial, and laser rim-and-face methods.

For example, a technician is aligning a pump/motor combination that has a 5″ coupling and operates at 1300 rpm. The vertical offset is 0.005″, the horizontal offset is 0.003″, the vertical angularity is 0.006″, and the horizontal angularity is 0.005″. Using the shaft alignment tolerances chart indicates that at 1300 rpm, the acceptable offset tolerance is 0.0038″. The chart also indicates that the angularity tolerance of a 10″ coupling operating at 1300 rpm is acceptable at 0.010″.

The angular measurements must be doubled to equate with a 10″ coupling because the equipment coupling is 5″. This gives a recorded reading of 0.012″ for vertical angularity and 0.010″ for horizontal angularity. The results of the recorded readings indicate that both horizontal readings are in tolerance and both vertical readings are out of tolerance.

SPM Instrument, Inc.

An alignment computer uses electronic indicators and the electronic reverse dial method to correct machine misalignment.

Shaft Alignment Tolerances*			
Offset (Thousandths/Inch)			
Speed†	Excellent	Acceptable	
0 – 999	0.0030	0.0050	
1000 – 1999	0.0020	0.0038	
2000 – 2999	0.0015	0.0025	
3000 – 3999	0.0008	0.0015	
4000 – 4999	0.0005	0.0010	
5000 – 5999	0.0004	0.0008	
Angularity (Measured as Gap Size at 10″)			
Speed†	Excellent	Acceptable	
0 – 999	0.0070	0.020	
1000 – 1999	0.0030	0.010	
2000 – 2999	0.0025	0.005	
3000 – 3999	0.0020	0.004	
4000 – 4999	0.0015	0.003	
5000 – 5999	0.0010	0.002	

* in in.
† in rpm

Figure 13-19. A shaft alignment tolerances chart indicates suggested tolerances by speed in thousandths per inch.

Horizontal Movements

Angular and offset movements in the horizontal plane are normally made as jack screws are screwed in as dial indicator movement is observed. To move the MTBS away 0.020″, a dial indicator is placed at the back edge of the machine base plate and directly in-line with the adjusting jack screw. The front jack screw is screwed in until the indicator registers a 0.020″ movement. **See Figure 13-20.** When only one indicator is used, a movement at one end of a machine changes the machine position at the other end, complicating accurate movement at both ends. To overcome possible confusion during horizontal movement, two indicators, one at each foot, are used to display exact front and back motor/pump unit movement.

Angular Movements

Angular movements are generally made by adjustments at two machine feet. However, it is necessary to adjust only one set of feet such as the machine front feet or back feet only. For example, if calculations call for the front feet to be raised 0.050″ and the back feet 0.025″, the total angularity position is corrected by raising the front feet 0.025″ (0.050″ – 0.025″ = 0.025″).

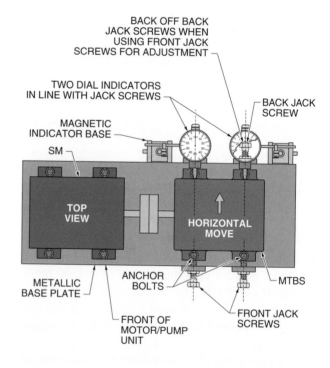

Figure 13-20. Angular and offset movements in the horizontal plane are normally made as jack screws are screwed in as dial indicator movement is observed.

To prevent compounding errors, new readings must be taken after each adjustment to the MTBS. To reduce the chances for error, always rotate dial indicators in one direction. Start back at the zero setting if movement direction has been reversed. Clock position movements (clockwise/counterclockwise) are those viewed from the MTBS toward the SM. Repeat dial indicator movements and recheck readings at least three times.

Angular measurements are easier to interpret when it is realized that each misalignment angle, in its own plane (vertical or horizontal), is the same whether it is measured off of the coupling face or the misalignment at the feet of the machine. Known readings of distance and gap may be used to determine angularity and gap at any other distance. This principle can be used to determine the shim thickness to eliminate any angle. The angle reading (gap) must always start at zero for this to occur. Indicator readings that begin at a number greater than zero must have the initial reading subtracted from the total needle movement to give the proper gap. **See Figure 13-21.**

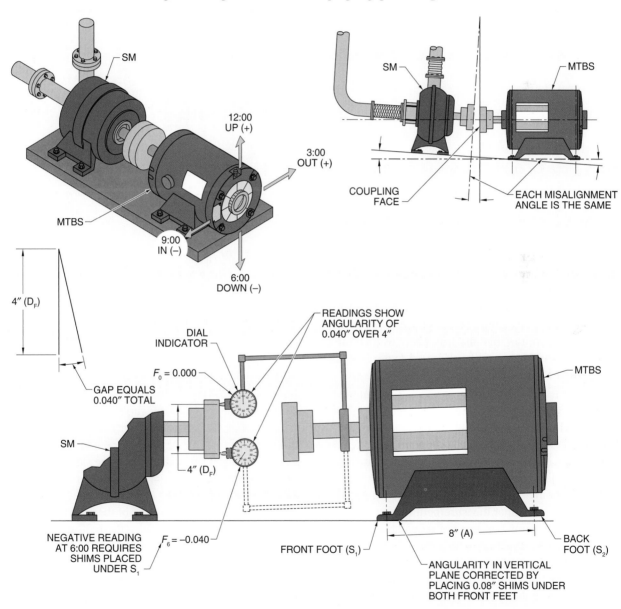

Figure 13-21. Each misalignment angle, in its own plane (vertical or horizontal), is the same whether it is measured off of the coupling face or the misalignment at the feet of the machine.

The gap at the desired distance is found by applying the formula:

$$G = \frac{g}{d} \times D$$

where

G = gap at desired distance (in in.)

g = known gap (in in.)

d = known distance from zero (in in.)

D = distance desired from zero (in in.)

For example, what is the gap at 8″ if the gap at 4″ from zero is 0.04″?

$$G = \frac{g}{d} \times D$$

$$G = \frac{0.04}{4} \times 8$$

$$G = 0.01 \times 8$$

$$G = \mathbf{0.08″}$$

Straightedge Method

The *straightedge alignment method* is a method of coupling alignment in which an item with an edge that is straight and smooth, such as a steel rule, feeler gauge, or taper gauge, is used to align couplings. A *taper gauge* (sometimes referred to as a gap gauge) is a flat, tapered strip of metal with graduations in thousandths of an inch or millimeters marked along its length. As the gauge is placed in a hole or gap, the reading on the gauge at the hole or gap edge is the diameter of the hole or gap at that point. A taper gauge used for this purpose is more reliable and less likely to give false trial-and-error readings like those of feeler gauges. **See Figure 13-22.**

Straightedge alignment is the oldest method used for measuring misalignment at couplings. This method, although easy to understand and perform, is highly inaccurate, produces no proper resolutions, and does not offer any repeatability. Any proper measuring with tolerances within a few thousandths of an inch must be read in thousandths of an inch and not fractions. Also, proper measurement is of no value if the readings are not consistent or repeatable. Another drawback is that the alignment is being measured on machined surfaces and the outside diameter, or OD, of many coupling hubs are not true with their bore, nor is the face of a hub perpendicular with the shaft. With the straight-edge method of alignment, coupling hub face and OD runout must be checked and compensated for in making corrective calculations. The straight-edge

method is, however, an excellent method for getting shafts roughly close for other alignment methods.

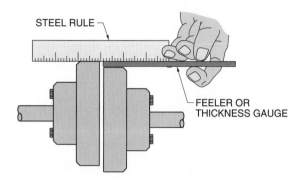

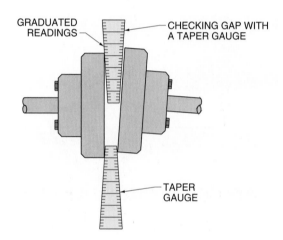

Figure 13-22. The straightedge alignment method uses a steel rule, feeler gauge, or taper gauge to align couplings.

To begin alignment, ensure that the MTBS is not bolt bound and the SM is firmly anchored. Check angular differences in the vertical plane (up and down angle) using a taper gauge. This measurement is made at the 12:00 and 6:00 positions on the shafts or couplings. Vertical or horizontal angularity corrections using shim stock must be performed at one location only, such as at the MTBS front feet or back feet. The opposing feet are used as the angle pivot point. The selection of the adjusted feet and the pivot feet is determined by the 6:00 reading. The front feet (S_1) are shimmed if the 6:00 reading is positive (greater than the 12:00 reading). The back feet are shimmed if the 6:00 reading is negative. The difference between the 12:00 and 6:00 readings is used to determine the thickness of the shim stock placed under the feet to eliminate the angle. **See Figure 13-23.** Shim stock thickness to eliminate angular misalignment in the vertical plane is based on the vertical angular gap (gap at 12:00 minus gap at 6:00), the diameter of the coupling, and the distance between front and back MTBS mounting holes. Shim stock thickness to eliminate angular misalignment in the vertical plane is found by applying the formula:

$$S = \frac{V_A}{2} \div \frac{D}{2} \times A$$

where

S = shim stock thickness (in in.)

V_A = vertical angular gap (in in.)

2 = constant

D = diameter of coupling (in in.)

A = distance between front and back MTBS mounting holes (in in.)

Straightedge Alignment

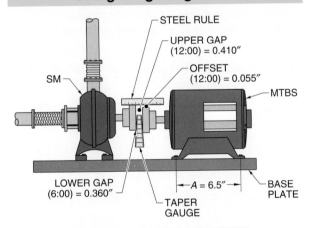

VERTICAL PLANE ALIGNMENT

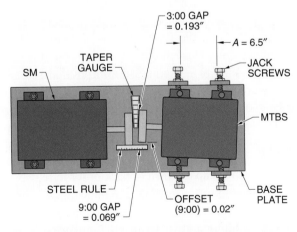

HORIZONTAL PLANE ALIGNMENT

Figure 13-23. Straightedge alignment is generally used for rough alignment prior to using a more precise alignment method.

For example, what is the shim stock thickness required to correct the angular misalignment in the vertical plane of a pump and motor assembly having a 5″ diameter coupling, a vertical angular gap of 0.050″ (0.410″ − 0.360″ = 0.050″), and an MTBS mounting hole distance of 6.5″?

$$S = \frac{V_A}{2} \div \frac{D}{2} \times A$$

$$S = \frac{0.050}{2} \div \frac{5}{2} \times 6.5$$

$$S = \frac{0.025}{2.5} \times 6.5$$

$$S = 0.01 \times 6.5$$

$$S = \mathbf{0.065''}$$

The misalignment in the vertical plane is adjusted by placing shims equal to 0.065″ under both back feet of the MTBS because the gap between the couplings is wider at 12:00 (0.410″) than at 6:00 (0.360″).

After the angular misalignment in the vertical plane has been corrected, in this case by placing 0.065″ shims under both back feet, the MTBS is checked for offset misalignment in the vertical plane (vertical offset). This is accomplished by laying a straightedge across the top of the shafts or couplings. With the straightedge held firmly and parallel against the highest shaft or coupling, feeler gauges are slid into the offset (if there is one) to determine the distance that the MTBS must be raised or lowered. Always check for shafts or couplings of unequal diameters.

For example, a straightedge placed on top of a 5″ coupling shows a 0.055″ offset between the straightedge and the coupling half connected to the MTBS. The MTBS must be raised 0.055″. After inserting shims equaling 0.055″ under all four feet and tightening mounting bolts, recheck the angular position in the vertical plane and the offset position in the vertical plane.

The angular position in the horizontal plane (side to side angle) and the offset position in the horizontal plane (side to side offset) are checked and corrected similar to the vertical plane adjustments. These movements are accomplished with the use of jack screws. The angular position in the horizontal plane is checked with the jack screws fingertight against the feet of the MTBS and the mounting bolts tight. The angular gap is checked with the taper gauge at the 3:00 and the 9:00 positions.

The adjustment to eliminate angular misalignment in the horizontal plane is based on the horizontal angular gap (gap at 3:00 minus gap at 9:00 or gap at 9:00 minus gap at 3:00), the diameter of the coupling, and the distance between the front and back MTBS mounting holes. Horizontal adjustment to eliminate angular misalignment in the horizontal plane is found by applying the formula:

$$S = \frac{H_A}{2} \div \frac{D}{2} \times A$$

where

S = shim stock thickness (in in.)

H_A = horizontal angular gap (in in.)

2 = constant

D = diameter of coupling (in in.)

A = distance between front and back MTBS mounting holes (in in.)

For example, what is the adjustment required to correct the angular misalignment in the horizontal plane of a pump and motor assembly having a 5″ diameter coupling, a horizontal angular gap of 0.124″ (0.193″ − 0.069″ = 0.124″), and a MTBS mounting hole distance of 6.5″?

$$S = \frac{H_A}{2} \div \frac{D}{2} \times A$$

$$S = \frac{0.124}{2} \div \frac{5}{2} \times 6.5$$

$$S = \frac{0.062}{2} \times 6.5$$

$$S = 0.0248 \times 6.5$$

$$S = \mathbf{0.161''}$$

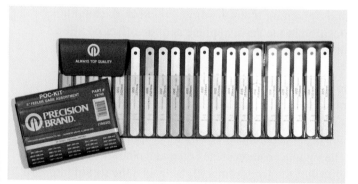

Precision Brand Products, Inc.

Feeler gauges are used to determine the distances that machines must be moved for correct alignment.

The horizontal angular adjustment of 0.161″ is at the back foot (S_2) because the gap is wider at the 3:00 position (0.193″) than the 9:00 position (0.069″).

Offset misalignment in the horizontal plane is corrected similar to correcting offset misalignment in the vertical plane. A straightedge and feeler gauge is placed at the 9:00 position and the offset is measured.

For example, a straightedge placed at the 9:00 position of a 5″ coupling shows a 0.020″ offset between the straightedge and the coupling half connected to the MTBS. The gap being between the straightedge and the MTBS coupling half at the 9:00 position requires the motor to be moved to the front a distance of 0.020″ by the jack screws. Recheck the angular position in the horizontal plane and the offset position in the horizontal plane after making any adjustments.

Straightedge alignment is not necessarily true to the shaft axis because straightedge alignment measures the condition of a coupling and shaft assembly along with machined surfaces of the coupling. Final results may easily be far from tolerance if this method is used as the total alignment method. Upon completion of alignment, release (unscrew) any pressure from jack screws.

Rim-and-Face Method

The *rim-and-face alignment method* is an alignment method in which the offset and angular gap of two shafts is determined using two dial indicators that measure the rim and face of a coupling. A *coupling rim* is the outside diameter surface of a coupling. A *coupling face* is the flat surface of a coupling half, facing the flat surface of the connecting coupling half.

A dial indicator measuring at the rim position measures offset directly under the indicator stem. Also, the difference in offset over the distance between two indicators is the angularity in thousandths per inch. Before shim size adjustments are determined, offset and angular gaps must be calculated and used to determine if a machine must be moved up or down or back or forth.

The rim-and-face method is the most widely used, most widely misused, and most troublesome of the precision alignment methods. It is misused by the technician who turns only one shaft to check alignment and troublesome because misuse creates a never ending search in trying to be within tolerance. This method also has additional error sources such as axial float and irregular coupling shapes. *Axial float* is the axial movement of a shaft due to bearing and bearing housing clearances.

Rim-and-face alignment may be accomplished by using the individual rim-and-face method or the combination rim-and-face method. The combination method is considerably faster and more accurate than the individual method.

Individual Rim-and-Face Alignment. The individual rim-and-face alignment method uses an indicator that is attached and used to measure the coupling face (angularity) and then repositioned to measure the coupling rim (offset). Rim and face readings must be taken with the coupling disconnected and both coupling halves rotated together. This is best accomplished when a mark is made on both coupling halves and kept in-line as the couplings are rotated. By rotating both couplings, shaft centerlines are measured, whereas rotating only one shaft measures one shaft in relation to the opposing coupling face or diameter.

Combination Rim-and-Face Alignment. The combination rim-and-face alignment method uses two dial indicators. One indicator reads the rim offset, and the other reads the face to measure the angularity. Both indicators are assembled using rods and couplings (hardware) and are assembled at the same end of the spanning rod. The two indicators measure the vertical and horizontal planes of the same shaft simultaneously.

Angular gap and offset information must be obtained before shim thickness and location can be determined. Angular gap and offset information used for shim placement is found by checking angular misalignment in the vertical plane (up and down), offset misalignment in the vertical plane (vertical offset), angular misalignment in the horizontal plane (side to side), and offset misalignment in the horizontal plane (side to side offset). The misalignment values are found by measuring with dial indicators. **See Figure 13-24.** Combination rim-and-face alignment is performed by applying the procedure:

1. Check for angular misalignment in the vertical plane (up and down).

Angular misalignment in the vertical plane is checked by measuring the face of the coupling at the 12:00 and 6:00 positions. The vertical angular gap equals the 6:00 reading minus the 12:00 reading if both total indicator readings are either positive or negative. The vertical angular gap equals the 6:00 reading plus the 12:00 reading if one reading is positive and the other reading is negative.

The shim stock thickness to adjust for angular misalignment in the vertical plane can be found once the vertical angular gap is determined. The shim stock thickness is based on the diameter traveled by the face indicator tip, the vertical angle gap, and the distance between front and back MTBS mounting holes. Shim stock thickness to eliminate angular misalignment in the vertical plane is found by applying the formula:

$$S = \frac{V_A}{2} \div \frac{D_F}{2} \times A$$

where

S = shim stock thickness (in in.)

V_A = vertical angular gap (in in.)

2 = constant

D_F = diameter traveled by the face indicator tip (in in.)

A = distance between front and back MTBS mounting holes (in in.)

2. Check for offset misalignment in vertical plane.

Offset misalignment in the vertical plane is checked by measuring the rim of the coupling at the 12:00 and 6:00 positions.

The shim stock thickness to adjust for offset misalignment in the vertical plane can be found once the vertical offset is determined. The shim stock thickness is based on the difference between the 12:00 and 6:00 rim readings minus the rod sag. Shim stock thickness to eliminate offset misalignment in the vertical plane is found by applying the formula:

$$V_0 = \frac{R_0 - R_6 - R_s}{2}$$

where

V_0 = offset in vertical plane (in in.)

R_0 = reading of rim at 12:00 (in in.)

R_6 = reading of rim at 6:00 (in in.)

RS = rod sag (in in.)

2 = constant

3. Check for angular misalignment in the horizontal plane (side to side).

Angular misalignment in the horizontal plane is checked by measuring the face of the coupling at the 3:00 and 9:00 positions. The horizontal angular gap equals the 9:00 reading minus the 3:00 reading if both indicator readings are either positive or negative. The horizontal angular gap equals the 9:00 reading plus the 3:00 reading if one reading is positive and the other negative.

Rim-and-Face Alignment

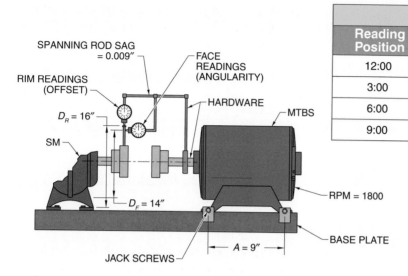

Dial Indicator Data		
Reading Position	Face Reading	Rim Reading
12:00	0.000	0.000
3:00	−0.006	−0.022
6:00	−0.016	−0.021
9:00	−0.010	+0.001

What are the proper shims to be placed under the feet of a pump/motor combination with readings as shown in Dial Indicator Data?

1. Check for angular misalignment in vertical plane.

Measure the face of the coupling at the 12:00 and 6:00 positions (−0.016″ + 0.000″ = −0.016″). Shim stock is placed beneath the front feet (S_1) because the 6:00 reading is negative.

$$S = \frac{V_A}{2} \div \frac{D_F}{2} \times A$$

$$S = \frac{-0.016}{2} \div \frac{14}{2} \times 9$$

$$S = \frac{-0.008}{7} \times 9$$

$$S = -0.001 \times 9$$

$$S = \mathbf{-0.009''}$$

2. Check for offset misalignment in the vertical plane.

Measure the rim of the coupling at the 12:00 and 6:00 positions (0.000″ − 0.021″ = −0.021″).

$$V_0 = \frac{R_0 - R_6 + R_S}{2}$$

$$V_0 = \frac{0.000 - 0.021 - 0.009}{2}$$

$$V_0 = \frac{-0.030}{2}$$

$$V_0 = \mathbf{-0.015''}$$

Shims are placed beneath the feet of the MTBS to raise the MTBS because the 6:00 reading is negative ($R_6 = -0.021''$). To compensate for vertical offset, shims equaling 0.015″ are placed under all four feet.

3. Check for angular misalignment in the horizontal plane.

Measure the face of the coupling at the 3:00 and 9:00 positions [−0.010″ − (−0.006″) = −0.004″].

$$S = \frac{H_A}{2} \div \frac{D_F}{2} \times A$$

$$S = \frac{-0.004}{2} \div \frac{14}{2} \times 9$$

$$S = \frac{-0.002}{7} \times 9$$

$$S = 0.00028 \times 9$$

$$S = \mathbf{0.0025''}$$

The angular gap of −0.004″ is adjusted at the front feet of the MTBS because the lesser gap reading is at 3:00. Adjust the front jack screws for an indicator movement of 0.0025″ and anchor the MTBS.

4. Check for offset misalignment in the horizontal plane.

Measure the rim of the coupling at the 3:00 and 9:00 positions (−0.022″ + 0.001″ = −0.021″).

$$H_0 = \frac{R_3 - R_9}{2}$$

$$H_0 = \frac{0.022 - 0.001}{2}$$

$$H_0 = \frac{0.021}{2}$$

$$H_0 = \mathbf{0.0105''}$$

Lightly snug the anchor bolts and make the 0.0105″ adjustment. At this stage, the machine should be rechecked and corrected until a minimum acceptable tolerance for the 1800 rpm motor is met (offset = 0.002″ to 0.003″ and angularity = 0.003″ to 0.010″).

Figure 13-24. Rim and face readings determine the offset and angle respectively of one shaft relative to another.

The horizontal adjustment to eliminate angular misalignment in the horizontal plane is based on the horizontal angular gap (gap at 3:00 minus gap at 9:00 or the gap at 9:00 minus the gap at 3:00), the diameter traveled by the face indicator tip, and the distance between the front and back MTBS mounting holes. Horizontal adjustment to eliminate angular misalignment in the horizontal plane is found by applying the formula:

$$S = \frac{H_A}{2} \div \frac{D_F}{2} \times A$$

where

S = shim stock thickness (in in.)

H_A = horizontal angular gap (in in.)

2 = constant

D_F = diameter traveled by the face indicator tip (in in.)

A = distance between the front and back MTBS mounting holes (in in.)

4. Check for offset misalignment in the horizontal plane (horizontal offset). Offset misalignment in the horizontal plane is checked by measuring the rim of the coupling at the 3:00 and 9:00 positions.

The shim stock thickness to adjust for offset misalignment in the horizontal plane can be found once the horizontal offset is determined. The shim stock thickness is based on the difference between the 3:00 and 9:00 rim readings. Rod sag deviation has no effect in the horizontal plane. Shim stock thickness to eliminate offset misalignment in the horizontal plane is found by applying the formula:

$$H_0 = \frac{R_3 - R_9}{2}$$

where

H_0 = offset in horizontal plane (in in.)

R_3 = reading of the rim at 3:00 (in in.)

R_9 = reading of the rim at 9:00 (in in.)

2 = constant

Reverse Dial Method

The *reverse dial method* is an alignment method that uses two dial indicators to take readings off of opposing sides of coupling rims, giving two sets of shaft runout readings. Since it is faster and more accurate than the rim-and-face method, the reverse dial method is not affected by axial float. Each indicator shows both angle and offset. Reverse dial indicator readings can be illustrated by a plotted layout. **See Figure 13-25.**

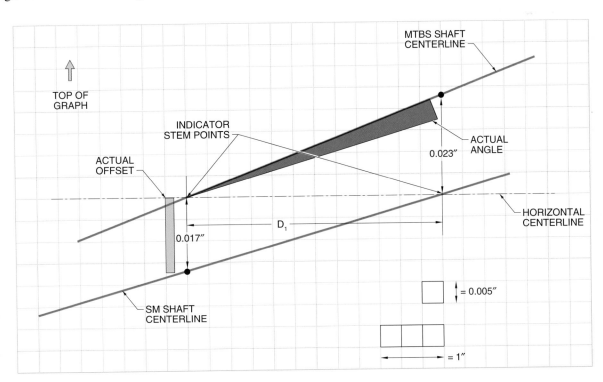

Figure 13-25. Plotting the location of one shaft in relation to the other using reverse dial readings provides information for the movement required for alignment.

Plots are laid out using graph paper. Each square represents a horizontal and vertical measurement. Horizontal measurements are in inches, providing a representative view of the overall machine dimensions. Vertical measurements are total indicator readings plotted in thousandths of an inch. Plotting may be done to view the top and side relative positions.

The plot shows the relative position of shaft centerlines and the indicator reading dimensions. The horizontal squares are 1″ per three-square division and the vertical squares are 0.005″ per division.

Plotting begins by drawing a horizontal centerline and placing a mark on the centerline that represents an indicator stem point. Count the appropriate number of squares and place a mark at the second indicator stem point. This distance represents the distance between the indicator stem points on the two shafts. To plot the MTBS shaft centerline, plot half of the MTBS reading toward the top of the graph from one indicator point and make a mark. For example, if the total MTBS reading is 0.046″, a mark is made 0.023″ toward the top of the graph from one of the indicator stem points (4½ squares).

Tech Tip

Plotting a graph of alignment measurements allows easy experimentation with various shim moves, produces an exact hard copy description of the amount of shim movement needed that can be stored in the plant maintenance files for future reference, and is less expensive than calculators and computers.

A line is drawn from this mark through the opposing indicator stem point. To plot the SM shaft centerline, plot half of the SM reading from the indicator point toward the bottom of the graph and make a mark. Draw a line from this mark through the opposing indicator point. These lines indicate that reverse dial readings give offset and angular positions. The alignment objective is to end up with both lines parallel. For example, if the total SM reading is 0.034″, a mark is made 0.017″ toward the bottom of the graph from the opposite indicator stem point. The difference between these lines should be adjusted for proper alignment.

Plotting provides a graphic illustration as well as an indication of the movements required for alignment. In this case, it is shown that any corrective move must be made by raising the SM or lowering the MTBS.

Exact movements and shim thicknesses are determined by calculating the angular and offset dimensions in the vertical and horizontal planes. **See Figure 13-26.** Reverse dial alignment is obtained by applying the procedure:

1. Correct vertical offsets.

Angular and offset conditions are corrected using the SM and MTBS offset readings. Corrections are first determined by adjusting the TIR to represent the shaft offset. The vertical shaft offset at the SM is found by applying the formula:

$$ST_V = \frac{S_0 - S_6 - RS_1}{2}$$

where
ST_V = vertical shaft offset at SM (in in.)
S_0 = SM indicator reading at 12:00 (in in.)
S_6 = SM indicator reading at 6:00 (in in.)
RS_1 = SM rod sag (in in.)
2 = constant

The vertical shaft offset at the MTBS is found by subtracting the MTBS rod sag and indicator reading at 6:00 from the MTBS indicator reading at 12:00 and dividing by 2. The vertical shaft offset at the MTBS is found by applying the formula:

$$M_V = \frac{M_0 - M_6 - RS_2}{2}$$

where
M_V = vertical shaft offset at MTBS (in in.)
M_0 = MTBS indicator reading at 12:00 (in in.)
M_6 = MTBS indicator reading at 6:00 (in in.)
RS_2 = MTBS rod sag (in in.)
2 = constant

The vertical shim corrections are calculated after the shaft offsets are determined. The vertical shim correction under both MTBS front feet (S_1) is found by applying the formula:

$$VS_1 = \left[(ST_V + M_V) \times \left(\frac{D_2}{D_1} \right) \right] - ST_V$$

where
VS_1 = vertical shim correction under both MTBS front feet (in in.)
ST_V = vertical shaft offset at SM (in in.)
M_V = vertical shaft offset at MTBS (in in.)
D_2 = distance between SM indicator and MTBS front feet (in in.)
D_1 = distance between indicators (in in.)

The vertical shim correction under both MTBS back feet (S_2) is found by applying the formula:

$$VS_1 = \left[\left(ST_V + M_V \right) \times \left(\frac{D_3}{D_1} \right) \right] - ST_V$$

where

VS_2 = vertical shim correction under both MTBS back feet (in in.)

ST_V = vertical shaft offset at SM (in in.)

M_V = vertical shaft offset at MTBS (in in.)

D_3 = distance between SM indicator and MTBS back feet (in in.)

D_1 = distance between indicators (in in.)

2. Correct horizontal offsets.

Angular and offset conditions are corrected using reverse dial offset readings. The horizontal offset TIR of both machines is adjusted to represent true offsets divided by 2. The horizontal shaft offset at the MTBS is found by applying the formula:

$$ST_H = \frac{S_9 - S_3 - RS_1}{2}$$

where

ST_H = horizontal shaft offset at MTBS (in in.)

S_9 = SM indicator reading at 9:00 (in in.)

S_3 = SM indicator reading at 3:00 (in in.)

RS_1 = SM rod sag (in in.)

2 = constant

The horizontal shaft offset at the SM is found by applying the formula:

$$M_H = \frac{M_9 - M_3 - RS_2}{2}$$

where

M_H = horizontal shaft offset at SM (in in.)

M_9 = MTBS indicator reading at 9:00 (in in.)

M_3 = MTBS indicator reading at 3:00 (in in.)

RS_2 = MTBS rod sag (in in.)

2 = constant

Horizontal shaft offsets are then used to calculate the side movement of the MTBS. The horizontal corrective movement of the MTBS front feet (S_1) is found by applying the formula:

$$HS_1 = \left[\left(ST_H + M_H \right) \times \left(\frac{D_2}{D_1} \right) \right] - ST_H$$

where

HS_1 = horizontal corrective movement at MTBS front feet (in in.)

ST_H = horizontal shaft offset at MTBS (in in.)

M_H = horizontal shaft offset at SM (in in.)

D_2 = distance between SM indicator and MTBS front feet (in in.)

D_1 = distance between indicators (in in.)

The horizontal corrective movement of the MTBS back feet (S_2) is found by applying the formula:

$$HS_2 = \left[\left(ST_H + M_H \right) \times \left(\frac{D_3}{D_1} \right) \right] - ST_H$$

where

HS_2 = horizontal corrective movement at MTBS back feet (in in.)

ST_H = horizontal shaft offset at MTBS (in in.)

M_H = horizontal shaft offset at SM (in in.)

D_3 = distance between SM indicator and MTBS back feet (in in.)

D_1 = distance between indicators (in in.)

Tech Tip

Motor couplings are rated according to the amount of torque they are designed for. Couplings are rated in pound-inches (lb-in) or pound-feet (lb-ft). The torque rating must be correct for the application to prevent the coupling from bending or breaking. A bent coupling causes misalignment and vibration, and a broken coupling prevents the motor from operating. Flexible couplings exist that allow a motor to operate a driven load while compensating for slight misalignments between the motor and load shaft.

Manufacturing & Maintenance Systems, Inc.

Alignment performed to the correct motor/pump tolerance increases the life of pump seals and bearings.

Reverse Dial Alignment

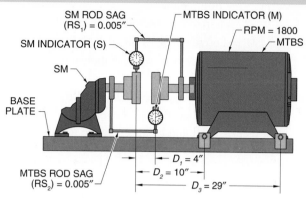

Dial Indicator Data		
Reading Position	SM Reading	MTBS Reading
12:00	0.000	0.000
3:00	+ 0.005	− 0.006
6:00	+ 0.035	− 0.046
9:00	+ 0.030	− 0.040

Note: Solve for values inside the parentheses first and work from the inside out.

What adjustments must be made to the MTBS using the reverse dial alignment method for proper alignment?

1. Correct vertical offsets.

$$ST_V = \frac{S_0 - S_6 - RS_1}{2}$$

$$ST_V = \frac{0.000 - 0.035 - 0.005}{2}$$

$$ST_V = \frac{-0.040}{2}$$

$$ST_V = -0.020''$$

$$M_V = \frac{M_0 - M_6 - RS_2}{2}$$

$$M_V = \frac{0.000 - (-0.046) - 0.005}{2}$$

$$M_V = \frac{-0.041}{2}$$

$$M_V = 0.0205''$$

$$VS_1 = \{(ST_V + M_V) \times \left(\frac{D_2}{D_1}\right)\} - ST_V$$

$$VS_1 = \{(-0.020 + 0.0205) \times \left(\frac{10}{4}\right)\} - (-0.020)$$

$$VS_1 = (0.0005 \times 2.5) + 0.020$$
$$VS_1 = 0.00125 + 0.020$$
$$VS_1 = \mathbf{0.02125}$$

$$VS_2 = \{(ST_V + M_V) \times \left(\frac{D_3}{D_1}\right)\} - ST_V$$

$$VS_2 = \{(-0.020) + 0.0205 \times \left(\frac{29}{4}\right)\} - (-0.020)$$

$$VS_2 = (0.0005 \times 7.25) + 0.020$$
$$VS_2 = 0.003625 + 0.020$$
$$VS_2 = \mathbf{0.023625''}$$

VS_1 and VS_2 results indicate that the MTBS front feet (S_1) are raised 0.021″ and the back feet (S_2) are raised 0.024″.

Vertical correction of the MTBS is not possible without raising the SM. Raise the SM by placing 0.100″ shims under all feet and start over.

2. Correct horizontal offsets.

$$ST_H = \frac{S_9 - S_3 - RS_1}{2}$$

$$ST_H = \frac{0.030 - 0.005 - 0.005}{2}$$

$$ST_H = \frac{0.020}{2}$$

$$ST_H = 0.010''$$

$$M_H = \frac{M_9 - M_3 - RS_2}{2}$$

$$M_H = \frac{-0.040 - (-0.006) - 0.005}{2}$$

$$M_H = \frac{-0.039}{2}$$

$$M_H = -0.0195''$$

$$HS_1 = \{(ST_H + M_H) \times \left(\frac{D_2}{D_1}\right)\} - ST_H$$

$$HS_1 = \{(0.010 + (-0.0195)) \times \left(\frac{10}{4}\right)\} - 0.010$$

$$HS_1 = (-0.0095 \times 2.5) - 0.010$$
$$HS_1 = -0.02375 - 0.010$$
$$HS_1 = \mathbf{-0.03375''}$$

$$HS_2 = \{(ST_H + M_H) \times \left(\frac{D_3}{D_1}\right)\} - ST_H$$

$$HS_2 = \{(0.010 + (-0.0195)) \times \left(\frac{29}{4}\right)\} - 0.010$$

$$HS_2 = (-0.0095 \times 7.25) - 0.010$$
$$HS_2 = -0.068875 - 0.010$$
$$HS_2 = \mathbf{-0.078875''}$$

Using two zeroed indicators at the back of the MTBS (one on the same plane as S_1 and one on the same plane as S_2) adjust the front feet (S_1) toward the technician by 0.034″ and the back feet (S_2) toward the technician by 0.079″. The large numbers indicate a very coarse adjustment.

Figure 13-26. Reverse dial alignment procedures offer net misalignment values at one setting from which shim placement can be determined.

Electronic Reverse Dial Method

The *electronic reverse dial method* is an alignment method that uses the reverse dial as a base method with the dial indicators replaced with electromechanical sensing devices. The electronic reverse dial method is supported by computer-aided electronic instrumentation. The sensing devices detect physical movement and convert the movement to an electrical signal, which is sent to a calculator. Before the calculator can process an adjustment response, it must have the electronic movement signal and certain physical dimensions, which must be entered by the technician. The calculator computes MTBS movements as its response. **See Figure 13-27.**

Electronic Reverse Dial Method

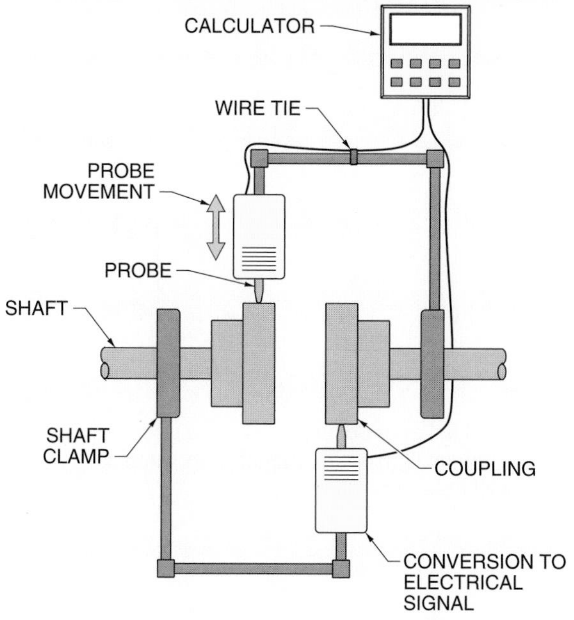

Figure 13-27. In the electronic reverse dial method, electromechanical aligning devices convert a mechanical movement into an electrical signal.

Inspect equipment condition before beginning any alignment procedure. Check motor and pump bases for breaks or cracks. All contact surfaces must be clean, smooth, and flat. Foundation and base plates must be in good condition. Check shaft, coupling, and bearing condition for any irregularities. Finally, organize tools, materials, and workplace. Anchor bolt condition, eccentric shafts, and soft foot must be checked and corrected before the alignment procedure begins.

The electronic reverse dial method requires that two sensing devices be used, each being assembled using rods and couplings and attached to a coupling or shaft. Installation is similar to using dial indicators where one sensor is opposite the other. Sensor wires must be secured with slack to prevent unwanted tugging forces.

Some manufacturers recommend that the coupling be disconnected while others suggest that they be connected. Always follow manufacturer's recommendations. However, any force, including coupling force, can create enough resistance to produce adverse responses.

The shafts are rotated together and readings are entered with the press of a button at the 12:00, 3:00, 6:00, and 9:00 positions. Some manufacturers recommend entering readings at 12:00, then forward (clockwise) to 3:00, back (counterclockwise) to 9:00, forward (clockwise) to 6:00, and back (counterclockwise) to 12:00. Check manufacturer's requirements. Also, add any necessary thermal expansion compensation information to the calculation.

In preparation for corrective horizontal movements, place two dial indicators against the farthest front and back feet of the MTBS and zero both indicators as a starting reference. Vertical and horizontal movements are made according to the calculator's indication.

Finally, repeat all measurements until shaft runout is within tolerance. Upon startup, vibration readings can be taken to compare previous readings with present readings to determine alignment condition and establish data in order to determine future progressing conditions.

Laser Rim-and-Face Method

The *laser rim-and-face alignment method* is an alignment method in which laser devices are placed opposite each other to measure alignment. The laser rim-and-face method is used when extreme accuracy and fast alignment are required. Even though the initial cost of laser equipment are higher than that of other methods, known and hidden payback costs are generally worth the extra expense. Known paybacks due to extreme accuracy are those items that are measurable, such as less emergency downtime, less need for extra or spare replacement inventories, and less utility costs. Energy savings on an accurately aligned machine are 7% to 12% per alignment over a marginally acceptable aligned machine. Hidden savings are those that are not measurable, such as the availability and utilization of power elsewhere or the increased morale within a smoother running facility.

Laser alignment devices operate using the rim-and-face method, with the dial indicator being replaced with a laser beam. The beam is directed to a 90° prism reflector, which is directed back to the sending unit where a receiving transducer (photo position detector) accepts the signal and converts it into an impulse for the calculator. **See Figure 13-28.** A benefit of using a laser beam is that there is no rod sag to calculate and the alignment is not affected by distance or axial float. The beam sensor is able to detect up and down movement. Offsets and angles are read and measured accurately when the return beam is detected at one position (12:00) and then detected at another position (6:00). The calculator determines offset as a direct send/receive reading and determines the angularity by measuring right triangles.

Once set up, laser alignment devices check soft foot, compensate for thermal expansion, indicate the movement of a machine during alignment, rapidly couple machines with multiple couplings, and determine shim placement. The ease and simplicity of making alignment moves is noted when all adjusting moves are being observed on a screen as they happen. A graphic display presents the condition of each foot when soft foot or angular soft foot is checked. Finally, all rechecking is completed and corrected in a matter of seconds.

An advantage of a laser alignment device is that measurements are not required to be read, recorded, and calculated for proper movements to be made. Also, the corrective values for the machine feet appear automatically in the computer display.

A laser alignment device requires care in handling to maintain its high level of calibration. Dropping or bumping may result in loss of calibration and alignment integrity. Care must also be used with lasers because steam, dust, and sunlight can adversely affect the laser beam.

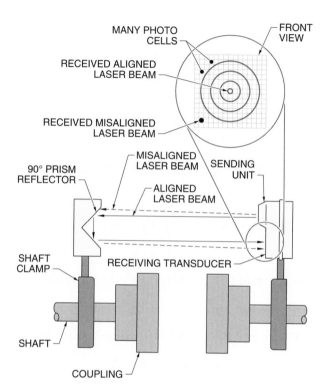

Figure 13-28. Laser accuracy is based upon being able to send a pinpoint light beam, reflect it, and send it to another location without deflection.

Review

1. What is the difference between a coupling and a flexible coupling?

2. What is the intended purpose of challenging a start switch?

3. How is it possible to ensure any transmission of an outside force does not affect the proper alignment of machines?

4. Define dowel effect and explain how it is corrected.

5. Describe proper shim stock.

6. Define soft foot and list its four conditions.

7. Explain the two methods of measuring soft foot.

8. Define thermal expansion.

9. Describe how accurate indicator rod sag is determined.

10. What is the choice of alignment method based on?

11. Describe a method for overcoming confusion during horizontal movement.

12. Explain the straightedge alignment method.

13. Why is the rim-and-face method typically the most troublesome of the precision alignment methods?

14. Explain how computer-aided electronic instruments support the electronic reverse dial method.

15. Explain how the laser rim-and-face alignment method provides known and hidden paybacks despite the high initial cost of laser equipment.

Digital Resources

ATPeResources.com/Quicklinks
Access Code 462409

CHAPTER 14

ELECTRICAL PRINCIPLES

An understanding of electrical principles is required for anyone working with electricity in the industrial field. The National Electrical Code® safeguards persons and property from hazards arising from the use of electricity. Electrical safety procedures and precautions should be practiced and proper protective equipment should be worn by all personnel working with electrical systems.

Salisbury

Chapter Objectives

- Describe atomic theory and its relation to electricity.
- Identify the electrical quantities of voltage, current, and resistance.
- Use Ohm's law to perform calculations involving voltage, current, and resistance.
- Use the power formula to perform calculations involving power, voltage, and current.
- Describe the three electrical circuit connections and how they operate.
- Describe the properties of magnetism as they relate to electricity.
- Explain the purpose of the National Electrical Code®.
- Explain the importance of electrical safety.

Key Terms

- electricity
- atom
- valence electron
- law of charges
- conductor
- insulator
- electron flow
- voltage
- polarity
- rectifier
- current
- direct current
- alternating current
- resistance
- Ohm's law
- power
- electrical circuit
- series connection
- parallel connection
- series/parallel connection
- magnetism
- magnetic flux lines
- induction
- electromagnet
- generator
- power distribution
- National Electrical Code®
- electrical shock
- authorized individual
- arc flash
- arc flash boundary
- grounding
- lockout
- tagout

ELECTRICAL THEORY

Electricity is a physical occurrence involving electric charges and their effects when the charges are in motion and at rest. Electrical components and circuits are designed to operate in a predetermined manner to safely produce light, heat, and rotary and linear motion; to transfer and store information; and to provide many other uses in industrial facilities. Electricity always follows basic scientific laws and principles. Understanding these basic laws and principles is the first step for anyone who wishes to work safely on any electrical system.

Atomic Theory

According to atomic theory, all matter is an organized collection of atoms. An *atom* is the smallest building block of matter that cannot be reduced into smaller units without changing its basic characteristics. The three fundamental particles contained in atoms are protons, neutrons, and electrons. A *proton* is a particle with a positive electrical charge. A *neutron* is a neutral particle with a mass approximately equal to that of a proton. An *electron* is a particle with a negative charge.

A *nucleus* is the heavy, dense center of an atom. Protons and neutrons combine to form the nucleus of an atom,

and electrons orbit around the nucleus. Electrons travel at such a high rate of speed in their orbits that they form shells around the nucleus. A *shell* is an orbiting layer of electrons in an atom. Atoms each have a different number of electrons and a different number of electron shells. Each shell can only hold a specific number of electrons. Shells are numbered innermost to outermost 1, 2, 3, 4, 5, 6, and 7 or lettered K, L, M, N, O, P, and Q. **See Figure 14-1.**

Valence Electrons. A *valence shell* is the outermost shell of an atom. A *valence electron* is an electron located in a valence shell. Valence electrons can be moved from atom to atom.

When atoms gain or lose electrons, an electrical charge is produced. When there are as many electrons as there are protons in an atom, the atom is electrically neutral. However, a positive (+) or negative (−) electrical charge can be produced when there is a different number of electrons compared to protons. A *positive charge* is an electrical charge produced when there are fewer electrons than protons in an atom. A *negative charge* is an electrical charge produced when there are more electrons than protons in an atom.

These electrical charges are further described by the law of charges. The *law of charges* states that opposite charges attract each other and like charges repel each other.

Atoms

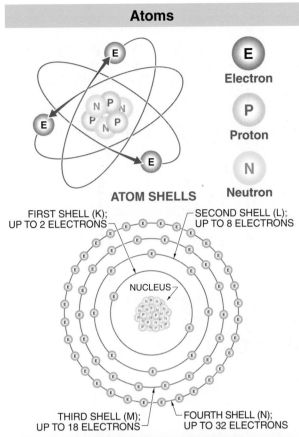

E — Electron

P — Proton

N — Neutron

ATOM SHELLS

FIRST SHELL (K); UP TO 2 ELECTRONS

SECOND SHELL (L); UP TO 8 ELECTRONS

NUCLEUS

THIRD SHELL (M); UP TO 18 ELECTRONS

FOURTH SHELL (N); UP TO 32 ELECTRONS

Electron Configuration

Shell Number	Shell Letter	Maximum Number of Electrons
1	K	2
2	L	8
3	M	18
4	N	32
5	O	50
6	P	72
7	Q	98

Figure 14-1. Atoms are composed of protons, neutrons, and electrons. Electrons form shells around the nucleus of an atom. Each shell can only hold a specific number of electrons.

See Figure 14-2. Therefore, a positively charged atom attracts a negatively charged atom; a positively charged atom repels another positively charged atom; and a negatively charged atom repels another negatively charged atom. All charged atoms exert a force on one another, even if they are not in physical contact.

Law of Charges

FORCE EXERTED

NEGATIVELY CHARGED PARTICLE

POSITIVELY CHARGED PARTICLE

OPPOSITE CHARGES ATTRACT

LIKE CHARGES REPEL

Figure 14-2. The law of charges states that opposite charges attract each other and like charges repel each other.

Safety Tip

Approximately 40% of the electricity generated in the United States is used in industry, 34% in residences, and 26% in commercial applications.

The electrical behavior of atoms varies according to the physical structure of the atom and the type of matter it composes. Most atoms do not have a full valence shell containing a maximum number of valence electrons. The number of valence electrons in the atom determines whether that atom allows electrons to easily move from atom to atom or whether that atom prevents the electrons from moving. When electrons move, it creates electricity.

A material that allows electrons to easily move from atom to atom can be used as a conductor. A *conductor* is material that has very little electrical resistance and permits electrons to move through it easily. For example, copper is a good conductor because copper atoms allow their valence electrons to move easily from atom to atom.

Another type of matter, such as rubber, does not allow electricity to easily pass through it and can act as an insulator. An *insulator* is a material that has a very high electrical resistance and resists the flow of electrons.

Electron Flow. *Electron flow* is the traveling of a displaced valence electron from one atom to another. Electron flow can be described using two different conventions. The two conventions are conventional current flow and electron current flow. *Conventional current flow* is a convention that shows current as flowing from positive to negative. *Electron current flow* is a convention that shows current as flowing from negative to positive. **See Figure 14-3.** Both conventions are still used today to describe electron flow. Conventional current flow is used more in the electrical field and by electrical engineers to aid in explaining electrical circuit properties. Conventional current flow is also used in the electrical field to assist in explaining the operation of solid-state electronic components. *Note:* The difference between the two conventions has no impact on how an electrical circuit or a solid-state component actually functions in the field. The operation, the output, and the calculations are the same no matter which convention is used.

Voltage

All electrical circuits must have a source of power to produce work. The source of power used depends on the application and the amount of power required. All sources of power produce a set voltage level or voltage range.

Voltage (E) is the amount of electrical pressure in a circuit. Voltage is measured in volts (V). Voltage is also known as electromotive force (EMF) or potential difference. Voltage can be produced by electromagnetism (generators), chemical action (batteries), light (photocells), heat

(thermocouples), pressure (piezoelectricity), or friction (static electricity). **See Figure 14-4.**

Voltage is either direct current (DC) or alternating current (AC) voltage. *DC voltage* is voltage that flows in one direction only. *AC voltage* is voltage that reverses its direction of flow at regular intervals. DC voltage is used in almost all portable equipment, such as automobiles, golf carts, flashlights, and cameras. AC voltage is used in residential, commercial, and industrial lighting and power distribution systems.

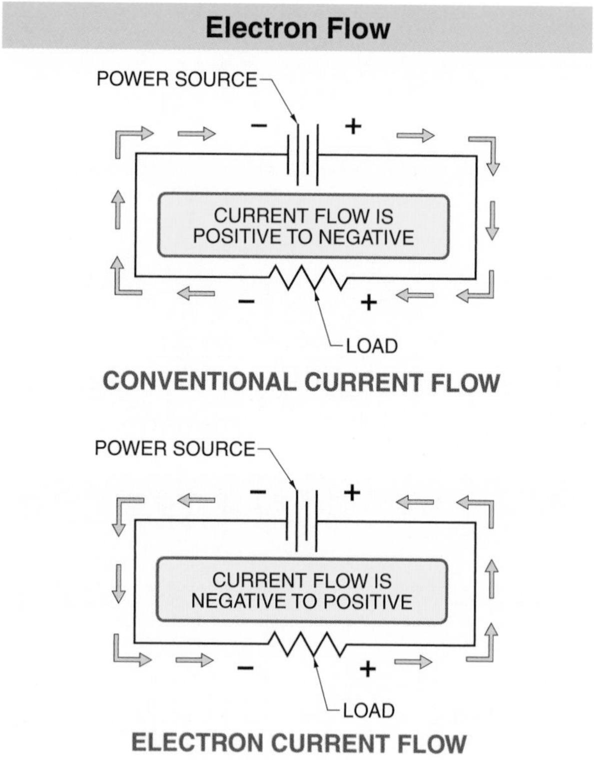

Figure 14-3. Conventional current flow is a convention that shows current as flowing from positive to negative.

All DC voltage sources have a positive and a negative terminal. The positive and negative terminals establish polarity in a circuit. *Polarity* is the positive (+) or negative (−) charge of an object. All points in a DC circuit have polarity. The most common power sources that directly produce DC voltage are batteries and photocells.

In addition to obtaining DC voltage directly from batteries and photocells, DC voltage is also obtained from a rectified AC voltage supply. A *rectifier* is a device that converts AC voltage to DC voltage by allowing the voltage and current to move in only one direction.

Voltage Sources

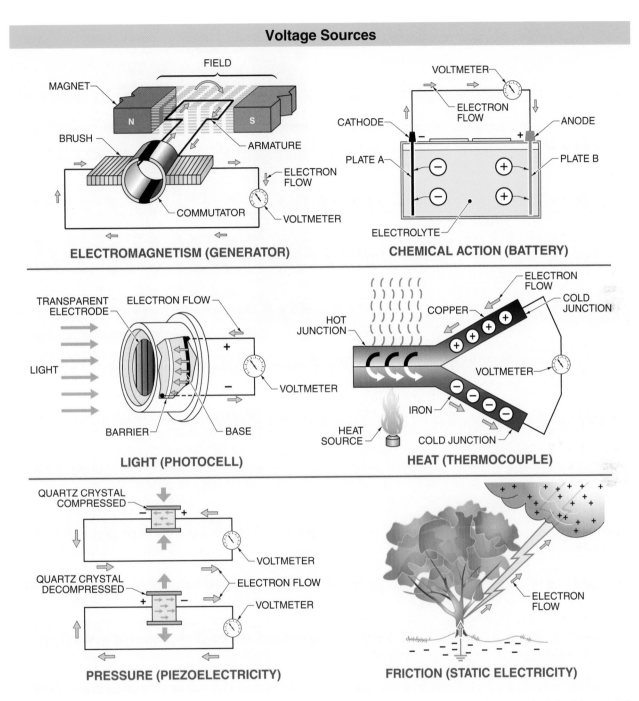

Figure 14-4. Voltage can be produced by electromagnetism (generators), chemical action (batteries), light (photocells), heat (thermocouples), pressure (piezoelectricity), or friction (static electricity).

DC voltage obtained from a rectified AC voltage supply varies from almost pure DC voltage to half-wave DC voltage. **See Figure 14-5.** Common DC voltage levels include 1.5 V, 3 V, 6 V, 9 V, 12 V, 24 V, 36 V, 125 V, 250 V, 600 V, 1200 V, 1500 V, and 3000 V.

Safety Tip

Transient voltage (voltage spike) is a temporary, unwanted voltage in an electrical circuit. Transient voltages can occur due to lightning strikes, unfiltered electrical equipment, or power being switched on and off excessively.

DC Voltage

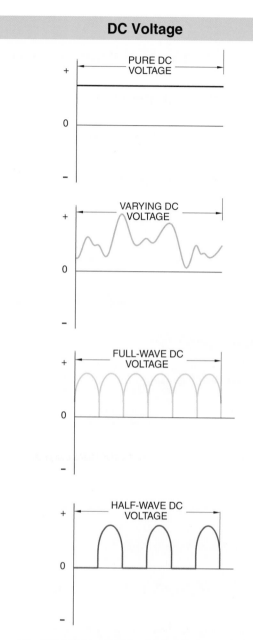

Standard DC Voltages

Device	Level (in V)
Flashlights, watches, etc.	1.5, 3
Toys, automobiles, trucks	6, 9, 12, 24, 36
Printing presses, small electric railway systems	125, 250, 600
Large electric railway systems	1200, 1500, 3000

Figure 14-5. DC voltage is voltage that flows in one direction only. DC voltage can be obtained by either a battery, a photocell, or a rectified AC voltage.

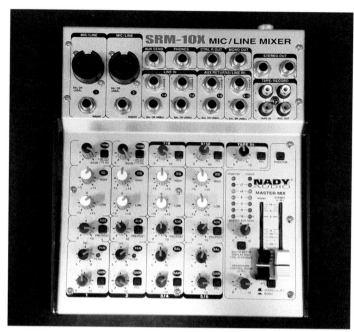

Because rectifiers convert AC power to DC power, they are commonly used inside the power supply of electric equipment.

AC voltage is the most common type of voltage used to produce work. AC voltage is produced by generators which produce AC sine waves as they rotate. An *AC sine wave* is the waveform formed by the smooth, repetitive oscillation of voltage or current. The wave reaches its peak positive value at 90°, returns to 0 V at 180°, increases to its peak negative value at 270°, and returns to 0 V at 360°. A *cycle* is one complete positive and one complete negative alternation of a wave form. An *alternation* is half of a cycle. A sine wave has one positive alternation and one negative alternation per cycle.

AC voltage is either single-phase (1φ) or three-phase (3φ). Single-phase AC voltage contains only one alternating voltage waveform. Three-phase AC voltage is a combination of three alternating voltage waveforms, each displaced 120 electrical degrees (one-third of a cycle) apart. Three-phase voltage is produced when three coils are simultaneously rotated in a generator. **See Figure 14-6.**

Almost any level of AC voltage is available. Low AC voltages (6 V to 24 V) are used for doorbells and security systems. Medium AC voltages (110 V to 120 V) are used in residential applications for lighting, heating, cooling, cooking, and operating motors. High AC voltages (208 V to 480 V) are used in industrial applications to convert raw materials into usable products, in addition to providing lighting, heating, and cooling for plant personnel.

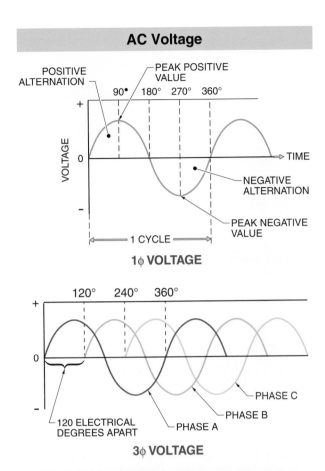

AC Voltage

1φ VOLTAGE

3φ VOLTAGE

Standard AC Voltages

Device	Level (in V)
Doorbells, security systems	6, 24
Most residential appliances (TVs, coffeemakers), lighting applications	110, 115, 120
Industrial motors, heating elements	208, 480

Figure 14-6. AC voltage is the most common type of voltage used to produce work. AC voltage can either be single-phase (1φ) or three-phase (3φ).

Current

Electrons flow through a circuit when a source of power is connected to a device that uses electricity. *Current (I)* is the flow of electrons through an electrical circuit. Current is measured in amperes (A). An *ampere* is a quantity of electrons passing a given point in one second. The more power a load requires, the larger the amount of current flow. **See Figure 14-7.**

Current can be direct or alternating. *Direct current (DC)* is current that flows in only one direction. Direct current flows in any circuit connected to a power supply producing a DC voltage. *Alternating current (AC)* is current that reverses its direction of flow at regular intervals. Alternating current flows in any circuit connected to a power supply producing an AC voltage.

Circuit Current Flow

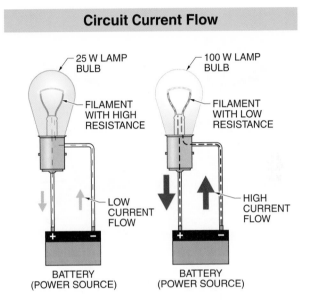

Figure 14-7. Electrons flow through a circuit when a source of power is connected to a device that uses electricity.

Resistance

Resistance (R) is the opposition to current flow. Resistance is measured in ohms. An *ohm* is the resistance of a conductor in which an electrical pressure of 1 V causes an electrical current of 1 A to flow. The Greek symbol omega (Ω) is used to represent ohms. Higher resistance measurements are expressed using prefixes, as in kilohms (k Ω) and megohms (M Ω).

Resistance limits the flow of current in an electrical circuit. The higher the resistance, the lower the current flow. Likewise, the lower the resistance, the higher the current flow. **See Figure 14-8.** Components designed as insulators, such as rubber or plastic, should have a very high resistance. Components designed as conductors, such as wires or switch contacts, should have a very low resistance. The resistance of insulators decreases when they are damaged by moisture or overheating. The resistance of conductors increases when they are damaged by burning or corrosion. Factors that affect the resistance of conductors are the size of the wire, the length of the wire, the conductor material, and temperature.

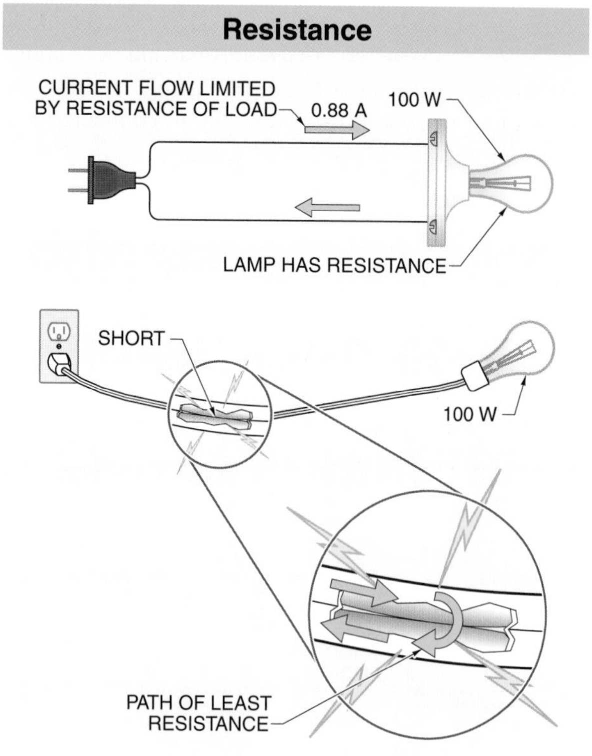

Resistance

CURRENT FLOW LIMITED
BY RESISTANCE OF LOAD— →0.88 A   100 W

LAMP HAS RESISTANCE

SHORT

100 W

PATH OF LEAST
RESISTANCE

Figure 14-8. Resistance limits the flow of current in an electrical circuit.

Conductors are insulated and rated according the American Wire Gauge (AWG) system. A conductor with a large cross-sectional area has less resistance than a conductor with a small cross-sectional area. A large conductor may also carry more current. The longer the conductor, the greater the resistance as well. Short conductors have less resistance than long conductors of the same size. Certain metal conductors can carry more current than others. For example, copper is a better conductor (less resistance) than aluminum and may carry more current for a given size. Temperature also affects resistance. For metals, the higher the temperature, the greater the resistance.

Ohm's Law

Ohm's law is the relationship between voltage (E), current (I), and resistance (R) in an electrical circuit. Ohm's law states that current in a circuit is proportional to the voltage and inversely proportional to the resistance. Using Ohm's law, any value in this relationship can be found when the other two are known. Ohm's law can be easily shown with a visual aid. **See Figure 14-9.**

Ohm's Law

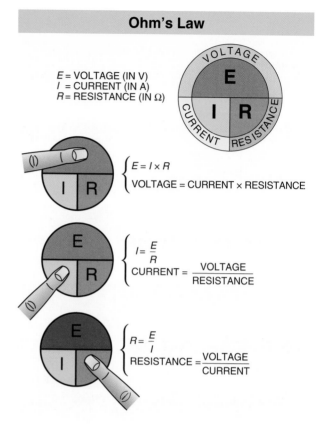

E = VOLTAGE (IN V)
I = CURRENT (IN A)
R = RESISTANCE (IN Ω)

$\begin{cases} E = I \times R \\ \text{VOLTAGE} = \text{CURRENT} \times \text{RESISTANCE} \end{cases}$

$\begin{cases} I = \dfrac{E}{R} \\ \text{CURRENT} = \dfrac{\text{VOLTAGE}}{\text{RESISTANCE}} \end{cases}$

$\begin{cases} R = \dfrac{E}{I} \\ \text{RESISTANCE} = \dfrac{\text{VOLTAGE}}{\text{CURRENT}} \end{cases}$

Figure 14-9. Ohm's law is the relationship between voltage, current, and resistance in an electrical circuit.

Calculating Voltage Using Ohm's Law. Ohm's law states that voltage (E) in a circuit is equal to current (I) multiplied by resistance (R). To calculate voltage using Ohm's law, apply the following formula:

$E = I \times R$

where

E = voltage (in V)

I = current (in A)

R = resistance (in Ω)

For example, what is the voltage in a circuit that delivers 2 A to an electric motor with a resistance of 60 Ω? **See Figure 14-10.**

$E = I \times R$

$E = 2 \times 60$

$E = \mathbf{120\ V}$

Calculating Current Using Ohm's Law. Ohm's law states that current (I) in a circuit is equal to voltage (E) divided by resistance (R). To calculate current using Ohm's law, apply the following formula:

$$I = \frac{E}{R}$$

where

I = current (in A)

E = voltage (in V)

R = resistance (in Ω)

For example, what is the current in a circuit with a 40 Ω light connected to a 120 V supply? **See Figure 14-11.**

$$I = \frac{E}{R}$$

$$I = \frac{120}{40}$$

$$I = \textbf{3 A}$$

Calculating Voltage

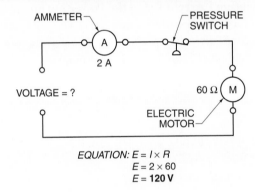

EQUATION: $E = I \times R$

$E = 2 \times 60$

$E = \textbf{120 V}$

Figure 14-10. Voltage can be calculated by multiplying the total resistance of a circuit by the total current of the circuit.

Calculating Current Using Ohm's Law

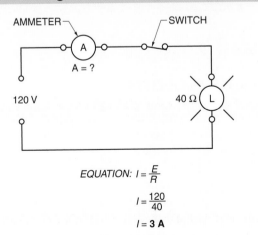

EQUATION: $I = \frac{E}{R}$

$I = \frac{120}{40}$

$I = \textbf{3 A}$

Figure 14-11. Current can be calculated by dividing the supply voltage by the total resistance of a circuit.

Calculating Resistance Using Ohm's Law. Ohm's law states that resistance (R) in a circuit is equal to voltage (E) divided by current (I). To calculate resistance using Ohm's law, apply the following formula:

$$R = \frac{E}{I}$$

where

R = resistance (in Ω)

E = voltage (in V)

I = current (in A)

For example, what is the resistance of a circuit in which an electric motor draws 4 A and is connected to a 120 V supply? **See Figure 14-12.**

$$R = \frac{E}{I}$$

$$R = \frac{120}{4}$$

$$R = \textbf{30 } \Omega$$

Calculating Resistance Using Ohm's Law

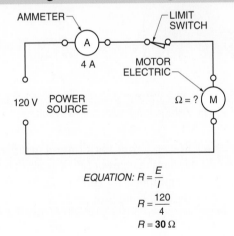

EQUATION: $R = \frac{E}{I}$

$R = \frac{120}{4}$

$R = \textbf{30 } \Omega$

Figure 14-12. Resistance can be calculated by dividing the supply voltage by the current in a circuit.

Power

Power is the rate of doing work or using energy. In electrical circuits, power is used by devices to perform work. Power is rated in watts (W) and can be measured in kilowatts (kW) or megawatts (MW). Power is equal to voltage (E) times current (I).

The *power formula* is a formula that shows the relationship between power (P), voltage (E), and current (I) in an electrical circuit. Any value in this relationship may be found using the power formula when the other two values are known. The power formula can be easily shown with a visual aid. **See Figure 14-13.**

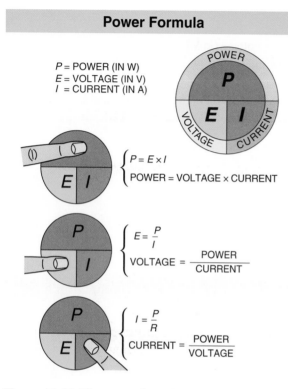

Figure 14-13. The power formula is the relationship between power, voltage, and current in a circuit.

Calculating Power Using the Power Formula. The power formula states that power (P) in a circuit is equal to voltage (E) multiplied by current (I). To calculate power, apply the following formula:

$$P = E \times I$$

where

P = power (in W)

E = voltage (in V)

I = current (in A)

For example, what is the power of a load that draws 5 A when connected to a 120 V supply?

$$P = E \times I$$
$$P = 120 \times 5$$
$$P = \mathbf{600\ W}$$

Calculating Voltage Using the Power Formula. The power formula states that voltage (E) in a circuit is equal to power (P) divided by current (I). To calculate voltage using the power formula, apply the following formula:

$$E = \frac{P}{I}$$

where

E = voltage (in V)

P = power (in W)

I = current (in A)

For example, what is the voltage in a circuit in which a 600 W load draws 5 A?

$$E = \frac{P}{I}$$
$$E = \frac{600}{5}$$
$$E = \mathbf{120\ V}$$

Calculating Current Using the Power Formula. The power formula states that current (I) in a circuit is equal to power (P) divided by voltage (E). To calculate current using the power formula, apply the following formula:

$$I = \frac{P}{E}$$

where

I = current (in A)

P = power (in W)

E = voltage (in V)

For example, what is the current in a circuit in which a 600 W load is connected to a 120 V supply?

$$I = \frac{P}{E}$$
$$I = \frac{600}{120}$$
$$I = \mathbf{5\ A}$$

ELECTRICAL CIRCUITS

An *electrical circuit* is the interconnection of conductors and electrical components through which current is designed to flow. Electrical circuits can consist of a power source, conductor(s), switch(es), and load(s). Electrical circuits can be either open or closed. An *open circuit* is an electrical circuit that has a gap or opening that does not allow current to flow. A *closed circuit* is an electrical circuit with a complete path that allows current to flow. Types of circuit connections include series, parallel, and series/parallel connections.

Series Circuits

Fuses, switches, loads, and other electrical components can be connected in series. A *series connection* is a circuit connection that has two or more components connected so there is only one path for current to flow. **See Figure 14-14.** Opening the circuit at any point stops the flow of current, such as when a fuse blows, a circuit breaker trips, or a switch or load opens.

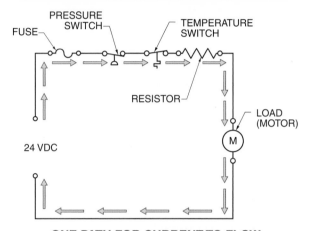

Series Circuits

ONE PATH FOR CURRENT TO FLOW

Figure 14-14. Series circuits can have as many components as required, but there is only one path for current to flow.

Resistance in Series Circuits. The total resistance in a circuit containing series-connected loads equals the sum of the resistances of all the loads. The resistance in the circuit increases if loads are added in series and decreases if loads are removed. To calculate the total resistance of a series circuit, apply the following formula:

$$R_T = R_1 + R_2 + R_3 + ...$$

where

R_T = total resistance (in Ω)

R_1 = resistance 1 (in Ω)

R_2 = resistance 2 (in Ω)

R_3 = resistance 3 (in Ω)

For example, what is the total resistance of a circuit that has 2 Ω, 4 Ω, and 6 Ω resistors connected in series?

$$R_T = R_1 + R_2 + R_3$$
$$R_T = 2 + 4 + 6$$
$$R_T = \mathbf{12\ \Omega}$$

Voltage in Series Circuits. The total voltage applied across loads connected in series is divided across the loads. Each load drops a set percentage of the applied voltage. The exact voltage drop across each load depends on the resistance of that load. The voltage drops across any two loads are the same if the resistance values are the same. To calculate total voltage of a series circuit when the voltage across each load is known or measured, apply the following formula:

$$E_T = E_1 + E_2 + E_3 + ...$$

where

E_T = total applied voltage (in V)

E_1 = voltage drop across load 1 (in V)

E_2 = voltage drop across load 2 (in V)

E_3 = voltage drop across load 3 (in V)

For example, what is the total applied voltage of a circuit containing 4 V, 8 V, and 12 V drops across three loads?

$$E_T = E_1 + E_2 + E_3$$
$$E_T = 4 + 8 + 12$$
$$E_T = \mathbf{24\ V}$$

Current in Series Circuits. The current in a circuit containing series-connected loads is the same throughout the circuit. The current in the circuit will decrease if the circuit resistance increases, and the current will increase if the circuit resistance decreases. To find total current in a series circuit, apply the following formula:

$$I_T = I_1 = I_2 = I_3 = ...$$

where

I_T = total circuit current (in A)

I_1 = current through load 1 (in A)

I_2 = current through load 2 (in A)

I_3 = current through load 3 (in A)

For example, what is the total current through a series circuit if the current measured at each load is 2 A?

$$I_T = I_1 = I_2 = I_3$$
$$I_T = 2 = 2 = 2$$
$$I_T = \mathbf{2\ A}$$

Parallel Circuits

Fuses, switches, loads, and other components can be connected in parallel. A *parallel connection* is a circuit connection that has two or more components connected so there is more than one path for current to flow. **See Figure 14-15.**

Parallel Circuits

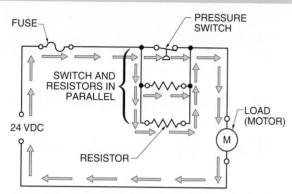

MORE THAN ONE PATH FOR CURRENT TO FLOW

Figure 14-15. Parallel circuits have multiple paths for current to flow.

Care must be taken when working with parallel circuits because current could be flowing in one part of the circuit even though another part of the circuit is OFF. Understanding and recognizing parallel-connected components and circuits enables a technician to take proper measurements, make circuit modifications, and troubleshoot the circuit.

Resistance in Parallel Circuits. The total resistance in a circuit containing parallel-connected loads is less than the smallest resistance value. The total resistance decreases if loads are added in parallel and increases if loads are removed. To calculate total resistance in a parallel circuit containing two resistors, apply the following formula:

$$R_T = \frac{R_1 \times R_2}{R_1 + R_2}$$

where

R_T = total resistance (in Ω)

R_1 = resistance 1 (in Ω)

R_2 = resistance 2 (in Ω)

For example, what is the total resistance in a circuit containing resistors of 16 Ω and 24 Ω connected in parallel?

$$R_T = \frac{R_1 \times R_2}{R_1 + R_2}$$

$$R_T = \frac{16 \times 24}{16 + 24}$$

$$R_T = \frac{384}{40}$$

$$R_T = \mathbf{9.6\ \Omega}$$

To calculate total resistance in a parallel circuit with three or more resistors, the following formula can be applied:

$$R_T = \frac{1}{\dfrac{1}{R_1}} + \frac{1}{\dfrac{1}{R_2}} + \frac{1}{\dfrac{1}{R_3}} + \cdots$$

where

R_T = total resistance (in Ω)

R_1 = resistance 1 (in Ω)

R_2 = resistance 2 (in Ω)

R_3 = resistance 3 (in Ω)

For example, what is the total resistance in a circuit containing resistors of 16 Ω, 24 Ω, and 48 Ω connected in parallel?

$$R_T = \frac{1}{\dfrac{1}{R_1}} + \frac{1}{\dfrac{1}{R_2}} + \frac{1}{\dfrac{1}{R_3}}$$

$$R_T = \frac{1}{\dfrac{1}{16}} + \frac{1}{\dfrac{1}{24}} + \frac{1}{\dfrac{1}{48}}$$

$$R_T = \frac{1}{0.06250} + \frac{1}{0.04167} + \frac{1}{0.02083}$$

$$R_T = \mathbf{8\ \Omega}$$

Voltage in Parallel Circuits. The voltage across each load is the same when loads are connected in parallel. The voltage across each load remains the same if parallel loads are added or removed. To find total voltage in a parallel circuit when the voltage across a load is known or measured, apply the following formula:

$$E_T = E_1 = E_2 = E_3 = \ldots$$

where

E_T = total applied voltage (in V)

E_1 = voltage across load 1 (in V)

E_2 = voltage across load 2 (in V)

E_3 = voltage across load 3 (in V)

For example, what is the total applied voltage if the voltage across three parallel-connected loads is 96 VDC?

$$E_T = E_1 = E_2 = E_3$$

$$E_T = 96 = 96 = 96$$

$$E_T = \mathbf{96\ V}$$

Current in Parallel Circuits. Total current in a circuit containing parallel-connected loads equals the sum of the current through all the loads. Total current increases if loads are added in parallel and decreases if loads are removed. To calculate total current in a parallel circuit, apply the following formula:

$$I_T = I_1 + I_2 + I_3 + ...$$

where

I_T = total circuit current (in A)

I_1 = current through load 1 (in A)

I_2 = current through load 2 (in A)

I_3 = current through load 3 (in A)

For example, what is the total current in a circuit containing three loads connected in parallel if the current through the three loads is 6 A, 4 A, and 2 A?

$$I_T = I_1 + I_2 + I_3$$
$$I_T = 6 + 4 + 2$$
$$I_T = \mathbf{12\ A}$$

Series/Parallel Circuits

Fuses, switches, loads, and other components can be connected in a series/parallel connection. A *series/parallel connection* is a circuit connection with a combination of series- and parallel-connected components. An example of a series/parallel connection is a DC compound motor. **See Figure 14-16.**

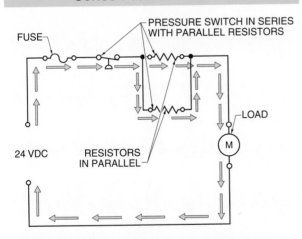

Series-Parallel Circuits

FUSE

PRESSURE SWITCH IN SERIES WITH PARALLEL RESISTORS

LOAD

M

24 VDC   RESISTORS IN PARALLEL

ONE FLOW PATH AND MORE THAN ONE PATH FOR CURRENT TO FLOW

Figure 14-16. A series/parallel connection is a circuit connection with a combination of series- and parallel-connected components.

Tech Tip

The average power consumption at any one-time for a common residence is approximately 3.33 kW. Thus, 1 megawatt (1 MW) provides enough power for approximately 300 homes.

Resistance in Series/Parallel Circuits. A series/parallel circuit may contain any number of individual resistors (loads) connected in any number of different series/parallel circuit combinations. A series/parallel combination is always equal to one combined total resistance value. The total resistance in a circuit containing series/parallel connected resistors equals the sum of the series loads and the equivalent resistance of the parallel combinations. To calculate total resistance in a series/parallel circuit that contains two resistors in series connected to two resistors in parallel, apply the following formula:

$$R_T = \left(\frac{R_{P1} \times R_{P2}}{R_{P1} + R_{P2}} \right) + R_{S1} + R_{S2}$$

where

R_T = total resistance (in Ω)

R_{P1} = parallel resistance 1 (in Ω)

R_{P2} = parallel resistance 2 (in Ω)

R_{S1} = series resistance 1 (in Ω)

R_{S2} = series resistance 2 (in Ω)

For example, what is the total resistance of a 150 Ω and 50 Ω resistor connected in parallel with a 25 Ω and 100 Ω resistor connected in series?

$$R_T = \left(\frac{R_{P1} \times R_{P2}}{R_{P1} + R_{P2}} \right) + R_{S1} + R_{S2}$$

$$R_T = \left(\frac{150 \times 50}{150 + 50} \right) + 25 + 100$$

$$R_T = \left(\frac{7500}{200} \right) + 125$$

$$R_T = 37.5 + 125$$

$$R_T = \mathbf{162.5\ \Omega}$$

Current in Series/Parallel Circuits. The total current in a series/parallel circuit and the current in individual parts of the circuit follow the same laws of current as in a series or parallel circuit. Current is the same in each series part of the series/parallel circuit, and current is also equal to the sum of each parallel combination in each parallel part of the series/parallel circuit.

Voltage in Series/Parallel Circuits. The total voltage applied across resistors (loads) connected in a series/parallel circuit is divided across the individual resistors (loads). The higher the resistance of any one resistor or equivalent parallel resistance, the higher the voltage drop. Likewise, the lower the resistance of any one resistor or equivalent parallel resistance, the lower the voltage drop.

MAGNETISM

A *magnet* is a device that attracts iron. *Magnetism* is a force that interacts with other magnets and ferromagnetic materials. A *ferromagnetic material* is a material, such as soft iron or steel, that is easily magnetized. *Magnetic flux lines* are the invisible lines of force that make up a magnetic field. The more dense the flux lines, the stronger the magnetic force. Flux is most dense at the ends of a magnet. For this reason, the magnetic force is strongest at the ends of a magnet. **See Figure 14-17.**

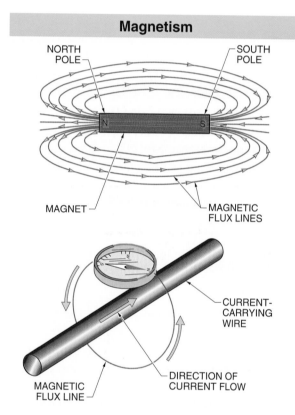

Magnetism

Figure 14-17. Magnetic flux lines are the invisible lines of force that make up a magnetic field.

All magnets and magnetic fields have a north pole and a south pole. Flux lines flow from north poles to south poles. The fundamental law of magnetism is that unlike poles attract each other and like poles repel each other. For example, if two magnetic forces are aligned, such as the north poles of two magnets, then the poles would repel each other. The same would be true if the south poles of two magnets were likewise aligned. The force of repulsion would increase if the two poles were moved closer together. But if a north pole of one magnet and the south pole of another magnet were aligned, they would attract each other.

Magnetism can also be a product of an electric current due to each electron within the current having its own line of magnetic force. Therefore, as current flows through a straight wire, each electron, grouped with other electrons, sets up a circular field of magnetism that surrounds the wire. The flux lines circling the wire are magnetic, which sets up magnetic fields along the length of the wire, creating an electromagnet. An *electromagnet* is a magnet created when electricity passes through a wire.

Induction is the process of causing electrons to align or uniformly join to create a magnetic or electrical force. Placing an iron bar in a strong magnetic field causes induction. Electrons join and the lines of force in the field combine to pass through the iron bar, causing the bar to become a magnet.

A coiled wire with an electric current passing through it has a magnetic field set up around the entire coil. Placing an iron core within the coil concentrates the flux lines and transfers the magnetism of the wire to the iron core by induction. The iron core added to a coil of wire increases the strength of the magnetic field. The iron core remains magnetized as long as current continues to flow through the coiled wire and ceases when the current is turned OFF. **See Figure 14-18.**

Electromagnetism Principles

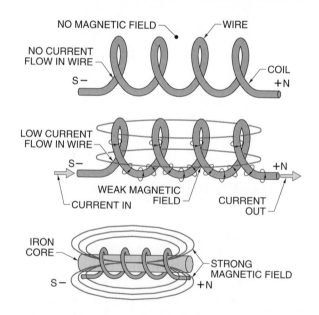

Figure 14-18. The magnetism created as electric current flows through a coil of wire can be strengthened by placing an iron core within the coil.

Electromagnetic Induction

Electromagnetic induction is the process by which voltage is induced in a wire when lines of force from a magnetic field move across the wire. This can be done by moving a wire through a magnetic field. **See Figure 14-19.** Increasing the speed of the wire movement through the magnetic field increases the voltage.

A *generator* is a machine that converts mechanical energy into electrical energy. A generator consists of a loop of wire, or an armature, which rotates between north and south magnetic poles. An *armature* is the movable part of a generator or motor. A flow of current is induced as the loop of wire cuts through the magnetic field between the poles. The direction of current flow is determined by the magnetic north pole and the direction of the wire movement through the magnetic field.

Alternating Current Generation. AC is generated, as a loop of wire (an armature) enters and leaves a magnetic field. **See Figure 14-20.** In position 1, the rotor is about to rotate in the clockwise direction. There is no current flow at this point because the rotor is not cutting any magnetic flux lines. As the rotor rotates from position 1 to position 2, the rotor begins to cut across the magnetic flux lines. The voltage in segments AB and CD increases as the rotor rotates. The maximum number of magnetic flux lines are cut when the rotor is in position 2. The induced voltage is greatest in this position.

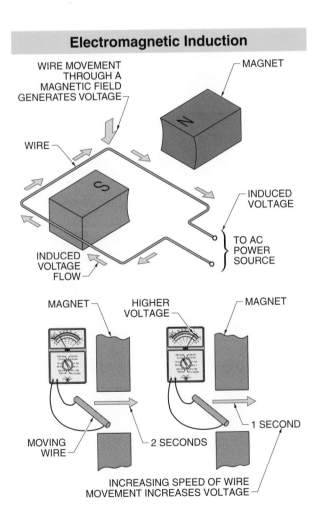

Figure 14-19. A voltage can be induced in a wire by electromagnetic induction when the wire is moved through a magnetic field.

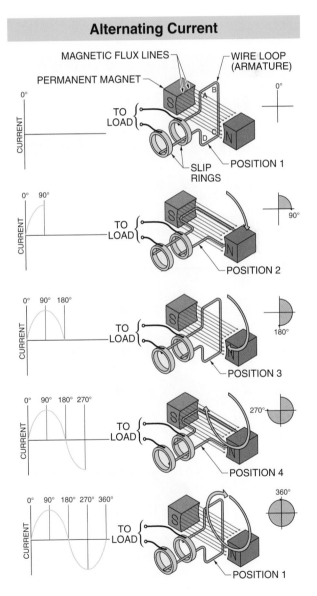

Figure 14-20. Voltage changes direction in a wire as it alternately passes through north and south magnetic poles.

From position 2 to position 3, the voltage decreases to zero because the rotor is cutting across fewer and fewer magnetic flux lines, and finally none at position 3. As the rotor continues to rotate from position 3 to position 4, the voltage increases but in the opposite direction. The voltage reaches a maximum negative value at position 4, and then returns to zero at position 1. Alternations of current (reversing polarity) continue as long as the loop rotates. Slip rings are used to connect the external load to the armature without interfering with its rotation.

Direct Current Generation. DC can be generated using electromagnetic induction. DC generators operate on the principle that when a coil of wire is rotated within a magnetic field, a voltage is induced in the coil.

The amount of voltage induced is determined by the rate at which the coil is rotated within the field. DC generators consist of field windings, an armature, a commutator, and brushes. **See Figure 14-21.**

The field windings of a DC generator are magnets used to produce the magnetic field. The armature (coil) rotates between the field windings. A *commutator* is a ring made of insulated segments that keep the armature windings in the correct polarity to interact with the main fields. The commutator is mounted on the same shaft as the armature and rotates with the shaft. A *brush* is the sliding contact that rides against the commutator segments and is used to connect the armature to the external circuit.

Direct Current Generation

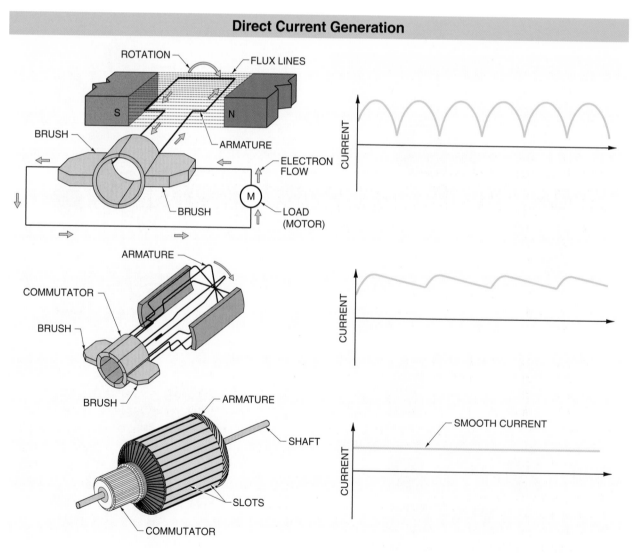

Figure 14-21. In a DC generator, the more armature windings added to the construction of an armature, the smoother the DC voltage output.

In DC generators, the armature consists of wires wound around a core of laminated iron sections. Each armature wire consists of two ends, with each end connected to one segment of the commutator. Each segment on the commutator is separated by an insulating material. As the armature rotates, voltage is induced in coil and sent through the armature to the commutator. The brushes are forced against the commutator as it rotates, which allows for current to flow into the connected electrical circuit. The resulting output voltage is pulsating (ripples) DC voltage. The more armature windings added to the construction of an armature, the smoother the DC voltage output.

The output voltage can be determined by the rotational speed of the armature. The amount of voltage is consistent if the number of rotations used to make the magnetic field remains constant. Increasing the speed of the armature increases the output voltage. Decreasing the rotational speed decreases the output voltage. DC voltages of 6 V, 12 V, 24 V, 36 V, 125 V, 250 V, and 600 V are typically used to drive loads.

Power Distribution

Power distribution is the process of delivering electrical power to where it is needed. Power distribution includes all parts of an electrical utility system from the power generating plant to the customer's service-entrance equipment. Power control, protection, transformation, and regulation must take place before any power is delivered. **See Figure 14-22.** The distribution system includes the following:

- Step-up transformers—The generated voltage is stepped up to a transmission voltage level. The transmission voltage level is usually between 12.47 kV and 245 kV.
- Power plant transmission lines—The 12.47 kV to 245 kV power plant transmission lines deliver power to the transmission substations.
- Transmission substations—The transmission substations transform the voltage to a lower primary (feeder) voltage. The primary voltage level is usually between 4.16 kV and 34.5 kV.
- Primary transmission lines—The 4.16 kV to 34.5 kV primary transmission lines deliver power to the distribution substations and heavy industry.
- Distribution substations—The distribution substations transform the voltage down to utilization voltages. Utilization voltage levels range from 480 V to 4.16 kV.

- Distribution lines—Distribution lines carry the power from the distribution substations along streets or rear-lot lines to the final step-down transformers.
- Final step-down transformers—The final step-down transformers transform the voltage to 480 V. The final step-down transformers may be installed on poles, grade-level pads, or in underground vaults. The secondary of the final step-down transformers is connected to service drops that deliver the power to the customer's service-entrance equipment.

A generator is commonly used as an emergency power supply when the primary power supply fails.

A typical power distribution system delivers power to industrial, commercial, and residential customers. In a power distribution system, standard voltage levels at fixed current ratings are delivered to set points, such as receptacles. These voltage levels are typically 110 V, 115 V, 120 V, 208 V, 220 V, 240 V, 277 V, 430 V, 440 V, 460 V, and 480 V. However, there is no such thing as a standard voltage or current level in an electrical circuit, because these levels are continuously being changed to meet circuit requirements. In a typical heavy industrial facility, the electricity is delivered directly from a transmission substation to an outside transformer vault.

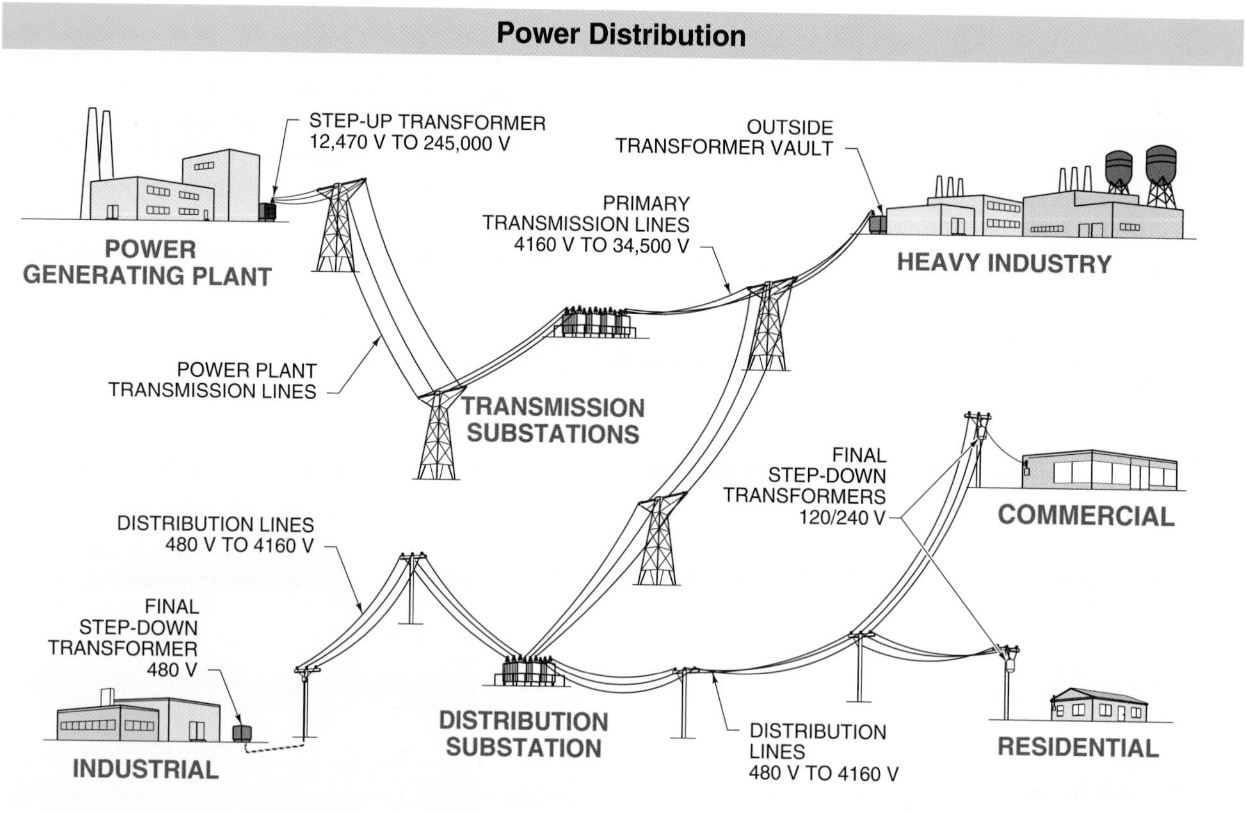

Power Distribution

Figure 14-22. High-voltage electricity from power plants is transmitted through electrical power lines and substations.

NATIONAL ELECTRICAL CODE®

The *National Electrical Code®* (NEC®) is published by the National Fire Protection Association. The purpose of the NEC® is the practical safeguarding of persons and property from hazards arising from the use of electricity. The NEC® is updated on a three-year cycle. It is adopted by governmental bodies that have legal jurisdiction over electrical installations and for use by insurance inspectors. The authority having jurisdiction (AHJ) is responsible for enforcing the NEC®. **See Figure 14-23.**

The NEC's® scope of coverage includes the following:
- Electrical conductors and equipment in structures, mobile homes, RVs, floating buildings, yards, carnivals, parking lots, industrial substations, and public and private buildings
- Installations of conductors and equipment connected to the electrical supply
- Installations of other outside conductors and equipment
- Installations of fiber-optic cable and raceways
- Installations in buildings used by an electric utility that are not part of the generating plant, substation, or control center

The NEC® does not cover the following:
- Ships, trains, aircraft, or automotive vehicles other than mobile homes and RVs
- Installations in mines
- Installations of communication equipment controlled by communications utilities and located outdoors
- Installations controlled by electric utilities for communications, metering, generation, control, transformation, transmission, or distribution of electrical energy

ELECTRICAL SAFETY

Improper electrical wiring or misuse of electricity can cause destruction of equipment, fire damage to property, and fatal incidents to personnel. Safe working habits are required when operating on an electrical circuit or component because the electric parts that are normally enclosed are exposed. Electrical safety standards should be practiced by all personnel working with electrical systems.

Figure 14-23. The authority having jurisdiction (AHJ) is responsible for enforcing the NEC®.

Electrical Shock

An *electrical shock* is a shock that results any time a body becomes part of an electrical circuit. Electrical shock effects range from a mild sensation to paralysis to death. Also, severe burns may occur where current enters and exits the body. The severity of an electrical shock depends on the amount of electric current in milliamps (mA) that flows through the body, the length of time the body is exposed to the current flow, the path the current takes through the body, and the physical size and condition of the body through which the current passes. **See Figure 14-24.**

When handling a victim of an electrical shock accident, apply the following procedures:

1. Break the circuit to free the person immediately and safely. Never touch any part of a victim's body when the victim is in contact with the circuit. When the circuit cannot be turned OFF, use any nonconducting device to free the person. Resist the temptation to touch the person when power is not turned OFF.
2. After the person is free from the circuit, send for help and determine if the person is breathing.

3. When there is no breathing or pulse, start CPR if trained to do so. Always get medical attention for a victim of electrical shock.
4. When the person is breathing and has a pulse, check for burns and cuts. Burns are caused by contact with a live circuit and can be found at the points where the electricity entered and exited the body. Treat the entrance and exit burns as thermal burns and get medical help immediately.

Electrical Shock Effects

Approximate Current*	Effect on Body†
Over 20	Causes severe muscular contractions, paralysis of breathing, heart convulsions
15–20	Painful shock; may be frozen or locked to point of electrical contact until circuit is de-energized
8–15	Painful shock; removal from contact point by natural reflexes
8 or less	Sensation of shock but probably not painful

100 mA — CURRENT IN 100 W LAMP CAN ELECTROCUTE 20 ADULTS
CURRENT
50 mA — HEART CONVULSIONS, USUALLY FATAL
15 mA–20 mA — PAINFUL SHOCK, INABILITY TO LET GO
0 mA–5 mA — SAFE VALUES
1mA
0 mA — NO SENSATION

* in mA
† effects vary depending on time, path, amount of exposure, and condition of body

Figure 14-24. Electrical shock is a condition that results any time a body becomes part of an electrical circuit.

The NFPA 70E® identifies three approach boundaries to protect against electrical shock. These boundaries are the limited approach boundary, the restricted approach boundary, and the prohibited approach boundary. Each boundary is viewed as a sphere, extending 360° around an exposed energized conductor or circuit part. The size of each boundary is based on the phase-to-phase nominal voltage of the energized conductor or circuit part. **See Figure 14-25.**

Approach Boundaries to Energized Parts for Shock Prevention				
Nominal System (Voltage, Range, Phase to Phase*)	Limited Approach Boundary		Restricted Approach Boundary (Allowing for Accidental Movement)	Prohibited Approach Boundary
	Exposed Movable Conductor	Exposed Fixed-Circuit Part		
less than 50	N/A	N/A	N/A	N/A
50 to 300	10'-0"	3'-6"	Avoid contact	Avoid contact
301 to 750	10'-0"	3'-6"	1'-0"	0'-1"
751 to 15,000	10'-0"	5'-0"	2'-2"	0'-7"

* in V

Figure 14-25. The limited approach, restricted approach, and prohibited approach boundaries protect personnel from electrical shock.

The *limited approach boundary* is the distance from an exposed energized conductor or circuit part at which a person can get an electric shock and is the closest distance an unqualified person can approach. The *restricted approach boundary* is the distance from an exposed energized conductor or circuit part where an increased risk of electric shock exists due to the close proximity of the person to the energized conductor or circuit part. The *prohibited approach boundary* is the distance from an exposed energized conductor or circuit part inside which any work performed is considered the same as making contact with the energized conductor or circuit part.

Only an authorized individual (qualified person) can cross the limited approach boundary. An *authorized individual* is a knowledgeable individual to whom the authority and responsibility to perform a specific task has been given. The authorized individual must wear appropriate PPE for shock and arc flash hazards. If unauthorized individuals are working near the limited approach boundary, they must be warned of the hazards and remain outside the limited approach boundary.

The amount of current that passes through a circuit depends on the voltage and resistance of the circuit. During an electrical shock, a person's body becomes part of the circuit. The resistance a person's body offers to the flow of current varies. Sweaty hands have less resistance than dry hands. A wet floor has less resistance than a dry floor. The lower the resistance, the greater the current flow. The greater the current flow, the greater the severity of shock.

Safety Tip

Per OSHA regulations, it is an employer's duty to furnish each employee with a place of employment that is free from recognized hazards that cause or are likely to cause death or serious physical harm.

Arc Flash and Arc Blast

An *electric arc* is a discharge of electric current across an air gap. Arcs are caused by excessive voltage ionizing an air gap between two conductors, or by accidental contact between two conductors and followed by reseparation. When an electric arc occurs, there is the possibility of an arc flash or an arc blast. An *arc flash* is an extremely high-temperature discharge produced by an electrical fault in the air. Arc flash temperatures reach 35,000°F. An *arc blast* is an explosion that occurs when the air surrounding electrical equipment becomes ionized and conductive. The threat of arc blast is greatest from electrical systems of 480 V and higher. Arc blasts are possible in systems of lesser voltage but are not likely to be as destructive as in a high-voltage system.

Tech Tip

There are no equations available to accurately calculate safe approach distance for an arc blast hazard. Developing these equations are goals of the IEEE/NFPA Joint Arc Flash Research Program.

Arc flash and arc blast are always a possibility when working with electrical equipment. A potential cause for arc flash and arc blast is improper test instrument and meter use. For example, an arc blast can occur by connecting an ammeter across two points of a circuit that is energized with a voltage higher than the rating of the meter. To prevent causing arc blast or arc flash, an electrical system needs to be de-energized, locked out, and tagged out prior to performing work. Only qualified electricians are allowed to work on energized circuits of 50 V or higher.

The *arc flash boundary* is the distance from an exposed energized conductor or circuit part where bare skin would receive the onset of a second-degree burn. **See Figure 14-26.** The arc flash boundary is dependent on the available short-circuit current, maximum total clearing time of the overcurrent protection device, the voltage of the circuit, and a standard factor that varies if the actual short-circuit current is known or not known. The arc flash boundary can be calculated using the equations in NFPA 70E Informative Annex D.

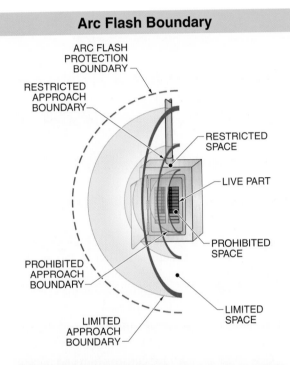

Arc Flash Boundary

Working Voltage*	Limited Approach Boundary Distance†
Up to 750	3
750 to 2000	4
2000 to 15,000	16
15,000 to 36,000	19
Over 36,000	Must wear protection

* in V
† in ft

Figure 14-26. An arc flash boundary is the distance from an exposed energized conductor or circuit part where bare skin would receive the onset of a second-degree burn.

Grounding

Electrical circuits are grounded to safeguard equipment and personnel against the hazards of electrical shock. Proper grounding of electrical tools, machines, equipment, and transmission systems is one of the most important factors in preventing hazardous conditions.

Grounding is the connection of all exposed non-current-carrying metal parts to the earth. Grounding provides a direct path for unwanted (fault) current to the earth without causing harm to persons or equipment. Grounding is accomplished by connecting the circuit to non-current-carrying metal parts, such as a metal underground pipe, a metal frame of a building, a concrete-encased electrode, or a ground ring. **See Figure 14-27.**

An unwanted current may exist because of insulation failure or if a current-carrying conductor makes contact with a non-current-carrying part of the system. In a properly grounded system, the unwanted current flow blows fuses or trips circuit breakers. Once the fuse is blown or circuit breaker is tripped, the circuit is open and no additional current flows.

Personal Protective Equipment

The use of personal protective equipment is required whenever work may occur on or near energized exposed electrical circuits. The National Fire Protection Association standard *NFPA 70E, Standard for Electrical Safety in the Workplace* addresses "electrical safety requirements for employee workplaces that are necessary for the safeguarding of employees in pursuit of gainful employment."

For maximum safety, personal protective equipment and safety requirements must be followed as specified in NFPA 70E®; OSHA Standard Part 1910 *Subpart 1—Personal Protective Equipment* (1910.132 to 1910.138); and other applicable safety mandates. Personal protective equipment includes arc-rated clothing, head protection, eye protection, ear protection, and hand protection. **See Figure 14-28.**

Arc-Rated Clothing. Sparks from an electrical circuit can cause a fire. Approved arc-rated clothing must be worn for protection from electrical arcs when performing certain operations on or near energized equipment or circuits. Arc-rated clothing must be kept clean and sanitary, and must be inspected prior to each use. Defective clothing must be removed from service immediately and replaced. Defective arc-rated clothing must be tagged "unsafe" and returned to a supervisor.

Head Protection. Head protection requires using a protective helmet. A *protective helmet* is a hard hat that is used in the workplace to prevent injury from the impact of falling and flying objects, and from electrical shock. Protective helmets resist penetration and absorb impact force.

Grounding Methods

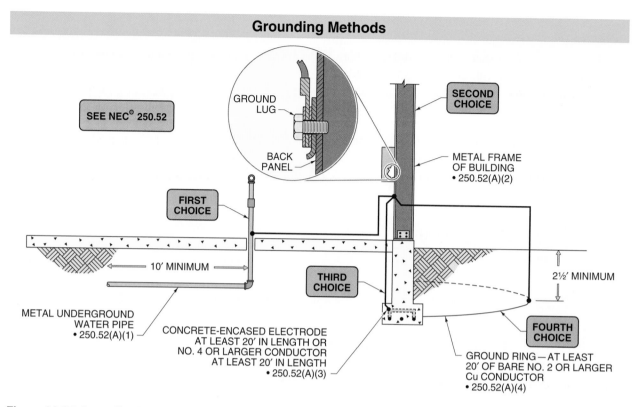

Figure 14-27. Grounding is accomplished by connecting a circuit to a metal underground pipe, a metal frame of a building, a concrete-encased electrode, or a ground ring.

Personal Protective Equipment

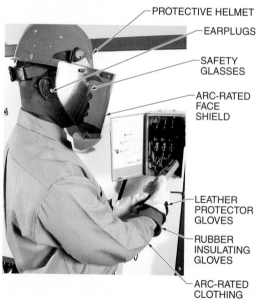

- PROTECTIVE HELMET
- EARPLUGS
- SAFETY GLASSES
- ARC-RATED FACE SHIELD
- LEATHER PROTECTOR GLOVES
- RUBBER INSULATING GLOVES
- ARC-RATED CLOTHING

Figure 14-28. The use of personal protective equipment is required whenever work occurs on or near energized exposed electrical circuits.

Protective helmet shells are made of durable, lightweight materials. A shock-absorbing lining consists of crown straps and a headband that keeps the shell away from the head to provide ventilation.

Protective helmets are identified by class of protection against specific hazardous conditions. Class A, B, and C helmets are used for construction and industrial applications. Class A protective helmets protect against low-voltage shocks and burns and impact hazards and are commonly used in construction and manufacturing facilities. Class B protective helmets protect against high-voltage shock and burns, impact hazards, and penetration by falling or flying objects. Class C protective helmets are manufactured with lighter materials yet provide adequate impact protection. **See Figure 14-29.**

Eye Protection. Eye protection must be worn to prevent eye or face injuries caused by flying particles, contact arcing, and radiant energy. Eye protection must comply with OSHA 29 CFR 1910.133, *Eye and Face Protection*. Eye protection standards are specified in ANSI Z87.1, *Occupational and Educational Personal Eye and Face Devices*. Eye protection includes safety glasses, arc-rated face shields, and arc-rated hoods. **See Figure 14-30.**

Protective Helmets

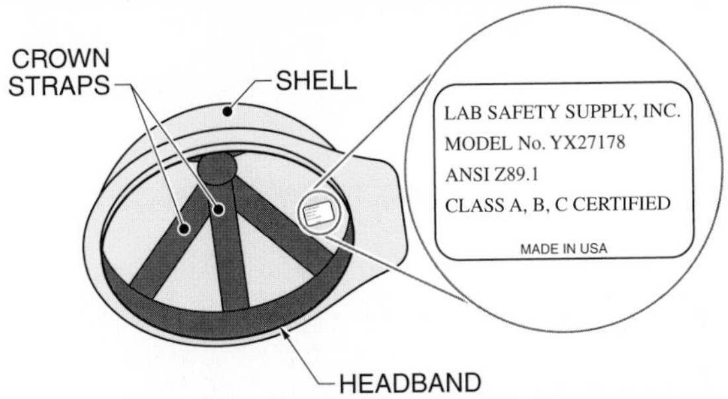

CROWN STRAPS

SHELL

LAB SAFETY SUPPLY, INC.
MODEL No. YX27178
ANSI Z89.1
CLASS A, B, C CERTIFIED

MADE IN USA

HEADBAND

Class	Use
A	General service, low-voltage protection
B	Utility service, high-voltage protection
C	Special service, no voltage protection

Lab Safety Supply, Inc.

Figure 14-29. Protective helmets are identified by class of protection against hazardous conditions as Class A, B, and C.

Safety glasses are an eye protection device with special impact-resistant glass or plastic lenses, a reinforced frame, and side shields. Plastic frames are designed to keep the lenses secured in the frame if an impact occurs and to minimize the shock hazard when working with electrical equipment. Side shields provide additional protection from flying objects. Tinted-lens safety glasses protect against low-voltage arc hazards.

An *arc-rated face shield* is an eye and face protection device that covers the entire face with a plastic shield and is used for protection from flying objects. Tinted face shields protect against low-voltage arc hazards. An *arc-rated hood* is an eye and face protection device that consists of a flame-resistant hood and face shield.

Safety glasses, arc-rated face shields, and arc-rated hoods must be properly maintained to provide protection and clear visibility. Lens cleaners are available that clean without risk of lens damage. Pitted, scratched, and crazed lenses reduce vision and may cause lenses to fail on impact. (Crazing is a defect caused by exposure to aggressive solvents, chemicals, or heat that leaves microscopic cracks within the lenses.)

Safety Tip

Over 3000 people are killed every year in the United States from electrical shock and over 85,000 injuries occur due to failure to properly control hazardous energy sources during maintenance.

Eye Protection

SIDE SHIELDS

IMPACT-RESISTANT LENS

REINFORCED FRAMES

SAFETY GLASSES

ARC-RATED FACE SHIELD

ADJUSTABLE HEADBAND

ARC FACE SHIELD

ARC-RATED HOOD

HOOD FACE SHIELD

ARC-RATED HOOD

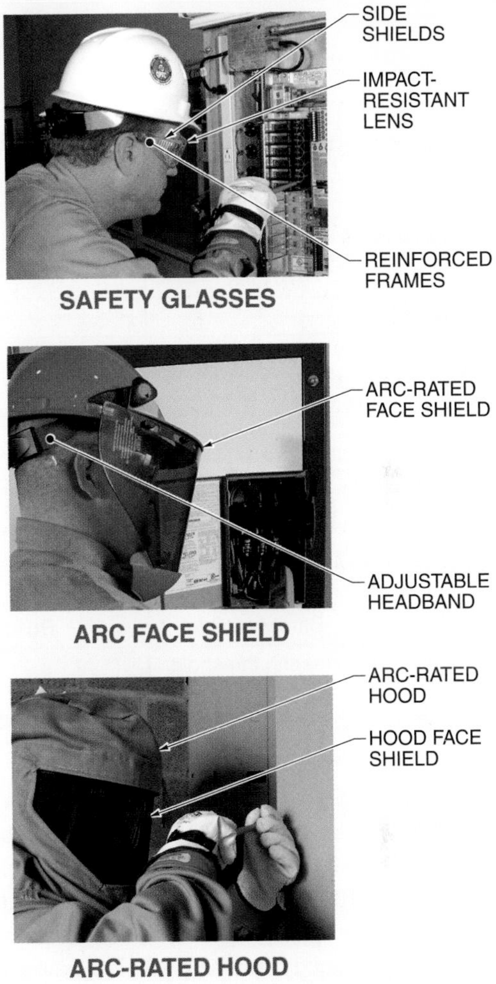

Figure 14-30. Eye protection includes safety glasses, arc-rated face shields, and arc-rated hoods.

Ear Protection. Ear protection devices are worn to limit the noise entering the ear and include earplugs and earmuffs. An earplug is made of moldable rubber, foam, or plastic and inserted into the ear canal. An earmuff is worn over the ears. A tight seal around an earmuff is required for proper protection.

Power tools and equipment can produce excessive noise levels. Technicians subjected to excessive noise levels may develop hearing loss over time. The severity of hearing loss depends on the intensity and duration of exposure. Noise intensity is expressed in decibels. A *decibel (dB)* is a unit of measure used to express the relative intensity of sound. **See Figure 14-31.** Ear protection is worn to prevent hearing loss.

Sound Levels			
Average Decibels (dB)	**Loudness**	**Examples**	**Exposure Duration**
140	Deafening	Jet airplane taking off, air raid siren, locomotive horn	—
130	Pain Threshold		2 min
120	Feeling Threshold		7 min
110	Uncomfortable		30 min
100	Very Loud	Chainsaw	2 hr
90	Noisy	Shouting, auto horn	4 hr
85			8 hr
80	Moderately loud	Vacuum cleaner	25.5 hr
70	Loud	Telephone ringing, loud talking	—
60	Moderate	Normal conversation	—
50	Quiet	Hair dryer	—
40	Moderately Quiet	Refrigerator running	—
30	Very Quiet	Quiet conversation, broadcast studio	—
20	Faint	Whispering	—
10	Barely audible	Rustling leaves, soundproof room, human breathing	—
0	Hearing threshold	Intolerably quiet	—

Figure 14-31. Industrial technicians subjected to excessive noise levels may develop hearing loss over time.

Ear protection devices are assigned a noise reduction rating (NRR) number based on the noise level reduced. For example, an NRR of 27 means that the noise level is reduced by 27 dB when the device is tested at the factory. To determine approximate noise reduction in the field, 7 dB is subtracted from the NRR. For example, an NRR of 27 provides a noise reduction of approximately 20 dB in the field.

Hand Protection. Hand protection consists of rubber insulating gloves and leather protectors worn to prevent injuries to hands caused by cuts or electrical shock. The primary purpose of rubber insulating gloves and leather protectors is to insulate hands and lower arms from possible contact with live conductors. Rubber insulating gloves offer a high resistance to current flow to help prevent an electrical shock, and the leather protectors protect the rubber glove and add additional insulation. Rubber insulating gloves are rated, labeled, and color-coded to indicate the maximum voltage at which the gloves offer adequate protection. **See Figure 14-32.**

WARNING: Rubber insulating gloves are designed for specific applications. Leather protector gloves are required for protecting rubber insulating gloves. Rubber insulating gloves must not be used alone. Serious injury or death can result from improper use of rubber insulating gloves, or from using outdated and/or the wrong type of rubber insulating gloves for an application.

Rubber Insulating Glove Rating*		
Class	**Maximum Use**	**Color Code**
00	500 V	Beige
0	1 kV (1000 V)	Red
1	7.5 kV (7500V)	White
2	17 kV (17,000 V)	Yellow
3	26.5 kV (26,500 V)	Green
4	36 kV (36,000 V)	Orange

* Refer to ASTM D—120-02 Standard Specification for Rubber Insulating Gloves.

Figure 14-32. Rubber insulating gloves are rated, labeled, and color-coded to indicate the maximum voltage at which the gloves offer adequate protection.

Lockout/Tagout

Electrical power must be removed when electrical equipment is inspected, serviced, repaired, or replaced. Power is removed and the equipment locked out and tagged out to ensure the safety of personnel working with the equipment.

Per OSHA standards, equipment must be locked out and tagged out before any preventive maintenance or service is performed. *Lockout* is the process of removing a source of power and installing a lock that prevents the power from being turned ON. *Tagout* is the process of placing a tag on a locked-out power source that indicates that the power may not be restored until the tag is removed.

A danger tag has the same importance and purpose as a lock and is used alone only when a lock does not fit the disconnect device. The danger tag shall be attached at the disconnect device with a tag tie or equivalent and shall have space for the worker's name, craft, and other required information. A danger tag must withstand the elements and expected atmosphere for as long as the tag remains in place. **See Figure 14-33.**

Lockout/Tagout

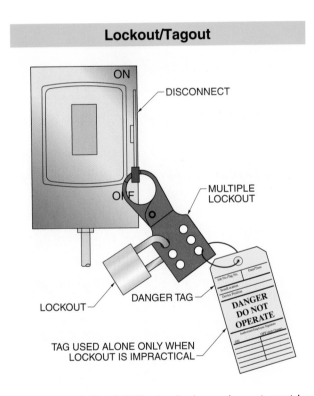

Figure 14-33. Per OSHA standards, equipment must be locked out and tagged out before any preventive maintenance or service is performed.

A lockout/tagout must be used when the following conditions exist:
- When servicing electrical equipment that does not require power to be ON to perform the service
- When removing or bypassing a machine guard or other safety device
- When the possibility exists of being injured or caught in moving machinery
- When clearing jammed equipment
- When the danger exists of being injured if equipment power is turned ON

Lockouts and tagouts do not by themselves remove power from a circuit. An approved procedure is followed when applying a lockout/tagout. Lockouts and tagouts are attached only after the equipment is turned OFF and tested to ensure that power is OFF. The lockout/tagout procedure is required for the safety of workers due to modern equipment hazards. OSHA provides a standard procedure for equipment lockout/tagout, as follows:

1. Prepare for machinery shutdown.
2. Shut down machinery or equipment.
3. Isolate machinery or equipment.
4. Apply lockout and/or tagout.
5. Release stored energy.
6. Verify isolation of machinery or equipment.

WARNING: Personnel should consult OSHA Standard 1910.147 — *The Control of Hazardous Energy (Lockout/Tagout)* for industry standards on lockout/tagout.

A lockout/tagout must not be removed by any person other than the person who installed it, except in an emergency. In an emergency, the lockout/tagout may be removed only by authorized personnel. The authorized personnel must follow approved procedures. A list of company rules and procedures are given to any person who may use a lockout/tagout.

When more than one electrician is required to perform a task on a piece of equipment, each electrician shall place a lockout/tagout on the energy isolating device(s). A multiple-lockout/tagout device, or hasp, must be used because energy-isolating devices typically cannot accept more than one lockout/tagout at one time. A *hasp* is a multiple-lockout/tagout device.

Lockout Devices. A *lockout device* is a lightweight enclosure that allows the lockout of a standard control device. Lockout devices are available in various shapes and sizes that allow for the lockout of ball valves, gate valves, and electrical plugs.

Lockout devices resist chemicals, cracking, abrasion, temperature changes, and are available in colors to match ANSI pipe colors. Lockout devices are sized to fit standard industry control-device sizes. **See Figure 14-34.**

Locks used to lock out a device may be color-coded and individually keyed. The locks are rust resistant and available with various-size shackles.

Lockout Devices

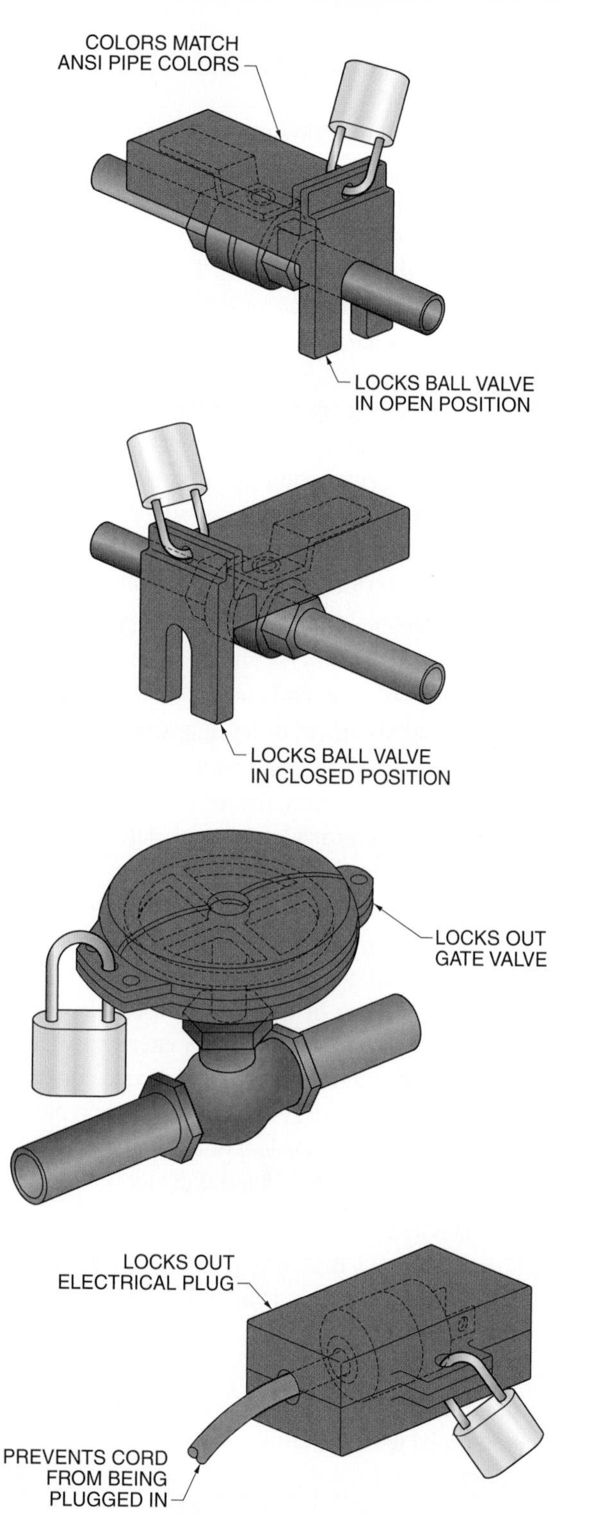

COLORS MATCH
ANSI PIPE COLORS

LOCKS BALL VALVE
IN OPEN POSITION

LOCKS BALL VALVE
IN CLOSED POSITION

LOCKS OUT
GATE VALVE

LOCKS OUT
ELECTRICAL PLUG

PREVENTS CORD
FROM BEING
PLUGGED IN

Figure 14-34. Lockout devices are sized to fit standard industry control-device sizes.

Danger tags provide additional lockout and warning information. Various danger tags are available. Danger tags may include warnings such as "Do Not Start" or "Do Not Operate," or may provide space to enter information such as the worker's name, the date, and the reason for the lockout. Tag ties must be strong enough to prevent accidental removal and must be self-locking and nonreusable.

Lockout/tagout kits are also available. A lockout/tagout kit contains items required to comply with OSHA lockout/tagout standards. Lockout/tagout kits contain reusable danger tags, tag ties, multiple-lockout devices, locks, magnetic signs, and information on lockout/tagout procedures. **See Figure 14-35.**

Lockout/Tagout Kits

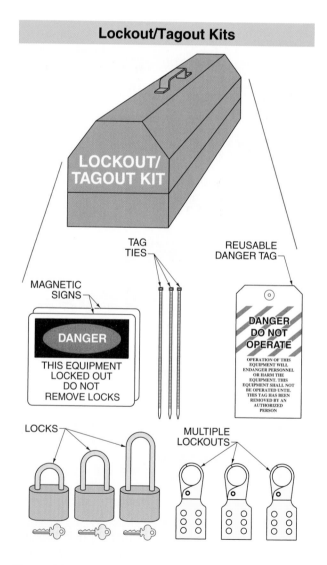

Figure 14-35. A lockout/tagout kit contains items required to comply with OSHA lockout/tagout standards.

Review

1. What are the three fundamental particles contained in an atom?

2. What is the law of charges?

3. What is the difference between a conductor and an insulator?

4. Define electron flow as well as the two conventions used to describe it.

5. Define voltage, current, and resistance and list the unit of measure for each.

6. What kind of electrical device converts AC voltage to DC voltage?

7. Conductors are insulated and rated according to what system?

8. Define power and describe the power formula.

9. List the three types of electric circuit connections.

10. Why are magnetic forces strongest at the ends of a magnet?

11. Define alternating current and explain how it is generated.

12. Define power distribution and list the steps in the distribution system.

13. What is the purpose of the National Electrical Code® (NEC®) and how often is it updated?

14. Describe the procedure when handling a victim of an electrical shock accident.

15. List the three types of approach boundaries.

Digital Resources

ATPeResources.com/Quicklinks
Access Code 462409

CHAPTER **15**

ELECTRICAL APPLICATIONS

Electrical systems are integral to the operation of an industrial facility. Mechanical technicians frequently encounter electrical circuits and devices integrated with other systems as well, such as fluid power and HVAC systems. Basic operational knowledge of common electrical devices and troubleshooting skills for electrical circuits is critical. Mechanical technicians should know how to use common electrical test instruments for testing and verification.

Baldor Electric Company

Chapter Objectives

- List the common electrical test tools and their basic functions and uses.

- Identify common electrical devices and their operating principles and procedures.

- Describe the different types of coil-operated devices.

- Compare three-phase motors and single-phase motors and their troubleshooting procedures.

- Describe variable-frequency drives and their function.

- Compare line diagrams and wiring diagrams.

Key Terms

- continuity
- continuity tester
- voltage tester
- multimeter
- transformer
- heating element
- switch
- fuse
- circuit breaker
- ground fault
- ground-fault circuit interrupter (GFCI)
- coil
- armature
- solenoid
- relay
- contact
- contactor
- motor starter
- holding contact
- induction motor
- stator
- rotor
- split-phase motor
- starting winding
- running winding
- capacitor-start motor
- capacitor
- capacitor start-and-run motor
- variable-frequency drive
- inverter
- line diagram
- wiring diagram

ELECTRICAL MEASURING DEVICES

A mechanical technician must often determine the electrical condition of a component within a system. This can involve determining if the component is receiving an electric current when required, if the conductors are able to carry the current, and if the component operates when electricity is present.

Continuity Testers

Continuity is the condition of a circuit that is closed (allows current flow). Continuity can be determined with a continuity tester or a multimeter. A *continuity tester* is a device that indicates if a circuit is open or closed. **See Figure 15-1.** An *open circuit* is an electrical circuit that has a gap or opening that does not allow current flow. A break in a conductor or an open switch causes an open circuit, preventing the flow of electricity. A *closed circuit* is an electrical circuit with a complete path that allows current to flow.

Continuity Tester Use. A continuity tester must be used only in an inoperative circuit with the power OFF. The tester supplies its own voltage and current by the use of batteries. The circuit is complete and the continuity tester lights when the tester leads are touching both ends of a continuous (unbroken) wire or circuit. The wire may be broken or disconnected if the tester does not light.

Continuity Testers

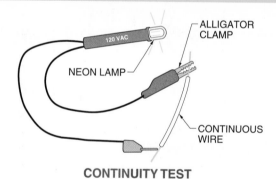

CONTINUITY TEST

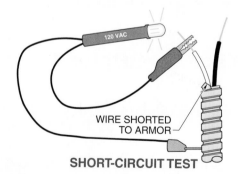

SHORT-CIRCUIT TEST

Figure 15-1. A continuity tester is a test instrument with a small light that indicates a continuous, unbroken path for current to flow.

Continuity testers are also used to check for short circuits between component wiring and the body of a component. This is accomplished by touching one tester lead to one of the wire ends and the other tester lead to the component body. A short-circuited component, meaning a bare spot in the wire insulation, allows the wire to come in contact with the component body, giving an indication of a short from the tester.

Voltage Testers

A *voltage tester* is a device that indicates approximate voltage level and type (AC or DC) by the movement of a pointer on a scale. **See Figure 15-2.** Basic voltage testers do not have a meter or digital display. A voltage tester should not be used on low-power circuits because it draws high current, and no voltage indication is given on low-power circuits. The range for most voltage testers is between 90 V and 600 V.

Voltage Testers

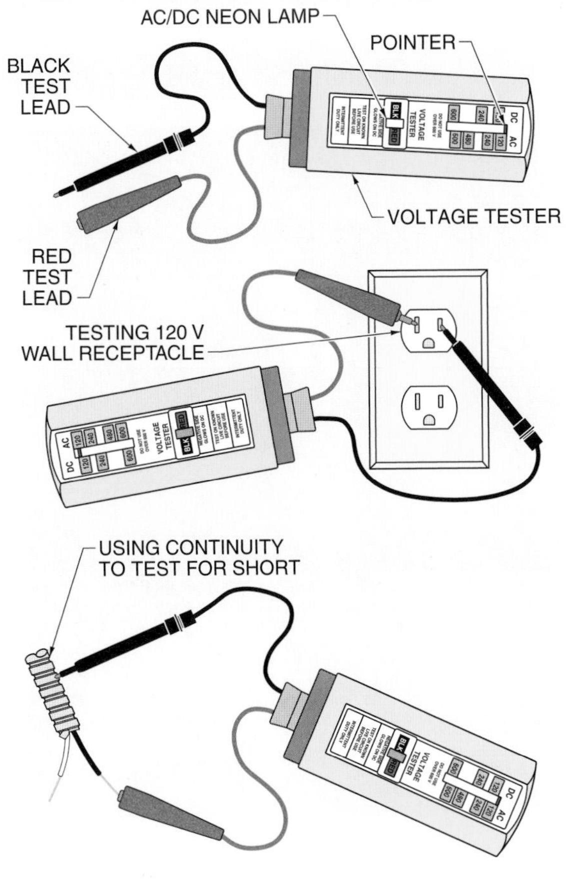

AC/DC NEON LAMP
POINTER
BLACK TEST LEAD
RED TEST LEAD
VOLTAGE TESTER

TESTING 120 V WALL RECEPTACLE

USING CONTINUITY TO TEST FOR SHORT

Figure 15-2. A voltage tester indicates the presence of AC or DC voltage and its approximate level.

Most voltage testers contain a solenoid that vibrates when connected to AC and does not vibrate when connected to DC, though the pointer indicates the tested voltage. Most voltage testers also include neon lights that determine if the voltage is AC or DC. Voltage testers are designed for intermittent use and should not be connected to a power supply for more than 15 sec. Simple voltage testers that are compact and rugged enough for mechanical technicians generally have the capability of being a voltage tester and a continuity tester.

Voltage Tester Use. The procedure for using a voltage tester is to connect one test probe to one side of a circuit and then connect the other test probe to the other side of the circuit. A reading is taken from the pointer. The white wire, when present, should be checked first to determine that it is the neutral conductor. A *neutral conductor* is a wire that carries current from one side of the load to ground. Connecting a voltage tester between the neutral wire and a grounded surface should give zero voltage. All other hot wires are then checked to ground or neutral and should read the supply voltage level. Caution must be taken when testing any voltage over 24 V.

WARNING: Ensure no part of the body contacts any part of a live circuit, including the metal contact points at the tip of a tester. Always assume that circuits are powered (hot) until they have been checked.

Multimeters

A *multimeter* is a portable test tool that is capable of measuring two or more electrical quantities. Multimeters may be analog or digital. **See Figure 15-3.** An analog multimeter indicates readings by the mechanical motion of a pointer. The value is read from right to left on the scale. Extremely high readings on the scale are to the far left and are known as overload (OL) readings. A digital multimeter (DMM) indicates readings as numerical values. Digital multimeters help reduce the chance of user error when taking readings by displaying exact values measured.

Multimeters have a function switch that enables the testing of various electrical quantities. Most multimeters can measure current (AC and DC), voltage (AC and DC), and resistance. Some multimeters can also measure capacitance, frequency, temperature, and other quantities. Attachments can be used with a digital multimeter to extend its measurement capabilities to current (with a clamp), temperature, pressure, light intensity, gas concentration, and other quantities. **See Figure 15-4.**

Multimeters

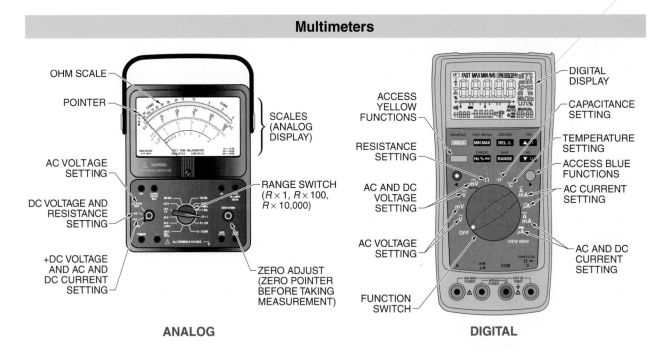

OHM SCALE

POINTER

SCALES (ANALOG DISPLAY)

AC VOLTAGE SETTING

DC VOLTAGE AND RESISTANCE SETTING

RANGE SWITCH ($R \times 1$, $R \times 100$, $R \times 10,000$)

+DC VOLTAGE AND AC AND DC CURRENT SETTING

ZERO ADJUST (ZERO POINTER BEFORE TAKING MEASUREMENT)

ANALOG

ACCESS YELLOW FUNCTIONS

RESISTANCE SETTING

AC AND DC VOLTAGE SETTING

AC VOLTAGE SETTING

FUNCTION SWITCH

DIGITAL DISPLAY

CAPACITANCE SETTING

TEMPERATURE SETTING

ACCESS BLUE FUNCTIONS

AC CURRENT SETTING

AC AND DC CURRENT SETTING

DIGITAL

Figure 15-3. A multimeter, either analog or digital, is capable of measuring multiple electrical quantities. Digital multimeters are particularly versatile because they can typically measure many different quantities and have additional supporting features.

Most digital multimeters also have special functions to make taking measurements easier, safer, and less prone to error. A meter with auto-ranging automatically chooses the most appropriate range in which to take the reading. The HOLD function records the currently displayed reading. This is used when technicians cannot see the display while taking a measurement. The MIN/MAX function records the minimum, maximum, and average values since the beginning of a measurement. The RELATIVE function establishes a baseline value and displays the current measurement as a positive or negative difference from that baseline. All of the functions can be used for any of the measured electrical quantities and for measurements from attachments. Also, these functions can be used in combination with each other.

Multimeters do not require an external power source because they use a battery to supply their own current and voltage. Some connections to live circuits can potentially damage a multimeter and its battery, though most are protected from incorrect use and damage by fuses in their circuitry. Technicians must be certain that a multimeter's batteries are in good condition before use.

The function and range switches must be set at the correct quantity and scale. Care must be taken to ensure that a multimeter is connected to a circuit correctly. Incorrect settings can make a measurement meaningless or unsafe.

Digital Multimeter Attachments

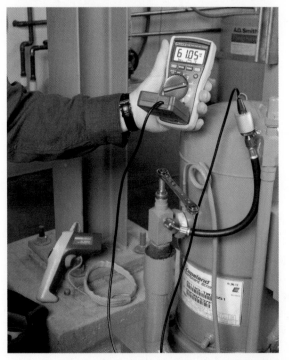

Fluke Corporation

Figure 15-4. Digital multimeter attachments are used to provide measurement capabilities beyond electrical quantities.

Measuring Resistance. Resistance is measured either by first checking that all power is OFF in the circuit being tested or by removing the component from the circuit. The black test lead is plugged into the negative jack (–), and the red test lead is plugged into the positive jack (+). Then the meter leads are touched across the component, and the resistance is displayed. **See Figure 15-5.** The meter should be turned OFF after measurements are taken to save battery life.

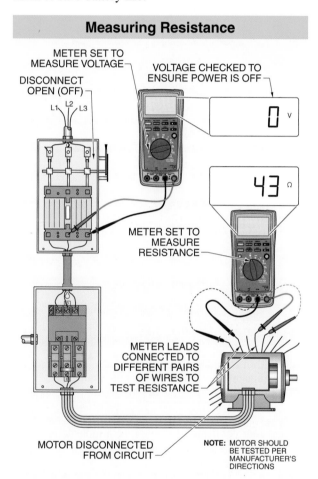

Measuring Resistance

METER SET TO MEASURE VOLTAGE

DISCONNECT OPEN (OFF)

L1 L2 L3

VOLTAGE CHECKED TO ENSURE POWER IS OFF

0 V

43 Ω

METER SET TO MEASURE RESISTANCE

METER LEADS CONNECTED TO DIFFERENT PAIRS OF WIRES TO TEST RESISTANCE

MOTOR DISCONNECTED FROM CIRCUIT

NOTE: MOTOR SHOULD BE TESTED PER MANUFACTURER'S DIRECTIONS

Figure 15-5. Resistance measurements are taken with the component removed from the circuit or the circuit de-energized.

Measuring Voltage. Voltage measurements are taken with the leads connected in parallel with the component or section of the circuit being tested. The circuit must be energized to be tested. **See Figure 15-6.** If multiple voltage measuring functions are available (such as VAC, VDC, or mV), the correct one must be selected. Then, the black test lead is plugged into the negative jack (–), and the red test lead is plugged into the positive jack (+). The best practice is to first test a known line circuit, ensuring that the multimeter is working properly.

Then, the test leads are touched to the parts of the circuit being tested and the measurement read. When testing many parts of a circuit, it is common practice, and safer, to clip the negative test lead onto a neutral conductor and move only the positive test lead to touch various points. This practice also helps trace voltage drop across components. When testing is complete, the multimeter is verified again by measuring the known live circuit. (This verify, test, and verify procedure is known as the three-point test method.)

Measuring Current. There are two types of current measurements—those taken with a clamp-on probe, or open fork-style probe, and those taken in-line. **See Figure 15-7.** Meters with a clamp-on or fork-style probe measure current in a closed circuit by measuring the strength of the magnetic field around a single conductor. These probes do not need to touch the conductor to measure current. Some meters are built around this type of probe, while including other measuring functions available through standard test lead jacks. To take this type of measurement, the selector switch is set to the correct function, and the jaws of the clamp or fork are held around the wire to be tested.

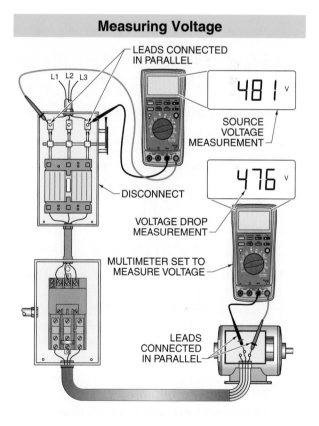

Measuring Voltage

LEADS CONNECTED IN PARALLEL

L1 L2 L3

481 V

SOURCE VOLTAGE MEASUREMENT

476 V

DISCONNECT

VOLTAGE DROP MEASUREMENT

MULTIMETER SET TO MEASURE VOLTAGE

LEADS CONNECTED IN PARALLEL

Figure 15-6. Voltage measurements are taken with the leads connected in parallel with the component or section of the circuit being tested.

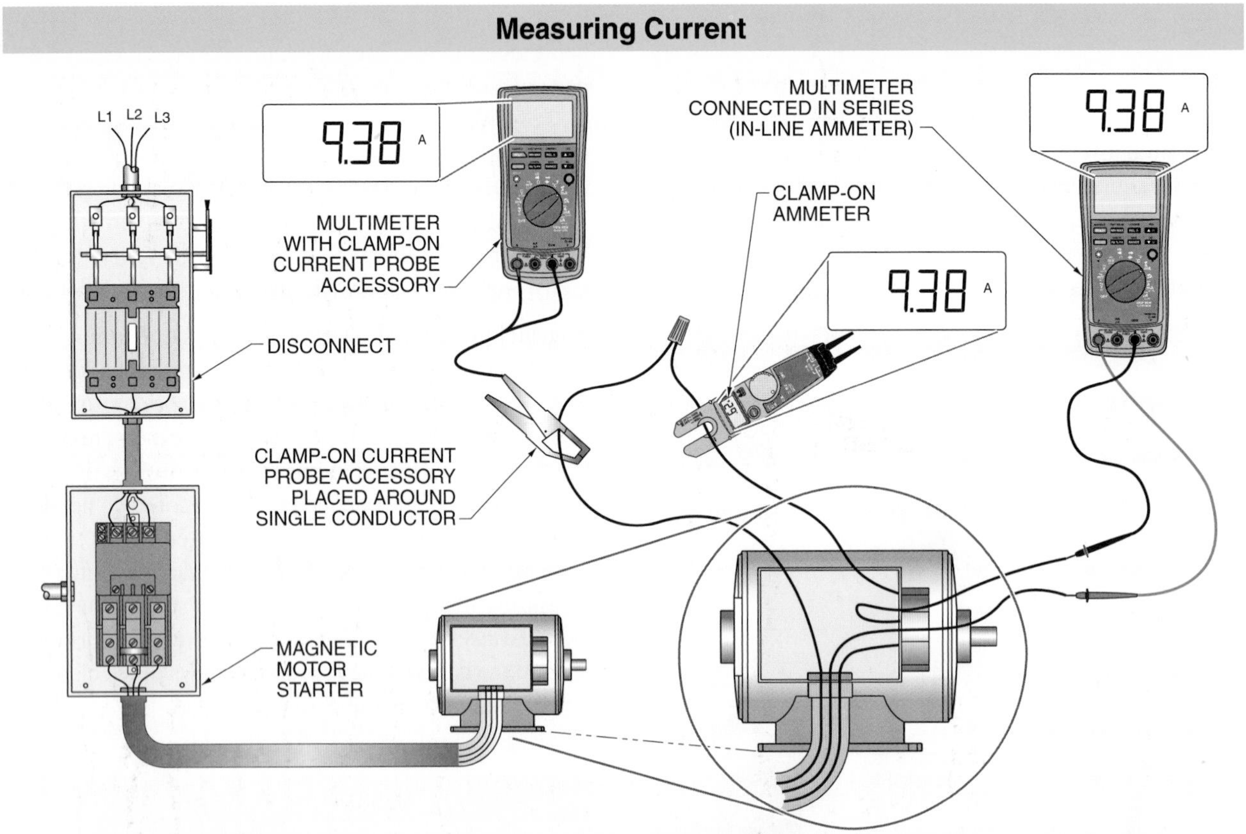

Measuring Current

Figure 15-7. Current is measured with a clamp-on ammeter, a multimeter connected in series, or a multimeter with a clamp-on current probe accessory.

Current Measurment Devices		
	Clamp-On Ammeter	In-Line Ammeter	DMM With Clamp-On Current Probe Accessory
Range	Up to 3000 A	Less than 10 A	Up to 1000 A
Procedure	Measurement does not require opening circuit	Measurement requires opening circuit	Measurement does not require opening circuit

Alternatively, a clamp-on current probe accessory can be connected to a meter through the standard jacks to add this capability to the meter. When using an accessory, the correct multimeter setting must be selected, which may or may not be a current type. The accessory should specify the correct setting, such as outputting 1 mV for each ampere measured. When reading the measurement, the displayed value must be adjusted as needed to represent the correct units.

In-line current measurements are made by opening the circuit and connecting the multimeter's test leads so that it closes the circuit. Current flows through the meter as it becomes part of the circuit. The black test lead is plugged into the negative jack (–), and the red test lead is plugged into a current jack. There may be more than one current jack, with one being for measurements up to 10 A and the other for measurements up to about 400 mA. Then, while the circuit is OFF, the test leads should be connected into an open in the circuit. When the circuit is turned ON again, the multimeter displays the in-line current.

This method of current measurement is more accurate but must be limited to circuits that can be easily opened and are known to have currents less than 10 A, which greatly limits its applicability in industrial settings. In-line current measurements also present greater opportunity for electrical shock.

ELECTRICAL DEVICES

An electrical circuit includes a source of electricity and a load. Any practical circuit also includes one or more devices to control the flow of electricity, such as switches. In addition, a circuit should also include a protection device (a fuse or circuit breaker) to ensure that the circuit operates within its electrical limits. A mechanical technician should be familiar with the most common examples of these types of electrical devices.

Source Voltage

Before troubleshooting an electrical device that is suspected of causing a problem, the mechanical technician should always test the source voltage to the device first. This ensures that a problem is not farther upstream in the circuit and helps focus the troubleshooting efforts in the right place. An electrical problem could be as simple as a switch upstream that is not in the ON position.

Source voltage is tested by measuring the voltage between each ungrounded (hot) conductor (if there is more than one) and between each hot conductor and ground. **See Figure 15-8.** Either a voltage tester or multimeter may be used, though a multimeter is more precise. Incoming voltage should be within ±10% of the nominal voltage value. A high- or low-voltage situation indicates a supply problem, and troubleshooting should be redirected there.

Conductors must be identified correctly. The grounded (neutral) conductor is grounded to earth at the distribution panel and is always white or gray. Ungrounded (hot) conductors are generally red or black, although colors other than red or black are sometimes used. The ground conductor may be either a separate wire (bare copper or with green insulation) within the conduit, cable, or raceway, or it may be the metal enclosure or conduit itself.

Proper grounding of the enclosure, conduit, or equipment frame should also be verified. When measuring voltage between hot conductors and grounded materials, such as enclosures or conduits, properly grounded equipment indicates the source voltage.

Transformers

A *transformer* is an electrical device that uses electromagnetism to change AC voltage from one level to another. A transformer consists of two looped conductors (windings) and an iron core. **See Figure 15-9.** The iron core is used as a magnetic coupling between the windings.

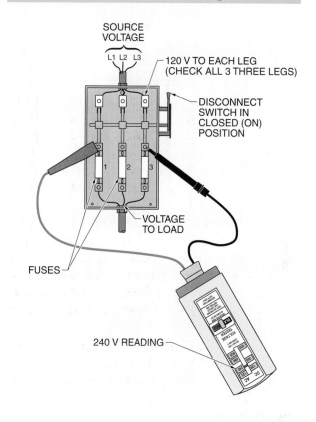

Testing Source Voltage

Figure 15-8. Before troubleshooting a malfunctioning electrical device, the source voltage to the device should be checked first to ensure that the problem is not upstream.

The two windings are referred to as the primary and secondary windings. The *primary winding* is the power input winding of a transformer and is connected to the incoming power supply. The *secondary winding* is the output or load winding of a transformer and is connected to the load. A magnetic field builds up around the primary winding when current passes through its conductor. As the magnetic field builds around the primary winding, current is induced in the secondary winding. This is done without any physical connection between the two windings.

A solid iron core creates unwanted eddy currents. An *eddy current* is an electric current that is generated and dissipated in a conductive material in the presence of an electromagnetic field. Eddy currents create an excessive amount of heat. Therefore, transformer cores are made from laminated (layered) sheets of iron, which reduces the amount of eddy currents.

Transformers

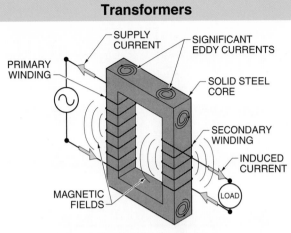

TRANSFORMER OPERATION

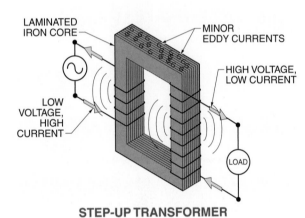

STEP-UP TRANSFORMER

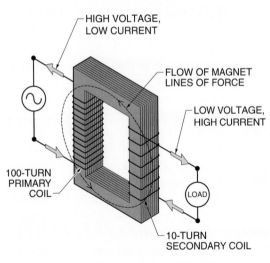

STEP-DOWN TRANSFORMER

Figure 15-9. In a transformer, one coil induces a voltage into another coil through magnetic fields. The ratio of the numbers of turns in each coil determines whether the voltage steps up or down.

The number of turns in each winding establishes whether the transformer steps up or steps down voltage. A transformer is a step-up transformer if it has more turns in the secondary winding than in the primary winding. The transformer is a step-down transformer if there are fewer turns in the secondary winding than the primary winding. The ratio of the number of turns of wire in the secondary winding to those in the primary winding determines how much the voltage is stepped up or down. For example, if a primary winding contains 100 turns and is connected to an input supply of 120 V, and the secondary winding has 50 turns (half of the primary), the output voltage to the load is 60 V (half the input voltage). In this case, the transformer is a step-down transformer with a 2:1 ratio.

Testing Transformers. The input and output voltages are checked if there appears to be a problem with a transformer. The transformer is good as far as voltages are concerned if the voltage is within ±10% of the nameplate rating. A multimeter set on the resistance setting may be used if a break in the wires or a short is suspected.

The meter leads are touched to the ends of each winding. Normally, the windings should have some resistance (the results being neither zero nor overload). Exact resistance values for each winding vary with transformer type and size, and they may be found in the manufacturer's information or by testing a known, good transformer of the same type and size. **See Figure 15-10.**

To check a transformer for a short, the meter leads should be touched between each of the secondary and primary wires, or between each of the secondary and primary wires and the core. The test lead must be touched to the metal of the core, not to the paint or varnish. An overload reading indicates that the transformer is operating properly. Any other reading indicates a short.

Heating Elements

A *heating element* is a conductor with an intentionally high resistance for producing heat when connected to an electrical power supply. Electric current flows through the conductor, making the heating element hot. Heating elements are used in electric ovens, dryers, and forced-air heating systems.

Safety Tip

The K rating should be considered when a power transformer is selected for an application. The higher the K rating, the better the transformer will be at handling nonlinear loads.

Testing Transformers

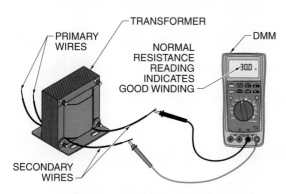

OPEN CIRCUITS IN WINDINGS

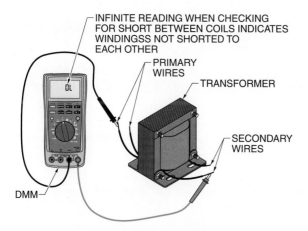

SHORT CIRCUIT BETWEEN
PRIMARY AND SECONDARY WINDINGS

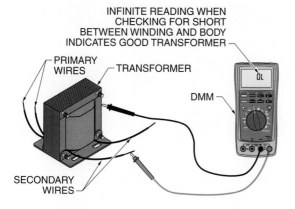

WINDINGS SHORTED TO CORE

Figure 15-10. A multimeter set to measure resistance can be used to test transformer windings.

Testing Heating Elements. The most common problem with heating elements is an open circuit, which is usually caused by overheating, high voltage, physical damage, or shorting to nearby metal. Heating elements are tested by measuring their resistance.

A good heating element has some resistance when tested. **See Figure 15-11.** A defective heating element with an open indicates an overload resistance when connected to a DMM set to measure resistance. A short in a heating element indicates 0 Ω resistance.

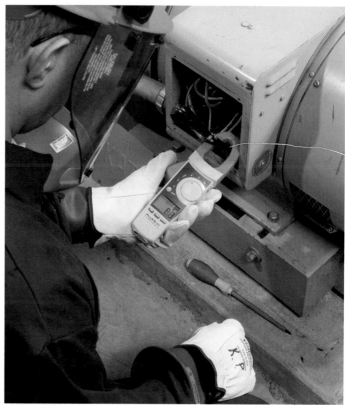

Fluke Corporation

Clamp meters are frequently used in troubleshooting, such as when measuring the current draw of a motor.

Switches

A *switch* is a device that starts or stops the flow of electrical energy in a circuit. An electrical system must contain devices that permit it to be safely turned ON or OFF, limit it in strength, or control its direction. Controlling the ON/OFF function of power is accomplished using a switch. **See Figure 15-12.** A switch is closed when it allows current to flow in a circuit. A switch is open when it prevents flow of current in a circuit.

Testing Heating Elements

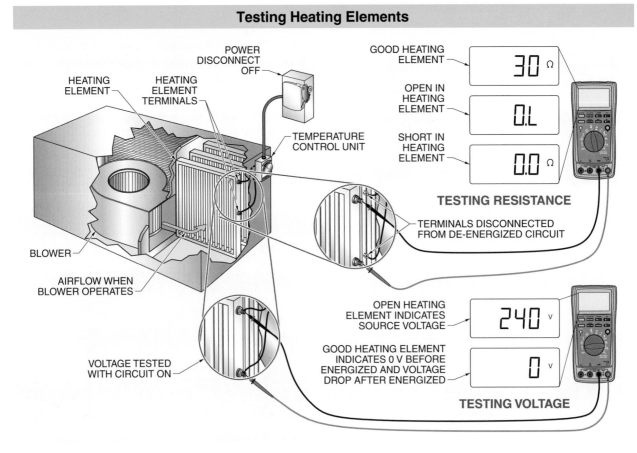

Figure 15-11. Heating elements are tested by measuring their resistance.

Switches

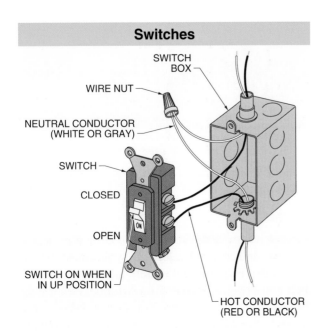

Figure 15-12. ON/OFF switches completely interrupt or allow current flow through a circuit.

Testing Switches. A switch is tested by either measuring voltage (when the switch is connected to a voltage supply) or testing for resistance/continuity (when the switch is removed from the circuit). If measuring voltage, the voltage tester or multimeter is connected to the load terminal of the switch and ground. **See Figure 15-13.** The source voltage should be indicated when the switch is in the ON position, and no voltage should be indicated when the switch is in the OFF position.

Testing a switch with a continuity tester (or a multimeter set to measure resistance) is performed with the switch removed from electrical energy. A continuity test ensures that the mechanisms within the switch are properly making or breaking contacts. One of the tester leads is connected to one of the switch terminals, and the other lead to the remaining terminal. A switch in good operating condition indicates continuity when the switch is in the ON position and no continuity when the switch is in the OFF position.

Testing Switches

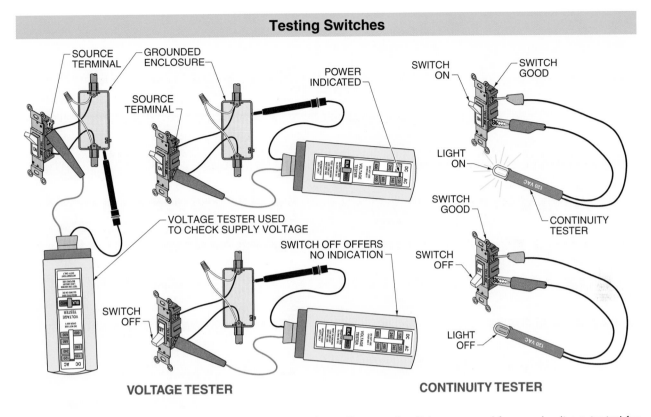

Figure 15-13. Switches in a circuit are tested by measuring voltage, and switches removed from a circuit are tested for continuity.

Fuses and Circuit Breakers

Fuses and circuit breakers automatically interrupt the flow of current in a circuit if the current exceeds the amount the circuit was designed to handle safely. A *fuse* is an overcurrent protection device with a fusible link that melts and opens a circuit in an overcurrent condition. **See Figure 15-14.** A *circuit breaker* is a manually operated switch with a mechanism that automatically opens the circuit if an overload condition occurs. Overcurrent conditions may be caused by an overloaded circuit that uses too many loads, or by a short circuit where a bare hot wire touches another hot wire or any metal that is grounded.

Excessive heat is generated in a conductor with too much current. This principle is used in safety devices such as fuses and circuit breakers, which are designed as an intentional weak point in a circuit. Fuses have a higher resistance than the rest of a circuit, so the fuse generates a lot of heat quickly and melts itself, opening the electrical circuit before damage is done to the conductor or other electrical devices.

Circuit breakers use a variety of methods for automatically opening a circuit. One design uses a bimetallic strip in which two dissimilar metals are bonded together. One metal has greater resistance, and the other metal has little resistance. During a high current flow, the difference between their resistances causes one metal to expand more than the other, which causes the strip to bend and separate the electrical contacts.

Testing Fuses and Circuit Breakers. Fuses or circuit breakers are tested for unintentional open circuits. They may be tested by measuring voltage while they are connected to supply power or by measuring resistance/continuity when they are removed from power.

If measuring voltage, a fuse or circuit breaker is tested between one supply side terminal and a terminal on the load side of another line. Never check between terminals of the same line when checking fuse or breaker condition. For example, voltage can be used to check the fuse condition of a 3ϕ circuit. Fuse 3 may be checked with one probe contacting the L1 or L2 supply terminals and the other probe on Fuse 3's load side terminal. Testing Fuse 2 is accomplished by placing one probe of the voltage tester on the L1 or L3 supply terminals and the other probe contacting the F2 load terminal. A good fuse or circuit breaker measures source voltage in this test.

Fuses and Circuit Breakers

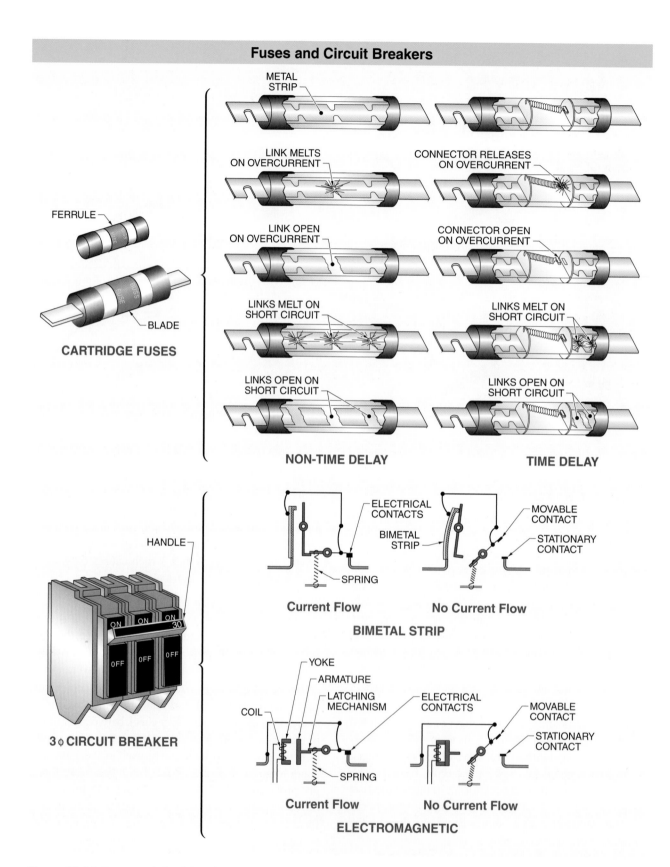

CARTRIDGE FUSES

3 φ CIRCUIT BREAKER

NON-TIME DELAY

TIME DELAY

BIMETAL STRIP

ELECTROMAGNETIC

Figure 15-14. Fuses and circuit breakers open a circuit when an overcurrent condition or short circuit occurs.

To test the continuity of a fuse or circuit breaker, it is first removed from the circuit. **See Figure 15-15.** Fuses are removed with a fuse puller. A *fuse puller* is a device made of a nonconductive material such as nylon that is used to grasp and remove cartridge fuses. With the fuse or breaker removed, place continuity tester leads across the fuse or breaker. The fuse or breaker is good if it has continuity, and it is bad if it does not. A circuit breaker must be in the ON position when testing for continuity.

Ground-Fault Circuit Interrupters (GFCIs)

A *ground fault* is an unintended current path between a ground and an ungrounded (hot) conductor. A ground fault may result from defective electrical equipment, improperly installed equipment, or through misuse of good equipment. Ground faults may damage equipment. More importantly, however, a ground fault may cause an electrical shock resulting in injury or death to any person who becomes part of the ground-fault circuit. **See Figure 15-16.**

Safety Tip

A neutral-to-ground connection must not be made in any subpanel, receptacle, or equipment because it creates the potential threat of electric shock from the ground wire.

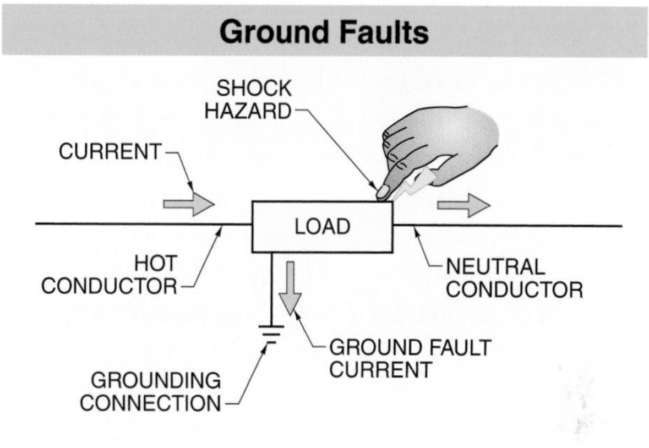

Ground Faults

Figure 15-16. A ground fault is an unwanted diversion of current to ground. This creates a shock hazard to anyone who may touch the associated load.

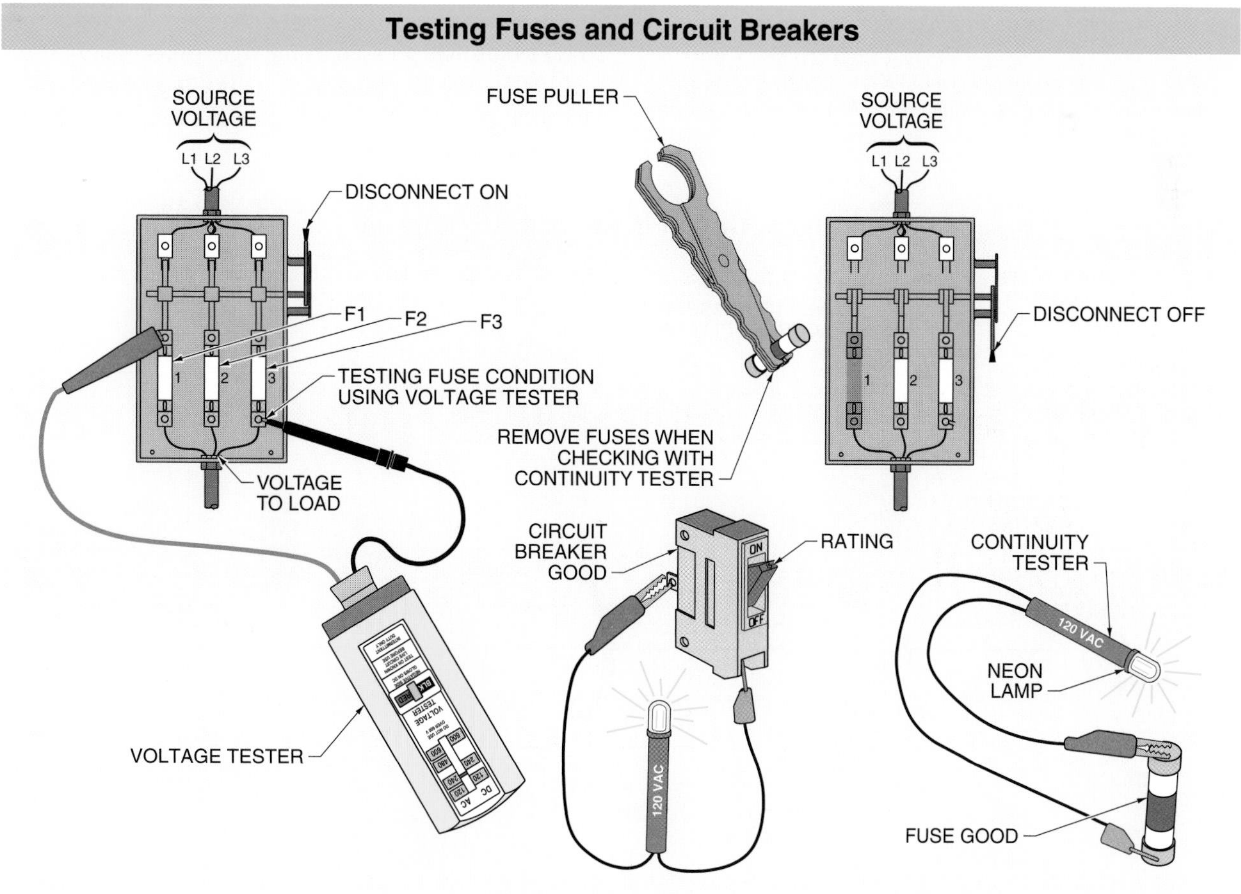

Testing Fuses and Circuit Breakers

Figure 15-15. Fuses and circuit breakers are tested by checking either the voltage across them or their continuity.

A *ground-fault circuit interrupter (GFCI)* is an electrical device that protects personnel by detecting potentially hazardous ground faults and quickly disconnecting power from the circuit. Any current over 8 mA is considered potentially dangerous, depending on the path the current takes, the amount of time exposed to the shock, and the physical condition of the person shocked. GFCIs are required in wet locations, such as kitchens, bathrooms, construction sites, marinas, and other areas in which a circuit may experience a ground fault.

A GFCI circuit compares the amount of current in the ungrounded (hot) conductor with the amount of current in the neutral conductor. **See Figure 15-17.** If the current in the neutral conductor becomes less than the current in the hot conductor, then the "missing" current is returning to the source through some other path. This is a ground-fault condition. A fault current as low as 4 mA to 6 mA will activate the GFCI and interrupt the circuit. Once activated, the fault condition is cleared, and the GFCI is manually reset before power is restored to the circuit.

Safety Tip

GFCIs have terminals marked LINE and LOAD. Incoming power lines to a GFCI should always be connected to the LINE terminals.

GFCI receptacles are typically wired to provide GFCI protection at all other receptacles installed downstream on the same circuit. GFCI circuit breakers, when installed in a load center or panelboard, provide GFCI protection and conventional circuit overcurrent protection for all branch circuit components connected to the circuit breaker. Plug-in GFCIs provide ground-fault protection for devices plugged into them. These plug-in devices are often used by personnel working with power tools in an area that does not include GFCI receptacles.

Testing GFCI Receptacles. GFCI receptacles include their own testing function. A button on the front is a switch that closes a short between the hot and neutral conductors. **See Figure 15-18.** A resistor is used to limit the current flow through the short, but the diverted current is enough that it should activate the sensing circuit and cause the receptacle's contacts to automatically open, shutting off all power to the receptacle and any connected loads.

To conduct a GFCI test, a load is connected to the receptacle, such as a light, and turned ON. The TEST button is pressed. If the light immediately shuts OFF, then the receptacle is functioning properly. The RESET button is pressed after the test to reset the sensing circuit.

Ground-Fault Circuit Interrupter (GFCI) Circuits

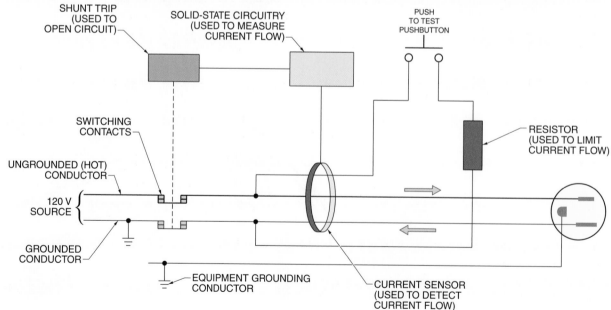

Figure 15-17. A ground-fault circuit interrupter (GFCI) circuit senses a ground fault and immediately opens contacts in the hot and neutral conductors.

Testing GFCI Receptacles

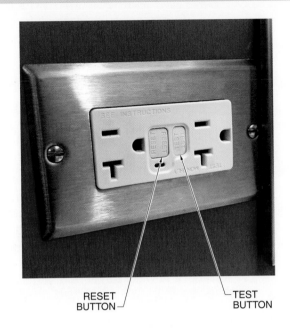

RESET BUTTON

TEST BUTTON

Figure 15-18. GFCI receptacles have TEST and RESET buttons for testing proper operation.

A similar test can be performed on any standard receptacle that is connected downstream of the GFCI receptacle and is, therefore, covered by the GFCI's ground-fault protection. A receptacle tester with an included GFCI test function is required. The tester is plugged into the standard receptacle, and the tester is activated. Just like the internal GFCI circuit, the tester shorts the hot and neutral conductors through a resistor, tripping the GFCI contacts. If successful, the tester's lights turn OFF. If the test fails, however, the problem could be either the GFCI receptacle or the wiring between it and the standard receptacle.

Coil-Operated Devices

Electromechanical devices convert electrical energy into mechanical energy. Often this is accomplished by creating a magnetic field in a coil and using the magnetic field to move a set of contacts or a metal plunger. A *coil* is a winding of insulated conductors arranged to produce a magnetic field. This magnetic field can be used to move an iron armature. An *armature* is the moving part of a coil-operated device. The moving armature can then be used as a source of mechanical energy.

Solenoids. A *solenoid* is a device that converts electrical energy into a linear mechanical force. Solenoids consist of a doughnut-shaped coil surrounding a movable iron plunger. **See Figure 15-19.** Current in the coil produces an electromagnetic field that moves the plunger. The plunger is attached to a valve or other movable device. The most common application of a solenoid is for opening and closing valves.

Relays. A *relay* is an electrical switch that is actuated by a separate circuit. An electromechanical relay is essentially an application of a solenoid in which the plunger is used to mechanically open or close a set of electrical contacts. **See Figure 15-20.** A *contact* is a conducting part of a switch to connect or disconnect a circuit or component. This way, a relay on one circuit can energize or de-energize another circuit.

Relays are the bridge between completely separate power circuits and control circuits. A *power circuit* is an electrical circuit that connects a load to main power lines. A *control circuit* is an electrical circuit consisting of switches and other control devices that is designed to operate a load in a power circuit in a particular way. The voltage of a control circuit is usually much lower than a power circuit in order to use lighter conductors and lower current devices, though it does not have to be.

For example, a relay in a low-voltage control circuit can open or close contacts in a high-voltage load circuit. The load in the control circuit is the relay coil. The relay coil becomes an electromagnet, causing the armature to move when the control circuit is energized. This closes the relay contacts to energize high-voltage loads, such as motors, lights, or heating elements.

Relays are commonly used to control heating elements. The control circuit and heating element (load) circuit are separate circuits. The control circuit operates at 24 V, 0.5 A. The heating element circuit operates at 460 V, 10 A. As temperature falls in a room, the NO contacts in the thermostat close the control circuit to energize the relay coil. Current flowing through the relay coil produces a magnetic field that closes the NO contacts in the high-voltage load circuit. The closed contacts complete the circuit, and power flows to the heating elements.

Tech Tip

A relay contact block may contain more contacts than required for a particular application. Unused relay contacts can be connected in parallel with used contacts to divide the current flow over both sets of contacts.

Solenoid Operation

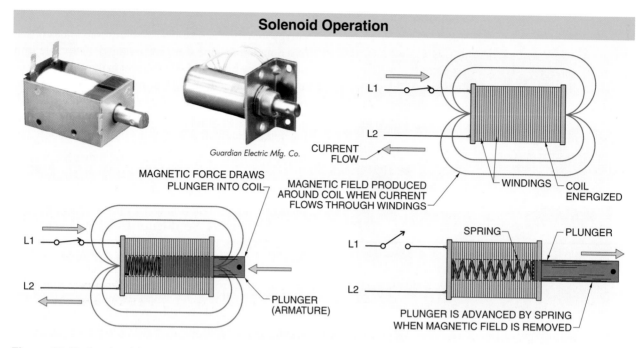

Guardian Electric Mfg. Co.

MAGNETIC FORCE DRAWS PLUNGER INTO COIL

MAGNETIC FIELD PRODUCED AROUND COIL WHEN CURRENT FLOWS THROUGH WINDINGS

CURRENT FLOW

WINDINGS

COIL ENERGIZED

L1

L2

PLUNGER (ARMATURE)

SPRING

PLUNGER

L1

L2

PLUNGER IS ADVANCED BY SPRING WHEN MAGNETIC FIELD IS REMOVED

Figure 15-19. A solenoid uses the magnetic field generated by a coil to move an iron plunger.

Relays

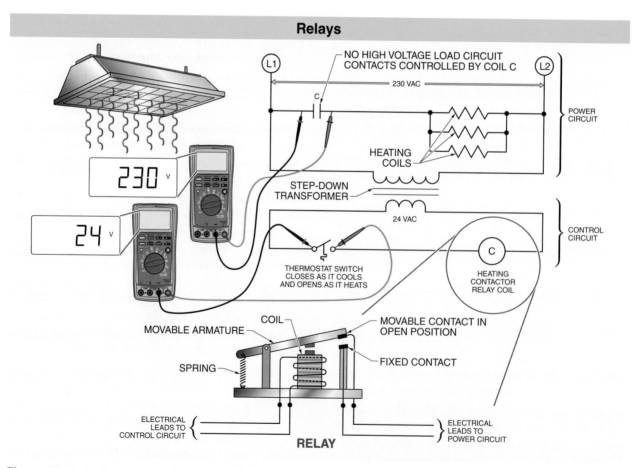

NO HIGH VOLTAGE LOAD CIRCUIT CONTACTS CONTROLLED BY COIL C

L1

L2

230 VAC

C

POWER CIRCUIT

HEATING COILS

STEP-DOWN TRANSFORMER

24 VAC

CONTROL CIRCUIT

230 V

24 V

THERMOSTAT SWITCH CLOSES AS IT COOLS AND OPENS AS IT HEATS

C

HEATING CONTACTOR RELAY COIL

MOVABLE CONTACT IN OPEN POSITION

MOVABLE ARMATURE

COIL

FIXED CONTACT

SPRING

ELECTRICAL LEADS TO CONTROL CIRCUIT

ELECTRICAL LEADS TO POWER CIRCUIT

RELAY

Figure 15-20. A relay opens or closes a set of electrical contacts to control a separate circuit.

Contactors and Motor Starters. Contactors and motor starters are types of heavy-duty relays that are specialized for certain types of circuits. **See Figure 15-21.** A *contactor* is a heavy-duty relay for controlling high-current loads. Contactors are commonly used for switching motors, heating elements, and large lighting circuits. A *motor starter* is a heavy-duty relay that includes motor overload protection. Overload protection is accomplished through the use of overload relays. An *overload relay* is a time-delay device that senses motor current temperature and disconnects the motor from the power supply if the current is excessive for a certain length of time.

Contactors and Motor Starters

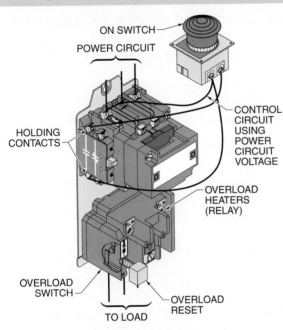

MOTOR STARTER

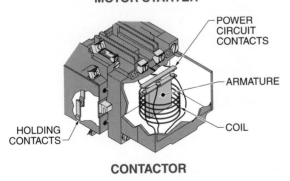

CONTACTOR

Figure 15-21. Contactors and motor starters are heavy-duty relays intended for controlling high-current loads. They also have additional features, such as holding contacts and overload relays.

Relay contacts are switched only as long as the control circuit is energized. If a pushbutton is used in a control circuit to energize a relay coil, then the relay coil de-energizes and the power contacts return to normal as soon as the pushbutton is released. For contactors and motor starters, however, it is usually desirable for the load to remain ON even when an ON pushbutton is released and until a separate OFF pushbutton is pressed. Continuous operation using a temporary current is accomplished through use of a holding contact. A *holding contact* is an auxiliary contact used to maintain current flow to a relay coil. Holding contacts are part of a contactor or motor starter and open and close with the power contacts. These contacts are wired into the control circuit. Once an ON pushbutton is pressed, holding contacts use the current of the power circuit to maintain the closed condition of the control circuit. Once an OFF pushbutton is pressed, the circuit is opened and current flow to the holding contact coil is removed, which causes the holding contacts and power contacts to open.

Testing Coils. To troubleshoot a coil-operated device, it must be removed from the circuit. Any covers must be removed to expose the coil in order to visually inspect it for burnt, broken, or frozen parts. The plunger must be free to move and not jammed. Most coil devices have ways to be manually operated.

The coil is then inspected further by testing its resistance with a multimeter. **See Figure 15-22.** The solenoid is good if the resistance is within ±15% of the normal coil or rated value. This information may be found in manufacturer's data or by testing a good solenoid of the same type and size. A low or zero reading indicates a short in the solenoid coil windings. An overload reading indicates that the coil has a broken wire and is open.

Testing Contactors and Motor Starters. Testing contactors and motor starters is done by first visually inspecting the contactor or starter for physical damage. The source voltage from the power circuit must be checked next. **See Figure 15-23.** Each hot line must be within ±10% of the load voltage rating. Also, each incoming power wire to ground must be checked for a blown fuse upstream. The voltage used by the control circuit and the control device must be checked. The control device can be a start/stop switch or pilot switch, such as a pressure switch, float switch, or flow switch.

On a motor starter, a technician tests each side of the overload for the presence of voltage. An overload may be tripped if there is supply voltage and power voltage but no output voltage.

Testing Coils

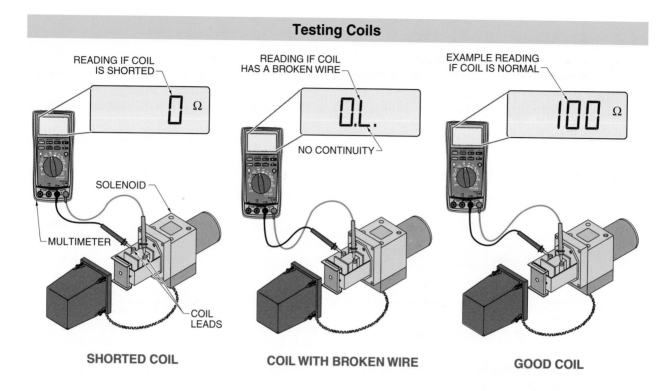

Figure 15-22. A coil, whether in a solenoid, a relay, a contactor, or a motor starter, can be tested by measuring its resistance.

Testing Contactors and Motor Starters

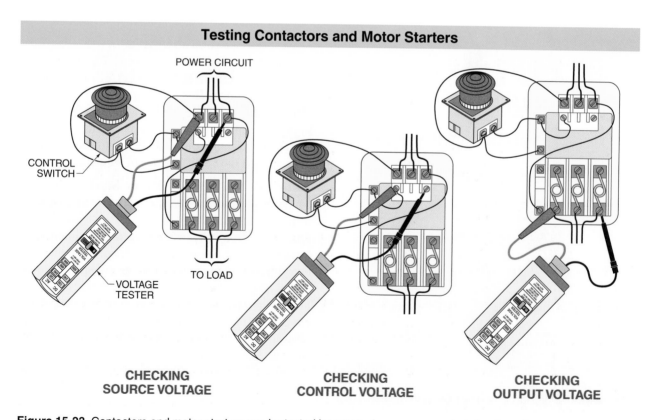

Figure 15-23. Contactors and motor starters can be tested by measuring source, control circuit, and output voltages.

A technician must not attempt to reset an overload switch unless authorized to do so. Overload switches protect a motor and normally trip when there is a high current condition. Other checks must be made by an authorized technician to determine the overload condition before resetting an overload switch.

Motors

Motors are very common and found in a wide variety of equipment. Large motors are usually found in industrial facilities and warrant individual maintenance and troubleshooting. Motors are broadly categorized as either three-phase or single-phase motors, with multiple operating designs in each category.

Three-Phase Motors. The most common industrial motor is the three-phase (3ϕ) induction motor. An *induction motor* is a motor that rotates due to the interaction between the magnetic fields of the stator and rotor. **See Figure 15-24.** A 3ϕ motor contains three coils of wire wound into a stator. A *stator* is a slotted circular iron frame that is the stationary part of an AC motor. The stator windings do not move and surround the rotor. A *rotor* is the rotating part of an AC motor. The rotor is mounted on the motor shaft, which is attached to the load the motor is powering.

Each coil in a 3ϕ motor is wound so that one half of the coil is opposite the other half of the coil in the stator. Each stator coil receives a single phase of electric current. Each phase lags or leads the other two phases to create a pulsating magnetic field that creates a rotating magnetic field in the stator. The rotating magnetic field passes through the aluminum or copper bars of the squirrel cage built into the rotor. This induces current in the squirrel cage bars, producing a magnetic field around each rotor bar. The rotating magnetic field of the stator pushes on the magnetic field around each rotor bar, causing the rotor to rotate because magnetic fields cannot cross or pass through each other. The torque produced by the interaction of the three magnetic fields starts the motor and keeps it rotating.

Single-Phase Motors. Single-phase (1ϕ) motors are seldom larger than 1 HP and are used for light loads. Most 1ϕ motors are induction motors that operate on the same principle as 3ϕ induction motors. **See Figure 15-25.** A *split-phase motor* is a 1ϕ AC motor that includes a starting winding and a running winding in the stator. A *starting winding* is a coil that is energized temporarily at startup to create the torque required to start a 1ϕ motor rotating.

The starting winding is wound out-of-phase with the running winding of the motor. A *running winding* is a coil that continues to operate a 1ϕ motor after it has started.

Three-Phase (3ϕ) Motors

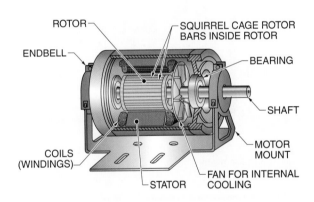

Figure 15-24. In a three-phase (3ϕ) induction motor, the rotating magnetic field of the stator pushes on the magnetic field around each rotor bar, causing the rotor to rotate.

At startup, both windings are energized and create two out-of-phase magnetic fields that produce the torque to start the motor rotating. A centrifugal switch opens when the motor reaches approximately 75% of full speed, disconnecting the starting winding from the circuit. **See Figure 15-26.** A *centrifugal switch* is a switch that opens when a rotor reaches a certain speed and reconnects when the rotor falls below that speed. This allows the motor to operate on the running winding only. The starting winding is not designed to carry current for long periods of time. The starting winding burns out quickly if the centrifugal switch does not open.

Split-Phase Motors

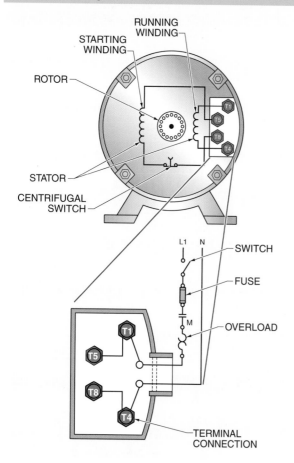

Figure 15-25. A split-phase motor has starting and running windings in the stator.

Baldor Electric Company

Motors are a common part of many industrial systems.

Centrifugal Switches

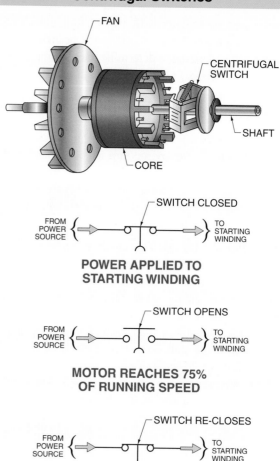

POWER APPLIED TO STARTING WINDING

MOTOR REACHES 75% OF RUNNING SPEED

MOTOR FALLS TO APPROXIMATELY 40% OF RUNNING SPEED

Figure 15-26. At approximately 75% speed, a split-phase motor's centrifugal switch opens, disconnecting the starting winding.

A capacitor-start motor is selected if a heavy load must be operated. A *capacitor-start motor* is a 1ϕ motor that has a capacitor in the starting winding. **See Figure 15-27.** A *capacitor* is a device that stores electrical energy in an electrostatic field. A capacitor consists of two metal plates separated by a dielectric (insulator). The capacitor causes the starting winding current to be nearly 90° out-of-phase with the running winding current. Starting torque is increased by adding a capacitor in series with the starting winding. A *capacitor start-and-run motor* is a 1ϕ motor that has capacitors in both the starting and running windings.

Capacitor Motors

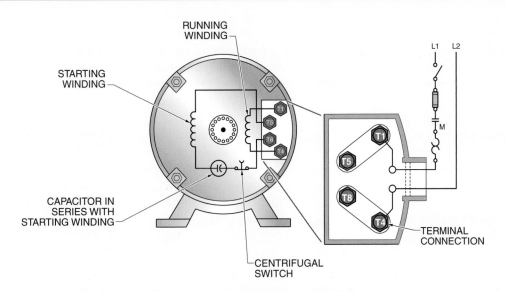

CAPACITOR-START MOTOR

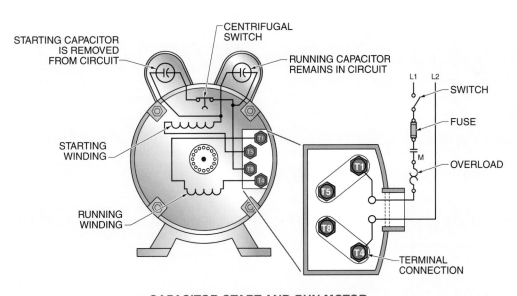

CAPACITOR START-AND-RUN MOTOR

Figure 15-27. Capacitor motors have capacitors in the starting and/or running windings, which increase the starting and/or running torque.

Troubleshooting Motors. Overheating is the most common cause of motor failure. Overheating melts the insulation on stator coil wires. The uninsulated wires touch, shorting the coils and destroying the motor. Overheating is often caused by overloading a motor, which causes excessive current flow in the stator coils. As a motor starts, it receives high levels of current (high inrush current) until the rotor begins to rotate. Normally, the high inrush current lasts a few seconds at startup. Most motors can handle brief overloads without damage. However, if the motor does not start or is completely stalled by a heavy load, the motor can overheat and fail quickly. Overheating can also be caused by starting and stopping or changing direction frequently.

If a motor is excessively hot, the current draw should be measured and compared to the motor nameplate data. **See Figure 15-28.** A motor is normally loaded if the motor current is between 95% and 105% of the rated current. Current above 105% of rated current causes overloading if continued for excessive periods of time. Tripped overloads on a motor starter are a strong indication of motor overloading.

Low voltage or high voltage can also cause overheating, so motor voltage should also be checked, especially if the current draw is high. Voltage should be within ±10% of the motor's required voltage. The voltage source and path should be checked if the voltage is not within the acceptable range.

If the input voltage is within the acceptable range, the motor bearings should be lubricated and the current draw checked. If current then falls to an acceptable level, the bearings are most likely damaged and should be replaced. If the current remains high after all the bearings are lubricated, then the motor is disconnected from its load and operated alone. If current returns to normal, the problem is in the load, which is inspected for other mechanical problems. The motor must be replaced if current remains high with the load disconnected because the motor coils may be damaged. A motor that is overloaded with a properly functioning load should be replaced with a motor of greater horsepower and better electrical efficiency.

Motor coils are tested using a DMM set to measure resistance. **See Figure 15-29.** The three individual motor coils should have the same resistance. They should not have continuity with each other. If they do, the insulation around some of the wires has melted, and bare wires from separate coils are touching. In addition, overheated motor windings are likely to be blackened and have a distinct burnt insulation smell. The three coils that are connected electrically in the center should have a different resistance reading from the three individual coils. There should be no continuity between the connected coils and the individual coils. The coil configuration varies for different motors. For this reason, manufacturer's literature must be consulted before conducting motor resistance tests.

The motor is damaged and must be rewound or replaced if one or more coils has an OL (infinite resistance) reading or the individual coils have continuity with the motor frame. Rewinding a motor involves the complete replacement of all the copper wire in the stator. It is sometimes economically feasible to rewind large motors; however, it is usually less expensive to replace small motors. The root cause of the damage must be determined to avoid having the same problem in the future.

A magnetic motor starter is operated manually when a motor is not working and the overload contacts are not tripped. This involves closing the power contacts in the starter, which connects the motor to its voltage source.

Measuring Motor Current Draw

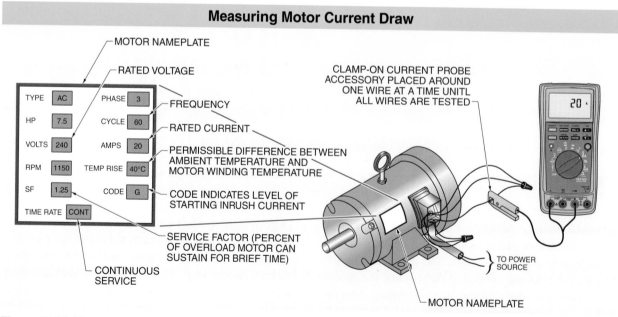

Figure 15-28. The current draw of a hot motor can be measured and compared to the motor nameplate data.

Testing 3ϕ Motor Coils

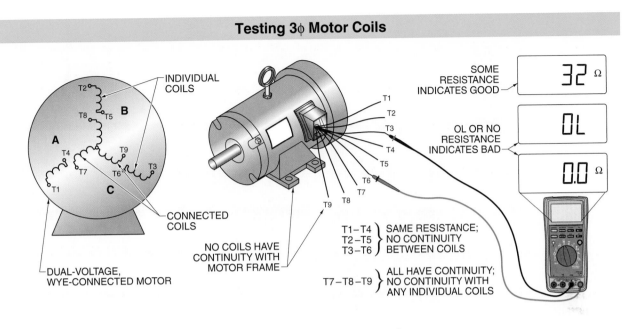

Figure 15-29. A 3ϕ motor's coils are tested using a DMM set to measure resistance.

This is accomplished by different means in different starters. Closing the power contacts should only be done if the circuit is properly fused and the auxiliary contacts are not used to control any other loads.

WARNING: Always use proper safety equipment and follow manufacturer's recommended safety procedures when working on or near energized magnetic motor starters.

Tech Tip

Three-phase motors can be used where only single-phase power exists because motor drives are available to deliver a three-phase (typically 230 VAC) output up to about 2 HP when connected to a 120 VAC or 230 VAC single-phase power source.

Single-phase motors are simpler to troubleshoot. They are usually operated directly by a switch, and the motor is protected by a fuse and/or overload contacts. If a 1ϕ motor does not start, the overload contact must first be reset and the supply voltage checked. If the motor has a capacitor for the starting winding, it can be tested using a DMM with the appropriate function. The capacitance of a capacitor should be within ±10% of its rated value. Some capacitors can be replaced if faulty. Otherwise, the motor is usually replaced without further testing since 1ϕ motors are generally small and inexpensive.

Variable-Frequency Drives

Magnetic motor starters can switch a motor ON or OFF but cannot control a motor's speed or operating characteristics. A *variable-frequency drive* is a motor controller that is used to change the speed of AC motors by changing the frequency of the source voltage. The frequency of a power supply determines the speed of AC motors. Since normal power line frequency is a constant 60 Hz (in the United States), it must be changed to increase or decrease the speed of the motor. In addition to controlling motor speed, variable-frequency drives can control motor acceleration time, deceleration time, motor torque, and motor braking.

A drive changes the frequency of the voltage applied to a motor by converting the incoming AC voltage to a DC voltage and inverting it back to an AC voltage that simulates the desired fundamental frequency. **See Figure 15-30.** An *inverter* is a device that changes DC voltage into AC voltage of any frequency. The *fundamental frequency* is the desired voltage frequency simulated by the varying ON/OFF pulses at a higher carrier frequency. The *carrier frequency* is the frequency of the short voltage pulses of varying length that simulate a lower fundamental frequency. The higher the carrier frequency, the more individual pulses there are to reproduce the fundamental frequency and the closer the output sine wave is to a pure fundamental frequency sine wave.

Variable-Frequency Drives

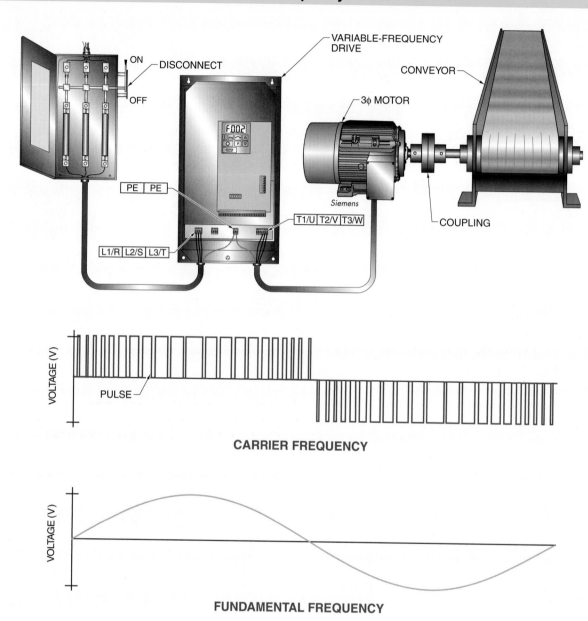

Figure 15-30. A variable-frequency drive changes the frequency of the voltage applied to a motor by converting the incoming AC voltage to a DC voltage and inverting it back to an AC voltage that simulates the desired fundamental frequency.

The carrier frequency of most variable-frequency drives ranges from 1 kHz to about 20 kHz. The carrier frequency can be changed to meet particular load requirements. Noise is noticeable in the 1 kHz to 2 kHz range because it is within the range of human hearing and is amplified by the motor. Higher carrier frequencies reduce heat in the motor because motors operate at cooler levels when the voltage more closely simulates a pure sine wave. However, the fast switching of the inverter section of a variable-frequency drive produces large voltage spikes that can gradually damage motor insulation. The motors controlled by variable-frequency drives often require additional cooling and higher quality insulation to resist overheating.

Troubleshooting Variable-Frequency Drives. Variable-frequency drives are one of the more complicated electrical devices. They contain a number of high-current electrical components and complex electronics. Thorough troubleshooting may require expertise beyond the level of the mechanical technician. However, there are some initial tests that can be used to identify common problems, or at least narrow down the possible causes, before calling in a specialized technician.

First, the motor being controlled by the variable-frequency drive must be put into a safe operating condition in case it starts or stops unexpectedly. This may require disconnecting its drive shaft from the mechanical assemblies. Then, the power supply to the drive should be checked for proper voltage.

Many variable-frequency drives have an electronic display or some other type of status indicator. **See Figure 15-31.** Some can even be connected to computers for loading operating programs. The drives should be checked for any fault codes. Fault codes may be numerical or abbreviated codes that require referencing the drive's manual for more information. Often, a fault code can identify a problem easily, such as a low-voltage condition. If possible, the drive settings can be checked against the facility's documentation for any inadvertent changes.

Voltage and current output from a drive to a motor can be measured and may help troubleshoot an operating problem. However, the high-frequency pulsing output produced by the drive can lead to false readings by many multimeters, so a more sophisticated meter with a low-pass filter feature may be required.

Some variable-frequency drives include test points for measuring the DC bus voltage used internally to convert the incoming power to motor output power. These measurements may indicate whether or not a problem is in the rectification section of the drive. The manufacturer's manual should provide more information on expected readings and what they mean.

ELECTRICAL CIRCUIT PRINTREADING

Technicians rely heavily on diagrams to show the interconnections between devices and power sources in electrical systems. Otherwise, it is difficult or impossible to understand the operations and interactions from studying the physical wiring alone.

The most common types of electrical diagrams are line diagrams and wiring diagrams. Line diagrams are commonly used for control circuits, and wiring diagrams are commonly used for power circuits. In fact, an electrical print may include both types of diagrams together, with the control transformer being the dividing point between the power and control parts of the circuit. **See Figure 15-32.**

Abbreviations and Symbols

Abbreviations and symbols are used for consistency and to conserve space on drawings. Symbols are preferred over abbreviations whenever possible because abbreviations are language-specific. Standards and industry organizations specify abbreviations and symbols used for a particular type of system, though variations and alternative symbols may also be used. A legend may be included on a print to describe all or some (the less common or nonstandard) abbreviations and symbols.

Most devices are labeled in some way on an electrical diagram, which is critical for keeping track of components in complicated circuits. Labels may be highly abbreviated or use whole words or phrases. Labeling generally helps the reader understand the function of each device or circuit, but too many labels can clutter the diagram and make it harder to follow.

Variable-Frequency Drive Display

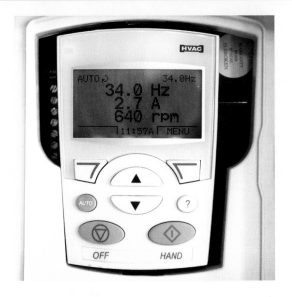

Figure 15-31. Many variable-frequency drives have a display that can show fault codes, which are a common part of troubleshooting drive malfunctions.

Wiring Diagrams and Line Diagrams

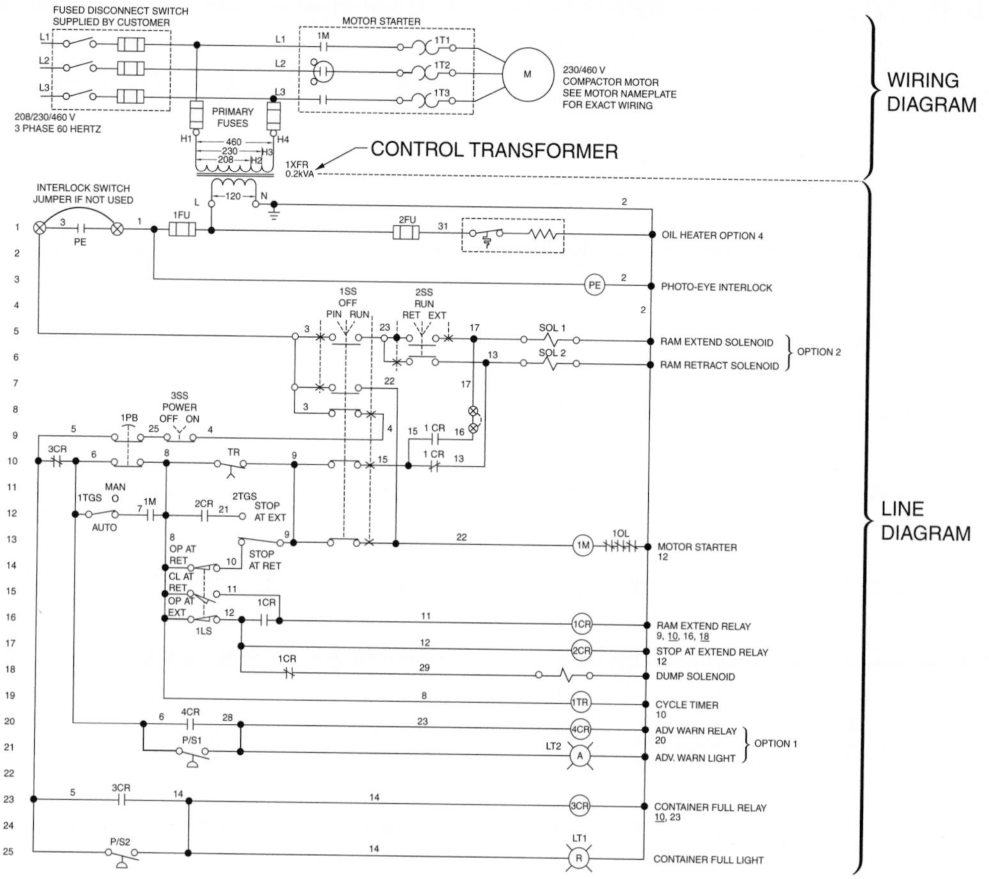

Figure 15-32. Electrical diagrams may include circuit information in both wiring diagram and line diagram formats.

Line Diagrams

A *line diagram* is a diagram that uses lines and graphic symbols to show the logic and operation of an electrical circuit. Like other electrical prints and diagrams, line diagrams show the electrical connections between input devices, loads, relays, meters, and other components. However, line diagrams follow a standard format that is particularly suited for describing the circuit's operation and logic. For this reason, line diagrams are used primarily for control circuits, the circuits that turn loads ON and OFF.

Line diagrams resemble a ladder, so they are also known as ladder diagrams. The power lines (L1 and L2 or L and N) are shown as vertical lines. The horizontal "rungs" between them each make a small circuit that has one or more inputs and one output. The arrangement of inputs in series or parallel clearly illustrates the circuit's control logic, such as AND (requiring both inputs) or OR (requiring only either input) logic.

Line Diagram Notations. Line diagrams may contain a variety of notations to help understand and troubleshoot the circuit. **See Figure 15-33.** Text notes are often included near the rungs to describe that individual circuit's role in the whole system. For example, the overflow switch on line 11 turns on a red light. Also, each rung has a line reference number, usually listed along the left side. A line reference number is an identification number assigned to each horizontal line.

Along the right side, cross-reference numbers contained within parentheses indicate where contacts controlled by a coil are located throughout the circuit. Normally open (NO) contacts have no underline, while normally closed (NC) contacts are underlined. For example, the relay coil is a load in one rung, but it has contacts that serve as inputs in other rungs. The coil's rung includes cross-reference numbers for each rung that contains one of the relay's contacts.

Line Diagram Notations

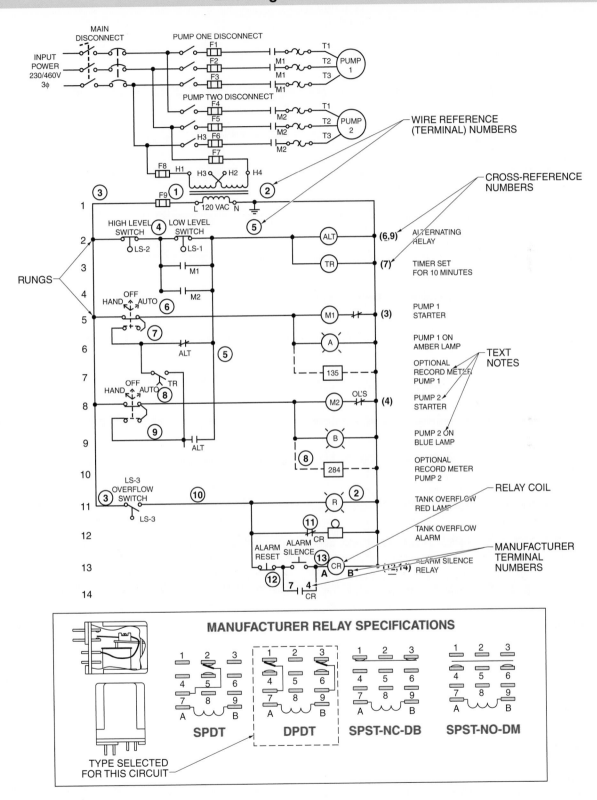

Figure 15-33. Line diagrams include reference information, which aids in wiring and troubleshooting.

Other sets of numbers on a line diagram include wire reference (terminal) numbers and manufacturer terminal numbers. When a terminal strip is used for the physical wiring, the terminal numbers should be included on the line or wiring diagram. All wire lines that are electrically connected to each other are assigned a single wire reference number. By convention, terminal 1 is always the ungrounded (hot) source conductor and terminal 2 is the grounded (neutral) source conductor. All other wire reference numbers are assigned from top to bottom and from left to right. Some method may be used to make these numbers distinct on the diagram, such as circling them.

Manufacturer terminal numbers are used only when a device has its own numbered wiring or terminal strip. These numbers are then included as needed on the line diagram adjacent to those connections.

Reading line diagrams is a critical skill for electrical troubleshooting. Fortunately, while the particular needs of an industrial process may call for a custom motor control circuit, most utilize simple and common circuits, or at least minor variations of them. Even seemingly complex diagrams can usually be broken down into simpler sections, which can be read and analyzed easily.

Common Line Diagram Circuits. A standard start-stop station used to control a single small- to medium-sized motor is one of the most common industrial control circuits. The control circuit can be represented in both line and wiring diagram form. **See Figure 15-34.** This control circuit contains a fuse, a stop pushbutton, a start pushbutton, a magnetic motor starter contact coil (M coil), a motor, and an overload (OL) contact. Pressing the start pushbutton energizes the magnetic motor starter contact coil (M), which switches multiple sets of contacts. The motor contacts in each of the three power phases close, which energizes the motor. At the same instant, the M contacts in the control circuit close, providing an alternate path for current flow (the holding circuit) that keeps the M coil energized when the start pushbutton is released. The fuse protects the control circuit from excessive current.

The M coil remains energized until the stop pushbutton is pressed, the overloads (all OLs) or fuse open due to excessive current flow, or when power to the circuit is lost. When the M coil is de-energized, opening all M contacts, the motor is de-energized. At the same instant, the control circuit M contacts open, releasing the holding circuit. This is an important safety consideration. The circuit will not re-energize unless the start pushbutton is pressed again.

Three-Wire Motor Starter Control

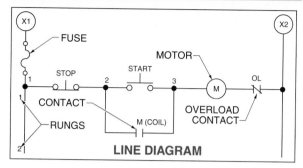

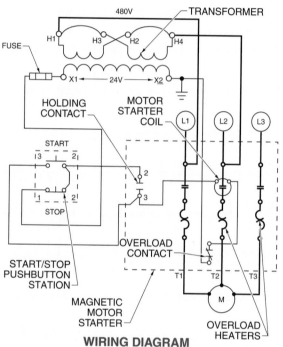

Figure 15-34. The most common motor control circuit is a three-wire, start-stop station circuit. The magnetic motor starter coil's auxiliary contacts are used to keep the circuit energized after the start button is released.

A simpler and more common circuit is the two-wire circuit. It is called a two-wire circuit because it does not use the terminal 3 connection point on the motor starter contacts to provide a holding circuit. **See Figure 15-35.** When a switch closes, the M coil energizes and the motor starts. When the switch opens, the M coil de-energizes and the motor stops. This circuit is usually used with automatic control switches (such as level, pressure, or temperature switches), which creates a safety concern because technicians working on the circuit could be hurt if the motor starts automatically. Such circuits must be locked-out carefully.

Common two-wire circuits are a liquid-level circuit that controls a pump motor; a pressure switch circuit that controls an air compressor motor; and a temperature switch circuit that controls a heating system.

Wiring Diagrams

A *wiring diagram* is a diagram that shows the electrical connections of a circuit. Some wiring diagrams show limited device layout information. For example, they show that the STOP and START pushbuttons of a motor control circuit are located in the same control station. However, there is no indication as to where the start-stop station is located. It could be twenty feet away with the wires running through conduit, or it could be built into the cover of the enclosure containing the motor starter. Therefore, wiring diagrams cannot be relied on for accurate location information.

Wiring diagrams are not made to clearly show how a circuit works as it can be difficult to follow the conductors on complicated diagrams. Therefore, wiring diagrams are typically used for simpler circuits, such as the power circuit connections for a 3ϕ motor. If a circuit is more complex, the circuit may be shown in line diagram form. Or the circuit may be divided into sections based on function and each section drawn separately, with extra space to clarify or annotate the connections.

Wiring Diagram Notations. A wiring diagram may include notes, labels, and callouts to identify various devices or subcircuits. Frequently, dashed lines are drawn around certain sections to identify that portion as a group. For example, a box may be drawn around the wiring that represents the internal connections of a magnetic motor starter. By noting the wiring that crosses the dashed lines, the technician can differentiate between the starter's internal and external connections. The wiring diagram may also indicate wiring information such as terminal numbers and wire color.

Wiring diagrams may use line thickness to differentiate between voltage levels. **See Figure 15-36.** For example, higher voltage power circuits (feeding motors and other large loads) are usually drawn in thick lines, while lower voltage control circuits are drawn in thin lines.

Two-Wire Motor Starter Control

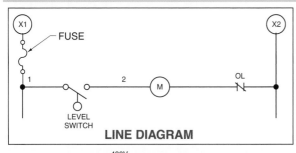

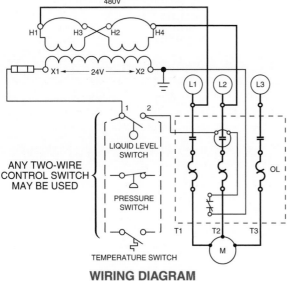

Figure 15-35. A two-wire motor control circuit is very simple but often used in ways that can be hazardous if it operates automatically.

Wiring Diagram Notations

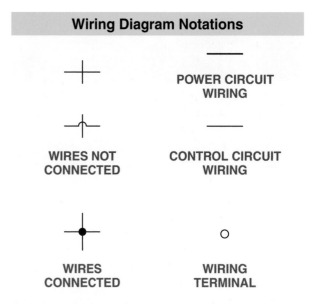

Figure 15-36. Wiring diagrams use line and dot conventions to indicate circuit type and connections.

Solid black dots at line intersections mean that the wires are connected electrically. It does not mean that the wires are spliced at this point, only that they are connected is some way, such as coming from the same terminal. Wires crossing in a diagram with no dot, or with one wire drawn as "jumping over" the other, indicate that there is no electrical connection.

Common Wiring Diagram Circuit. The wiring diagram of the standard start-stop station shows all the wiring of the circuit. **See Figure 15-37.** The motor starter and its wiring are shown in thick lines to identify them as part of the power circuit. The power circuit shows the 3ϕ power supply connecting to the magnetic motor starter at the L (line) connections. When the motor starter contacts are closed, the current then passes through the large motor contacts and the overloads to the T connections on the motor.

The two, thick lines running from L1 and L2 on the starter supply the high-voltage primary (H1 and H4) connections of the transformer. The voltage coming out of the secondary side, X1 and X2, is 24 V. The low-voltage wiring is shown in thin lines. There is a fuse in the wire leading from X1 to the stop button, labelled wire reference (terminal) number 1. The M contacts are labeled 2 and 3. These connection numbers are important when locating loose or damaged wires.

Schneider Electric USA, Inc.

Many motor control circuits incorporate the holding contacts on the motor starter. With standard wire reference numbering, these contacts are designated as terminals 2 and 3 on the starter.

Wiring Diagrams

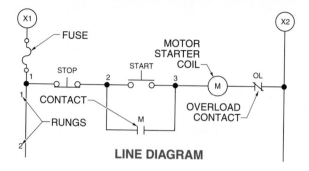

LINE DIAGRAM

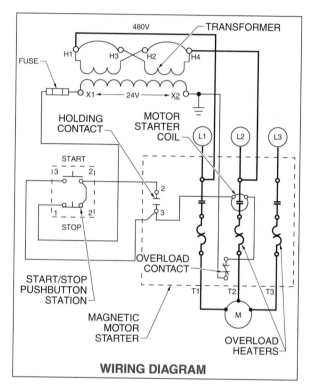

WIRING DIAGRAM

Figure 15-37. A wiring diagram usually includes the control circuit and the power circuit. The control circuit, however, is typically not as easily interpreted in this format.

Review

1. What is the difference between an open circuit and a closed circuit?

2. What are some of the common special functions on digital multimeters?

3. What are the two types of current measurements?

4. What should always be verified before troubleshooting a malfunctioning electrical device?

5. What is the basic operating principle of a transformer?

6. How are heating elements tested?

7. What is the purpose of fuses and circuit breakers?

8. How are fuses and circuit breakers tested?

9. How does a ground fault circuit interrupter (GFCI) work?

10. What are the differences between coils, relays, contactors, and motor starters?

11. What are the causes and consequences of overheating in a motor?

12. How does a variable-frequency drive produce the output power needed to run a motor at a variety of speeds?

13. How do some variable-frequency drives provide information on status or faults?

14. What is a line diagram and what is its primary advantage?

15. What are the four types of numerical notations on line diagrams?

Digital Resources

ATPeResources.com/Quicklinks
Access Code 462409

CHAPTER 16

HYDRAULIC PRINCIPLES

Hydraulics is the branch of science that deals with the practical application of water or other liquids at rest or in motion. Hydrostatics is the study of liquids at rest and the forces exerted on them or by them. Hydrodynamics is the study of the forces exerted on a solid body by the motion or pressure of a fluid. A liquid is a fluid that can flow readily and assume the shape of its container. Fluid flow is the movement of fluid caused by a difference in pressure between two points. In a hydraulic system, fluid flow is produced by the action of a pump and is expressed as a measurement of gallons per minute or liters per minute.

Chapter Objectives

- Define hydraulics and the two major divisions of hydraulics.

- Identify the different types of pressure.

- Identify the terms related to fluid flow.

- Define mechanical advantage.

- Describe the different types of energy involved with the mechanics of hydrostatics.

Key Terms

- hydraulics
- hydrostatics
- hydrodynamics
- fluid
- pressure
- force
- head
- flow
- lift
- friction

- viscosity
- volume
- capacity
- velocity
- mechanical advantage
- energy
- efficiency
- horsepower
- torque

HYDRAULICS

Hydraulics is the branch of science that deals with the practical application of water or other liquids at rest or in motion. The two major divisions of hydraulics are hydrostatics and hydrodynamics.

Hydrostatics

Hydrostatics is the study of liquids at rest and the forces exerted on them or by them. *Equilibrium* is the condition when all forces and torques are balanced by equal and opposite forces and torques. Most hydraulic systems apply hydrostatic principles. For example, the liquid in an automobile's hydraulic braking system is at rest and the pressure throughout the system is in equilibrium. **See Figure 16-1.** The brake system is activated by applying pressure to the foot pedal. The liquid in the system transmits the applied force from the foot pedal to the slave cylinder piston. The slave cylinder piston transmits the force to the brake pad, which applies pressure to the brake drum. The pressure of the liquid is equal in all parts of the system, but higher than the pressure of the liquid when the system is at rest.

Hydrostatics

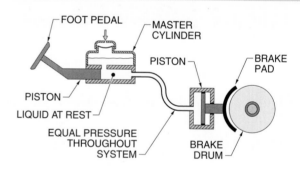

BRAKE SYSTEM AT REST

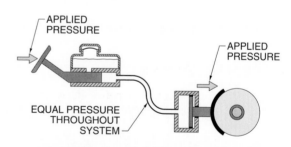

BRAKE SYSTEM ACTIVATED

Figure 16-1. Hydrostatics is the study of liquids at rest and the forces exerted on them or by them.

Atlas Technologies, Inc.

Hydraulics provides the force required for the stamping and fabrication of most products.

Hydrodynamics

Hydrodynamics is the study of the forces exerted on a solid body by the motion or pressure of a liquid. For example, liquids are transferred through a nonpositive displacement pump by centrifugal force. A *nonpositive displacement pump* is a pump that is not sealed between its inlet and outlet. *Centrifugal force* is the outward force produced by a rotating object. **See Figure 16-2.** The liquid is forced to the discharge (outlet) port by rotating impeller vanes. The output of the pump may be reduced or completely blocked if the pressure in the discharge circuit is increased because there is no positive displacement of liquid.

hollowing tree trunks and connecting them together. The ancient Romans, Greeks, and Egyptians used lead, copper, or bronze plumbing for conveying water. About 500 B.C., Persian aqueducts using metal plumbing were tunneled through mountains. Around 100 B.C., hydraulic machines, such as the water-lifting machine and the Archimedes water-screw, were developed.

The water-lifting machine (water wheel) was a device equipped with paddles which raised water by the force of current from a stream. The Archimedes water-screw was another device used to raise water from a stream or lake up to an irrigation ditch. The Archimedes water-screw consisted of a wood core with layers of pitch-covered wood strips attached to form a spiral. **See Figure 16-3.** This assembly was covered to create a spiral tube. With one end of the tube lowered into the water, the complete assembly was rotated, allowing water to hydrostatically work its way up the screw.

Devices applying hydrodynamic principles appeared around 1500 A.D., when the piston concept was used to pump water to the top of a 40′ Roman aqueduct. The Ramelli quadruple suction pump used a water wheel to drive a wooden peg gear mechanism. **See Figure 16-4.** The peg gear mechanism drove a worm gear that was connected to a rotating crank. The rotating crank was attached to a reciprocating piston (suction pump) that would raise water with each rotation of the crank.

Hydrodynamics

DISCHARGE (OUTLET) PORT

ROTATING IMPELLER BLADES

FLUID FLOW

DIRECTION OF ROTATION

LIQUID-FILLED CELLS

NON-POSITIVE DISPLACEMENT PUMP

Figure 16-2. Hydrodynamics is the study of the forces exerted on a solid body by the motion or pressure of a liquid.

Hydraulic System History

Early hydraulic systems consisted of diverting streams for village irrigation and water supply and digging wells. In prehistoric Europe, plumbing was formed by

Archimedes Water-Screw

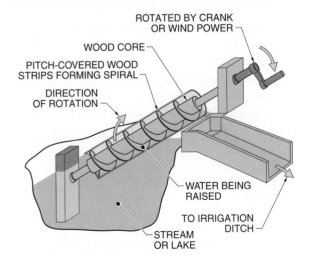

ROTATED BY CRANK OR WIND POWER

WOOD CORE

PITCH-COVERED WOOD STRIPS FORMING SPIRAL

DIRECTION OF ROTATION

WATER BEING RAISED

TO IRRIGATION DITCH

STREAM OR LAKE

Figure 16-3. The Archimedes water-screw hydrostatically raised water as the device was rotated.

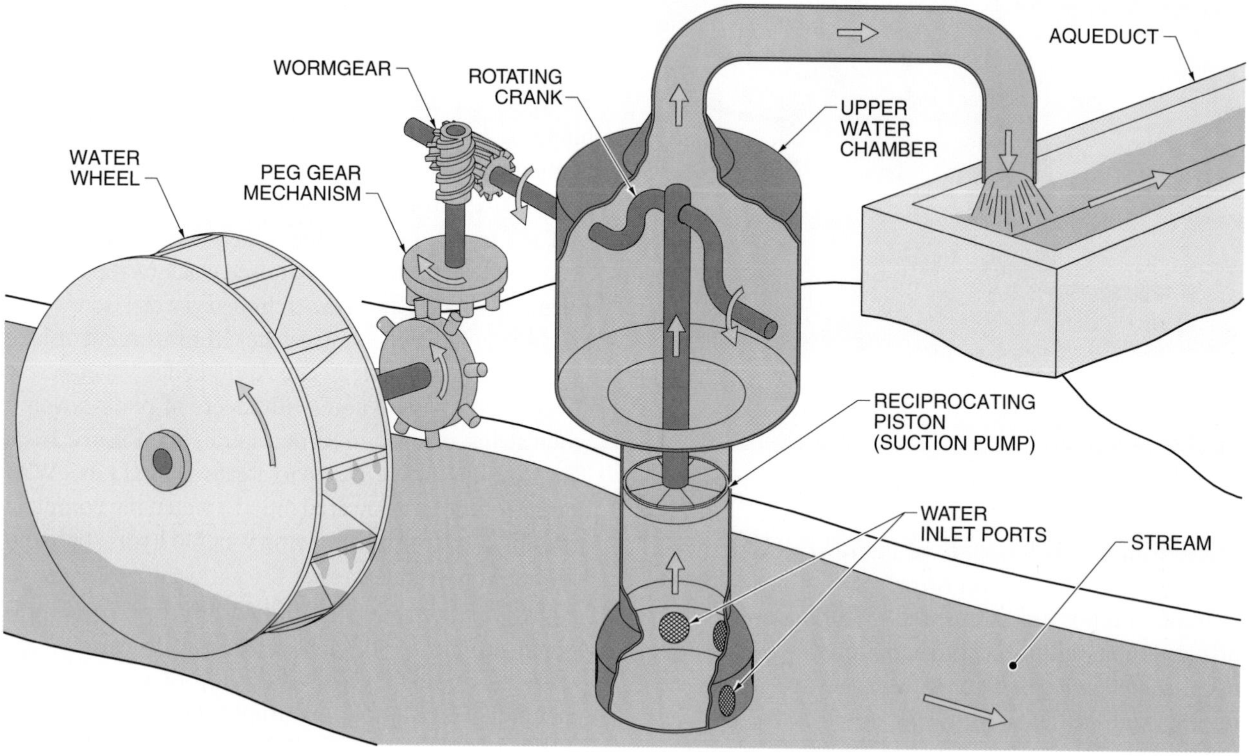

Figure 16-4. Devices applying hydrodynamic principles appeared around 1500 A.D., when the piston concept was used to pump water to the top of a 40′ Roman aqueduct.

FLUID CHARACTERISTICS

In hydraulics, the term "fluid" refers to gases as well as liquids. A *fluid* is a substance that tends to flow or conform to the outline of its container (such as a liquid or a gas). Fluids yield easily to pressure. A *liquid* is a fluid that can flow readily and assume the shape of its container. Liqids have no independent shape but do have a definite volume. Liquids do not expand indefinitely and are only slightly compressible. A *gas* is a fluid that has neither independent shape nor volume and tends to expand indefinitely. Oxygen, hydrogen, etc. are gases.

Liquids make convenient fluids for transmitting force because they are not highly compressible like gases. The term "fluid" is used in reference to a liquid because liquids are specifically used in hydraulic systems. Work produced in a hydraulic system is dependent on the pressure and flow of the fluid in the system.

Pressure

Pressure is the measure of force per unit area. In 1653, French scientist Blaise Pascal realized that enclosed fluids under pressure follow a definite law. Pascal's law states that pressure at any one point in a static liquid is the same in every direction and acts with equal force on equal areas. *Force* is an interaction that tends to change the state of rest or motion of an object. However, it was not until the 20th century that fluid power became a means of energy transmission.

Pressure is expressed as atmospheric, gauge, and absolute. *Atmospheric pressure* is the force exerted by the weight of the atmosphere on the Earth's surface. The weight of the atmosphere, acting over a height of several hundred thousand feet above the Earth's surface, varies slightly with weather conditions. The weight of the atmosphere at sea level is 14.7 pounds per square inch absolute (psia). Atmospheric pressure is also expressed in inches of mercury absolute (in. Hg abs) and is measured with a mercury barometer. A *mercury barometer* is an instrument that measures atmospheric pressure using a column of mercury. **See Figure 16-5.**

Tech Tip

Mineral-base oil is the most widely used hydraulic fluid. It has excellent lubricating properties, does not cause rusting, dissipates heat readily, and can be cleaned easily by mechanical filtration and gravity separation.

Pressure

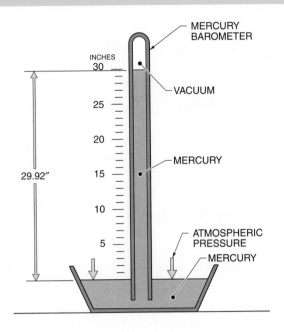

Pressure Equivalents				
Atmospheres	PSIA	PSIG or PSI	in. Hg ABS	in.Hg
3	44.7	29.4	89.76	—
2	29.4	14.7	59.84	—
1	14.7	0	29.92	0
—	10	—	20	10
—	5	—	10	20
—	0	—	0	29.92

Figure 16-5. A mercury barometer is an instrument that measures atmospheric pressure using a column of mercury.

A mercury barometer consists of a glass tube that is closed on one end and completely filled with mercury. The tube is inverted and the open end is submerged in a dish of mercury. A vacuum is created at the top of the tube as the mercury tries to run out of the tube. *Vacuum* is a pressure lower than atmospheric pressure. The pressure of the atmosphere on the mercury in the open dish prevents the mercury in the tube from running out of the tube. The height of the mercury in the tube corresponds to the pressure of the atmosphere on the mercury in the open dish.

A mercury barometer is commonly calibrated in inches of mercury (in. Hg). At sea level, the atmosphere can support 29.92″ Hg in the tube. A barometric pressure of 29.92″ Hg equals one atmosphere or 14.7 psi. Pressures above one atmosphere are generally expressed in psi and pressures below one atmosphere are generally expressed in in. Hg. Minute pressure changes are expressed in inches of water column (in. WC). Atmospheric pressure at sea level should be able to hold water in a column 406.91″ (33.9′) high because atmospheric pressure is able to hold mercury in a column 29.92″ high and water is 13.6 times lighter than mercury.

Gauge pressure is pressure above atmospheric pressure that is used to express pressures inside a closed system.

Gauge pressure assumes that atmospheric pressure is zero (0 psi). Most pressures are measured as gauge pressure unless otherwise specified. Gauge pressure is expressed in pounds per square inch gauge (psig) or psi.

Absolute pressure is pressure above a perfect vacuum. Absolute pressure is the sum of gauge pressure plus atmospheric pressure. Absolute pressure is expressed in pounds per square inch absolute (psia).

Pressure outside a closed system (such as normal air pressure) is expressed in pounds per square inch absolute. The difference between gauge pressure and absolute pressure is the pressure of the atmosphere at sea level at standard conditions (14.7 psia). A pressure gauge reads 0 psig at normal atmospheric pressure. To find absolute pressure when gauge pressure is known, the atmospheric pressure of 14.7 psia is added to the gauge pressure. Absolute pressure is found by applying the formula:

$$psia = psig + 14.7$$
where
$psia$ = pounds per square inch absolute
$psig$ = pounds per square inch gauge
14.7 = constant (atmospheric pressure at standard conditions)

For example, what is the absolute pressure in a system when a pressure gauge reads 100 psig?

$psia = psig + 14.7$

$psia = 100 + 14.7$

$psia = \textbf{114.7 psia}$

Pressure, other than atmospheric pressure, is considered to be artificial and is produced to transfer or amplify force in hydraulic systems. This transferred or amplified force is used to do work such as lifting a car with a hydraulic jack, running a conveyor with a hydraulic motor, or shaping steel into car or truck components.

Area, force, and pressure are the basis of all hydraulic systems. The force exerted by a liquid is based on the size of the area on which the liquid pressure is applied. In hydraulic systems, this area usually refers to the face of a piston, which is circular in shape. Area is always expressed in square units, such as sq in. or sq mm.

A circle with a diameter the same as a square has less area. The area of a circle is exactly 78.54% of the area of a square with the same measurement. **See Figure 16-6.** The area of a circle is found by applying the formula:

$A = 0.7854 \times D^2$

where

A = area (in sq in.)

0.7854 = constant

D^2 = diameter squared

For example, what is the area of a circle with a diameter of 3″?

$A = 0.7854 \times D^2$

$A = 0.7854 \times (3 \times 3)$

$A = 0.7854 \times 9$

$A = \textbf{7.069 sq in.}$

The area of a piston can be found if the force and pressure applied to a cylinder are known. The applied pressure on a piston can be found if the amount of force and the piston area are known. Also, the force produced by a piston can be found if the area and pressure applied to a piston are known. Two of the values must be known to find the unknown value.

The relationship between force, pressure, and area can be recalled using a force, pressure, and area formula pyramid. By covering the letter of the unknown value, the formula for finding the solution is shown.

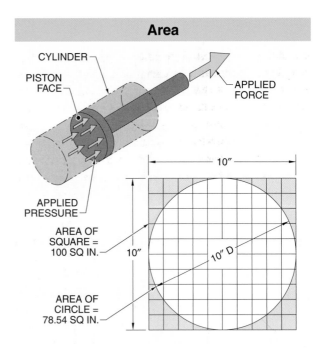

Area

Figure 16-6. In hydraulic systems, area usually refers to the face of a piston, which is in circular shape.

See Figure 16-7. For example, to find area, covering A indicates that F is divided by P. To find pressure, covering P indicates F is divided by A. To find force, covering F indicates P is multiplied by A.

For example, what is the area of a piston face with 5000 lb of force exerting 250 psi?

$A = \dfrac{F}{P}$

$A = \dfrac{5000}{250}$

$A = \textbf{20 sq in.}$

Force and diameter must be known to calculate the required pressure of a cylinder. The area of a piston is calculated from the diameter and then used to find the pressure in a cylinder.

For example, how much pressure is required to move a 5000 lb force with a 4″ D piston?

1. Find area of piston face.

$A = 0.7854 \times D^2$

$A = 0.7854 \times (4 \times 4)$

$A = 0.7854 \times 16$

$A = 12.566$ sq in.

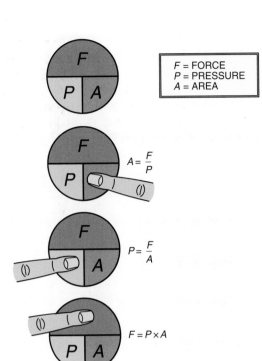

F = FORCE
P = PRESSURE
A = AREA

$$A = \frac{F}{P}$$

$$P = \frac{F}{A}$$

$$F = P \times A$$

Figure 16-7. The force, pressure, or area of a piston can be calculated if any two values are known.

2. Find required pressure.

$$P = \frac{F}{A}$$

$$P = \frac{5000}{12.566}$$

$$P = \textbf{397.899 psi}$$

Pressure and diameter must be known when calculating the required force of a cylinder. The area of a piston is calculated using the diameter of the piston and then the area is used to find the required force.

For example, how much force does 450 psi produce in a 4″ D piston?

1. Find area of piston face.

$$A = 0.7854 \times D^2$$

$$A = 0.7854 \times (4 \times 4)$$

$$A = 0.7854 \times 16$$

$$A = 12.566 \text{ sq in.}$$

2. Find required force.

$$F = P \times A$$

$$F = 450 \times 12.566$$

$$F = \textbf{5654.7 lb}$$

Head Pressure. *Head* is the difference in the level of a liquid (fluid) between two points. Head is expressed in feet. *Head pressure* is the pressure created by fluid stacked on top of itself. **See Figure 16-8.**

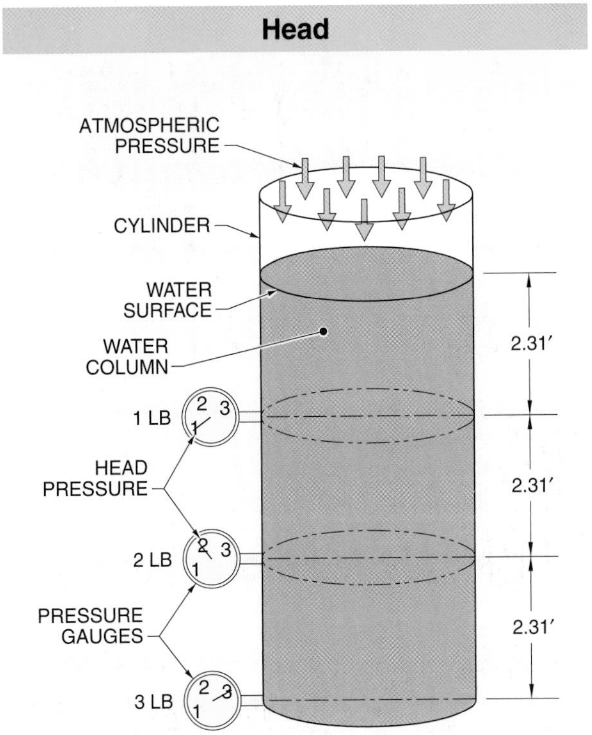

Head

ATMOSPHERIC PRESSURE

CYLINDER

WATER SURFACE

WATER COLUMN

1 LB

HEAD PRESSURE

2 LB

PRESSURE GAUGES

3 LB

2.31′

2.31′

2.31′

Figure 16-8. Head pressure is the pressure created by fluid stacked on top of itself.

In an open cylinder, the pressure of the fluid at any depth in the cylinder is proportional to the height of the column of fluid. The pressure in a column of fluid is determined using the column's height and the fluid's weight, not the shape of the vessel. The pressure at the same level in each vessel is identical if the pressure surrounding the different-shaped vessels is the same and the fluid in each vessel is the same. The pressure of the fluid at any level in a vessel is based on the height of the fluid above that level and is the same at that level regardless of the shape of the vessel. **See Figure 16-9.**

Tech Tip

The head pressure at any point in a container is directly proportional to the density of the fluid and the depth below the surface of the fluid.

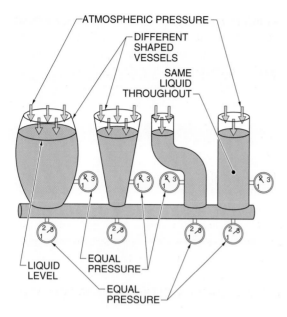

Figure 16-9. The pressure of the fluid at any level in a vessel is based on the height of the fluid above that level and is the same at the level regardless of the shape of the vessel.

The pressure at the base of a column of fluid is calculated by multiplying the weight of the fluid by its height. The weight of a fluid is obtained from a Fluid Weights/Temperature Standards table. **See Appendix.** The pressure of a fluid in a cylinder is found by applying the formula:

$$P = w \times h$$

where

P = pressure at base (in psi)

w = weight of fluid (in lb/cu in. from Fluid Weights/Temperature Standards table)

h = height (in in.)

For example, what is the pressure at the base of a 72″ D cylindrical vessel that contains 96″ of water? *Note:* The weight of water (in lb/cu in.) equals 0.0361 at 39°F (from Fluid Weights/Temperature Standards table).

$$P = w \times h$$
$$P = 0.0361 \times 96$$
$$P = \textbf{3.466 psi}$$

In a hydraulic system, head pressure is the energy or pressure that supplies a hydraulic pump. Atmospheric pressure and head pressure combine to feed the suction (intake) line connecting a hydraulic pump to a reservoir.

Head is classified as static or dynamic. *Static head* is the height of a fluid above a given point in a column at rest. *Static head pressure* is a pressure created by the weight of a fluid. Static head pressure is potential energy. The pressure of water per foot of static head is calculated by using 0.0361 lb/cu in. or 2.31′ head of water for each psi. **See Figure 16-10.**

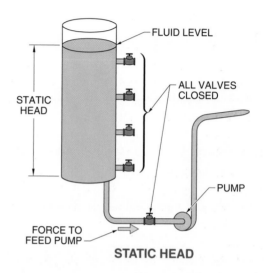

STATIC HEAD

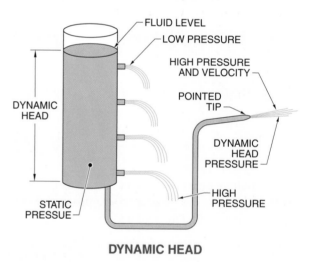

DYNAMIC HEAD

Figure 16-10. Static head is the height of a fluid above a given point in a column at rest. Dynamic head is the head of fluid in motion.

For example, a storage tank located on the second floor of a building contains 10′ of water. The tank feeds a pump 14′ below on the first floor. The total static head from the surface of the water to the pump below is 24′. Therefore, the theoretical head pressure is 10.397 psi (0.0361 × 12″ × 24′ = 10.397 psi). Static head pressure provides pressure to move a fluid when a port or valve is opened.

Dynamic head is the head of fluid in motion. Dynamic head represents the pressure necessary to force a fluid from a given point to a given height. *Dynamic head pressure* is the pressure and velocity produced by a fluid in motion. Dynamic head pressure results when a valve is opened and fluid is allowed an open flow. Dynamic head pressure may be used to direct an open flow of fluid. For example, dynamic head pressure was used in early prospecting days to wash away the sides of mountains to retrieve gold. This was accomplished by piping water from higher lakes and using dynamic head pressure to produce a high pressure and high velocity.

Lift Pressure. *Lift* is the height at which atmospheric pressure forces a fluid above the elevation of its supply source. **See Figure 16-11.** A pipe with one end in fluid and the other end open to the atmosphere is in equilibrium. Atmospheric pressure lifts (pushes) the liquid in the pipe when a pump is placed on the end of the pipe open to the atmosphere and a vacuum is drawn. With respect to pump operation, lift is the height measured from the elevation of the supply source to the center of the pump's inlet port.

The maximum height a fluid at a standard temperature of 62°F can be lifted is determined by the barometric pressure. Temperature standards are established by agencies such as the American National Standards Institute (ANSI) and the International Organization for Standardization (ISO) to create national and international uniformity. Temperature standards for fluids are required due to the fluctuation of fluid volume at different temperatures.

Static lift is the height to which atmospheric pressure causes a column of fluid to rise above the supply to restore equilibrium. The weight of a column of fluid required to create equilibrium is equal to atmospheric pressure. For example, when an elevated pump is turned ON, the pump removes air from its plumbing, creating a partial vacuum. The fluid then rises to a height that is determined by atmospheric pressure. Atmospheric pressure essentially lifts the fluid to a height of equilibrium, or the balance between the atmosphere's pressure and the water's weight. **See Figure 16-12.**

Tech Tip

The energy applied to a fluid by a pump goes either into the production of usable pressure or velocity in the fluid or into friction losses.

Lift

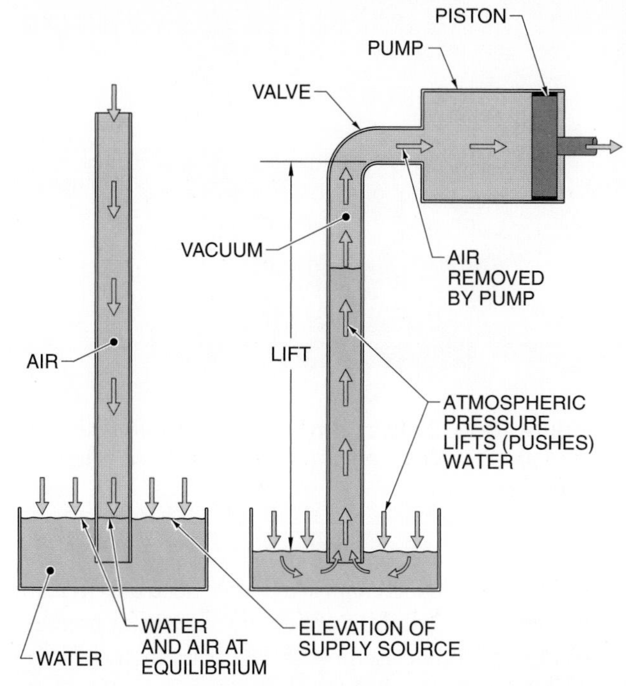

Altitude above Sea Level*	Barometer Reading†	Atmospheric Pressure‡	Theoretical Lift at Standard Temperature of 62°F*
0	29.92	14.7	34
1000	28.8	14.2	33
2000	27.7	13.6	31.5
3000	26.7	13.1	30.2
4000	25.7	12.6	29.1
5000	24.7	12.1	28
6000	23.8	11.7	27
7000	22.9	11.2	26
8000	22.1	10.8	25
9000	21.2	10.4	24
10,000	20.4	10.0	23

* in ft
† in in. Hg
‡ in psi

Figure 16-11. Lift is the height at which atmospheric pressure forces a liquid above the elevation of its supply source.

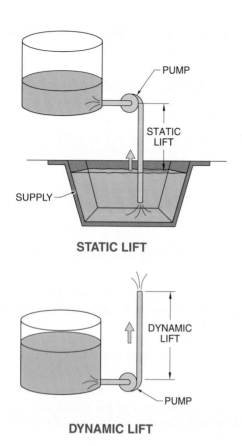

STATIC LIFT

DYNAMIC LIFT

Figure 16-12. Static lift is the height to which atmospheric pressure causes a column of the fluid to rise above the supply to restore equilibrium. Dynamic lift is the lift of fluid in motion.

Dynamic lift is the lift of fluid in motion. Dynamic lift represents the pressure necessary to lift a fluid from a given point to a given height. The dynamic lift and distance a fluid can be raised vary due to pump imperfections and pipe friction. Therefore, the limit of actual pump lift ranges to approximately 25′. A pump can force a fluid to greater heights depending on the force exerted on the fluid.

Practical dynamic lift is considerably less than practical static lift because of the friction within piping lengths, piping sizes, number of fittings such as elbows or valves, and because of the fact that most installations are higher than sea level. The practical lift in a pump moving water generally falls in the 20′ to 25′ range.

Total column is the fluid head plus lift. Total column may be static or dynamic. **See Figure 16-13.** *Static total column* is static head plus static lift. Static columns are determined from pressures created by a fluid at rest.

A pump removing water from a retention pond to supply an overhead cooling tower is an example of a static total column. The pump, which is mounted above the pond, assists in lifting the water up to its center (static lift), and at the same time supports the column of water up to the cooling tower (static head). The sum of a pump's static lift and static head is the static total column.

Total Column

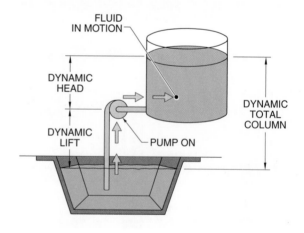

Figure 16-13. Static or dynamic columns are determined from pressures created by a fluid at rest or in motion.

Dynamic total column is dynamic head plus dynamic lift. Dynamic columns are determined from pressures created by a fluid in motion. Dynamic total column is the total column of fluid in motion and represents the pressure from the total column plus frictional resistance.

Eaton Corporation

Fixed displacement gear pumps produce fluid flow in a hydraulic system from 6.4 gpm to 24.1 gpm and operate at a pressure of 3000 psi continuously.

Fluid Flow

Fluid flow is the movement of fluid caused by a difference in pressure between two points. In a hydraulic system, fluid flow is produced by the action of a pump and is expressed as a measurement of gallons per minute (gpm) or liters per minute (lpm). Fluid flow in a hydraulic system is affected by friction and the viscosity of the fluid. Fluid flow is based on the volume and capacity of the system and the velocity of the fluid in the system. Fluid flow also affects the speed of a hydraulic system.

In a hydraulic system with flowing fluid, pressure is caused by total resistance to the fluid flow from a pump. Pressure results only where there is resistance to flow. Resistance to flow is comprised of friction throughout the system and actuator loads. A pressure change that occurs to a fluid due to its flow is generally expressed in psi.

Friction. Friction is generated throughout a hydraulic system between the piping wall and the fluid, and within the fluid as fluid molecules slide by one another. The faster a fluid flows, the greater the friction. Any friction generated becomes a resistance to fluid flow. Pressure must be increased to overcome the friction. Each component in a hydraulic system offers resistance and reduction of available working pressure. **See Figure 16-14.**

Tech Tip

A fluid vaporizes if an attempt is made to lift the fluid more than the distance atmospheric pressure is capable of raising it.

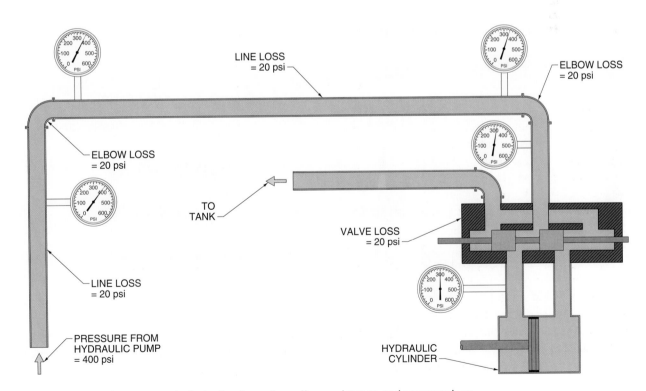

LINE LOSS = 20 psi

ELBOW LOSS = 20 psi

ELBOW LOSS = 20 psi

TO TANK

VALVE LOSS = 20 psi

LINE LOSS = 20 psi

PRESSURE FROM HYDRAULIC PUMP = 400 psi

HYDRAULIC CYLINDER

Figure 16-14. Each component of a hydraulic system offers resistance and pressure loss.

A fluid flows because of a difference in pressure. The pressure of a moving fluid is always higher upstream. *Pressure drop* is the pressure differential between upstream and downstream fluid flow caused by resistance. The pressure developed in a hydraulic system is designed to be used as hydraulic leverage. Pressure and fluid flow rate are independent of each other, but both assist in the output. Pressure provides the force and flow rate is used to provide speed. Flow rate is expressed in gpm and is typically determined by the capacity of the pump.

Fluids follow the path of least resistance. For example, a hydraulic system consists of two equal-diameter cylinders with a load of 500 lb on Cylinder A and 200 lb on Cylinder B. Cylinder B moves to the end of its travel before Cylinder A begins to move because of the reduced resistance produced by the lighter weight. **See Figure 16-15.**

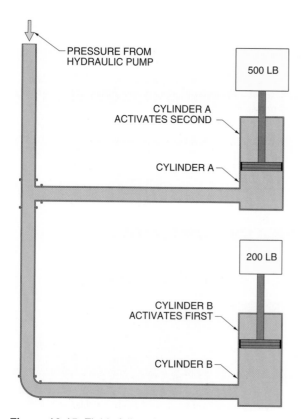

Figure 16-15. Fluids follow the path of least resistance.

Viscosity. *Viscosity* is the measure of a fluid's resistance to flow. Fluids that flow with difficulty have a high viscosity. Fluids that are thin and flow easily have a low viscosity. For example, cold honey has a high viscosity and water has a low viscosity.

Viscosity is determined under laboratory conditions by measuring the time required for a specific amount of a fluid at a specific temperature to flow through a specific size orifice. Viscosity is measured in Saybolt Universal Seconds (SUS) using a Saybolt viscometer. **See Figure 16-16.**

A *Saybolt viscometer* is a test instrument used to measure fluid viscosity. Upon reaching the proper temperature, a cork is pulled, allowing 60 mL of test fluid to flow out of the cylinder while being timed with a stopwatch. The measured time is the SUS. The Society of Automotive Engineers (SAE) has established standard numbers for oil viscosity readings.

For example, an SAE 10 oil at 130°F placed in a Saybolt viscometer takes between 90 sec and 120 sec to empty. An oil that is thicker (more viscous) and at the same temperature takes longer to empty the viscometer. This oil has a higher SAE number such as SAE 30, which takes between 185 sec and 255 sec to empty the viscometer.

The *viscosity index* is a scale used to show the magnitude of viscosity changes in lubrication oils with changes in temperature. The viscosity index indicates the relative change in SUS readings. Desirable oils are those that have a high viscosity index (relatively low SUS reading change). Oils with a low viscosity index register a large change in SUS readings as temperatures change. Oils with a high viscosity index change slightly as the temperature of the fluid changes.

Saybolt Viscometer

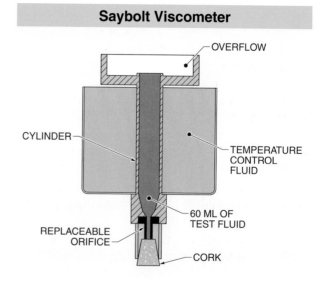

Figure 16-16. A Saybolt viscometer is an instrument used to measure the viscosity of a fluid.

High-viscosity fluids lead to high internal resistance (friction). This results in high resistance to flow through hydraulic components, creating slow component movement. High resistance increases power consumption, creating a considerable pressure drop throughout the system. Also, fluid temperatures rise when friction is high. Increased fluid temperatures can cause fluid breakdown and damage to pumps and seals.

Fluid viscosity that is too low can be equally harmful. Slippage and leaking may occur, producing an increase in wear because of a reduced lubricating fluid film between mechanical parts. *Slippage* is the internal leaking of fluid from a pump's outlet to a pump's inlet. Slippage occurs between the gear teeth and the housing and along the sides of the gears in a gear pump. A *gear pump* is a positive-displacement pump containing intermeshing gears that force the fluid from the pump. *Displacement* is the volume of fluid moved during each revolution of a pump's shaft. A *positive-displacement pump* is a pump that delivers a definite volume of fluid for each cycle of the pump at any resistance encountered. Slippage is desirable in limited amounts to lubricate moving parts. However, as pressures increase beyond the pump's pressure rating, slippage increases rapidly.

A difference exists between actual pump output and absolute pump output (zero slippage) because slippage indicates a decrease in pump output and fluid flow. *Volumetric efficiency* is the percentage of actual pump output compared to the pump output if there were no slippage. A typical positive-displacement pump in good condition has a volumetric efficiency of approximately 85% because some slippage is required for pump lubrication.

Volume. *Volume* is the size of a three-dimensional space or object as measured in cubic units. **See Figure 16-17.** Regardless of the shape of an object, volume is always expressed in cubic units such as cubic inches, cubic feet, cubic millimeters, or cubic meters. The volume of an object can be found by calculating the area of the object and multiplying that area by the height or length. For example, the volume of a cylinder can be calculated by applying the following procedure:

1. Find area of cylinder.

   $A = 0.7854 \times D^2$

   where

   A = area (in sq units)

   0.7854 = constant

   D^2 = diameter squared

2. Find volume of cylinder.

   $V = A \times l$

   where

   V = volume (in cu units)

   A = area (in sq units)

   l = length (in units)

Volume

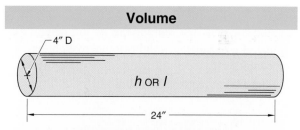

Calculating Area and Volume of a Cylinder

1. Find area of cylinder.	2. Find volume of cylinder.
$A = 0.7854 \times D^2$	$V = A \times h$
$A = 0.7854 \times (4 \times 4)$	$V = 12.566 \times 24$
$A = 0.7854 \times 16$	$V = \textbf{301.584 cu in.}$
$A = 12.566$ sq in.	

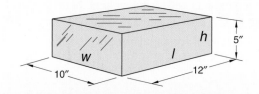

Calculating Area and Volume of a Rectangular Solid

1. Find area of rectangular solid.	2. Find volume of rectangular solid.
$A = l \times w$	$V = A \times h$
$A = 12 \times 10$	$V = 120 \times 5$
$A = 120$ sq in.	$V = \textbf{600 cu in.}$

Figure 16-17. Volume is the size of a three-dimensional space or object as measured in cubic units.

For example, what is the volume of a 4″ D cylinder that is 24″ long?

1. Find area of cylinder.

$A = 0.7854 \times D^2$

$A = 0.7854 \times (4 \times 4)$

$A = 0.7854 \times 16$

$A = 12.566$ sq in.

2. Find volume of cylinder.

$V = A \times l$

$V = 12.566 \times 24$

$V = 301.584$ cu in.

Note: The formula for finding the volume of a cylinder may also be expressed as $V = 0.7854 \times D^2 \times l$.

The volume of a rectangular solid is found by calculating the area and multiplying by the height. The volume of a rectangular solid is found by applying the procedure:

1. Find area of rectangular solid.

$A = l \times w$

where

A = area (in sq units)

l = length (in units)

w = width (in units)

2. Find volume of rectangular solid.

$V = A \times h$

where

V = volume (in cu units)

A = area (in sq units)

h = height (in units)

For example, what is the volume of a 12″ long, 10″ wide, and 5″ high rectangular solid?

1. Find area of rectangular solid.

$A = l \times w$

$A = 12 \times 10$

$A = 120$ sq in.

2. Find volume of rectangular solid.

$V = A \times h$

$V = 120 \times 5$

$V = \mathbf{600}$ **cu in.**

Capacity. *Capacity* is the ability to hold or contain something. Capacity is expressed in cubic units and is calculated from a container's volume. Fluids are measured in ounces, pints, quarts, gallons, liters, etc. based on the size of their containers. **See Appendix.** Fluid measurements can also be expressed in cubic units

(cu in., cu ft, etc.) because fluids occupy three dimensions. For example, one gallon of fluid equals 231 cu in.

The quantity of fluid required to fill a specific volume is determined by calculating the volume and dividing by 231. Capacity of a cylinder is found by applying the procedure:

1. Find area of cylinder.

$A = 0.7854 \times D^2$

2. Find volume of cylinder.

$V = A \times l$

3. Find capacity of cylinder.

$C = \dfrac{V}{231}$

where

C = capacity (in gal.)

V = volume (in cu in.)

231 = constant (cu in. of fluid per gallon)

For example, what is the capacity of a 4″ D cylinder that has a 24″ stroke?

1. Find area of cylinder.

$A = 0.7854 \times D^2$

$A = 0.7854 \times (4 \times 4)$

$A = 0.7854 \times 16$

$A = 12.566$ sq in.

2. Find volume of cylinder.

$V = A \times l$

$V = 12.566 \times 24$

$V = 301.584$ cu in.

3. Find capacity of cylinder.

$C = \dfrac{V}{231}$

$C = \dfrac{301.584}{231}$

$C = \mathbf{1.306}$ **gal.**

Less hydraulic fluid is required to retract a piston than is required to extend a piston. This is due to the piston rod taking up part of the cylinder volume (reduced capacity). The volume that the piston rod occupies must be subtracted from the total volume of the cylinder when determining the volume of fluid that a cylinder displaces when retracting. **See Figure 16-18.** The capacity of a cylinder when retracting is found by applying the procedure:

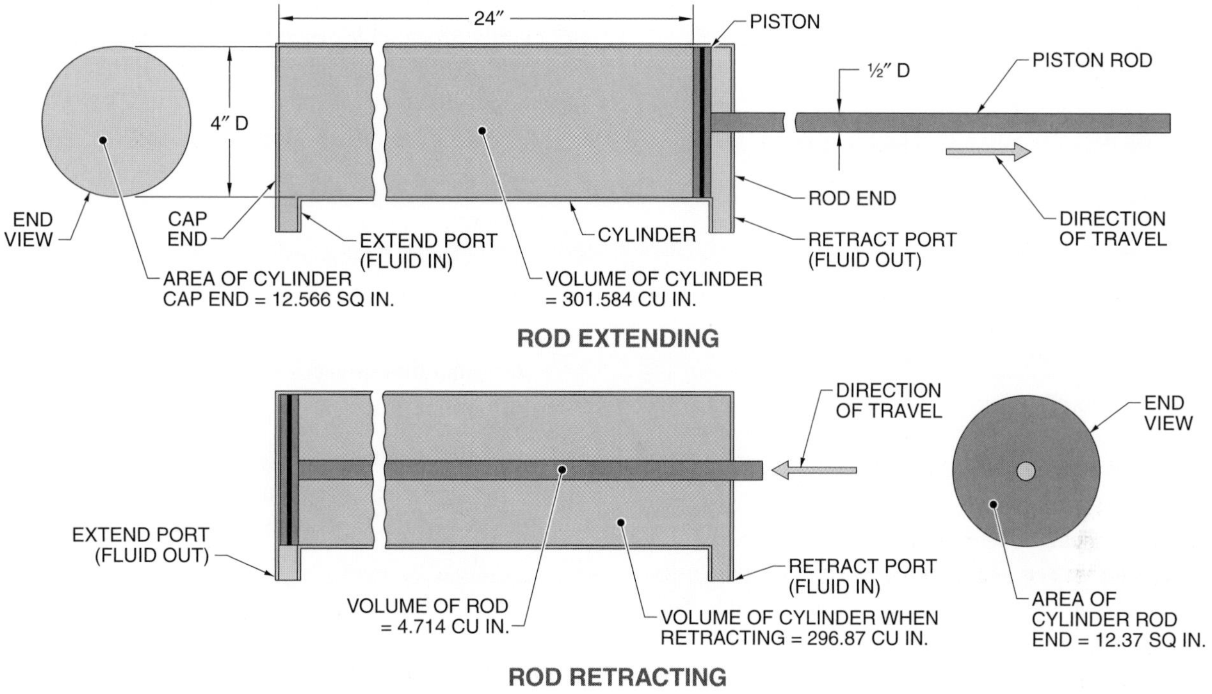

ROD EXTENDING

ROD RETRACTING

Figure 16-18. Less fluid is required to retract a piston than to extend a piston due to the piston rod occupying a part of the cylinder volume.

1. Find area of piston.
   $$A_p = 0.7854 \times D^2$$
2. Find volume of cylinder.
   $$V_c = A_p \times l_c$$
   where
   V_c = volume of cylinder (in cu units)
   A_p = area of piston (in sq units)
   l_c = length of cylinder (in units)
3. Find area of rod.
   $$A_r = 0.7854 \times D^2$$
4. Find volume of rod.
   $$V_r = A_r \times l_r$$
   where
   V_r = volume of rod (in cu units)
   A_r = area of rod (in sq units)
   l_r = length of rod (in units)
5. Find volume of cylinder when retracting.
   $$V_{cr} = V_c - V_r$$
   where
   V_{cr} = volume of cylinder when retracting (in cu units)
   V_c = volume of cylinder (in cu units)
   V_r = volume of rod (in cu units)

6. Find capacity of cylinder when retracting.
   $$C = \frac{V_{cr}}{231}$$

   For example, what is the capacity of a 4″ D hydraulic cylinder when retracting with a 24″ stroke and a ½″ piston rod?

   1. Find area of piston.
      $$A_p = 0.7854 \times D^2$$
      $$A_p = 0.7854 \times (4 \times 4)$$
      $$A_p = 0.7854 \times 16$$
      $$A_p = 12.566 \text{ sq in.}$$

   2. Find volume of cylinder.
      $$V_c = A_p \times l_c$$
      $$V_c = 12.566 \times 24$$
      $$V_c = 301.584 \text{ cu in.}$$

   3. Find area of rod.
      $$A_r = 0.7854 \times D^2$$
      $$A_r = 0.7854 \times (0.5 \times 0.5)$$
      $$A_r = 0.7854 \times 0.25$$
      $$A_r = 0.1964 \text{ sq in.}$$

4. Find volume of rod.

$$V_r = A_r \times l_r$$
$$V_r = 0.1964 \times 24$$
$$V_r = 4.714 \text{ cu in.}$$

5. Find volume of cylinder when retracting.

$$V_{cr} = V_c - V_r$$
$$V_{cr} = 301.584 - 4.714$$
$$V_{cr} = 296.87 \text{ cu in.}$$

6. Find capacity of cylinder when retracting.

$$C = \frac{V_{cr}}{231}$$
$$C = \frac{296.87}{231}$$
$$C = \textbf{1.285 gal.}$$

Velocity. *Velocity* is the distance a fluid travels in a specified time. **See Figure 16-19.** Velocity generally means the change of position of a fluid particle during a certain time interval. This may be represented as distance in feet per second (ft/sec).

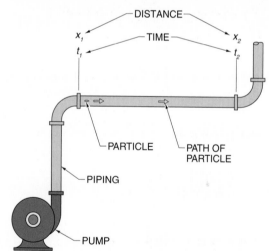

Velocity

Figure 16-19. Velocity is the distance a fluid travels in a specified time.

Velocity is measured as a vector. A *vector* is a quantity that has a magnitude and direction. A vector is commonly represented by a line segment whose length represents its magnitude and whose orientation represents its direction.

The velocity of a fluid particle is determined by subtracting its initial position from its final position and dividing by the value of the initial time subtracted from the final time. Velocity is found by applying the formula:

$$v = \frac{x_2 - x_1}{t_2 - t_1}$$

where

v = velocity (in ft/sec)
x_2 = final position (in ft)
x_1 = initial position (in ft)
t_2 = final time (in sec)
t_1 = initial time (in sec)

For example, what is the velocity of a fluid particle in a hydraulic system that is at point x_1 at 9:33:54 AM, and after traveling 50′ reaches point x_2 at 9:34:30 AM?

$$v = \frac{x_2 - x_1}{t_2 - t_1}$$
$$v = \frac{50 - 0}{9:34:30 - 9:33:54}$$
$$v = \frac{50}{36}$$
$$v = \textbf{1.389 ft/sec}$$

Tech Tip

The velocity of the hydraulic fluid in a system should not exceed recommended values because turbulent conditions result with loss of pressure and excessive heating.

Exercise caution around swinging arms and booms because anything that is supported by fluid pressure can fail if a hose breaks.

The velocity of a fluid varies from one moment to another as its speed or direction of flow changes. *Acceleration* is an increase in speed. Acceleration of a fluid is determined as its change in velocity per unit of time.

The symbol delta (Δ) is generally used to indicate a change. Acceleration is given in units of ft/sec² because velocity is measured in ft/sec and time is measured in sec. Acceleration, like velocity, is constantly changing within a hydraulic system. Pipes of various diameters, elbows, valves, and other components all affect the velocity and acceleration of the fluid within a hydraulic system. Acceleration of a fluid is found by applying the formula:

$$a = \frac{\Delta v}{\Delta t}$$

where

a = acceleration (in ft/sec²)

Δv = average velocity during Δt (in ft/sec)

Δt = time interval elapsed in traveled distance (in sec)

For example, what is the acceleration between measuring points when the fluid flow within a hydraulic system has an initial velocity of 15 ft/sec and changes to 30 ft/sec in 8 sec?

$$a = \frac{\Delta v}{\Delta t}$$

$$a = \frac{30 - 15}{8}$$

$$a = \frac{15}{8}$$

$$a = \textbf{1.875 ft/sec}^2$$

Flow rate is the volume of fluid flow per minute. A fluid in motion is always flowing, but its rate of flow may change. Fluid velocity depends on the rate of flow in gallons per minute (gpm) and the cross-sectional area of a pipe or component.

The velocity of a fluid increases at any restriction in a pipe or component if the flow rate remains the same in the system. Common restrictions include valves, elbows, pipes, reducers, etc. Also, the velocity of a fluid decreases as the cross-sectional area of a pipe or component increases. **See Figure 16-20.**

The law of conservation of matter states that the mass or volumetric flow rate of an incompressible fluid through a pipe is constant at every point in the pipe. The velocity of a fluid must increase at any restriction if there

are no leaks in the system and the flow rate remains constant. The velocity increases four times to maintain a constant rate of flow if a pipe diameter is changed to one-half of its original size. Velocity of a fluid in a pipe is found by applying the formula:

$$v = \frac{l_2}{\frac{A \times l_1}{231} \times \frac{60}{Q}}$$

where

v = velocity (in ft/sec)

l_2 = length of pipe (in ft)

A = cross-sectional area of pipe (in sq in.)

l_1 = length of pipe (in in.)

231 = constant (cu in. of fluid per gallon)

Q = flow rate (in gpm)

60 = constant (sec in 1 min)

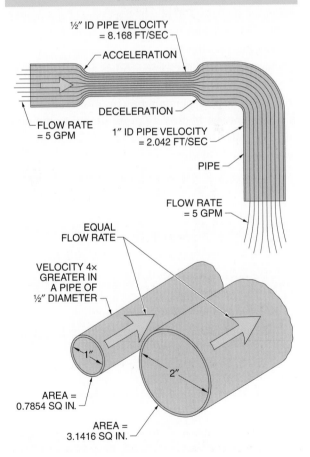

Figure 16-20. The velocity of a fluid increases at any restriction in a pipe or component if the flow rate remains the same in the system.

For example, what is the velocity of a fluid having a flow rate of 5 gpm through a 12″ section of 1″ D pipe?

$$v = \dfrac{l_2}{\dfrac{A \times l_1}{231} \times \dfrac{60}{Q}}$$

$$v = \dfrac{1}{\dfrac{0.7854 \times 12}{231} \times \dfrac{60}{5}}$$

$$v = \dfrac{1}{\dfrac{9.4248}{231} \times 12}$$

$$v = \dfrac{1}{0.0408 \times 12}$$

$$v = \dfrac{1}{0.4896}$$

$$v = \textbf{2.042 ft/sec}$$

Speed. The speed of a cylinder rod is determined by the volume of the cylinder, and the fluid flow rate (gpm). To determine the speed at which a cylinder rod moves, the flow rate at which hydraulic fluid is directed into the cylinder must be known.

The speed of a cylinder rod is independent of pressure. The speed of rod extension is usually expressed in inches per minute (in./min). The speed of rod extension is directly proportional to the flow rate. Cylinder rod extension speed is calculated by applying the formula:

$$s = 231 \times \dfrac{Q}{0.7854 \times D^2}$$

where

s = speed of extension (in in./min)

231 = constant (cu in. of fluid per gallon)

Q = flow rate (in gpm)

0.7854 = constant

D^2 = diameter of cylinder squared

For example, what is the rod speed of a 5″ D cylinder supplied by a 5 gpm pump?

$$s = 231 \times \dfrac{Q}{0.7854 \times D^2}$$

$$s = 231 \times \dfrac{5}{0.7854 \times 5 \times 5}$$

$$s = 231 \times \dfrac{5}{19.64}$$

$$s = 231 \times 0.255$$

$$s = \textbf{58.9 in./min}$$

Two methods of increasing the speed at which a load (or cylinder rod) in a hydraulic system moves are by using a smaller diameter cylinder or by increasing the rate of fluid flow to the cylinder. A small diameter cylinder produces an increase in speed and a decrease in the applied force as compared to a larger cylinder. Two cylinders of different diameters having the same length have different fluid capacities and, if both receive the same rate of fluid flow, the rate of travel and pressure output are different.

MECHANICAL ADVANTAGE

Mechanical advantage is the ratio of the output force to the input force of a device. Mechanical advantage is achieved when an applied input force is multiplied, resulting in a larger output force. **See Figure 16-21.** Devices that produce mechanical advantage include levers, block and tackles, gears, etc.

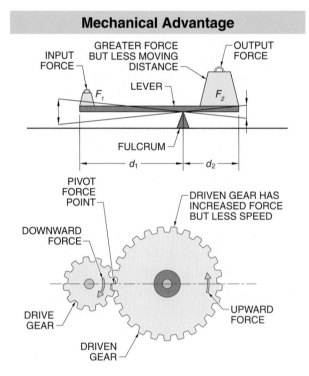

Figure 16-21. Mechanical advantage is the advantage gained by the use of a mechanism in transmitting force.

When using a lever, mechanical advantage results from a force applied a certain distance from a fulcrum. A *fulcrum* is a support on which a lever turns or pivots. In determining the force needed to balance a lever/fulcrum mechanism, the input force must be farther from the fulcrum than the output force or must have an input force equal to or greater than the output force. The force required to overcome a output force is calculated by applying the formula:

$$F_1 = \frac{F_2 \times d_2}{d_1}$$

where

F_1 = input force (in lb)

F_2 = output force (in lb)

d_1 = distance between input force and fulcrum (in ft)

d_2 = distance between output force and fulcrum (in ft)

For example, what is the input force, placed 15′ from the fulcrum, required to lift a output force of 800 lb placed 1½′ from the fulcrum?

$$F_1 = \frac{F_2 \times d_2}{d_1}$$

$$F_1 = \frac{800 \times 1.5}{15}$$

$$F_1 = \frac{1200}{15}$$

$$F_1 = \textbf{80 lb}$$

According to Pascal's law, pressure exerted on an enclosed fluid is transmitted undiminished in every direction. This is demonstrated by a fluid-filled bottle. As the cork is pressed further into the bottle, the pressure throughout the bottle increases until the incompressible fluid bursts the bottle. **See Figure 16-22.** The bottle bursts because the force applied to one area (the cork) is equal to the pressure multiplied by the larger area (the body of the bottle). The resulting force within a vessel is a product of the input force and the output pressure area divided by the input pressure area. Resulting force within a vessel is found by applying the formula:

$$F_2 = F_1 \times \frac{A_2}{A_1}$$

where

F_2 = resulting force (in lb)

F_1 = input force (in lb)

A_2 = area of output pressure (in sq in.)

A_1 = area of input pressure (in sq in.)

For example, what is the force placed on a bottle with a cork area of .4418 sq in. $(0.7854 \times 0.75 \times 0.75 = 0.4418)$ and a bottle surface area of 70.686 sq in. (3″ D × 3.1416 × 7.5″ h = 70.686 sq in.) when 5 lb is applied to the cork?

$$F_2 = F_1 \times \frac{A_2}{A_1}$$

$$F_2 = 5 \times \frac{70.686}{0.4418}$$

$$F_2 = 5 \times 159.99$$

$$F_2 = \textbf{800 lb}$$

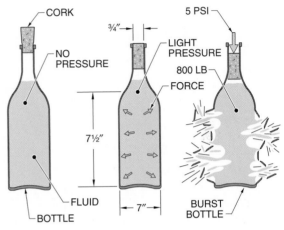

Figure 16-22. Pressure exerted on an enclosed fluid is transmitted undiminished in every direction.

Fluids are well suited for being transmitted through pipes, hoses, and passages because of these force characteristics. This force is energy, which can produce movement, work, or leverage when applied to a hydraulic application. For example, interconnected hydraulic cylinders of different diameters produce hydraulic leverage in a typical car jack. **See Figure 16-23.**

The output pressure of two interconnected cylinders is found by calculating the area of both cylinders, dividing the area of the output cylinder by the area of the input cylinder, and multiplying the result by the input force. Output pressure of two interconnected cylinders is found by applying the procedure:

1. Find area of input piston.

$$A_1 = 0.7854 \times D_1^2$$

2. Find area of output piston.

$$A_2 = 0.7854 \times D_2^2$$

3. Find output piston force.

$$F_2 = F_1 \times \frac{A_2}{A_1}$$

For example, what is the force of a 3″ D output piston interconnected to a ½″ D input piston if a 50 lb force is applied by the input piston?

1. Find area of input piston.
$$A_1 = 0.7854 \times D_1{}^2$$
$$A_1 = 0.7854 \times (0.5 \times 0.5)$$
$$A_1 = 0.7854 \times 0.25$$
$$A_1 = 0.0196 \text{ sq in.}$$

2. Find area of output piston.
$$A_2 = 0.7854 \times D_2{}^2$$
$$A_2 = 0.7854 \times (3 \times 3)$$
$$A_2 = 0.7854 \times 9$$
$$A_2 = 7.069 \text{ sq in.}$$

3. Find output piston force.
$$F_2 = F_1 \times \frac{A_2}{A_1}$$
$$F_2 = 50 \times \frac{7.069}{0.196}$$
$$F_2 = 50 \times 36.066$$
$$F_2 = \mathbf{1803.3 \text{ lb}}$$

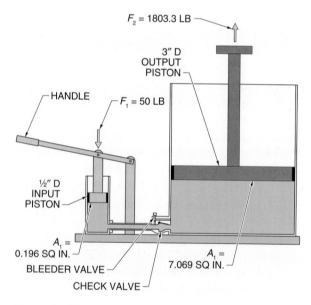

$F_2 = 1803.3$ LB

3″ D
OUTPUT
PISTON

HANDLE

$F_1 = 50$ LB

½″ D
INPUT
PISTON

$A_1 =$
0.196 SQ IN.

$A_1 =$
7.069 SQ IN.

BLEEDER VALVE

CHECK VALVE

Figure 16-23. Interconnected hydraulic cylinders of different diameters produce hydraulic leverage in a typical car jack.

At times, hydraulic system pressure must be determined before calculating either the input force or the output force. This may be required when determining the input force required to produce a given output force with given size cylinders. The required input force is determined by calculating the area of the output cylinder, calculating the pressure in the system, and determining the input force

based on the system pressure and area of the input cylinder. Input force is found by applying the procedure:

1. Find area of output piston.
$$A_2 = 0.7854 \times D_2{}^2$$

2. Find pressure in system.
$$P = \frac{F_2}{A_2}$$

3. Find area of input piston.
$$A_1 = 0.7854 \times D_1{}^2$$

4. Find input force required.
$$F_1 = P \times A_1$$

For example, what is the necessary input force on a 3″ D piston if a static load of 5000 lb being lifted by a 10″ piston stalls due to loss of input force?

1. Find area of output piston.
$$A_2 = 0.7854 \times D_2{}^2$$
$$A_2 = 0.7854 \times (10 \times 10)$$
$$A_2 = 0.7854 \times 100$$
$$A_2 = 78.54 \text{ sq in.}$$

2. Find pressure in system.
$$P = \frac{F_2}{A_2}$$
$$P = \frac{5000}{78.54}$$
$$P = 63.662 \text{ psi}$$

3. Find area of input piston.
$$A_1 = 0.7854 \times D_1{}^2$$
$$A_1 = 0.7854 \times (3 \times 3)$$
$$A_1 = 0.7854 \times 9$$
$$A_1 = 7.069 \text{ sq in.}$$

4. Find input force required.
$$F_1 = P \times A_1$$
$$F_1 = 63.662 \times 7.069$$
$$F_1 = \mathbf{450.027 \text{ lb}}$$

ENERGY AND WORK

The mechanics of hydrostatics, where the flow of fluid within an enclosed system is used to do work, is based on the theory of the conservation of energy. *Energy* is a measure of the ability to do work. The theory of the conservation of energy states that the total energy of a fluid at any point in a system is equal to the total energy of the fluid at another point, unless work has been done by the fluid on some external component.

Total energy is a measure of a fluid's ability to do work. In hydrostatics, total energy is the sum of static energy, kinetic energy, heat energy, and pressure energy. **See Figure 16-24.**

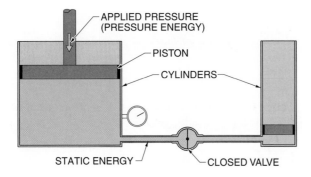

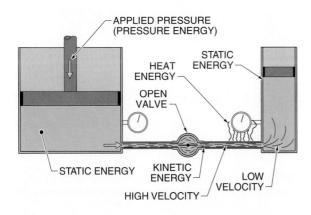

Figure 16-24. In hydrostatics, total energy is the sum of static energy, kinetic energy, heat energy, and pressure energy.

Static energy is the ability of a fluid to do work using the height and weight of the fluid above some reference point. Static energy is stored energy ready to be used. In a hydraulic system, static energy is transformed into kinetic energy when a valve is opened, allowing fluid to flow. This flow causes velocity, acceleration, and the ability to do work.

Kinetic energy is the energy of motion. Any moving object, such as a fluid in a hydraulic system, has kinetic energy. As a fluid flows through a system and through a hydraulic motor, it is kinetic energy. However, when the fluid enters a cylinder to do work, flow and velocity decrease and the system's kinetic

energy is changed to static energy. Kinetic energy can be changed into heat energy because of its movement (friction and pressure).

Heat energy is the ability to do work using the heat stored or built up in a fluid. Heat energy cannot be harnessed or used in a hydraulic system. Once a portion of kinetic energy is converted to heat energy, it is lost energy.

Pressure energy is the ability to do work through the pressure of a fluid. Energy exists in various forms and has the ability to change from one form to another. Pressure energy begins the moment pressure is applied at the beginning of a system. Pressure energy can be produced by manual force, such as a foot brake or car jack, or by the use of a pump. A pump, however, only creates fluid flow and does not add to the pressure until there is a resistance to the flow.

Pressure energy is introduced when resistance is met by a hydraulic pump. The transmission of energy throughout a hydraulic system begins at a pump motor as electrical energy and is converted by and at the motor into mechanical energy. The motor's mechanical energy is transferred to the hydraulic pump, which supplies kinetic energy to the system, which is ultimately converted back to mechanical energy as work. **See Figure 16-25.**

Efficiency

As energy is transmitted through a hydraulic system, it is reduced by friction, heat, resistance, and slippage. The degree to which energy is reduced is a measure of a system's efficiency. *Efficiency* is a measure of a component's or system's useful output energy compared to its input energy. Efficiency is expressed as a percentage. When new, the natural slippage within a hydraulic pump reduces its efficiency by as much as 15%. Electric motors are typically 85% efficient. No electrical, hydraulic, pneumatic, or mechanical system is 100% efficient. Total efficiency of more than one energy component in a system is found by applying the formula:

$$Eff_T = Eff_1 \times Eff_2 \times ... \times 100$$
where
Eff_T = total efficiency (in %)
Eff_1 = efficiency of component 1
Eff_2 = efficiency of component 2
100 = constant (to convert to percent)

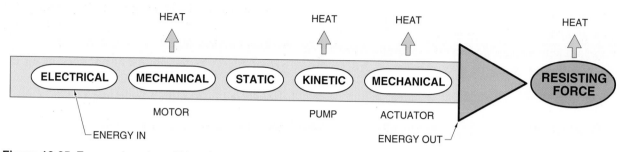

Figure 16-25. Energy changing within a hydrostatic system greatly reduces its overall efficiency.

For example, what is the total efficiency of a system power unit containing a motor listed as 85% efficient and a hydraulic pump listed as 90% efficient?

$$Eff_T = Eff_1 \times Eff_2 \times 100$$
$$Eff_T = 0.85 \times 0.90 \times 100$$
$$Eff_T = \textbf{76.5\%}$$

Energy that changes form from hydraulic to mechanical represents work. *Work* is the energy used when a force is exerted over a distance. Work is expressed in pound-feet (lb-ft). Work is found by applying the formula:

$$W = F \times d$$

where

W = work (in lb-ft)

F = force (in lb)

d = distance (in ft)

For example, how much work is performed by a forklift exerting a 3000 lb force over a vertical lift distance of 9'?

$$W = F \times d$$
$$W = 3000 \times 9$$
$$W = \textbf{27,000 lb-ft}$$

Work may also be expressed by the amount of power required. *Power* is the rate of doing work or using energy. Power is found by applying the formula:

$$P = \frac{F \times d}{t}$$

where

P = power (in lb-ft/time)

F = force (in lb)

d = distance (in ft or in.)

t = time (in sec, min, or hr)

For example, how much power is required to move a 3000 lb force 9″ in 8 sec?

$$P = \frac{F \times d}{t}$$
$$P = \frac{3000 \times 9}{8}$$
$$P = \frac{27,000}{8}$$
$$P = \textbf{3375 lb-ft/sec}$$

Tech Tip

Select a 25% larger cylinder and a 25% higher system pressure than is mathematically required to move the load when determining cylinder size and system pressure.

Snorkel

Rough terrain hydraulic scissor lifts have articulating rear axles for better traction and are available with 4-wheel drive that allows operation on up to 40% grades.

Horsepower

Mechanical energy is often expressed in horsepower (HP). One horsepower is the amount of energy required to lift 33,000 lb 1′ in 1 min. One horsepower equals 550 ft lb/sec. **See Figure 16-26.** Mechanical horsepower is found by applying the formula:

$$HP = \frac{F \times d}{550 \times t}$$

where

HP = horsepower
F = force (in lb)
d = distance (in ft)
550 = constant
t = time (in sec)

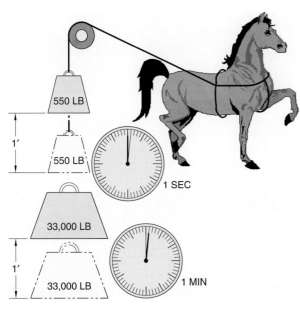

Figure 16-26. One horsepower is the amount of energy required to lift 33,000 lb 1′ in 1 min.

For example, what is the horsepower required to lift 3000 lb 9′ in 8 sec?

$$HP = \frac{F \times d}{550 \times t}$$

$$HP = \frac{3000 \times 9}{550 \times 8}$$

$$HP = \frac{27,000}{4400}$$

$$HP = \textbf{6.136 HP}$$

Horsepower in a hydraulic system is used to calculate the rate at which a system is doing work. To calculate hydraulic horsepower, pressure (in psi) and flow rate (in gpm) are used instead of ft, lb, and sec to determine mechanical HP. Also, hydraulic horsepower formulas use a conversion factor of 0.000583, which indicates the relationship between ft, lb, psi, and gpm. Hydraulic horsepower is found by applying the formula:

$$HP = P \times Q \times 0.000583$$
where
HP = horsepower
P = pressure (in psi)
Q = flow rate (in gpm)
0.000583 = constant

For example, what horsepower is needed in a hydraulic system to deliver 10 gpm at 800 psi?

$$HP = P \times Q \times 0.000583$$
$$HP = 800 \times 10 \times 0.000583$$
$$HP = \textbf{4.664 HP}$$

Torque

Torque is the twisting (rotational) force of a shaft. **See Figure 16-27.** The twisting effort at the shaft causes the shaft to rotate. The presence of torque indicates that there is a force present, even without rotation.

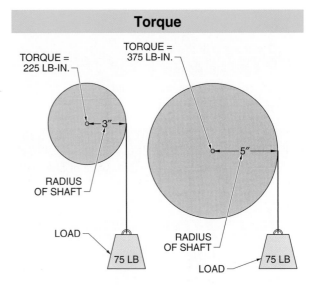

Figure 16-27. Torque is the twisting (rotational) force of a shaft.

Torque is measured at a distance from the motor shaft center. The farther a force is from the shaft's center, the greater its torque. Torque is normally measured in lb-in. and is equal to the product of its force (lb) times the distance from the shaft's center. Torque is found by applying the formula:

$T = F \times d$

where

T = torque (in lb-in.)

F = force (in lb)

d = distance from the shaft center (in in.)

For example, what is the torque required to overcome a 75 lb force connected 3″ from the motor shaft's center?

$T = F \times d$

$T = 75 \times 3$

$T = $ **225 lb-in.**

A motor with a large shaft or pulley would have to apply a greater torque. For example, if the distance between the shaft center and point of force is 5″, the torque required is 375 lb-in. (75 × 5 = 375 lb-in.).

Torque applied by a hydraulic motor can also be calculated by replacing force and distance with pressure (in psi) and hydraulic motor displacement per revolution, divided by 2π. The torque developed by a hydraulic motor is found by applying the formula:

$$T = \frac{P \times d}{2\pi}$$

where

T = torque (in lb-in.)

P = pressure (in psi)

d = motor displacement (in cu in.)

π = constant (3.1416)

For example, what is the available torque delivered by a hydraulic motor with a displacement of 2.146 cu in. per revolution and an applied pressure of 500 psi?

$$T = \frac{P \times d}{2\pi}$$

$$T = \frac{500 \times 2.146}{2 \times 3.1416}$$

$$T = \frac{1073}{6.283}$$

$T = $ **170.773 lb-in**

Cincinnati Milacron

Torque requirements on many machine processes vary with the type of material being machined.

The torque a hydraulic motor develops depends on its applied pressure and displacement. In most cases, if the available delivered torque is not enough, the pressure is increased and in some cases, depending on the motor type, the displacement can be increased. Either of these may be necessary to start a hydraulic motor and overcome breakaway, starting, and running torque.

Breakaway torque is the initial energy required to get a nonmoving load to turn. *Starting torque* is the energy required to start a load turning after it has been broken away from a standstill. *Running torque* is the energy that a motor develops to keep a load turning.

Changing displacement may allow for greater breakaway or starting torque, but it also has an adverse effect on the system's speed and operating pressure. An increase in displacement decreases the motor's speed and decreases the operating pressure. Decreasing displacement increases the speed of the motor and also increases the effect on operating pressure. **See Figure 16-28.**

Effect on Motor Output from System Changes*			
Change	Speed	Effect on Operating Pressure	Available Torque
Increase Displacement	Decreases	Decreases	Increases
Decrease Displacement	Increases	Increases	Decreases
Increase Pressure Setting	No effect	No effect	Increases
Decrease Pressure Setting	No effect	No effect	Decreases
Increase gpm	Increases	No effect	No effect
Decrease gpm	Decreases	No effect	No effect

* effects on changes assume working loads remain the same

Figure 16-28. Hydraulic systems may be adjusted to allow for higher torque requirements, but operating conditions may also change.

Increasing pressure or flow rate may be used to increase a system's working force or speed. Increasing pressure increases cylinder output force. Increasing flow rate increases cylinder speed. The output force of a hydraulic motor is determined by the amount of pressure acting on the area of its rotating parts. The output force of a hydraulic cylinder is determined by the amount of pressure acting on the area of its piston. The output speed of a hydraulic motor is determined by the rate at which the fluid flows through the motor. The output speed of a hydraulic cylinder is determined by how quickly the flow rate fills the volume ahead of the cylinder's piston.

Review

1. What is work produced in a hydraulic system dependent on?

2. What does Pascal's law state?

3. Describe the difference between gauge pressure and absolute pressure.

4. Define dynamic lift.

5. List two reasons why practical dynamic lift is considerably less than practical static lift.

6. Describe the relationship between fluid flow and friction in a hydraulic system.

7. What does the velocity index indicate?

8. Define pressure drop.

9. Why does it take less hydraulic fluid to retract a piston than it takes to extend a piston?

10. Define mechanical advantage.

11. Describe the theory of the conservation of energy.

12. List the four events that reduce energy as it is transmitted through a hydraulic system.

13. What does the presence of torque indicate?

14. What is horsepower in a hydraulic system used to calculate?

15. What can increasing the pressure and flow rate do to a hydraulic system and cylinder?

Atlas Technologies Inc.

CHAPTER 17

HYDRAULIC APPLICATIONS

Hydraulic circuits consist of controlling the movement of a contained liquid. Hydraulic diagrams explain, demonstrate, or clarify the relationship or functions between hydraulic components. Any hydraulic circuit must contain hydraulic fluid, a reservoir, piping, a pump, valves, and actuators. The hydraulic circuit application, complexity, and power requirements dictate the type and number of components used. Every energy source must be identified, understood, and disabled prior to working on a hydraulic system. Always follow the equipment manufacturer's recommendations when servicing hydraulic equipment and circuits.

Chapter Objectives

- Identify the three basic types of hydraulic diagrams.
- Explain how hydraulic fluid affects a hydraulic circuit.
- Describe the functions of reservoirs.
- List the different types of piping devices used in a hydraulic circuit.
- Explain the functions of pumps used in a hydraulic circuit.
- Identify the different types of valves used in a hydraulic circuit.
- Describe the functions of the parts of hydraulic cylinders and motors.
- List the different types of accumulators.
- Explain the importance of hydraulic circuit maintenance.
- Explain the importance of hydraulic circuit troubleshooting.
- Explain hydraulic circuit component testing.

Key Terms

- hydraulics
- hydraulic diagram
- fluid
- additive
- viscosity
- oxidation
- strainer
- filter
- contaminant
- reservoir
- sight gauge
- breather cap
- heat exchanger
- hose
- pipe
- tube
- flared fitting
- displacement
- gear pump
- vane pump
- piston pump
- pressure-relief valve
- sequence valve
- pressure-reducing valve
- pressure gauge
- directional control valve
- check valve
- flow control valve
- hydraulic actuator
- hydraulic cylinder
- hydraulic motor
- accumulator

HYDRAULIC CIRCUITRY

Hydraulics is the branch of science that deals with the practical application of water or other liquids at rest or in motion. A *liquid* is a fluid that can flow readily and assume the shape of its container. Hydraulic circuits control the movement of a contained liquid. A *hydraulic circuit* is a closed path through which hydraulic fluid flows or may flow. Basic hydraulic circuits include the storing of hydraulic fluid, a method of controlling its flow, and devices that transfer force.

Hydraulic Diagrams

A *hydraulic diagram* is the layout, plan, or sketch of a hydraulic circuit and is designed to explain, demonstrate, or clarify the relationship between or functions of hydraulic components. Hydraulic diagrams are used to illustrate how a system develops. For example, a diagram can show many reservoirs, even though there may be only one in the actual circuit. This is done as a convenience for the technician, to show the function and connection of each major component, and to prevent the overlapping of lines. **See Figure 17-1.** The three basic hydraulic diagrams are pictorial, cutaway, and graphic.

Pictorial Diagrams. A *pictorial diagram* is a diagram that uses drawings or pictures to show the relationship of each component in a circuit. **See Figure 17-2.** Pictorial diagrams generally use single lines to show the elements of a circuit. The components are shown using simple outlines to indicate their relative position and appearance in a circuit.

Hyster Company

Industrial forklift trucks typically have hydraulic circuits as part of the lifting mechanism.

Hydraulic Diagrams

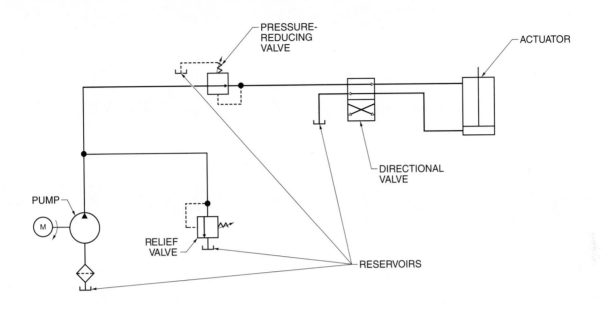

Figure 17-1. Hydraulic diagrams show functions and connections of each major component and prevent the overlapping of lines.

Pictorial Diagrams

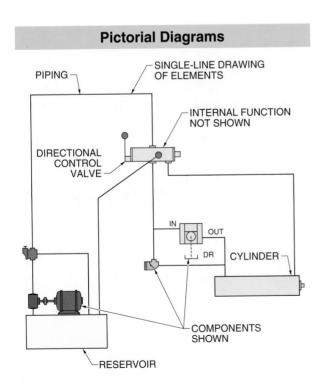

Figure 17-2. A pictorial diagram is a diagram that uses drawings or pictures to show the relationship of each component in a circuit.

Pictorial diagrams show the component's purpose within a circuit but do not provide the internal function or specific information about the component. For example, an outline of a directional control valve may be illustrated in a pictorial diagram, but its type is not defined.

Cutaway Diagrams. A *cutaway diagram* is a diagram that shows the internal details of components. **See Figure 17-3.** Cutaway diagrams provide more detail than pictorial diagrams. Cutaway diagrams consist of double-line drawings of the circuit components showing their operation and internal positions. A cutaway diagram provides an excellent understanding of simple circuits. Cutaway diagrams may be color-coded to show direction of flow or pressure of the fluid in the piping. Cutaway diagram color coding is as follows:

- Red indicates fluid flowing at system operating pressure or highest working pressure. *System operating pressure* is the pressure of a fluid after the pump until the flow is reduced, metered, or returned to the reservoir.
- Yellow indicates controlled flow by a metering device or lowest working pressure. *Controlled flow* is the fluid flow after a flow control device has reduced the flow rate of the fluid.

Cutaway Diagrams

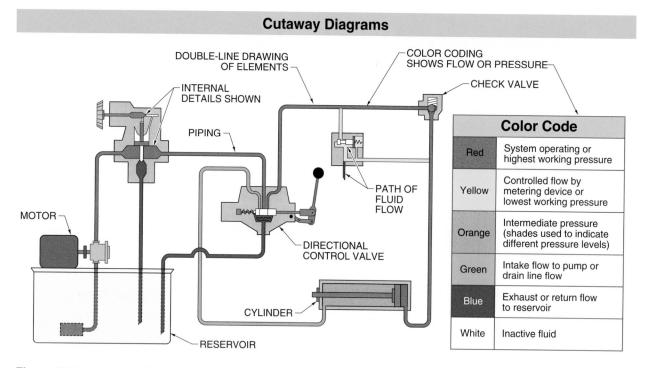

Figure 17-3. A cutaway diagram is a diagram showing the internal details of components and the path of fluid flow.

Color Code	
Red	System operating or highest working pressure
Yellow	Controlled flow by metering device or lowest working pressure
Orange	Intermediate pressure (shades used to indicate different pressure levels)
Green	Intake flow to pump or drain line flow
Blue	Exhaust or return flow to reservoir
White	Inactive fluid

- Orange indicates intermediate pressure that is lower than system operating pressure.
- Green indicates intake flow to pump or drain line flow. *Intake flow* is the fluid flow from the reservoir, through the filters, to the pump.
- Blue indicates exhaust or return flow to the reservoir. *Exhaust flow* is the fluid flow from the actuator, back through the valve, to the reservoir.
- White indicates inactive fluid (reservoir fluid).

A disadvantage of a cutaway diagram is that a considerable amount of space is required to show a system consisting of more than the minimum basic components. Also, cutaway diagrams do not indicate some elements such as type or direction of rotation of a pump or motor.

Graphic Diagrams. A *graphic diagram* is a drawing that uses simple line shapes (symbols) with interconnecting lines to represent the function of each component in a circuit. **See Figure 17-4.** Graphic diagrams are used when designing and troubleshooting fluid power circuits, because the connecting lines and symbols are used to explain how a circuit works. A *symbol* is a graphic element that indicates a particular device, etc. Graphic symbols simplify the explanation of a circuit and can be used by individuals that speak different languages

because a person does not have to speak a particular language to understand them. This promotes a universal understanding of hydraulic fluid power circuits.

Graphic Diagrams

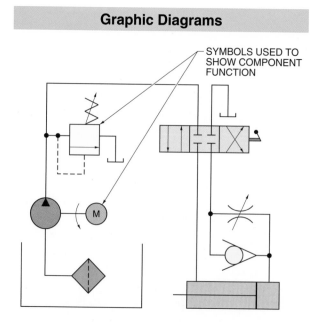

Figure 17-4. A graphic diagram is a diagram that uses symbols with interconnecting lines to represent the function of each component in a circuit.

The American National Standards Institute (ANSI), in collaboration with the American Society of Mechanical Engineers (ASME), has adopted a standard regarding symbols for fluid power diagrams, ANSI/ASME Y32.10-1967, *Graphic Symbols for Fluid Power Diagrams.* **See Appendix.** This standard illustrates the basic fluid power symbols and describes the principles on which the symbols are based.

Graphic symbols show flow paths, connections, and functions of components. They are not used to indicate the rate of flow or to offer pressure settings. Indications of rate of flow or pressure settings must be added to the graphic diagram. Also, graphic diagrams do not give the actual position of a component in the system. The components are generally positioned to show the flow of the system and how each component is related to the others.

Little written explanation is required in a graphic diagram because standard symbols and lines are used. For example, hydraulic circuit graphic diagrams use four different lines, with each representing a working pipe. The four lines are solid, dashed, dotted, and center lines. A solid line represents a main pipe, outline, shaft, or conductor. This main pipe, outline, shaft, or conductor is essentially the working element. A dashed line represents pilot piping for controlling a component's function. A dotted line represents exhaust or drain piping. A center line shows the outline of an enclosure.

Graphic symbols present considerable information in a small space and may contain information about the flow of fluid. **See Figure 17-5.** For example, the simplified symbol for a check valve shows a ball being held against a seat, indicating the direction of flow through that part of the line. Also, some symbols closely resemble the actual component. For example, the symbol for a cylinder is similar to a cutaway view of a cylinder, and the symbol for a pressure gauge looks much like the face of a gauge with its indicating needle.

Graphic diagrams can be used to understand the working mechanism of a hydraulic circuit by showing how and in which direction fluid flows through the circuit. Tracing fluid flow indicates the operation and location of each component used. Troubleshooting most modern hydraulic equipment can be difficult and time-consuming if an operating diagram is not available, due to the grouping of valves within manifolds or the complexity of the circuit. Without a graphic diagram, dismantling equipment can be required to determine the function of components within the circuit.

Hydraulic circuits are used with crawler cranes to help move large amounts of heavy materials.

Components and their parts are represented by shapes such as circles, triangles, squares, and rectangles. Circles generally represent a component that is round, such as a gauge. Circles also represent rotary devices, such as pumps or motors. Triangles generally represent direction of fluid flow. Triangles that are completely shaded represent liquid flow within a hydraulic circuit. Triangles that are unshaded represent gas flow within a pneumatic system. Triangles are used to distinguish between pumps and motors. For example, a pump (circle) having an unshaded triangle pointing out is an air compressor. A circle with a solid triangle pointing in the direction of circuit fluid flow is a hydraulic motor. Squares or rectangles generally represent valves and may be grouped together to show multiple internal functions of a valve such as that of a directional control valve.

Arrows are used in graphic diagrams to indicate an adjustable or variable component or to show shaft rotation on the near side of the shaft. A component that may be adjusted or varied is represented by an arrow passing through the symbol at approximately a 45° angle. For example, an angled arrow passing through four zigzag lines indicates adjustable spring pressure. An angled arrow passing through a circle represents an adjustable pump or motor. An arrow parallel to the short side of a symbol, within the symbol, indicates that the component is pressure-compensated. The arrow within the symbol is actually a piston that is activated by line or pilot pressure. An example of pressure compensation is that of a pressure-relief valve where the arrow (piston) within the symbol does not line up with the circuit lines until enough pressure from a pilot line allows the internal arrow to overcome spring pressure and align the internal passage with the circuit lines.

Safety Tip

Oil-soaked clothes must be stored or discarded properly because they are toxic and can also catch on fire from a match, cigarette, sparks, or any open flame.

HYDRAULIC CIRCUIT COMPONENTS

Hydraulic circuit components are used in a wide range of combinations for different applications. The circuit application, complexity, and power requirements dictate the type and number of components used. Any hydraulic circuit must contain six essential elements: hydraulic fluid to transmit force and motion, a reservoir (tank) to store the fluid, piping to transport the fluid through the circuit, a pump to move the fluid, valves to control the pressure and direction of the fluid, and actuators to convert hydraulic force into mechanical force. **See Figure 17-6.**

Graphic Symbols

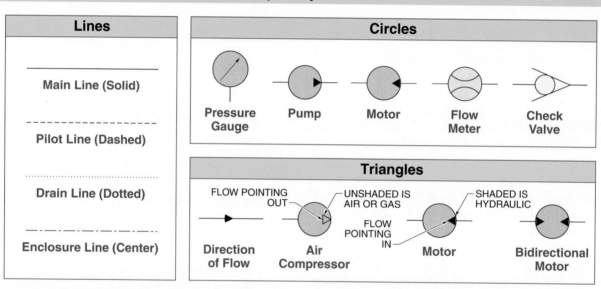

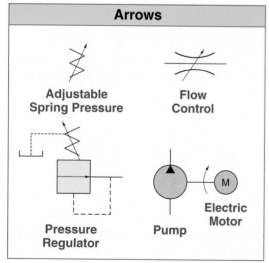

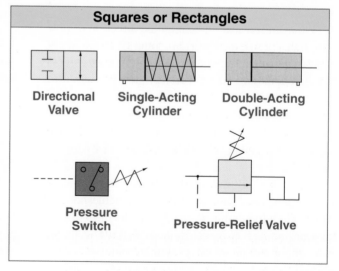

Figure 17-5. Graphic symbols present considerable information in a small space and may contain information about the flow of fluid.

Hydraulic Circuit Components

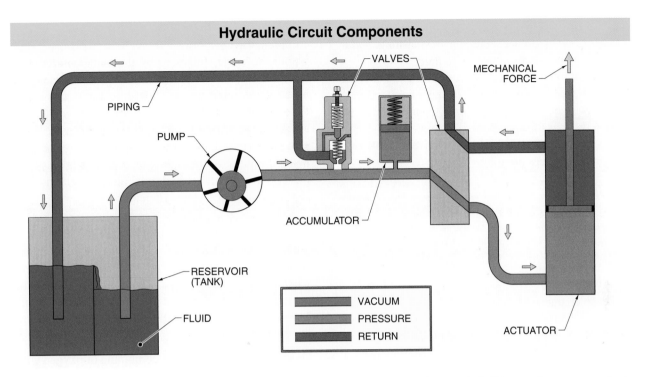

Figure 17-6. Hydraulic circuits require six elements to draw fluid into a system (vacuum), build operating pressure to do work, and return the fluid back to storage.

Hydraulic Fluid

Hydraulic fluids are liquids used to transmit force and are used to do work in hydraulic machinery and equipment. In addition, hydraulic fluids conduct heat away from metal surfaces and lubricate moving parts. Hydraulic fluids are used to create force because they can be applied instantly throughout a system, allow for control at different locations, increase or decrease force, and change direction.

Hydraulic fluids are used in a variety of machines such as industrial production equipment, forklifts, excavators, backhoes, bulldozers, aircraft, and automobiles. Automobile power steering systems, transmissions, and power brakes all require hydraulic fluids. Automobile brake fluid is specially made to have a high boiling point, a low freezing point, and hygroscopic abilities in order to absorb water.

Today, most hydraulic fluids are based on mineral oil base stocks. Natural oils, such as canola oil, are used as base stocks for hydraulic fluids where biodegradability and renewable sources are required. Other base stocks are used for fire resistance and extreme temperature applications. Some examples include glycol, esters, organophosphate ester, polyalphaolefin, propylene glycol, and silicone oils.

Biodegradable Hydraulic Fluids. Environmentally sensitive applications, such as marine dredges or farm tractors, in which there is the risk of an oil spill from a ruptured oil line benefit from using biodegradable hydraulic fluids with a base of canola oil. Typically, oils of this type are blended per ASTM standard D6006-11, *Standard Guide for Assessing Biodegradability of Hydraulic Fluids,* and are available as ISO 32, ISO 46, and ISO 68 specification oils.

Transmission of Energy. Since hydraulic fluid only compresses 0.5% for every 1000 psi applied to it, the transmission of energy through fluids is almost as efficient as through solids. Pascal's law, stating that pressure is distributed undiminished throughout a closed container, can be applied during the transmission of energy in a hydraulic circuit. Hydraulic fluids must have the following characteristics to be effective:

- Hydraulic fluids must lubricate by offering a substantial film of fluid even when subjected to high heat.
- Hydraulic fluids must have sufficient viscosity to resist leakage. Leakage results in loss of pump efficiency, loss of pressure, and generated heat.
- Hydraulic fluids must resist oxidation, thereby preventing the reactions of oil products to form gum, varnish, and sludge.

- Hydraulic fluids must resist or depress foaming caused by turbulence, agitation, or splashing. *Foaming* is excessive air in hydraulic fluid.
- Hydraulic fluids must resist rust, corrosion, and pitting caused by the chemical (usually acid) union of iron or steel with oxygen. *Pitting* is localized corrosion that has the appearance of cavities (pits).
- Hydraulic fluids must remain relatively stable over a broad temperature range.

Tech Tip

Water was the original hydraulic fluid, dating back to ancient Egypt. Beginning in the 1920s, mineral oil began being used as a base stock due to its inherent lubrication properties and ability to be used at temperatures above the boiling point of water.

These characteristics of hydraulic fluids are developed by special compounding of refined oil and various additives. A large number of compounded fluids are available due to the wide variety of materials used in hydraulic circuits, such as seals, rings, or flexible hoses. Care must be taken to ensure that proper fluids are used with compatible components. Only the compounded fluids specified by the manufacturer should be used.

Additives. An *additive* is a chemical compound added to a hydraulic fluid to change its properties. Additives protect hydraulic fluid in addition to protecting components. Hydraulic fluids are protected so they are free of destructive chemicals, resistant to foaming, and, in some cases, fire-resistant. Additives are also included to speed demulsification and stabilize viscosity. *Demulsification* is the act of separating water and oil quickly. *Emulsification* is the act of mixing oil and water. In most hydraulic circuits, water is damaging and must be rapidly separated from the hydraulic fluid.

Viscosity is the measurement of the resistance of a fluid's molecules to move past each other. The viscosity of a fluid must match the application. Hydraulic fluids without the proper viscosity cannot transmit power satisfactorily. Heavy fluids do not flow properly and cause a slow, sluggish operation. Light fluids leak around components and do not lubricate properly. Improper hydraulic fluid must be drained and replaced according to the manufacturer recommendations. Used hydraulic fluid must be disposed of according to state and local municipal codes.

A fluid's resistance to foaming is increased by adding antifoaming additives. Hydraulic fluids allowed to foam from air entering the system cause spongy component movement and higher-than-normal temperatures.

Overheating can damage hydraulic fluid and circuit components. Operating temperatures should be kept within the range suggested by the manufacturer. The reservoir and circuit components should be clean so heat can dissipate easily. Kinked lines can also cause excessive heat buildup. The relief valves should be set at the recommended level because excessive pressure generates additional heat. To prevent overheating hydraulic fluid, overspeeding or overloading a system must be avoided. Also, hydraulic fluid coolers must be kept clean and operating efficiently.

Fire-resistant fluids are sometimes required in environments where a flash point, fire point, or auto-ignition point exists. The *flash point* is the temperature at which oil gives off enough vapor to ignite briefly when touched with a flame. The *fire point* is the temperature at which oil ignites when touched with a flame. The *autoignition point* is the temperature at which oil ignites by itself.

Oxidation. *Oxidation* is the combining of oxygen with oil, which breaks down the basic oil composition. Oxidation greatly reduces the service life of a hydraulic fluid because oxygen readily combines with the hydrogen and carbon that make up the oil. The oxidation process creates resins, which are converted to varnish and gum and settle out as sludge. This harmful process also produces an acidic and highly corrosive fluid. Sludge and varnishes are considered contaminants and are by-products of already-destroyed oil. The oil must then be changed and flushed. A major additive to hydraulic fluids is one that prevents oxidation of the fluid.

Oxidation begins when hydraulic fluid reaches a high temperature while in the presence of air. The higher the temperature, the greater the oxidation. Oxidation rates double for approximately each 20°F increase in temperature. Boiling burnt hydraulic oils separates certain resins from the fluid. These resins create varnish and acid, which are two corrosive contaminants of a hydraulic circuit. Scalelike varnish results when these resins touch hot metal components. Filter-clogging sludge is formed as these particles drop off. Formed and baked resins in a system create an acidic condition in the fluid. This acid attacks and dissolves the metal it contacts.

The acidic condition of hydraulic fluid may be checked with the use of litmus paper. *Litmus paper* is a color-changing, acid-sensitive paper that is impregnated with lichens. *Lichen* is a fungus normally seen as a growth on tree trunks or rocks. Lichens turn from blue to a reddish color when submerged in acidic oil.

The presence of moisture and oxygen causes iron to rust. *Rust* is a form of oxidation in which metal oxides are chemically combined with water to form a reddish-brown scale on ferrous metal. A *ferrous metal* is a metal containing iron. The prevention of rust depends on an oils ability to form a film on metal surfaces, which prevents the metal from coming in contact with air or water.

Strainers and Filters. Most contaminants come from wear and tear and improper maintenance. In addition, contaminants enter a hydraulic system from the air surrounding the system and must be filtered or strained before and during circuit operation. A *strainer* is a fine metal screen that blocks contaminant particles. A *filter* is a device containing a porous substance through which a fluid can pass but particulate matter cannot.

The major maintenance function of any hydraulic fluid is keeping the fluid clean. Particle buildup interferes with lubrication by blocking flow or by rubbing or scraping against moving parts. Particle buildup also interferes with the cooling process by making heat transfer difficult. As heated fluid returns to the reservoir, the heat is normally given up to the walls and baffles. Particle buildup on the walls and baffles acts as an insulator, preventing cooling.

A *contaminant* is a substance that causes harm or damage to that with which it comes in contact. Particle contamination is a major source of failure for hydraulic circuits. Particle size is categorized in three different sizes: larger than internal clearances, same size as internal clearances, and smaller than internal clearances. Although particles larger than or the same size as internal clearances can cause damage, they are generally removed through proper filtration before damage occurs. Particles smaller than internal clearances are the most damaging over an extended period because these particles are highly abrasive, flow freely through a circuit, and cause rapid wear. If allowed to build up, these particles become silt and sludge and destroy all close-fitting components. Changing hydraulic fluid per manufacturer recommendations prevents destruction caused by these small particles.

Strainers are made of fine mesh wire screening elements wrapped around a metal frame. Strainers are used because screening is not as fine as a filter and offers less resistance to fluid flow. Most strainers can be cleaned periodically, while filters, made of porous materials, absorb particles from flowing fluids and must be replaced.

Strainer screens are rated in mesh and filters are rated in microns. *Mesh* is the number of horizontal and vertical threads per square inch. A *micron* (µm) is a unit of measure equal to one millionth of a meter, or 0.000039″. Most strainers remove particles above 100 mesh, while filters remove particles above 3µm. **See Figure 17-7.** Filter performance is based on the amount of particulate matter that can be removed from a hydraulic fluid.

Strainers remove particles with a straight flow path through one layer of material. Materials used for strainers are generally cloth thread, metallic thread, or perforated metal. The threads are laid in equal amounts vertically and horizontally with the amount of threads being counted per square inch. For example, a 200 mesh strainer has 200 vertical threads and 200 horizontal threads per square inch.

Certain mesh sizes produce certain pore (opening) sizes because of the accurate control of the strainer manufacturing process. A 200 mesh strainer has a pore size of 74µm. Strainer mesh is given an absolute rating because the pore size is accurately controlled. An *absolute rating* is an indication of the largest opening in a strainer element. Therefore, 74µm is the largest particle that can pass through a 200 mesh strainer. The higher the mesh number, the smaller the opening.

Filters do not have standard openings or pore sizes. Instead, filter elements remove contaminants by forcing the hydraulic fluid to flow through a thick, threaded material that does not have a straight flow path. Contaminants are trapped between the intertwining threads of the filter element. Eventually, the filter element must be replaced because trapped contaminants build up in the threads.

Hydraulic filter assemblies may be fitted with a pressure gauge that indicates when the filter needs to be replaced.

Relative Micron Particle Sizes

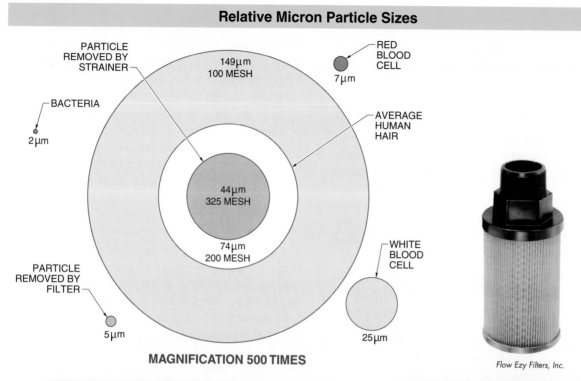

Flow Ezy Filters, Inc.

Relative Sizes	
Lower limit of visibility (naked eye)	40 μm
White blood cells	25 μm
Red blood cells	7 μm
Bacteria (cocci)	2 μm

Linear Equivalents		
1 in.	25.4 mm	25,400 μm
1 mm	0.0394 in	1000 μm
1 μm	25,400 of an in.	0.001 mm
1 μm	3.94×10^{-5} in.	0.000039 in

Screen Sizes		
Meshes per Linear Inch	Opening in Inches	Opening in μm
52.36	0.0117	297
72.45	0.0083	210
101.01	0.0059	149
142.86	0.0041	105
200.00	0.0029	74
270.26	0.0021	53
323.00	0.0017	44
–	0.00039	10
–	0.000019	0.5

Figure 17-7. Very small particles are removed from hydraulic fluids using strainers and filters.

Ideally, a filter should be placed in the hydraulic line before every component. Placing a filter in the hydraulic line before every component is not economical, so filters of various types and ratings are placed strategically in a hydraulic circuit to offer the best filtering capability and most economical results. Three basic locations for filters used in a hydraulic circuit are the suction strainer, pressure filter, and return-line filter. **See Figure 17-8.**

A *suction strainer* is a coarse filter attached to a pump inlet. Suction strainers can be installed inside a reservoir at the fluid inlet of a circuit and placed in-line between the reservoir and the pump or installed on the exterior of a reservoir. In-reservoir-type strainers are difficult to maintain because the strainer is inside the reservoir and cannot be seen. Exterior-mounted strainers offer ease of element change and have color-coded indicators that show the condition of the element.

Filters

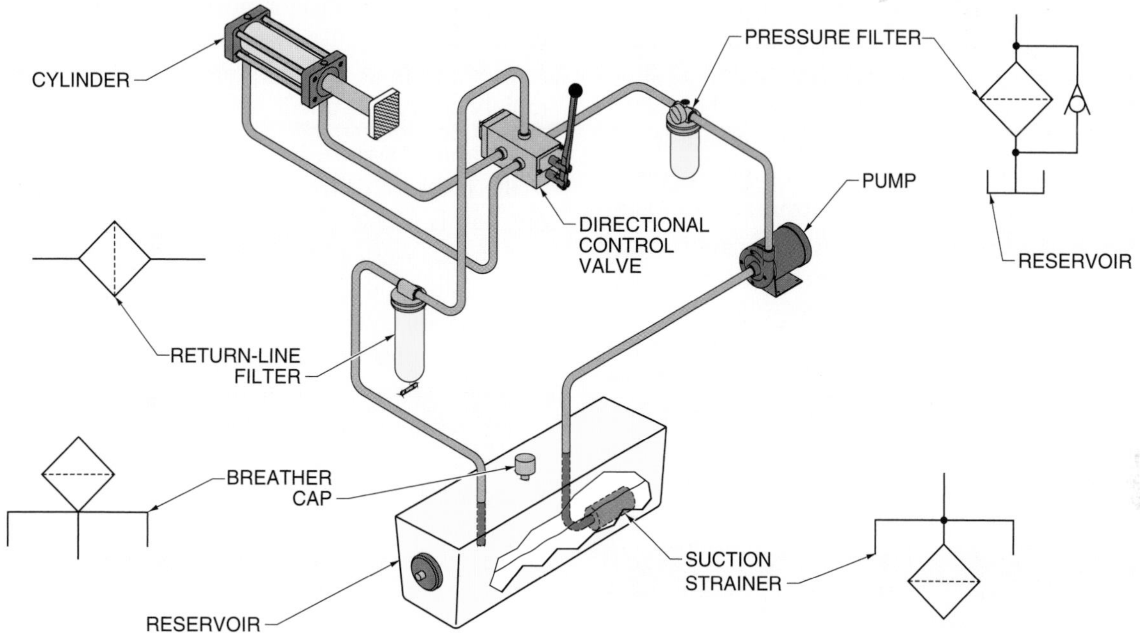

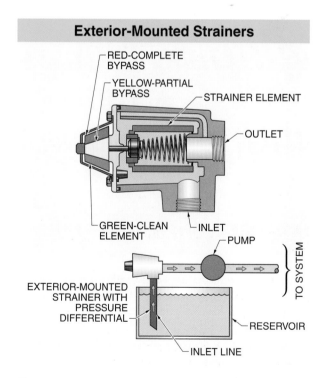

Figure 17-8. The suction strainer, pressure filter, and return-line filter are the three basic locations for filters found in a hydraulic circuit.

Exterior-mounted strainers have two types of indicators: those linked directly to a bypass valve and those that react to pressure differentials. Indicators linked to a bypass valve indicate valve movement by showing actual valve position. Pressure differential indicators are color coded to show the condition of the element. As the indicator rotates, it shows green for a clean element, yellow for partial bypass, and red for complete bypass. **See Figure 17-9.** Both indicator types are designed to be either visual, electronic, or both. Electronic indicators can be used to activate a warning bell or light or to shut down a machine until it is serviced.

Poor maintenance of the suction strainer starves the pump and prevents the lubrication of the pump, which causes it to generate high temperatures and contaminates. Suction strainers range from 25μm to 235μm.

A *pressure filter* is a very fine filter placed after a pump for protection of circuit components. Pressure filters are placed between the pump and components or between individual components in a circuit. Pressure filters placed between components have the advantage of filtering out particles introduced upstream by a deteriorating component.

Exterior-Mounted Strainers

Figure 17-9. Exterior-mounted strainers with pressure differential indicators are color coded to show the condition of the element.

Some pressure filters are capable of handling bidirectional flow. Bidirectional flow filters may be used between a directional control valve and the actuator. Pressure filters can filter out very fine particles because system pressure is used to push the hydraulic fluid through the minute openings. Particles can be pushed through the element or the element may collapse or tear if a filter becomes contaminated and is not equipped with a bypass. Pressure filters range from 5μm to 40μm.

A *return-line filter* is a filter positioned in a circuit just before the reservoir. Return-line filters filter hydraulic fluids before the fluid is returned to the reservoir and do not operate under pressure. These fluids may be finely filtered using normal operating back pressure. Poor maintenance increases back pressure and adversely affects circuit components. Also, return-line filters deliver clean fluid to a dirty reservoir if the reservoir is not properly maintained.

DoALL Company

Sight gauges on fluid reservoirs indicate fluid level with a quick glance.

Some filters are designed to allow hydraulic fluids to bypass the filter when a difference in specific pressure is sensed. In this case, a relief or bypass valve allows full fluid flow across the filter if proper filter maintenance is not performed and pressures increase above normal.

In some units, indicators are included in the filters to show their condition. Filter-clogging conditions may be indicated by dial indicator movement, a light or buzzer, or an equipment OFF button. Regardless of the hydraulic filter or fluid used, lack of a proper maintenance program rapidly destroys hydraulic equipment.

Allowable Pressure Drop. *Allowable pressure drop* is the pressure differential across a filter. Hydraulic pressure is usable energy but can be wasted by pressure drops. Wasted energy decreases efficiency and is costly. Initial and terminal pressure differential is energy consumption and must be kept as low as possible. For example, a pressure differential of 80 psid at 20 gpm consumes (wastes) 1 HP at the pump.

Filter elements have a collapse or destroy rating under certain pressures, construction particulars, and media strength. Filter manufacturers determine allowable pressure drop value to be less than the collapse rating of the element. Maximum drop value (terminal pressure drop) is typically 45 psi.

Filter selection from charts and tables is based on flow and pressure drop and may be provided by the filter manufacturers. Filter type, size, and use are typically part of hydraulic machine specifications.

Tech Tip

> *Always cap or plug open lines or connectors when installing or removing components to reduce the possibility of contaminants entering a system.*

Reservoirs

A *reservoir* is a container for storing fluid in a hydraulic circuit. The primary purpose of a reservoir is to provide storage space for the fluid required by the circuit. The reservoir capacity should normally be two to three times the volume of fluid pumped through the system in one minute. In addition to fluid storage, a reservoir also prevents fluid contamination, helps with fluid/air separation, and maintains safe fluid temperatures.

Reservoirs are constructed with a dished bottom to allow for drainage, a sight gauge, a breather cap, baffle plate(s), and return, drain, and suction lines. Reservoirs may also be equipped with a strainer to prevent fluid contamination, clean-out covers or removable tops, a drain plug, and permanent chip magnets. **See Figure 17-10.**

Reservoirs

RETURN LINE

DRAIN LINE

BREATHER CAP

SIGHT GAUGE

RESERVOIR SYMBOL

SUCTION LINE

BAFFLE PLATE

DISHED BOTTOM

STRAINER

DRAIN PLUG

PERMANENT CHIP MAGNET

CLEAN-OUT COVER

Figure 17-10. A reservoir is a container for storing fluid in a hydraulic system.

A *sight gauge* is a device used to visually inspect the hydraulic fluid level in a reservoir. A sight gauge can also be used to send a signal to a warning device if the hydraulic fluid level drops below the recommended operating level.

A *breather cap* is a device that allows atmospheric pressure to push the fluid up to the pump. Also, the fluid level is constantly rising and falling when the circuit is operating. A breather cap roughly filters the dirt-laden air, which enters and exits the reservoir as the system actuator is filled and exhausted. Breather caps become plugged and cause improper pump operation if they are not cleaned regularly. Breather caps are not used and the reservoirs are pressurized when reservoirs are used in applications in unclean or corrosive environments.

Baffle plates block the returning fluid from going directly to the suction line. The baffle plates are placed between the return and suction line ports in the reservoir to help settle the flow of fluid and allow contaminants or moisture to drop out and air bubbles to rise to the top. Additionally, baffle plates are used as heat exchangers to help reduce the temperature of the returning fluid.

To prevent fluid foaming and aeration as fluid returns to the reservoir, reservoir lines are extended below the fluid level, usually 2″ from the tank bottom. In many cases, lines terminating in the reservoir are cut at a 45° angle to prevent bottoming in the tank and are positioned so that fluid flow is directed toward the tank wall. This positioning promotes cooling and keeps the high-temperature fluid away from the pump inlet line.

A *clean-out cover* is a device used to access a reservoir when it requires cleaning from the buildup of solid materials. Two cleanout covers may often be located on opposite sides of the baffle. This allows both sides of the reservoir to be thoroughly cleaned.

A *drain plug* is a threaded device that is installed at the lowest point of a reservoir to allow for the removal of hydraulic fluid or draining of accumulated moisture. A *permanent chip magnet* is a magnetic device placed in a reservoir to attract and hold ferrous metal particles that have contaminated the system but have not been recovered by system filters or strainers. Permanent chip magnets are typically placed in the reservoir in the form of magnetic drain plugs or magnetic rings that are placed near the strainer.

To dissipate heat, reservoirs are narrow and deep as opposed to short and wide. This offers a large exterior surface to contact ambient air for cooling. For best operation, reservoir fluids should not exceed 160°F to 180°F. Generally, reservoirs dissipate about 70% of the heat generated within the system with the remaining 30% being radiated from the components or plumbing. The amount of heat dissipated also depends on the temperature difference between the reservoir surface and ambient air. Hydraulic equipment installed too close to furnaces or ovens may reach or surpass critical temperature levels.

The Gates Rubber Company

Hydraulic hoses should be enclosed in protective sleeves when subject to rubbing and should be installed with a bending radius of greater than six times the inside diameter.

Heat Exchangers. A *heat exchanger* is a device that transfers heat through a conducting wall from one fluid to another. Heat produced in an operating circuit is radiated away by the reservoir baffle plate and into the air surrounding the system components. This heat transfer is generally sufficient for a hydraulic circuit. However, a reservoir may not be capable of dissipating enough of the heat produced. In such cases, heat exchangers are placed in the hydraulic lines to remove excess and damaging heat. Heat exchangers are either air-cooled or water-cooled devices. **See Figure 17-11.**

Air-cooled heat exchangers (coolers) operate by pumping the hot circuit fluid through tubes attached to sheet metal fins. Cooling of the hot fluid is accomplished through the use of a blower, which blows air over the tube and fins. Air-cooled heat exchangers are similar to automobile radiators.

When water is available for a hydraulic system, a water-cooled heat exchanger is the preferred choice because it is more efficient than an air-cooled heat exchanger and is less affected by ambient temperatures. Water-cooled heat exchangers are typically located on the low-pressure return line of a hydraulic circuit so that their maximum operating pressure is 100 psi, with the possibility that pressures can go up to 200 psi. The return line is less likely to develop a leak because of its continuous high-volume, low-pressure fluid flow. Water-cooled heat exchangers must be protected from hydraulic fluid pressure surges by using a relief valve.

Water-cooled heat exchangers operate by pumping hot system fluid through a shell and over tubes containing circulated cool water. Heat from the hydraulic fluid is transmitted through the cooling tubes to the water by means of conduction. The circulating water carries away unwanted heat from the system fluid. This type of heat exchanger may be reversed to warm system fluids. By circulating warm water through the tubes, system fluid within the shell can be heated.

Piping

Hoses, pipes, and tubing are the basic piping devices used to connect components and to conduct fluid in a hydraulic circuit. Hydraulic circuit piping must be leakproof and strong enough to withstand required temperatures, vibrations, and pressures. Proper materials and procedures must be used to prevent excess restriction, turbulence, leakage, or dangerous situations.

Hoses. A *hose* is a flexible tube for carrying fluids under pressure. Hoses are fabricated in layers for use in high-pressure or extra-high-pressure hydraulic circuits. High-pressure hoses are capable of withstanding pressures up to 5000 psi. Hoses generally consist of an inner layer of soft synthetic rubber that is compatible with hydraulic fluids, two or more layers of multiple wire braid reinforcement, a layer of cotton braid, and a rubber cover. Extra-high-pressure hoses contain four or more wire braid layers. **See Figure 17-12.**

Hoses are installed to avoid twists and sharp bends. The bending radius of flexible hose must be greater than six times the inside diameter. Protective sleeves must encase any hose that is subject to rubbing. Hoses must not be excessively long or excessively short. Hoses that are excessively long have more internal resistance. Hoses tend to decrease in length when pressurized, so a hose that is excessively short, without any bend or flex, will fail prematurely.

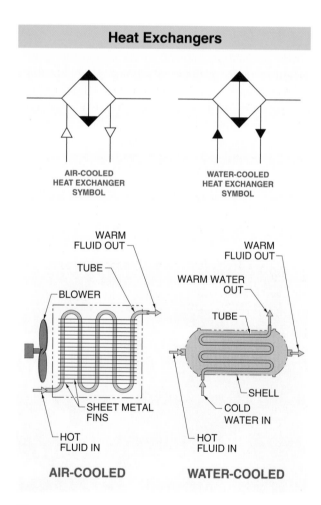

Heat Exchangers

AIR-COOLED
HEAT EXCHANGER
SYMBOL

WATER-COOLED
HEAT EXCHANGER
SYMBOL

WARM
FLUID OUT
TUBE
BLOWER
SHEET METAL
FINS
HOT
FLUID IN

AIR-COOLED

WARM
FLUID OUT
WARM WATER
OUT
TUBE
SHELL
COLD
WATER IN
HOT
FLUID IN

WATER-COOLED

Figure 17-11. Heat exchangers help to lower the temperature of hydraulic fluids when other methods are not sufficient.

Hoses

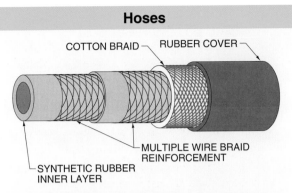

Hose Installation

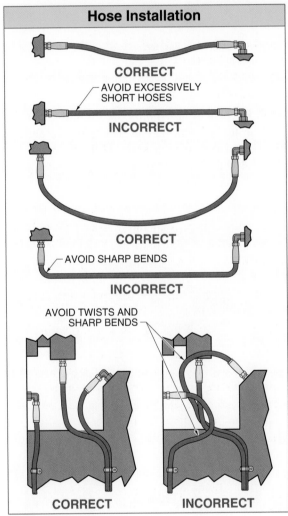

Figure 17-12. Hoses are fabricated in layers for use in high-pressure hydraulic circuits.

Pipes. A *pipe* is a hollow cylinder of metal or other material of substantial wall thickness. Pipe wall thickness is normally thick enough that the pipe may be threaded. Originally, pipe was manufactured with one wall thickness and its size was the actual inside diameter. This changed due to an increase in the strength requirements of pipe. Wall thicknesses were increased, which reduced the inside diameter (ID) of the pipe, leaving the outside diameter (OD) unchanged.

Pipe is designated according to its nominal size and wall thickness. Presently, the nominal pipe size indicates the thread size for connections. A standard ½″ pipe has an ID of 0.622″ and is classified as Schedule 40 pipe. An extra-heavy ½″ pipe is classified as Schedule 80 pipe with an ID of 0.546″. A double-extra-heavy pipe has an ID of 0.252″ (approximately Schedule 160). The actual inside diameter varies with piping, but the actual outside diameter remains constant for any given size pipe. **See Figure 17-13. See Appendix.**

Pipe is generally used for permanent installations. This is because pipe threads are tapered, and once a connection is broken, it must be tightened further to reseal. Each time a pipe is retightened, its length is changed. In some cases, replacing a pipe with a slightly longer pipe may be required.

Pipe

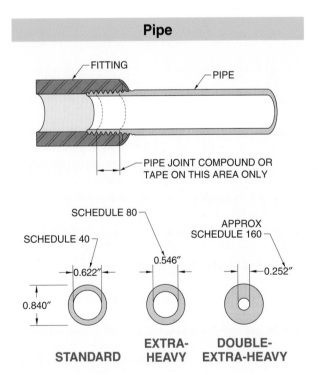

Figure 17-13. Pipe is designated according to its nominal size and wall thickness.

Proper installation procedures ensure the permanency of pipe fittings. Two-thirds of the threaded area should be covered with a pipe joint compound, such as Teflon® tape or paste. This is applied to the middle portion of the male thread to prevent any compound from entering and contaminating the system. Overtightening threads may cause premature leaking or undue stress.

Hydraulic lines and fittings should be made of steel, with the exception of flexible hoses. Galvanized pipe must not be used due to the possibility of metal flaking. Also, the zinc used for galvanizing reacts adversely with certain hydraulic fluid additives. Copper tubing should not be used because it reacts to hydraulic fluids. In addition to its reaction to the fluids, copper tends to harden and crack under the heat and vibrations of hydraulic circuits.

Tubing. A *tube* is a thin-walled, seamless or seamed, hollow cylinder. Tubing is soldered, welded, or formed for compression sealing because the wall thickness is usually too thin for threading. Tubing used for hydraulic circuits must always be seamless. As a general rule, tubing used in hydraulic circuits may be readily bent to a radius equal to five or six times the nominal (outside) diameter of the tubing. Carbon steel tubing offers distinct advantages over pipes when used in hydraulic circuits. One minor disadvantage is that tubing is more expensive than pipe. Advantages of tubing include the following:

• Tubing requires fewer connections than pipe because it can be bent. Tubing also absorbs vibrations better than pipe because of its flexibility.
• Tubing connections make every joint a union, permitting faster assembly and disassembly without the need for joint compound or tape. A *union* is a fitting used to connect or disconnect two tubes that cannot be turned.
• Tubing is lighter in weight than pipe and has a smoother inner surface, which produces less friction and less pressure loss.

Hydraulic tubing is connected using fittings. Tubing fittings may be flared or flareless fittings.

Tech Tip

High-performance hydraulic hose assemblies are constructed to withstand heated, pressurized oil. High-performance hydraulic hose assemblies are typically constructed of four layers, with nylon used as the inner layer, two layers of steel wire encased in braided wire as the second and third layers, and a polyurethane covering as the fourth (outer) layer.

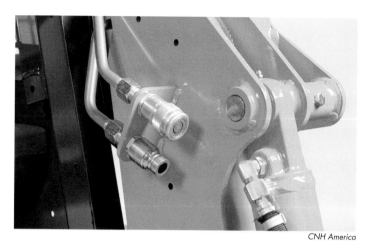

CNH America

Tubing should never be assembled in a straight line but instead should be bent to reduce vibration

Flared Fittings. A *flared fitting* is a fitting that is connected to tubing with an end that is spread outward. The body of the fitting is screwed tightly against the flared end of the tubing. Proper flaring provides a firm, leakproof connection. **See Figure 17-14.**

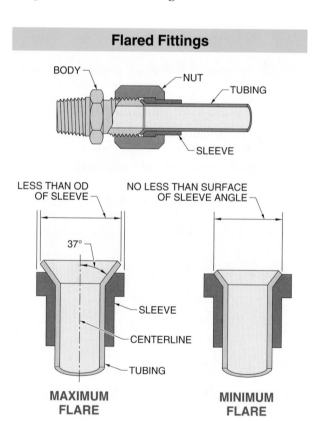

Figure 17-14. A flared fitting is a fitting that is connected to tubing with an end that is spread outward.

Flared fittings generally consist of a body, sleeve, and nut. A seal is made when the flared tubing is pressed against the angular seat of the body by the sleeve. The angles of the body, sleeve, and flared tubing ensure a good seal when the tubing, which is the softest of the three pieces, is pressed into the body. The standard flare angle for hydraulic tubing fittings is 37° from the centerline. The flare extends to cover the total angular surface of the sleeve, but not beyond the sleeve's outside diameter. The flare seats firmly and positively between the sleeve and the body when tubing is flared properly and tubing nuts are tightened securely. Flares that are too short do not provide enough mating area to prevent leaks, and flares that are too long hang up during assembly. Clean, square tubing cuts are achieved with a tube cutter. Hacksaws produce rough cuts that generally are not square.

Flared tubing connections may also have a flare angle of 45°. A 45° flare angle is used for low-pressure applications such as pneumatic, refrigeration, or automotive applications and is not to be used for high-pressure hydraulic circuits.

Incorrect flares may appear to assemble satisfactorily and may even pass initial pressure tests. They are not, however, reliable for continuous service. All tubing flares must conform to the sleeve and body used to join tubing sections.

Flared Joint Tightening. A positive seal is vital to prevent fluid loss, keep out contaminates, and maintain hydraulic circuit pressure. A positive seal does not allow the slightest amount of fluid to pass and is normally compressed between two rigid parts. A nonpositive seal allows a certain amount of leakage, which provides a lubricating film between surfaces. An example of a nonpositive seal is a piston and an O-ring moving within a cylinder.

Tightening a flared fitting is a positive seal and is accomplished by using the proper torque. Undertightening or overtightening a flared fitting nut is avoided by using a torque wrench with proper torque settings or by manually turning the fitting nut while observing witness marks applied to the sealing nut and body. **See Figure 17-15.**

Witness marks may be used when a torque wrench is not available. The joint is assembled with the nut bottomed out and tightened to fingertight. A line is marked, using a felt marker, lengthwise on a flat on the body and onto the corresponding flat of the nut. The nut is rotated with a wrench until a determined number of flats on the body have been passed. The number of flats passed is

based on the size of the fitting. For example, a size 8, ½″ flared fitting should be rotated 2 flats after fingertight. This fitting may also be tightened to 200 lb-in. to 300 lb-in. using a torque wrench.

Flared Joint Tightening

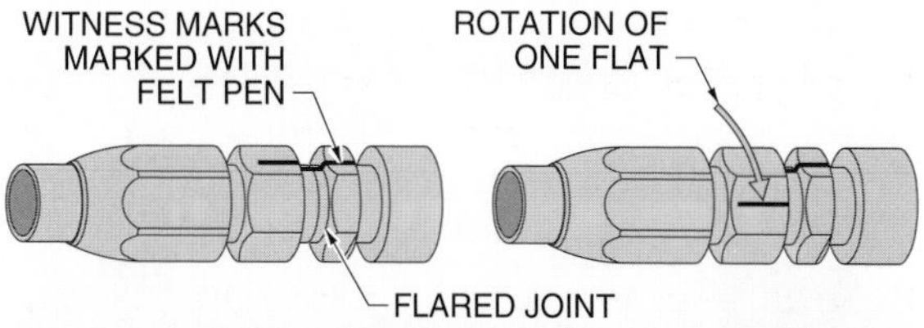

Flared Joint Tightening Specifications

Nominal Tube Size*	Fitting Size	No. of Flats Rotated	Torque†
¼	4	2½	10 – 120
⅜	6	2	100 – 200
½	8	2	200 – 300
⅝	10	1¾	300 – 400
¾	12	1	500 – 700
1	16	⅞	700 – 1000
1¼	20	⅞	1000 – 1300

* in in.
† in lb-in.

Figure 17-15. A flared fitting is tightened using a torque wrench or by turning the fitting nut while observing witness marks.

Tubing should never be assembled in a straight line. Bending tubing for assembly reduces vibration strains and compensates for thermal expansion. A gradual bend is preferred over elbow fittings because elbow fittings have sharp turns with high resistance to flow. Tubing must be bent with the correct radius and without kinks, wrinkles, or flattened bends. The bending radius should be greater than four times the tubing ID. Tubing must also be properly supported to minimize the stresses of vibration. **See Figure 17-16.**

Impact Flaring Method. The *impact flaring method* is a basic flaring method in which a flaring tool is inserted into the tubing end and hammered into the tubing until the tubing end is spread (flared) as required. Flaring tool kits consist of a split female die, a tubing clamp, and a variety of different-sized flaring tools. **See Figure 17-17.**

Tube Installation

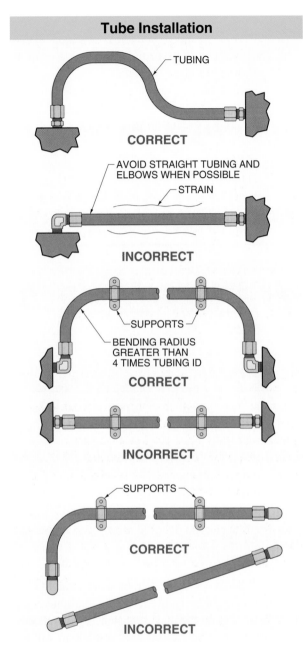

Figure 17-16. Bending tubing for assembly reduces vibration strains and compensates for thermal expansion.

Tubing is flared using the impact method by applying the following procedure:

1. Insert the split female die into the tubing clamp and slide it to the end, enabling the die to be spread open.
2. Insert the tubing into the appropriate die hole with the tubing end approximately $\frac{1}{32}''$ above the top surface of the die.
3. Place the tubing clamp directly over the tubing end and tighten.

4. Hold the tubing with one hand while hammering the flaring tool. This hand takes up the concussion from the blows and is used to feel the thud of the flaring tool when bottomed out.
5. Disassemble the parts and check the flare when the bottomed out thud of the flare tool is felt. Reinsert the tube $\frac{1}{32}''$ above the die surface and repeat steps 3 and 4 if a wider flare is required.

Impact Flaring Method

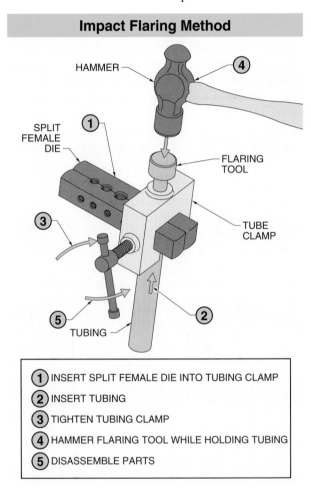

Figure 17-17. The impact flaring method uses a flaring tool that is inserted into the tubing end and hammered into the tubing until the tubing end is spread (flared) as required.

Flareless Fittings. A *flareless (compression) fitting* is a fitting that seals and grips by manual adjustable deformation. Flareless fittings are designed for thick wall tubes that are not suitable for flaring. Flareless fittings create a seal with a ferrule. A *ferrule* is a metal sleeve used for joining one piece of tube to another. The ferrule cuts into and compresses the tube when the nut is tightened onto the body. The nut is tightened a full turn after the completed assembly is fingertight. **See Figure 17-18.**

Flareless (Compression) Fittings

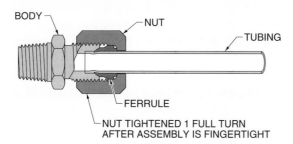

Figure 17-18. A flareless (compression) fitting uses a ferrule which cuts into and compresses the tube when the nut is tightened onto the body.

Pumps

A *pump* is a mechanical device that causes fluid to flow. Hydraulic pumps do not create energy, they convert the energy of a prime mover into hydraulic energy. A *prime mover* is a device that supplies rotating mechanical energy to a fluid power circuit. The two main types of prime movers used in fluid power circuits are electric motors and internal combustion engines. **See Figure 17-19.** Pumps are rated by their manufacturer with a pressure and fluid output rating in revolutions per minute (rpm). Pressure is produced by resistance against fluid flow.

Prime Movers

ELECTRIC MOTOR

INTERNAL COMBUSTION ENGINE

Figure 17-19. The two main types of prime movers used in fluid power systems are electric motors and internal combustion engines.

Boeing Commercial Airplane Group

Many aircrafts contain hydraulic systems for the control of flaps, stabilizers, landing gear, and brakes.

Displacement. *Displacement* is the volume of fluid moved during each revolution of a pump's shaft. Displacement is rated in cubic inches per revolution and is usually found using the pump model number provided by the manufacturer. Generally, pumps used in hydraulic circuits are positive-displacement pumps. *Positive displacement* is the movement of a fixed volume of fluid with each revolution of a pump shaft.

Displacement per minute represents the amount of hydraulic fluid a pump displaces in one minute. The amount is based on the displacement of the pump (in cubic inches per revolution) multiplied by the revolutions per minute of the prime mover. It is calculated using the following formula:

$$Din^3/min = rpm \times D$$

where

Din^3/min = displacement per minute

rpm = revolutions per minute of prime mover

D = pump volumetric displacement (in cu in.)

For example, what is the displacement of a pump with revolutions of 1.94 cu in. that has a prime mover that operates at 1120 rpm?

$$Din^3/min = rpm \times P_r$$
$$Din^3/min = 1120 \times 1.94$$
$$Din^3/min = \textbf{2173 cu in./min}$$

Flow Rate. *Flow rate* is the volume of fluid flow per minute. Flow rate is measured in gallons per minute and is typically observed with a flow meter attached to a hydraulic circuit after the outlet port of the pump. **See Figure 17-20.** Flow rate is calculated by applying the following formula:

$$Q = \frac{D \times rpm}{231}$$

where

Q = flow rate (in gpm)

D = pump volumetric displacement (in cu in.)

rpm = revolutions per minute

231 = constant (cu in. in one gal.)

For example, what is the flow rate of a pump that has a displacement of 4 cu in. and operates at 1725 rpm?

$$Q = \frac{D \times rpm}{231}$$

$$Q = \frac{4 \times 1725}{231}$$

$$Q = \frac{6900}{231}$$

$$Q = \textbf{29.87 gpm}$$

Flow Rate

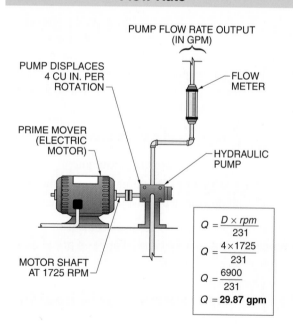

Figure 17-20. Flow rate is the amount of hydraulic fluid that passes a given point in one minute.

Positive-Displacement Pump Operation. All positive-displacement pumps operate similarly to create fluid flow. All hydraulic positive-displacement pumps have similar parts such as a shaft, pump housing, inlet port, and outlet port. **See Figure 17-21.** A positive-displacement pump operates in the following four basic steps:

1. The vacuum in a pump is created when the pump rotates and an increased volume is created at its inlet. The pressure in the reservoir is at atmospheric pressure, which is higher than the vacuum created at the inlet. Atmospheric pressure forces the fluid to flow from the reservoir into the inlet of the pump.
2. Once the fluid enters the pump through the inlet, the pump traps the fluid through a sealing method. A *seal* is a device in contact between two components that contains pressure and prevents leakage. The fluid travels through the pump towards the outlet side of the pump.
3. At the outlet side of the pump, the sealed chamber opens and releases the fluid into the pump outlet port.
4. The sealed chamber closes, preventing fluid in the outlet side of the pump from slipping to the inlet side of the pump.

Hydraulic pumps must be constructed to high-quality standards, used properly, and maintained using a scheduled maintenance program because pump efficiency is directly related to pump cleanliness. Positive displacement pumps used in hydraulic circuits are gear, vane, or piston pumps.

Gear Pumps. A *gear pump* is a positive-displacement pump containing intermeshing gears that force the fluid from the pump. One gear is driven by the prime mover (external gear pump). Gear pumps are the most widely used hydraulic pumps because of their simple design and ease of repair. **See Figure 17-22.**

Gear pumps develop fluid flow by carrying hydraulic fluid between the teeth of the two closely meshed gears and their housing. The housing includes an inlet port and an outlet port. The housing encloses the pumping cells between gear teeth. Side pressure plates act as replaceable wear plates. The two gears that develop flow are the drive gear and the driven (idler) gear. The drive gear is powered by a drive shaft that, when rotated, turns the driven gear.

Safety Tip

Relieve all hydraulic pressure before working on pressurized hydraulic lines or components by using the manual bypass or bleeder screw before disconnecting the parts of the system.

Positive-Displacement Pump Operation

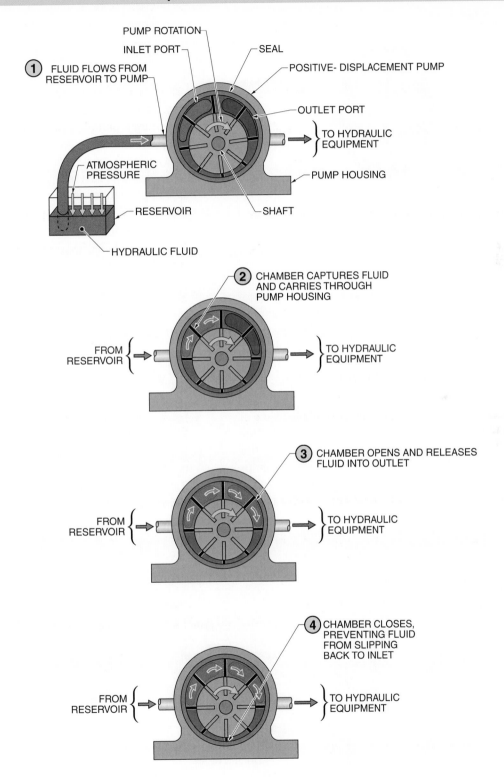

Figure 17-21. All positive-displacement pumps follow the same basic operational steps to create fluid flow.

Gear Pumps

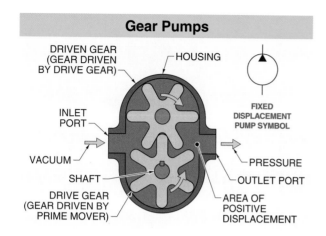

Figure 17-22. A gear pump contains two gears within a housing that are rotated to produce fluid flow.

As the gear teeth move apart at the inlet side of the pump, a partial vacuum is created. Pushed by atmospheric pressure, hydraulic fluid then fills the unfolding gear-tooth chamber voids. The chambers carry the hydraulic fluid around the outside of the gears, where it is then forced out (displaced) as the teeth begin to mesh together at the outlet side of the pump.

Hydraulic fluid that remains in the chamber voids as the gears mesh together and creates a high degree of pressure that is relieved by machined notches in the side plates. External gear pumps are available as single or double versions. Double versions have two single pumps that share a common inlet port and input shaft, with each pump having its own outlet port. Double pumps

save space and installation costs and can serve either a single circuit with great volume or two separate circuits.

Gear pumps can sometimes be repaired in the field. Gear pump assemblies consist of major parts (a frame, gears, housing, and a shaft) and minor parts (O-rings, backup rings, and seals) that may be repaired, replaced, or refurbished. Many gear pump manufacturers supply step-by-step procedures for the assembly and disassembly of their pumps. Gear pump manufacturers also supply part numbers and/or descriptions for every part that may need to be replaced. **See Figure 17-23.** For example, the two gears in a gear pump can be replaced by applying the following procedure:

1. Remove the pump from its motor coupling by removing the bolts.
2. Remove all bolts from the front of the pump that hold the pump assembly together.
3. Disassemble the pump by detaching the front plate assembly from the pump. Remove the two gears from the body housing. Detach the body housing from the back plate assembly. Place all parts in a straight line in the exact order that they were disassembled.
4. Inspect all seals that connect the three main sections together. If the seals are good, proceed to step 5. If the seals are worn or damaged, replace them.
5. Inspect the ball bearings for wear and lubrication. Replace or regrease as required.
6. Replace the two gears with new gears in the order that they have been laid out.
7. Reassemble the gear pump by working backwards from step 6.

Gear Pump Assemblies

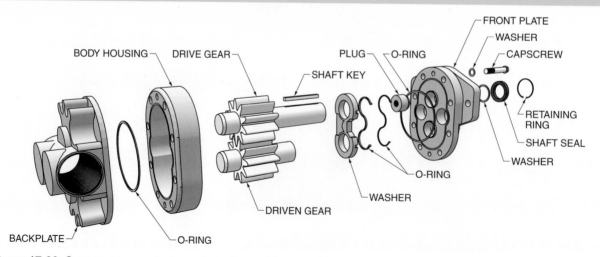

Figure 17-23. Gear pump manufacturers typically provide assembly diagrams for each specific pump.

External Gear Pumps. An *external gear pump* is a gear pump that consists of two externally toothed gears that form a seal within the pump housing. External gear pumps have two equally sized gears, the drive gear and driven gear, which rotate to cause fluid to flow into the circuit. **See Figure 17-24.** External gear pumps operate in the following four basic steps:

1. As the drive gear rotates, a vacuum is created on the inlet side of the pump as the gear teeth pull apart. Atmospheric pressure pushes the fluid from the reservoir into the inlet.
2. As both gears rotate away from the inlet, the gear teeth trap the fluid and force it between the gear teeth and the internal wall toward the pump outlet.
3. As the gear teeth reach the outlet, the fluid is forced out of the pump by decreasing volume.
4. As the gear teeth mesh together, they form a seal that does not allow the fluid to flow back into the inlet, forcing the fluid through the outlet port and into the hydraulic circuit.

Gear pumps consist of two meshing gears enclosed in a close-fitting housing.

External Gear Pump Operation

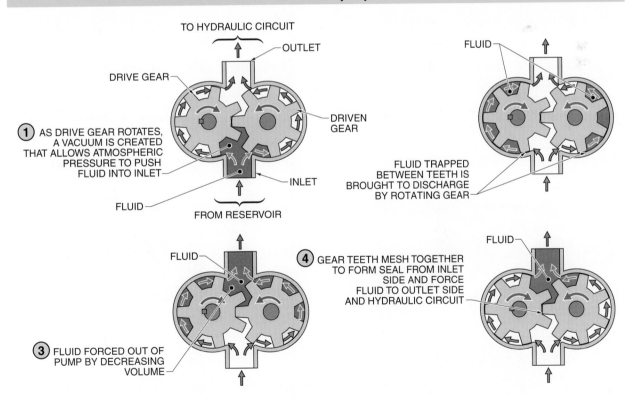

Figure 17-24. An external gear pump consists of meshing gears that form a seal with the pump housing and operate similar to the four basic steps of a positive-displacement pump.

Internal Gear Pumps. An *internal gear pump* is a gear pump that consists of a small external drive gear mounted inside a large internal spur gear (ring gear). The two gears rotate in the same direction. **See Figure 17-25.** A crescent seal separates the low- and high-pressure areas of the pump. A *crescent seal* is a crescent-moon-shaped seal between the gears and between the inlet and outlet sides of an internal gear pump. Because there are only two moving parts, internal gear pumps are reliable and easy to maintain. Internal gear pumps operate in the following three basic steps:

1. As the motor rotates the external gear, the gear teeth unmesh from the internal gear, creating a vacuum. Atmospheric pressure pushes the fluid from the reservoir into the inlet of the pump.

2. The fluid becomes trapped in the cavities of the unmeshed gears. As the gears rotate, the crescent seal separates the internal gear and the external gear. The fluid continues to move as the two gears continue to rotate.

3. As the two gears reach the end of the crescent seal, the gears begin to mesh, decreasing the volume. The decrease in volume forces the fluid out of the cavities between the teeth and causes the fluid to flow through the outlet of the pump and into the hydraulic circuit.

Tech Tip

In addition to hydraulic oil, internal gear pumps are used to pump fuel oil, lube oil, resins, polymers, alcohols, solvents, asphalt, bitumen, tar, polyurethane foam, paint, inks, pigments, soaps, surfactants, and glycol.

Continental Hydraulics

Directional control valves are available with up to 5 actuators (solenoid, air, cam, oil, and lever) and 12 spool options for use with pressures up to 4600 psi.

Internal Gear Pump Operation

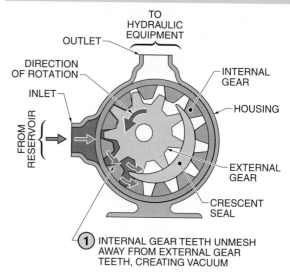

TO HYDRAULIC EQUIPMENT
OUTLET
DIRECTION OF ROTATION
INLET
FROM RESERVOIR
INTERNAL GEAR
HOUSING
EXTERNAL GEAR
CRESCENT SEAL

① INTERNAL GEAR TEETH UNMESH AWAY FROM EXTERNAL GEAR TEETH, CREATING VACUUM

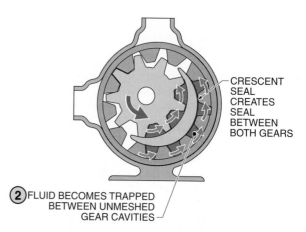

CRESCENT SEAL CREATES SEAL BETWEEN BOTH GEARS

② FLUID BECOMES TRAPPED BETWEEN UNMESHED GEAR CAVITIES

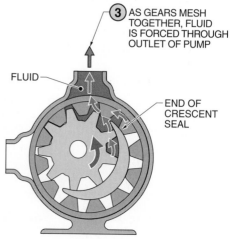

③ AS GEARS MESH TOGETHER, FLUID IS FORCED THROUGH OUTLET OF PUMP

FLUID

END OF CRESCENT SEAL

Figure 17-25. An internal gear pump consists of a small external drive gear mounted inside a large internal gear.

The gears used in a gear pump are generally spur gears. A *spur gear* is a gear that has straight teeth that are parallel to the shaft axes. For quieter operation and increased performance, helical or herringbone gears are used. A *helical gear* is a gear with teeth that are cut at an angle to its axis of rotation. A *herringbone gear* is a double helical gear that contains a right- and left-handed helix. **See Figure 17-26.**

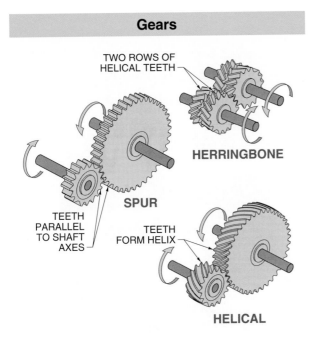

Gears

TWO ROWS OF HELICAL TEETH

HERRINGBONE

SPUR

TEETH PARALLEL TO SHAFT AXES

TEETH FORM HELIX

HELICAL

Figure 17-26. The gears used in a gear pump are generally spur gears. For quieter operation and increased performance, helical or herringbone gears are used.

Vane Pumps. A *vane pump* is a pump that contains vanes in an offset rotor. Vane pumps produce fluid flow as the rotor and vanes rotate. **See Figure 17-27.** As the rotor rotates, a vacuum is produced at the inlet allowing atmospheric pressure to push the fluid and fill the voids between the vanes. Centrifugal force, springs, and/or pressure under the vanes hold them firmly against a cam ring to form a positive seal between the tip of the vanes and the cam ring.

A *cam ring* is a metal ring that provides an area for fluid flow and a surface against which the vanes ride. The fluid enters and is discharged from the pump through a port plate. A *port plate* is a device that contains ports that connect the pump's internal inlet and discharge areas to the pump housing inlet and outlet ports. The inlet of the port plate is connected to the inlet of the pump housing. The outlet of the port plate is connected to the outlet of the pump housing.

As the rotor continues to rotate and fluid is carried forward, the vanes begin to retract as the volume of the void produced between the off-centered rotor and the cam ring is reduced. The fluid is forced out of the port plate outlet at increased pressure. Vane pumps remain efficient throughout their use due to the movement of the vanes, which compensates for wear.

Vane pumps are classified as unbalanced or balanced. An *unbalanced-vane pump* is a vane pump that has one set of internal ports and produces a pumping action in the chambers on one side of the rotor and shaft. The rotor is off-center from the cam ring, and each vane creates a pumping action once in each revolution. The pump is unbalanced because low inlet pressure is applied on one side of the rotor and high discharge pressure is applied on the other side of the rotor.

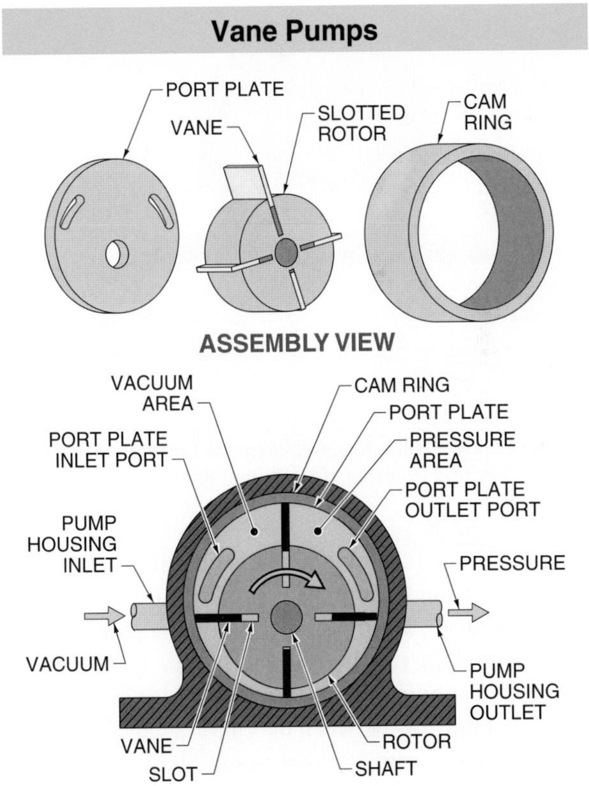

Vane Pumps

PORT PLATE
VANE
SLOTTED ROTOR
CAM RING

ASSEMBLY VIEW

VACUUM AREA
PORT PLATE INLET PORT
PUMP HOUSING INLET
VACUUM
VANE SLOT
CAM RING
PORT PLATE
PRESSURE AREA
PORT PLATE OUTLET PORT
PRESSURE
PUMP HOUSING OUTLET
ROTOR
SHAFT

Figure 17-27. A vane pump contains vanes in the slots of a rotor that produce fluid flow as the rotor rotates.

Tech Tip

Many manufacturers provide detailed assembly and disassembly instructions for vane pumps that make it possible to repair, replace, or refurbish any part of a vane pump.

A *balanced-vane pump* is a vane pump that has two sets of internal ports and produces a pumping action in chambers on both sides of the rotor and shaft. **See Figure 17-28.** In a balanced-vane pump, the rotor is centered inside the cam ring. The vanes create a pumping action twice in one revolution because the two inlet and two outlet ports on the port plate are 180° apart. Both inlet ports are connected together and both outlet ports are connected together so each leads to one respective port in the pump housing. The pump is balanced because low inlet pressure and high discharge pressure are applied to both sides of the rotor.

Balanced-Vane Pumps

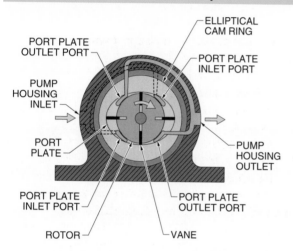

Figure 17-28. A balanced-vane pump has two sets of internal ports and contains an elliptical cam ring.

The volume of a vane pump is determined by how far the rotor and cam ring are offset. Changing the offset between the rotor and cam ring changes the volume of fluid supplied by the pump. The offset is changed by moving the cam ring to reduce or increase the size of the area between the cam ring and the rotor. Displacement control is performed by the turn of an external hand wheel or by a pressure compensator. A *pressure compensator* is a displacement control that alters displacement in response to pressure changes in a system.

Safety Tip

Bourdon tube pressure gauges can be damaged by pressure surges in the hydraulic system. To limit the damage, bourdon tube gauges may be oil filled.

A *pressure-compensated vane pump* is a vane pump equipped with a spring on the low-displacement side of the cam ring. The spring pressure is adjustable and determines the amount of fluid pressure required to create cam movement. The cam ring centers and pumping ceases when the pressure acting on the inner wall of the cam ring is high enough to overcome the spring force. In many cases, a pressure-compensated pump is used to limit circuit pressure. As a result, there is little fluid heating or wasted horsepower. **See Figure 17-29.**

Pressure-Compensated Vane Pumps

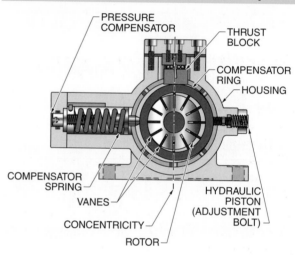

Figure 17-29. A pressure-compensated vane pump is a pump that automatically adjusts the amount of volume it displaces per rotation by centering the rotor when the pressure in the systems starts to build.

At full compensation, the displacement of a pump is zero plus minor internal leakage. All pressure-compensated pumps require the case oil to be drained when fully compensated because of the internal leakage. The hydraulic fluid left in the pump during zero displacement continues to rise in temperature if there is no fluid flow.

Many manufacturers provide detailed assembly and disassembly instructions for vane pumps that make it possible to repair, replace, or refurbish any part of a vane pump. It is also common for vane pump assemblies to have rebuild kits that can be ordered and kept on-site for quick repair. **See Figure 17-30.** For example, a cam ring in a vane pump is replaced by applying the following procedure:

1. Remove the pump from its foot bracket by removing the bolts.
2. Remove four bolts from the back of the pump.
3. Disassemble the pump into three sections. One section is the back housing with O-rings. The second section is the cartridge assembly. The third section is the front housing with the shaft.
4. Place all parts in a straight line in the exact order that they were disassembled.
5. Inspect all the seals that connect the three main sections together. If the seals are good, continue to step 6. If the seals are worn or damaged, replace them.
6. Inspect the ball bearings for wear and lubrication. Replace or regrease as required.
7. Remove two screws holding the vane cartridge together.
8. Place all parts from the vane cartridge in a straight line in the exact order that they were disassembled.
9. Remove the cam ring and the rotor, making sure that the vanes do not fall out of the rotor.
10. Remove all vanes from the rotor and inspect for premature wear.
11. Replace the damaged cam ring with a new cam ring.
12. Reassemble the vane pump by working backwards from step 11.

Piston Pumps. A *piston pump* is a hydraulic pump in which fluid flow is produced by reciprocating pistons. A piston pump consists of a piston block, pistons with shoes, a valve plate, a swash plate, and a drive shaft.

Industrial ferrous metal balers are hydraulically powered and can process up to 75 tons of material per hour.

A *swash plate* is an angled plate in contact with the piston heads that causes the pistons in the cylinders of a pump to extend and retract. **See Figure 17-31.** The pistons are connected to the swash plate by shoes. As the piston block rotates, the pistons slide over the valve plate. The valve plate contains two crescent-shaped ports. One port is connected to the inlet of the pump and the other is connected to the outlet. The pistons reciprocate, drawing fluid from the crescent-shaped inlet port due to the angle of the swash plate. At midrotation, the pistons are completely extended and filled with fluid. As the piston block continues to rotate, the cylinders begin to retract and force the fluid out of the crescent-shaped outlet port.

Vane Pump Assemblies

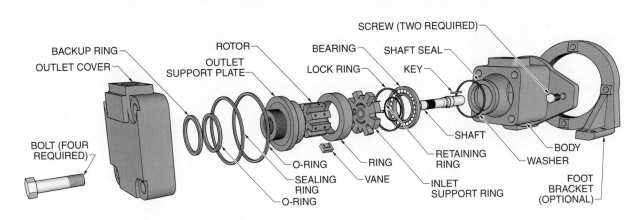

Figure 17-30. Vane pump manufacturers typically provide assembly diagrams for each specific pump.

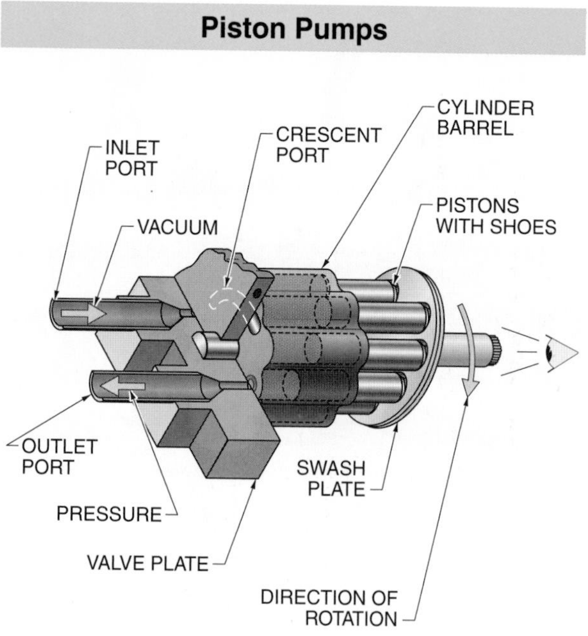

Piston Pumps

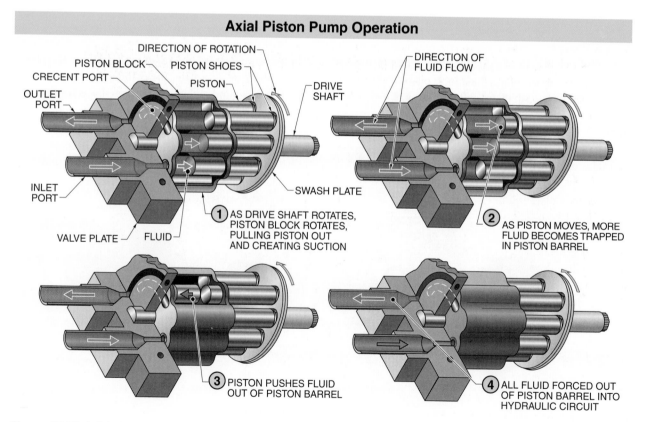

Figure 17-31. The pumping action of a piston is caused by reciprocating pistons within the cylinder barrel as the swash plate rotates.

Displacement of piston pumps is determined by the size and number of pistons and the stroke length. Stroke length in a variable-displacement pump is determined by the angle of the swash plates. Increasing or decreasing the piston stroke is done by pivoting the movable yoke. Pivoting the yoke allows for angle change of the swash plate. Positioning the yoke and its angle can be controlled by load-sensing control. Types of piston pumps include axial, variable-displacement, bent-axis, and radial piston pumps.

An *axial piston pump* is a piston pump that consists of pistons in a rotating piston block parallel to the drive shaft. **See Figure 17-32.** Axial piston pumps operate in the following four basic steps:

1. As the drive shaft rotates, the piston block rotates in the same direction, pulling a piston out and creating suction.
2. As the piston moves through the first half of the pump, it pulls more fluid in.
3. When the piston reaches the halfway point of a cycle (180°), the piston pushes fluid out of the piston barrel.
4. As the piston completes a 360° cycle, all fluid is pushed out of the piston barrel.

Axial Piston Pump Operation

Figure 17-32. Axial piston pumps consist of a number of pistons, a piston block, piston shoes, a swash plate, and a drive shaft and operate with four basic steps.

A *variable-displacement piston pump* is a piston pump in which the angle of the swash plate can be varied. **See Figure 17-33.** A variable-displacement piston pump works under the same principles as an axial piston pump with the exception of the variability of the angle of the swash plate. When the angle of the swash plate is varied, it changes the distance a piston pulls into the barrel. This causes the piston to allow more or less hydraulic fluid into its barrel, varying the flow rate of the pump.

The most common method to vary the angle of a swash plate is through internal pilot pressure. As the pressure in the hydraulic circuit begins to reach the set pressure of the pump, pressure from the internal pilot lines begins to push on the pilot valve attached to the swash plate.

Maximum pilot pressure is set with a setscrew that adjusts a control spring. As the swash plate begins to move, the distance the pistons pull back into the barrel changes. When the swash plate is vertical, there is no fluid flow produced by the pump. However, the prime mover still rotates at the same revolutions per minute, which saves energy because there is no load on the motor and there is no energy wasted from fluid moving through the pressure-relief valve.

A *bent-axis piston pump* is a piston pump in which the pistons and cylinders are at an angle to the drive shaft and thrust plate. Bent-axis piston pumps operate similarly to axial piston pumps, but rather than the swash plate being at an angle (offset), the pistons and piston block are at an angle (offset). **See Figure 17-34.** The angle at which the pistons and piston block are offset determines the amount of hydraulic fluid that each piston can take in. Thus, the angle at which the pistons and piston block are set determines the amount of fluid flow.

Bent-axis piston pumps can be either fixed or variable. Fixed bent-axis piston pumps work by rotation from the prime mover that the angled piston is attached to. As they rotate, the pistons extend and retract, creating fluid flow. Variable bent-axis piston pumps work by adjusting the angle at which the pistons and the piston block sit.

Tech Tip

The pistons in an axial piston pump reciprocate parallel to the centerline of the drive shaft of the piston block. Rotary shaft motion is converted into axial reciprocating motion. Most axial piston pumps contain multiple pistons and use check valves or port plates to direct fluid flow from its inlet to its outlet.

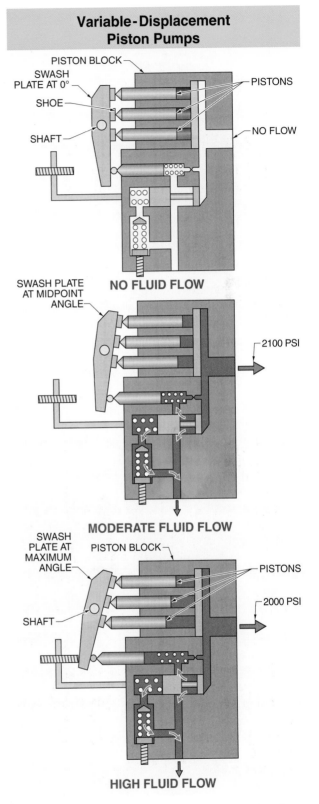

Variable-Displacement Piston Pumps

PISTON BLOCK
SWASH PLATE AT 0°
SHOE
SHAFT
PISTONS
NO FLOW

NO FLUID FLOW

SWASH PLATE AT MIDPOINT ANGLE
2100 PSI

MODERATE FLUID FLOW

SWASH PLATE AT MAXIMUM ANGLE
PISTON BLOCK
PISTONS
2000 PSI
SHAFT

HIGH FLUID FLOW

Figure 17-33. A variable-displacement piston pump has a swash plate at an angle that can be varied, thereby varying the amount of fluid flow.

Bent-Axis Piston Pumps

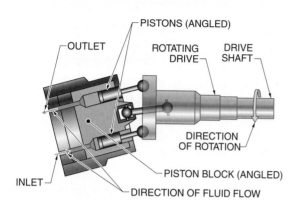

Figure 17-34. Bent-axis piston pumps operate in the same manner as an axial piston pump, but rather than the swash plate being at an angle, the pistons and piston block are at an angle.

Hydraulic pumps are used to provide pressures required in concrete compression testers.

A *radial piston pump* is a piston pump that consists of a cylinder barrel, pistons with shoes, a ring, and a valve block located perpendicular to the pump shaft. **See Figure 17-35.** Radial piston pumps are high-pressure hydraulic pumps capable of operating at 10,000 psi. Radial piston pumps are used because of the design of their pistons and barrel, which allow for a short stroke.

Radial Piston Pumps

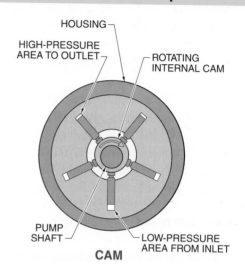

CAM

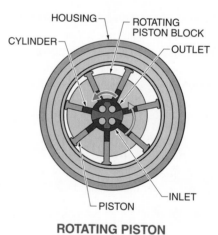

ROTATING PISTON

Figure 17-35. Radial piston pumps consist of reciprocating pistons in cylinders and can be classified as cam or rotating piston pumps.

Tech Tip

Certain hydraulic flow control valves are constructed of high-strength steel and have the ability to withstand pressures up to 3000 psi and temperatures up to 180°F.

Radial piston pumps are classified as cam or rotating piston pumps. In a cam pump, a rotating internal cam moves the pistons in cylinders. The cam is shaped to push the pistons out during one half of the cam rotation and allow the pistons to retract during the other half. There are also variable pressure-compensated radial piston pumps.

A variable pressure-compensated radial piston pump works similar to a variable axial piston pump by adjusting the stroke of the pistons to adjust the amount of fluid flow as pressure increases. The amount of fluid flow is adjusted by centering the cam ring and controlling the distance that the pistons extend and retract. In a rotating piston pump, pistons are housed in a rotating piston block that is offset inside the pump housing and rotates around a fixed shaft. Fluid enters the pump inlet as the pistons extend and is discharged from the pump outlet as the pistons retract.

Radial piston pumps operate on the same basic principles as axial piston pumps but are built with the pistons lying flat and facing inward toward the shaft. The inlet and outlet are located close to the shaft, and the piston block is off-center inside the cam ring. As the shaft rotates, the pistons extend and retract to complete the four basic operational steps of positive-displacement pumps.

Schematic Symbols. Hydraulic pump schematic symbols are used to determine general information about the pump used in a circuit. **See Figure 17-36.** While hydraulic pump schematic symbols do not provide direct information on pump type, such as piston, gear, or vane, they do provide information on whether the pump is unidirectional or bidirectional by using arrows. These symbols also provide information on whether the pump is fixed or variable.

Load-Sensing Control. *Load-sensing control* is a method of controlling variable-displacement pump output according to the amount of pressure required at load use. Varying the output of a pump as the workload varies can significantly reduce power and energy that is otherwise wasted. Controlling pump output consumes the least horsepower while generating the least heat. Load-sensing valves perform control by allowing power when required by the load or reducing pump output power when there is no load.

Hydraulic Pump Schematic Symbols

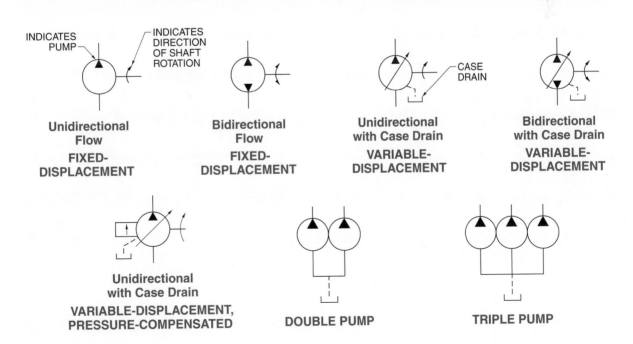

Figure 17-36. Hydraulic pump schematic symbols are used to determine general information about the type of pump used in a system.

Load-sensing controls operate by means of flow differential across an orifice, with the flow change caused by the pressure or absence of a workload (actuator). **See Figure 17-37.** As the workload is being used (flow), the position of the pump yoke is offering maximum displacement. When the workload is paused (no flow through the actuator), greater pressure is placed on the valve spool, overcoming spring pressure, which allows fluid flow to the pump's yoke. Pressure at the pump's yoke decreases the pump displacement (output) and reduces its flow, thereby protecting the system. When work (actuator) resumes, pressure to the compensator piston is decreased, the spring closes off flow to the yoke, and pump displacement maximizes.

Piston pumps are used when variable displacement is required. Variable displacement allows the control of fluid from a no-flow condition up to full flow. Regulating flow is accomplished by adjusting the swash plate from a no-angle position to a full-angle position. The pistons do not reciprocate if the swash plate is not at an angle. The pistons reciprocate, drawing fluid from the inlet and discharging it at the outlet, if the swash plate is at an angle. The displacement varies according to the angle of the swash plate.

Cavitation. *Cavitation* is the process in which microscopic gas bubbles expand in a vacuum and suddenly implode when entering a pressurized area. *Implosion* is an inward bursting. Cavitation occurs when the inlet of a pump is restricted. An indication of pump cavitation is a high shrieking sound or a sound similar to loose marbles or ball bearings in the pump. Cavitation is normally created when the suction line is damaged, plugged, or collapsed. Cavitation may also be caused by an increase in pump revolutions per minute that requires more hydraulic fluid than the circuit piping allows, hydraulic fluids with an increased viscosity due to lower ambient temperatures, and an increase in the viscosity of a hydraulic fluid in a circuit when the circuit has a long suction line.

As the pump pulls against a hydraulic fluid that does not flow, a greater vacuum is created. Any microscopic air or gas within the fluid expands. Expanded bubbles on the inlet side collapse rapidly on the outlet side of the pump. The small but tremendous implosions can cause great damage to pump parts. Theoretically, an air bubble exposed to 5000 psi hydraulic circuit cavitation may create an implosion pressure of 75,000 psi and travel at a speed of 600 fps to 4000 fps. **See Figure 17-38.**

Load-Sensing Control Operation

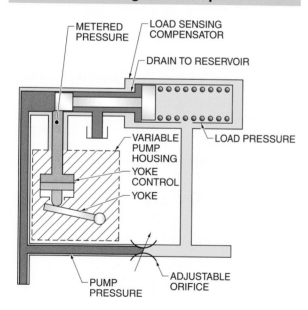

Figure 17-37. Load-sensing control is a method of controlling variable-displacement pump output according to the amount of pressure required at load use.

Cavitation

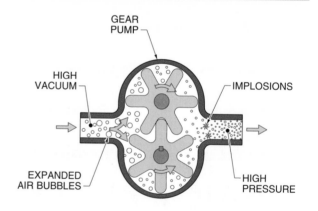

Figure 17-38. Cavitation occurs when gas bubbles expand in a vacuum and implode when entering a pressurized area.

Safety Tip

Hydraulic fluid becomes hot during use. Be careful not to burn hands on hot hydraulic fluid. Always let a system cool before beginning service procedures.

Pseudocavitation is artificial cavitation caused by air being allowed into the pump suction line. Pseudocavitation is caused by low reservoir fluid, contaminated fluid, or leaking pump suction lines. Pseudocavitation is indicated by an unchanged or lower pump intake vacuum. Cavitation and pseudocavitation are both affected by air in the prepump area, and both can be damaging. Cavitation is caused by low inlet pressure and is a change in size of the air molecules normally found in hydraulic fluids. Pseudocavitation is air being introduced into the fluid. Upon pump discharge, the pseudocavitation air is carried into the circuit, and the cavitation air molecules implode.

Valves

A *valve* is a device that controls the pressure, direction, or rate of fluid flow. Hydraulic circuit components, including valves, are equipped with one or more external openings (ports), which allow the flow of fluid to and from the component. Each external port is a primary or secondary port. A *primary port* is the source or inlet port. A primary port may be labeled with a P for primary or pressure. A *secondary port* is an external passage that allows fluid flow to other components. Secondary ports may be labeled A, B, or T. A- and B-labeled ports are ports that lead to other pressure components. A T-labeled (tank-labeled) port is a port that leads to the reservoir. Basic hydraulic valves operate by moving elements that open or block fluid passages that are connected to other components. Hydraulic circuit valves are grouped by their function. They may be pressure, directional, or flow control valves.

Pressure Control. Pressure control in a hydraulic circuit is concerned with maintaining or reducing (regulating) system pressure to operate a circuit. Pressure control valves include pressure-relief, sequence, and pressure-reducing valves.

Hydraulic circuit pressure is maintained by the use of a pressure-relief valve. A *pressure-relief valve* is a valve that sets a maximum operating pressure level for a circuit to protect the circuit from overpressure. Pressure-relief valves are normally closed valves that require higher-than-spring pressure to open. In a pressure-relief valve, pressure on a ball or poppet overcomes spring pressure, allowing fluid to flow. The inlet (primary) port is connected to circuit pressure and the discharge (secondary) port is connected to the reservoir. Ball or poppet movement is controlled by a predetermined pressure level. The pressure level, or spring pressure, is usually varied by screw adjustment. **See Figure 17-39.**

Pressure-Relief Valves

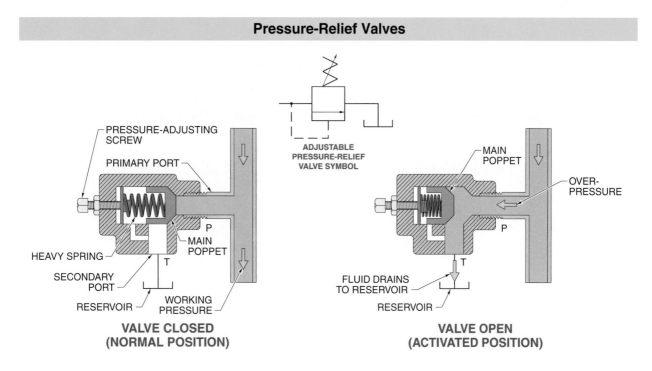

VALVE CLOSED (NORMAL POSITION)

VALVE OPEN (ACTIVATED POSITION)

Figure 17-39. A pressure-relief valve limits the maximum pressure in a hydraulic circuit.

Hydraulic valves may be direct-acting or pilot-operated. A *direct-acting valve* is a valve that is activated or directly moved by fluid pressure from the primary port. For example, in a pressure-relief valve, the spool or poppet is directly activated by an increase in circuit or upstream pressure. **See Figure 17-40.**

A *pilot-operated valve* is a valve that is actuated by hydraulic fluid in the line that is otherwise sent back to the reservoir. *Pilot operation* is the controlling of the function of a valve using system pressure or pressure supplied by an external (pilot) source. A *pilot line* is a passage used to carry fluid to control a valve. A pilot line is not used to power an actuator. Pilot lines may be externally plumbed to transfer the flow of fluid from another component or may be passages that are machined within a component.

Pilot-operated pressure-relief valves are used as circuit pressure overload protection and are also used for circuit operating pressure regulation. In a pilot-operated pressure-relief valve, pilot pressure is sensed through a control orifice by the pilot poppet. As circuit pressure builds, the pilot poppet opens, allowing the fluid to flow to the reservoir. This reduces the pressure of the outlet side of the main poppet, causing it to open and allowing greater fluid flow to the reservoir.

A pressure-relief valve that is normally closed can be situated between two linear actuators (cylinders) and used as a sequence valve. A *sequence valve* is a pressure-operated valve that diverts flow to a secondary actuator while holding pressure on the primary actuator at a predetermined minimum value after the primary actuator completes its travel. A *sequence* is the order of a series of operations or movements.

A clamp and stamp circuit is an example of a circuit in which a sequence valve can be used to control the sequence of circuit operations. A sequence valve is positioned in the circuit just ahead of the stamp cylinder. The pressure setting does not allow the main poppet to shift, preventing the primary port and secondary port from being connected until the set pressure has been reached. **See Figure 17-41.**

Fluid flow to the clamp cylinder extends the clamp cylinder while the stamp cylinder does not move. Pressure in the circuit continues to build after the clamp cylinder is fully extended. The sequence valve main poppet opens, allowing fluid flow to the stamp cylinder when the adjusting screw pressure is reached. This allows a part to be clamped and stamped in the correct sequence.

Valve Actuation Methods

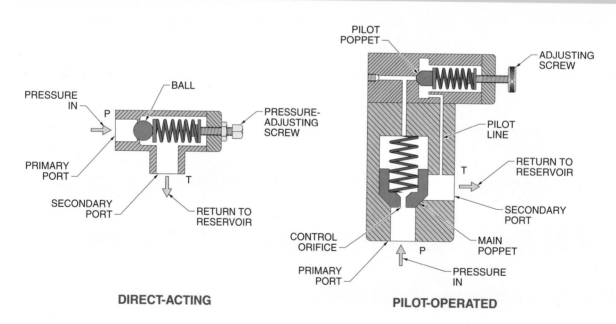

DIRECT-ACTING

PILOT-OPERATED

Figure 17-40. Hydraulic valves may be direct-acting (actuated by fluid pressure from the primary port) or pilot-operated (actuated by fluid in the line that is otherwise sent back to the reservoir).

Sequence Circuits

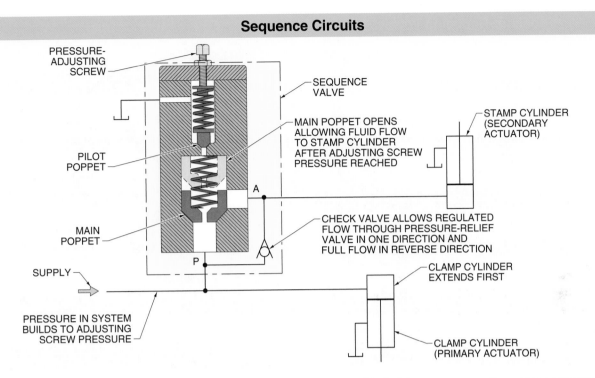

Figure 17-41. A sequence circuit diverts flow to a secondary actuator while holding pressure on the primary actuator at a predetermined minimum value after the primary actuator completes its travel.

System pressure set by pressure-relief valves may not always be sufficient to operate multiple actuators. A pressure-reducing valve is used where each actuator or circuit may require a lower pressure than the circuit's operating pressure. A *pressure-reducing valve* is a valve that limits the maximum pressure at its outlet, regardless of the inlet pressure. Pressure-reducing valves are normally open and may be direct-acting or pilot-operated. Pressure-reducing valves, which sense pressure from their secondary port, are normally direct-acting. However, pressure in another part of a circuit can be sensed and pilot pressure can be used to operate a pressure-reducing valve by means of an external pilot line. **See Figure 17-42.** For example, a pressure-reducing valve can be used when a metal workpiece must be stamped three times and each stamping operation must be set at a different pressure.

Pressure-reducing valves operate by pressure being sensed at their secondary port. The spool is moved off its normal position, reducing or blocking working pressure when higher-than-system pressure is reached. Excess fluid flow is diverted to the reservoir through the drain port. Only enough flow is passed to the outlet to maintain the preset pressure. A light flow is sent to the reservoir through the drain port if the valve closes completely. This prevents pressure from building up in the circuit. Pressure levels are maintained by a pressure adjusting screw.

Pressure intensity is measured by a pressure gauge when adjustments are made to pressure-control components. A *pressure gauge* is a device that measures the intensity of a force applied to a hydraulic fluid. Pressure gauges are required for adjusting control valves to within proper or required values, determining the forces exerted by a cylinder, or determining the torque produced by a hydraulic motor.

Pressure-Reducing Valves

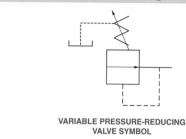

VARIABLE PRESSURE-REDUCING
VALVE SYMBOL

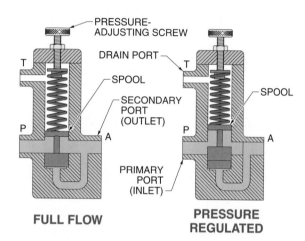

FULL FLOW

PRESSURE
REGULATED

Figure 17-42. Pressure-reducing valves are used to regulate pressure in one leg of a circuit or individual component pressure.

Many gauges used for measuring high pressure use a Bourdon tube. A *Bourdon tube* is the pressure-sensing element inside a mechanical pressure gauge that consists of a circular stainless steel or bronze tube that is flattened to make it flexible. A Bourdon tube is oval or elliptical in its cross-sectional area and is bent in the shape of the letter C. One end of the Bourdon tube is fixed to a frame where the hydraulic fluid enters. The other end is closed and free to move. As the hydraulic fluid pressure inside the tube changes, the elliptical cross section changes and the free end of the Bourdon tube tends to straighten. This actuates a linkage to a pointer gear, which moves the pointer to indicate the pressure on a scale. **See Figure 17-43.**

Directional Control. A *directional control valve* is a valve whose primary function is to direct or prevent flow through selected passages. A directional control valve allows fluid to be directed to actuators and other system components at the appropriate time and valve port. Directional control valves include check, two-way, three-way, and four-way valves.

Pressure Gauges

PRESSURE GAUGE

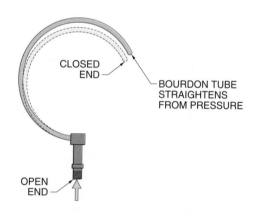

BOURDON TUBE OPERATION

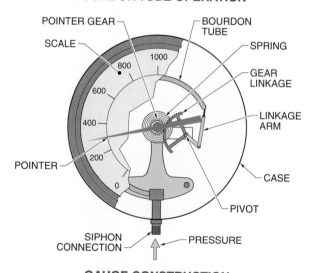

GAUGE CONSTRUCTION

Figure 17-43. Pressure gauges use a Bourdon tube to measure pressures.

A *check valve* is a valve that allows flow in only one direction. Check valves are normally closed and may be direct-acting or pilot-operated. A direct-acting check valve consists of a valve body, spring, and ball or poppet. The valve body contains an inlet (primary) and outlet (secondary) port. The spring holds the ball or poppet in one position. The ball or poppet blocks fluid flow when held against the seat or allows fluid flow when pushed off its seat as the inlet pressure rises high enough to overcome the spring pressure. The spring and fluid pressure forces the ball or poppet to seat, preventing fluid flow if fluid attempts to flow in the reverse direction. **See Figure 17-44.**

The flow through a check valve in both directions may be accomplished with the use of pilot operation. The external pilot supply may be hydraulic or pneumatic. Pilot-operated check valves operate normally as a check valve. Pilot pressure is needed at the pilot poppet when reverse fluid flow is required. The check valve poppet is unseated, allowing reverse flow when sufficient pilot pressure is produced at the pilot port. The pilot poppet and the main poppet control the flow of fluid.

Safety Tip

Prolonged high-velocity fluid pressure in ball check valves can cause ball-to-spring damage severe enough to result in significant fluid leaks. Alternative check valves are recommended for heavy-duty hydraulic applications.

Directional control valves are described by their number of ways and spool positions. The movement of the spool determines which way the fluid flows. A *way* is a route that fluid can take through a valve. For example, a check valve is referred to as a one-way valve because fluid flow is routed in only one direction. Ways may connect more than one port.

Atlas Technologies, Inc.

Hydraulic presses used in manufacturing stamping plants contain accumulators for the storage of the pressurized hydraulic fluid and to absorb circuit shock.

Check Valves

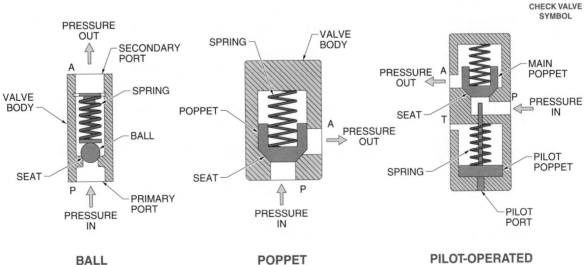

Figure 17-44. Check valves are normally closed and allow fluid flow in one direction.

A *two-way directional control valve* is a valve that has two main ports that allow or stop the flow of fluid. Two-way valves are used as shutoff, check, and quick-exhaust valves. A *three-way directional control valve* is a valve that has three main ports that allow or stop fluid flow or exhaust. **See Figure 17-45.** Three-way valves are used to control single-acting cylinders, fill-and-drain tanks, and nonreversible fluid power motors. A *four-way directional control valve* is a valve that has four main ports that change fluid flow from one port to another. Four-way valves are used to control the direction of double-acting cylinders and reversible fluid motors.

Two-way, three-way, and four-way directional control valves are constructed similarly with some minor changes. The basic parts of a directional control valve are the body and the internal spool. The body has a series of holes (ports) that are usually labeled A, B, P, and T. Ports labeled A and B are passage ports to and from an actuator. Ports labeled P are pump or pressure ports. Ports labeled T are tank ports. There are many different flow paths used to connect these ports. **See Figure 17-46.**

The spool must be in a specific position within the valve body to allow a specific flow direction. A *position* is the specific location of a spool within a valve which determines the direction of fluid flow through the valve. Most directional control valves are either two-way, three-way, or four-way valves and may have two or three positions. Symbols are used to show the various valves. The symbols indicate each valve position as an envelope or one square. Arrows are used to show fluid flow and indicate the number of ways in a valve. When the spool in the valve shifts, another envelope shows the change in fluid flow.

A four-way, two-position directional control valve is used to control the fluid flow in a typical extend and retract circuit. In this circuit, the spool directs flow from Port P to Port A when it is shifted left. **See Figure 17-47.** This allows fluid to flow to the cap end of the cylinder, forcing the piston rod to extend. At the same time, fluid is discharged from the rod end of the cylinder through Port B. The fluid passes through the spool and out Port T to the reservoir.

Fluid flows from Port P to Port B when the spool is shifted right. This allows fluid to flow to the rod end of the cylinder, forcing the piston to retract. At the same time, fluid is discharged from the cap end of the cylinder through Port A, through the spool, and out Port T to the reservoir. The four-way, two-position valve has a total of four passageways and two positions for the spool.

Spools in a directional control valve have high and low areas. Low areas allow fluid flow. High areas (lands) block fluid flow. Spool lands vary according to valve design and function. Some lands are wide and some narrow. Some spools have two lands, while others have four.

Two-Position, Three-Way Directional Control Valves

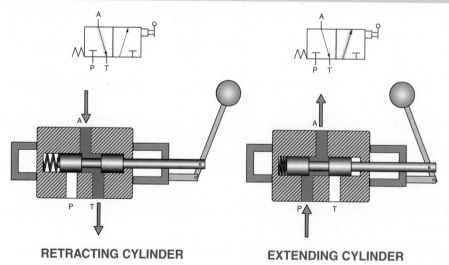

RETRACTING CYLINDER   EXTENDING CYLINDER

Figure 17-45. A two-position, three-way directional control valve has two positions, one to allow fluid flow from the cylinder port to the tank port and the other to allow fluid flow from the pump to the cylinder port.

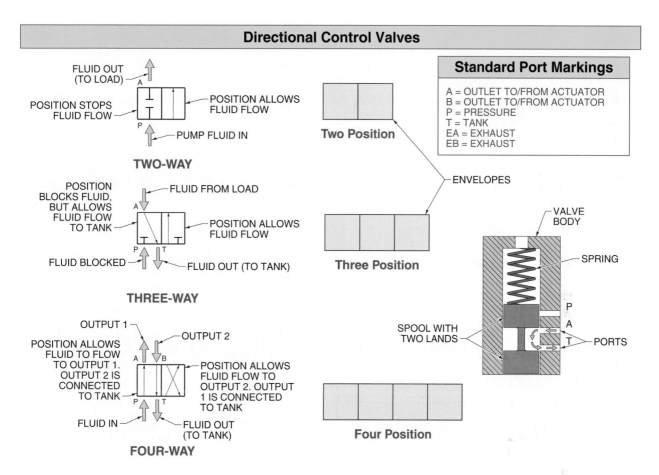

Figure 17-46. Directional control valve symbols describe fluid ways and spool position.

Spool fit is very precise. The clearance between a spool and valve body ranges from 8μ to 10μ. Although this clearance allows for oil leakage (used for sealing and lubrication), the spool land would prohibit even a red blood cell from passing between ports. For this reason, most spools are not interchangeable and must be handled with care.

Three-position directional control valves have a center position. The actuator motion is controlled by the left or right positions of the valve spool. The center position satisfies the circuits requirements. Many different valve spool center configurations are available. The most common is the tandem center configuration. The tandem center configuration blocks ports A and B and connects ports P and T. This allows the actuator to remain in its last placed position with the pump flow returning to the reservoir without going through the pressure-relief valve. In one tandem design, the center of the spool is drilled out, allowing a connection between ports P and T through the core of the spool.

The Gates Rubber Company

A backhoe uses extend and retract circuits to operate the boom, bucket, crowd, and stabilizing cylinders. The cylinders are double-acting to give full force in both directions.

Extend and Retract Circuits

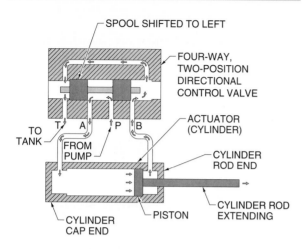

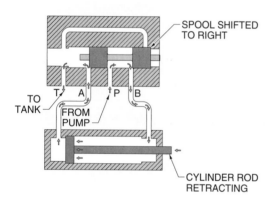

Figure 17-47. A four-way, two-position directional control valve is used to control the fluid in a typical extend and retract circuit.

Prince Manufacturing Corporation

Depending on the application, some directional control valves used in hydraulic systems can have more than one lever.

An example of a tandem-center directional control valve is that of an excavator. When the excavator bucket is lifted off the floor and lifting stops, the bucket remains in position until the valve is directed to raise or lower the bucket. Fluid flows through the spool's core from Port P to Port T while the bucket is immobile and the hand lever is in the center position. This allows fluid to return to the reservoir without loading the pump. **See Figure 17-48.**

The spool can be held in the center position by hydraulic pressure or springs. Spring centering is the most common. Three position valves are normally centered by equal spring pressure located at both ends.

Directional control valves must have a means to change the position of the valve spool. A *valve actuator* is a device that changes the position of a valve spool. Valve actuators may be manual, mechanical, pneumatic, hydraulic, or electrical. Manual valve actuators include levers, pushbuttons, or foot pedals. Mechanical valve actuators include cams depressing a plunger connected to the spool. Pneumatic or hydraulic valve actuators include air or fluid pilot operation. A common means of positioning spools is electrically with the use of a solenoid. Each method of valve actuation is shown by a symbol for the easy identification of the valve actuation method. **See Figure 17-49.**

System or pilot pressure is often used in combination with solenoids because many directional control valves require more pressure to operate the spool than a solenoid can produce. Typical pilot operation consists of a piggyback design where a smaller, solenoid-operated, directional control valve is positioned on top of the main valve. Flow from the small valve is directed as pilot pressure to either side of the main directional control valve spool when shifting is required. **See Figure 17-50.**

Flow Control. Flow control in a hydraulic circuit is primarily used to regulate speed. Flow control may be accomplished by using a variable-displacement pump or a flow control valve. A *variable-displacement pump* is a pump in which the displacement per cycle can be varied.

Tech Tip

Push-type solenoids are most effective when positioning the spools of smaller valves. When electricity is applied to a solenoid, its coil creates a magnetic field strong enough to attract the armature into it. The armature pushes on the spool while it is pulled into the magnetic field and allows the spool to shift properly.

Tandem-Center Positions

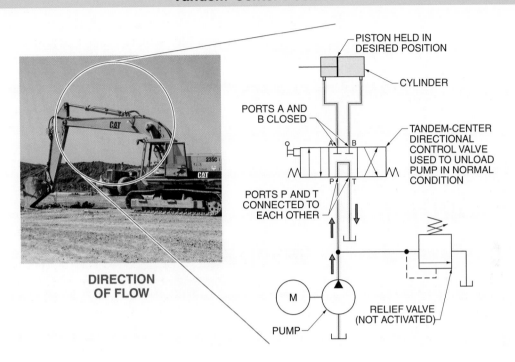

Figure 17-48. Tandem-center positions are used to hold a cylinder in the desired position while allowing fluid flow to be directed back to the tank without the need to activate a relief valve.

Valve Actuators

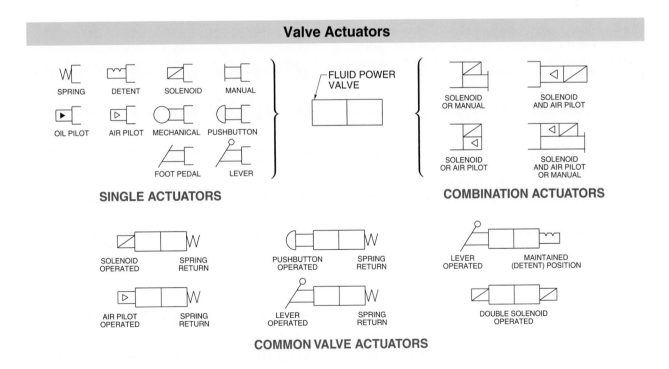

Figure 17-49. Valve spool shifting is accomplished through the use of valve actuators.

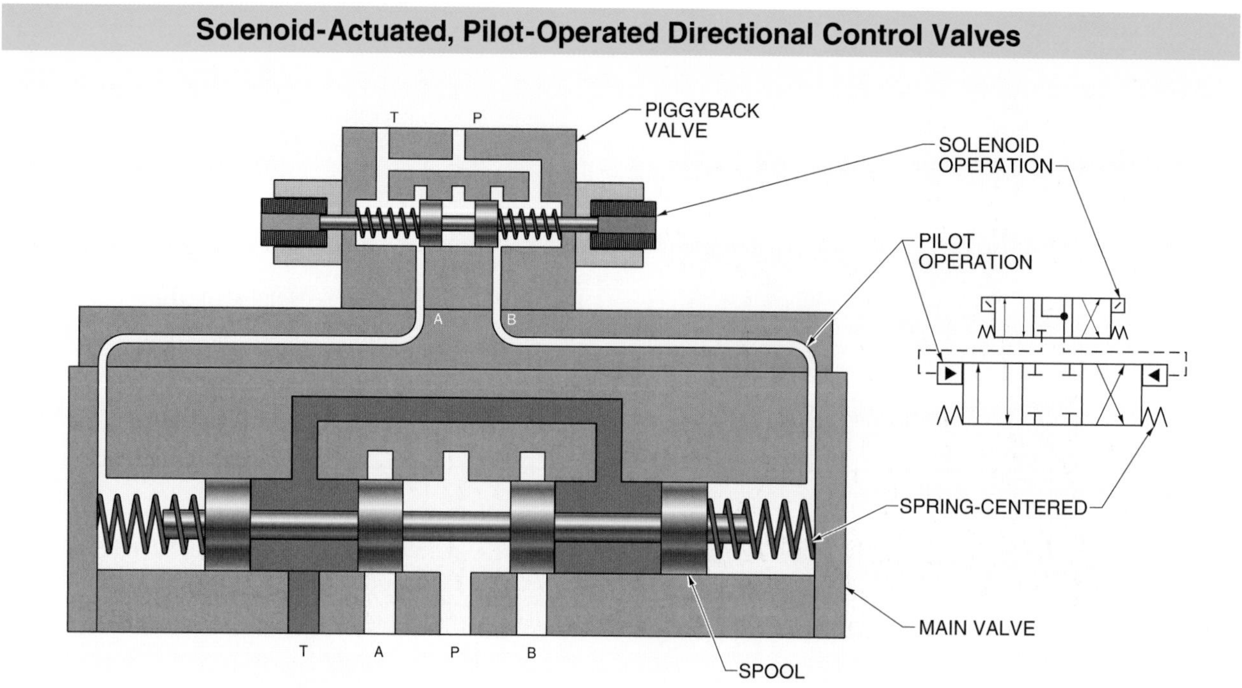

Solenoid-Actuated, Pilot-Operated Directional Control Valves

Figure 17-50. Pilot-operated directional control valves are often used in combination with solenoids in order to produce the pressure necessary to operate the spool.

A *flow control valve* is a valve whose primary function is to regulate the rate of fluid flow. By regulating the rate of fluid flow, the flow control valve becomes a resistor and increases hydraulic fluid pressure upstream from the valve. The increase in pressure opens the pressure-relief valve to allow excess fluid to return to the reservoir and reduced flow continues to a branch circuit or actuator. In some cases, the excess fluid is used in another circuit rather than being sent to the reservoir. Controlling the flow of fluid in a circuit is accomplished by using an orifice or needle valve. **See Figure 17-51.**

A *restrictive check valve* is a check valve with a specific size hole drilled through its center. Factory preset orifices are sized to control a flow rate at a specific inlet pressure. Fluid flow increases if the pressure increases at any factory preset orifice.

A *gate valve* is a two-position valve that has an internal gate that slides over the opening through which fluid flows. Gate valves are generally used for full flow or no flow operation and are not designed for restricting fluid flow. Fluid flows in a straight path through the valve, which offers very little pressure drop in the circuit when fully open. Vibration and wear occur when a gate valve is used in a partially open position. Any restricting of flow by a gate valve should be of very coarse metering.

A *globe valve* is an infinite-position valve that has a disk that is raised or lowered over a port through which fluid flows. Globe valves do not have a straight through path for fluid flow. Fluid flow makes two 90° turns when flowing through the valve. The opening between the seat and disk is controlled to meter or throttle the fluid flow from zero to full flow. *Metering* is regulating the amount or rate of fluid flow. *Throttling* is permitting the passing of a regulated flow. The flow is regulated in one direction only because of the two 90° turns in the globe valve's flow path. The flow direction through a globe valve is generally indicated by an arrow on the side of the valve's housing.

A *needle valve* is an infinite-position valve that has a narrow tapered stem (needle) positioned in line with a tapered hole or orifice. An *orifice* is a precisely sized hole through which fluid flows. The size of the orifice controls the flow rate by creating a pressure drop. The remaining pump flow, which is not passed through the orifice, is either dumped to the reservoir or used in another circuit. Needle valves offer precise flow control because of their cone-shaped needle and seat and the fine threaded adjusting stem. The fine threaded adjusting stem offers a very gradual change in orifice size.

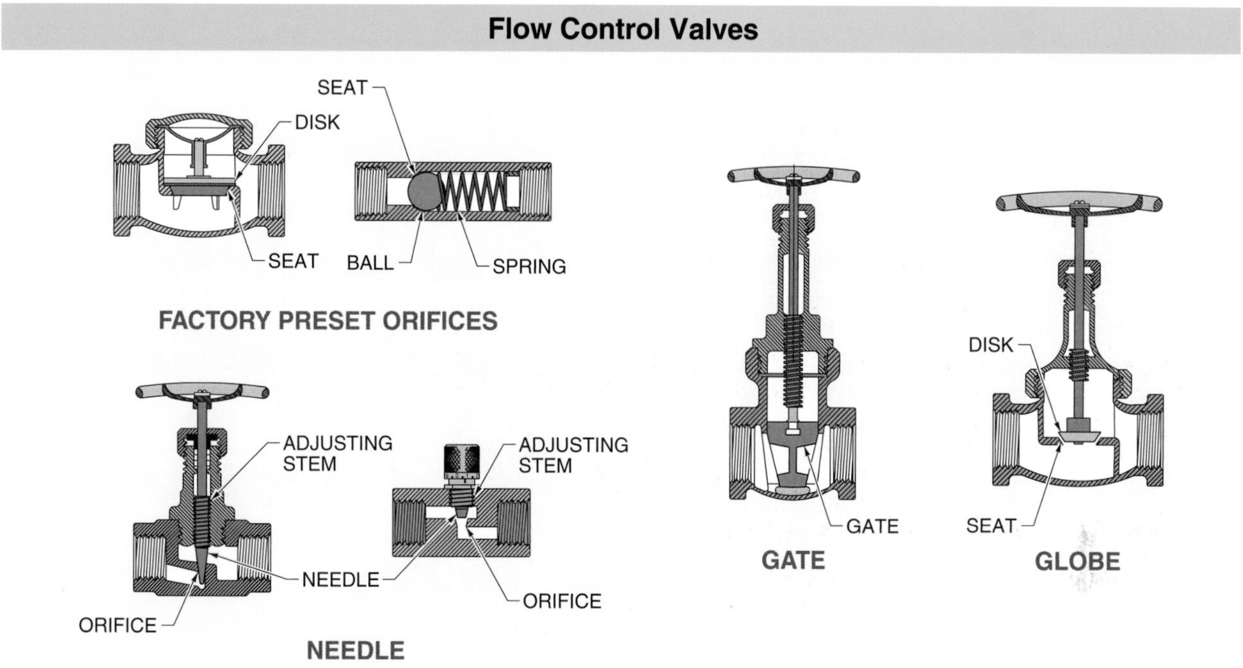

Flow Control Valves

Figure 17-51. A restrictive check valve or needle valve may be used to control the flow of fluid in a hydraulic circuit.

Flow control valves are available with built-in check valves for applications that require metered flow in only one direction. In one direction, the check valve directs fluid flow through the needle valve. Fluid passes freely through the check valve when the flow is reversed. An arrow on the side of the needle/check valve combination indicates the direction of controlled flow. The greater the pressure differential across a needle valve, the greater the flow rate through the valve. Thus, any change in pressure before or after a flow control valve affects the flow through the valve, resulting in a change in actuator speed. For example, an increase in the load on an actuator lowers the pressure differential and reduces the flow through the valve. This reduces the speed of the actuator. To increase the actuator speed, the flow control valve is partially opened. The speed of the actuator can become too great if the load on the actuator is reduced. This scenario can be corrected by using a pressure-compensated flow control valve.

A *pressure-compensated flow control valve* is a needle valve that makes allowances for pressure changes before or after the orifice through the use of a spring and spool. The needle valve's adjustment knob provides a controlled orifice, but any change in pressure is compensated for by the spool. Stable flow occurs when the spool has created a restriction equivalent to the balance of inlet and outlet pressure forces assisted by the spring and controlled orifice. **See Figure 17-52.**

Flow controls in a hydraulic circuit are normally used to control actuator speeds. The three basic methods of actuator flow control include meter-in (metering the flow at the actuator inlet), meter-out (metering the flow at the actuator discharge), and bleed-off (metering a portion of the inlet flow to the reservoir.

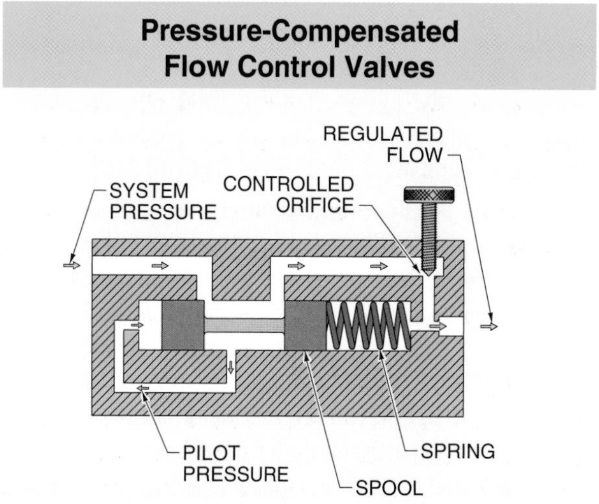

Pressure-Compensated Flow Control Valves

Figure 17-52. A pressure-compensated flow control valve is used to provide constant actuator speed with varying loads.

Metering the flow at the actuator inlet controls the amount of hydraulic fluid going into the actuator. This provides a very accurate and constant movement of the actuator when the load resistance is continually present, such as when a load is required to move at a controlled speed or a vertical load is being lifted. Metering the flow at the actuator's discharge prevents a jerky movement when the load is not constant. The use of either inlet or discharge flow controls requires a check valve to allow full return flow.

Metering a portion of the inlet flow is sometimes called bleed-off flow control and is not as accurate as the discharge flow control method. Bleed-off flow control is accomplished by placing a tee in the actuator's inlet plumbing. The flow control valve is installed at the tee between the supply line to the actuator and the reservoir. The advantage of this method is that less work is done by the pump and pressure-relief valve due to the excess hydraulic fluid being metered back to the reservoir. The disadvantage is that the metered or set orifice flow is to the reservoir, which offers the same problem as inlet metering when the load is not constant.

Actuators

A *hydraulic actuator* is a device that converts hydraulic energy into mechanical energy. Hydraulic actuators include cylinders and motors.

Cylinders. A *hydraulic cylinder* is a device that converts hydraulic energy into straight-line (linear) mechanical energy. A basic hydraulic cylinder consists of a cylinder body, a piston, a piston rod, and seals. The end through which the rod protrudes is the rod end, and the opposite end is the cap end. Fluid ports are located in the rod and cap ends. As the cylinder rod reciprocates, the rod is supported by a steady bearing which also holds the rod seal and wiper seal in place. **See Figure 17-53.**

Cylinders are classified as single- or double-acting. A *single-acting cylinder* is a cylinder in which fluid pressure moves the piston in only one direction. A small rod end port allows atmospheric air in and out of the rod end of the cylinder. The piston and piston rod extend as hydraulic fluid is pumped into the cap end port. The piston rod retracts when fluid in the cap end of the cylinder is released to the reservoir. The rod in a single-acting cylinder retracts either by gravity, spring, or some other mechanical force. An example of a single-acting cylinder is that of most lifting cylinders on a forklift. **See Figure 17-54.**

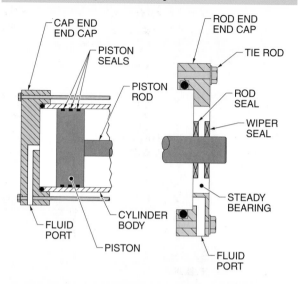

Hydraulic Cylinders

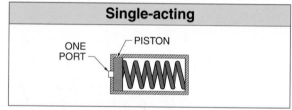

Single-acting

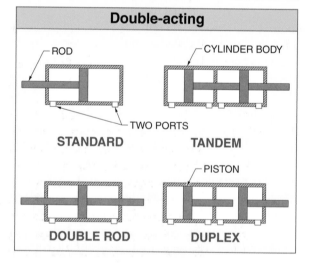

Double-acting

Figure 17-53. A hydraulic cylinder is an actuator that converts hydraulic energy into linear mechanical energy.

Safety Tip

Excessive wear on wiper seals can cause erratic movements in hydraulic cylinders. Regular inspection and replacement of wiper seals can prevent contaminants from entering the cylinder and negatively influencing performance.

Single-Acting, Spring-Return Cylinders

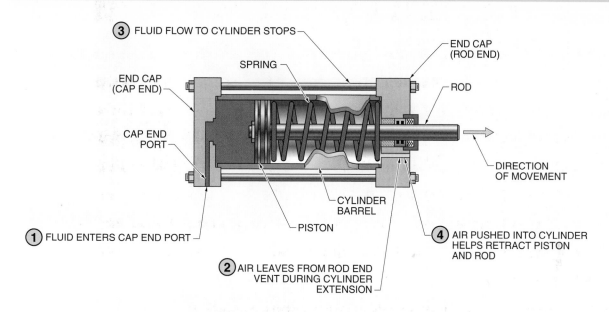

③ FLUID FLOW TO CYLINDER STOPS

SPRING

END CAP
(CAP END)

END CAP
(ROD END)

ROD

CAP END
PORT

DIRECTION
OF MOVEMENT

① FLUID ENTERS CAP END PORT

CYLINDER
BARREL

PISTON

**④ AIR PUSHED INTO CYLINDER
HELPS RETRACT PISTON
AND ROD**

**② AIR LEAVES FROM ROD END
VENT DURING CYLINDER
EXTENSION**

Figure 17-54. A single-acting, spring-return cylinder uses fluid flow for extension and a spring for retraction.

A *double-acting cylinder* is a cylinder that requires fluid flow for extending and retracting. A double-acting cylinder has ports and fluid at each end of the cylinder (both sides of the piston). The piston rod extends when fluid is pumped into the cap end port. The piston rod retracts with the fluid in the cap end of the cylinder and returns to the reservoir when fluid is pumped into the rod end port. Double-acting cylinders include tandem, duplex, and double rod cylinders. The cylinder used in an application is based on the requirements of the application.

Seals, an integral part of a cylinder, create positive contact across the piston and the rod, allowing for maximum pressure and preventing leakage. Seals must be well-lubricated, contaminate-free, and without any nicks or damage to provide maximum contact. They should be used on smooth, true, and unmarred surfaces.

Seals may be static or dynamic. A *static seal* is a seal used as a gasket to seal nonmoving parts. Static seals are used where there is contact between two parts but no motion, such as between stationary items taken apart and reassembled. A *dynamic seal* is a seal used between moving parts to prevent leakage or contamination. For example, dynamic seals are used on pistons and piston rods to allow the piston and rod to slide inside the cylinder. Seals may be positive or nonpositive. A *positive seal* is a seal that does not allow the slightest amount of fluid to pass. A *nonpositive seal* is a seal that allows a minute amount of fluid through to provide lubrication between surfaces. Seals include O-ring, Quad-ring®, lip, compression, and packing seals. **See Figure 17-55.**

A double-acting cylinder is able to extend or retract the piston rod by applying hydraulic fluid pressure to either side of the piston.

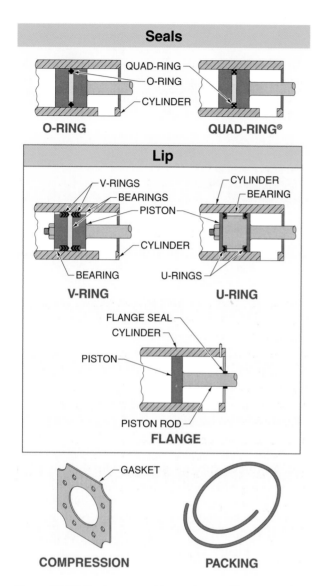

Figure 17-55. Seals are used to create a positive contact between moving parts to provide pressure and prevent leakage.

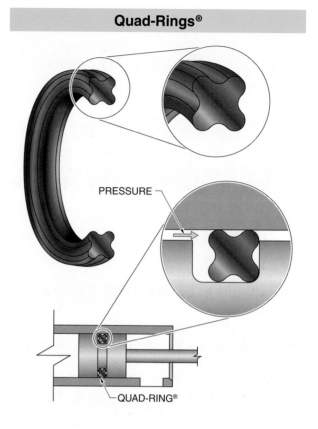

Figure 17-56. When pressure is applied to a Quad-ring®, it creates a dynamic seal by pressing against one side of the groove.

O-rings are the most commonly used seal in mechanical assemblies. In hydraulics, however, the pressures reached in a circuit cause an O-ring to extrude. O-rings are only used in hydraulics for special static applications. Quad-rings® are similar to O-rings. A *Quad-ring®* is a molded synthetic rubber seal having a basically square cross-sectional shape. Quad-rings® look like they are made of four O-rings. During use and under pressure, Quad-rings® offer a dynamic seal by being pressed (forced) against one side of their groove. This pressure forces the seal outward and against the sealing surface. Increased pressure on quad-rings results in greater sealing forces. **See Figure 17-56.**

Quad-rings® require a back-up ring for pressures over 1500 psi because of the disfiguration of the seal under pressure. The back-up ring prevents the resilient rubber of the Quad-ring® from being extruded out of the groove. Quad-rings® have twice the sealing power with greater resistance to rolling, twisting, or extruding. *Resilience* is the capability of a material to regain its original shape after being bent, stretched, or compressed.

A *lip seal* is a seal that is made of a resilient material that has a sealing edge formed into a lip. Lip seals are made of leather, rubber, or synthetic material. Lip seals use the hydraulic fluid's pressure on its lip to form a seal. As pressures increase, the seal tries to expand at the lip, creating a tighter seal. Lip seals are made in various shapes for various pressure duties. Lip seals are dynamic seals used for sealing rotating or reciprocating shafts, pistons, cylinder rod ends, and pump shafts. Typical lip seals are the V-ring, U-ring, and flange seal.

A *V-ring seal* is a lip seal shaped like the letter V. V-ring seals are dynamic seals used in severe operating condition applications. V-ring seal materials include impregnated and treated leather, rubber, and asbestos. The advantages of V-ring seals are that dissimilar seals of different materials may be used in combination to provide the best pressure, wear, and friction characteristics for an application.

A *U-ring seal* is a lip seal shaped like the letter U. U-ring seals are dynamic seals used in reciprocating and rotating applications. System pressure increases the sealing force on the lips of the seal. U-ring seal materials include impregnated and treated leather, rubber, and other elastic compounds.

Flange seals are generally used as cylinder rod seals instead of piston seals. Flange seal materials include impregnated and treated leather, rubber, and other elastic compounds.

Compression seals are static seals commonly referred to as gaskets. A *gasket* is a seal used between machined parts or around pipe joints to prevent the escape of fluids. A gasketed joint is sealed by the molding of the gasket material into the imperfections of the mating surfaces of the joint. Gasket material may be rubber, leather, synthetic, or metal. Metal is used by itself or in combination with other softer materials. For example, in some cases, a sheet of brass or copper may be soft enough to form a proper seal when sandwiched between two harder metals.

Packing is a bulk deformable material or one or more mating deformable elements reshaped by manually adjustable compression. Packing is used where some form of motion occurs between rigid members of an assembly. Packing must only be deformed enough to allow or throttle leakage between the moving and stationary parts. This leakage becomes a lubricant and coolant for the packing. On large applications, the leakage rate may be as high as 10 drops per minute. On small or light applications, one drop per minute is sufficient.

Packing must be pliable enough to provide a radial seal when axially compressed. Packing material is made of braided, woven, or twisted cotton or flax. Solid lubricants, such as graphite or mica, are added to the packing to protect the moving parts. Packing must be adjusted frequently to compensate for wear.

In addition to soft metals, seal materials also include plastics, elastoplastics, and elastomers. These types of seal materials have different pressure ratings, operating temperature ranges, and hydraulic fluid compatibilities.

Plastic seals can be made from polytetrafluoroethylene. *Polytetrafluoroethylene (PTFE)*, commonly referred to as Teflon®, is a low-friction plastic material used as sealing in low-pressure hydraulic circuits. It can operate in high temperatures and has good compatibility with hydraulic fluids.

Elastoplastic seals can be made from polyurethane. *Polyurethane* is a hard, chemical-resistant plastic used as a sealing material in hydraulic circuits. Polyurethane is harder and stronger than other types of plastics but has less elasticity and flexibility than elastomers. It is suitable for high-pressure applications and has good resistance to abrasion. However, it can be destroyed by exposure to water at temperatures above 140°F and therefore cannot be used in any hydraulic circuits with water-based fluids.

Although not very costly to purchase, hydraulic seals can be expensive when coupled with the loss of production and downtime. Analyzing the condition of a seal during a repair can sometimes indicate the cause of a major breakdown and future prevention. Clues may indicate whether the breakdown occurred as the result of chemical breakdown, heat damage, contamination, or improper workmanship.

Chemical breakdown is generally the result of the wrong kind or lack of certain additives in the hydraulic fluid. Failure caused by chemical breakdown may cause the seal to exhibit swelling, shrinkage, discoloration, softening, and/or loss of the seal lip. If chemical breakdown is suspected, hydraulic fluid can be analyzed for content, followed by a complete system flush before the replacement of the hydraulic fluid.

In addition to replacing worn or damaged seal materials, hydraulic rods must be regularly maintained for proper operation.

Heat damage is a physical degradation of a seal material and can be caused by use of the wrong seal size or material, high surrounding temperatures, high dynamic friction, or excessive circuit loads. Excessive circuit loads can result in high compression. This can damage lip-sealing properties. A heat-damaged seal exhibits a hard and brittle material with cracks and a broken or chipped body or lip parts. In some applications, worn seals can increase the operating temperature of the circuit enough to create heat damage to other hydraulic seals within the circuit.

Correcting a heat problem requires locating the point in the circuit where excessive heat is being generated. If a proactive maintenance program is being used, circuit temperature records can indicate an area with temperature increases. If a damaged seal was recently replaced, the possibility of an incorrect seal must be considered. After repair and during normal operation, heat levels must be monitored at different locations to observe for unusual heat rise at some other location, possibly identifying the cause of present seal damage.

A circuit can be contaminated by elements of damage or wear from within the circuit or elements of dirt, grime, and moisture that have worked their way into the circuit. Elements of damage or wear within a circuit can be caused by improper maintenance of strainers and filters. Contaminants can enter a hydraulic circuit through a breather in a harsh environment or through the rod-end seal of a hydraulic cylinder.

Rod-end seals are primarily meant to contain a pressurized hydraulic fluid and to wipe the rod clean at each stroke. However, if the retracting-rod piston pressure is lower in the cylinder than on the extending side of the piston, a negative pressure could draw in air, moisture, and dirt through the rod-end seal. In this case, the fluid demand on the rod side of the cylinder is exceeding the volume of fluid supplied by the pump. Negative pressure at the rod end can be the cause of excessive loads on the end of the piston as it returns. These are usually overhead, gravity, or pressure-return loads.

Correcting a contamination problem caused by elements entering through the breather can sometimes be done by replacing a faulty breather, conditioning the environment around the unit, pressurizing the reservoir, or changing the breather type.

Contamination drawn in through the rod end can be the result of a restriction between the cylinder and the directional control valve and not allowing full flow of fluid on the return stroke.

Lack of a clean work area can prematurely ruin seals as they are being installed. Other work-habit causes of contamination or damage to seals include improper handling of a seal before and during installation, creating cuts, scratches, or nicks; installation of the wrong seal; installing a seal backwards; and not using enough lubrication to ease installation and prevent scratching of the seal surface.

Hydraulic Motors. A *hydraulic motor* is a device that converts hydraulic energy into rotary mechanical energy. Hydraulic motors are often referred to as rotary actuators. Hydraulic motor construction is similar to the construction of hydraulic pumps. Hydraulic motors include gear, vane, and piston models. The transformation of hydraulic energy to rotary mechanical energy is a reversal of the pumping function. **See Figure 17-57.**

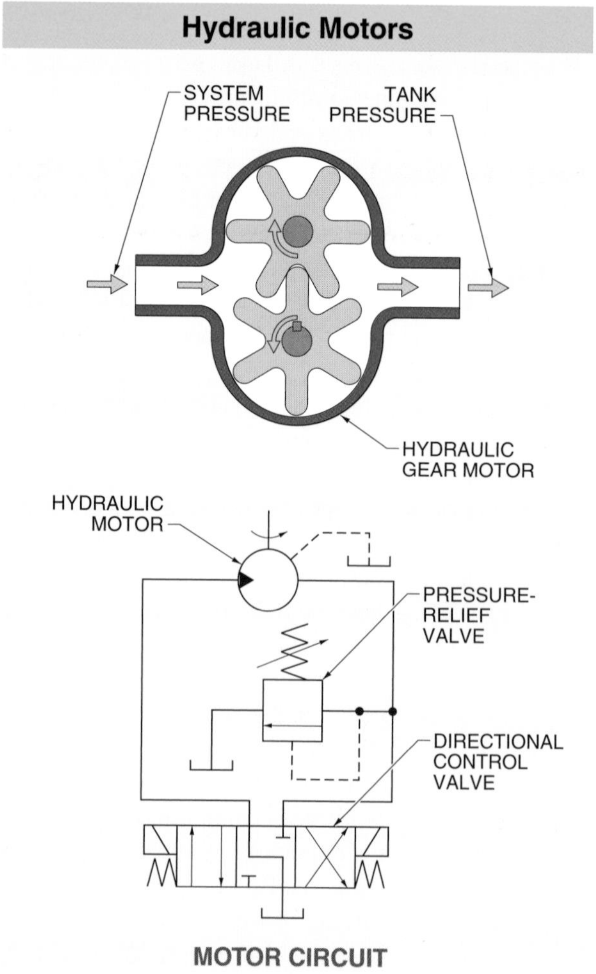

Figure 17-57. A hydraulic motor is a device that converts hydraulic energy into rotary mechanical energy.

Rotary mechanical energy is produced in a gear motor when hydraulic fluid at high pressure is forced against the teeth of the upper and lower gears, causing them to rotate. Similarly, hydraulic fluid at high pressure forced against the vanes of a hydraulic vane motor causes the shaft to rotate. The difference between a vane motor and a gear motor is that the vanes in a vane motor need to be extended by spring action. The vanes are generally extended by hydraulic fluid pressure after motor torque is developed. The type of hydraulic motor selected depends on its application and amount of torque needed.

Accumulators

An *accumulator* is a container in which hydraulic fluid is stored under pressure. Accumulators store hydraulic fluid under pressure (potential energy) until it is needed. Accumulators also maintain circuit pressure, develop circuit flow, and absorb circuit shock. Accumulators may also maintain circuit pressure in an emergency if a pump fails. This allows a circuit to complete cycling before shutdown occurs. Accumulators may be spring-loaded, weight-loaded, or hydro-pneumatic. **See Figure 17-58.**

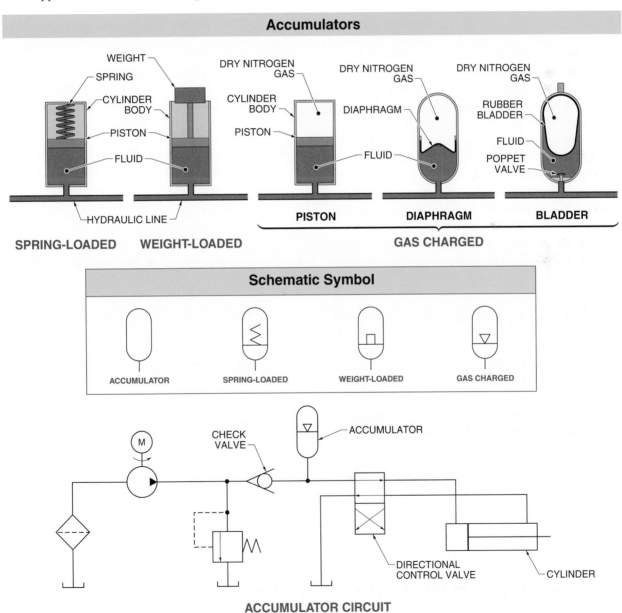

Figure 17-58. Accumulators store hydraulic fluids under pressure for use at a later time.

WARNING: All accumulators are extremely dangerous when their energy is uncontrolled. For this reason, all accumulator energy must be released or blocked before performing repairs.

A *spring-loaded accumulator* is an accumulator that applies force to a fluid by means of a spring. Spring-loaded accumulators consist of a cylinder body, piston, and spring. The piston rides between the spring and the hydraulic fluid inside the cylinder body. Circuit pressure and the compression rate of the spring determine the amount of stored energy. In many cases, a mechanical stop at the spring area is used to prevent excessive pressure from overcompressing and damaging the spring. Most spring-loaded accumulators are adjustable, allowing for varied amounts of stored fluid pressure.

A *weight-loaded accumulator* is an accumulator that applies force to a hydraulic fluid by means of heavy weights. The weights are generally iron or concrete and offer a constant pressure throughout the piston stroke. A weight-loaded accumulator may cause excessive pressure surges in a circuit when the accumulator is quickly discharged and suddenly stopped because of the constant pressure of the heavy weights.

A *gas charged accumulator* is an accumulator that uses compressed gas over hydraulic fluid to store energy. Dry nitrogen is used as the gas in hydro-pneumatic accumulators because air-oil vapors are explosive. Gas charged accumulators are divided into categories according to the method used to separate the gas from the hydraulic fluid. The methods include piston, diaphragm, and bladder.

A *piston gas charged accumulator* is an accumulator with a floating piston acting as a barrier between the gas and hydraulic fluid. Gas occupies the volume of space above the piston and is pressurized as circuit hydraulic fluid enters and occupies the space below the piston. The gas pressure equals the circuit pressure when the accumulator is pressurized.

A *diaphragm gas charged accumulator* is an accumulator with a flexible diaphragm separating the gas and hydraulic fluid. Diaphragm gas charged accumulators are generally small and lightweight with capacities up to 1 gal. They are constructed of two steel hemispheres bolted together with a flexible, dish-shaped, rubber diaphragm clamped between them. The top half of the sphere is pressurized (precharged) with gas. Hydraulic fluid supplied by circuit pressure is applied to the bottom hemisphere, compressing the gas. The gas acts as a spring against the diaphragm as fluid pressures equalize the pressure in each hemisphere.

A *bladder gas charged accumulator* is an accumulator consisting of a seamless steel shell, a rubber bladder (bag) with a gas valve, and a poppet valve. The steel shell is cylindrical in shape and rounded at both ends. A large opening at the bottom of the shell is used to insert the bladder. A small opening at the top of the shell is used for the bladder's gas valve. The one-piece, pear-shaped bladder is molded of synthetic rubber. It includes a molded gas valve, which is fastened to the inside upper end of the steel shell by a locknut. The bottom end of the shell is equipped with a poppet valve at the discharge port. This valve closes off the port when the accumulator is fully discharged, preventing the bladder from being squeezed into the discharge port opening.

Accumulators are precharged while empty of hydraulic fluid. *Precharge pressure* is the pressure of the compressed gas in an accumulator prior to the admission of hydraulic fluid. The higher the precharge pressure, the less fluid the accumulator can hold. Precharge pressures vary with each application. Typically, an application consists of the circuit pressure range and the volume of fluid required in that range. Precharge pressures should never be less than ⅓ of the maximum circuit pressure.

HYDRAULIC CIRCUIT MAINTENANCE

Special precautions must be followed when servicing powered equipment. Every energy source must be identified, understood, and disabled prior to working on a machine. All energy is categorized as kinetic or potential. *Kinetic energy* is the energy of motion. For example, a saw blade or grinding wheel in operation has kinetic energy. *Potential energy* is stored energy a body has due to its position, chemical state, or condition. For example, hydraulic accumulators, raised loads, or equipment counterweights have potential energy.

Controlling an energy source that is potentially dangerous is accomplished by using an energy-isolating device. An *energy-isolating device* is a device that prevents the transmission or release of energy. A potentially dangerous energy source may be electrical, mechanical, hydraulic, pneumatic, chemical, thermal, or any other energy source that could cause injury to personnel.

WARNING: The heat and pressures of hydraulic circuits can cause severe burns and injury.

Energy-isolating devices include manually operated electrical circuit breakers, disconnect switches, slip blinds, line valves, blocks, and similar devices that prevent the transmission or release of energy.

Pushbuttons, selector switches, and other circuit control devices are not energy-isolating devices. The following four basic steps are used to control hazardous energy in a hydraulic circuit:

1. Prepare to completely disable the circuit through identification of all energy sources.
2. Isolate the equipment by turning OFF all switches (including main disconnects), closing all necessary valves, and disconnecting, capping, or blocking auxiliary energy sources such as accumulators. **See Figure 17-59.**
3. Apply lockout/tagout/blockout devices to energy-isolating devices.
4. Verify that all necessary energy sources are isolated and locked. This is accomplished by attempting circuit startup. All area personnel must be warned of the startup attempt and cleared to safety. Return switches and controls to their OFF position after verification has been satisfied.

Dangerous Pressures

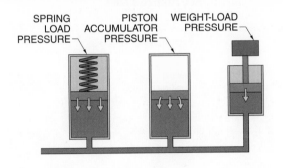

Figure 17-59. Dangerous pressures must be disconnected, capped, or blocked from accumulators before servicing a hydraulic circuit.

In many hydraulic circuit repair operations, the machinery requires testing and must be energized before other work is performed. This can be safely accomplished by using the following five basic steps:

1. Verify that all personnel are at a safe distance.
2. Remove tools and materials from the machinery.
3. Remove necessary energy-isolation devices and reenergize machine functions following completion of safety procedures.
4. Perform test or tryout.
5. Return all energy sources to isolation, including the purging of necessary systems. Replace locks and tags as required.

Cleanliness is the key to successful troubleshooting and repair of hydraulic circuits. Practices that must be followed to maintain a clean circuit include the following:

- Adhere to a preventive maintenance program for fluid, filters, and strainers.
- Use only clean equipment to prevent contamination when replacing or changing hydraulic fluid.
- Clean and cover fluid containers and store in a clean, dry area when a maintenance project is complete.
- Replace packing or seals before obvious replacement is necessary.
- Maintain clean hydraulic equipment. Keep dust and dirt buildup down by removing fluids from equipment surfaces.

Tech Tip

Absorbing hydraulic shock is one of the most important operations of a properly functioning accumulator. Accumulators help soften abrupt pressure surges dealt to hydraulic systems.

Fluid Maintenance

The equipment manufacturer's recommendations must always be followed when servicing hydraulic equipment and circuits. Contaminants such as dirt, sand, scale, and other foreign materials are often found in new circuits when installed. These contaminants must be flushed from the circuit after an initial break-in period. Also, the manufacturer's recommendations must always be checked as to whether cylinders should be extended during a fluid change.

When flushing contaminants from a circuit, the fluid is removed according to the equipment manufacturer's recommendations. Any hoses that need draining are disconnected. As much fluid as possible is removed from the circuit. Any leftover contaminants could cause early equipment failure. Levers are shifted to release trapped fluids. All fitting surfaces are cleaned with a dry, lint-free cloth before reassembling. All exterior surfaces are cleaned with a cleaning solvent after plugging any breather or dipstick holes to prevent the introduction of contaminants to the interior of the reservoir. Any filters or breathers are replaced or cleaned before refilling the reservoir. The equipment is run at idle for a few minutes to warm the circuit upon completion of hydraulic fluid change. The actuators are operated until foaming stops, and then hydraulic fluid is added to the full level. Finally, the hydraulic circuit is checked for leaks.

HYDRAULIC SYSTEM TROUBLESHOOTING

Today's hydraulic system components are becoming increasingly more advanced and complex. However, many basic component malfunctions may not require increased repair costs or production downtime. A knowledge of proper hydraulic system operation, applying a systematic approach, and using the process of elimination can help formulate an accurate diagnosis when troubleshooting a hydraulic system. Basic troubleshooting procedures can help solve many problems before resorting to sophisticated hydraulic component testers or a specialist.

All troubleshooting procedures begin with a knowledge of proper hydraulic circuit operation. In hydraulic circuits, malfunctions are related to problems with hydraulic fluid pressure, flow, or direction. All interlinked components ensure that one or all of these three characteristics are operating within normal levels based on the circuit design. Troubleshooting procedures are implemented to isolate or identify a malfunction involving one of these three hydraulic fluid characteristics.

In most cases, a malfunction of one component can affect the operation of another. This is why it is critical to follow a standard systematic approach to troubleshooting. When a problem occurs in a hydraulic system, the first questions to ask are the following:

- When did the problem first occur?
- Did the problem get worse gradually or was it sudden?
- When was the last time the system was serviced?
- What was the last maintenance procedure performed?
- Does the date of the last procedure coincide with the starting date of the problem?

A gradual decline in hydraulic system performance typically indicates a loss or degradation of hydraulic fluid and its negative effect on system components. An abrupt malfunction typically indicates a control device failure, such as the failure of a solenoid or switch, or damage to internal or external component parts. And finally, if the timing of a failure and maintenance coincide, the malfunction may be due to the result of an error, such as the replacement of the wrong viscosity fluid, an inaccurate breather replacement, or the loose installation of set screws.

Hydraulic Circuit Troubleshooting Safety

Many safety concerns arise when troubleshooting hydraulic systems. Industrial technicians are required to identify, understand, and disable every energy source prior to working on a hydraulic circuit. Proper safety procedures must be followed, personal protective equipment must be worn, and safety hazards must be identified in order to prevent injury or death. Safety hazards when troubleshooting hydraulic systems include the following:

- High-pressure fluids
- Pressurized pipes and hoses, especially when disconnecting them
- Supported loads or accumulator conditions that are ignored or overlooked
- Electrical devices that are not locked out or tagged out
- High-temperature fluids

Troubleshooting Using Human Senses

Human senses can be used to troubleshoot hydraulic system failures and malfunctions. Of the five human senses, the three senses of sight, touch, and hearing can be used to obtain information when performing troubleshooting procedures. The sense of taste is not applicable in hydraulic system troubleshooting, and the sense of smell can only be used to detect burnt hydraulic fluid or components. **See Figure 17-60.**

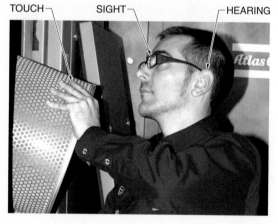

Human Senses

TOUCH — SIGHT — HEARING

Atlas Copco

Figure 17-60. The senses of sight, touch, and hearing can be used to identify hydraulic circuit malfunctions.

Sight. Many malfunctions can be observed. Recognizing that a problem exists generally begins with observing erratic, slow, or no movement in a component, such as the actuator; unusual gauge readings; worn or broken external components; leaking fluids; or smoking or burnt solenoids.

External observations can indicate an internal component malfunction. For example, a slower-than-normal movement of a hydraulic cylinder can indicate a problem with fluid flow, as fluid flow determines hydraulic cylinder response and speed.

Touch. Touch can be helpful when identifying damaged or overheated components and when detecting abnormal vibrations. However, using touch to identify hydraulic system malfunctions can be dangerous. Care must be taken when temperatures are high enough to cause burns or when the pressure of a leak is high enough to cause serious personal injury. If extreme heat is presumed, then temperature measurement devices must be used.

Hearing. An abnormal sound in a hydraulic system is usually the first indication of a component malfunction. By focusing on an abnormal sound, the technician can find the general location of the malfunctioning component while the system continues to operate. The technician can then schedule a repair check and avoid an unscheduled system shutdown.

Abnormal sounds can be caused by worn external or internal components, clogged inlet strainers, pump cavitation, air in the hydraulic fluid, improper valve settings, contaminated filters, or worn or misaligned couplings. In hydraulic systems, abnormal sounds are often caused by cavitation. Cavitation occurs when air infuses with the hydraulic fluid. The air compresses and decompresses, creating a sputter. Symptoms of cavitation may include erratic hydraulic cylinder movement or foaming of the hydraulic fluid.

Hydraulic Circuit Troubleshooting Procedures

When troubleshooting a hydraulic system, the goal is to identify the malfunctioning component in the shortest time possible. When troubleshooting, it is helpful to understand the operation of all components, be able to read hydraulic diagrams, and recognize trouble signals. Troubleshooting procedures can involve checking electrically controlled hydraulic devices and checking the temperature of hydraulic components.

Troubleshooting Fluid Power Electrical Control Components. When a malfunction occurs in an electrically controlled hydraulic device, it must be determined if the malfunction involves an electrical component or connection, or the hydraulic device itself. Electrically controlled hydraulic devices, such as directional control valves, have solenoids that actuate to control the direction of the hydraulic fluid. A light-emitting diode (LED) is used as part of a solenoid. If the LED is lit, then the solenoid is operating, and the fault would be in the directional control valve.

If the LED is not lit, the solenoid may not be receiving power or the LED may be damaged. A digital multimeter (DMM) can be used to check for voltage in the solenoid circuit. **See Figure 17-61.** Voltage tests are performed by applying the following procedure:

1. Verify that the electrical supply is providing power to the solenoid circuit.
2. Refer to the ladder diagram of the circuit to determine which switch must be closed in order to activate the solenoid.
3. Perform a voltage test across solenoid contacts to verify that the solenoid is receiving voltage.
4. Perform a voltage test across the switch contacts to verify that the switch is closed and the solenoid is receiving power.

Troubleshooting Hydraulic Circuits Using Infrared Thermometers. Temperature is an important indicator of the condition of a hydraulic system device when performing both quality control checks and hydraulic system troubleshooting. With the use of an infrared thermometer, a fault can be promptly located.

Checking the temperature of individual components using an infrared thermometer can indicate a component with a possible malfunction. For example, if temperature measurements of an actuator are significantly high, this may indicate that the cylinder hydraulic fluid is under pressure and could be bypassing the actuator piston seal. **See Figure 17-62.**

Heat is generated when hydraulic fluid moves from an area of high pressure to an area of low pressure without performing useful work. If a component in a hydraulic system is several degrees hotter than the rest of the system, internal leakage or blockage could be present. For example, warmer-than-normal return fluid may indicate a malfunctioning directional control valve.

Hydraulic fluid temperatures above 180°F will damage piston seals and accelerate degradation of the hydraulic fluid. Also, fluid temperature that is too high reduces the viscosity of the fluid below the optimum value for the hydraulic system. When viscosity decreases, increased internal leakage occurs, which increases the components' heat load and gradually damages hydraulic system components.

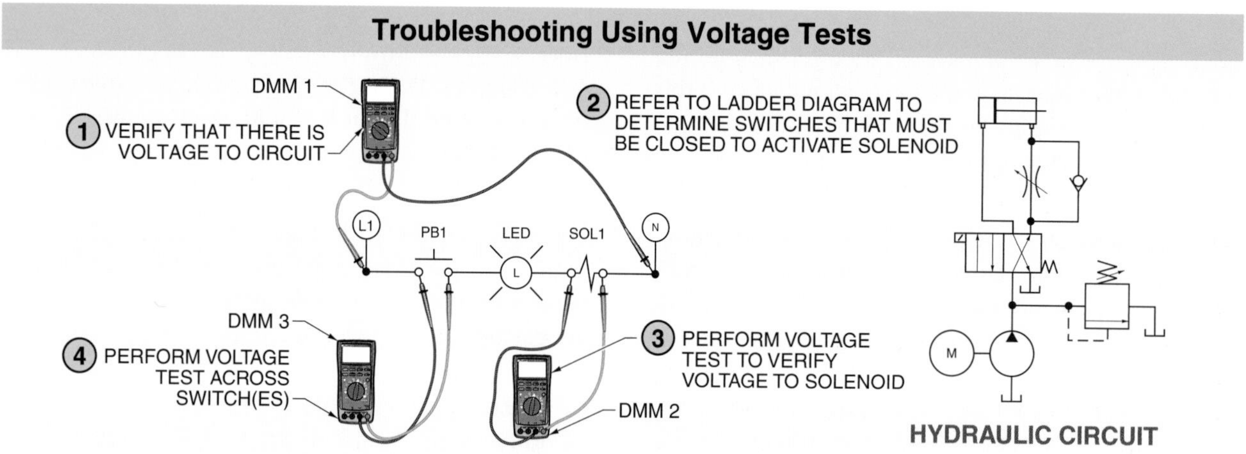

Troubleshooting Using Voltage Tests

(1) VERIFY THAT THERE IS VOLTAGE TO CIRCUIT

DMM 1

(2) REFER TO LADDER DIAGRAM TO DETERMINE SWITCHES THAT MUST BE CLOSED TO ACTIVATE SOLENOID

L1   PB1   LED   SOL1   N

DMM 3

(4) PERFORM VOLTAGE TEST ACROSS SWITCH(ES)

(3) PERFORM VOLTAGE TEST TO VERIFY VOLTAGE TO SOLENOID

DMM 2

HYDRAULIC CIRCUIT

Figure 17-61. Troubleshooting an electrically controlled hydraulic device can be performed by taking voltage readings with a DMM.

Temperature Measurements Using Infrared Thermometers

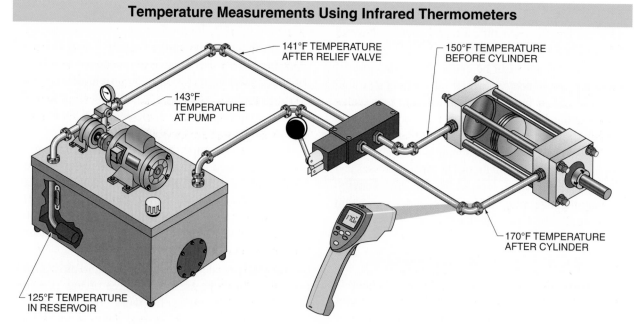

141°F TEMPERATURE AFTER RELIEF VALVE

150°F TEMPERATURE BEFORE CYLINDER

143°F TEMPERATURE AT PUMP

170°F TEMPERATURE AFTER CYLINDER

125°F TEMPERATURE IN RESERVOIR

Figure 17-62. Checking the temperature of individual components using an infrared thermometer can detect a component with a possible malfunction.

HYDRAULIC SYSTEM COMPONENT TESTING

A simple hydraulic system consists of a pump, a relief valve, a directional control valve, and an actuator (either a hydraulic cylinder or a fluid power motor). When a hydraulic system is contaminated, failures, malfunctions, and component degradation occur. All components must be kept clean through regular oil and filter maintenance.

When an actuator operates slowly, erratically, in reverse, or not at all, it typically indicates a failure or malfunction in the circuit. Unless the problem is obvious, such as burnt, broken, or leaking components, testing begins at the actuator. Components should not be tested under pressure. Components tested include hydraulic cylinders, directional control valves, suction strainers, pumps, and relief valves.

Hydraulic Cylinder Testing

Hydraulic cylinder operation can be affected by mechanical binds, improper fluid flow, piston seal leakage, or inaccurate timing control. A systematic approach should be followed when testing a hydraulic cylinder.

Hydraulic cylinders are first checked for external damage, such as bent shafts, misaligned attachments, or damaged walls. External damage can cause mechanical binds in hydraulic cylinders.

Hydraulic cylinders are then checked for internal leakage across the piston seal. To check for internal leakage, the return pipe or hose at the cylinder and at the directional control valve is first disconnected. Then the open directional control valve port is plugged to prevent any back pressure from the tank return. The system is started, and the directional control valve is used to run the piston toward the disconnected end. The valve is left stalled in this position while under pressure. Significant hydraulic fluid leakage indicates a broken or worn seal. (A small amount of leakage across the seal can be expected with pistons having metal ring seals.) **See Figure 17-63.**

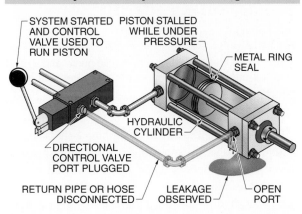

Figure 17-63. A systematic approach should be used when testing a hydraulic cylinder for mechanical binds, improper fluid flow, seal leakage, or inaccurate timing control.

After testing for leakage across the piston seal, the return pipe or hose is reattached and the test repeated using the directional control valve to return the piston to the other end of the hydraulic cylinder. If there is no sputtering in the cylinder and no significant amount of fluid leakage, the hydraulic cylinder is acceptable for use. Significant hydraulic fluid leakage indicates a broken or worn seal, and sputtering can indicate a damaged hydraulic cylinder wall.

To determine the condition of a hydraulic cylinder malfunctioning in one direction but not the other, the supply pipe or hose is switched with the return pipe or hose at the directional control valve. If the fault has switched sides, the problem involves the directional control valve. However, if the fault has not changed, the problem involves the hydraulic cylinder.

Directional Control Valve Testing

Directional control valves regulate fluid flow. A malfunctioning directional control valve can have adverse effects to the hydraulic cylinder it controls.

Testing a directional control valve for proper operation is accomplished by first disconnecting the hydraulic cylinder pipe or hose ports from the valve. Pressure gauges are then installed in each open port on the valve. The return port is then uncoupled so any leakage can be observed. Then the hydraulic circuit is started and gauge pressures monitored while shifting the control valve to a working position. The test is repeated by shifting the valve to other working positions. If the correct pressures are indicated by both gauges, then the directional control valve is operating properly.

Testing a directional control valve can help detect leakage, binding spools, or a broken return spring due to improper spool position, incorrect pressure, or hydraulic fluid leakage at the return port. **See Figure 17-64.**

Suction Strainer Testing

Approximately 75% of hydraulic failures and malfunctions occur due to inadequate fluid maintenance, which leads to low oil levels and contaminated oil. Contaminated oil directly affects the performance of a suction strainer, which can cause failure to an entire hydraulic circuit. Contaminated oil passed through suction strainers can initially cause increased pump sputter, reduction of hydraulic fluid flow, or reduced hydraulic fluid pressure.

Suction strainer testing can be accomplished by first removing the strainer from the reservoir. Once removed, the strainer is inspected for contaminates or degradation, such as clogging debris and damage or openings in the wire mesh. If damage exists, the strainer must be replaced. If the strainer is not damaged, then the wire mesh is washed in a solvent compatible with the circuit's hydraulic fluid. For example, kerosene can be used for strainers operating with petroleum-based hydraulic fluid. **CAUTION:** Do not use gasoline or other flammable or explosive solvents. Finally, after the suction strainer is cleaned of contaminants, it is then reinserted into the reservoir.

Directional Control Valve Testing

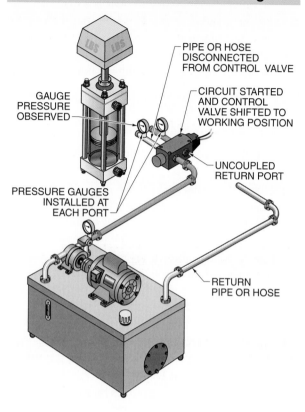

Figure 17-64. Testing the working pressure of a directional control valve can help detect leakage, a binding spool, or a broken return spring.

Suction strainers must be examined and cleaned at every oil change. The oil level is checked often, with hydraulic cylinders extended. If the suction strainers are blocked, then the hydraulic fluid causes insufficient pump suction, which increases the passing of contaminants into the circuit. Contaminants can obstruct ports and orifices and cause blockages around moving components. For example, contaminants can keep an AC solenoid armature from completely closing, causing the solenoid coil to burn out.

Pump and Relief Valve Testing

The purpose of both a pump and a relief valve is to generate fluid flow and develop full hydraulic circuit pressure. If starting the pump and tightening the adjustment screw on the relief valve develops full circuit pressure, then the pump and relief valve are operating properly. However, if full pressure cannot be developed, a test must be conducted to indicate which component is malfunctioning.

A pump and a relief valve can be tested by first isolating them from the rest of the hydraulic circuit. Isolation includes removing, capping, and/or plugging upstream ports and components from the valve. When isolated from the rest of the circuit, hydraulic fluid circulates from the suction strainer, then to the pump, then through the relief valve, then to a pressure gauge, and then back to the suction strainer in the reservoir. **See Figure 17-65.**

Pump and Relief Valve Testing

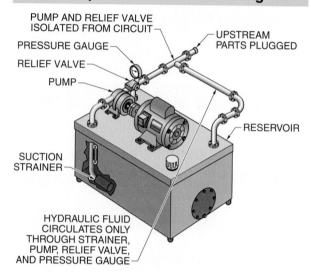

Figure 17-65. Before testing a hydraulic pump and a relief valve, all other circuit components must be isolated or plugged.

Pump malfunctions include slipping belts, sheared shaft pins or keys, broken couplings, or pump slippage. Pump slippage occurs when hydraulic fluid passes through the internal parts of a pump. Pump slippage causes the hydraulic fluid temperature to rise much higher than the reservoir fluid temperature. In normal operating conditions, the pump case runs about 20°F higher than the reservoir temperature. Pump slippage is detected through the use of a flow meter or by observing fluid flow. Fluid flow can be observed by attaching a short length of hose to the discharge port of the pressure gauge and positioning the hose above the reservoir filler opening. **See Figure 17-66.**

Safety Tip

When replacing a directional control valve, it is important to know the flow path for each position so that the replaced control valve will operate in the same manner.

Observing Fluid Flow

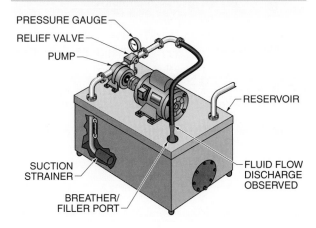

PRESSURE GAUGE

RELIEF VALVE

PUMP

RESERVOIR

SUCTION STRAINER

FLUID FLOW DISCHARGE OBSERVED

BREATHER/ FILLER PORT

Figure 17-66. Observing fluid flow and pressure at the relief valve can indicate the condition of both the hydraulic pump and the valve.

When the hydraulic circuit is started, the relief valve screw is tightened and then loosened to observe the discharge of fluid into the reservoir. If there is a full stream of hydraulic fluid when the relief screw is loosened, then the relief valve is operating properly.

Pump slippage would not allow pressure to build as the relief valve screw is tightened. The condition of the relief valve can be determined by observing the discharge fluid into the reservoir. If the volume of hydraulic fluid does not substantially decrease as the relief valve adjustment screw is tightened and gauge pressure does not increase, then the relief valve is malfunctioning and needs to be cleaned or replaced.

Tech Tip

The viscosity index is used to indicate the effects that temperature can have on the viscosity of a hydraulic fluid. The greater the viscosity index value of a hydraulic fluid, the less of an effect temperature has on the viscosity of the fluid.

Review

1. Define hydraulic diagram and explain what it is designed to do.

2. How do graphic symbols promote a universal understanding of fluid power systems?

3. Identify two environmentally sensitive applications that benefit from biodegradable hydraulic fluid use.

4. Define oxidation and explain how its process reduces the service life of hydraulic fluid.

5. Define pressure filter.

6. In addition to fluid storage, what are three functions of a reservoir?

7. How are pumps rated?

8. Define gear pump.

9. Define and describe the differences between an unbalanced-vane pump and a balanced-vane pump.

10. Define piston pump and list its major components.

11. List the four basic steps of axial piston pump operation.

12. Define cavitation.

13. List and describe the various roles of a pressure gauge.

14. Aside from storing hydraulic fluid under pressure, what are four advantages of an accumulator?

15. What are the five steps that must be taken in order to maintain cleanliness while troubleshooting and repairing hydraulic circuits?

Digital Resources

ATPeResources.com/Quicklinks
Access Code 462409

CHAPTER 18

PNEUMATIC PRINCIPLES

Pneumatics is the branch of science that deals with the transmission of energy using a gas. Pneumatic systems are based on fluid (gas) theory, which states that a fluid can flow, has no definite shape, and is susceptible to an increase in volume with an increase in temperature. The physical characteristics of gases are affected by pressure, volume, and temperature. Useful pneumatic pressure is produced by compressing atmospheric air and pushing it into a tank for later use. Contaminants of a compressed air system includes particulates, oil, and water. Contaminants must be removed to prevent damage to a compressed air system.

Humphrey Products Company

Chapter Objectives

- Describe the uses of pneumatic systems.
- Identify the properties and characteristics of gas.
- List the different types of gas laws.
- Discuss the effects of compression in pneumatic systems.
- Explain how air temperature and moisture content affect pneumatic systems.
- Explain how to remove contaminants from compressed air.

Key Terms

- pneumatics
- pneumatic system
- thermal expansion
- pressure
- atmospheric pressure
- vacuum
- gauge pressure
- absolute pressure
- volume
- gas laws
- compression
- multistage compression
- reciprocating compressor
- humidity
- relative humidity
- intake filter
- condensation
- dew point
- moisture separator
- pressure drop
- coalescing filter
- desiccant dryer
- adsorption
- refrigerant dryer

PNEUMATICS

Pneumatics is the branch of science that deals with the transmission of energy using a gas. A *pneumatic system* is a combination of components that controls energy through the use of a pressurized gas within an enclosed circuit. A pneumatic system uses gas, such as air or dry nitrogen, for power. Pneumatic systems use compressed (pressurized) air or other gas to transmit or control power.

Pneumatic System History

One of the earliest forms of pneumatics was the bellows. A *bellows* is a device that draws air in through a flapper valve when expanded and expels the air through a nozzle when contracted. A bellows consists of two wooden handles and a flexible leather cover. It was used to start or increase the heat of a fire through the use of compressed air. **See Figure 18-1.**

Today, pneumatic systems are used in the entertainment industry to operate air motors for ski lift drives and air cylinders for moving concert backdrops. In the medical field, doctors and dentists use nonsparking pneumatic drills and saws because of the flammability of oxygen, a widely used gas. Also, car washes use air motors instead of electric motors for rotating scrub brushes to prevent electrical shorts due to the high water content within the working area.

Bellows

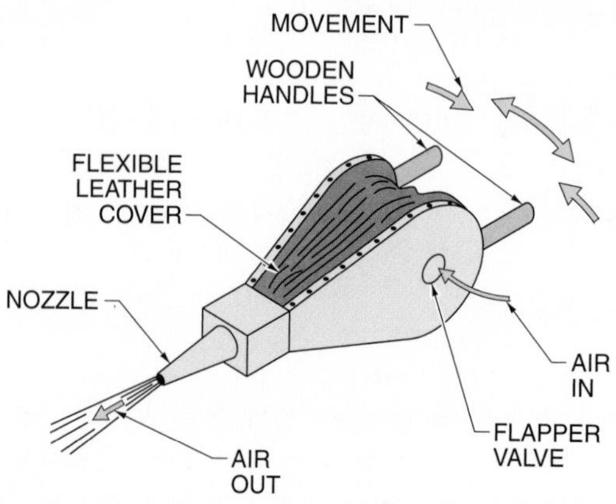

Figure 18-1. An early use of pneumatics was the bellows, which was used to start and increase the heat of a fire.

The manufacturing industry, however, is the largest user of pneumatic systems. Pneumatic systems in manufacturing include the use of air motors to move conveyors, rotate assembly fixtures, act as vibrators for parts feeding, rotate fluid mixing or agitating shafts, etc. Air cylinders are used to clamp and hold parts for machining or assembly, close or open safety guards, operate cutting blades, inject ink in a printer, etc.

Pneumatic systems have replaced many hydraulic systems because pneumatic systems are cleaner. Pneumatic logic systems have been used for electronic circuit replacement because overheating is seldom a problem in pneumatic circuits. In addition, pneumatic circuits are not likely to cause an explosion in hazardous environments. Pneumatic systems are based on fluid (gas) theory, which states that a fluid can flow, has no definite shape, and is susceptible to an increase in volume with an increase in temperature. Unlike molecules of a solid or liquid, air molecules are not freely attracted to each other and are easily compressible.

GAS CHARACTERISTICS

All substances are made up of atoms. An *atom* is the smallest building block of matter that cannot be divided into smaller units without changing its basic character. Atoms combine to form molecules. For example, an oxygen molecule (O_2) is made up of two atoms of oxygen. Water (H_2O) is made up of two atoms of hydrogen and one atom of oxygen.

Molecules are always in motion, and this motion creates heat. Without molecular motion, there is no heat. Heating a substance is the same as increasing the motion of its molecules. Increasing the motion of molecules sends them farther apart, thus creating thermal expansion. *Thermal expansion* is the dimensional change of a substance due to a change in temperature. For example, a metal bar expands when heat increases its molecular motion, causing its molecules to move farther apart. Cooling a substance slows molecular motion and decreases the distance between molecules.

Some molecules have a strong attraction to each other and are known as solids. Others have an attraction to each other but require some freedom. These are known as liquids. Molecules that move quickly and freely are known as gases. **See Figure 18-2.**

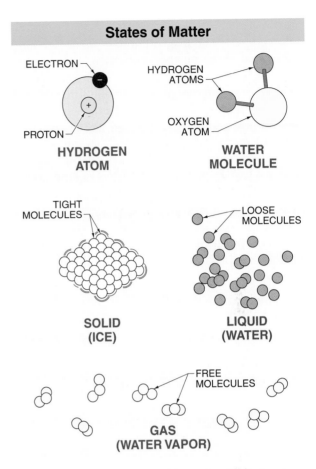

States of Matter

Figure 18-2. The closeness of the molecules that make up a substance determines whether the substance is a solid, liquid, or gas.

Although gas molecules are far apart, they can be pushed closer together, allowing gas to be compressed. Also, because gas molecules are always moving, they are constantly bumping each other and the walls of the container they occupy. This motion allows gas to expand to fill the volume and shape of its container. A balloon remains inflated because of the constant impact of gas molecules striking the balloon wall. When a gas is contained, any force applied to that gas is transmitted equally throughout its container.

Pressure

Pressure is the measure of force per unit area. Pressure increases as gas molecules are forced closer together through compression. Air is a mixture of gases that has weight and creates a pressure on Earth's surface through compression. The pressure exerted on Earth's surface varies with altitude, temperature, and humidity.

The force acting on a unit area is generally a unit of weight. For example, the weight of the air molecules that make up the atmosphere is measured in a 1 sq in. area and is referred to as atmospheric pressure. *Atmospheric pressure* is the force exerted by the weight of the atmosphere on Earth's surface. Atmospheric pressure at sea level is equal to about 14.7 pounds per square inch absolute (psia). At higher altitudes, there is less air, so the atmosphere exerts less weight on each square inch of Earth's surface. Atmospheric pressure decreases at higher altitudes. The pressure of air in Denver, Colorado, (5280′ above sea level) is 12.2 psia and the pressure of air on top of Mt. Everest (29,002′ above sea level) is approximately 5 psia. **See Figure 18-3.**

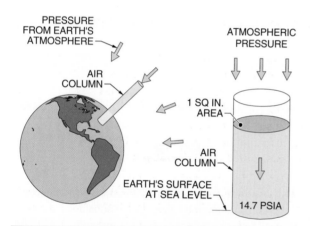

Atmospheric Pressure		
Pressure*	Location	Elevation†
5.0	Mt. Everest	29,002
11.1	Mexico City	7556
12.2	Denver, CO	5280
14.7	Sea Level	0

* in psia
† in ft

Figure 18-3. Atmospheric pressure at sea level is equal to about 14.7 pounds per square inch absolute and decreases at higher altitudes.

The total pressure of compressed air is equal to the number of pounds that are applied per a given unit area. The variables of force and area are used to calculate the amount of pressure acting on a surface. To calculate pressure, the force (pounds) applied is divided by the given unit area (square inches). For example, if a 100 lb force is applied to an area of 5 sq in., the resulting pressure is 20 psia (100 ÷ 5 = 20 psia).

The total force of compressed air is equal to the amount of pressure acting on a given unit area. Force is calculated by the amount of pressure (pounds per square inch absolute) multiplied by the given unit area (square inches). For example, air applying 1 psia on a 5 sq in. surface exerts a total force of 5 lb (1 × 5 = 5 lb).

Pascal's law states that pressure applied to a confined static fluid (liquid or gas) is transmitted with equal intensity throughout the fluid. This means that because a fluid takes the shape of its container and is compressed (under pressure), pressure is the same at any point, regardless of the size or shape of the container.

A slight change in pressure at any point of the container is instantly transmitted throughout the container. For example, a 50′ long pressurized hose becomes a gas station announcing bell when it is plugged at one end, connected to a bell at the other, and laid across a driveway. As the weight of a car drives over the hose, pressure is increased and is instantly transmitted to activate the bell. Pneumatic forces can also be transmitted over considerable distances with little loss. For example, a mile-long freight train uses one common pneumatic system to operate the air brakes of each car.

Measuring Pressure

Pneumatic systems use gases to transmit and control energy to produce mechanical work. The power of a pneumatic system is pressure (psi). Normally, pressure (gauge pressure) is any pressure greater than atmospheric pressure. A *vacuum* is any pressure lower than atmospheric pressure.

Atmospheric pressure (pressure at sea level) is only a reference level and is used when determining more specific types of pressure measurements. The two most commonly used pressure measurements are gauge and absolute pressure.

Gauge pressure is the pressure above atmospheric pressure that is used to express pressures inside a closed system. A typical pressure gauge measures pressures above the surrounding atmosphere. Any pressure below 0 psig is a vacuum. Gauge pressure is expressed in pounds per square inch gauge (psig) or psi.

Tech Tip

Liquids and gases expand when heated. Any excessive pressure within a system creates undue strain on seals and packing. The amount of gas expansion can be calculated by applying Charles' law.

A pressure gauge reads 0 psig at normal atmospheric pressure. For example, a simple plunger pressure gauge used to check tire pressures indicates 0 when not being used even though its plunger has 14.7 psia on both sides (at sea level). When a pressure of 30 lb is indicated when checking an inflated tire with a pressure gauge, the actual pressure within the tire is equal to 44.7 psia (gauge pressure reading plus atmospheric pressure). Gauge pressure shows the numerical value of the difference between atmospheric pressure and absolute pressure. **See Figure 18-4.**

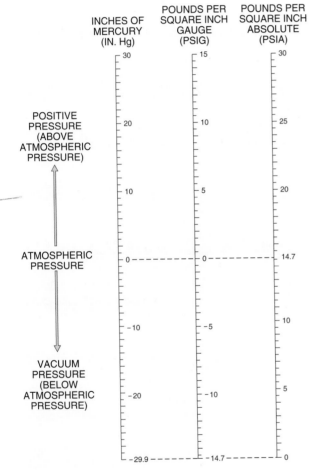

Figure 18-4. Gauge pressure shows the numerical value of the difference between atmospheric pressure and absolute pressure.

Absolute pressure is pressure above a perfect vacuum. Absolute pressure is the sum of gauge pressure plus atmospheric pressure. Absolute pressure is expressed in pounds per square inch absolute (psia).

Pressure outside a closed system (such as normal air pressure) is expressed in pounds per square inch

absolute. The difference between gauge pressure and absolute pressure is the pressure of the atmosphere at sea level at standard conditions (14.7 psia).

Tech Tip

Industrial pneumatic systems are designed to operate with minimal frictional resistance and working pressures of approximately 90 psi.

Volume

Any substance consisting of molecules, such as solids, liquids, or gases, is considered to have volume. *Volume* is the size of a three-dimensional space or object as measured in cubic units. Liquids and solids have definite volumes that do not vary significantly when compressed. Gases have indefinite volumes that vary significantly when compressed.

Volume values in other units of measure must be converted to cubic units for calculations, etc. A conversion table is used to determine the conversion constant required to convert one unit of measure to another. **See Appendix.** For example, the volume of a 2000 gal. (capacity) vessel is divided by the conversion constant of 7.48 to convert gallons to cubic feet. The 2000 gal. vessel has a volume of 267.4 cu ft (2000 ÷ 7.48 = 267.4). To convert cubic feet to gallons, divide cubic feet by 0.1337. For example, a 100 cu ft vessel can hold 747.9 gal. (100 ÷ 0.1337 = 747.9 gal.). The volume of a cube measuring $1' \times 1' \times 1'$ is 1 cu ft. A cubic foot of air may be contained in a cube, cylinder, or any other shaped vessel. The actual volume of air in a container is the same as the volume of the space within the container. **See Figure 18-5.**

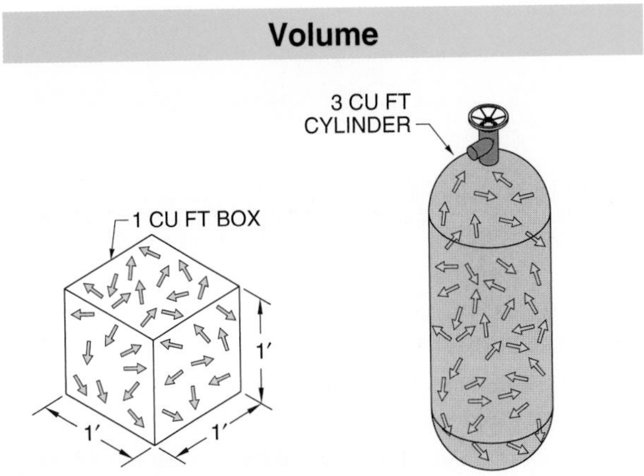

Volume

Figure 18-5. The actual volume of air in a container is the same as the volume of the space within the container.

Gas Laws

The physical characteristics of gases are affected by pressure, volume, and temperature. The relationships between each property are established as gas law equations. *Gas laws* are the relationships between the volume, pressure, and temperature of a gas. Gas laws are used to determine the change in volume, pressure, or temperature of a gas. The behavior of a gas is affected by the characteristics of the gas and the interaction between these characteristics. Gas law equations assume the gas to be under perfect (ideal) conditions.

Boyle's Law. Boyle's law is concerned with the compression of gas. Boyle's law states that the volume of a given quantity of gas varies inversely with the pressure as long as the temperature remains constant. Two vessels or cylinders of the same size may contain air of different pressures because air is compressible. When the volume of a gas is reduced by one half, the pressure of the gas doubles. When air is forced to occupy a smaller volume its pressure increases. **See Figure 18-6.**

Boyle's Law

NOTE: TEMPERATURE REMAINS CONSTANT

V_1 = 60 CU FT

GAS

P_1 = 20 PSIA

EXTERNAL FORCE

V_2 = 30 CU FT

PRESSURE INCREASES

DECREASED VOLUME

P_2 = 40 PSIA

COMPRESSED GAS

Figure 18-6. The volume of a given quantity of gas varies inversely with the pressure as long as the temperature remains constant.

Boyle's law is used when temperature changes are not a factor, such as in determining the consumption of air by an air cylinder or determining the volume/pressure capacity of a vessel. The final pressure is determined by multiplying the initial pressure by the initial volume of the gas and dividing by the final volume. For calculating purposes, pressures must be absolute values or converted from gauge pressure to absolute pressure at sea level. Final pressure is determined by applying the formula:

$$P_2 = \frac{P_1 \times V_1}{V_2}$$

where
P_2 = final pressure (in psia)
P_1 = initial pressure (in psia)
V_1 = initial volume (in cubic units)
V_2 = final volume (in cubic units)

For example, what is the final pressure of 60 cu ft of air at 20 psia when compressed to 30 cu ft?

$$P_2 = \frac{P_1 \times V_1}{V_2}$$

$$P_2 = \frac{20 \times 60}{30}$$

$$P_2 = \frac{1200}{30}$$

$$P_2 = \textbf{40 psia} = \textbf{25.3 psi}$$

Clippard Instrument Laboratory, Inc.

A cylinder rod retracts when a gas (air) at an increased pressure fills the volume in the cylinder around the piston rod.

In addition, when the initial and final pressures are known along with the initial volume, the final volume may be determined by multiplying the initial pressure by the initial volume and dividing by the final pressure. Final volume is determined by applying the formula:

$$V_2 = \frac{P_1 \times V_1}{P_2}$$

where
V_2 = final volume (in cubic units)
P_1 = initial pressure (in psia)
V_1 = initial volume (in cubic units)
P_2 = final pressure (in psia)

For example, what is the final volume of 60 cu ft of air at 20 psia when compressed to 40 psia?

$$V_2 = \frac{P_1 \times V_1}{P_2}$$

$$V_2 = \frac{20 \times 60}{40}$$

$$V_2 = \frac{1200}{40}$$

$$V_2 = \textbf{30 cu ft}$$

Tech Tip

In a pneumatic system, energy is stored and distributed in a potential state (compressed air). Useful work results from a pneumatic system when the compressed air is allowed to convert its potential energy into kinetic energy.

Charles' Law. Boyle's law remains practical as long as temperatures do not change. However, temperatures increase when air is compressed. Charles' law states that the volume of a given mass of gas is directly proportional to its absolute temperature provided the pressure remains constant. *Absolute temperature* is the temperature on a scale that begins with absolute zero. *Absolute zero* is the temperature at which substances possess no heat.

The absolute temperature scale was first determined in 1872 by William Rankine, and is referred to as the Rankine scale (°R or °abs). Molecules are still in motion at 0°F. However, molecules do not move at 0°R. A comparison of Rankine and Fahrenheit scales shows that the temperature in degrees Rankine is always 460° greater than the temperature in degrees Fahrenheit. **See Figure 18-7.**

Degrees Fahrenheit is converted to degrees Rankine by adding 460° to the Fahrenheit temperature. Degrees Fahrenheit is converted to degrees Rankine by applying the formula:

$$°R = 460 + °F$$

where

$°R$ = degrees Rankine

460 = constant

$°F$ = degrees Fahrenheit

For example, what is the Rankine equivalent of 96°F?

$$°R = 460 + °F$$

$$°R = 460 + 96$$

$$°R = \textbf{556°R}$$

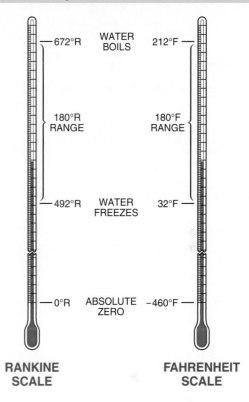

Fahrenheit/Rankine Temperature Conversion

672°R — WATER BOILS — 212°F

180°R RANGE — 180°F RANGE

492°R — WATER FREEZES — 32°F

0°R — ABSOLUTE ZERO — –460°F

RANKINE SCALE   **FAHRENHEIT SCALE**

Figure 18-7. Absolute temperature (°R) is always 460° greater than the temperature in degrees Fahrenheit (°F).

According to Charles' law, if the temperature of a gas is increased, the volume increases proportionately as long as the pressure does not change. **See Figure 18-8.** Variations of the Charles' law equation are used to determine a change in volume or temperature. Final volume is found by multiplying initial volume by the final temperature and dividing by the initial temperature. Final volume is found by applying the formula:

$$V_2 = \frac{V_1 \times T_2}{T_1}$$

where

V_2 = final volume (in cubic units)

V_1 = initial volume (in cubic units)

T_2 = final temperature (in °R)

T_1 = initial temperature (in °R)

For example, what is the final volume of a gas that occupies 40 cu ft at 60°F when the temperature is increased to 90°F?

Charles' Law

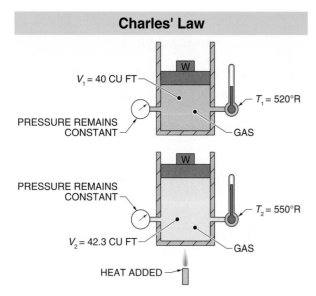

Figure 18-8. The volume of a given mass of gas in directly proportional to its absolute temperature provided the pressure remains constant.

1. Convert initial temperature to °R.
$$°R = 460 + °F$$
$$°R = 460 + 60$$
$$°R = 520°R$$

2. Convert final temperature to °R.
$$°R = 460 + °F$$
$$°R = 460 + 90$$
$$°R = 550°R$$

3. Calculate final volume.
$$V_2 = \frac{V_1 \times T_2}{T_1}$$
$$V_2 = \frac{40 \times 550}{520}$$
$$V_2 = \frac{22,000}{520}$$
$$V_2 = \textbf{42.3 cu ft}$$

In addition, final temperature is found by multiplying the final volume by the initial temperature and dividing by the initial volume. Final temperature is found by applying the formula:

$$T_2 = \frac{V_2 \times T_1}{V_1}$$

where

T_2 = final temperature (in °R)
V_2 = final volume (in cubic units)
T_1 = initial temperature (in °R)
V_1 = initial volume (in cubic units)

For example, what is the final temperature of 10 cu ft of gas at 492°R that is increased to 14 cu ft?

$$T_2 = \frac{V_2 \times T_1}{V_1}$$
$$T_2 = \frac{14 \times 492}{10}$$
$$T_2 = \frac{6888}{10}$$
$$T_2 = \textbf{688.8°R = 288.8°F}$$

Gay-Lussac's Law. Gay-Lussac's law states that if the volume of a given gas is held constant, the pressure exerted by the gas is directly proportional to its absolute temperature. **See Figure 18-9.** Gay-Lussac's law is used to determine the pressure based on an increase in temperature. Final pressure is determined by multiplying the initial pressure by the final temperature and dividing by the initial temperature. Final pressure is found by applying the formula:

$$P_2 = \frac{P_1 \times T_2}{T_1}$$

where

P_2 = final pressure (in psia)
P_1 = initial pressure (in psia)
T_2 = final temperature (in °R)
T_1 = initial temperature (in °R)

Gay-Lussac's Law

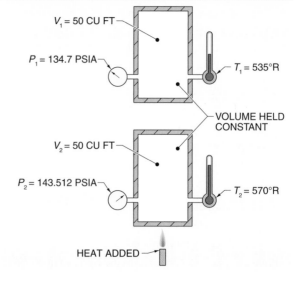

Figure 18-9. The pressure exerted by a gas is directly proportional to its absolute temperature if the volume of the gas is held constant.

Advanced Assembly Automation Inc.

Automated industrial systems use pneumatics for applications such as pneumatic wrenches, air clamps, ejection molding machines, and automatic gates and doors.

For example, what is the final pressure in a 50 cu ft vessel holding air at 120 psig at 75°F if the temperature is increased to 110°F?

1. Convert initial pressure to absolute pressure.

$$psia = 14.7 + psig$$
$$psia = 14.7 + 120$$
$$psia = 134.7 \text{ psia}$$

2. Convert initial temperature to °R.

$$°R = 460 + °F$$
$$°R = 460 + 75$$
$$°R = 535°R$$

3. Convert final temperature to °R.

$$°R = 460 + °F$$
$$°R = 460 + 110$$
$$°R = 570°R$$

4. Calculate final pressure.

$$P_2 = \frac{P_1 \times T_2}{T_1}$$

$$P_2 = \frac{1347 \times 570}{535}$$

$$P_2 = \frac{76,779}{535}$$

$$P_2 = \textbf{143.512 psia} = \textbf{128.812 psi}$$

Combined Gas Law. The relationship between pressure, volume, and temperature is best determined when Boyle's, Charles', and Gay-Lussac's laws are combined. The combined gas law covers all the variables regarding the relationships between the pressure, volume, and temperature of a gas. Final pressure is found by applying the formula:

$$P_2 = \frac{P_1 \times V_1}{T_1} \times \frac{T_2}{V_2}$$

where

P_2 = final pressure (in psia)
P_1 = initial pressure (in psia)
V_1 = initial volume (in cubic units)
T_1 = initial temperature (in °R)
T_2 = final temperature (in °R)
V_2 = final volume (in cubic units)

For example, what is the final pressure in a 40 cu in. cylinder containing air at 100 psig and 100°F when compressed to 20 cu in. while heated to 150°F?

1. Convert initial gauge pressure to absolute pressure.

$$psia = 14.7 + psig$$
$$psia = 14.7 + 100$$
$$psia = 114.7 \text{ psia}$$

2. Convert initial temperature to °R.

$$°R = 460 + °F$$
$$°R = 460 + 100$$
$$°R = 560°R$$

3. Convert final temperature to °R.

$$°R = 460 + °F$$
$$°R = 460 + 150$$
$$°R = 610°R$$

4. Calculate final pressure.

$$P_2 = \frac{P_1 \times V_1}{T_1} \times \frac{T_2}{V_2}$$

$$P_2 = \frac{114.7 \times 40}{560} \times \frac{610}{20}$$

$$P_2 = \frac{4588}{560} \times 30.5$$

$$P_2 = 8.193 \times 30.5$$

$$P_2 = \textbf{249.886 psia} = \textbf{235.186 psi}$$

The combined gas law equation may be rearranged to determine any value when the others are known. Final volume is found by applying the formula:

$$V_2 = \frac{P_1 \times V_1}{T_1} \times \frac{T_2}{P_2}$$

where

V_2 = final volume (in cubic units)

P_1 = initial pressure (in psia)

V_1 = initial volume (in cubic units)

T_1 = initial temperature (in °R)

T_2 = final temperature (in °R)

P_2 = final pressure (in psia)

For example, what is the final volume of air if 40 cu ft of air at 12 psia is compressed to 42 psia and the compressor suction temperature is 60°F and the discharge temperature is 180°F?

1. Convert initial temperature to °R.

°R = 460 + °F

°R = 460 + 60

°R = 520°R

2. Convert final temperature to °R.

°R = 460 + °F

°R = 460 + 180

°R = 640°R

3. Calculate final volume.

$$V_2 = \frac{P_1 \times V_1}{T_1} \times \frac{T_2}{P_2}$$

$$V_2 = \frac{12 \times 40}{520} \times \frac{640}{42}$$

$$V_2 = \frac{480}{520} \times 15.238$$

$$V_2 = 0.923 \times 15.238$$

$$V_2 = \textbf{14.065 cu ft}$$

Final temperature may be found when initial and final pressure and volume and initial temperature are known. Final temperature is found by applying the formula:

$$T_2 = \frac{P_2 \times V_2}{P_1 \times V_1} \times T_1$$

where

T_2 = final temperature (in °R)

P_2 = final pressure (in psia)

V_2 = final volume (in cubic units)

P_1 = initial pressure (in psia)

V_1 = initial volume (in cubic units)

T_1 = initial temperature (in °R)

For example, what is the final temperature of 80 cu ft of air at 65°F and 14.7 psia when compressed to 40 cu ft at 40 psia?

1. Convert initial temperature to °R.

°R = 460 + °F

°R = 460 + 65

°R = 525°R

2. Calculate final temperature.

$$T_2 = \frac{P_2 \times V_2}{P_1 \times V_1} \times T_1$$

$$T_2 = \frac{40 \times 40}{14.7 \times 80} \times 525$$

$$T_2 = \frac{1600}{1176} \times 525$$

$$T_2 = 1.36 \times 525$$

$$T_2 = \textbf{714°R}$$

Saylor-Beall Manufacturing Company

Air compressors may be powered by electric motors or gasoline engines for use in a variety of locations.

COMPRESSION

Useful pneumatic pressure is produced by first compressing quantities of atmospheric air and then pushing it into a tank (receiver) for later use. As a piston retracts, it draws air (suction) through the inlet port at one pressure. As the piston extends, it decreases (compresses) the volume of air, which increases the pressure of the discharged air.

The difference between the suction (inlet) pressure and the discharge pressure is the ratio of compression (R_c). The ratio of compression is 3 when a compressor triples the absolute inlet pressure. Ratio of compression is calculated by dividing the absolute discharge pressure by the absolute inlet pressure. Ratio of compression is found by applying the formula:

$$R_c = \frac{P_2}{P_1}$$

where

R_c = ratio of compression
P_2 = final pressure (in psia)
P_1 = initial pressure (in psia)

For example, what is the ratio of compression if a compressor inlet pressure is 1.5 psi vacuum and the discharge pressure is 40 psig?

1. Convert final pressure to absolute.
$psia = 14.7 + psig$
$psia = 14.7 + 40$
$psia = 54.7$ psia

2. Convert initial pressure to absolute.
$psia = 14.7 + psig$
$psia = 14.7 + -1.5$
$psia = 13.2$ psia

3. Calculate ratio of compression.

$$R_c = \frac{P_2}{P_1}$$

$$R_c = \frac{54.7}{13.2}$$

$$R_c = \mathbf{4.14}$$

Air temperature increases as a piston extends and the air molecules are forced closer together (compressed). This happens because as the compressor forces gas molecules closer together, it also increases the collisions of the molecules, thus increasing their temperature. **See Figure 18-10.** The temperature increase is based on the inlet temperature and the ratio of compression. Increasing the ratio of compression increases the temperature of the discharged air.

The amount of energy required to compress a quantity of air to a given pressure depends on the rate that heat is dissipated. Higher compression temperatures require greater amounts of energy (horsepower) to accomplish the same task as those having lower temperatures. Greater horsepower is required to meet output demands if discharge temperatures are excessive. The inlet air temperature or ratio of compression may be reduced to reduce the need for greater horsepower and to reduce the discharge temperature. The ratio of compression can be decreased by reducing the discharge pressure, increasing the inlet pressure, or both.

Compression

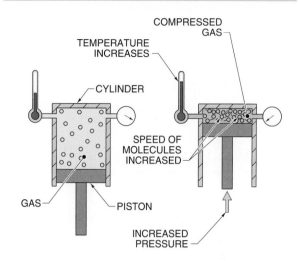

Figure 18-10. Compressing gas molecules increases the collisions of the molecules, thus increasing their temperature.

Heat generated during compression is controlled by limiting the pressure of the compressed air or by cooling. Lowering the suction temperature increases the standard volume of air processed and decreases the discharge temperature if the ratio of compression value and inlet pressure remain the same.

Multistage Compression

Air must be compressed in two or more steps (stages) for a reciprocating compressor to obtain pressures over 100 psi. A *reciprocating compressor* is a device that compresses gas by means of a piston that moves back and forth in a cylinder. In multistage compression, the air moves from one pumping chamber to another with each chamber receiving and discharging a higher pressure. **See Figure 18-11.**

Multistage compression is required when the ratio of compression is greater than 6. When a compressing unit is required to have a total ratio of compression of 18, compression is usually accomplished in three stages. For example, the upper limit of a two-stage compressor at sea level with each stage at a ratio of compression of 4 is 220.5 psi.

Safety Tip

Compressed air must not be used for cleaning purposes except where reduced to less than 30 psi and then only with effective chip guarding and personal protective equipment.

Multistage Compression

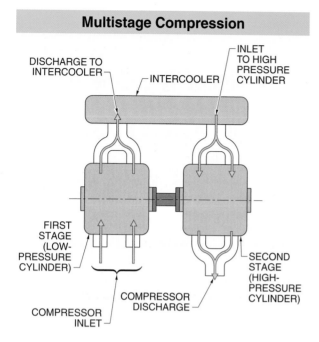

Figure 18-11. In multistage compression, the air moves from one pumping chamber to another with each chamber receiving and discharging a higher pressure.

Intercooling

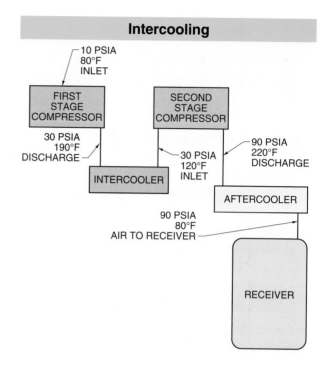

Figure 18-12. Intercooling removes a portion of the heat of compression as air is fed from one compression stage to another.

Compression Intercooling

Heat generated in a cylinder due to compression of a reciprocating compressor is removed by lubrication, conduction through the cylinder walls, and intercooling. *Intercooling* is the process of removing a portion of the heat of compression as the air is fed from one compression stage to another. Intercooling is generally performed by a tube and shell heat exchanger (intercooler). An *intercooler* is a heat exchanger between the discharge of one compression stage and the inlet of the next compression stage. A tube and shell intercooler uses cool water to dissipate heat. **See Figure 18-12.**

An *aftercooler* is a heat exchanger for cooling the discharge from a compressor. Aftercoolers control the amount of water vapor in a compressed air system by condensing the water vapor into liquid form. In a distribution or process manufacturing system, liquid water can cause significant damage to the equipment that uses compressed air.

Intercooling is used to reduce the temperature of the compressed air and reduce the volume of the air before it reaches the next stage. Also, for every 5°F absorbed at the intercooler, approximately 1% of horsepower is saved. Without intercooling, multistage compression does not allow an overall reduction in discharge temperature of compressed air.

AIR TEMPERATURE AND MOISTURE CONTENT

Atmospheric air contains water in some form or another. The water may be solid (snow or ice crystals), visible (clouds or fog), or invisible (water vapor). Invisible vapor makes up most of the moisture in atmospheric air. *Humidity* is the amount of moisture in the air. Humidity comes from water that has evaporated into the air.

Water in a pneumatic system is a natural occurrence because the air that a compressor takes in contains significant amounts of water from the atmosphere. It takes 7.8 cu ft of atmospheric air to produce 1.0 cu ft of compressed air at 100 psig. This means that after compression there is 7.8 times more moisture in the compressed air as compared to the atmospheric air.

Air generally contains less moisture than it is capable of holding. However, as the moisture level in air rises to its maximum, the air becomes saturated. *Saturated air* is air that holds as much moisture as it is capable of holding. The amount of moisture air is capable of holding is also greatly affected by the temperature of the air. The higher the air temperature, the greater the amount of moisture it is able to hold.

Pressure gauges are used to monitor air pressure readings in most pneumatic systems.

At standard pressure and temperature, the weight of 1 cu ft of air equals 0.076 lb. An average room ($12' \times 13' \times 8' = 1248$ cu ft) contains 94.848 lb of air (1248 cu ft × 0.076 lb = 94.848 lb).

Air at saturation and at a temperature of 80°F is capable of holding about 0.022 lb of moisture per pound of dry air. The air in an average room (1248 cu ft) at saturation is capable of holding 2.086 lb of water vapor (94.848 lb × 0.022 lb = 2.086 lb). Moisture in the air is more commonly expressed in grains. **See Figure 18-13.** There are 150 grains (gr) of moisture per pound of saturated air at 80°F. Therefore, an average room (1248 cu ft) contains 14,227.2 gr of moisture (150 gr × 94.848 lb = 14,227.2 gr).

Saturated Air Moisture Holding Properties			
Dew Point*	Pounds of Moisture†	Grains of Moisture†	Grains of Moisture‡
100	0.044	300	19.766
90	0.03	210	14.790
80	0.022	150	10.934
70	0.015	110	7.980
60	0.011	76	5.745
50	0.0075	53	4.076
40	0.005	36	2.749
30	0.0033	24	1.935
20	0.002	15	1.235
10	0.0014	9	0.776
0	0.0008	5.5	0.481

* in °F
† per lb of dry air
‡ per cu ft of dry air

Figure 18-13. The total amount of moisture that air is capable of holding varies based on the temperature of the air.

Air that is at 100°F is saturated when it contains 19.766 gr of moisture per cu ft. Air that is at 60°F is saturated when it contains only 5.745 gr of moisture per cu ft. This is why cooling the air at the compressor reduces the moisture content in the receiver.

Relative humidity is the percentage of moisture contained in air compared to the maximum amount of moisture (saturation) it is capable of holding. For example, if air at 80°F contains 4.374 gr of moisture per cu ft, the air is at 40% relative humidity, because 4.374 gr is only 40% of the 10.934 gr saturation level. The relative humidity of this air increases when it is cooled because the air is capable of holding less moisture, even though the total grains of moisture does not change. Similarly, if this air is heated, the relative humidity decreases because the air is capable of holding more moisture. **See Figure 18-14.**

AIR CONTAMINANTS

Contaminants of a compressed air system include particulates, oil, and water. Each contaminant damages a pneumatic system in a different way. A *particulate* is a fine solid particle that remains individually dispersed in a gas. Particulates in a system grind away metal surfaces, lodge between moving parts, or plug orifices. Oil is a contaminant when it is carried over from the compressor. Oil carried over from the compressor is too thick for air line conditioning. Thick compressor oil gets thicker as it travels through a system, picking up dust, dirt, and other solids. The result is plugged orifices or gummy, sticky moving parts. Water is a pneumatic system's most obvious contaminant. Water in a system blocks passageways, slows actuators, rusts metal components, and mixes with other contaminants.

Contaminant Removal

Removing particulates begins with a pneumatic system's intake filter. An *intake filter* is a filter that removes solids from the free air at the compressor inlet port. *Free air* is air at atmospheric pressure and ambient temperature. The condition of the free air at the compressor normally determines the type or method of intake filter used. Excessive moisture or corrosive gases may require either a purifying filter or extra piping to reach cleaner air from a remote area. Intake filters must be cleaned or replaced periodically.

Moisture Held By Air*									
Temperature†	Percent Saturation								
	10	20	30	40	50	60	70	80	90	100
−10	0.028	0.057	0.086	0.114	0.142	0.171	0.200	0.228	0.256	0.285
0	0.048	0.096	0.144	0.192	0.240	0.289	0.337	0.385	0.433	0.481
10	0.078	0.155	0.233	0.210	0.388	0.466	0.543	0.621	0.698	0.776
20	0.124	0.247	0.370	0.494	0.618	0.741	0.864	0.988	1.112	1.235
30	0.194	0.387	0.580	0.774	0.968	1.161	1.354	1.548	1.742	1.935
40	0.285	0.570	0.855	1.140	1.424	1.709	1.994	2.279	2.564	2.749
50	0.408	0.815	1.223	1.630	2.038	2.446	2.853	3.261	3.668	4.076
60	0.574	1.149	1.724	2.298	2.872	3.447	4.022	4.596	5.170	5.745
70	0.798	1.596	2.394	2.192	3.990	5.788	5.586	6.384	7.182	7.980
80	1.093	2.187	3.280	4.374	5.467	6.560	7.654	8.747	9.841	10.934
90	1.479	2.958	4.437	5.916	7.395	8.874	10.353	11.832	13.311	14.790
100	1.977	3.953	5.930	7.906	9.883	11.860	13.836	15.813	17.789	19.766

* in grains
† in °F

Figure 18-14. The percentage of moisture contained in air compared to the maximum amount of moisture (saturation) it is capable of holding varies based on the temperature of the air.

Conditioning Compressed Air. Contaminant removal in a compressed air system is accomplished by conditioning the air. This requires the ambient air to be filtered, compressed, and cooled for moisture separation before it enters the receiver. Cooling the compressed air is accomplished by use of an aftercooler. An aftercooler is a heat exchanger that cools air that has been compressed. Aftercoolers generally use water as the cooling medium. Aftercoolers remove up to 80% of the moisture by means of condensation. **See Figure 18-15.**

Condensation is the change in state from a gas or vapor to a liquid. The ability of air to hold water vapor decreases when the air is cooled. The point at which water vapor begins to change to a liquid is referred to as dew point (dp). *Dew point* is the temperature to which air must be cooled in order for the moisture in the air to begin condensing. Dew point is also known as saturation temperature. An example of condensation is the moisture that condenses on a glass of water with ice in it. The outside surface of the glass lowers the temperature of the surrounding atmospheric air enough to cause the water vapor in the air to condense (turn to liquid) on the surface of the glass.

The greatest volume of water and oil is removed at the discharge of the aftercooler by means of a moisture separator. A *moisture separator* is a device that separates a large percentage of water from the cooled air through a series of plates or baffles. A moisture separator is located at the discharge port of an aftercooler. The plates or baffles are shaped and placed in such a way that air entering the separator is directed into a swirling motion. Centrifugal force causes large contaminants (mostly water) to strike the wall of the separator and flow to the bottom, where they are drained.

Conditioning Circuit Air. The remaining contaminants, such as oil carry-over, water vapor, and particulates, are removed in sequence to maximize the remaining removal equipment. Filters are required in each circuit to remove particulates or liquids that remain in the system or are introduced by the system. For example, particulates may be introduced from wear of moving metal components. Oil is introduced as oil carry-over. *Oil carry-over* is the released lubricating oil from the walls of the compressor cylinder and piston. Oil carry-over travels with the compressed air into the system as a vapor. As compressor cylinder walls or piston seals (rings) wear, more oil is carried over into the system.

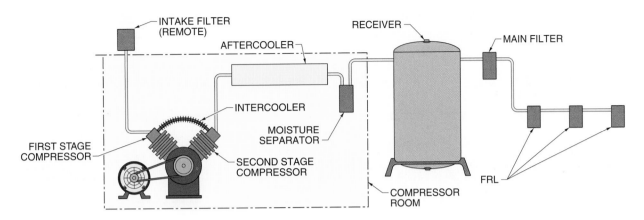

Figure 18-15. Compressor intake air is conditioned by removing particulates from the free air with an intake filter and removing 80% of its moisture with an aftercooler.

A basic pneumatic circuit contains a general purpose filter, regulator, and lubricator. The filter removes contaminants before the air enters the regulator. The regulator adjusts system pressure to the required circuit pressure. The lubricator releases or injects the proper protective lubricant into the system. These three components, when used in combination, are commonly referred to as an FRL (filter/regulator/lubricator).

Air in a pneumatic system must always be conditioned through filtering. The filtering components must be of the correct type and placed properly in the system to maximize their filtering functions. Component placement for conditioning compressed air requires removing contaminants in sequence. **See Figure 18-16.**

Dust and dirt are removed using a general purpose or particulate filter. These filters also remove some liquids, but their major function is to remove solids. General purpose filters remove condensed (liquid) water, oil, and solid particles as small as 5 microns (μ), or 0.0002″. General purpose filters allow most water and oil vapors, as well as submicron solid particles, to pass. Due to the significant oil, water, and particle collection during filtering, general purpose filter elements must be changed often. A general purpose filter is always used in a pneumatic system because it removes the solids that would otherwise clog a more expensive filter. Solid particles create an increase in flow resistance. For this reason, the life of a general purpose filter is determined by the quantity of solids retained, not by the amount of liquid visible.

The coarsest grade of filter element that satisfactorily protects the system should be used. This grade provides the longest element life without a sizable pressure drop. *Pressure drop* is the pressure differential between upstream and downstream fluid flow caused by resistance. Resistance creating a pressure drop may be caused by friction between the air and its plumbing, intentional creation of resistance by placing an orifice (restriction) in a system, or blockage of flow created by contaminants.

Pressure drops on new filters should not exceed 2 psi. In some cases, multiple filters can be used in series. The beginning filter having the coarsest element is followed by a filter with a finer element. The success of filter elements to remove solids at this stage determines the life (and expense) of any downstream liquid removing elements.

Safety Tip

All air receivers shall be constructed in accordance with the **ASME Boiler and Pressure Vessel Code, Section VIII, Edition 1968** *and installed so that all drains are easily accessible. Under no circumstances shall an air receiver be buried underground or located in an inaccessible place.*

Removing Oil And Water. Liquid-removing filters, such as coalescing filters, become plugged and ineffective prematurely if solids are allowed to continue downstream. A *coalescing filter* is a device that removes submicron solids and vapors of oil or water by uniting very small droplets into larger droplets.

Coalescing filters are ideal filters for removing oil and water from a system. Fine liquid droplets are continuously trapped in the element. The droplets grow in size and emerge on the outside surface of the element to flow to the filter drain. Coalescing elements function at their original efficiency even when saturated and continue functioning well until restricted by particulates. When coalescing filters become plugged, they must be discarded and replaced. **See Figure 18-17.**

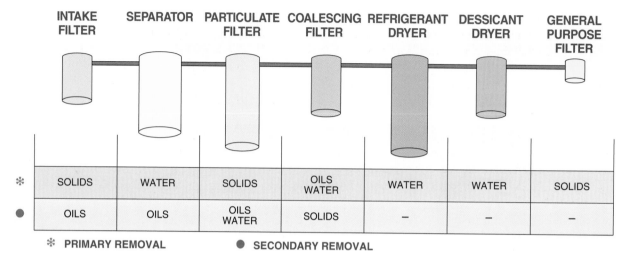

	INTAKE FILTER	SEPARATOR	PARTICULATE FILTER	COALESCING FILTER	REFRIGERANT DRYER	DESSICANT DRYER	GENERAL PURPOSE FILTER
✱	SOLIDS	WATER	SOLIDS	OILS WATER	WATER	WATER	SOLIDS
●	OILS	OILS	OILS WATER	SOLIDS	–	–	–

✱ **PRIMARY REMOVAL**   ● **SECONDARY REMOVAL**

Figure 18-16. Filtering components must be of the right type and placed properly in the system to maximize the filtering functions.

Even with excellent prefiltering, some solids make their way to the coalescing element, shortening its life. With proper maintenance, coalescing filter element life should be approximately 2000 operating hours. Oil not removed by a coalescing filter and allowed to continue downstream enters actuators to become a gummy, sticky resistance producing early actuator failure.

The temperature of ambient air must be considered when placing equipment that conditions compressed air. Coalescing filters are most effective when placed in the coolest location of the system.

Dryers. Pneumatic technology is used extensively in automatic processing equipment. Automatic processing equipment such as instrumentation, measuring devices, controllers, etc. are some of the devices used in automatic production operations. *Instrumentation* is the area of industry that deals with the measurement, evaluation, and control of process variables. A *process variable* is any characteristic that changes its value during any operation within the process. Process variables include temperature, pressure, flow, force, etc.

To be dependable and effective, most automatic processing equipment depends on clean, dry air. General purpose and coalescing filters are not completely effective in removing all water vapor because of the ability of water to change form at different temperatures. In many cases, proper installation and placement of coalescing filters offers sufficient protection. However, a dryer is used when water vapor may contaminate a sensitive device.

A *dryer* is a device that dries air through cooling and condensing. Dryers leave air dry enough for applications

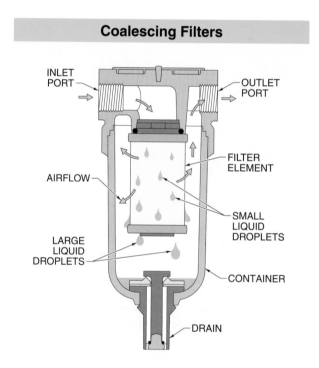

Coalescing Filters

Figure 18-17. A coalescing filter is a device that removes submicron solids and vapors of oil or water by uniting very small droplets into larger droplets.

such as instrumentation, air logic, etc. *Dry air* is air free of water vapor or oil droplets. Water vapor is not always visible. Most water vapor droplets are in the $0.5\mu - 2\mu$ range, and the smallest size droplet visible is about 15μ. Dryers remove water vapor from the air using desiccants or refrigeration.

A desiccant dryer is used to remove invisible water vapor when maximum drying is required. A *desiccant dryer* is a device that removes water vapor by adsorption. *Adsorption* is the adhesion of a gas or liquid to the surface of a porous material. A desiccant dryer removes water vapor using material such as silica gel or alumina. Adsorption offers the capability of removing 99.9% of the water vapor in the air. The desiccant material adsorbs water and becomes saturated and ineffective. For this reason, two cylinders may be interconnected to allow for the heating, evaporation, and regeneration of used desiccant. **See Figure 18-18.** While one side is drying compressed air, the other is being reactivated by use of an embedded heating coil or dry air being passed through the desiccant. The desiccant used as drying material can also adsorb any oil present in the system, leaving the dryer contaminated and ineffective.

A *refrigerant dryer* is a device designed to lower the temperature of the compressed air to 35°F. The cooling provided by a refrigeration system causes the water in the air to condense by lowering the relative humidity and the dew point of the air. The condensed liquids are drained automatically. Generally, the cold dry air flows through a heat exchanger to precool the incoming air. **See Figure 18-19.**

Dryers are normally placed in the coolest downstream location to lessen the energy required to lower the air temperature and condense the moisture in the air. To prevent contamination, dryers must be used for water removal only. Oil that is not removed upstream from the dryer builds up on the dryer tube walls. The oil buildup acts as an insulator, which reduces the dryer capacity. Desiccant dryers are most effective when placed after coalescing filters and closest to the point of air use. The air entering a desiccant dryer should have a maximum temperature of 100°F.

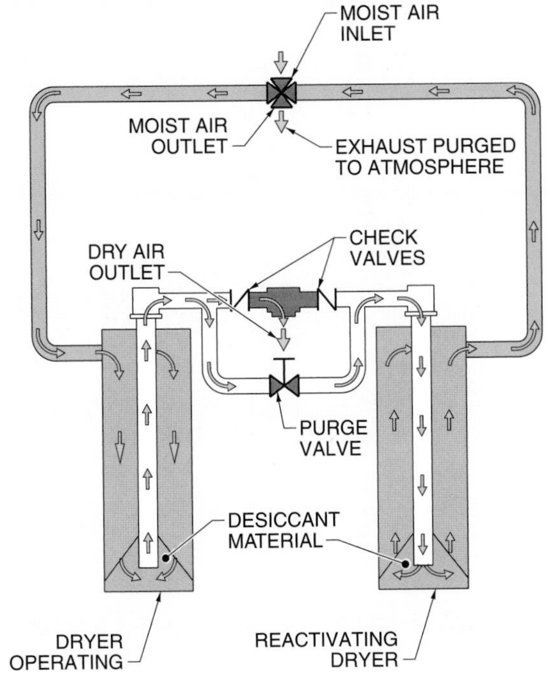

Figure 18-18. A desiccant dryer is a device that removes water vapor by adsorption.

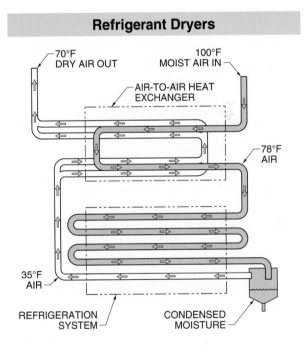

Figure 18-19. A refrigerant dryer is a device designed to lower the temperature of the compressed air to 35°F using a refrigeration system.

Review

1. Define pneumatic system.

2. How does the attraction of molecules differ among solids, liquids, and gases?

3. List three factors that cause the pressure on the Earth's surface to vary.

4. Define Boyle's law and explain how two cylinders of the same size are able to contain air of different pressures.

5. Define Charles' law and explain the difference between the Rankine scale and the Fahrenheit scale.

6. What does the combined gas law cover?

7. Explain why air temperature increases as a piston extends and air molecules are forced closer together (compressed).

8. What role does intercooling play in multistage compression?

9. Define relative humidity.

10. How do particulates, oil, and water act as contaminants in a pneumatic system?

11. Why does air condense on the outside surface of a glass of water with ice in it?

12. List two reasons why resistance could create a pressure drop.

13. Define coalescing filter.

14. What is a refrigerant dryer designed to do?

15. List two reasons why dryers are normally placed in the coolest downstream location.

Digital Resources

ATPeResources.com/Quicklinks
Access Code 462409

CHAPTER 19

PNEUMATIC APPLICATIONS

A pneumatic system transmits and controls energy through the use of pressurized gas within an enclosed circuit. A pneumatic system consists of a compressor, receiver, pressure switch, piping, check valve, receiver safety valve, pressure gauge, and pneumatic circuit. A pneumatic circuit is a combination of air operated components that are connected to perform work. Pneumatic circuit components include check valves, filters, lubricators, pressure valves, directional control valves, flow control valves, and actuators. Pneumatic logic elements are miniature air valves used as switching devices to provide decision making signals in a pneumatic circuit.

ARO Fluid Products Div., Ingersoll-Rand

Chapter Objectives

- Explain the benefits of pneumatic circuitry.
- List the different types of air compressors used in pneumatic systems.
- Explain the importance of pressure control in a compressor used in a pneumatic system.
- Define safety relief valve.
- Explain why pneumatic system piping must be installed.
- Define check valve.
- List the components used in the conditioning, controlling, and directing of air in pneumatic circuits.
- Describe the different types of actuator devices.
- Identify pneumatic logic.

Key Terms

- pneumatic system
- pneumatic circuit
- graphic diagram
- symbol
- air compressor
- positive displacement compressor
- piston compressor
- helical screw compressor
- vane compressor
- pressure compensator
- pressure switch
- safety relief valve
- check valve
- filter
- lubricator
- pressure regulator
- flow control valve
- air cylinder
- seal
- o-ring
- lip seal
- air motor
- logic
- pneumatic logic element

PNEUMATIC CIRCUITRY

A *pneumatic system* is a system that transmits and controls energy through the use of a pressurized gas within an enclosed circuit. Pneumatic systems compress, store, and provide clean and safe air for a pneumatic circuit. A *pneumatic circuit* is a combination of air-operated components that are connected to perform specific work. The use of pneumatic circuits on industrial assembly lines has increased greatly since the development of mass production. Pneumatic circuits are being used today where hydraulic circuits were used in the past because pneumatic circuits offer benefits over hydraulic circuits. Pneumatic circuit benefits include the following:

- easy air storage and use in remote locations because air is compressible
- provides potential energy without the use of electricity
- cleaner than hydraulic circuits
- economical because initial costs are relatively low for equipment and spare parts
- overheating is generally not a problem
- high reliability due to fewer moving parts
- electrical shock and spark-free controls, enabling use in wet or explosive locations
- pneumatic logic components that can be used in place of electrical switches, relays, resistors, and timers

Graphic Diagrams

Pneumatic system and circuit understanding is clarified by using graphic diagrams. A *graphic diagram* is a drawing that uses simple line shapes (symbols) with interconnecting lines to represent the function of each component in a system or circuit. A *symbol* is a graphic element which indicates a particular device. To simplify a graphic diagram, each device or component is given a symbol in place of a full drawing or picture. The path along the connecting lines from any one of the symbols (components) to any other component can be traced to determine the system or circuit operation. **See Figure 19-1.**

Safety Tip

Pneumatic system hose fittings must be tightened securely because a whipping hose can damage equipment and cause injury to personnel.

504

Graphic Diagrams

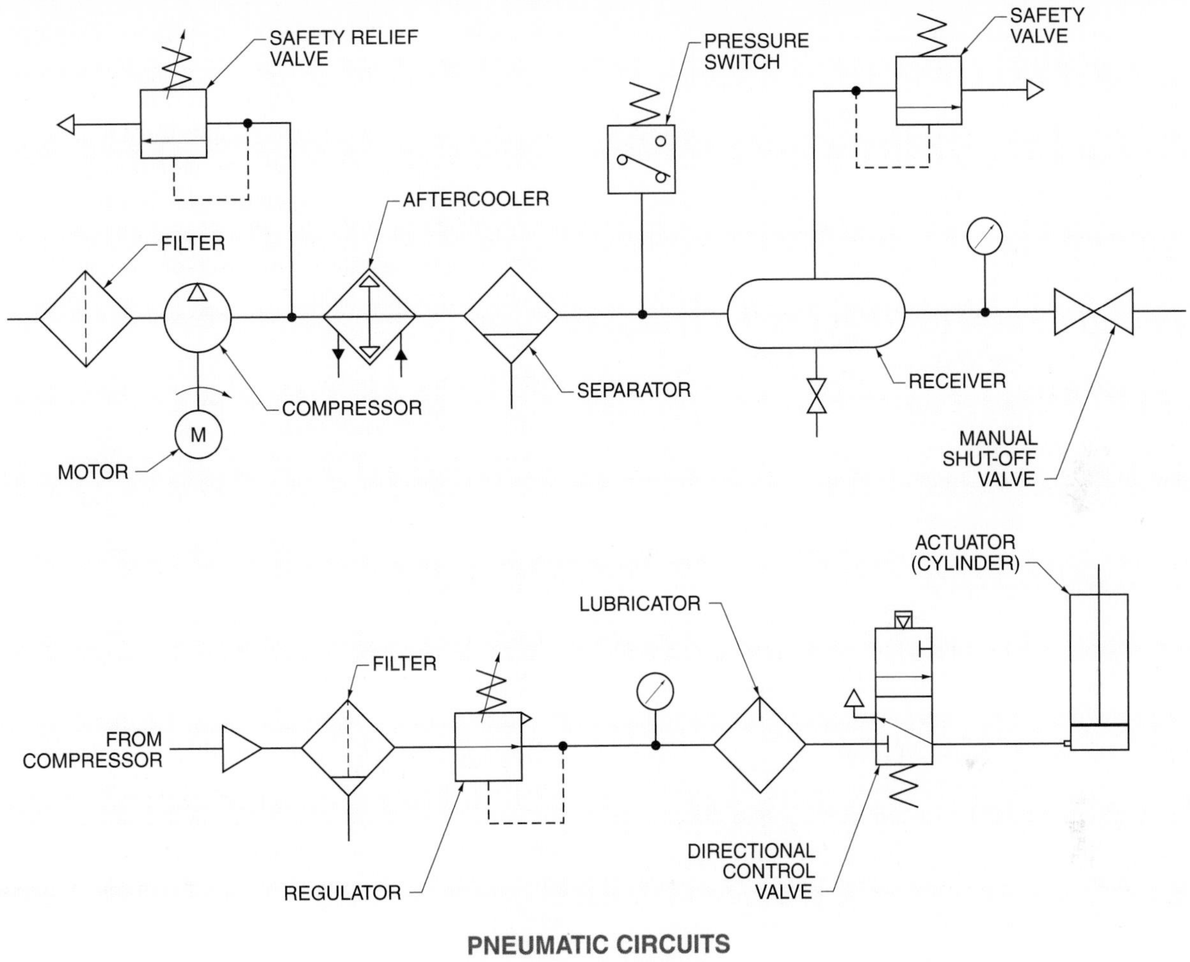

PNEUMATIC CIRCUITS

Figure 19-1. A graphic diagram uses symbols and interconnecting lines to represent the function of each component in a system circuit.

The graphic diagram of a pneumatic system begins at the atmospheric air input into the system. This is normally at a filter or breather symbol that is located upstream from the compressor. The diagram shows all components used to compress and store air for use in a pneumatic circuit. The components include the breather, compressor, safety relief valve, aftercooler, separator, pressure switch, receiver, and manual shut-off valve. A pneumatic system ends at the shut-off valve downstream from the receiver.

The graphic diagram of a pneumatic circuit shows the circuit components beginning with the compressor, which indicates the circuit's beginning and direction of airflow. The compressor symbol is followed by a filter, a regulator, a lubricator, a directional control valve, and an actuator (cylinder).

Complex circuit graphic diagrams may show origination and direction of airflow by the use of an arrow or an "S" (supply). The piping connecting the components in a pneumatic graphic diagram is traced to determine the circuit operation. For example, the circuit piping is traced beginning at the supply, through the pilot-operated 4-way valve, and on to the actuator. This reveals that pressure is being applied to the rod end port of the actuator. **See Figure 19-2.**

The solenoid must be activated to shift the 3-way valve to allow pilot air to shift the 4-way valve to allow pressure to flow to the cap end of the actuator. The actuator piston extends when the 4-way valve is shifted by pilot pressure. Both valves return to their original position by spring pressure when power is removed from the solenoid.

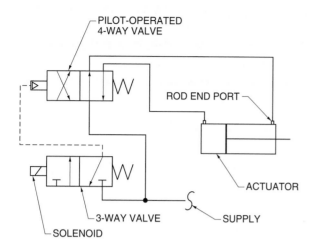

PILOT-OPERATED
4-WAY VALVE

ROD END PORT

ACTUATOR

3-WAY VALVE

SUPPLY

SOLENOID

Figure 19-2. The piping connecting the components in a pneumatic graphic diagram is traced to determine the circuit operation.

PNEUMATIC SYSTEMS

A pneumatic system consists of a compressor to compress air, a receiver (tank) to store the compressed air, a pressure switch to shut down the compressor when the preset pressure has been met or to activate unloaders, piping to transfer air within the system, a check valve to prevent compressed air from backing into the compressor, a receiver safety valve to prevent dangerous overpressure, and a pressure gauge to observe system pressure. **See Figure 19-3.**

Air Compressors

An *air compressor* is a device that takes air from the atmosphere and compresses it to increase its pressure. Compressing air for storage is accomplished using a positive displacement compressor, which increases the air's pressure by reducing its volume in a confined space. A *positive displacement compressor* is a compressor that compresses a fixed quantity of air with each cycle. Compressors generally have fixed operating speeds and constant pumping rates. Positive displacement compressors include piston, helical screw, and vane compressors.

Tech Tip

Compressor air intake filters must be kept clean because a dirty intake filter decreases a compressor's efficiency and performance.

Piston Compressors. A *piston compressor* is a compressor in which air is compressed by reciprocating pistons. To *reciprocate* is to move forward and backward alternately. Piston compressors, also referred to as reciprocating compressors, may be single-stage or multistage. A *single-stage compressor* is a compressor that uses one piston to compress air in a single stroke before it is discharged. A *multistage compressor* is a compressor that uses two or three cylinders, each with a progressively smaller diameter, to produce progressively higher pressures.

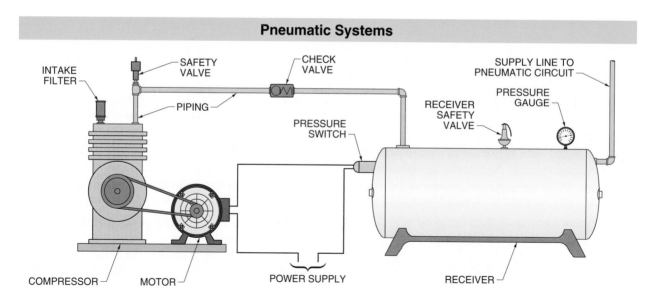

Pneumatic Systems

INTAKE
FILTER

SAFETY
VALVE

CHECK
VALVE

SUPPLY LINE TO
PNEUMATIC CIRCUIT

PIPING

PRESSURE
GAUGE

RECEIVER
SAFETY
VALVE

PRESSURE
SWITCH

COMPRESSOR

MOTOR

POWER SUPPLY

RECEIVER

Figure 19-3. A basic pneumatic system consists of a compressor, receiver, pressure switch, piping, check valve, receiver safety valve, pressure gauge, and a circuit to perform the work required.

Piston compressors consist of a crankcase, cylinder(s), crankshaft, connecting rod(s), piston(s), piston rings, and an inlet and outlet valve. **See Figure 19-4.** The reciprocating motion of the piston fills the cylinder and compresses the air with each alternation. Pistons are generally driven in a reciprocating motion by a crankshaft. A *crankshaft* is a shaft that has one or more eccentric surfaces that produce a reciprocating motion when the shaft is rotated. An *eccentric surface* is a surface that has a different center than the center of the crankshaft. A section of a crankshaft that centers on a different axis than the shaft is said to be eccentric or to have offset. The crankshaft is connected to the piston by the connecting rod. A *connecting rod* is the rod that connects the crankshaft to the piston. A crankshaft may be driven by a motor, a gasoline engine, or another prime mover.

Piston Compressors

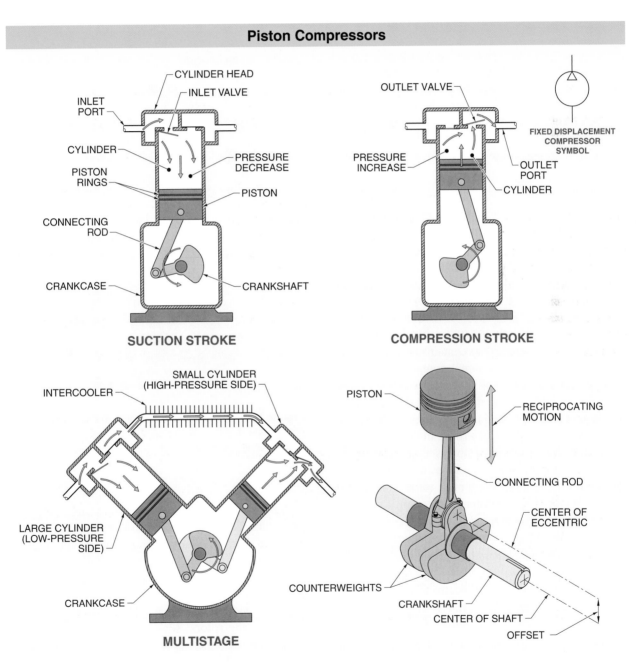

Figure 19-4. Air is compressed as a reciprocating piston draws air in on one stroke and pushes it out under pressure on its alternating stroke.

As the crankshaft rotates, the piston reciprocates within the cylinder. An increasing volume is produced in the cylinder as the crankshaft pulls the piston downward. At the end of the stroke, the cylinder is filled with air and the intake valve closes. The piston compresses the air as it moves upward. The discharge (outlet) valve opens to force the air to the next stage or to a receiver when the air pressure reaches a high pressure.

Staging is required to move compressed air to higher pressures. *Staging* is the process of dividing the total pressure among two cylinders by feeding the outlet from the first large (low-pressure) cylinder into the inlet of a second small (high-pressure) cylinder. The air compressed by the first cylinder is boosted to a higher pressure. A connecting intercooler is required between each stage to increase compressor efficiency and to lower the air temperature at the outlet of the first stage.

Helical Screw Compressors. A *helical screw compressor* is a compressor that contains meshing screw-like helical rotors that compress air as they turn. The meshing rotors (screws) draw air in at one end of a close-fitting chamber. The air is forced along the rotors as they rotate. The air trapped between the rotors is compressed as the size of the cavities between the rotors is progressively reduced. Airflow is positive and continuous because the air is constantly being drawn in and forced axially along the rotors. **See Figure 19-5.**

Helical screw compressors contain one, two, or three rotors. Two-rotor helical screw compressors are the most common in industry. A two-rotor helical screw compressor contains one male and one female rotor, which generally have four and six lobes, respectively. A *lobe* is the screw helix of a rotor. In the operation of a two-rotor helical screw compressor, air entering the chamber is compressed as the lobes of the male rotor mesh with the lobes of the female rotor. This pushes the trapped air along the rotors and compresses it. Air is progressively compressed until the lobes pass the outlet port, discharging the compressed air.

Helical screw compressors may have dry or oil-flooded compressing mechanisms. Dry screw compressing mechanisms compress air without the lobes of the rotors contacting each other as they rotate. The constant and close separation is accomplished by both rotors being rotated by a set of timing gears that are driven by the prime mover. Prime movers normally rotate the rotors at speeds between 3000 rpm and 12,000 rpm. At these relatively high speeds, the rotors turn freely with a

carefully controlled clearance between the rotors and the housing. The clearance is protected by a light film of oil. Lubrication is not a major factor and screw life is long because there is no contact between the meshed rotors. Dry screw mechanisms are used in applications where oil-free air is required, such as instrumentation, paint spraying, clean rooms, etc.

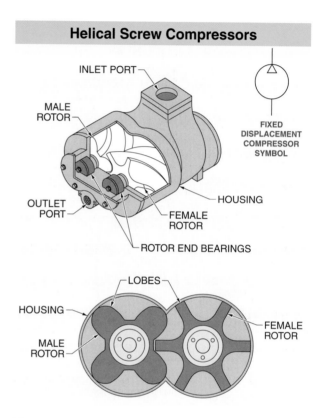

Figure 19-5. Air is compressed in a helical screw compressor as the size of the cavities between the rotors is progressively reduced.

Oil-flooded compressing mechanisms compress air with the lobes of the rotors contacting each other. The male rotor is driven by the prime mover and meshes with and drives the female rotor. This driving motion causes surface contact between the rotors, which must be well-lubricated. Lubrication is accomplished by injecting oil through internal passages to the rotors and rotor end bearings. **See Figure 19-6.** The oil bath lubricates the rotors, seals the rotor clearances for high-compression efficiency, and absorbs the heat of compression. This results in low outlet air temperature. The oil is separated from the compressed air, filtered, cooled, and returned to the compressor for reuse.

Oil-Flooded Compressing Mechanism

SIGHT FLOW
OIL RETURN LINE

AIR INLET

HELICAL SCREW
COMPRESSOR

OIL
COOLER

CHECK
VALVE

AIR/OIL
OUTLET

OIL
SUMP

AIR/OIL
SEPARATOR

OIL
FILTER

UNLOADER
SOLENOID
VALVE

AIR
RECEIVER

CONTROL
CENTER

AIR PRESSURE
SWITCH

	AIR
	AIR/OIL
	OIL

Figure 19-6. Oil-flooded compressing mechanisms require oil injected through internal passages to the rotors and rotor end bearings.

Gast Manufacturing Company

Rotary vane compressors are available in oilless or lubricated models and are used in food processing equipment and air bearings.

Vane Compressors. A *vane compressor* is a positive-displacement compressor that has multiple vanes located in an offset rotor. The vanes form a seal as they are forced against the cam ring. The offset of the rotor in the cam ring produces different distances between the rotor and cam ring at different points inside the compressor. As the rotor rotates, its offset position allows the vanes to slide out and draw air from the inlet port. As the rotor continues to rotate, the volume between the vanes and the cam ring decreases, pushing the vanes into their slots

in the rotor. The decreasing volume compresses the air and forces it out of the outlet port. Compressor vanes are normally made of high-temperature metal and are spring-loaded to ensure outward sliding and contact with the housing. **See Figure 19-7.** Vane compressors may be either single-stage (up to 50 psi) or two-stage (50 psi to 125 psi).

Vane Compressors

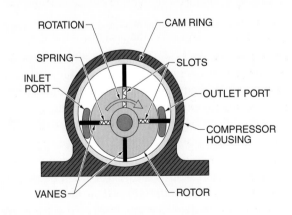

ROTATION

CAM RING

SPRING

SLOTS

INLET
PORT

OUTLET PORT

COMPRESSOR
HOUSING

VANES

ROTOR

Figure 19-7. Vane compressors use sliding vanes to trap inlet air and compress it as the volume between the vanes decreases.

Pressure Control

The compressor in a pneumatic system must be able to maintain the required (set) level of pressure at all times. Extensive damage occurs to the receiver, motor, or compressor if the compressor is allowed to continue compressing air beyond the system's high-pressure limit.

A pressure compensator is used to avoid damage to a system using a vane compressor. A *pressure compensator* is a displacement control that alters displacement in response to pressure changes in a system. A pressure-compensated vane compressor consists of a pressure ring, a pressure-adjustment spring, and a thrust block. As the pressure builds within the compressor to the set level, the pressure ring is forced to a center position, where compression ceases. The pressure level is set by a pressure compensator adjustment screw, which adjusts the spring force required to center the pressure ring. A pressure switch or unloading valve is used to prevent damage to a piston or helical screw compressor. **See Figure 19-8.**

A *pressure switch* is a device that senses a high- or low-pressure condition and relays an electrical signal to turn the compressor motor ON or OFF. The motor is shut OFF when the receiver pressure reaches its preset maximum. The motor does not restart until the preset minimum pressure is reached. Pressure switch operation works well in applications involving intermittent air demand or low air consumption.

Excessive stopping and starting of a compressor motor overheats and burns up the motor. Generally, a compressor motor should not be cycled ON and OFF more than three times per hour. A system with an unloader must be used if the air consumption is high enough to create excessive stopping and starting.

An *unloading valve* is a device that senses a high-pressure condition and removes the compression energy. An unloading valve is placed in the compressor system to allow the motor to continue to run even after the high-pressure setting has been reached. Releasing the compression energy is accomplished by closing the compressor's inlet valve to prevent flow through the compression chamber, holding the inlet valve open. Each unloading valve is operated by pilot pressure from the receiver to a piston within the unloading valve and allows the prime mover to remain operational without building pressure. Unloading

valves are used in applications where the air demand is high, constant, or both.

Safety Relief Valves

Pressure developed by a compressor is designed to be regulated by the system's pressure control system. However, in emergencies such as the failure of the pressure control system, pressure may build to a dangerous level. To relieve unsafe overpressure, safety relief valves are placed at the compressor and receiver. A *safety relief valve* is a device that prevents excessive pressure from building up by venting air to the atmosphere.

Safety relief valves operate as normally closed valves with spring-loaded poppets. The poppet is moved off its seat when the force of the air on the poppet becomes greater than the spring force. The undesirably high air pressure is exhausted through the valve vent port. **See Figure 19-9.**

Safety relief valves are designed strictly for safety and are not intended for frequent operation. However, safety valves do require periodic maintenance. Maintenance of safety valves consists of verifying that the valve can move freely. On most valves, this is done by moving the test lever or by pulling a ring to unseat the poppet. The valve vent port normally remains clean from the air's exhausting velocity if regular testing begins when the valve is new. However, a dirty and untested valve may leak after being tested for the first time if it is allowed to sit and accumulate dust, dirt, and oil.

Tech Tip

The oil used in a compressor is not the same oil used to lubricate pneumatic valves. If air is allowed to carry compressor oil vapor to a valve, the valve could varnish, causing it to stick. Oil supplied by a lubricator is transported by gravity and the working air to various points in a system.

Pneumatic System Piping

Piping in a pneumatic system begins at the compressor and runs to each component so energy can be transmitted by use of the compressed air. As the air is used throughout the system, it is returned (exhausted) to the atmosphere. For this reason, pneumatic system piping does not return to the beginning stage (compressor) like hydraulic system piping. Pneumatic system piping must be installed to minimize pressure loss and air leaks and to provide liquid drainage.

Pressure Control

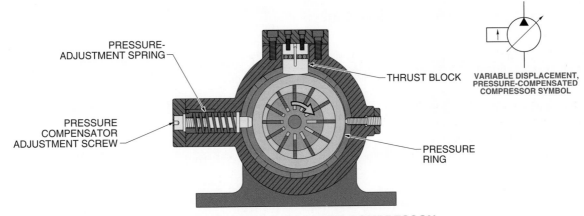

PRESSURE-COMPENSATED VANE COMPRESSON

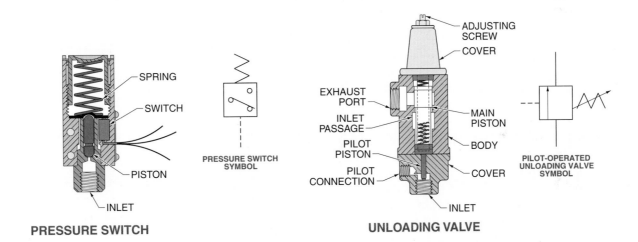

PRESSURE SWITCH

UNLOADING VALVE

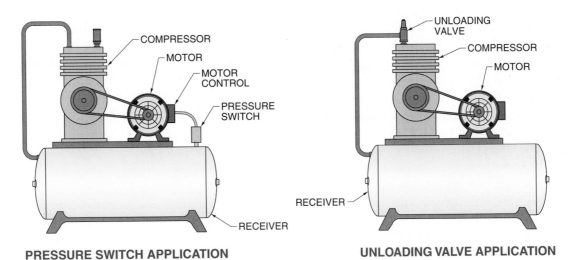

PRESSURE SWITCH APPLICATION

UNLOADING VALVE APPLICATION

Figure 19-8. Continuous compressor pressure buildup is regulated by a pressure compensator, pressure switch, or unloading valve.

Safety Relief Valves

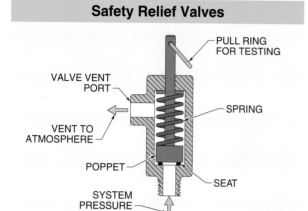

Figure 19-9. Safety relief valves operate when system pressure builds high enough to overcome spring pressure, which pushes a poppet off its seat.

Pressure Loss. Pressure loss in a pneumatic system is a result of the resistance created within the system and circuit components, the work load demands, and the size and length of pipe and pipe fittings. Pressure loss is also created during system operation by moving controls such as valves. Pressure loss occurs only while air is moving through the piping. The maximum pressure drop in a system should be less than 10% over the system operating pressure. Pressure loss figures are taken from pressure loss constants tables when determining total system pressure loss resulting from friction through lengths of pipe, pipe fittings, and other components such as valves. Pressure loss constants tables are also used to determine if the system has the required capacity. **See Appendix.**

Calculating the total pressure drop in a pneumatic circuit from piping, fittings, and other components is determined by calculating the individual pressure drops in each pipe, fitting, and component and adding these values together. The individual pressure drops in pipe, fittings, and components are calculated from the flow rate, pressure, and constant based on the component. Pressure loss in a component is found by applying the formula:

$$\Delta P = \frac{CQ^2}{1000} \times \frac{14.7}{14.7 + P}$$

where

ΔP = pressure drop (in psi)

C = constant (from pressure loss constants table)

Q = air flow rate (scfm)

14.7 = constant (atmospheric pressure)

1000 = constant

P = working pressure (in psi)

For example, what is the pressure drop in a pneumatic system having a working pressure of 100 psi, an air flow rate of 50 scfm, and containing 100′ of 1″ Schedule 40 pipe, one 1″ gate valve, three 1″ 90° elbows, and one 1″ 50µ filter?

1. Calculate pressure drop for 1″ pipe. *Note:* C value for 1″ pipe = 1.66 (from pipe pressure loss constants table).

$$\Delta P = \frac{CQ^2}{1000} \times \frac{14.7}{14.7 + P}$$

$$\Delta P = \frac{1.66 \times 50^2}{1000} \times \frac{14.7}{14.7 + 100}$$

$$\Delta P = \frac{1.66 \times 2500}{1000} \times \frac{14.7}{114.7}$$

$$\Delta P = \frac{4150}{1000} \times 0.128$$

$$\Delta P = 4.15 \times 0.128$$

$$\Delta P = \mathbf{0.531\ psi}$$

2. Calculate pressure drop for 1″ gate valve. *Note:* C value for 1″ gate valve = 0.018 (from pipe fitting pressure loss constants table).

$$\Delta P = \frac{CQ^2}{1000} \times \frac{14.7}{14.7 + P}$$

$$\Delta P = \frac{0.018 \times 50^2}{1000} \times \frac{14.7}{14.7 + 100}$$

$$\Delta P = \frac{0.018 \times 2500}{1000} \times \frac{14.7}{114.7}$$

$$\Delta P = \frac{45}{1000} \times 0.128$$

$$\Delta P = 0.045 \times 0.128$$

$$\Delta P = \mathbf{0.006\ psi}$$

3. Calculate pressure drop for three 90° 1″ elbows. *Note:* C value for one 90° 1″ elbow = 0.043 (from pipe fitting pressure loss constants table). Total pressure drop for three 90° elbows = 0.129 (0.043 × 3 elbows = 0.129).

$$\Delta P = \frac{CQ^2}{1000} \times \frac{14.7}{14.7 + P}$$

$$\Delta P = \frac{0.129 \times 50^2}{1000} \times \frac{14.7}{14.7 + 100}$$

$$\Delta P = \frac{0.129 \times 2500}{1000} \times \frac{14.7}{114.7}$$

$$\Delta P = \frac{322.5}{1000} \times 0.128$$

$$\Delta P = 0.323 \times 0.128$$

$$\Delta P = \mathbf{0.041\ psi}$$

4. Calculate pressure drop for a 1″ 50μ filter. *Note:* C value for a 1″ 50μ filter = 0.20 (from filter pressure loss constants table).

$$\Delta P = \frac{CQ^2}{1000} \times \frac{14.7}{14.7 + P}$$

$$\Delta P = \frac{0.20 \times 50^2}{1000} \times \frac{14.7}{14.7 + 100}$$

$$\Delta P = \frac{0.20 \times 2500}{1000} \times \frac{14.7}{114.7}$$

$$\Delta P = \frac{500}{1000} \times 0.128$$

$$\Delta P = 0.0 \times 0.128$$

$$\Delta P = \mathbf{0.064\ psi}$$

5. Find total system pressure drop by adding each pressure drop from piping and components.

$$\Delta P = 0.531 + 0.006 + 0.041 + 0.064$$

$$\Delta P = \mathbf{0.642\ psi}$$

Selecting slightly oversized pipe adds a safety margin and allows for additional demands on a system. Also, to ensure that an adequate supply of air is available, pressure loss may be calculated when the purchase of additional equipment is anticipated.

Clippard Instrument Laboratory, Inc.

Needle valves are used in pneumatic systems to control actuator speeds by the adjustment of a tapered needle.

Leaks. Leaks in a pneumatic system are very costly, and most leaks in a facility go unnoticed. In most cases, large savings are realized when a company takes the time to look, listen, and document major leaks within the facility. Leaks that cannot be heard may be located by brushing soapy water on each threaded or compression fitting. The presence of bubbles indicates a leak. Repairs are scheduled starting with the most major leak. Cost savings from fixing air leaks can be substantial. For example, a 1/32″ hole in a system having an initial pressure of 100 psig loses 1.6 standard cubic feet per minute (scfm). The annual cost savings from repairing this hole, based on $0.25/1000 scfm, is $210.24 (1.60 × 60 × 24 × 365 = 840,960 cu ft per year. [840,960 ÷ 1000] × 0.25 = $210.24). **See Figure 19-10.**

Pneumatic System Air Loss*

Initial Pressure†	Hole Size‡				
	1/64	1/32	1/16	1/8	1/4
40	0.195	0.775	3.10	12.4	50.0
60	0.265	1.05	4.25	17.0	68.0
80	0.335	1.35	5.35	21.5	85.5
100	0.405	1.60	6.50	26.0	104.0
125	0.495	2.0	7.90	31.5	126.0

* in scfm
† in psig
‡ in in.

Air Leak Costs*

Hole Size†	Annual Cost‡
1/64	53.22
1/32	210.24
1/16	854.10
1/8	3416.40
1/4	13,665.60

* @100 psig based on $0.25/1000 scfm, 24 hr/day, 52 weeks/year
† in in.
‡ in dollars

Figure 19-10. In most cases, large savings are realized by fixing major leaks within a facility.

Liquid Drainage. Draining liquid is generally accomplished by installing the main header at a downward pitch of 1″ for every 10′. A *main header* is the main air supply line that runs (generally overhead) between the receiver and the circuits in a pneumatic system. The main header should terminate with a moisture water leg and drain valve. To prevent moisture from draining into a circuit branch line, each branch line is connected to the top of the main header or branch header. Moisture water legs should also be placed at locations in the system where the header passes through areas of low temperatures because water drops out through condensation at these locations. **See Figure 19-11.**

Liquid Drainage

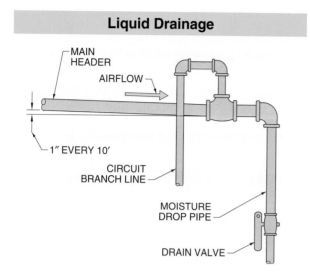

Figure 19-11. Headers must have the proper slope so that moisture that condenses within the header flows to a moisture water leg and drain valve.

Pipe Installation. A pneumatic system graphic diagram shows the layout of pneumatic system piping. The graphic diagram shows the main header from the receiver and the feeder lines that supply individual circuits. The graphic diagram also shows the location of the various components used to condition (filter, regulate, lubricate, etc.) the air in the individual circuits. Pneumatic system piping should be installed in a two-way loop around a plant whenever possible. The two-way loop provides two paths for airflow. This prevents significant pressure drop in the circuits at the end of the main header. **See Figure 19-12.**

Piping Materials. Standard piping is generally used for pneumatic system main headers and feeder lines. Pipe material and connections are normally determined by the working load. Generally, pipe over 6″ in diameter has welded joints, pipe between 3″ and 6″ has bolted flange connections, and pipe under 3″ is threaded. Rigid feeder lines should be connected to the main header or feeder lines as close to the point of use as possible.

A thread-lubricating material must be used to prevent leaks when threaded connections are used. Thread-lubricating material should be placed on the male fitting only. The first two lead-in threads must be left bare to allow for thread starting and to prevent contamination from pipe dope or Teflon tape.

Copper, nylon, or plastic tubing may be used to pipe individual circuits. Copper tubing is used because it resists abrasion and heat damage. Nylon or plastic tubing is normally used for circuits because it is easy to work with, can be cut to length with a sharp knife and installed quickly and easily, and, in most cases, can be used with operating pressures up to 200 psi. Also, the cost of nylon or plastic tubing is considerably less than the cost of copper tubing.

Check Valves

A *check valve* is a valve that allows flow in only one direction. A check valve is used in a pneumatic system to prevent the flow of stored compressed air in the receiver from flowing back into the compressor.

A check valve consists of a body with a primary (inlet) port and a secondary (outlet) port. A ball or poppet is held against the inlet port by a spring. Air pressure at the inlet port that is greater than the spring pressure moves the ball or poppet off of its seat, thus allowing airflow. The ball or poppet is held against the inlet port by spring and air pressure if air attempts to flow in the opposite direction. **See Figure 19-13.**

An example of a check valve application is a hand tire pump used to pump air into bicycle tires. Air is free to flow through the hose and into the tire when the handle is pushed down. The check valve prevents air from flowing out of the tire when the handle is lifted to draw air into the cylinder.

Saylor-Beall Manufacturing Company

Air compressors are used in water pumping station installations to supply compressed air for air tools, valve actuation, and air motors.

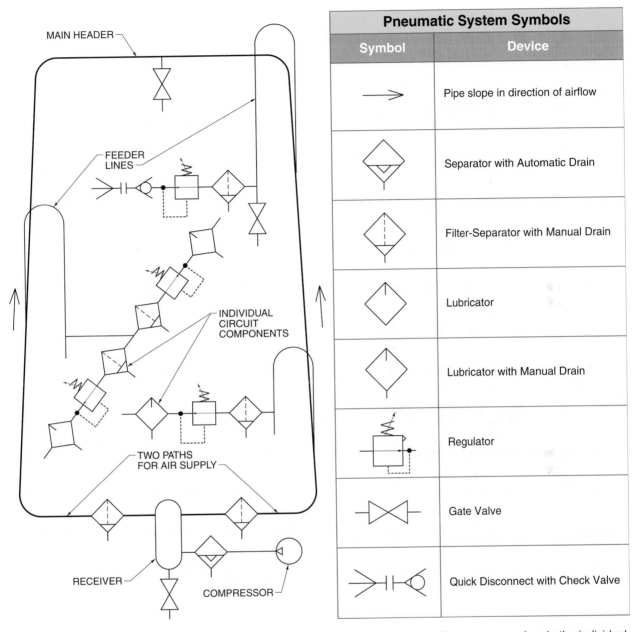

Pneumatic System Symbols	
Symbol	**Device**
→	Pipe slope in direction of airflow
	Separator with Automatic Drain
	Filter-Separator with Manual Drain
	Lubricator
	Lubricator with Manual Drain
	Regulator
	Gate Valve
	Quick Disconnect with Check Valve

Figure 19-12. A two-way looped pneumatic system piping installation prevents significant pressure drop in the individual circuits.

PNEUMATIC CIRCUIT COMPONENTS

Pneumatic circuit components use stored compressed air from a pneumatic system to support the work process or to perform the actual work. Work process supporting components condition, control, or direct the air. The symbols for most of the components used in a pneumatic circuit are similar to those used in hydraulic circuits. Actuators perform the actual work. An *actuator* is a device that transforms fluid energy into linear or rotary mechanical force.

Conditioning Compressed Air

Conditioning compressed air in a pneumatic system begins at the compressor, where the air is filtered and cooled and moisture is removed. Cooling and removing the moisture at this point in a pneumatic system is the minimum required for proper compressed air storage. Additional air conditioning is required at each circuit in the system. System air conditioning consists of filtering and lubricating the air according to the requirements of the individual circuits. To be effective, air conditioning devices must be of the proper type and size and must be installed and maintained correctly.

Check Valves

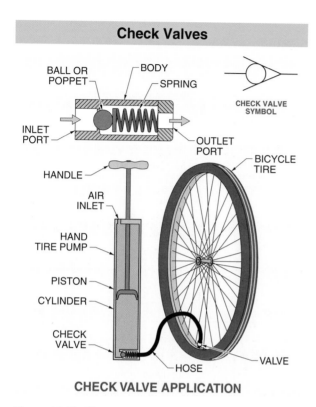

Figure 19-13. Check valves are used in a pneumatic system to allow air to flow in one direction.

Filters. A *filter* is a device containing a porous substance through which a fluid can pass but particulate matter cannot. **See Figure 19-14.** Ideally, filters, or a combination of filters, should remove all solids and liquids from the air. In many pneumatic circuits, a particulate filter, which removes solid particles of 5µ and smaller, is placed just ahead of a coalescing filter. The coalescing filter removes oil and moisture. A screen filter should be placed before the particulate filter if the air is extra dirty and the particulate filter requires frequent changing. Filters must be changed or cleaned per manufacturer specifications.

Tech Tip

An intake filter acts as a silencer for air as it rushes into a compressor. Filters are required in a pneumatic system because polluted air may also contain undesirable gases which, when mixed with the moisture found in a pneumatic system, could be corrosive.

Filters

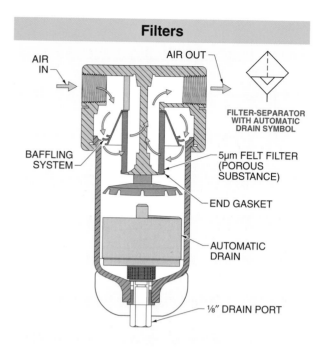

Figure 19-14. A filter contains a porous substance through which air can pass but particulate matter cannot.

Lubricators. A *lubricator* is a device that injects atomized oil into the air sent to pneumatic components. Proper lubrication of moving components is generally accomplished by atomizing light oil into the air stream. Lubricators use an orifice to create a pressure differential in the device. The pressure differential creates a siphon effect on the feeder tube. The siphon effect draws the oil up to the drip tube, where it is dripped into the air stream. **See Figure 19-15.** Oil flow is metered by an oil adjustment screw. The oil adjustment screw controls the number of oil drops that are released to be atomized into the system air.

Oil flow must be regulated because under- or over-lubrication may be a problem. Depending on the actuator requirements, begin by using 1 drop for every 10 cfm. This may be regulated up or down based on actuator requirements. Generally, a light-grade oil is satisfactory for lubricator atomizing. The grade of oil selected must atomize properly for the correct lubrication of the components.

Lubricators should be placed downstream from the filter and should be as close as possible to the components being lubricated. Lubricators should be placed no more than 10′ from the lubricated components because atomized oil begins to drop out of the air beyond 10′.

Lubricators

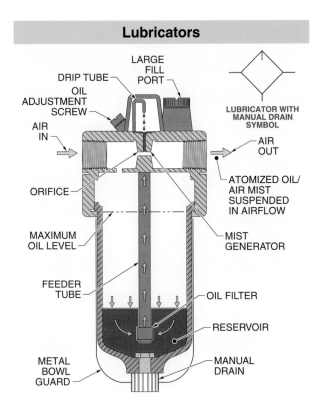

Figure 19-15. A lubricator is a device that injects atomized oil into the air sent to pneumatic components.

Controlling and Directing Air

Pneumatic circuit controls are valves designed to control the pressure, direction, and flow of air through a circuit. Pneumatic circuit valves include pressure regulators, directional control valves, and flow control valves. Also, special valves are used as switching devices to control time or sequence of events. Control valves are operated by system or circuit pressure or by the exhaust from another valve. The operation of one valve from the exhaust of another is a form of pilot operation.

Pressure Regulators. A *pressure regulator* is a valve that restricts and/or blocks downstream airflow. Pressure regulators (pressure-reducing valves) are used in a pneumatic circuit to provide a constant and proper air pressure to pneumatic components. Pressure regulators generally control circuit pressures from 0 psi to 150 psi, depending on the circuit's maximum pressure and application. Pressure regulators operate on the pressure differential between the downstream pressure, the regulating spring force, and the upstream pressure. The upstream pressure and the regulating spring force

equal the downstream pressure. A pressure regulator always adjusts to a balanced pressure (equilibrium). Pressure regulators may be diaphragm or piston design. **See Figure 19-16.**

Pressure Regulators

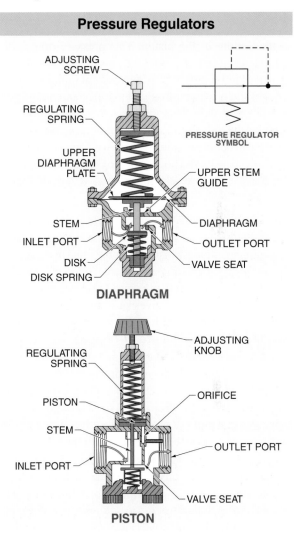

Figure 19-16. Pressure regulators provide constant and proper pressure to pneumatic components.

A diaphragm pressure regulator uses a metallic (bronze) or nylon reinforced rubber diaphragm to sense a pressure differential between a regulating spring and a disc spring. The regulating spring exerts a force on the upper side of the diaphragm. The diaphragm is connected to the disc by a stem. An increase in outlet pressure increases the upward force on the diaphragm. This closes or reduces the flow through the regulator passageway. A decrease in outlet pressure reduces the upward pressure on the bottom of the diaphragm. This opens or increases the flow through the regulator passageway.

Piston regulators use a piston to sense the pressure differential between the valve's inlet and outlet pressures. The regulator body acts as a cylinder for the piston. Pressure from the regulating spring forces the stem to open the valve seat and allow airflow. Downstream pressure is sensed through the orifice and is applied to the bottom of the piston. This pressure offsets the regulating spring and inlet air pressure to maintain a constant outlet pressure. A decrease in outlet pressure reduces the pressure through the orifice and on the bottom of the piston. This increases the spring pressure, opening the valve seat, allowing increased airflow. An increase in outlet pressure increases the pressure through the orifice and on the bottom of the piston. This reduces the spring pressure, closing the valve seat, allowing reduced airflow.

Directional Control Valves. Directional control valves direct the flow of air to an actuator or another valve in a pneumatic circuit. Most directional control valves are 2-way, 3-way, or 4-way valves. A *way* is a flow path through a valve. Two-way valves have two main ports for airflow. Three-way valves have three main ports for airflow. Four-way valves have four main ports (possibly five) for airflow. **See Figure 19-17.**

Directional control valves are placed in different positions to start, stop, or change the direction of fluid flow. A *position* is the specific location of a spool within a valve which determines the direction of fluid flow through the valve. A 2-position valve can be placed in two positions and a 3-position valve can be placed in three positions.

A 3-way, 2-position, solenoid-operated, spring-return directional control valve may be used to activate a single-acting cylinder. The valve directs air to the cap end of the cylinder when the solenoid is energized. De-activating the solenoid causes the spring to shift the valve spool, exhausting the air in the cylinder to the atmosphere. **See Figure 19-18.**

A 3-way, 3-position, manually-operated, spring-centered directional control valve may be used to control a single-acting cylinder. The left position extends the cylinder and right position retracts the cylinder. The center position allows the cylinder to be stopped and held in any position between fully extended and fully retracted.

Tech Tip

A constant air leak at the vent hole of a piston regulator is an indication of damaged or worn piston seals that require replacement.

Valve Ways

Standard Port Markings

P = PRESSURE
E = EXHAUST
A = OUTLET TO/FROM ACTUATOR
B = OUTLET TO/FROM ACTUATOR
EA = EXHAUST
EB = EXHAUST

AIR OUT (TO LOAD) — A

POSITION STOPS AIRFLOW — POSITION ALLOWS AIRFLOW

P

AIR IN

2-WAY

POSITION BLOCKS COMPRESSOR AIRFLOW BUT ALLOWS EXHAUST AIRFLOW — AIR FROM LOAD

A — POSITION ALLOWS COMPRESSOR AIRFLOW

P   E

BLOCKED — AIR OUT (EXHAUST)

3-WAY

OUTPUT 1 — OUTPUT 2

POSITION ALLOWS COMPRESSOR AIRFLOW TO OUTPUT 1. OUTPUT 2 IS CONNECTED TO EXHAUST — A   B — POSITION ALLOWS COMPRESSOR AIRFLOW TO OUTPUT 2. OUTPUT 1 IS CONNECTED TO EXHAUST

P   E

AIR IN — AIR OUT (EXHAUST)

4-WAY

Figure 19-17. Directional control valves have 2, 3, or 4 ways to direct the flow of air to an actuator or another valve in a pneumatic circuit.

Clippard Instrument Laboratory, Inc.

The shifting of the spool in a directional control valve determines the ports to which fluid flows.

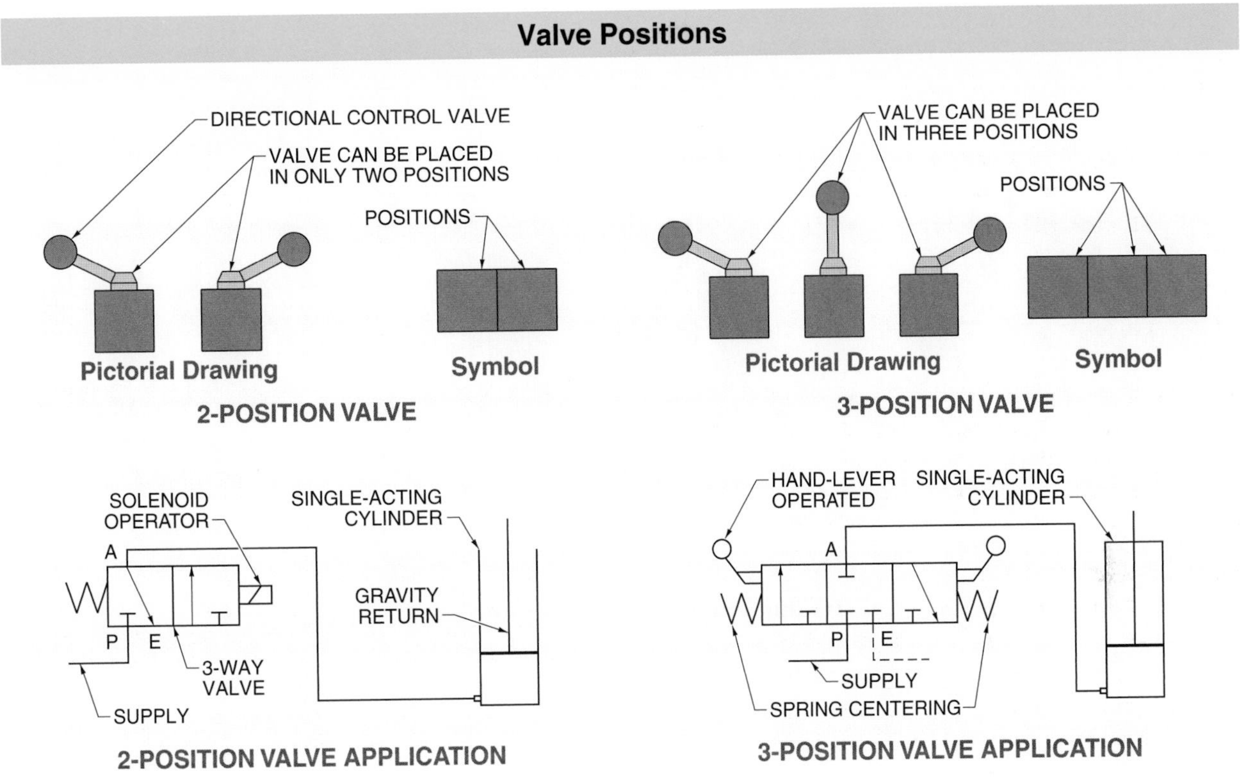

Valve Positions

Figure 19-18. Directional control valves are placed in different positions to start, stop, or change the direction of fluid flow.

Directional control valves may be operated electrically, mechanically, manually, or by pilot operation and may be normally open or normally closed. Normally open valves allow flow between the inlet and outlet ports when the valve operator is not energized. Normally closed valves require the valve operator to open a path between the inlet and outlet ports.

Most directional control valves are 2- or 3-position valves. Two-position valves have two positions in which the spool can be placed. These positions are referred to as the extreme positions. Three-position valves have a center (neutral) position in addition to the two extreme positions. The neutral position is generally the deactivated position, where the internal spool is normally centered by spring action on both ends of the spool.

The neutral position produces various functions based on the design of the valve spool. For example, one neutral function has all ports blocked. This function, known as the closed center supply, allows for infinite positioning of a cylinder. In this case, the cylinder remains in its last actuated position when the operator is first activated and deactivated because the air is not allowed to exhaust.

Another neutral function is the open center supply design. In the open center supply design, the neutral ports from the supply to the outputs are open. This allows the supply pressure to hold the outputs in their actuated positions. In the open center exhaust design, the neutral ports from the outputs are to the exhausts and are open, allowing the cylinder ports to exhaust while blocking the inlet port. **See Figure 19-19.**

Pneumatic circuits using 4-way directional control valves are the most common in industry. A 4-way, 2-position, hand-lever operated, spring-return directional control valve may be used to activate a double-acting cylinder in both directions. In this circuit, the hand-lever operator fully extends or retracts the piston rod depending on the position of the operator. **See Figure 19-20.**

A 4-way, 3-position, hand-lever operated, spring-centered directional control valve may be used with a double-acting cylinder for infinite positioning. As the hand lever is operated, the piston rod extends or retracts. The piston remains in its present location whenever the lever is released and the springs center the valve spool. This use is similar to positioning forks on a forklift.

Valve Center Positions

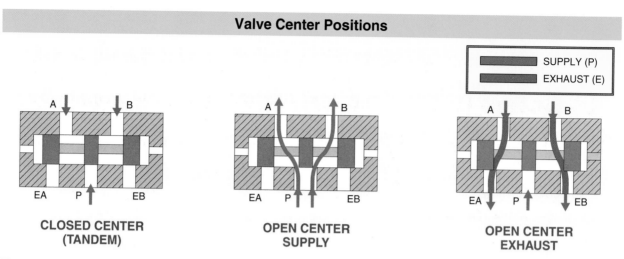

Figure 19-19. Center positions of directional control valves are designed to exhaust, block, or allow inlet or exhaust to travel to another component.

Four-Way Valve Applications

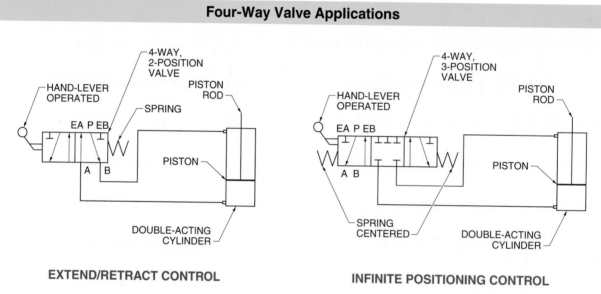

Figure 19-20. Pneumatic circuits using 4-way, 5 ported directional control valves are commonly used in industry to control the operation of double-acting cylinders.

Directional control valves may be controlled electrically by a solenoid. A *solenoid* is a device that converts electrical energy into a linear, mechanical force. The mechanical force in a solenoid is created by a magnetic field that is set up by the flow of electric current through a coil of wire. In pneumatic circuits, solenoids are used to allow or prevent (open/close) airflow in 2-way valves or control the position of the spool in 3-way valves. **See Figure 19-21.**

In 2-way valves, solenoids are used to control the operation of a plunger to open or close ports. This produces a flow or no flow condition. In 3-way valves,

solenoids may directly move the spool of the main valve (solenoid-operated) or move the spool of a pilot valve (pilot-operated). In a 3-way solenoid-operated valve, an electrical signal to the solenoid pushes the solenoid rod, which shifts the main valve spool. This controls the flow of air to the outlet ports of the main valve. In a 3-way, solenoid-controlled, pilot-operated valve, an electrical signal to the solenoid pushes the solenoid rod, which shifts the spool of a pilot valve attached to the main valve. The movement of the pilot valve spool allows airflow to shift the spool of the main valve.

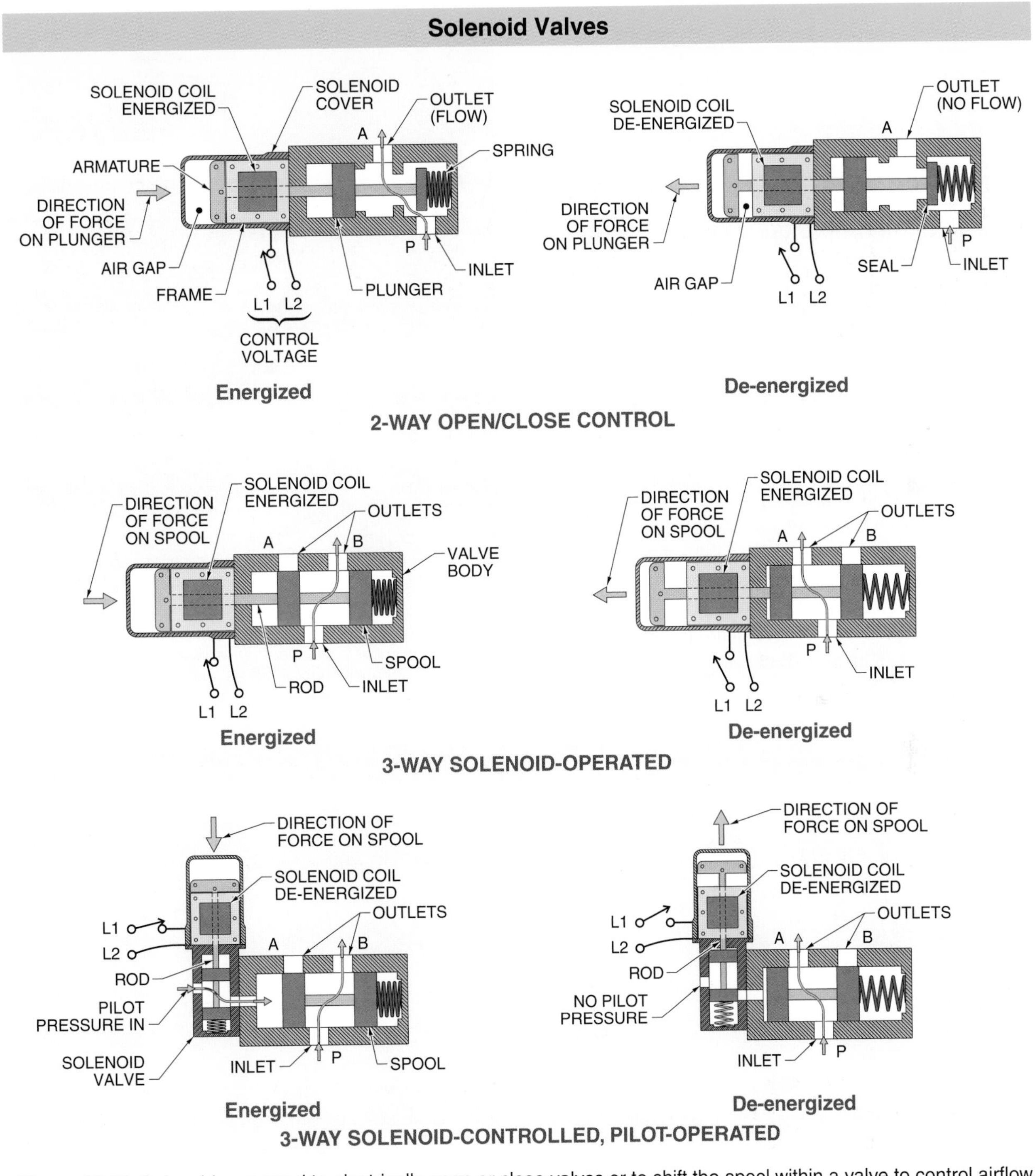

Solenoid Valves

Figure 19-21. Solenoids are used to electrically open or close valves or to shift the spool within a valve to control airflow.

Flow Control Valves. A *flow control valve* is a valve whose primary function is to regulate the rate of fluid flow. Flow control valves, sometimes referred to as needle valves, are normally used for metering airflow to control motor speed, cylinder piston speed, or valve spool shifting speed (for timing). **See Figure 19-22.**

Cylinder piston movement or motor speed is controlled precisely and smoothly if the airflow is regulated at the exhaust rather than at the inlet. Regulating exhaust air is used on cylinders under light or no load where the volume of air supplied is less than the amount required to rapidly and smoothly move the piston. Heavy loads may be regulated at the inlet or outlet port as long as rapid speed is not required. Also, a flow control valve can be coupled with a check valve to give regulated flow in one direction and full flow in the reverse direction.

Actuators

An actuator transforms fluid energy into linear or rotary mechanical force. An air cylinder produces linear mechanical force. An air motor produces rotary mechanical force.

Air Cylinders. An *air cylinder* is a device that converts compressed air energy into linear mechanical energy. Pneumatic cylinders operate by air pressure and flow acting on a piston. Work performed is a product of the area of the cylinder bore and the air pressure.

Cylinders are classified as single-acting or double-acting and are manufactured in a variety of diameters, stroke lengths, and mounting arrangements. A *single-acting cylinder* is a cylinder in which fluid pressure moves the piston in only one direction. The piston is returned by spring or gravity force. A *double-acting cylinder* is a cylinder that requires fluid flow for extending and retracting. The major parts of an air cylinder are the cylinder body, ends, piston, piston rod, and seals. **See Figure 19-23.**

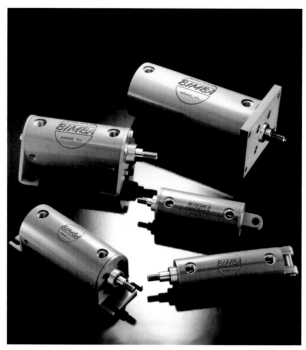

Bimba Manufacturing Company

Double-wall cylinders use easy-to-assemble bolt-on mounting kits to convert one basic cylinder into various National Fluid Power Association (NFPA) mounting styles, such as a front flange, side lugs, pivot, end lugs, and clevis.

Flow Control Valves

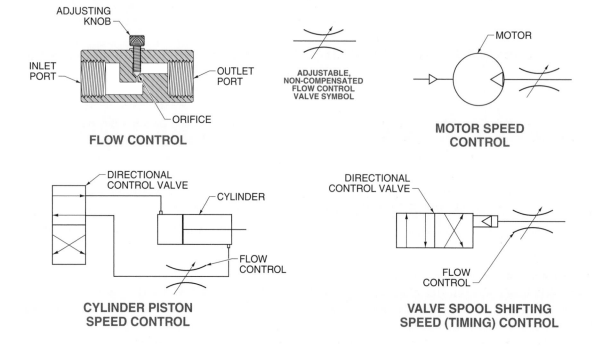

Figure 19-22. Flow control valves use a fine threaded adjusting screw to precisely meter the flow of air within a circuit.

Air Cylinders

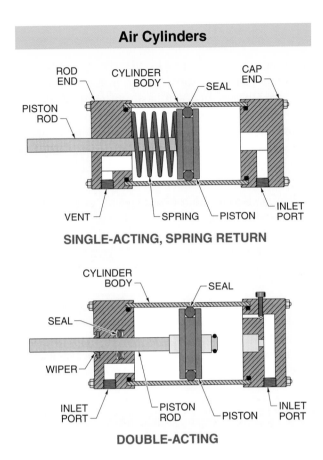

SINGLE-ACTING, SPRING RETURN

DOUBLE-ACTING

Figure 19-23. Air cylinders are actuators that are operated by the force of compressed air in one (single-acting) or two (double-acting) directions.

Air cylinders use various seals to prevent air pressure loss and to protect the internal cylinder parts from outside contaminants. A *seal* is a device in contact between two components that contains pressure and prevents leakage. Seals may be static or dynamic. A *static seal* is a seal used as a gasket to seal nonmoving parts. They are used between two stationary parts that may be taken apart and reassembled. A *dynamic seal* is a seal used between moving parts that prevents leakage or contamination. Dynamic seals are used on pistons and piston rods to allow the piston and rod to slide inside the cylinder. Seals include O-rings, lip seals, wipers, and packing. Seal material may be Teflon, nylon, leather, or rubber. **See Figure 19-24.**

O-rings are the most commonly used seal in pneumatic applications. An *O-ring* is a molded synthetic rubber ring having a round cross section. O-rings are used as static and dynamic seals and may be used in some high-pressure operations. O-rings used in dynamic applications depend on the smoothness of the moving

parts and the closeness of their fit for the best service life. O-rings should have a 10% compression between the cylinder and piston groove walls when installed. As pressure builds, the O-ring becomes distorted in an attempt to completely fill all voids at one end of its groove. This forced distortion becomes the seal under dynamic conditions.

Cylinder Seals

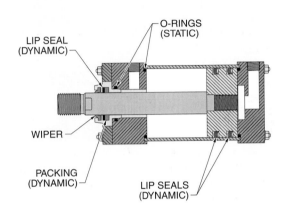

Figure 19-24. Seals are used in cylinders to prevent leakage between moving parts (dynamic seals) or prevent leakage between two immovable parts (static seal).

Pressure that becomes excessive may distort the O-ring enough to squeeze it out of its groove and into any void between the piston and cylinder wall. This may be prevented by using a backup ring between the O-ring and the wall of the piston groove. A backup ring supports the O-ring's distortion and must be installed on the side of the O-ring receiving the least pressure. Backup rings must be installed on both sides of the O-ring if the O-ring receives high pressure in both directions. **See Figure 19-25.**

High-pressure dynamic forces within a cylinder are best contained by using lip seals. A *lip seal* is a seal that is made of a resilient material that has a sealing edge formed into a lip. Lip seals are specifically designed for reciprocating motion and form a tighter seal as pressure increases. A lip seal uses the air's pressure on its lip to form a seal. Lip seals include V-ring and cup seals.

Tech Tip

Abruptly stopping an air cylinder during high speed operation can damage the cap end. To prevent this damage, a cushion is often installed.

Backup Rings

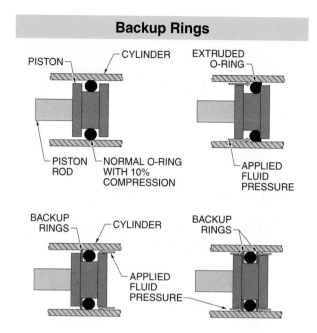

Figure 19-25. Backup rings support O-rings during compression.

A *V-ring seal* is a lip seal shaped like the letter V. V-ring seals are dynamic seals used in high-pressure and severe operating condition applications. A *cup seal* is a lip seal whose lip forms the shape of a cup. Cup seals are used specifically as a piston seal and may be used as a seal for single-acting cylinders. Two cup seals may be used back-to-back in double-acting cylinders. **See Figure 19-26.**

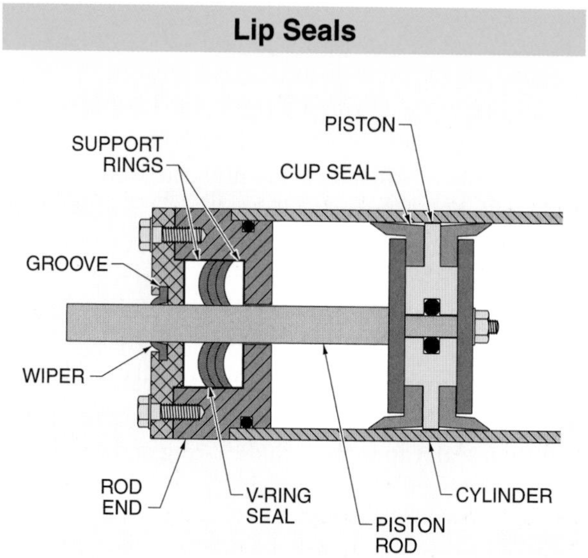

Lip Seals

Figure 19-26. Lip seals are designed for reciprocating motion and form a tighter seal as pressure increases.

A *wiper* is a seal designed to prevent foreign abrasive or corrosive material from entering a cylinder. Wipers are designed with a lip to wipe the rod clean of foreign materials with each stroke of the piston rod. Wipers are normally installed with a slip fit into a machined groove on the outermost portion of the rod end. Wipers protect the end sealing material in addition to removing contamination from the rod. Wipers are made of metal or synthetic material and are not designed to seal against pressure.

Packing is a bulk deformable material or one or more mating deformable elements reshaped by manually adjustable compression. Packing seals the piston rod to prevent air from escaping around the rod. Packing uses various designs and materials to seal in the cylinder pressure.

A *piston cushioning device* is a device within a cylinder that provides a gradual deceleration of the piston as it nears the end of its stroke. Piston cushioning devices help to reduce the shock produced when a piston reaches the end of its stroke. **See Figure 19-27.**

Piston Cushioning

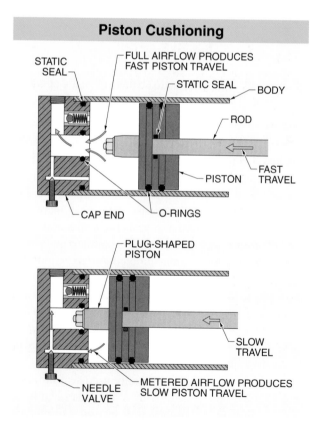

Figure 19-27. Piston cushioning devices are used at the end of a piston stroke to prevent possible damage of the load or cylinder component by slowing piston movement.

A piston cushioning device consists of a plug-shaped piston and a slightly larger bore on the inside of the cylinder end. The piston cushioning device may be seal-less, have a V-ring or O-ring seal on the plug, or have an O-ring set in a machined groove in the rod end. The plug may be on one or both sides of the piston, depending on whether the cylinder is cushioned at one or both ends. The cushioning device operates by trapping a volume of air and compressing it as the piston approaches the end of its stroke. The air is trapped between the piston and the cylinder end as the plug enters the bore. Needle valves installed in the cylinder end(s) are adjusted to allow the trapped air to be metered through the port at the desired rate. This adjustment determines the intensity of the piston cushioning.

High-pressure punching machines or high-pressure forming tools require increased actuator output force. In most cases, increasing the supply pressure to the actuator is sufficient to increase the actuator output force. An intensifier may be added to a circuit if a pressure greater than the supply system pressure is required. An *intensifier (booster)* is a device that converts low-pressure fluid power into high-pressure fluid power. An intensifier uses the difference in the areas of the two cylinders to increase output pressure without an increase in input pressure. Intensifiers normally consist of a large surface area piston (operating piston) connected to a small surface area piston (ram). The pressure exerted by an intensifier is determined by dividing the area of the operating piston by the product of the area of the ram and the system operating pressure. **See Figure 19-28.** Intensifier pressure is found by applying the formula:

$$P_o = \frac{A_c}{A_R} \times P_I$$

where

P_O = outlet pressure (in psi)

A_C = area of operating piston (in sq in.)

A_R = area of ram (in sq in.)

P_I = inlet pressure (in psi)

Tech Tip

Intensifiers enable high force levels to be produced in a low-pressure pneumatic system. Intensifiers are mounted close to the work cylinder so high pressure is in confined to the small part of the circuit between the intensifier and the work cylinder. A pressure relief valve in the high pressure portion of the intensifier circuit helps prevent excessive pressures.

Intensifiers

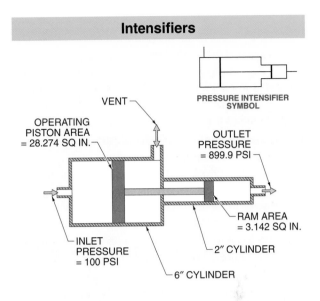

Figure 19-28. An intensifier converts low-pressure fluid power into high-pressure fluid power by using the difference in areas of two pistons to increase output pressure without an increase in input pressure.

For example, what is the outlet pressure produced by a 6″ D operating piston and a 2″ D ram operating at 100 psi? *Note:* the area of the 6″ operating piston = 28.274 sq in. and the area of the 2″ ram = 3.142 sq in.

$$P_o = \frac{A_c}{A_R} \times P_I$$

$$P_o = \frac{28.274}{3.142} \times 100$$

$$P_o = 8.999 \times 100$$

$$P_o = \textbf{899.99 psi}$$

Air Motors. An *air motor* is an air-driven device that converts fluid energy into rotary mechanical energy. In many cases, air motors are selected over electric motors because air motors are two to four times lighter than a direct replacement electric motor. Also, air motors can stall for an indefinite period of time without overheating or burning up. Air motors can be reversed without any strain or shock and rarely break down suddenly if maintained. Air motors normally wear slowly with a gradual reduction in power, allowing for scheduled maintenance. Disadvantages of air motors are that air motors are less efficient than electric motors, air motors slow down as the work load increases, and supplying sufficient air for operating an air motor may be a problem.

The most popular air motor is the vane air motor. A *vane air motor* is an air motor that contains a rotor with vanes that are rotated by compressed air. Vane air motors are simple in design, available in a wide range of sizes, and are easily maintained. Most vane air motors are connected to a gear train because the rotor and vanes rotate at a high speed. The gear train develops greater output torque by reducing the motor speed. Vane air motors are popular for pump motors, portable tools, mixing motors, and air-operated hoists because they are able to produce different speeds or torque from the same input pressure. Rotary vane motors are available in sizes up to 10 HP and capable of speeds of up to 15,000 rpm at operating pressures of 100 psi.

Vane air motor design is similar to a vane compressor. Vane motors develop torque by the air pressure acting on the exposed surfaces of the vanes. The vanes slide in and out of the rotor, which is connected to the drive shaft. **See Figure 19-29.** As the rotor rotates, the vanes follow the surface of the housing due to centrifugal force. Good lubrication is required because of the constant sliding of the vanes and housing. Lubrication may be provided by an in-line lubricator, which injects atomized oil into the air stream. Recommended oil usage for a vane air motor is one drop per minute for every 50 cfm to 75 cfm of airflow. In many cases, air motor manufacturers provide lubrication and flow rate charts.

Air motors are available in lubricated and oilless models for use in mixing equipment, conveyor drives, pump drives, hoists, and winches.

PNEUMATIC LOGIC

Logic is the determination of an output based on the state(s) of a system's input(s). Logic is used to describe electrical or pneumatic switching (switching logic) because of the varied number of functions available. Pneumatic logic is referred to as a binary system. A *binary system* is a system that has two values, such as one and zero. In pneumatic logic, ones and zeros each have the coded meaning of pressure or no pressure. Binary functions are logical functions that may be considered ON or OFF, normally open or normally closed (NO or NC), or 1 or 0. In pneumatic logic terms, a 3-way directional control valve performs the same logical function as a light switch (ON or OFF).

Signal, Decision, and Action

All pneumatic circuits are designed based on how energy is used to perform work. The design should include a method of signaling the elements in the circuit to start their function followed by the elements making their decisions in causing an action. An *element* is a logic device that is capable of making a 0 or 1 output decision based on its input. Similar to electrical switching devices, pneumatic logic elements are the decision maker within the three basic control divisions of signal, decision, and action.

A *signal* in pneumatic logic is a condition that initiates a start or stop of fluid flow by opening or closing a valve. A signal component is a start/stop switch, relief

Vane Air Motors

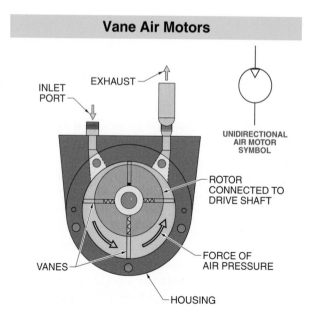

INLET PORT

EXHAUST

UNIDIRECTIONAL AIR MOTOR SYMBOL

ROTOR CONNECTED TO DRIVE SHAFT

FORCE OF AIR PRESSURE

VANES

HOUSING

Figure 19-29. A vane air motor is an air motor that contains a rotor with vanes that are rotated by compressed air.

valve, directional control valve, pressure switch, flow switch, etc. All signals rely on another condition to occur. This condition may be manual, mechanical, or automatic. For example, a manual condition occurs when a person operates a palm button or foot switch.

A mechanical condition occurs when a limit valve is mechanically operated. Limit valves are the most common pneumatic logic control. A *limit valve* is a mechanically actuated 3-way valve that is used to either monitor motion or measure position of an object. Limit valves generally sense an object by the use of a lever with a ball or roller at its tip. For example, a container on a moving conveyor makes contact with the switching mechanism of a limit valve, thereby sending a signal. A limit valve signal is 1 or 0, where 1 is when the valve is actuated (pressure), and 0 is when the valve is released (no pressure). An example of an automatic signal is that of a flow control switch or pressure switch that automatically produces an input when flow or pressure is detected or met.

A *decision* is a judgment or conclusion reached or given. A decision is the selection of the action or work to be accomplished based on an input. The decision process selects, sorts, and redirects the input information to a directional control valve, which causes an action to take place. An *action* is the work of an actuator or a pilot operator, which becomes the input for another section of a control circuit. The action produced by a pneumatic logic control circuit generally results in the pilot operation of a directional control valve. In some circuits, pneumatic logic controls operate a pneumatic/electric switch. The valve or switch is used to activate an actuator or become a signal in another circuit.

Pneumatic logic controls make their decision based on input signals received and relay the order for action. In many cases, the action that ultimately occurs changes the original input signal. In other words, the completion of the final action can signal a reversal of circuit operation. For example, when a cylinder receives the order to extend, there must also be an order to retract.

Pneumatic Logic Elements

A *pneumatic logic element* is a miniature air valve used as a switching device to provide decision making signals in a pneumatic circuit. A pneumatic logic element accepts input signals, makes logical decisions based on the input signals, and provides an output signal. The output signal is used to power output devices. Pneumatic logic elements are similar to electrical

relays in that they provide an output based on any input information. Pneumatic logic elements are static in nature and only require low air pressure for operation (generally between 75 psi and 90 psi) because they have no continuous airflow. The air supplied to logic elements should be filtered to remove particulates and moisture, regulated, and unlubricated. Particulate filtration should be 40µ or less.

Honeywell's MICRO SWITCH Division

A limit switch, which may be a pneumatic or electric switch, is mechanically operated when an object moves the limit switch lever.

Pneumatic logic elements are approximately one or two cubic inches in size and are considered miniature pneumatic switches. Each element is attached to a manifold, eliminating the need for plumbing between switches. A *manifold* is a device that contains passageways that enable one input signal to be divided into several output signals. Piping into the manifold comes from components such as door switches (for safety), palm switches (for activation), limit switches (for detecting product movement), etc. Piping out of the manifold is sent to pilot-operated directional control valves, which activate components such as clamping cylinders, air motors, drills, etc.

The National Fluid Power Association (NFPA) has designated symbols for pneumatic logic elements. These symbols are NFPA's standardization and are used in diagramming pneumatic logic controls. These symbols can be compared to similar electrical switching logic, electrical relay logic, or hydraulic controls. **See Appendix.**

Pneumatic logic elements were designed for operation sequencing, automated production, and controlling certain machine functions. Pneumatic logic elements make the decisions as to the work and order of work to be done. The output from a single element or combination of elements can provide a decision (pilot signal) required of a directional control valve that determines the machine's action. The basic logic elements used in pneumatic circuits include the AND, OR, and NOT elements.

AND. An *AND logic element* is a logic element that provides a logic level 1 only if all inputs are at logic level 1. An AND logic element has two or more inputs and one output. The output supplies pressure (ON or 1) only if all of its inputs have pressure. The output supplies no pressure (OFF or 0) if one or more input has no pressure. **See Figure 19-30.**

In an AND logic element, the output is 0 if both inputs are 0 and if only one input is 0. The output is 1 if both inputs are 1. Output decisions based on various inputs are shown using a truth table. A *truth table* is a table that lists the output condition of a logic element or combination of logic elements for every possible input condition. On logic elements, the A and B ports are the input ports, and the C port is the output port.

An example of AND logic use is a safety circuit on a punch press. In this circuit, the operator is required to press two palm buttons (switches) to activate the press. The operator's left hand presses one switch and the right hand is required to press the other switch before the press activates. This safety circuit keeps both hands out of a machine. An AND logic installation is similar to using two 3-way directional control valves that use manual palm buttons as actuators. In this application, valve 1

receives supply air at its inlet. The outlet of valve 1 is connected to the inlet of valve 2. The outlet of valve 2 is connected to the inlet of the actuator. As the palm button on valve 1 is pressed and held down, supply air is sent to valve 2. As the palm button on valve 2 is pressed and held down, supply air is sent to the actuator for circuit operation. Activating either valve by itself does not send air pressure to the actuator.

OR. An *OR logic element* is a logic element that provides a logic level 1 if one or more inputs are at logic level 1. An OR logic element has two or more inputs and one output. The output is 1 if any one or more input(s) are 1. The output is 0 if all inputs are 0. **See Figure 19-31.**

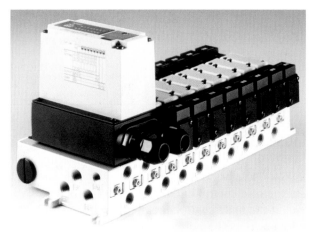

Humphrey Products Company

A relay system features manifold-mounted plug-in solenoid valves that can be changed rapidly for circuit modification and use a single power supply.

AND Logic Elements

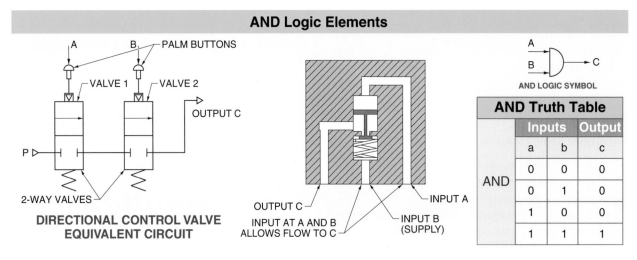

Figure 19-30. An AND logic element requires two inputs to produce an output signal.

OR Logic Elements

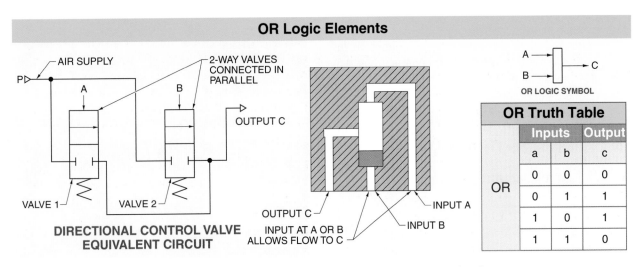

Figure 19-31. An OR logic element requires one or more inputs to produce an output signal.

An example of OR logic use is a circuit in which two switches at different locations operate the same cylinder or motor. Actuation of one switch or the other can start or stop the actuator. Two or more 2-way NC directional control valves connected in parallel may be used to show the operation of OR logic. The supply air is connected to the inlet of each valve and the output of each valve is connected to the actuator or device to be controlled. Operating either valve activates the actuator.

NOT. A *NOT logic element* is a logic element that provides an output that is the opposite of the input. A NOT logic element has one input, one output, and one supply. The supply normally flows through the element until an input signal stops the air flow. The output is 1 if the input is 0. The output is 0 if the input is 1. **See Figure 19-32.**

NOT logic applications include machine safety circuits. For example, a NOT logic element may be used for a safety block-out on a punch press. A valve is activated, which allows airflow to the input of the NOT element when a safety block is shifted under the ram on a press. This element does not allow air flow to the palm buttons, which prevents accidental ram operation. The NOT truth table shows that any signal (pressure) to port A removes any output at C. An equivalent circuit is a normally open valve. The supply is present as long as the valve is not actuated.

Tech Tip

The life expectancy of pneumatic logic control systems is 50 to 100 times longer than corresponding electrical devices such as high-maintenance switches.

NOT Logic Elements

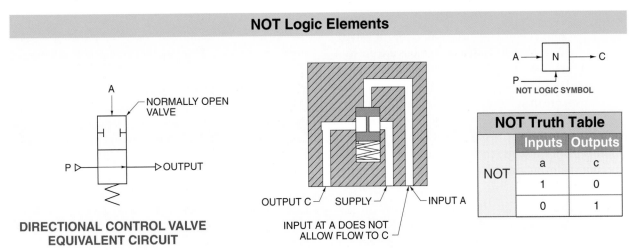

Figure 19-32. A NOT logic element provides an output that is the opposite of the input.

Logic Element Combinations

AND, OR, and NOT elements may be connected in combination to form additional logic combinations. An example of using an AND, OR, and NOT combination in a circuit is a safety anti-tie down, two-hand, palm button circuit. An anti-tie down circuit requires both hands of a machine operator to activate two palm buttons to operate a machine. **See Figure 19-33.**

Logic Element Combination

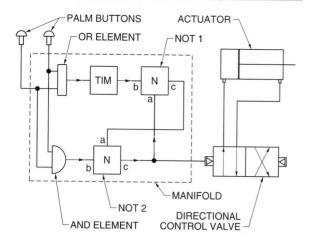

Figure 19-33. AND, OR, and NOT logic elements may be combined in a safety anti-tie down, two-hand, palm button circuit.

The anti-tie down circuit uses a timer to prevent either palm button from being tied down and defeating the safety feature. This circuit uses one OR, one AND, two NOT, and a timing element (TIM.). This combination consists of two inputs (palm buttons) and one output (pilot control to directional control valve). The machine activates if both palm buttons are pressed and held simultaneously. However, the timer prevents an output if one pushbutton is pressed before the other (normally more than $\frac{1}{10}$ of a second). The elements are arranged so that the output signal is 1 only when both palm buttons are 1 (pressed). The output is 0 as soon as either palm button is released. The output remains 0 until both palm buttons are released and pressed again.

Tech Tip

In actual industrial applications, pneumatic logic control systems often respond twice as fast as electrical controls.

Atlas Technologies, Inc.

Two-hand anti-tie down circuits are used for safety on industrial punch presses to prevent an operator's hands from being inside the press when it operates.

The output of the AND element becomes 1 and continues through NOT 2 to supply the output signal if both palm button inputs become 1 simultaneously. The pressure at the output of NOT 2 is also the input of NOT 1, preventing flow through NOT 1 after the timed period. If either palm button is held down without the other, the flow of air through the timer after its timed period continues through NOT 1 because there is no pressure at its A port. The flow from NOT 1 flows to the A port of NOT 2, holding NOT 2 closed and not allowing an output signal. To reactivate palm button operation, both palm buttons must be released, allowing the elements to exhaust. This also prevents double-tripping of a machine.

AND, OR, and NOT logic elements are the basic logic elements. Other logic elements include NAND, NOR, and flip-flop elements. As elements are added to a circuit, the complexity of the total circuit increases.

The NAND element is a combination of the NOT and AND elements. The NOR element is a combination of the NOT and OR elements. The flip-flop element is an element with two inputs and two outputs. When a signal is applied to one input of a flip-flop, a corresponding output is turned ON and the other output is turned OFF. In a flip-flop, one output is always ON and the other is OFF.

Electric and pneumatic controls are often used in machines designed for production manufacturing.

Memory and timer elements are also used to produce pneumatic logic. Memory elements are capable of memorizing information for recall at a later time. Memory decisions are the 1 and 0 (ON and OFF, pressure and no pressure) produced by other elements, except that once a decision is set, the decision is retained until it is reset. Timer elements (differentiators) are created by restricting the flow of the input signal into a timing chamber (accumulator). The time it takes to build enough pressure for an output signal is the time delay.

Pneumatic logic elements may be combined to form simple or complex circuits. Pneumatic logic elements are being used more often as pneumatic circuits and are easier and safer to troubleshoot than electronic circuits and are more dependable as long as the supply air is clean, dry, and properly maintained. One drawback is that pneumatic logic circuits occupy more space than electronic circuits.

Review

1. List at least four benefits a pneumatic circuit provides that a hydraulic circuit does not.

2. List eight different components used to compress and store air for use in a pneumatic circuit.

3. Explain how a positive displacement compressor is used to compress air for storage.

4. Define vane compressor.

5. Why does pneumatic system piping not return to the beginning stage like hydraulic system piping does?

6. List at least two causes of pressure loss in a pneumatic system.

7. Why is a check valve used in a pneumatic system?

8. Describe the role of the oil adjustment screw within a lubricator.

9. Define pressure regulator.

10. What is the functional difference between normally open valves and normally closed valves?

11. What is the primary function of a flow control valve?

12. List five major parts of an air cylinder.

13. List the advantages and disadvantages of using air motors instead of electric motors.

14. What is the difference between an AND logic element and an OR logic element?

15. Describe what is shown in the NOT truth table.

 Digital Resources

ATPeResources.com/Quicklinks
Access Code 462409

CHAPTER 20

PREVENTIVE MAINTENANCE PROGRAMS

A successful preventive maintenance program utilizes regularly scheduled evaluation of critical equipment, machinery, and systems. Preventive maintenance programs help detect potential problems, and maintenance tasks can be immediately scheduled to prevent deterioration in system operating condition. A successful maintenance program includes time-based maintenance tasks and corrective maintenance to provide support for all plant production or manufacturing systems. Preventive maintenance programs can be either manual or computer-based systems.

Atlas Copco

Chapter Objectives

- List the different types of maintenance.

- Explain ways to record and organize maintenance information.

- Explain the importance of developing a preventive maintenance program.

- Identify the information used in a preventive maintenance system.

- List the different types of testing methods available for preventive maintenance.

- Explain power quality analysis.

Key Terms

- maintenance
- preventive maintenance
- predictive maintenance
- manual preventive maintenance program
- preventive maintenance system
- computerized maintenance management system
- plant survey
- inventory control
- barcode
- infrared thermography
- thermal imager
- thermal signature
- ferrography
- ultrasonic analysis
- strobe tachometer
- electrical power system

PREVENTIVE MAINTENANCE

Maintenance is the planning and action to minimize and prevent equipment breakdowns and lost production time. If a problem occurs, maintenance personnel generally have a prepared and organized plan to return operations to normal in as short a time as possible. Maintenance is categorized as routine, emergency, and preventive.

Routine maintenance consists of maintenance procedures that involve servicing operating equipment on a scheduled basis. Routine maintenance typically consists of activities such as checking and adjusting oil levels, lubricating moving parts, replacing filters, inspecting V belts, and replacing or sharpening cutting devices.

Emergency maintenance consists of maintenance procedures that involve reacting to operating equipment breakdowns. These breakdowns must be corrected immediately and for certain facilities may require the assistance of outside contractors.

Preventive maintenance (PM) consists of maintenance procedures that involve scheduled inspections of, adjustments to, and repairs to equipment to verify that equipment is in proper working order. Repairs include the replacement of worn parts prior to failure based on known useful life spans or observed conditions. PM is the combination of technical and administrative measures taken to maintain production operating equipment and ensure that it is capable of performing at its peak operating condition. Technical and administrative measures include equipment identification for PM, equipment maintenance requirements, and accurate maintenance records.

While some facilities allow their equipment to operate continuously without regular maintenance, culminating in a catastrophic failure, other facilities overhaul machinery and routinely rebuild equipment at regularly scheduled time intervals. A PM program helps anticipate potential problems and monitor equipment closely for increased maintenance requirements. Original equipment manufacturers (OEMs) provide specifications and operator's manuals, which are designed for helping end users keep their machinery in peak operating condition. **See Figure 20-1.**

Predictive maintenance (PDM) is the monitoring of wear conditions and equipment operating characteristics for comparison against a predetermined tolerance to predict potential malfunctions or failures. In certain types of facilities, PDM practices are sometimes used in conjunction with PM practices to analyze equipment that has been through regular PM cycles. PDM is basically a series of operating equipment inspections coupled with administrative data collecting and scheduling. Inspections involve a combination of measuring conditions and subjective (look, listen, feel, and smell) inspections. Administrative data collecting and scheduling, such as equipment identification, equipment information, equipment maintenance requirements, and maintenance records, are used to record and organize maintenance information.

Operator's Manuals

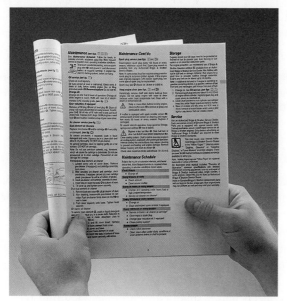

Briggs & Stratton Corporation

Figure 20-1. Original equipment manufacturers supply operator's manuals for each specific piece of equipment that offer safety functions as well as maintenance specifications.

Equipment Identification

Every piece of equipment within an operating facility requires some type of maintenance. Identifying each piece of equipment is required in order to track the type and frequency of maintenance and its cost. All equipment must have its own identification number.

Identification of equipment is generally done by OEM asset number, or a facility may create its own internal identification system. It may be more convenient to identify equipment by zone or department. For example, a filling machine in a bottling department for a liquid soap manufacturer can be identified as "B-FI-1" (for "bottling-filling-machine 1"). Once the specific equipment identification is made, a master list of equipment, which includes each identification number and piece of equipment, is prepared. **See Figure 20-2.**

Equipment Identification

I.D. #	Description	I.D. #	Description
FILLING EQUIPMENT: MASTER LIST			
1.	B-FI-1 Bottling Filling Machine 1	25.	
2.	B-FI-2 Bottling Filling Machine 2	26.	
3.	B-FI-3 Bottling Filling Machine 3	27.	
4.	B-CP-1 Bottling Capping Machine 1	28.	
5.	B-CP-2 Bottling Capping Machine 2	29.	
6.	B-CP-3 Bottling Capping Machine 3	30.	
7.	LS-CV-1 Liquid Soap Conveyor 1	31.	
8.	LS-CV-2 Liquid Soap Conveyor 2	32.	
9.	LS-CV-3 Liquid Soap Conveyor 3	33.	

Figure 20-2. A master list identifying each piece of operating equipment for a specific facility may be organized using company asset numbers and an internal numbering system.

Properly compiling the required information for equipment is accomplished by gathering and recording information on each piece of equipment. Information gathered may be from the OEM or operator's manuals, purchase records, supplier information, or equipment nameplates. **See Figure 20-3.**

Equipment Information Sources

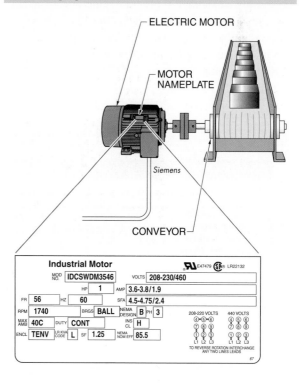

Figure 20-3. Accurate equipment information is typically found on the equipment nameplate.

Equipment Maintenance Requirements

OEMs may provide recommended maintenance procedures and schedules for the equipment they manufacture. This information may also be found in operating manuals, through direct contact with the supplier, on manufacturer websites, or in manufacturer catalogs. For example, equipment or product suppliers can suggest the type of lubricant, lubrication intervals, and amount of lubricant to use. Emergency maintenance situations also determine or affect routine scheduling. **See Figure 20-4.**

Maintenance Records

As maintenance work is performed, records of specific tasks performed, parts required, suppliers used, and costs incurred must be kept. This information can be used to determine the frequency of PM. When emergency maintenance situations arise, this information can be used to make decisions about repairing or replacing equipment or changing work schedules. **See Figure 20-5.**

MANUAL PREVENTIVE MAINTENANCE PROGRAMS

Developing a manual PM program requires establishing goals, creating organized maintenance plans, prioritizing production areas, cataloging operating equipment, prioritizing equipment, planning PM schedules, and employee training.

Maintenance Records

MAINTENANCE RECORD			
Equipment ID: GC 101	Description: Input Conveyor		
Date	Repair Work Performed:	Workers:	Cost:
6/19/2005	Replaced drive bearings	Smith	325.00
1/24/2007	Installed new drive chain	Gordon	550.00
10/10/2009	Replaced conveyor pads	Smith	850.00

Figure 20-5. Maintenance records for each piece of equipment offer the vital historical information required when developing a PM program.

Manual PM programs are typically used in small-capacity manufacturing facilities that do not have multiple operations or many employees. For example, a machine shop with less than thirty employees that produces small batch custom work is classified as a small-capacity facility, while an automotive assembly plant with several hundred employees is classified as a large-capacity facility. A manual PM program introduced into a small-capacity manufacturing facility allows introduction of a computerized maintenance management system (CMMS) program at a later date.

Equipment Maintenance Requirements

MAINTENANCE SCHEDULE						
Equipment ID: GC 101				Description INPUT CONVEYOR		
		Daily	Weekly	Monthly	6 Months	12 Months
1.	Check Grease Bearings			X		
2.	Check Grearbox Oil			X		
3.	Change Gearbox Oil					X
4.	Grease Motor Bearings					
5.	Check V-Belts		X			
6.	Oil Chains		X			
7.	Check Leg Belt			X		
8.	Check Leg Cups					
9.	Check Roto Guard/Oil/Belt					
10.	Check Head Pulley					
11.	Check Grad Chain Paddles					
12.	Check Hanger Bearings					
13.	Check Air Filter					

Figure 20-4. A list of maintenance requirements for a specific piece of equipment is frequently referenced and updated after each scheduled maintenance procedure.

Establishing Goals

Established goals must be realistic, specific, and measurable to be effective. Goals must have detailed, planned steps and required actions. Employees that are expected to implement the plans to achieve specific goals must understand the required expectations. To be realistic, a goal must represent an attainable objective. A *realistic goal* is a goal that can be measured in terms of achievement and performance and for which sufficient means of implementation are possessed. A *specific goal* is a goal that incorporates an action plan that outlines how the goal is to be achieved and a performance measurement that provides goal evaluation. The performance measurement in a goal is often a date or specific length of time, but it could be any objective that can be used to determine when a specific goal has been attained.

Creating Organized PM Plans

Creating organized PM plans requires detailing the planning and action steps required to develop a program. Initial planning of an organized PM plan may include the best methods to gain plan acceptance from company management, required equipment for production operations, required training, and specific uses for equipment.

Prioritizing Production Areas

Prioritizing the best production areas to implement a PM plan may require selecting an area that has the most downtime, requires the most maintenance, or can increase production output with minimal changes. For example, an area can be chosen that has more problems than other production areas, and input can be gathered from individuals, such as machine operators and maintenance personnel, who work in the area. This input can be used to troubleshoot and solve each problem until each problem area is addressed.

Cataloging Operating Equipment

To catalog operating equipment, a list of all equipment in the production area requiring a PM plan is created. Then, for each separate piece of equipment, a list of relevant information is created. Relevant information may include the following:

- equipment name and description
- date placed in service
- manufacturer name, part number, and serial number
- maintenance history
- downtime record
- safety considerations
- effects on productivity
- tasks required for PM

Prioritizing Equipment for PM

Once pieces of equipment have PM programs, they can then be prioritized in order of most requirement to least requirement of PM. All possible maintenance information should then be gathered, including any manufacturer's PM suggestions, air pressure settings, electrical measurements, lubrication type and frequency, or any function vital to equipment operation.

Planning PM Schedules

When planning a PM schedule, it must be determined how often specific inspections and measurements need to be taken from equipment. OEM recommendations usually have the best information for these tasks. Generally, because equipment can differ in use and have varying operating parameters (speeds, pressures, temperature, etc.), a schedule for each particular piece of equipment must be developed.

PM scheduling also requires a method of communicating with the technician. This is normally accomplished by means of a work order. A *work order* is informational documentation created for communication between a technician and maintenance personnel or other relevant parties. Work orders typically include information such as job type, scheduling, parts required for repair or maintenance, and tasks required. PM scheduling is used to make decisions on the type of information that must be collected on a work order, the party responsible for generating the work order, and the details that must be included on the work order. PM scheduling also requires electronic data entry into the internal system.

Tech Tip

Typically, a PM program must be presented to company management. This is one reason why a manual PM program is started on a small scale and records of decreased downtime or increased production are tracked. This information can be used to present a return on investment (ROI) from a PM program.

Employee Training

Ensuring that PM procedures are implemented as intended is accomplished through on-the-job training of technicians. Technicians must be trained to have an understanding of the program, why it is to be used, and the benefits it provides. Employee training can be accomplished in a classroom setting through various examples and applications. Initial employee training should occur frequently, such as weekly or bimonthly, and then taper off, such as once a quarter, as the program is established. Employee training should also include revisions to the program as they become necessary.

COMPUTERIZED MAINTENANCE MANAGEMENT PROGRAMS

Modern maintenance operations require more than just basic PM activities designed to prevent major problems. A *preventive maintenance (PM) system* is a system used to record and organize maintenance information, which is then used to make the decisions required to maintain the equipment in a facility. **See Figure 20-6.** Information in a PM system includes equipment data, maintenance costs, consumables, time on task, and breakdown resolutions. With a consistent, accurate flow of operation information, PM systems can help increase efficiency, reduce costs, and minimize health and safety problems. PM systems can also be used to document compliance with environmental, health, and safety regulations. The PM system may include the entire facility or only the departments that are expected to see the greatest benefit from improved maintenance practices.

Although PM systems can be implemented as hard copy (paper-based) systems, many organizations use computerized maintenance management systems. A *computerized maintenance management system (CMMS)* is a software package that organizes PM information and automatically generates reports, work orders, and other data for implementing and improving future maintenance activities. **See Figure 20-7.** CMMS software provides quick access to maintenance information, issues and tracks work orders, determines the costs of maintenance activities, schedules maintenance work, manages maintenance inventories, and assists in troubleshooting.

CMMSs also help analyze the frequency and type of PM work and make adjustments in order to keep equipment in peak operating condition with minimal cost. Excessive PM work increases maintenance costs, but inadequate PM work also results in high maintenance costs due to an increase in breakdowns. Data compiled from a PM system is also used for assessing plant performance, equipment service life, energy costs, equipment purchase requirements, insurance costs, and personnel and plant budget decisions. CMMSs require plant surveys, plant documentation, program implementation, and inventory control.

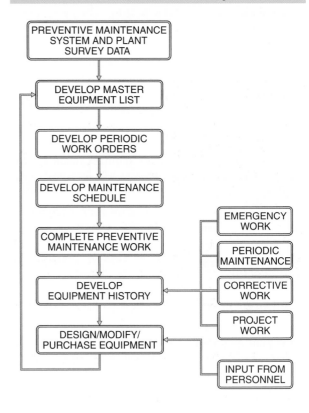

Preventive Maintenance Systems

Figure 20-6. A preventive maintenance system is used to record and organize maintenance information, which is then used to make the decisions required to maintain the facility and equipment.

Plant Surveys

A plant survey is the first step in implementing a PM system. A *plant survey* is a complete inventory and condition assessment of a facility's equipment and structure. Data from the plant survey is entered into the PM system to create a master file for each piece of equipment. A master file lists the equipment manufacturer, vendor, serial and model numbers, identifying codes, parts suppliers, equipment location, and entire service history.

Computerized Maintenance Management System (CMMS)

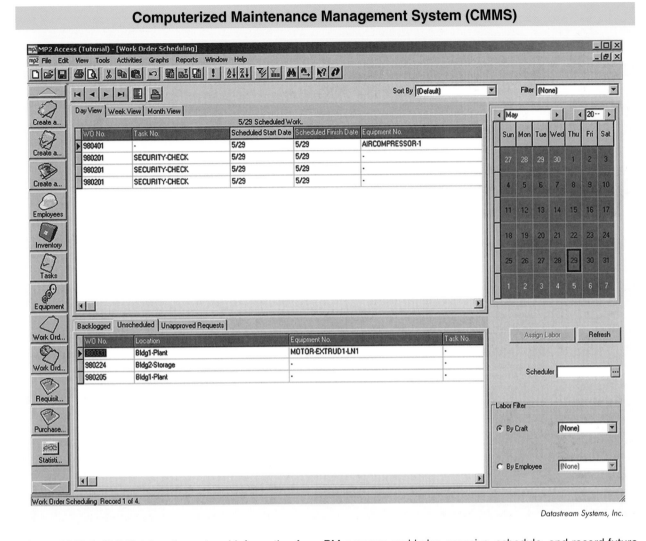

Datastream Systems, Inc.

Figure 20-7. A CMMS takes the entered information for a PM program and helps organize, schedule, and record future maintenance activities.

A plant survey is completed by in-house personnel or outside contractors. As part of the plant survey, the equipment is carefully inspected and analyzed. **See Figure 20-8.** The equipment condition is noted along with any repairs required to return the equipment to peak performance. These repairs become the first corrective work orders in the new PM system.

A plant survey typically uses hard copy forms to gather data for electronic entry into a computer system. In addition to hard copy forms, information can also be entered directly into a mobile device and transferred into the main system at a later time.

A master equipment file expands with newer information as the equipment is maintained and repaired. The data compiled is used to identify possible problems when troubleshooting and quantify the equipment's reliability.

The equipment reliability and total cost of operation information then influences future purchasing decisions.

Plant Documentation

All equipment documentation is gathered while assembling the master equipment files for the plant survey. This includes construction prints, wiring diagrams and schematics, replacement parts data, and installation, operating, maintenance, and troubleshooting manuals. For example, all industrial equipment comes with electrical drawings that show how the equipment connects to building systems and process controls. **See Figure 20-9.** System documentation may be a mixture of paper-based documents or electronic files. Maintenance personnel must be able to locate information quickly in either paper or electronic formats.

Plant Surveys and Master Equipment Files

PLANT SURVEY FORM

EQUIPMENT IDENTIFICATION: Furnace #1 - FUR0001.00

LOCATION: 4247 Piedmont Building 2, Floor-1 Room 23 CL15F

DATE OF PURCHASE: 08/17

VENDOR: Lewis Systems, Inc.

1862 Erie St. Cleveland, OH

VENDOR PHONE: 216-555-1340

MANUFACTURER: Brown Boveri, Inc.

MODEL #:   IT6P

SERIAL #:   OP2810C2B

EQUIPMENT DATA: (List all parts and part numbers important to the operation of the equipment.)

PLANT SURVEY FORM

Description:	Furnace #1	Voltage:	440	Warranty ID:	IT6P882
Asset ID:	FUR0001.00	Amperage:	600	Warranty Date:	02/17
Asset Type:	Furnace Systems	Wattage:	264000		
Parent ID:		Phase:	3	YTD Labor Hr:	45.00
Priority:	8 Active:☒	Elec Line:	10	YTD Downtime:	22.00
Manufacturer:	Brown Boveri, Inc.	Air Area:	COMP 6		
Model:	IT6P			TD Labor Hr:	740.00
Serial Number:	OP2810C2B			TD Downtime:	493.00
Vendor:	Lewis Systems, Inc.	Counter UOM:			
Vendor Address:	1862 Erie St.	Current Counter:	1	YTD Labor Cost:	724.00
	Cleveland, OH 55117	Counter Rollover:	0	YTD Misc Cost:	1231.22
Vendor Phone:	216-555-1340	Meter UOM:		YTD Part Cost:	7701.19
Asset Tag:	00509	Current Meter:	1234550	Total:	9656.41
Location:	4247 Piedmont Building 2				
	Floor-1 Room 23 CL15F			TD Labor Cost:	9620.60
Department ID:		Meter Rollover:	0	TD Misc Cost:	5631.80
Cost Center:	Fixed Asset Repair	Purchase Date:	02/17	TD Part Cost:	14,942.00
Supervisor:	Jones, Fred	Install Date:	08/05	Total:	30,194.40
		Retire Date:			
		Install Cost:	97000		
		Replacement Cost:	97000		

Comment: Manufact. warranty extremely strict. Document all hours worked and parts used.

Report Totals:	YTD Labor Hr: 45.00	YTD Labor Cost: 724.00	TD Labor Cost: 9620.60
	YTD Downtime: 22.00	YTD Misc Cost: 1231.22	TD Misc Cost: 5631.80
Assets: 1	TD Labor Hr: 740.00	YTD Part Cost: 7701.19	TD Part Cost: 14,942.00
	TD Downtime: 493.00	Total: 9656.41	Total: 30,194.40

MASTER EQUIPMENT FILE

DPSI (DP Solutions, Inc.)

Figure 20-8. A plant survey and data from the PM system are combined to create a master equipment file for each piece of equipment.

Plant Documentation

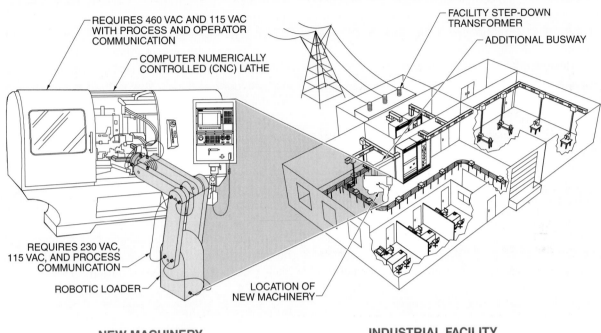

REQUIRES 460 VAC AND 115 VAC
WITH PROCESS AND OPERATOR
COMMUNICATION

COMPUTER NUMERICALLY
CONTROLLED (CNC) LATHE

REQUIRES 230 VAC,
115 VAC, AND PROCESS
COMMUNICATION

ROBOTIC LOADER

FACILITY STEP-DOWN
TRANSFORMER

ADDITIONAL BUSWAY

LOCATION OF
NEW MACHINERY

NEW MACHINERY   **INDUSTRIAL FACILITY**

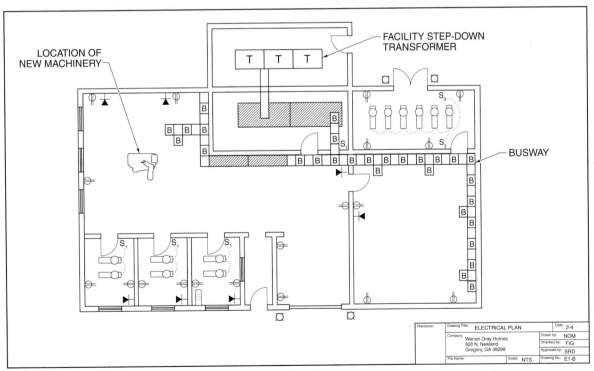

LOCATION OF
NEW MACHINERY

FACILITY STEP-DOWN
TRANSFORMER

BUSWAY

Revisions:	Drawing Title:	ELECTRICAL PLAN	Date: 2-4
	Company: Warren Gray Homes 826 N. Newland Gregory, GA 38299		Drawn by: NOM
			Checked by: FIG
			Approved by: SRD
	File Name:	Scale: NTS	Drawing No.: E1-B

DRAWING

Figure 20-9. Industrial equipment comes with plant documentation such as drawings that show how the equipment connects to building systems and process controls.

Hard copy documents require two copies. A reference copy, which is often the original document, is stored in a secure and permanent location. The working copy is used for daily tasks. The reference copy is used to make a new working copy if the working copy is lost or damaged. Any changes must be noted on both sets of documents.

Electronic filing systems are becoming standard for storing maintenance information, as they are easily accessed. For example, electronic files of large, complicated drawings are sometimes easier to use than traditional hard copy drawings because specific sections can be enlarged on a computer screen to display details. The files are stored on shared computers or networked hard drives so that they are available to all maintenance personnel. Although electronic files are often considered to be a reference copy, some organizations print hard copies to record changes and to use as backup copies in the event of computer network problems.

Tech Tip

Some CMMSs can be integrated with building automation systems. This allows the maintenance system to access the building information being generated and shared by the building automation system controllers. This system information can be used for maintenance applications, such as ensuring optimal system settings, monitoring energy use, and tracking equipment operating time.

Program Implementation

Once all the necessary data is gathered, the amount and type of PM work is determined. Working from OEM recommendations, maintenance and inspection routines are developed and anticipated costs are calculated. Maintenance costs and OEM warranties determine the maximum amount of work a specific piece of equipment requires. **See Figure 20-10.**

It is undesirable to incur a high cost on maintenance for a piece of equipment when an entire replacement unit can have a low cost. Exceptions to this include when the unit is critical for operations, affects health or safety issues, or cannot be quickly replaced due to extensive lead time from the OEM. For example, most electric motors receive little or no PM, but an electric motor that is critical to production can receive regular insulation testing to anticipate its possible failure so that it can be replaced prior to failure.

Changes to the developed routines are made as required to assign the appropriate amount of PM to each piece of equipment. Work is then scheduled and assigned. For example, routine lubrication and inspections might be assigned to equipment operators, while work requiring more skill is assigned to qualified maintenance personnel. Efficient scheduling generates a consistent workload and ensures that equipment is properly prepared for use.

Maintenance personnel periodically review maintenance routines against actual costs incurred, adjusting tasks as required to balance equipment availability and costs. For example, if a piece of equipment is performing well, the scheduled maintenance interval can be lengthened. The breakdown rate of each piece of equipment is monitored and the frequency of work is changed if necessary. Other factors affecting the type and frequency of PM are energy usage and equipment output.

Inventory Control

Inventory control is the organization and management of commonly used consumables, vendors and suppliers, and purchasing records in a PM system. Each type of consumable is assigned a unique part number that is used to track its quantity on hand and associate the consumable with data such as the supplier, equipment application, purchase date, and cost. The most effective use of an inventory control system is one that integrates this information with a CMMS.

A *barcode* is a series of vertical lines and spaces of varying widths that is used to represent data. **See Figure 20-11.** To expedite data entry and tracking, a part number label is often printed in a computer code that can be read automatically with an image scanning device, such as an optical barcode reader. Barcodes have been used for this purpose since the 1970s, because they are inexpensive and reliable. Barcodes can be scanned quickly and easily with an optical barcode reader, which inputs the identification number into software.

Scanning barcodes as parts are entered and removed from inventory allows the computer to keep up-to-date records of available parts. When quantities of parts fall to certain levels, the system is triggered and generates a purchase order for the amount of parts required. This helps maintain an adequate inventory to support operations without excessive parts and materials.

Preventive Maintenance Program Implementation

Datastream 7i - Microsoft Internet Explorer

Datastream 7i

START CENTER   MY ACCOUNT   HELP   ABOUT   LOGOUT

Work   Materials   Equipment   Purchasing   Operations   Administration

Asset: 203000   Boiler Pump Feed

Organization: ORG1
Dept: FAC1
Status: Installed

List View | Record View | Comments | Events | Costs | PM Schedule | Structure | **Warranties** | Add Tab

Warranty	Description	Coverage Type	Active	Duration	Duration UOM	Threshold	Threshold UOM	Sta
B-G-24M	Bell & Gosset 24 Month Warranty	Calendar	☑	720		60		09/
B-G-24M	Bell & Gosset 24 Month Warranty	Usage	☑		50,000		500	

Add Warranty Coverage | Delete Warranty Coverage

Warranty Details

Warranty: B-G-24M   Bell & Gosset 24 Month Warranty

Coverage Type: Usage   **Active:** ☑

Duration: **Duration UOM:** 50,000 HUR

Threshold: **Threshold UOM:** 500

Start Date: **Starting Usage:** 0

Expiration Date: **Expiration Usage:** 50,000

Last Value: 100

(Submit) (Clear)

Datastream Systems, Inc.

Figure 20-10. Entering information about equipment warranties into a PM system helps determine the most cost-effective solution to equipment breakdowns.

Barcodes

978-0-8269-3712-4

90000

9 780826 937124

Figure 20-11. Barcodes are often used for inventory control, which requires the organization and management of parts commonly used for maintenance tasks.

PREVENTIVE MAINTENANCE TESTS

PM is the scheduled work required to keep equipment in peak operating condition. PM includes basic tasks such as inspecting, cleaning, tightening, lubricating, and replacing consumable parts. PM tests include infrared thermography, ferrography, ultrasonic analysis, strobe tachometer tests, and periodic inspections and measurements.

Infrared Thermography

Infrared thermography is the science of using electronic optical devices to detect and measure radiation and correlating the radiation level to surface temperature. Radiation is the movement of heat that occurs as radiant energy (electromagnetic waves) without a direct medium of transfer. Radiant energy measurement is valuable because overheating equipment is often an indication of impending failure.

A *thermal imager* is a device that detects heat levels in the infrared-wavelength spectrum without making direct contact with the target. Thermal imagers operate by focusing a lens on a target. The thermal imager detects radiation given off by the target and displays it on a liquid crystal display (LCD) screen. In addition to producing a still or video image, thermal imagers digitally display temperature readings. **See Figure 20-12.**

Thermal Imagers

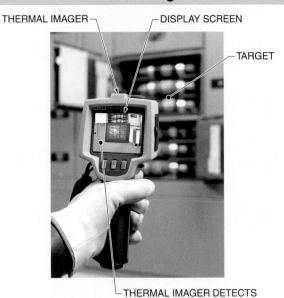

THERMAL IMAGER —   — DISPLAY SCREEN

— TARGET

THERMAL IMAGER DETECTS INFRARED RADIATION GIVEN OFF BY TARGET

Fluke Corporation

Figure 20-12. A thermal imager is a device that detects heat levels in the infrared-wavelength spectrum without making direct contact with the target.

A *baseline inspection* is an inspection intended to establish a reference point using equipment operating under normal conditions and in good working order. A *trending inspection* is an inspection performed after a baseline inspection to provide images for comparison. Monitoring trends over time often provides diagnostic and predictive information. This allows a technician to compare any differences or similarities that may be indicative of equipment performance.

Baseline and trending inspections should be scheduled at a frequency determined by consequence of failure and the condition of the equipment. By monitoring trends, PM can be performed to reduce unplanned downtime and costly failures.

Thermal Signatures. A *thermal signature* is a false-color picture of the infrared energy emitted from an object. Hot surfaces are displayed with a lighter color than cool surfaces. The lighter the color of the image displayed, the hotter the surface of the object being tested. The thermal signature of a normally operating piece of equipment can be used to evaluate the operation of a suspect piece of equipment. Any differences resulting from a change in equipment condition are then noted. **See Figure 20-13.** The primary benefit of infrared thermography is that tests can be performed quickly and without destruction to equipment. Also, since thermal imagers do not require contact with an object, they can be used while the equipment or component is in operation.

Thermal Signatures

VISIBLE LIGHT IMAGE

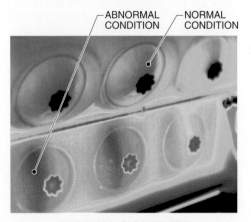

— ABNORMAL   — NORMAL
CONDITION   CONDITION

THERMAL SIGNATURE

Fluke Corporation

Figure 20-13. A thermal signature is a false-color picture of the infrared energy being emitted from an object.

When excessive heat is detected the image is analyzed and further testing conducted. The cause of overheating is located and remedied before a dangerous situation develops. Special training may be required to operate thermal imagers and interpret the results. Thermal imagers can be used to analyze a wide range of items including electric motor couplings, windows and doors, ceilings, walls, and roofs, and storage tanks. **See Figure 20-14.**

Safety Tip

Prior to performing a thermal inspection, the maintenance technician should perform a "walk-down" of the planned inspection route to visually inspect for possible safety issues. Safety issues should be evident because all thermal imager targets radiate energy that is measurable on the infrared spectrum. As a target's heat increases, it radiates more energy. Very hot targets radiate enough energy to be seen by the human eye.

Thermal Imager Applications

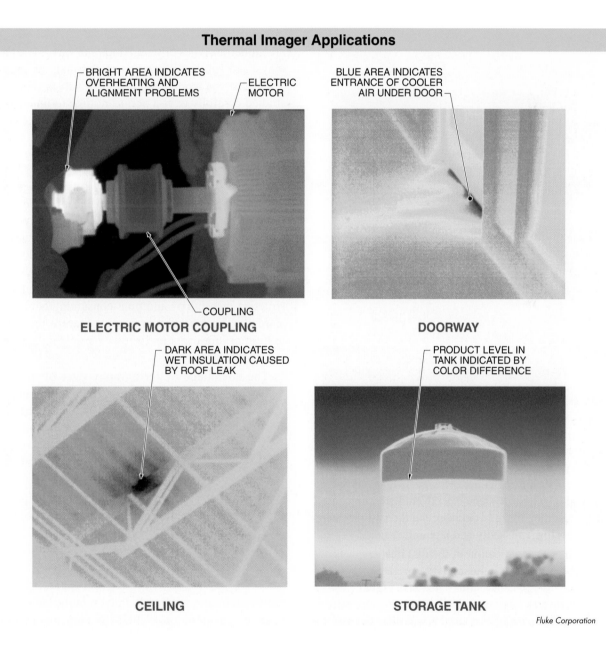

BRIGHT AREA INDICATES OVERHEATING AND ALIGNMENT PROBLEMS

ELECTRIC MOTOR

COUPLING

ELECTRIC MOTOR COUPLING

BLUE AREA INDICATES ENTRANCE OF COOLER AIR UNDER DOOR

DOORWAY

DARK AREA INDICATES WET INSULATION CAUSED BY ROOF LEAK

CEILING

PRODUCT LEVEL IN TANK INDICATED BY COLOR DIFFERENCE

STORAGE TANK

Fluke Corporation

Figure 20-14. Thermal imagers can be used to indicate electric motor coupling alignment problems, doorway leaks, roof leaks, and storage tank product levels.

Thermal imagers are used to analyze electric motors because motors are susceptible to heat-related failure. For example, motor misalignment or imbalance typically results in overheating of the motor coupling and/or bearings. Motor bearings that are warmer than normal may indicate a possible problem that should be investigated further. Similarly, motor couplings and shaft bearings that are operating normally should exhibit thermal signatures that are close to the ambient air temperature. **See Figure 20-15.**

Thermal Signature Analysis

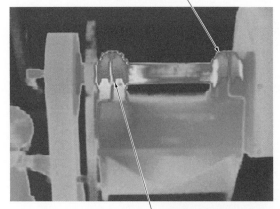

NORMAL SIGNATURE
(BEARING NEAR AMBIENT
TEMPERATURE)

ABNORMAL SIGNATURE
(BEARING WARMER
THAN AMBIENT
TEMPERATURE)

Fluke Corporation

Figure 20-15. A thermal signature of a motor that shows the motor bearings to be warmer than normal may indicate a possible problem that should be investigated further.

All types of thermal insulation can be evaluated by visually inspecting the target area for variations in the surface thermal signature. However, many types of thermal insulation systems are covered with unpainted metal cladding that can decrease the thermal signature due to its low emissivity and high reflectance.

A common application of thermal imagers is determining the levels of solids, liquids, or gases in vessels such as storage tanks and silos. **See Figure 20-16.** Although most vessels typically have instrumentation to indicate the level of material inside them, the data is often inaccurate because of improper instrumentation function. At other times, the data is accurate but must be independently confirmed.

A skilled technician can often determine the level of material in an uninsulated tank. Where insulation is present, it may take longer for thermal signatures to appear or some enhancement may be required. Material levels in a vessel can be enhanced by applying heat or inducing cooling with evaporation. For example, briefly spraying water on a tank and waiting a few minutes for the exterior surface of the tank to change temperature is often enough to reveal the level of material in the tank.

Steam traps and most valves exhibit temperature differences across them when working properly. There are numerous types of steam traps and valves and each may have subtle differences in thermal signatures. Therefore, the thermal signatures of valves should be studied carefully over a period of time to understand how the valves normally function.

Ferrography

Ferrography is wear particle analysis using diagnostic techniques to evaluate the condition of interacting lubricated parts or components. Ferrography is an easily performed inspection method for determining the health of a system and providing early failure detection. The use of ferrography to evaluate a system's condition helps eliminate the need for potentially destructive inspections and time-consuming equipment teardown. Ferrography is used to analyze the debris in system lubricants and shows particle size, shape, color, and quantity.

The advantage of ferrography over other types of PM tests is its capacity to detect a broad range of types and sizes ($0.1\mu m$–$500\mu m$) of wear particles. Ferrographic analysis encompasses wear (metallic and nonmetallic), contaminant (crystals, water, and organic and inorganic compounds), and lubricant (friction polymers) monitoring.

Typical wear problems identified by ferrography include gear teeth wear through excessive load or speed, misalignment, fractures, rolling contact failure, water contamination in oil or poor lubricant condition, oil additive depletion, outside contaminants such as sand or dust, cam shaft and cylinder wall failure, oil filter failure, and many others. Regular monitoring of wear particle concentration (WPC) can alert maintenance technicians of problems earlier than other types of tests, which can help maintenance managers schedule regular maintenance and have replacement parts available.

Liquid Levels in Tanks

TANK

WATER LEVEL

WATER STORAGE TANK

THERMAL IMAGER DISPLAYS THE LEVEL OF WATER IN TANK BASED ON DIFFERENCES IN TEMPERATURE BETWEEN AIR AND WATER INSIDE

Fluke Corporation

Figure 20-16. A common application of thermal imagers is determining the levels of solids, liquids, or gases in vessels such as storage tanks and silos.

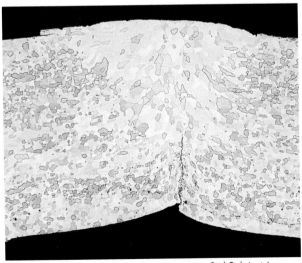

Stork Technimet, Inc.

Ultrasonic testing is used to locate internal faults in structural and welded components that are not visible to the naked eye.

Ferrography requires taking an oil sample from the equipment to be tested and depositing it on a glass slide that is positioned near a magnet. The magnet draws the metallic particles onto the slide for viewing under a microscope. Because of the skill and equipment required, ferrography typically requires the use of a laboratory that specializes in such analysis. **See Figure 20-17.**

Ultrasonic Analysis

Ultrasonic analysis is a nondestructive test method in which mechanical vibrational waves are used to test the integrity of a solid material. Operating mechanical equipment produces a normal sound signature. As components develop internal faults such as cracks or voids, a change in the normal sound signature occurs. Handheld or portable ultrasonic test units are typically used to perform ultrasonic analysis.

Ferrography

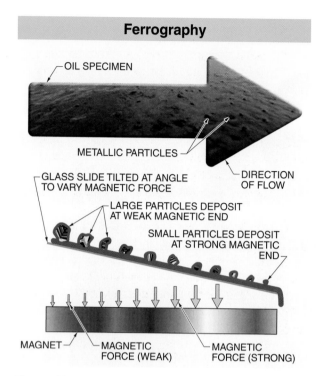

Figure 20-17. Ferrography requires taking an oil sample from the equipment to be tested and depositing it on a glass slide that is positioned near a magnet.

A test probe attached to the test unit is placed in contact with the component under test and transmits a signal that is displayed on the screen of the test unit. The size of the waveform on the screen indicates the presence and location of an internal fault. **See Figure 20-18.** This change can be noted as a shift in intensity on the screen and/or as a qualitative sound change that can be heard through headphones and recorded for further analysis.

As changes begin to occur in mechanical equipment, the subtle, directional nature of ultrasound allows these potential warning signals to be detected before actual failure, often before they are detected by vibration or infrared methods. For example, when a bearing begins to wear, large spikes are produced in its ultrasonic signal caused by flat spots or scratches on the bearing race. The spikes are heard as pops or crackles through the headset. Once the ultrasound produced by the bearing begins to indicate these characteristics, replacement of the bearing can be planned during normal production shutdown.

Strobe Tachometer Testing

A *strobe tachometer* is a test tool that measures the rotational speed of an object by use of a flashing (strobe) light.

A strobe tachometer measures speed by synchronizing the flash rate of a light with the speed of the moving object. Strobe tachometers can take speed measurements through glass, which eliminates direct exposure to hazardous areas. Strobe tachometers are also used for analysis of motion and vibration. Strobe tachometers measure speeds from 20 rpm to approximately 100,000 rpm, or 20 ft/min to approximately 12,500 ft/min.

Ultrasonic Analysis

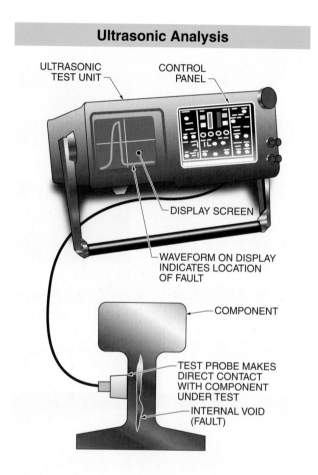

Figure 20-18. Handheld or portable ultrasonic test units are typically used to perform ultrasonic analysis.

Strobe Tachometer Measurement Procedures. Before taking any measurements using a strobe tachometer, the tachometer must be checked to ensure it is designed to take measurements on the object being tested. The operator's manual should be referred to for all measurement precautions, limitations, and procedures. The required personal protective equipment must always be worn and all safety rules must be followed when taking the measurement.

See Figure 20-19. To measure the speed of an object with a strobe tachometer, apply the following procedure:

1. Turn the tachometer ON.
2. Set the mode switch to the required rpm, ft/min, or m/min units.
3. Set tachometer range. Adjust strobe flash rate.
4. Ensure there is no danger of any part of the human body coming in contact with moving objects.
5. Align the visible light beam with the object being measured and reduce flash rate to freeze image.
6. Read the speed displayed on the meter.
7. Record all measurements.
8. Turn the tachometer OFF.

Periodic Inspections and Measurements

Even when a system has been fully inspected and adjusted for optimal efficiency, it must be regularly monitored for additional problems that may occur.

Equipment gradually ages, and even if a unit was operating acceptably during the previous inspection, its condition may deteriorate to a point where maintenance or replacement is required. Therefore, major equipment and motors must be inspected periodically.

A periodic inspection schedule should be implemented so that problems can be found quickly and equipment can be repaired before significant time and energy is wasted. Many PM tests may become part of a regular maintenance program. **See Figure 20-20.**

Continuous energy-use monitoring may be beneficial for some systems, such as those that have large electrical loads. Data logging meters can be used on a continual basis to record the power and energy parameters of specific equipment. Alternatively, electrical sensors can be permanently installed into panels or enclosures. The data from these sensors is sent to a computer system via a communication network for future monitoring, review, and analysis.

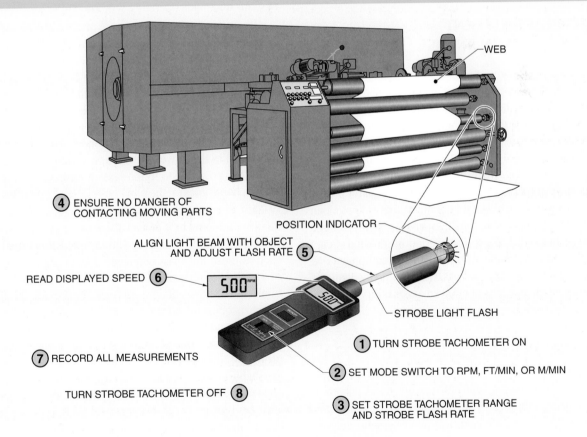

Strobe Tachometer

④ ENSURE NO DANGER OF CONTACTING MOVING PARTS

POSITION INDICATOR

ALIGN LIGHT BEAM WITH OBJECT AND ADJUST FLASH RATE ⑤

READ DISPLAYED SPEED ⑥ **500** RPM

STROBE LIGHT FLASH

① TURN STROBE TACHOMETER ON

⑦ RECORD ALL MEASUREMENTS

② SET MODE SWITCH TO RPM, FT/MIN, OR M/MIN

TURN STROBE TACHOMETER OFF ⑧

③ SET STROBE TACHOMETER RANGE AND STROBE FLASH RATE

WEB

Figure 20-19. Strobe tachometer are used for rotational speed measurements.

Periodic Inspections

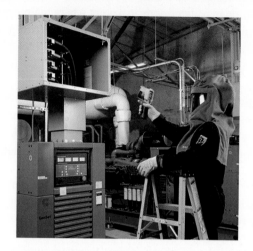

Fluke Corporation

Figure 20-20. A periodic inspection schedule should be implemented so that problems can be found quickly and equipment can be repaired before significant time and energy is wasted.

ELECTRICAL EQUIPMENT PREVENTIVE MAINTENANCE

PM keeps equipment in peak operating condition. Electrical equipment must be kept clean, dry, and cool. The interior and exterior of electrical equipment must be kept clean to prevent dirt from creating paths for short circuits. Moisture lowers resistance and damages insulation, causing short circuits. However, the main cause of electrical equipment failure is overheating. Electrical equipment must have proper ventilation and be kept clean for proper heat dissipation. Electrical equipment that is kept clean and dry with its connections tight and its supply voltage within manufacturer's recommendations provides years of efficient and reliable service. Electrical equipment PM requires power quality analysis. **Caution:** Only qualified persons, such as licensed electricians or utility workers, should perform work on electrical systems.

Power Quality Analysis

An *electrical power system* is a system that produces, transmits, distributes, and delivers electrical power in order to operate electrical loads designed for connection to the system. An electrical power system may be small and simple or large and complex. For example, a portable generator used by company electricians or by an emergency crew is a small, self-contained electrical power system. A large utility company serving a metropolitan area is a large electrical power system.

Regardless of the size of an electrical power system, power that allows the loads to operate satisfactorily must be supplied. Damage to electrical equipment occurs when electrical power is not supplied at the proper level (voltage), amount (current), type (single-phase or three-phase and AC or DC), or condition (purity).

When electrical power is properly supplied to a load, the load should operate for years without a problem. When problems occur with a load, the load must be serviced or replaced. Servicing or replacing a load is only a short-term solution when a problem exists with the electrical power system. Problems with electrical power systems or loads result in damage to production equipment, costly downtime, and/or safety hazards.

The quality of the incoming power must be tested to ensure proper system and load operation before a load is connected to a system or when a load is serviced. Power quality must also be tested as part of a PM program. Electrical system PM is performed to keep electrical systems and loads operating with little or no downtime.

When troubleshooting power quality problems, an electrician must identify the source of the problem. The problem may come from an outside source such as a lightning strike on the incoming power lines of the facility. The electrical distribution system of the facility or building can be the source of problems because of improper grounding and/or undersized conductors. Power quality problems may also be caused by one or all of the loads connected to the power distribution system. Loads connected to a power distribution system can cause problems such as harmonics on the power lines. **See Figure 20-21.**

Measurements must be taken to troubleshoot power quality problems in a power distribution system. Some measurements are taken and interpreted immediately, such as those for phase sequence, low or high voltages, or overloaded circuits. Other measurements, such as transient voltages, voltage sags, or voltage swells, require time to acquire an understanding of the system and circuit problems and for acquiring usable data. Acquiring measurements over time is preferred because power quality problems often occur intermittently and may only be detectable at certain times of the day or during a certain sequence of production operations. Power quality problems are found using test instruments such as oscilloscopes, clamp-on ammeters, phase detectors, or multimeters or by using specialized test instruments such as harmonic analyzers and power quality analyzers. **See Figure 20-22.**

Power Problem Sources

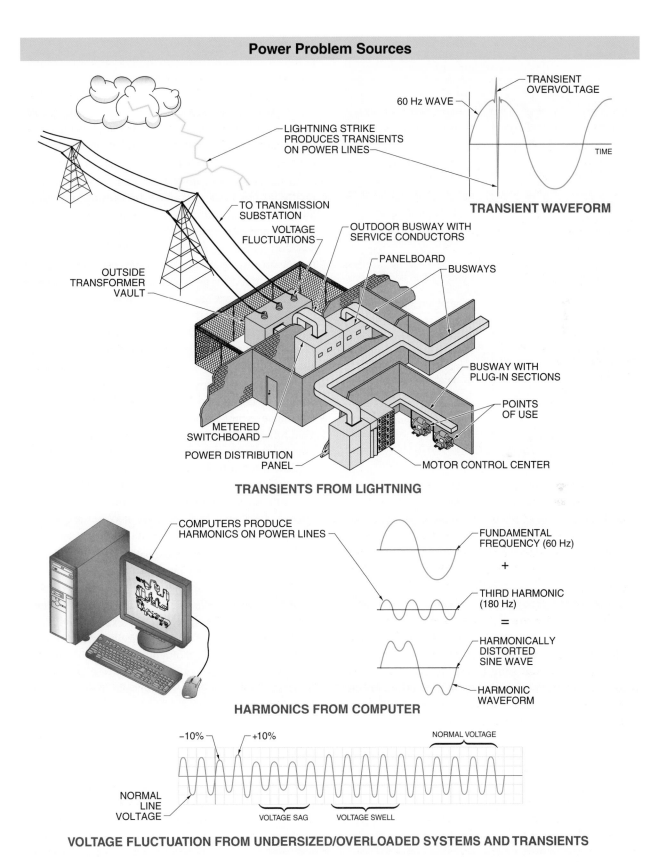

TRANSIENT WAVEFORM

TRANSIENTS FROM LIGHTNING

HARMONICS FROM COMPUTER

VOLTAGE FLUCTUATION FROM UNDERSIZED/OVERLOADED SYSTEMS AND TRANSIENTS

Figure 20-21. Power-related problems result from causes outside a facility or from loads connected inside a facility.

Power Quality Problem Test Instruments	
Power Quality Problem	**Test Instrument Used to Detect Problem**
Improper phase sequence, three-phase lines	Phase detector tester
Phase unbalance	Three-phase power analyzer meter
Voltage unbalance, three-phase lines	Digital multimeter
Single phasing	Clamp-on ammeter
Current unbalance, three-phase lines	Clamp-on ammeter
Overcurrent problems	Clamp-on ammeter
Overheated equipment	Temperature meter
Transients	Power analyzer meter
Harmonics	Harmonic meter – power analyzer meter
Power interruptions	Voltmeter with MIN MAX – power analyzer meter
Voltage problems (sags/swells)	Voltmeter with MIN MAX – power analyzer meter
Power factor	Power analyzer meter
Noise	Oscilloscope – power analyzer meter

Figure 20-22. Power quality problems are found using test instruments such as oscilloscopes, clamp-on ammeters, multimeters, temperature meters, and three-phase power quality analyzers.

Loose connections that create intermittent open pathways in an electrical system are the most common type of electrical problem. Connections can become loose from heating, cooling, and vibration. Loose connections cause high resistance, which leads to dangerous overheating and open circuits. Busway connections and other electrical equipment should be inspected according to manufacturer instructions. **WARNING:** The tightness of each electrical connection should be tested at least annually. The integrity of electrical connections should only be tested on circuits that are deenergized and locked and tagged out.

Conductor (wire) insulation must be inspected annually for cracking, darkening, brittleness, or wearing away of the insulation on moving equipment. Chemicals, oil, and dirt cause insulation to deteriorate quickly. The correct type of insulation must be selected for use in the various operating environments. For example, insulation rated for high temperatures must be used on equipment that operates at high temperatures. The size of the conductor and its insulation quality must comply with provisions of the NEC®. A clamp-on ammeter set to measure current can be used to verify that conductors are not carrying excess current.

Voltage drop across contacts should be tested annually for correct operation and to ensure that high resistance is not present at the contacts. Equipment used infrequently, such as disconnects, overloads, and circuit breakers, should be operated at least annually to prevent them from becoming stuck in one position. The mechanical operation of relays, solenoids, and magnetic starters should be inspected quarterly. All springs, interlocks, and mechanical stops should be in their proper position.

Noise, such as humming or chattering, is an indication of mechanical problems interfering with the closing of contacts. Changes in operating sounds, loudness, and/or frequency can indicate developing electrical problems. Equipment should be inspected for excessive vibration and high temperatures. Overheated insulation has a distinct odor and is an indication of a potential serious problem. All equipment enclosures must be inspected annually for cleanliness, physical damage, and solid mounting.

Review

1. Define maintenance.

2. What does PM consist of?

3. Define predictive maintenance (PDM).

4. What does developing a manual PM program require?

5. Define PM system.

6. What are the benefits of a CMMS?

7. What are some of the items included in the master file of a plant survey?

8. Why is it important to scan barcodes as parts are entered and removed from inventory?

9. Define infrared thermography.

10. What is the primary benefit of using infrared thermography as a PM tool?

11. What is a thermal signature?

12. Define ferrography and describe how tests are performed.

13. What is used to test the integrity of a solid material during ultrasonic analysis?

14. How does a strobe tachometer measure speed?

15. List several test instruments that can be used to find power quality problems.

Digital Resources

ATPeResources.com/Quicklinks
Access Code 462409

Appendix

MATH REVIEW

FORMULAS

An *equation* is a means of showing that two numbers or two groups of numbers are equal to the same amount. **See Figure A-1.**

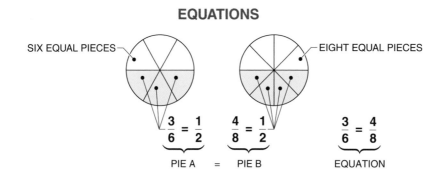

Figure A-1.

A *formula* is a mathematical equation that contains a fact, rule, or principle. Formulas are used in all trade areas. They are used to find the area of plane figures such as circles, triangles, quadrilaterals, etc. Italic letters are used in formulas to represent values (amounts).

For example, $a + b = c$ is a formula. **See Figure A-2.** In a formula, any number or letter may be transposed from left to right or from right to left of the equal sign. When transposed, the sign of the number or letter is changed to the opposite sign.

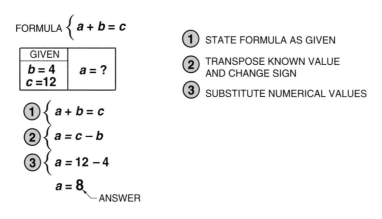

Figure A-2.

The sign is always in front of the number or letter of which it is a part. For example, in the formula $a + b = c$, if $b = 4$ and $c = 12$, the value of a is found by changing the formula to $a = c - b$, or $a = 12 - 4$. A formula can be changed to solve for any unknown value if the other values are known. Subscript letters or numbers may be used in formulas to distinguish between similar dimensions of different objects. For example, V_c may indicate the volume of a cylinder and V_t may indicate the volume of a tank.

PLANE FIGURES

A *plane figure* is a flat figure with no depth. All plane figures are composed of straight or curved lines. Plane figures include circles, triangles, quadrilaterals, and polygons. The area of a plane figure is measured in square units such as square inches, square feet, square millimeters, square meters, etc.

LINES

A *line* is the boundary of a surface. **See Figure A-3.** Lines are measured in linear units such as inches, feet, yards, millimeters, meters, etc.

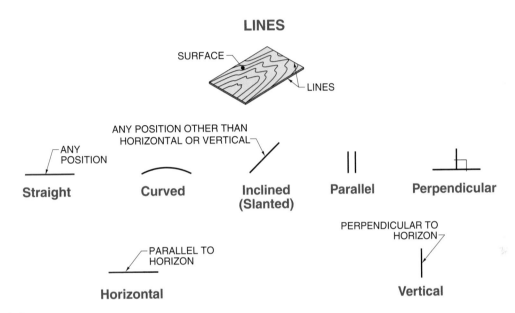

Figure A-3.

A *straight line* is the shortest distance between two points. It is commonly referred to as a line. A *curved line* is a line that continually changes direction. It is commonly referred to as a curve.

All lines may be drawn in any position, unless they are horizontal or vertical. An *inclined line* is a line that is slanted. It is neither horizontal nor vertical. *Parallel lines* are two or more lines that remain the same distance apart. The symbol for parallel lines is ‖.

A *perpendicular line* is a line that makes a 90° angle with another line. The symbol for perpendicular is ⊥. A *horizontal line* is a line that is parallel to the horizon. It may be referred to as a level line.

A *vertical line* is a line that is perpendicular to the horizon. It is often referred to as a plumb line. *Plumb* is an exact verticality (determined by a plumb bob and line) with the Earth's surface.

ANGLES

An *angle* is the intersection of two lines or sides. **See Figure A-4.** The *vertex* is the point of intersection of the sides of an angle. To identify angles, letters are placed at the end of each side and at the vertex.

ANGLES

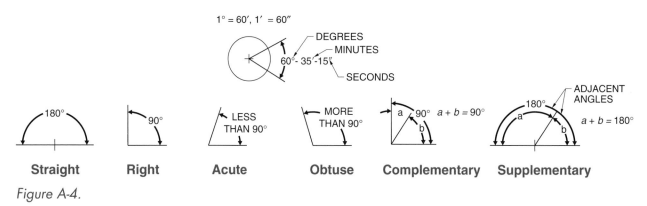

Figure A-4.

When referring to an angle, the vertex letter is read second. The angle symbol ($\angle$) is used to indicate an angle. The way the two sides of an angle intersect determines the size and type of angle.

Angles are measured in degrees, minutes, and seconds. The symbol for degrees is $°$, the symbol for minutes is $'$, the symbol for seconds is $''$. There are 360° in a circle (one revolution). There are 60′ in one degree and 60″ in one minute. For example, an angle might contain 112°-30′-12″.

A *straight angle* is an angle formed by two lines that intersect to form a straight line and measures exactly 180°. It is one-half of a revolution, or $^{360}\!/_2 = 180°$.

A *right angle* is an angle formed by two perpendicular lines and measures exactly 90°. It is one-fourth of a revolution, or $^{360}\!/_4 = 90°$.

An *acute angle* is an angle that contains less than 90°. An *obtuse angle* is an angle exceeding 90° but less than 180°. For example, a 45° angle is an acute angle, and a 135° angle is an obtuse angle.

Lines can intersect to create more than one angle. *Complementary angles* are two angles formed by three lines in which the sum of the two angles equals 90°. Each complementary angle is an acute angle. For example, a 30° angle and a 60° angle are acute angles that are complementary angles.

To find the complementary angle of a known acute angle, subtract the known angle from 90. For example, to find the complementary angle of a 40° angle, subtract 40 from 90 (90 – 40 = 50). The complementary angle to a 40° angle is a 50° angle.

Supplementary angles are two angles formed by three lines in which the sum of the two angles equals 180°. For example, a 45° angle and a 135° angle are supplementary angles.

To find the supplementary angle of a known angle, subtract the known angle from 180. For example, to find the supplementary angle of a 70° angle, subtract 70 from 180 (180 – 70 = 110). The supplementary angle to a 70° angle is a 110° angle.

Adjacent angles are angles that have the same vertex and one side in common. Adjacent angles are formed when two or more lines intersect.

For example, when two straight lines intersect, four angles and four sets of adjacent angles are formed. The sum of adjacent angles that form a straight line equals 180°. The two angles opposite each other when two straight lines intersect are equal.

Linear Measure

Linear measure is the measurement of length. It is used to find the one-dimensional length of an object. It measures distances such as how far, how long, etc. The common units used for linear measure are the inch ($''$) in the English system and the millimeter (mm) in the metric system.

Area

Area is the number of unit squares equal to the surface of an object. For example, a standard size sheet of plywood is 4′ × 8′. It contains an area of 32 sq ft (4 × 8 = 32 sq ft).

Area is expressed in square inches, square feet, and other units of measure. A square inch measures 1″ × 1″ or its equivalent. A square foot contains 144 sq in. (12″ × 12″ = 144 sq in.). The area of any plane figure can be determined by applying the proper formula. **See Figure A-5.**

AREA

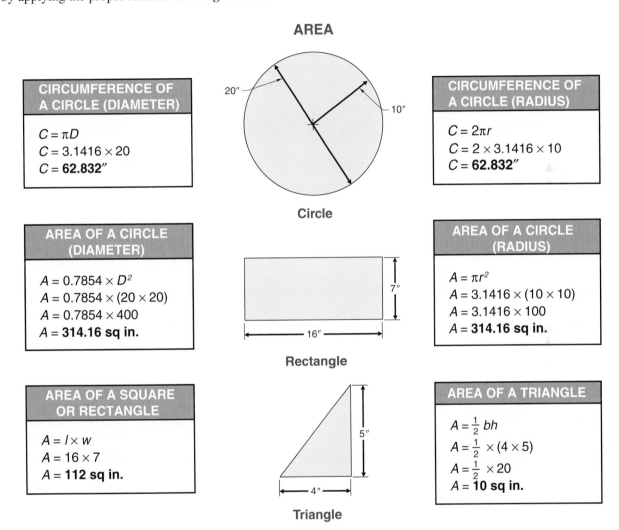

CIRCUMFERENCE OF A CIRCLE (DIAMETER)

$C = \pi D$
$C = 3.1416 \times 20$
$C = \mathbf{62.832''}$

CIRCUMFERENCE OF A CIRCLE (RADIUS)

$C = 2\pi r$
$C = 2 \times 3.1416 \times 10$
$C = \mathbf{62.832''}$

Circle

AREA OF A CIRCLE (DIAMETER)

$A = 0.7854 \times D^2$
$A = 0.7854 \times (20 \times 20)$
$A = 0.7854 \times 400$
$A = \mathbf{314.16\ sq\ in.}$

AREA OF A CIRCLE (RADIUS)

$A = \pi r^2$
$A = 3.1416 \times (10 \times 10)$
$A = 3.1416 \times 100$
$A = \mathbf{314.16\ sq\ in.}$

Rectangle

AREA OF A SQUARE OR RECTANGLE

$A = l \times w$
$A = 16 \times 7$
$A = \mathbf{112\ sq\ in.}$

AREA OF A TRIANGLE

$A = \frac{1}{2}\ bh$
$A = \frac{1}{2} \times (4 \times 5)$
$A = \frac{1}{2} \times 20$
$A = \mathbf{10\ sq\ in.}$

Triangle

Figure A-5.

CIRCLES

A *circle* is a plane figure generated about a centerpoint. **See Figure A-6.** All circles contain 360°. The *circumference* is the boundary of a circle.

The *diameter* is the distance from circumference to circumference through the centerpoint. The *centerpoint* is the point a circle or arc is drawn around.

An *arc* is a portion of the circumference. The *radius* is the distance from the centerpoint to the circumference. It is one-half the length of the diameter.

CIRCLES

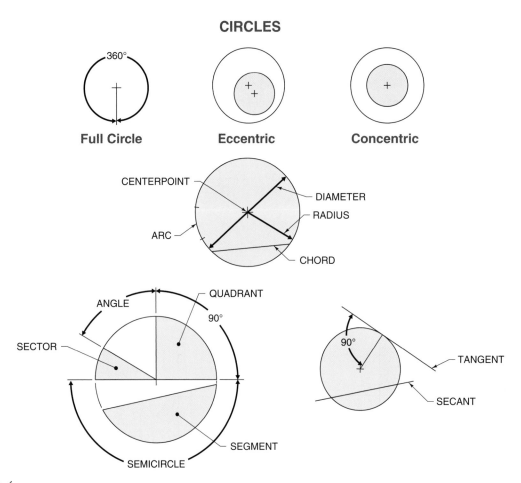

Full Circle   **Eccentric**   **Concentric**

Figure A-6.

A *chord* is a line from circumference to circumference not through the centerpoint. A *quadrant* is one-fourth of a circle containing 90°.

A *sector* is a pie-shaped piece of a circle. A *segment* is the portion of a circle set off by a chord. A *semicircle* is one-half of a circle containing 180°.

Concentric circles are two or more circles with different diameters but the same centerpoint. *Eccentric circles* are two or more circles with different diameters and different centerpoints.

A *tangent* is a straight line touching the curve of the circumference at only one point. A tangent is perpendicular to the radius. A *secant* is a straight line touching the circumference at two points.

Circumference of a Circle (Diameter). When the diameter is known, the circumference of a circle is found by applying the formula:

$$C = \pi D$$

where

C = circumference

π = 3.1416

D = diameter

For example, what is the circumference of a 20″ diameter circle?

$$C = \pi D$$
$$C = 3.1416 \times 20$$
$$C = \textbf{62.832″}$$

Circumference of a Circle (Radius). When the radius is known, the circumference of a circle is found by applying the formula:

$C = 2\pi r$

where

C = circumference

2 = constant

π = 3.1416

r = radius

For example, what is the circumference of a 10″ radius circle?

$C = 2\pi r$

$C = 2 \times 3.1416 \times 10$

$C = \mathbf{62.832″}$

Area of a Circle (Diameter). When the diameter is known, the area of a circle is found by applying the formula:

$A = 0.7854 \times D^2$

where

A = area

0.7854 = constant

D^2 = diameter squared

For example, what is the area of a 28″ diameter circle?

$A = 0.7854 \times D^2$

$A = 0.7854 \times (28 \times 28)$

$A = 0.7854 \times 784$

$A = \mathbf{615.754 \ sq \ in.}$

Area of a Circle (Radius). When the radius is known, the area of a circle is found by applying the formula:

$A = \pi r^2$

where

A = area

π = 3.1416

r^2 = radius squared

For example, what is the area of a 14″ radius circle?

$A = \pi r^2$

$A = 3.1416 \times (14 \times 14)$

$A = 3.1416 \times 196$

$A = \mathbf{615.754 \ sq \ in.}$

Triangles

A *triangle* is a three-sided polygon with three interior angles. The sum of the three angles of a triangle is always 180°. The sign (Δ) indicates a triangle. **See Figure A-7.** The *altitude* of a triangle is the perpendicular dimension from the vertex to the base. The *base* of a triangle is the side upon which the triangle stands. Any side can be taken as the base.

The angles of a triangle are named by uppercase letters. The sides of a triangle are named by lowercase letters. For example, a triangle may be named ΔABC and contain sides d, e, and f.

TRIANGLES

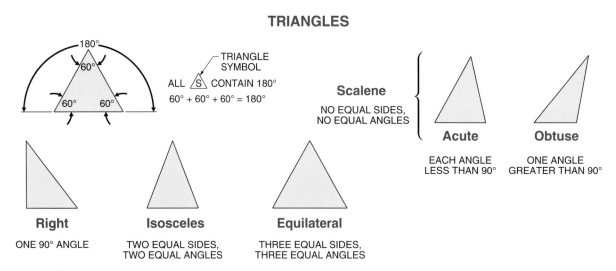

Figure A-7.

The different kinds of triangles are right triangles, isosceles triangles, equilateral triangles, and scalene triangles. A *right triangle* is a triangle that contains one 90° angle. An *isosceles triangle* is a triangle that contains two equal angles and two equal sides. An *equilateral triangle* is a triangle that has three equal angles and three equal sides. Each angle of an equilateral triangle is 60°. A *scalene triangle* is a triangle that has no equal angles or equal sides. A scalene triangle may be acute or obtuse. An *acute triangle* is a scalene triangle with each angle less than 90°. An *obtuse triangle* is a scalene triangle with one angle greater than 90°.

Area of a Triangle. The area of a triangle is found by applying the formula:

$A = \frac{1}{2}bh$
where
A = area
$\frac{1}{2}$ = constant
b = base
h = height

For example, what is the area of a triangle with a 10″ base and a 12″ height?

$A = \frac{1}{2}bh$
$A = \frac{1}{2} \times (10 \times 12)$
$A = \frac{1}{2} \times 120$
$A = \textbf{60 sq in.}$

Pythagorean Theorem. The *Pythagorean theorem* states that the square of the hypotenuse of a right triangle is equal to the sum of the squares of the other two sides. The *hypotenuse* is the side of a right triangle opposite the right angle.

A right triangle is said to have a 3-4-5 relationship and often is used for laying out right angles and checking corners for squareness. To check a corner for squareness, measure 3′ along one side and 4′ along the other side. These two points measure 5′ apart when the corner is square. **See Figure A-8.**

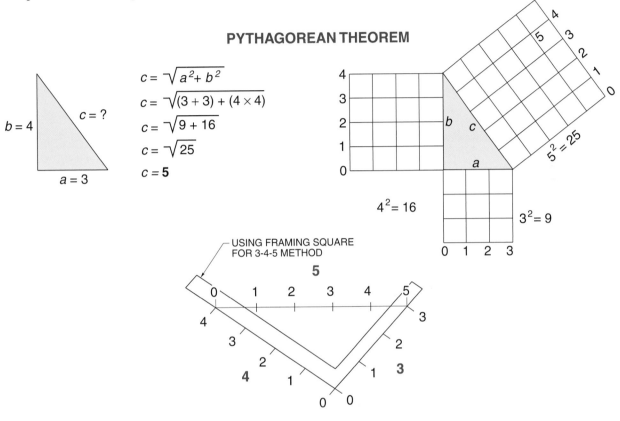

Laying Out and Checking Right Angles

Figure A-8.

The length of the hypotenuse of a right triangle is found by applying the formula:

$$c = \sqrt{a^2 + b^2}$$

where

c = length of hypotenuse

a^2 = length of one side squared

b^2 = length of other side squared

For example, what is the length of the hypotenuse of a triangle having sides of 3′ and 4′?

$$c = \sqrt{a^2 + b^2}$$

$$c = \sqrt{(3 \times 3) + (4 \times 4)}$$

$$c = \sqrt{9 + 16}$$

$$c = \sqrt{25}$$

$$c = \mathbf{5'}$$

Regular Polygons

A *polygon* is a many-sided plane figure. All polygons are bound by straight lines. A *regular polygon* is a polygon with equal sides and equal angles. An *irregular polygon* has unequal sides and unequal angles. Polygons are named according to their number of sides. Typical polygons include the triangle (three sides), quadrilateral (four sides), pentagon (five sides), hexagon (six sides), and octagon (eight sides). **See Figure A-9.**

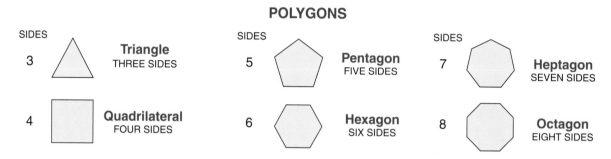

Figure A-9.

Quadrilaterals

A *quadrilateral* is a four-sided polygon with four interior angles. The sum of the four angles of a quadrilateral is always 360°. The kinds of quadrilaterals are squares, rectangles, rhombuses, rhomboids, trapezoids, and trapeziums. **See Figure A-10.**

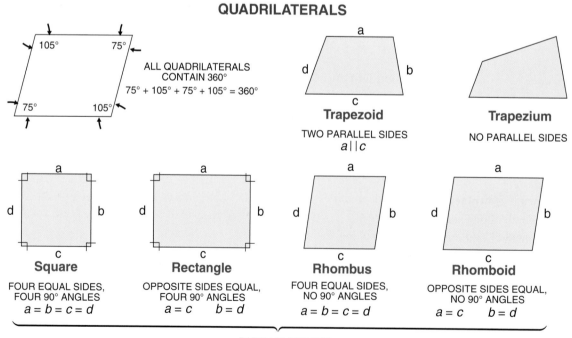

Figure A-10

A *square* is a quadrilateral with all sides equal and four 90° angles. A *rectangle* is a quadrilateral with opposite sides equal and four 90° angles. A *rhombus* is a quadrilateral with all sides equal and no 90° angles. A *rhomboid* is a quadrilateral with opposite sides equal and no 90° angles.

The square, rectangle, rhombus, and rhomboid are parallelograms. A *parallelogram* is a four-sided plane figure with opposite sides parallel and equal.

A *trapezoid* is a quadrilateral with two sides parallel. A *trapezium* is a quadrilateral with no sides parallel. Trapezoids and trapeziums are not parallelograms because all opposite sides are not parallel.

Area of a Square or Rectangle. The area of a square or the area of a rectangle is found by applying the formula:

$A = l \times w$

where

A = area

l = length

w = width

For example, what is the area of a 22′-0″ × 16′-0″ storage room?

$A = l \times w$

$A = 22 \times 16$

$A = $ **352 sq ft**

SOLIDS

A *polyhedron* is any of a variety of solids bound by plane surfaces (faces). A *regular solid* (polyhedon) is any of a variety of solids with faces that are regular polygons (equal sides). An *irregular polyhedron* is any of a variety of solids with faces that are irregular polygons (unequal sides).

Solids have length, height, and depth. The five regular solids are the tetrahedron, hexahedron, octahedron, dodecahedron, and icosahedron. Other common solids are prisms, cylinders, pyramids, cones, and spheres. Less common solids include the torus and ellipsoid. **See Figure A-11.**

Regular Solids

A *tetrahedron* is a regular solid of four triangles. A *hexahedron* is a regular solid of six squares. It is commonly referred to as a cube. An *octahedron* is a regular solid of eight triangles. A *dodecahedron* is a regular solid of twelve pentagons. An *icosahedron* is a regular solid of twenty triangles.

Prisms

A *prism* is a solid with two bases that are parallel and identical polygons. *Bases* are the ends of a prism. The three or more sides of a prism are parallelograms. **See Figure A-12.** A prism can be triangular, rectangular, pentagonal, hexagonal, octagonal, etc., according to the shape of its bases.

Lateral faces are the sides of a prism. There are as many of these lateral faces as there are sides in one of the bases.

The *altitude* of a prism is the perpendicular distance between the two bases. When the bases are perpendicular to the faces, the altitude equals the edge of a lateral face.

A *right prism* is a prism with lateral faces perpendicular to the bases. An *oblique prism* is a prism with lateral faces not perpendicular to the bases.

A *parallelepiped* is a prism with bases that are parallelograms. A *right parallelepiped* is a prism with all edges perpendicular to the bases. A *rectangular parallelepiped* is a prism with bases and faces that are all rectangles.

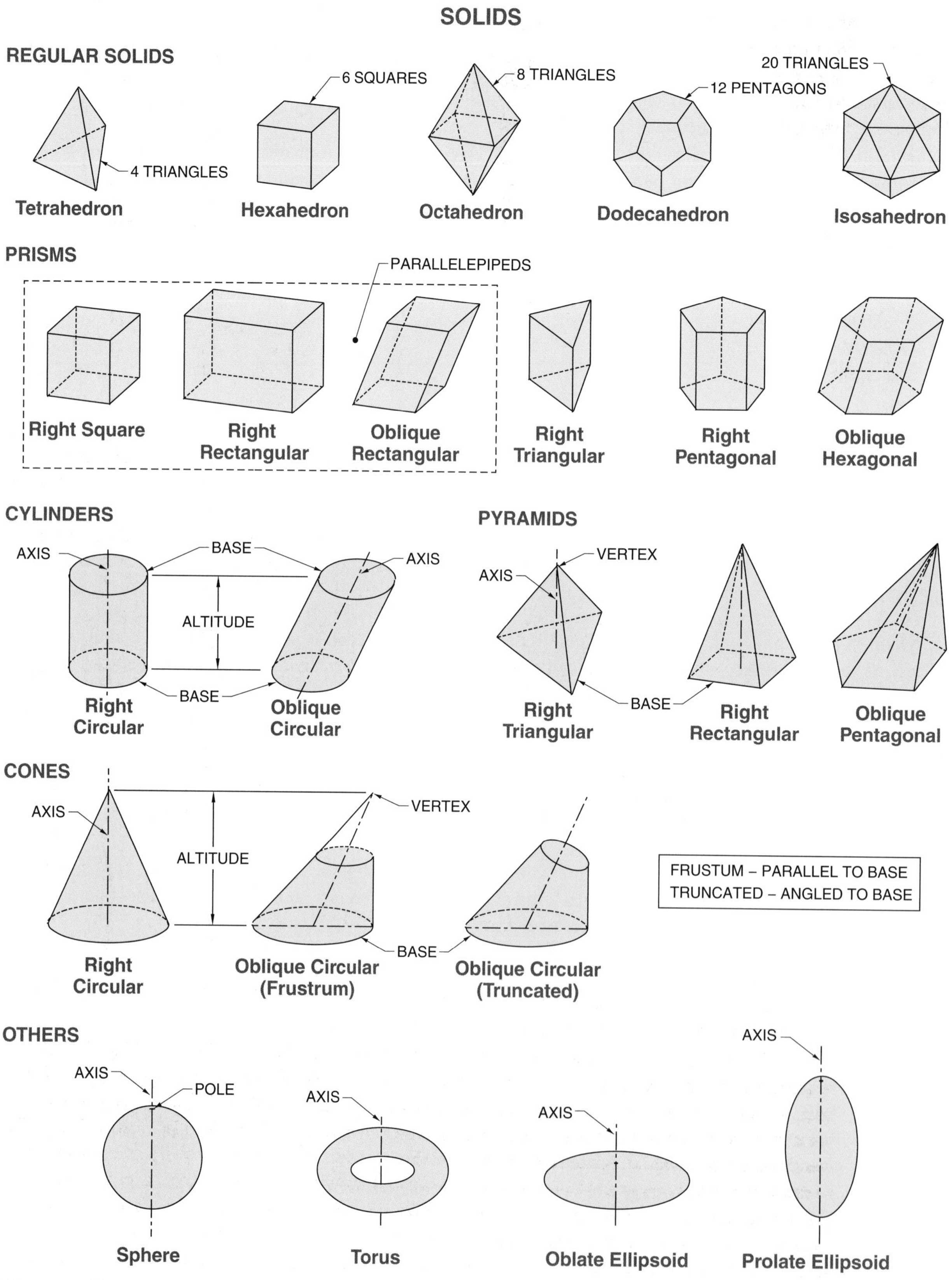

SOLIDS

REGULAR SOLIDS

Tetrahedron — 4 TRIANGLES

Hexahedron — 6 SQUARES

Octahedron — 8 TRIANGLES

Dodecahedron — 12 PENTAGONS

Isosahedron — 20 TRIANGLES

PRISMS

PARALLELEPIPEDS

Right Square

Right Rectangular

Oblique Rectangular

Right Triangular

Right Pentagonal

Oblique Hexagonal

CYLINDERS

AXIS — BASE — AXIS — ALTITUDE — BASE

Right Circular

Oblique Circular

PYRAMIDS

AXIS — VERTEX — BASE

Right Triangular

Right Rectangular

Oblique Pentagonal

CONES

AXIS — VERTEX — ALTITUDE — BASE

Right Circular

Oblique Circular (Frustrum)

Oblique Circular (Truncated)

FRUSTUM – PARALLEL TO BASE
TRUNCATED – ANGLED TO BASE

OTHERS

AXIS — POLE

AXIS

AXIS

AXIS

Sphere

Torus

Oblate Ellipsoid

Prolate Ellipsoid

Figure A-11

PRISMS

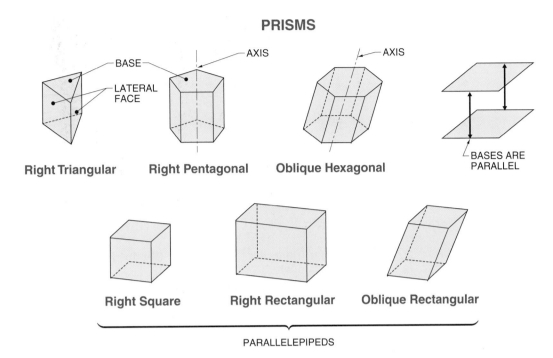

Figure A-12

Cylinders

A *cylinder* is a solid generated by a straight line moving in contact with a curve and remaining parallel to the axis and its previous position. Each position of the straight line forms an element of the cylinder.

A *right cylinder* is a cylinder with the axis perpendicular to the base. An *oblique cylinder* is a cylinder with the axis not perpendicular to the base. **See Figure A-13.**

CYLINDERS

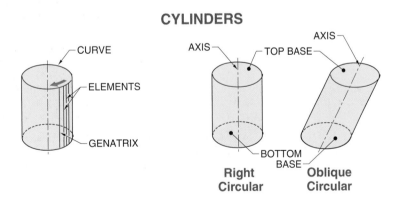

Figure A-13

Pyramids

A *pyramid* is a solid with a base that is a polygon and sides that are triangles. The *vertex* is the common point of the triangular sides that form the pyramid. **See Figure A-14.**

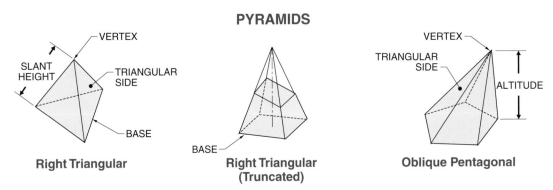

PYRAMIDS

Right Triangular

Right Triangular (Truncated)

Oblique Pentagonal

Figure A-14

The *altitude* of a pyramid is the perpendicular distance from the vertex to the base. Pyramids are named according to the kind of polygon forming the base, such as triangular, quadrangular, pentagonal, and hexagonal.

A *regular pyramid* has a base that is a regular polygon and a vertex that is perpendicular to the center of the base. The *slant height* is the distance from the base to the vertex parallel to a side. It is the altitude of one of the triangles that forms the sides.

Cones

A *cone* is a solid generated by a straight line moving in contact with a curve and passing through the vertex. Cones have a circular base and a surface that tapers from the base to the vertex.

The altitude of a cone is the perpendicular distance from the vertex to the base. The slant height is the distance from the vertex to any point on the circumference of the base. **See Figure A-15.**

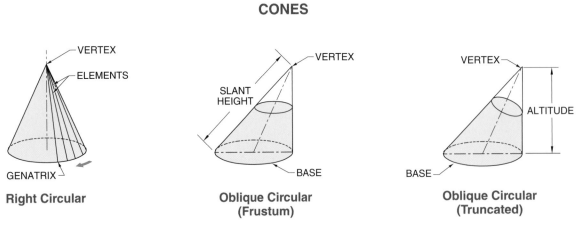

CONES

Right Circular

Oblique Circular (Frustum)

Oblique Circular (Truncated)

Figure A-15

Conic Sections. A *conic section* is a curve produced by a plane intersecting a right circular cone. A *right circular cone* is a cone with the axis located at a 90° angle to the circular base. The four conic sections are the circle, ellipse, parabola, and hyperbola. **See Figure A-16.**

CONIC SECTIONS

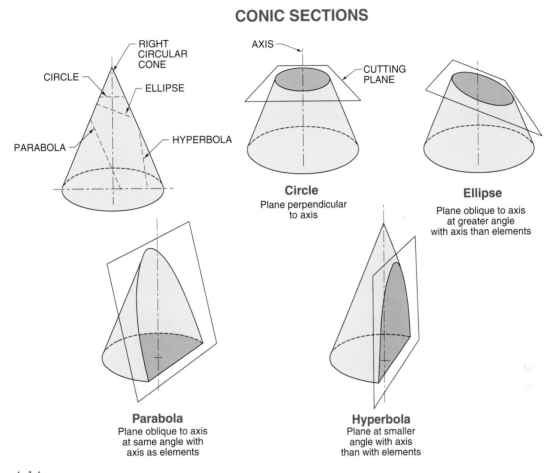

Figure A-16

A *circle* is a plane figure formed by a cutting plane perpendicular to the axis of a cone. An *ellipse* is a plane figure formed by a cutting plane oblique to the axis of a cone but at a greater angle with the axis than with the elements of the cone.

A *parabola* is a plane figure formed by a cutting plane oblique to the axis and parallel to the elements of the cone. A *hyperbola* is a plane figure formed by a cutting plane that has a smaller angle with the axis than with the elements of the cone.

Frustums. A *frustum* of a pyramid or cone is the remaining portion of a pyramid or cone with a cutting plane passed parallel to the base. A truncated pyramid or cone is the remaining portion of a pyramid or cone with the cutting plane passed not parallel to the base. **See Figure A-17**

FRUSTUMS AND TRUNCATIONS

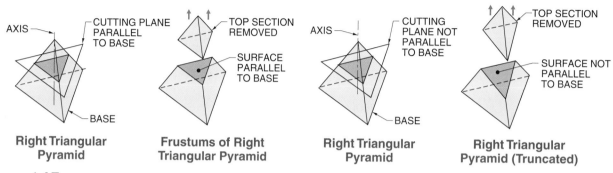

Right Triangular Pyramid **Frustums of Right Triangular Pyramid** **Right Triangular Pyramid** **Right Triangular Pyramid (Truncated)**

Figure A-17

Spheres

A *sphere* is a solid generated by a circle revolving about one of its axes. All points on the surface are an equal distance from the center of the sphere. **See Figure A-18.**

SPHERES

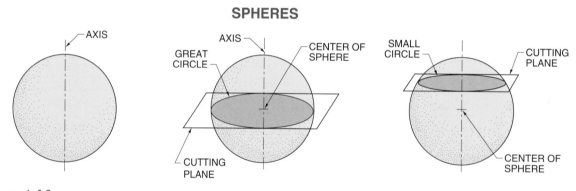

Figure A-18

A *great circle* is the circle formed by passing a cutting plane through the center of a sphere. A *small circle* is the circle formed by passing a cutting plane through a sphere but not through the center. The circumference of a sphere is equal to the circumference of a great circle.

VOLUME

Volume is the size of a three-dimensional space or object as measured in cubic units. For example, the volume of a standard size concrete block is 1024 cu in. (8″ × 8″ × 16″ = 1024 cu in.).

Volume is expressed in cubic inches, cubic feet, cubic yards, and other units of cubic measure. A cubic inch measures 1″ × 1″ × 1″ or its equivalent.

A cubic foot contains 1728 cu in. (12″ × 12″ × 12″ = 1728 cu in.). A cubic yard contains 27 cu ft (3′ × 3′ × 3′ = 27 cu ft).

Finding Volume

The volume of a solid figure can be determined by applying the proper formula. **See Figure A-19.**

Volume of a Rectangular Solid. The volume of a rectangular solid is found by applying the formula:

$V = l \times w \times h$

where

V = volume

l = length

w = width

h = height

VOLUME

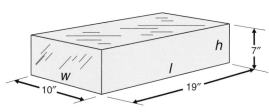

VOLUME OF A RECTANGULAR SOLID

$V = l \times w \times h$

$V = 19 \times 10 \times 7$

$V = 1330$ cu in.

$V = \frac{1330}{1728} =$ **0.770 cu ft**

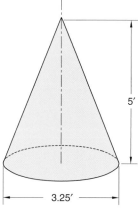

VOLUME OF A CONE

1. Solve for area of base

$A_b = 0.7854 \times D^2$

$A_b = 0.7854 \times (3.25 \times 3.25)$

$A_b = 0.7854 \times 10.563$

$A_b =$ **8.296 sq ft**

2. Solve for volume

$V = \frac{A_b \, a}{3}$

$V = \frac{8.296 \times 5}{3}$

$V = \frac{41.48}{3}$

$V =$ **13.826 cu ft**

VOLUME OF A SPHERE (DIAMETER)

$V = \frac{\pi D^3}{6}$

$V = \frac{3.1416 \times 7^3}{6}$

$V = \frac{3.1416 \times 343}{6}$

$V = \frac{1077.569}{6}$

$V =$ **179.595 cu ft**

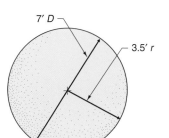

VOLUME OF A SPHERE (RADIUS)

$V = \frac{4\pi \, r^3}{3}$

$V = \frac{4 \times 3.1416 \times 3.5^3}{3}$

$V = \frac{4 \times 3.1416 \times 42.875}{3}$

$V = \frac{538.784}{3}$

$V =$ **179.595 cu ft**

VOLUME OF A CYLINDER (DIAMETER)

$V = 0.7854 \times D^2 \times h$

$V = 0.7854 \times (16 \times 16) \times 60$

$V = 0.7854 \times 256 \times 60$

$V = 12,063.744$ cu in.

$V = \frac{12,063.744}{1728}$

$V =$ **6.981 cu ft**

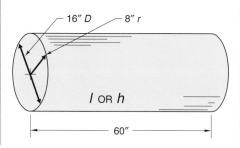

VOLUME OF A CYLINDER (RADIUS)

$V = \pi \, r^2 \times l$

$V = 3.1416 \times (8 \times 8) \times 60$

$V = 3.1416 \times 64 \times 60$

$V = 12,063.744$ cu in.

$V = \frac{12,063.744}{1728}$

$V =$ **6.981 cu ft**

Figure A-19

For example, what is the volume of a 24″ × 12″ × 8″ rectangular solid?

$V = l \times w \times h$

$V = 24 \times 12 \times 8$

$V = $ 2304 cu in.

Volume of a Cone. The volume of a cone is found by first solving for the area of the base and then solving for volume. The area of the base is found by applying the formula:

$A_b = 0.7854 \times D^2$

where

A_b = area of base

0.7854 = constant

D^2 = diameter squared

The volume of the cone is then found by applying the formula:

$$V = \frac{A_b a}{3}$$

where

V = volume

A_b = area of base

a = altitude

3 = constant

For example, what is the volume of a cone that has a 14″ diameter and a 35″ altitude?

1. Solve for the area of the base.

$A_b = 0.7854 \times D^2$

$A_b = 0.7854 \times (14 \times 14)$

$A_b = 0.7854 \times 196$

$A_b = 153.938$ sq in.

2. Solve for the volume.

$$V = \frac{A_b a}{3}$$

$$V = \frac{153.938 \times 35}{3}$$

$$V = \frac{5387.83}{3}$$

$V = $ 1795.943 cu in.

Volume of a Sphere (Diameter). When the diameter is known, the volume of a sphere is found by applying the formula:

$$V = \frac{\pi D^3}{6}$$

where

V = volume

π = 3.1416

D^3 = diameter cubed

6 = constant

For example, what is the volume of a sphere that is 4'-0" in diameter?

$$V = \frac{\pi D^3}{6}$$

$$V = \frac{3.1416 \times 4^3}{6}$$

$$V = \frac{3.1416 \times 64}{6}$$

$$V = \frac{201.062}{6}$$

$$V = \textbf{33.510 cu ft}$$

Volume of a Sphere (Radius). When the radius is known, the volume of a sphere is found by applying the formula:

$$V = \frac{4\pi r^3}{3}$$

where

V = volume

4 = constant

π = 3.1416

r^3 = radius cubed

3 = constant

For example, what is the volume of a sphere that has a 2'-0" radius?

$$V = \frac{4\pi r^3}{3}$$

$$V = \frac{4 \times 3.1416 \times 2^3}{6}$$

$$V = \frac{4 \times 3.1416 \times 8}{6}$$

$$V = \frac{100.531}{6}$$

$$V = \textbf{33.510 cu ft}$$

Volume of a Cylinder (Diameter). When the diameter is known, the volume of a cylinder is found by applying the formula:

$V = 0.7854 \times D^2 \times h$

where

V = volume

0.7854 = constant

D^2 = diameter squared

h = height

For example, what is the volume of a tank that is 4′-0″ in diameter and 12′-0″ long?

$V = 0.7854 \times D^2 \times h$

$V = 0.7854 \times (4 \times 4) \times 12$

$V = 0.7854 \times 16 \times 12$

$V = \mathbf{150.797\ cu\ ft}$

Volume of a Cylinder (Radius). When the radius is known, the volume of a cylinder is found by applying the formula:

$V = \pi r^2 \times l$

where

V = volume

π = 3.1416

r^2 = radius squared

l = length

For example, what is the volume of a tank that has a 2′-0″ radius and is 12′-0″ long?

$V = \pi r^2 \times l$

$V = 3.1416 \times (2 \times 2) \times 12$

$V = 3.1416 \times 4 \times 12$

$V = \mathbf{150.797\ cu\ ft}$

English System

	Unit	Abbr	Equivalents
LENGTH	mile	mi	5280′, 320 rd, 1760 yd
	rod	rd	5.50 yd, 16.5′
	yard	yd	3′, 36″
	foot	ft *or* ′	12″, 0.333 yd
	inch	in. *or* ″	0.083′, 0.028 yd
AREA $A = l \times w$	square mile	sq mi *or* mi²	640 A, 102,400 sq rd
	acre	A	4840 sq yd, 43,560 sq ft
	square rod	sq rd *or* rd²	30.25 sq yd, 0.00625 A
	square yard	sq yd *or* yd²	1296 sq in., 9 sq ft
	square foot	sq ft *or* ft²	144 sq in., 0.111 sq yd
	square inch	sq in. *or* in²	0.0069 sq ft, 0.00077 sq yd
VOLUME $V = l \times w \times h$	cubic yard	cu yd *or* yd³	27 cu ft, 46,656 cu in.
	cubic foot	cu ft *or* ft³	1728 cu in., 0.0370 cu yd
	cubic inch	cu in. *or* in³	0.00058 cu ft, 0.000021 cu yd
CAPACITY WATER, FUEL, ETC. — *US liquid measure*	gallon	gal.	4 qt (231 cu in.)
	quart	qt	2 pt (57.75 cu in.)
	pint	pt	4 gi (28.875 cu in.)
	gill	gi	4 fl oz (7.219 cu in.)
	fluid ounce	fl oz	8 fl dr (1.805 cu in.)
	fluid dram	fl dr	60 min (0.226 cu in.)
	minim	min	⅛ fl dr (0.003760 cu in.)
VEGETABLES, GRAIN, ETC. — *US dry measure*	bushel	bu	4 pk (2150.42 cu in.)
	peck	pk	8 qt (537.605 cu in.)
	quart	qt	2 pt (67.201 cu in.)
	pint	pt	½ qt (33.600 cu in.)
DRUGS — *British imperial liquid and dry measure*	bushel	bu	4 pk (2219.36 cu in.)
	peck	pk	2 gal. (554.84 cu in.)
	gallon	gal.	4 qt (277.420 cu in.)
	quart	qt	2 pt (69.355 cu in.)
	pint	pt	4 gi (34.678 cu in.)
	gill	gi	5 fl oz (8.669 cu in.)
	fluid ounce	fl oz	8 fl dr (1.7339 cu in.)
	fluid dram	fl dr	60 min (0.216734 cu in.)
	minim	min	1/60 fl dr (0.003612 cu in.)
MASS AND WEIGHT COAL, GRAIN, ETC. — *Avoirdupois*	ton		2000 lb
	short ton	t	2000 lb
	long ton		2240 lb
	pound	lb *or* #	16 oz, 7000 gr
	ounce	oz	16 dr, 437.5 gr
	dram	dr	27.344 gr, 0.0625 oz
	grain	gr	0.037 dr, 0.002286 oz
GOLD, SILVER, ETC. — *Troy*	pound	lb	12 oz, 240 dwt, 5760 gr
	ounce	oz	20 dwt, 480 gr
	pennyweight	dwt *or* pwt	24 gr, 0.05 oz
	grain	gr	0.042 dwt, 0.002083 oz
DRUGS — *Apothecaries'*	pound	lb ap	12 oz, 5760 gr
	ounce	oz ap	8 dr ap, 480 gr
	dram	dr ap	3 s ap, 60 gr
	scruple	s ap	20 gr, 0.333 dr ap
	grain	gr	0.05 s, 0.002083 oz, 0.0166 dr ap

Metric System

LENGTH	Unit	Abbreviation	Number of Base Units
	kilometer	km	1000
	hectometer	hm	100
	dekameter	dam	10
	meter*	m	1
	decimeter	dm	0.1
	centimeter	cm	0.01
	millimeter	mm	0.001
AREA $A = l \times w$	square kilometer	sq km or km^2	1,000,000
	hectare	ha	10,000
	are	a	100
	square centimeter	sq cm or cm^2	0.0001
VOLUME $V = l \times w \times h$	cubic centimeter	cu cm, cm^{32}, or cc	0.000001
	cubic decimeter	dm^3	0.001
	cubic meter*	m^3	1
CAPACITY WATER, FUEL, ETC. VEGETABLES, GRAIN, ETC. DRUGS	kiloliter	kl	1000
	hectoliter	hl	100
	dekaliter	dal	10
	liter*	l	1
	cubic decimeter	dm^3	1
	deciliter	dl	0.10
	centiliter	cl	0.01
	milliliter	ml	0.001
MASS AND WEIGHT COAL, GRAIN, ETC. GOLD, SILVER, ETC. DRUGS	metric ton	t	1,000,000
	kilogram	kg	1000
	hectogram	hg	100
	dekagram	dag	10
	gram*	g	1
	decigram	dg	0.10
	centigram	cg	0.01
	milligram	mg	0.001

* base units

Prefixes			
Multiples and Submultiples	**Prefixes**	**Symbols**	**Meaning**
$1{,}000{,}000{,}000{,}000 = 10^{12}$	tera	T	trillion
$1{,}000{,}000{,}000 = 10^{9}$	giga	G	billion
$1{,}000{,}000 = 10^{6}$	mega	M	million
$1000 = 10^{3}$	kilo	k	thousand
$100 = 10^{2}$	hecto	h	hundred
$10 = 10^{1}$	deka	d	ten
Unit $1 = 10^{0}$			
$0.1 = 10^{-1}$	deci	d	tenth
$0.01 = 10^{-2}$	centi	c	hundredth
$0.001 = 10^{-3}$	milli	m	thousandth
$0.000001 = 10^{-6}$	micro	μ	millionth
$0.000000001 = 10^{-9}$	nano	n	billionth
$0.000000000001 = 10^{-12}$	pico	p	trillionth

Conversion Table												
Initial Units	**Final Units**											
	giga	**mega**	**kilo**	**hecto**	**deka**	**base**	**deci**	**centi**	**milli**	**micro**	**nano**	**pico**
giga		3R	6R	7R	8R	9R	10R	11R	12R	15R	18R	21R
mega	3L		3R	4R	5R	6R	7R	8R	9R	12R	15R	18R
kilo	6L	3L		1R	2R	3R	4R	5R	6R	9R	12R	15R
hecto	7L	4L	1L		1R	2R	3R	4R	5R	8R	11R	14R
deka	8L	5L	2L	1L		1R	2R	3R	4R	7R	10R	13R
base	9L	6L	3L	2L	1L		1R	2R	3R	6R	9R	12R
deci	10L	7L	4L	3L	2L	1L		1R	2R	5R	8R	11R
centi	11L	8L	5L	4L	3L	2L	1L		1R	4R	7R	10R
milli	12L	9L	6L	5L	4L	3L	2L	1L		3R	6R	9R
micro	15L	12L	9L	8L	7L	6L	5L	4L	3L		3R	6R
nano	18L	15L	12L	11L	10L	9L	8L	7L	6L	3L		3R
pico	21L	18L	15L	14L	13L	12L	11L	10L	9L	6L	3L	

R = move the decimal point to the right

L = move the decimal point to the left

English to Metric Equivalents

LENGTH

A = l x w

V = l x w x h

	Unit	Metric Equivalent
LENGTH 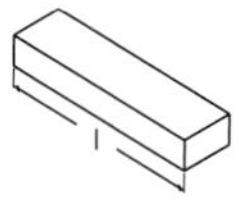	mile	1.609 km
	rod	5.029 m
	yard	0.9144 m
	foot	30.48 cm
	inch	2.54 cm
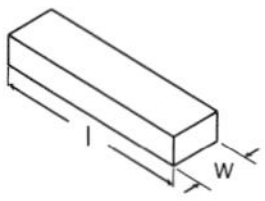 **AREA**	square mile	2.590 k^2
	acre	0.405 hectacre, 4047 m^2
	square rod	25.293 m^2
	square yard	0.836 m^2
	square foot	0.093 m^2
	square inch	6.452 cm^2
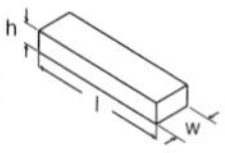 **VOLUME**	cubic yard	0.765 m^3
	cubic foot	0.028 m^3
	cubic inch	16.387 cm^3
 **CAPACITY** — US liquid measure	gallon	3.785 l
	quart	0.946 l
	pint	0.473 l
	gill	118.294 ml
	fluid ounce	29.573 ml
	fluid dram	3.697 ml
	minim	0.061610 ml
 US dry measure	bushel	35.239 l
	peck	8.810
	quart	1.101 l
	pint	0.551 l
 British imperial liquid and dry measure	bushel	0.036^3m
	peck	0.0091^3m
	gallon	4.546 l
	quart	1.136 l
	pint	568.26 cm^3
	gill	142.066 cm^3
	fluid ounce	28.412 cm^3
	fluid dram	3.5516 cm^3
	minim	0.059194^3cm
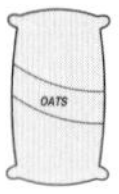 **MASS AND WEIGHT** — Avoirdupois	short ton	0.907 t
	long ton	1.016 t
	pound	0.454 kg
	ounce	28.350 g
	dram	1.772 g
	grain	0.0648 g
 Troy	pound	0.373 kg
	ounce	31.103 g
	pennyweight	1.555 g
	grain	0.0648 g
 Apothecaries'	pound	0.373 kg
	ounce	31.103 g
	dram	3.888 g
	scruple	1.296 g
	grain	0.0648 g

CAPACITY

WATER, FUEL, ETC.

VEGETABLES, GRAIN, ETC.

DRUGS

MASS AND WEIGHT

COAL, GRAIN, ETC.

GOLD, SILVER, ETC.

DRUGS

Metric to English Equivalents

LENGTH

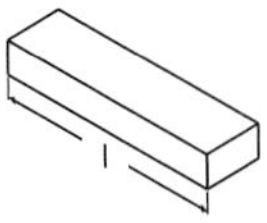

Unit	English Equivalent
kilometer	0.62 mi
hectometer	109.36 yd
dekameter	32.81′
meter	39.37″
decimeter	3.94″
centimeter	0.39 ″
millimeter	0.039″

AREA

$A = l \times w$

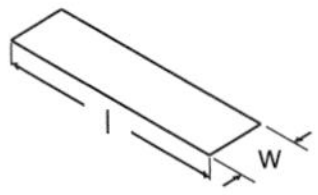

Unit	English Equivalent
square kilometer	0.3861 sq mi
hectacre	2.47 A
are	119.60 sq yd
square centimeter	0.155 sq in.

VOLUME

$V = l \times w \times h$

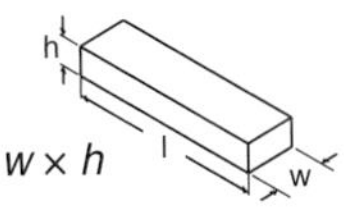

Unit	English Equivalent
cubic centimeter	0.061 cu in.
cubic decimeter	61.023 cu in.
cubic meter	1.307 cu yd

CAPACITY

WATER, FUEL, ETC.

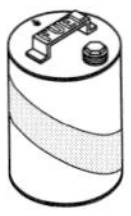

VEGETABLES, GRAIN, ETC.

DRUGS

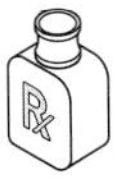

Unit	*cubic*	*dry*	*liquid*
kiloliter	1.31 cu yd		
hectoliter	3.53 cu ft	2.84 bu	
dekaliter	0.35 cu ft	1.14 pk	2.64 gal.
liter	61.02 cu in.	0.908 qt	1.057 qt
cubic decimeter	61.02 cu in.	0.908 qt	1.057 qt
deciliter	6.1 cu in.	0.18 pt	0.21 pt
centiliter	0.61 cu in.		338 fl oz
milliliter	0.061 cu in.		0.27 fl dr

MASS AND WEIGHT

COAL, GRAIN, ETC.

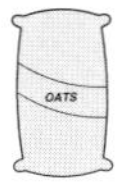

GOLD, SILVER, ETC.

DRUGS

Unit	English Equivalent
metric ton	1.102 t
kilogram	2.2046 lb
hectogram	3.527 oz
dekagram	0.353 oz
gram	0.035 oz
decigram	1.543 gr
centigram	0.154 gr
milligram	0.015 gr

Stock Material Weight*

Material	Weight	Material	Weight	Material	Weight
Metals		Chestnut	30	Granite	172
Aluminum, bronze	481	Cypress, southern	32	Greenstone, trap	187
Aluminum, cast-hammered	165	Douglas fir	34	Gypsum, alabaster	159
Antimony	416	Elm, American	35	Limestone	160
Arsenic	358	Hemlock, eastern, western	28	Magnesite	187
Bismuth	608	Hickory	53	Marble	168
Brass, cast-rolled	534	Larch, western	36	Phosphate rock, apatite	200
Chromium	428	Maple, red, black	38 – 40	Pumice, natural	40
Cobalt	552	Oak	51	Quartz, flint	165
Copper, cast-rolled	556	Pine, white, yellow, western	27 – 28	Sandstone, bluestone	147
Gold, cast-hammered	1205	Poplar, yellow	28	Slate, shale	172
Iron, cast, pig	450	Redwood	30	Soapstone, talc	169
Iron, slag	172	Spruce	28	**Bituminous Substances**	
Iron, wrought	485	Tamarack	37	Asphaltum	81
Lead	706	Walnut	39 – 40	Coal, anthracite	97
Magnesium	109	**Liquids**		Coal, bituminous	84
Manganese	456	Acids, muriatic, 40%	75	Coal, coke	75
Mercury	848	Acids, nitric, 91%	94	Coal, lignite	78
Molybdenum	562	Acids, sulphuric, 87%	112	Graphite	131
Nickel	545	Alcohol, 100%	49	Paraffin	56
Platinum, cast-hammered	1330	Gasoline	42	Petroleum, crude	55
Silver, cast-hammered	656	Lye, soda, 66%	106	Petroleum, refined	50
Steel	490	Oils	58	Pitch	69
Tin, cast-hammered	459	Petroleum	55	Tar, bituminous	75
Tungsten	1180	Water, 4°C	62	**Brick Masonry**	
Vanadium	350	Water, ice	56	Common brick	120
Zinc, cast-rolled	440	Water, seawater	64	Pressed brick	140
Solids		Water, snow, fresh fallen	8	Soft brick	100
Carbon, amorphous, graphitic	129	**Gases**		**Concrete**	
Cork	15	Air, 0°C	0.08071	Cement, cinder, etc.	100
Ebony	76	Ammonia	0.0478	Cement, slag, etc.	130
Fats	58	Carbon dioxide	0.1234	Cement, stone, sand	144
Glass, common, plate	160	Carbon monoxide	0.0781	**Building Material**	
Glass, crystal	184	Gas, natural	0.038 – 0.039	Ashes, cinders	40 – 45
Phosphorous, white	114	Hydrogen	0.00559	Cement, Portland, loose	90
Resins, rosin, amber	67	Nitrogen	0.0784	Cement, Portland, set	183
Rubber	58	Oxygen	0.0892	Lime, gypsum, loose	65 – 75
Silicon	155	**Minerals**		Mortar, set	103
Sulphur, Amorphous	128	Asbestos	153	Slags, bank, screenings	98 – 117
Wax	60	Basalt	184	Slags, bank slag	67 – 72
Timber, US Seasoned		Bauxite	159	Slags, machine slag	96
Ash, white	41	Borax	109	**Earth**	
Beech	44	Chalk	137	Clay, damp, plastic	110
Birch, yellow	43	Clay	137	Dry, packed	95
Cedar, white, red	22 – 23	Dolomite	181	Mud, packed	115

* in lb/cu ft

Twist Drill Fractional, Number, and Letter Sizes

Drill No.	Frac	Deci	Drill No.	Frac	Deci	Drill No.	Frac	Deci	Drill No.	Frac	Deci
80	—	0.0135	42	—	0.0935	7	—	0.201	X	—	0.397
79	—	0.0145	—	3/32	0.0938	—	13/64	0.203	Y	—	0.404
—	1/64	0.0156				6	—	0.204			
78	—	0.0160	41	—	0.0960	5	—	0.206	—	13/32	0.406
77	—	0.0180	40	—	0.0980	4	—	0.209	Z	—	0.413
			39	—	0.0995				—	27/64	0.422
76	—	0.0200	38	—	0.1015	3	—	0.213	—	7/16	0.438
75	—	0.0210	37	—	0.1040	—	7/32	0.219	—	29/64	0.453
74	—	0.0225				2	—	0.221			
73	—	0.0240	36	—	0.1065	1	—	0.228	—	15/32	0.469
72	—	0.0250	—	7/64	0.1094	A	—	0.234	—	31/64	0.484
			35	—	0.1100				—	1/2	0.500
71	—	0.0260	34	—	0.1110	—	15/64	0.234	—	33/64	0.516
70	—	0.0280	33	—	0.1130	B	—	0.238	—	17/32	0.531
69	—	0.0292				C	—	0.242			
68	—	0.0310	32	—	0.116	D	—	0.246	—	35/64	0.547
—	1/32	0.0313	31	—	0.120	—	1/4	0.250	—	9/16	0.562
			—	1/8	0.125				—	37/64	0.578
67	—	0.0320	30	—	0.129	E	—	0.250	—	19/32	0.594
66	—	0.0330	29	—	0.136	F	—	0.257	—	39/64	0.609
65	—	0.0350				G	—	0.261			
64	—	0.0360	—	9/64	0.140	—	17/64	0.266	—	5/8	0.625
63	—	0.0370	28	—	0.141	H	—	0.266	—	41/64	0.641
			27	—	0.144				—	21/32	0.656
62	—	0.0380	26	—	0.147	I	—	0.272	—	43/64	0.672
61	—	0.0390	25	—	0.150	J	—	0.277	—	11/16	0.688
60	—	0.0400				—	9/32	0.281			
59	—	0.0410	24	—	0.152	K	—	0.281	—	45/64	0.703
58	—	0.0420	23	—	0.154	L	—	0.290	—	23/32	0.719
			—	5/32	0.156				—	47/64	0.734
57	—	0.0430	22	—	0.157	M	—	0.295	—	3/4	0.750
56	—	0.0465	21	—	0.159	—	19/64	0.2297	—	49/64	0.766
—	3/64	0.0469				N	—	0.302			
55	—	0.0520	20	—	0.161	—	5/16	0.313	—	25/32	0.781
54	—	0.0550	19	—	0.166	O	—	0.316	—	51/64	0.797
			18	—	0.170				—	13/16	0.813
53	—	0.0595	—	11/64	0.172	P	—	0.323	—	53/64	0.828
—	1/16	0.0625	17	—	0.173	—	21/64	0.328	—	27/32	0.844
52	—	0.0635				Q	—	0.332			
51	—	0.0670				R	—	0.339			
50	—	0.0700	16	—	0.177	—	11/32	0.344	—	55/64	0.859
			15	—	0.180				—	7/8	0.875
49	—	0.0730	14	—	0.182	S	—	0.348	—	57/64	0.891
48	—	0.0760	13	—	0.185	T	—	0.358	—	29/32	0.906
—	5/64	0.0781	—	3/16	0.188	—	23/64	0.359	—	59/64	0.922
47	—	0.0785				U	—	0.368			
46	—	0.0810	12	—	0.189	—	3/8	0.375	—	15/16	0.938
			11	—	0.191				—	61/64	0.953
45	—	0.0820	10	—	0.194	V	—	0.377	—	31/32	0.969
44	—	0.0860	9	—	0.196	W	—	0.386	—	63/64	0.984
43	—	0.0890	8	—	0.199	—	25/64	0.391	—	1	1.000

Metric Screw Threads

Coarse (general purpose)		Fine	
Nom Size & Thd Pitch	Tap Drill Dia (mm)	Nom Size & Thd Pitch	Tap Drill Dia (mm)
M1.6 × 0.35	1.25	—	—
M1.8 × 0.35	1.45	—	—
M2 × 0.4	1.60	—	—
M2.2 × 0.45	1.75	—	—
M2.5 × 0.45	2.05	—	—
M3 × 0.5	2.50	—	—
M3.5 × 0.6	2.90	—	—
M4 × 0.7	3.30	—	—
M4.5 × 0.75	3.75	—	—
M5 × .8	4.20	—	—
M6.3 × 1	5.30	—	—
M7 × 1	6.00	—	—
M8 × 1.25	6.80	M8 × 1	7.00
M9 × 1.25	7.75	—	—
M10 × 1.5	8.50	M10 × 1.25	8.75
M11 × 1.5	9.50	—	—
M12 × 1.75	10.30	M12 × .25	10.50
M14 × 2	12.00	M14 × 1.5	12.50
M16 × 2	14.00	M16 × 1.5	14.50
M18 × 2.5	15.50	M18 × 1.5	16.50
M20 × 2.5	17.50	M20 × 1.5	18.50
M22 × 2.5	19.50	M22 × 1.5	20.50
M24 × 3	21.00	M24 × 2	22.00
M27 × 3	24.00	M27 × 2	25.00
M30 × 3.5	26.50	M30 × 2	28.00
M33 × 3.5	29.50	M30 × 2	31.00
M36 × 4	32.00	M36 × 3	33.00
M39 × 4	35.00	M39 × 3	36.00
M42 × 4.5	37.50	M42 × 3	39.00
M45 × 4.5	40.50	M45 × 3	42.00
M48 × 5	43.00	M48 × 3	45.00
M52 × 5	47.00	M52 × 3	49.00
M56 × 5.5	50.50	M56 × 4	52.00
M60 × 5.5	54.50	M60 × 4	56.00
M64 × 6	58.00	M64 × 4	60.00
M68 × 6	62.00	M68 × 4	64.00
M72 × 6	66.00	—	—
M80 × 6	74.00	—	—
M90 × 6	84.00	—	—
M100 × 6	94.00	—	—

Drilled Hole Tolerances

Drill Size	Tolerance*	
	Plus	Minus
0.0135 (No. 80) − 0.185 (No. 13)	0.003	0.002
0.1875 − 0.246 (D)	0.004	0.002
0.250 (E) − 0.750	0.005	0.002
0.756 − 1.000	0.007	0.003
1.0156 − 2.000	0.010	0.004
2.0312 − 3.500	0.015	0.005

* generally accepted tolerances for good practice

Unified and American Standard Screw Thread Pitches and Recommended Tap Drill Sizes*

AMERICAN NATIONAL COARSE STANDARD THREAD (N.C.) FORMERLY US STANDARD					AMERICAN NATIONAL FINE STANDARD THREAD (N.F.) FORMERLY S.A.E. THREAD				
Sizes	Threads per Inch	Outside Diameter of Screw	Tap Drill Sizes	Decimal Equiva-lent of Drill	Sizes	Threads per Inch	Outside Diameter of Screw	Tap Drill Sizes	Decimal Equiva-lent of Drill
1	64	0.073	53	0.0595	0	80	0.060	$3/64$	0.0469
2	56	0.086	50	0.0700	1	72	0.073	53	0.0595
3	48	0.099	47	0.0785	2	64	0.086	50	0.0700
4	40	0.112	43	0.0890	3	56	0.099	45	0.0820
5	40	0.125	38	0.1015	4	48	0.112	42	0.0935
6	32	0.138	36	0.1065	5	44	0.125	37	0.1040
8	32	0.164	29	0.1360	6	40	0.138	33	0.1130
10	24	0.190	25	0.1495	8	36	0.164	29	0.1360
12	24	0.216	16	0.1770	10	32	0.190	21	0.1590
$1/4$	20	0.250	7	0.2010	12	28	0.216	14	0.1820
$5/16$	18	0.3125	F	0.2570	$1/4$	28	0.250	3	0.2130
$3/8$	16	0.375	$5/16$	0.3125	$5/16$	24	0.3125	I	0.2720
$7/16$	14	0.4375	U	0.3680	$3/8$	24	0.375	Q	0.3320
$1/2$	13	0.500	$27/64$	0.4219	$7/16$	20	0.4375	$25/64$	0.3906
$9/16$	12	0.5625	$31/64$	0.4843	$1/2$	20	0.500	$29/64$	0.4531
$5/8$	11	0.625	$17/32$	0.5312	$9/16$	18	0.5625	0.5062	0.5062
$3/4$	10	0.750	$21/32$	0.6562	$5/8$	18	0.625	0.5687	0.5687
$7/8$	9	0.875	$49/64$	0.7656	$3/4$	16	0.750	$11/16$	0.6875
1	8	1.000	$7/8$	0.875	$7/8$	14	0.875	0.8020	0.8020
$1\tfrac{1}{8}$	7	1.125	$63/64$	0.9843	1	14	1.000	0.9274	0.9274
$1\tfrac{1}{4}$	7	1.250	$1\,7/64$	1.1093	$1\tfrac{1}{8}$	12	1.125	$1\,3/64$	1.0468
					$1\tfrac{1}{4}$	12	1.250	$1\,11/64$	1.1718

* Courtesy of South Bend Lathe Works. Table is based on 75% thread depth.

Regular Nut Eyebolts

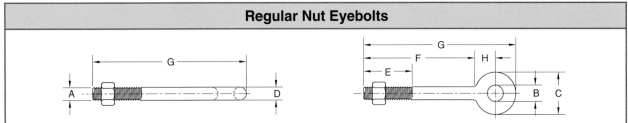

Shank Diameter and Length*	G-291 Stock No. Galv	Working Load Limit†	Weight per 100†	Dimensions*							
				A	B	C	D	E	F	G	H
¼ × 2	1043230	650	6.00	0.25	0.50	1.00	0.25	1.50	2.00	3.06	0.56
¼ × 4	1043258	650	13.50	0.25	0.50	1.00	0.25	2.50	4.00	5.06	0.56
⁵⁄₁₆ × 2¼	1043276	1200	18.75	0.31	0.62	1.25	0.31	1.50	2.25	3.56	0.69
⁵⁄₁₆ × 4¼	1043294	1200	25.00	0.31	0.62	1.25	0.31	2.50	4.25	5.56	0.69
⅜ × 2½	1043310	1550	24.33	0.38	0.75	1.50	0.38	1.50	2.50	4.12	0.88
⅜ × 4½	1043338	1550	37.50	0.38	0.75	1.50	0.38	2.50	4.50	6.12	0.88
⅜ × 6	1043356	1550	43.75	0.38	0.75	1.50	0.38	2.50	6.00	7.62	0.88
½ × 3¼	1043374	2600	50.00	0.50	1.00	2.00	0.50	1.50	3.25	5.38	1.12
½ × 6	1043392	2600	62.50	0.50	1.00	2.00	0.50	3.00	6.00	8.12	1.12
½ × 8	1043418	2600	75.00	0.50	1.00	2.00	0.50	3.00	8.00	10.12	1.12
½ × 10	1043436	2600	88.00	0.50	1.00	2.00	0.50	3.00	10.00	12.12	1.12
½ × 12	1043454	2600	100.00	0.50	1.00	2.00	0.50	3.00	12.00	14.12	1.12
⅝ × 4	1043472	5200	101.25	0.62	1.25	2.50	0.62	2.00	4.00	6.69	1.44
⅝ × 6	1043490	5200	120.00	0.62	1.25	2.50	0.62	3.00	6.00	8.69	1.44
⅝ × 8	1043515	5200	131.00	0.62	1.25	2.50	0.62	3.00	8.00	10.69	1.44
⅝ × 10	1043533	5200	162.50	0.62	1.25	2.50	0.62	3.00	10.00	12.69	1.44
⅝ × 12	1043551	5200	175.00	0.62	1.25	2.50	0.62	4.00	12.00	14.69	1.44
¾ × 4½	1043579	7200	185.90	0.75	1.50	3.00	0.75	2.00	4.50	7.69	1.69
¾ × 6	1043597	7200	180.00	0.75	1.50	3.00	0.75	3.00	6.00	9.19	1.69
¾ × 8	1043613	7200	200.00	0.75	1.50	3.00	0.75	3.00	8.00	11.19	1.69
¾ × 10	1043631	7200	237.50	0.75	1.50	3.00	0.75	3.00	10.00	13.19	1.69
¾ × 12	1043659	7200	251.94	0.75	1.50	3.00	0.75	4.00	12.00	15.19	1.69
¾ × 15	1043677	7200	300.00	0.75	1.50	3.00	0.75	5.00	15.00	18.19	1.69
⅞ × 5	1043695	10,600	275.00	0.88	1.75	3.50	0.88	2.50	5.00	8.75	2.00
⅞ × 8	1043711	10,600	325.00	0.88	1.75	3.50	0.88	4.00	8.00	11.75	2.00
⅞ × 12	1043739	10,600	400.00	0.88	1.75	3.50	0.88	4.00	12.00	15.75	2.00
1 × 6	1043757	13,300	425.00	1.00	2.00	4.00	1.00	3.00	6.00	10.31	2.31
1 × 9	1043775	13,300	452.00	1.00	2.00	4.00	1.00	4.00	9.00	13.31	2.31
1 × 12	1043793	13,300	550.00	1.00	2.00	4.00	1.00	4.00	12.00	16.31	2.31
1 × 18	1043819	13,300	650.00	1.00	2.00	4.00	1.00	7.00	18.00	22.31	2.31
1¼ × 8	1043837	21,000	750.00	1.25	2.50	5.00	1.25	4.00	8.00	13.38	2.88
1¼ × 12	1043855	21,000	900.00	1.25	2.50	5.00	1.25	4.00	12.00	17.38	2.88
1¼ × 20	1043873	21,000	1150.00	1.25	2.50	5.00	1.25	6.00	20.00	25.38	2.88

* in in.
† in lb

Shoulder Nut Eyebolts

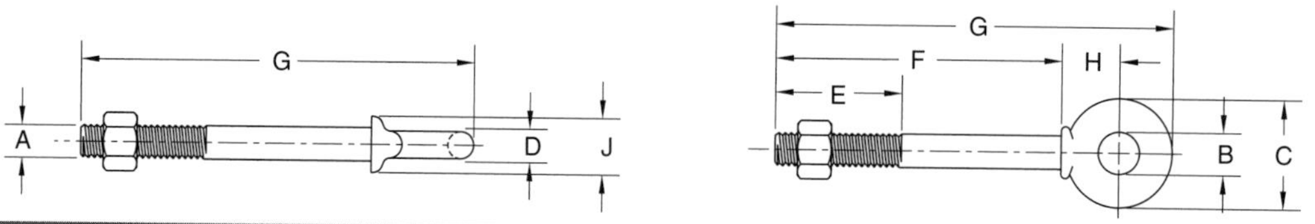

Shank Diameter and Length*	G-277 Stock No. Galv	Working Load Limit†	Weight per 100†	Dimensions*								
				A	B	C	D	E	F	G	H	J
¼ × 2	1045014	650	6.61	0.25	0.50	0.88	0.19	1.50	2.00	2.94	0.50	0.47
¼ × 4	1045032	650	8.61	0.25	0.50	0.88	0.19	2.50	4.00	4.94	0.50	0.47
⁵⁄₁₆ × 2¼	1045050	1200	12.50	0.31	0.62	1.12	0.25	1.50	2.25	3.50	0.69	0.56
⁵⁄₁₆ × 4¼	1045078	1200	18.75	0.31	0.62	1.12	0.25	2.50	4.25	5.50	0.69	0.56
⅜ × 2½	1045096	1550	19.00	0.38	0.75	1.38	0.31	1.50	2.50	3.97	0.78	0.66
⅜ × 4½	1045112	1550	31.58	0.38	0.75	1.38	0.31	2.50	4.50	5.97	0.78	0.66
½ × 3¼	1045130	2600	37.50	0.50	1.00	1.75	0.38	1.50	3.25	5.12	1.00	0.91
½ × 6	1045158	2600	56.25	0.50	1.00	1.75	0.38	3.00	6.00	7.88	1.00	0.91
⅝ × 4	1045176	5200	75.00	0.62	1.25	2.25	0.50	2.00	4.00	6.44	1.31	1.12
⅝ × 6	1045194	5200	100.25	0.62	1.25	2.25	0.50	3.00	6.00	8.44	1.31	1.12
¾ × 4½	1045210	7200	125.00	0.75	1.50	2.75	0.62	2.00	4.50	7.44	1.56	1.38
¾ × 6	1045238	7200	150.00	0.75	1.50	2.75	0.62	3.00	6.00	8.94	1.56	1.38
⅞ × 5	1045256	10,650	225.00	0.88	1.75	3.25	0.75	2.50	5.00	8.46	1.84	1.56
1 × 6	1045292	10,650	375.00	1.00	2.00	3.75	0.88	3.00	6.00	9.97	2.09	1.81
1 × 9	1045318	13,300	429.00	1.00	2.00	3.75	0.88	4.00	9.00	12.97	2.09	1.81
1¼ × 8	1045336	13,300	650.00	1.25	2.50	4.50	1.00	4.00	8.00	12.72	2.47	2.28
1¼ × 12	1045354	21,000	775.00	1.25	2.50	4.50	1.00	4.00	12.00	16.72	2.47	2.28
1½ × 15	1045372	24,000	1425.00	1.50	3.00	5.50	1.25	6.00	15.00	20.75	3.00	2.75

* in in.
† in lb

Machinery Eyebolts

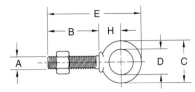

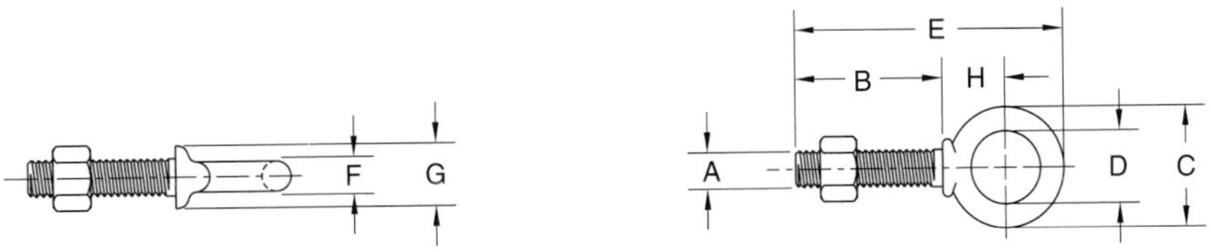

Shank Diameter and Length*	Stock No.	Working Load Limit†	Weight per 100†	Dimensions*							
				A	B	C	D	E	F	G	H
¼ × 1	9900182	650	3.20	0.25	1.00	0.88	0.50	1.94	0.19	0.47	0.50
5⁄16 × 1⅛	9900191	1200	6.20	0.31	1.13	1.12	0.62	2.38	0.25	0.56	0.69
⅜ × 1¼	9900208	1550	12.50	0.38	1.25	1.38	0.75	2.72	0.31	0.66	0.78
½ × 1½	9900217	2600	25.00	0.50	1.50	1.75	1.00	3.38	0.38	0.91	1.00
⅝ × 1¾	9900226	5200	50.00	0.63	1.75	2.25	1.25	4.19	0.50	1.12	1.31
¾ × 2	9900235	7200	87.50	0.75	2.00	2.75	1.50	4.94	0.62	1.38	1.56
⅞ × 2¼	9900244	10,600	150.00	0.88	2.25	3.25	1.75	5.72	0.75	1.56	1.84
1 × 2½	9900253	13,300	218.00	1.00	2.50	3.75	2.00	6.47	0.88	1.81	2.09
1¼ × 3	9900262	21,000	380.00	1.25	3.00	4.50	2.50	7.72	1.00	2.28	2.47
1½ × 3½	9900271	24,000	700.00	1.50	3.50	5.00	3.00	9.25	1.25	2.75	3.00

* in in.
† in lb

Sling Angle Loss Factors

Angle from Horizontal*	Loss Factor
90	1.000
85	0.996
80	0.985
75	0.966
70	0.940
65	0.906
60	0.866
55	0.819
50	0.766
45	0.707
40	0.643
35	0.574
30	0.500

* in degrees

Choker Sling Loss Factors

Choke Angle*	Loss Factor
120 – 180	0.75
90 – 119	0.65
60 – 89	0.55
30 – 59	0.40

* in degrees

Filter Pressure Loss Constants

Filter Length*	Micron Size	Pipe Size*						
		1/8	1/4	3/4	1	1 1/4	1 1/2	2
3.5	5	115.0	55.0	—	—	—	—	—
	25	112.0	49.0	—	—	—	—	—
	100	92.0	41.0	—	—	—	—	—
14	5	—	—	0.47	0.34	0.34	0.34	—
	25	—	—	0.34	0.23	0.20	0.20	—
	50	—	—	0.32	0.20	0.19	0.19	—
	75	—	—	0.32	0.20	0.19	0.19	—
17	25	—	—	—	—	—	0.05	0.028
	50	—	—	—	—	—	0.036	0.020
	75	—	—	—	—	—	0.032	0.018

* in in.

Pipe Pressure Loss Constants (100' of Schedule 40 Pipe)

Pipe Size*	Constant (C)
1/8	2300.00
1/4	450.00
3/8	91.00
1/2	26.40
3/4	5.93
1	1.66
1 1/4	0.40
1 1/2	0.174
2	0.046
2 1/2	0.018
3	0.006

* in in.

Pipe Fitting Pressure Loss Constants

Pipe Fitting	Pipe Size*								
	1/8	1/4	3/8	1/2	3/4	1	1 1/4	1 1/2	2
45° Elbow	8.30	2.20	0.53	0.21	0.059	0.021	0.007	0.004	0.001
90° Elbow	15.40	4.09	1.09	0.42	0.12	0.043	0.014	0.007	0.002
Gate Valve	6.7	1.76	0.47	0.18	0.05	0.018	0.006	0.003	0.001
Globe Valve	175.3	46.40	12.70	4.75	1.36	0.48	0.16	0.08	0.03
Tee-Side Flow	31.0	8.14	2.37	0.81	0.24	0.08	0.03	0.014	0.005
Tee-Run Flow	10.4	2.74	0.80	0.26	0.08	0.03	0.009	0.004	0.002

* in in.

Pipe							
		Inside Diameter (BW Gauge)			Nominal Wall Thickness		
Nominal ID*	OD (BW Gauge)	Standard	Extra-Heavy	Double Extra-Heavy	Schedule 40	Schedule 60	Schedule 80
1/8	0.405	0.269	0.215	—	0.068	0.095	—
1/4	0.540	0.364	0.302	—	0.088	0.119	—
3/8	0.675	0.493	0.423	—	0.091	0.126	—
1/2	0.840	0.622	0.546	0.252	0.109	0.147	0.294
3/4	1.050	0.824	0.742	0.434	0.113	0.154	0.308
1	1.315	1.049	0.957	0.599	0.133	0.179	0.358
1 1/4	1.660	1.380	1.278	0.896	0.140	0.191	0.382
1 1/2	1.900	1.610	1.500	1.100	0.145	0.200	0.400
2	2.375	2.067	1.939	1.503	0.154	0.218	0.436
2 1/2	2.875	2.469	2.323	1.771	0.203	0.276	0.552
3	3.500	3.068	2.900	2.300	0.216	0.300	0.600
3 1/2	4.000	3.548	3.364	2.728	0.226	0.318	—
4	4.500	4.026	3.826	3.152	0.237	0.337	0.674
5	5.563	5.047	4.813	4.063	0.258	0.375	0.750
6	6.625	6.065	5.761	4.897	0.280	0.432	0.864
8	8.625	7.981	7.625	6.875	0.322	0.500	0.875
10	10.750	10.020	9.750	8.750	0.365	0.500	—
12	12.750	12.000	11.750	10.750	0.406	0.500	—

* in in.

Fluid Weights/Temperature Standards		
Fluid	Weight*	Temperature†
Air	4.33×10^{-5}	20°C/68°F @ 29.92 in. Hg
Gasoline	0.0237 – 0.0249	20°C/68°F
Kerosene	0.0296	20°C/68°F
Mercury	0.49116	0°C/32°F
Oil, fuel	0.0036 – 0.0353	15°C/59°F
Oil, lubricating	0.0307 – 0.0318	15°C/59°F
Seawater	0.0370	15°C/59°F
Water	0.0361	4°C/39°F

* in lb/cu in.
† laboratory temperature under which numerical values are defined

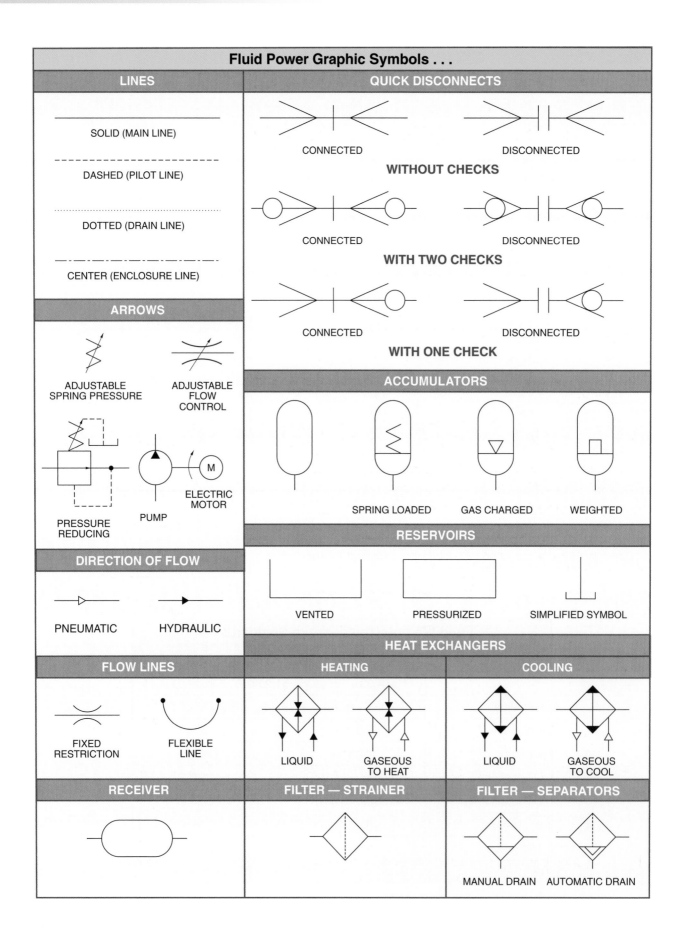

Fluid Power Graphic Symbols . . .

LINES

SOLID (MAIN LINE)

DASHED (PILOT LINE)

DOTTED (DRAIN LINE)

CENTER (ENCLOSURE LINE)

ARROWS

ADJUSTABLE
SPRING PRESSURE

ADJUSTABLE
FLOW
CONTROL

PRESSURE
REDUCING

PUMP

ELECTRIC
MOTOR

DIRECTION OF FLOW

PNEUMATIC   HYDRAULIC

FLOW LINES

FIXED
RESTRICTION

FLEXIBLE
LINE

RECEIVER

QUICK DISCONNECTS

CONNECTED   DISCONNECTED

WITHOUT CHECKS

CONNECTED   DISCONNECTED

WITH TWO CHECKS

CONNECTED   DISCONNECTED

WITH ONE CHECK

ACCUMULATORS

SPRING LOADED   GAS CHARGED   WEIGHTED

RESERVOIRS

VENTED   PRESSURIZED   SIMPLIFIED SYMBOL

HEAT EXCHANGERS

HEATING	COOLING
LIQUID GASEOUS TO HEAT	LIQUID GASEOUS TO COOL

FILTER — STRAINER

FILTER — SEPARATORS

MANUAL DRAIN   AUTOMATIC DRAIN

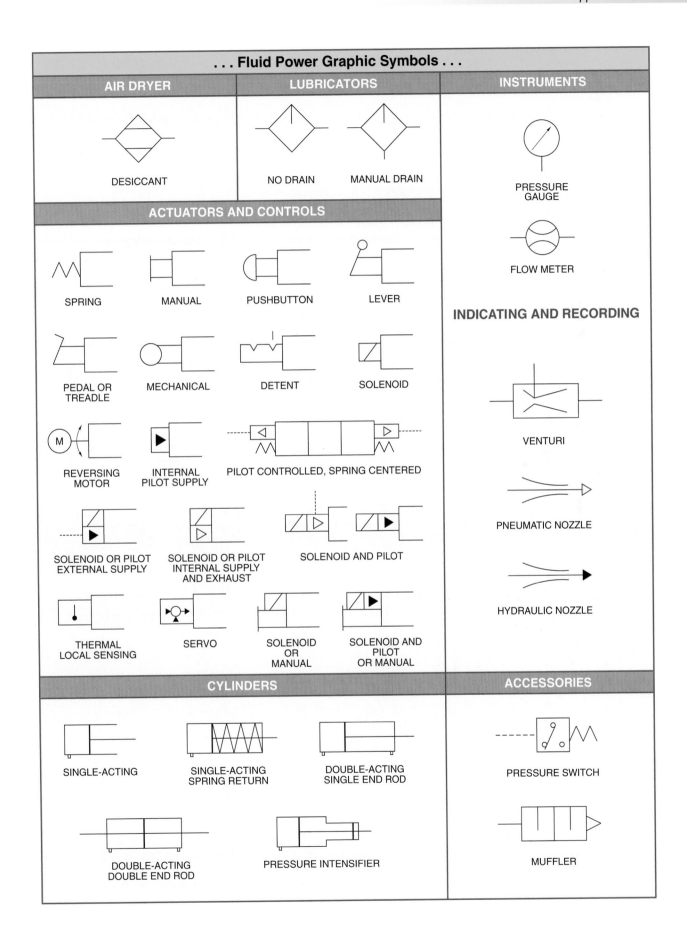

. . . Fluid Power Graphic Symbols . . .

AIR DRYER

DESICCANT

LUBRICATORS

NO DRAIN MANUAL DRAIN

INSTRUMENTS

PRESSURE GAUGE

FLOW METER

ACTUATORS AND CONTROLS

SPRING MANUAL PUSHBUTTON LEVER

PEDAL OR TREADLE MECHANICAL DETENT SOLENOID

REVERSING MOTOR INTERNAL PILOT SUPPLY PILOT CONTROLLED, SPRING CENTERED

SOLENOID OR PILOT EXTERNAL SUPPLY SOLENOID OR PILOT INTERNAL SUPPLY AND EXHAUST SOLENOID AND PILOT

THERMAL LOCAL SENSING SERVO SOLENOID OR MANUAL SOLENOID AND PILOT OR MANUAL

INDICATING AND RECORDING

VENTURI

PNEUMATIC NOZZLE

HYDRAULIC NOZZLE

CYLINDERS

SINGLE-ACTING SINGLE-ACTING SPRING RETURN DOUBLE-ACTING SINGLE END ROD

DOUBLE-ACTING DOUBLE END ROD PRESSURE INTENSIFIER

ACCESSORIES

PRESSURE SWITCH

MUFFLER

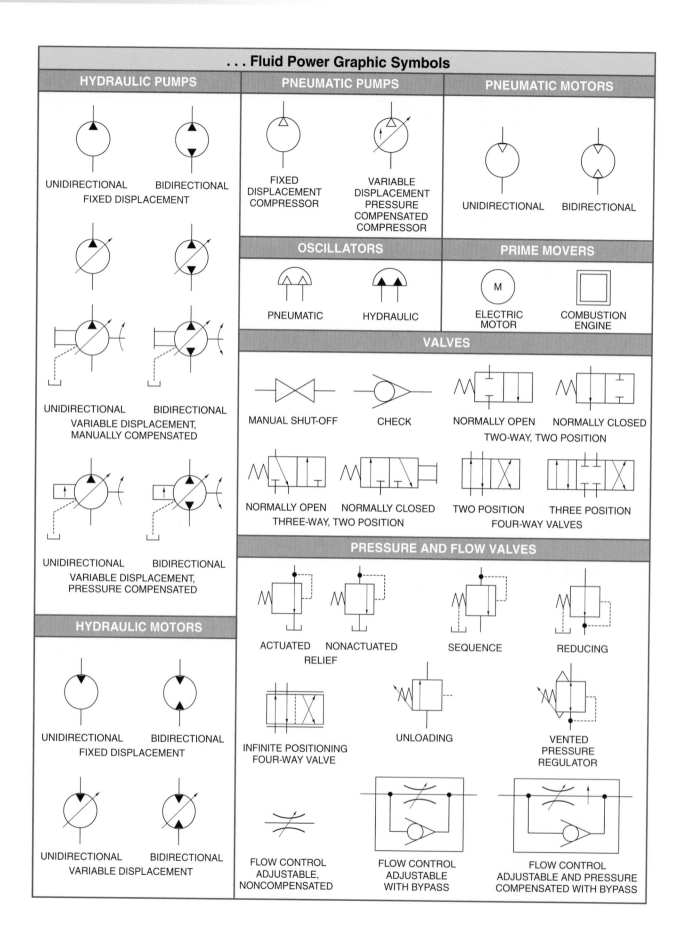

... Fluid Power Graphic Symbols

HYDRAULIC PUMPS

UNIDIRECTIONAL   BIDIRECTIONAL
FIXED DISPLACEMENT

UNIDIRECTIONAL   BIDIRECTIONAL
VARIABLE DISPLACEMENT,
MANUALLY COMPENSATED

UNIDIRECTIONAL   BIDIRECTIONAL
VARIABLE DISPLACEMENT,
PRESSURE COMPENSATED

HYDRAULIC MOTORS

UNIDIRECTIONAL   BIDIRECTIONAL
FIXED DISPLACEMENT

UNIDIRECTIONAL   BIDIRECTIONAL
VARIABLE DISPLACEMENT

PNEUMATIC PUMPS

FIXED
DISPLACEMENT
COMPRESSOR

VARIABLE
DISPLACEMENT
PRESSURE
COMPENSATED
COMPRESSOR

OSCILLATORS

PNEUMATIC   HYDRAULIC

PNEUMATIC MOTORS

UNIDIRECTIONAL   BIDIRECTIONAL

PRIME MOVERS

M
ELECTRIC
MOTOR

COMBUSTION
ENGINE

VALVES

MANUAL SHUT-OFF   CHECK   NORMALLY OPEN   NORMALLY CLOSED
TWO-WAY, TWO POSITION

NORMALLY OPEN   NORMALLY CLOSED
THREE-WAY, TWO POSITION

TWO POSITION   THREE POSITION
FOUR-WAY VALVES

PRESSURE AND FLOW VALVES

ACTUATED   NONACTUATED
RELIEF

SEQUENCE   REDUCING

INFINITE POSITIONING
FOUR-WAY VALVE

UNLOADING

VENTED
PRESSURE
REGULATOR

FLOW CONTROL
ADJUSTABLE,
NONCOMPENSATED

FLOW CONTROL
ADJUSTABLE
WITH BYPASS

FLOW CONTROL
ADJUSTABLE AND PRESSURE
COMPENSATED WITH BYPASS

Logic Symbols

LOGIC ELEMENT	AND	OR	NOT	NAND	NOR
LOGIC ELEMENT FUNCTION	OUTPUT IF ALL CONTROL INPUT SIGNALS ARE ON	OUTPUT IF ANY CONTROL INPUT SIGNAL IS ON	OUTPUT IF SINGLE CONTROL INPUT SIGNAL IS OFF	OUTPUT IF ALL CONTROL INPUT SIGNALS ARE ON	OUTPUT IF ANY CONTROL INPUT SIGNAL IS ON
MIL-STD-806B AND ELECTRONIC LOGIC SYMBOL					
ELECTRICAL RELAY LOGIC SYMBOL					
ELECTRICAL SWITCH LOGIC SYMBOL					
ASA (JIC) VALVING SYMBOL					
ARO PNEUMATIC LOGIC SYMBOL					
NFPA STANDARD			SUPPLY	SUPPLY	SUPPLY
BOOLEAN ALGEBRA SYMBOL	$(\ )\cdot(\ )$	$(\ )+(\ )$	$\overline{(\ )}$	$\overline{(\ )\cdot(\ )}$	$\overline{(\ )+(\ )}$
FLUIDIC DEVICE TURBULENCE AMPLIFIER					

Timing Belt Standard Pitch Lengths and Tolerances*										
Belt Length Designation	Pitch Length	Number of Teeth for Standard Lengths			Belt Length Designation	Pitch Length	Number of Teeth for Standard Lengths				
		MXL (0.080)	XL (0.200)	L (0.375)			XL (0.200)	L (0.375)	H (0.500)	XH (0.875)	XXH (1.250)
36	3.600	45	—	—	230	23.000	115	—	—	—	—
40	4.000	50	—	—	240	24.000	120	64	48	—	—
44	4.400	55	—	—	250	25.000	125	—	—	—	—
48	4.800	60	—	—	255	25.500	—	68	—	—	—
56	5.600	70	—	—	260	26.000	130	—	—	—	—
60	6.000	75	30	—	270	27.000	—	72	54	—	—
64	6.400	80	—	—	285	28.500	—	76	—	—	—
70	7.000	—	35	—	300	30.000	—	80	60	—	—
72	7.200	90	—	—	322	32.250	—	86	—	—	—
80	8.000	100	40	—	330	33.000	—	—	66	—	—
88	8.800	110	—	—	345	34.500	—	92	—	—	—
90	9.000	—	45	—	360	36.000	—	—	72	—	—
100	10.000	125	50	—	367	36.750	—	98	—	—	—
110	11.000	—	55	—	390	39.000	—	104	78	—	—
112	11.200	140	—	—	420	42.000	—	112	84	—	—
120	12.000	—	60	—	450	45.000	—	120	90	—	—
124	12.375	—	—	33	480	48.000	—	128	96	—	—
124	12.400	155	—	—	507	50.750	—	—	—	58	—
130	13.000	—	65	—	510	51.000	—	136	102	—	—
140	14.000	175	70	—	540	54.000	—	144	108	—	—
150	15.000	—	75	40	560	56.000	—	—	—	64	—
160	16.000	200	80	—	570	57.000	—	—	114	—	—
170	17.000	—	85	—	600	60.000	—	160	120	—	—
180	18.000	225	90	—	630	63.000	—	—	126	72	—
187	18.750	—	—	50	660	66.000	—	—	132	—	—
190	19.000	—	95	—	700	70.000	—	—	140	80	56
200	20.000	250	100	—	750	75.000	—	—	150	—	—
210	21.000	—	105	56	770	77.000	—	—	—	88	—
220	22.000	—	110	—	800	80.000	—	—	160	—	64
225	22.500	—	—	60	840	84.000	—	—	—	96	—

* in in.

Timing Belt Standard Widths and Tolerances*					
Belt Section	**Standard Belt Widths**		**Tolerances on Width for Pitch Lengths**		
	Designation	**Dimensions**	**Up to and Including 33″**	**Over 33″ up to and Including 66″**	**Over 66″**
MXL (0.080)	012	0.12	+0.02	—	—
	019	0.19	−0.03		
	025	0.25	+0.02		
XL (0.200)	025	0.25	−0.03	—	—
	037	0.38	+0.03		
L (0.375)	050	0.50	−0.03	+0.03 −0.05	—
	075	0.75	+0.03		
	100	1.00	−0.03		
H (0.500)	075	0.75	+0.03 −0.03	+0.03 −0.05	+0.03 −0.05
	100	1.00			
	150	1.50			
	200	2.00	+0.03	+0.05	+0.05
			−0.05	−0.05	−0.06
	300	3.00	+0.05	+0.06	+0.06
			−0.06	−0.06	−0.08
XH (0.875)	200	2.00	—	+0.19 −0.19	+0.19 −0.19
	300	3.00			
	400	4.00			
XXH (1.250)	200	2.00	—	—	+0.19 −0.19
	300	3.00			
	400	4.00			
	500	5.00			

* in in.

Allowable Tight Side Tension for AA Section V-Belts*

Belt Speed†	Pulley Effective Diameter‡							
	3.0	3.5	4.0	4.5	5.0	5.5	6.0	6.5
200	30	46	57	66	73	79	83	88
400	23	38	49	58	65	71	76	80
600	18	33	44	53	60	66	71	75
800	14	30	41	50	57	63	67	72
1000	12	27	38	47	54	60	65	69
1200	9	24	36	45	52	57	62	66
1400	7	22	34	42	49	55	60	64
1600	5	20	32	40	47	53	58	62
1800	3	18	30	38	46	51	56	60
2000	1	16	28	37	44	50	54	58
2200	—	15	26	35	42	48	53	57
2400	—	13	24	33	40	46	51	55
2600	—	11	23	31	39	44	49	53
2800	—	9	21	30	37	43	47	51
3000	—	8	19	28	35	41	46	50
3200	—	6	17	26	33	39	44	48
3400	—	4	16	24	31	37	42	46
3600	—	2	14	23	30	35	40	44
3800	—	1	12	21	28	34	38	43
4000	—	—	10	19	26	32	37	41
4200	—	—	8	17	24	30	35	39
4400	—	—	6	15	22	28	33	37
4600	—	—	4	13	20	26	31	35
4800	—	—	2	11	18	24	29	33
5000	—	—	—	9	16	22	27	31
5200	—	—	—	7	14	20	24	28
5400	—	—	—	4	12	17	22	26
5600	—	—	—	2	9	15	20	24
5800	—	—	—	—	7	13	18	22

* in lb-ft
† in fpm
‡ in in.

Allowable Tight Side Tension for BB Section V-Belts*									
Belt Speed†	Pulley Effective Diameter‡								
	5.0	5.5	6.0	6.5	7.0	7.5	8.0	8.5	9.0
200	81	93	103	111	119	125	130	135	140
400	69	81	91	99	107	113	118	123	128
600	61	74	84	92	99	106	111	116	121
800	56	68	78	87	94	101	106	111	115
1000	52	64	74	83	90	96	102	107	111
1200	48	60	71	79	86	93	98	103	107
1400	45	57	67	76	83	89	95	100	104
1600	42	54	64	73	80	86	92	97	101
1800	39	51	61	70	77	84	89	94	98
2000	36	49	59	67	74	81	86	91	96
2200	34	46	56	64	72	78	84	89	93
2400	31	43	53	62	69	75	81	86	90
2600	29	41	51	59	67	73	78	83	88
2800	26	38	48	57	64	70	76	81	85
3000	23	35	45	54	61	68	73	78	82
3200	21	33	43	51	59	65	70	75	80
3400	18	30	40	49	56	62	68	73	77
3600	15	27	37	46	53	59	65	70	74
3800	12	24	35	43	50	57	62	67	71
4000	9	22	32	40	47	54	59	64	69
4200	7	19	29	37	45	51	56	61	66
4400	4	16	26	34	42	48	53	58	63
4600	1	13	23	31	39	45	50	55	60
4800	—	10	20	28	35	42	47	52	57
5000	—	6	16	25	32	39	44	49	53
5200	—	3	13	22	29	35	41	46	50
5400	—	—	10	18	26	32	38	42	47
5600	—	—	6	15	22	29	34	39	43
5800	—	—	3	11	19	25	31	36	40

* in lb-ft
† in fpm
‡ in in.

Allowable Tight Side Tension for CC Section V-Belts*

Belt Speed[†]	Pulley Effective Diameter[‡]								
	7.0	8.0	9.0	10.0	11.0	12.0	13.0	14.0	15.0
200	121	158	186	207	228	244	257	268	278
400	99	135	164	187	206	221	234	246	256
600	85	122	151	173	192	208	221	232	242
800	75	112	141	164	182	198	211	222	232
1000	67	104	133	155	174	190	203	214	224
1200	60	97	126	149	167	183	196	207	217
1400	54	91	120	142	161	177	190	201	211
1600	48	85	114	137	155	171	184	196	205
1800	43	80	108	131	150	166	179	190	200
2000	38	75	103	126	145	160	174	185	195
2200	33	70	98	121	140	155	169	180	190
2400	28	65	93	116	135	150	164	175	185
2600	23	60	88	111	130	145	159	170	180
2800	18	55	83	106	125	140	154	165	175
3000	13	50	78	101	120	135	149	160	170
3200	8	45	73	96	115	130	144	155	165
3400	3	39	68	91	110	125	138	150	160
3600	—	34	63	86	104	120	133	145	154
3800	—	29	58	80	99	115	128	139	149
4000	—	24	52	75	94	109	123	134	144
4200	—	18	47	70	88	104	117	128	138
4400	—	12	41	64	83	98	112	123	133
4600	—	7	35	58	77	93	106	117	127
4800	—	1	29	52	71	87	100	111	121
5000	—	—	23	46	65	81	94	105	115
5200	—	—	17	40	59	75	88	99	109
5400	—	—	11	34	53	68	81	93	103
5600	—	—	5	27	46	62	75	86	96
5800	—	—	—	21	40	55	68	80	90

* in lb-ft
† in fpm
‡ in in.

Allowable Tight Side Tension for DD Section V-Belts*									
Belt Speed†	Pulley Effective Diameter‡								
	12.0	13.0	14.0	15.0	16.0	17.0	18.0	19.0	20.0
200	243	293	336	373	405	434	459	482	503
400	195	245	288	325	358	386	412	434	455
600	167	217	259	297	329	358	383	406	426
800	146	196	239	276	308	337	362	385	405
1000	129	179	222	259	291	320	345	368	389
1200	114	164	207	244	277	305	331	353	374
1400	101	151	194	231	263	292	318	340	361
1600	89	139	182	219	251	280	305	328	349
1800	78	128	170	207	240	269	294	317	337
2000	67	117	159	196	229	258	283	306	326
2200	56	106	149	186	218	247	272	295	316
2400	45	95	138	175	208	236	262	284	305
2600	35	85	128	165	197	226	251	274	294
2800	24	74	117	154	187	215	241	263	284
3000	14	64	106	144	176	205	230	253	273
3200	3	53	96	133	165	194	219	242	263
3400	—	42	85	122	155	183	209	231	252
3600	—	31	74	111	144	172	198	220	241
3800	—	20	63	100	132	161	186	209	230
4000	—	9	51	89	121	150	175	198	218
4200	—	—	40	77	109	138	163	186	207
4400	—	—	28	65	97	126	152	174	195
4600	—	—	16	53	85	114	139	162	183
4800	—	—	3	40	73	102	127	150	170
5000	—	—	—	28	60	89	114	137	158
5200	—	—	—	15	47	76	101	124	145
5400	—	—	—	1	34	62	88	111	131
5600	—	—	—	—	20	49	74	97	118
5800	—	—	—	—	6	35	60	83	104

* in lb-ft
† in fpm
‡ in in.

Vibration Characteristics

Problem	Frequency	Amplitude	Remarks/Comments
Bent Shaft	Normally 1× rpm. Often 2× and 3× rpm.	High amplitude in axial direction	A bent shaft is indicated by a phase difference between two bearings of the same machine.
Angular Misalignment	Normally 1× and 2× rpm. Often 3× rpm.	High amplitude in axial direction	Angular misalignment is indicated by a large phase difference between two bearings of a direct-coupled machine.
Offset Misalignment	1× and 2× rpm	High amplitude in radial direction	Normally produces a radial vibration at a frequency of 2× rpm.
Misaligned bearing on shaft	Normally 1× and 2× rpm. Often 3× rpm.	High amplitude in axial direction	Vibration eliminated by correctly installing bearing.
Mechanical Looseness	2× rpm	Often erratic	May be accompanied by unbalance and/or misalignment. May result from loose mounting bolts or excessive bearing clearance.
Plain Bearings	1× rpm	High amplitude in horizontal direction. Often high amplitude in vertical direction.	Vibration normally due to excessive bearing clearance, looseness, or lubrication problems.
Oil Whirl	Slightly less than ½× rpm	Often quite severe	Oil whirl vibration drives shaft into whirling path around bearing. Causes include excessive bearing wear, an increase in lube oil pressure, or a change of oil frequency.
Unbalance	1× rpm	High amplitude in radial direction	Unbalance is the most common cause of vibration.
Gear Tooth Wear	Very high. Gear teeth × rpm.	High amplitude at gear mesh frequency	Tooth wear indicated by excitation of gear natural frequency and includes sidebands at running speed of bad gear. Also indicated by excitation at gear mesh frequency.
Gear Eccentricity/ Backlash	1× gear rpm	Moderately high amplitude sidebends at gear mesh frequency	Improper backlash normally excites gear natural frequency and gear mesh frequency.
Gear Misalignment	1×, 2×, and 3× gear mesh frequency	Low amplitude at 1× gear mesh frequency. High amplitude at 2× and 3× gear mesh frequency.	Gear misalignment normally excites second order gear mesh frequency harmonics and includes sidebands at running speed.
Broken Gear Tooth	1× rpm of gear	High amplitude	Broken gear tooth excites gear natural frequency with sidebands at gear running speed.
Worn or Loose V-Belts	Less than motor or driven machine rpm	Normally unsteady amplitude readings	Worn or loose belts often produce high frequency vibration and noise such as a chirp or squeal.
V-Belt Pulley Misalignment	1× rpm	High amplitude vibration in axial direction	V-belt pulley misalignment normally produces highest axial vibrations at the fan rpm.
Eccentric V-Belt Pulleys	1× rpm of eccentric pulley	High amplitude vibration in-line with belts	Largest vibration occurs in the direction of belt tension.
Beat Vibration	Continually changing	Continually changing	Belt vibration occurs as beats or pulses occurring at regular intervals.

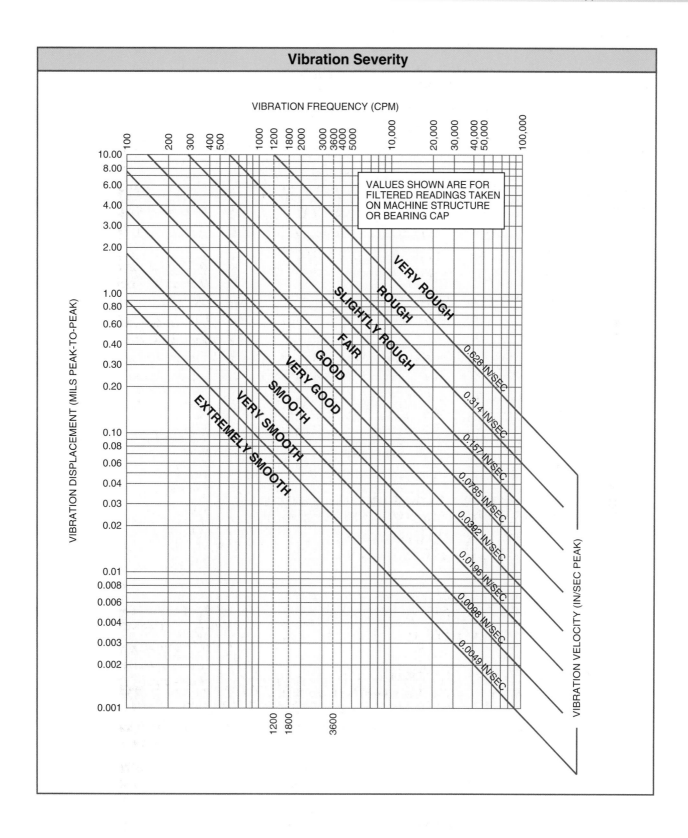

Abbreviations . . .

A

Absolute	ABS
Actual	ACT
Adapter	ADPT
Addendum	ADD
Adjust	ADJ
Advance	ADV
Allowance	ALLOW
Alloy	ALY
Altitude	ALT
Aluminum	AL
American Standard	AMER STD
American Wire Gauge	AWG
Amount	AMT
Anneal	ANL
Apparatus	APPAR
Approved	APP
Approximate	APPROX
Arc Weld	ARC/W
Area	A
Arrangement	ARR
Assemble	ASSEM
Assembly	ASSY
Authorized	AUTH
Auxiliary	AUX

B

Babbitt	BAB
Back-feed	BF
Back Pressure	BP
Ball Bearing	BB
Base Line	BL
Base Plate	BP
Bearing	BRG
Benchmark	BM
Bending Moment	M
Between	BET
Bevel	BEV
Bill of Materials	B/M
Bolt Circle	BC
Both Faces	BF
Both Sides	BS
Both Ways	BW
Bottom	BOT
Bottom Chord	BC
Bracket	BRKT
Brass	BRS
Brazing	BRZG
Break	BRK
Brinell Hardness	BH
British Standard	BR STD
Broach	BRO
Bronze	BRZ
Bushing	BUSH

C

Cabinet	CAB
Cadmium Plate	CD PL
Capacity	CAP
Cap Screw	CAP SCR
Carbon	C
Carburize	CARB
Carriage	CRG
Case Harden	CH

Cast	C
Cast Iron	CI
Cast-Iron Pipe	CIP
Cast Steel	CS
Casting	CSTG
Castle Nut	CAS NUT
Center	CTR
Centerline	CL
Center of Gravity	CG
Center Punch	CP
Ceramic	CER
Chamfer	CH or CHAM
Channel	CHAN
Chrome Molybdenum	CR MOLY
Chromium Plate	CR PL
Chrome Vanadium	CR VAN
Circle	CIR
Circular Pitch	CP
Circumference	CIRC
Clearance	CL
Clockwise	CW
Coated	CTD
Cold Drawn	CD
Cold Drawn Steel	CDS
Cold Finish	CF
Cold Punched	CP
Cold Rolled Steel	CRS
Concentric	CONC
Copper Plate	COP PL
Corrosion Resistant	CRE
Corrosion Resistant Steel	CRES
Cotter	COT
Counterclockwise	CCW
Counterbore	CB or CBORE
Counterdrill	CD or CDRILL
Countersink	CSK or CSINK
Countersink Other Side	CSKO
Coupling	CPLG
Cross Section	XSECT
Cubic	CU
Cubic Foot	CU FT
Cubic Inch	CU IN
Cylinder	CYL

D

Decimal	DEC
Dedendum	DED
Depth	DP or DEEP
Deep Drawn	DD
Degree	DEG
Density	D
Design	DSGN
Detail	DET
Diagonal	DIAG
Diagram	DIAG
Diameter	DIA
Diametral Pitch	DP
Dimension	DIM
Dovetail	DVTL
Dowel	DWL
Drafting	DFTG
Draftsman	DFTSMN

Drawing	DWG
Drill	DR
Drive	DR
Drop Forge	DF
Duplicate	DUP

E

Each	EA
Eccentric	ECC
Electric	ELEC
Elongation	ELONG
Enclose	ENCL
Engineer	ENGR
Envelope	ENV
Equipment	EQUIP
Equivalent	EQUIV
Existing	EXIST
Extension	EXT
Extrude	EXTR

F

Fabricate	FAB
Far Side	FS
Feet	FT
Feet Per Minute	FPM
Feet Per Second	FPS
Figure	FIG
Fillet	FIL
Finish	FIN
Finish All Over	FAO
Fitting	FTG
Fixture	FIX
Flange	FLG
Flashing	FL
Flat	F
Flat Head	FH
Flexible	FLEX
Forged Steel	FST
Forging	FORG
Forward	FWD
Foundry	FDRY
Fractional	FRAC
Furnish	FURN

G

Gauge	GA
Galvanize	GALV
Galvanized Iron	GI
Galvanized Steel	GS
Gasket	GSKT
General	GEN
Glass	GL
Grade	GR
Grind	GRD or GND
Groove	GRV

H

Half-Round	½RD
Hard	H
Hard Drawn	HD
Harden	HDN
Hardware	HDW
Head	HD
Headless	HDLS

. . . Abbreviations . . .

Heat	HT	Military	MIL	**Q**			
Heat Treat	HT TR	Millimeter	MM	Quadrant	QUAD		
Heavy	HVY	Minimum	MIN	Quality	QUAL		
Height	HGT	Minute	MIN	Quantity	QTY		
Hexagon	HEX	Miscellaneous	MISC	Quarter Round	¼RD		
High-Speed	HS	Mold Line	ML				
High-Speed Steel	HSS	Molded	MLD	**R**			
High-Tensile Cast Iron	HTCI	Molding	MLDG	Radial	RAD		
High-Tensile Steel	HTS	Morse Taper	MOR T	Radians	RAD		
Horizontal	HORIZ	Mounted	MTD	Radius	R		
Hot Rolled	HR	Mounting	MTG	Ream	RM		
Hot Rolled Steel	HRS	Multiple	MULT	Reassemble	REASM		

I

				Received	RECD
		N		Rectangle	RECT
Impregnate	IMPG	National	NATL	Reference	REF
Inch	IN.	Near Face	NF	Reference Line	REF L
Inches Per Minute	IPM	Near Side	NS	Reinforce	REINF
Indicate	IND	New British Standard		Relief	REL
Inside Diameter	ID	(Imperial Wire		Remove	REM
Install	INSTL	Gauge)	NBS	Require	REQ
Internal	INT	Nipple	NIP	Required	REQD
International Pipe		Nominal	NOM	Return	RET
Standard	IPS	Normal	NORM	Reverse	REV
Intersect	INT	Not To Scale	NTS	Revolutions Per	
Iron	I	Number	NO.	Minute	RPM
Irregular	IRREG			Right Hand	RH
		O		Rivet	RIV
J		Octagon	OCT	Rockwell Hardness	RH
Joint	JT	On Center	OC	Roller Bearing	RB
Junction	JCT	Opening	OPNG	Root Diameter	RD
		Opposite	OPP	Root Mean Square	RMS
K		Original	ORIG	Round	RD
Key	K	Outside Diameter	OD		
Keyseat	KST	Overall	OA	**S**	
Keyway	KWY			Schedule	SCH
Knockout	KO	**P**		Schematic	SCHEM
		Pair	PR	Screw	SCR
L		Parallel	PAR	Secondary	SEC
Laboratory	LAB	Part	PT	Section	SECT
Laminate	LAM	Patent	PAT	Semi-Finished	SF
Lateral	LAT	Pattern	PATT	Set Screw	SS
Left Hand	LH	Permanent	PERM	Shaft	SFT
Length	LG	Perpendicular	PERP	Sheet	SH
Limit	LIM	Phenolic	PHEN	Shop Order	SO
Linear	LIN	Pitch	P	Shoulder	SHLD
Locate	LOC	Pitch Circle	PC	Side	S
Low-Speed	LS	Pitch Diameter	PD	Similar	SIM
Lubricate	LUB	Plate	PL	Sketch	SK
		Point	PT	Sleeve	SLV
M		Position	POS	Sleeve Bearing	SB
Machine	MACH	Pound	LB	Slotted	SLOT
Machine Steel	MS	Pounds Per Square		Socket	SOC
Malleable	MALL	Inch	PSI	Space	SP
Malleable	Iron MI	Precast	PRCST	Special Treatment	
Manual	MAN	Prefabricated	PREFAB	Steel	STS
Manufacture	MFR	Preferred	PFD	Specification	SPEC
Manufactured	MFD	Primary	PRIM	Speed	SP
Manufacturing	MFG	Production	PROD	Spherical	SPHER
Material	MAT or MATL	Profile	PF	Spotface	SF or SFACE
Material List	ML	Project	PROJ	Spring	SPG
Maximum	MAX	Proposed	PROP	Square	SQ
Mechanical	MECH	Punch	PCH	Stainless	STN
Metal	MET			Stainless Steel	SST or SS
Micrometer	MIC			Standard	STD

. . . Abbreviations

Steel	STL
Stock	STK
Straight	STR
Stress Anneal	SA
Structural	STR
Supplement	SUPP
Supply	SUP
Surface	SURF
Symbol	SYM
Symmetrical	SYM
Synthetic	SYN
T	
Tachometer	TACH
Tangent	TAN
Taper	TPR
Technical	TECH
Tee	T
Teeth	T
Teeth Per Inch	TPI
Temperature	TEMP
Template	TEMP
Tensile Strength	TS
Tension	TENS
Thick	THK
Thread	THD
Threads Per Inch	TPI
Through	THRU
Tolerance	TOL
Tool Steel	TS
Total	TOT
Total Indicator Reading	TIR
Tubing	TUB
Typical	TYP
U	
United States Gauge	USG
United States Standard	USS
V	
Velocity	V
Vertical	VERT
Vibrate	VIB
Volume	VOL
W	
Washer	WASH
Weight	WT
Wheel Base	WB
Width	W
Wire	W
With	W/
Without	W/O
Woodruff	WDF
Wrought	WRT
Wrought Iron	WI

Printreading Symbols

Meaning	Symbol	Meaning	Symbol
Straightness	—	Projected tolerance zone	Ⓟ
Flatness	▱	Diameter	⌀
Circularity	○	Basic dimension	[50]
Cylindricity	⌭	Reference dimension	(50)
Profile of a line	⌒	Conical taper	⌲
Profile of a surface	⌓	Taper	⌳
All around	↙⊖	Counterbore/spotface	⌴
Angularity	∠	Countersink	⌵
Perpendicularity	⊥	Depth/deep	↧
Parallelism	//	Square	□
Position	⌖	Dimension not to scale	1̲5̲
Concentricity	◎	Arc length	⌒105
Symmetry	⌯	Radius	R
Circular runout	*↗	Spherical radius	SR
Total runout	*⌰	Spherical diameter	S⌀
Maximum material condition	Ⓜ	Between	*↔
Least material condition	Ⓛ	Statistical tolerance	⟨ST⟩

* may be filled or not filled

Ohm's Law

V = VOLTAGE (IN V)
I = CURRENT (IN A)
R = RESISTANCE (IN Ω)

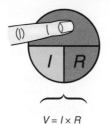

$$V = I \times R$$

VOLTAGE =
CURRENT × RESISTANCE

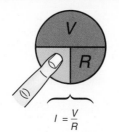

$$I = \frac{V}{R}$$

CURRENT= $\dfrac{\text{VOLTAGE}}{\text{RESISTANCE}}$

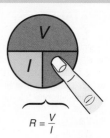

$$R = \frac{V}{I}$$

RESISTANCE = $\dfrac{\text{VOLTAGE}}{\text{CURRENT}}$

Power Formula

P = POWER (IN W)
V = VOLTAGE (IN V)
I = CURRENT (IN A)

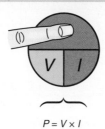

$$P = V \times I$$

POWER =
VOLTAGE × CURRENT

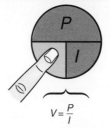

$$V = \frac{P}{I}$$

VOLTAGE = $\dfrac{\text{POWER}}{\text{CURRENT}}$

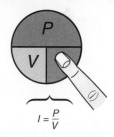

$$I = \frac{P}{V}$$

CURRENT = $\dfrac{\text{POWER}}{\text{VOLTAGE}}$

Parallel Circuit Calculations

RESISTANCE

$$R_T = \frac{R_1 \times R_2}{R_1 + R_2}$$

where
R_T = total resistance (in Ω)
R_1 = resistance 1 in (in Ω)
R_2 = resistance 2 in (in Ω)

VOLTAGE

$$V_T = V_1 = V_2 = \ldots$$

where
V_T = total applied voltage (in V)
V_1 = voltage drop across load 1 in (in V)
V_2 = voltage drop across load 2 in (in V) V)

CURRENT

$$I_T = I_1 + I_2 + I_3 + \ldots$$

where
I_T = total circuit current (in A)
I_1 = current through load 1 (in A)
I_2 = current through load 2 (in A)
I_3 = current through load 3 (in A)

Series Circuit Calculations

RESISTANCE

$$R_T = R_1 + R_2 + R_3 + \ldots$$

where
R_T = total resistance (in Ω)
R_1 = resistance 1 in (in Ω)
R_2 = resistance 2 in (in Ω)
R_3 = resistance 3 in (in Ω)

VOLTAGE

$$V_T = V_1 + V_2 + V_3 + \ldots$$

where
V_T = total applied voltage (in V)
V_1 = voltage drop across load 1 in (in V)
V_2 = voltage drop across load 2 in (in V)
V_3 = voltage drop across load 3 (in V)

CURRENT

$$I_T = I_1 = I_2 = I_3 = \ldots$$

where
I_T = total circuit current (in A)
I_1 = current through load 1 (in A)
I_2 = current through load 2 (in A)
I_3 = current through load 3 (in A)

Power Formulas – 1ϕ, 3ϕ					
Phase	To Find	Use Formula	Example		
			Given	Find	Solution
1ϕ	I	$I = \dfrac{VA}{V}$	32,000 VA, 240 V	I	$I = \dfrac{VA}{V}$ $I = \dfrac{32{,}000 \text{ VA}}{240 \text{ V}}$ $I = \textbf{133 A}$
1ϕ	VA	$VA = I \times V$	100 A, 240 V	AV	$VA = I \times V$ $VA = 100 \text{ A} \times 240 \text{ V}$ $VA = \textbf{24,000 VA}$
1ϕ	V	$V = \dfrac{VA}{I}$	42,000 VA, 350 A	V	$V = \dfrac{VA}{I}$ $V = \dfrac{42{,}000 \text{ VA}}{350 \text{ A}}$ $V = \textbf{120 V}$
3ϕ	I	$I = \dfrac{VA}{V \times \sqrt{3}}$	72,000 VA, 208 V	I	$I = \dfrac{VA}{V \times \sqrt{3}}$ $I = \dfrac{72{,}000 \text{ VA}}{360 \text{ V}}$ $I = \textbf{200 A}$
3ϕ	VA	$VA = I \times V \times \sqrt{3}$	2 A, 240 V	VA	$VA = I \times V \times \sqrt{3}$ $VA = 2 \times 416$ $VA = \textbf{831 VA}$

Glossary

A

abrasion: The removal or displacement of material due to the pressure of hard particles.

abrasive wear: Wear caused by small, hard particles.

absolute pressure: Pressure above a perfect vacuum.

absolute rating: An indication of the largest opening in a strainer element.

absolute temperature: The temperature on a scale that begins with absolute zero.

absolute zero: The temperature at which substances possess no heat.

acceleration: An increase in speed.

accelerometer transducer: A device constructed of quartz crystal material that produces an electric current when compressed.

accumulator: A container in which fluid is stored under pressure.

accuracy: The degree to which a measurement conforms to a specific standard.

acid: A sour substance with a pH value less than 7.

AC sine wave: The waveform formed by the smooth, repetitive oscillation of voltage or current.

action: The work of an actuator or a pilot operator, which becomes the input for another section of a control circuit.

actuator: A device that transforms fluid energy into linear or rotary mechanical force.

acute angle: An angle that contains less than 90°.

acute triangle: A scalene triangle with each angle less than 90°.

AC voltage: Voltage that reverses its direction of flow at regular intervals.

additive: A chemical compound added to a hydraulic fluid to change its properties.

adhesive: A substance that is used to bond materials together at the surface.

adhesive tape: A strip of material coated on one or both sides with an adhesive and used for the purpose of fastening.

adhesive wear: *See* galling.

adjacent angles: Angles that have the same vertex and one side in common.

adsorption: The adhesion of a gas or liquid to the surface of a porous material.

aftercooler: A heat exchanger for cooling the discharge from a compressor.

air cylinder: A device that converts compressed air energy into linear mechanical energy.

air motor: An air driven device that converts fluid energy into rotary mechanical energy.

alignment: The location (within tolerance) of one axis of a coupled machine shaft relative to that of another.

alkali: A bitter substance with a pH value greater than 7.

allowable pressure drop: The pressure differential across a filter.

alternating current (AC): Current that reverses its direction of flow at regular intervals.

alternation: One half of a cycle.

altitude: **1.** The perpendicular dimension from the vertex to the base of a triangle. **2.** The perpendicular distance between the two bases of a prism. **3.** The perpendicular distance from the vertex to the base of a pyramid.

ambient temperature: The temperature of the air surrounding a piece of equipment.

ampere: A quantity of electrons passing a given point in one second.

analog multimeter: A multimeter that can measure two or more electrical quantities and display the measured quantities along calibrated scales using a needle.

anchoring: Any means of fastening a mechanism securely to a base or foundation.

AND logic element: A logic element that provides a logic level 1 only if all inputs are at logic level 1.

angle: 1. A geometric figure formed by two lines extending from the same point. **2.** The intersection of two lines or sides.

angular contact bearing: A rolling contact bearing designed to carry both heavy axial (thrust) loads and radial loads.

angular misalignment: 1. In flexible belt drives, a condition where two shafts are parallel but at different angles with the horizontal plane. **2.** In motor couplings, a condition where one shaft is at an angle to the other shaft.

angular soft foot: A condition that exists when one machine foot is bent and not on the same plane as the other feet.

anvil: The fixed measuring surface of a micrometer.

application drawing: A type of drawing that shows the use of a specific piece of equipment or product in an application.

arc: A portion of the circumference.

arc blast: An explosion that occurs when the air surrounding electrical equipment becomes ionized and conductive.

arc flash: An extremely high-temperature discharge produced by an electrical fault in the air.

arc flash boundary: The distance from an exposed energized conductor or circuit part where bare skin would receive the onset of a second-degree burn.

architect's scale: A triangular scale used to draw objects to a particular size.

arc-rated face shield: An eye and face protection device that covers the entire face with a plastic shield and is used for protection from flying objects.

arc-rated hood: An eye and face protection device that consists of a flame-resistant hood and face shield.

arc welding: A welding process that uses the electrical resistance of an arc bridging a gap to generate the heat necessary for melting a workpiece.

area: The number of unit squares equal to the surface of an object.

armature: 1. The moveable part of a generator or motor. **2.** The moving part of a coil-operated device.

arrowhead: A symbol that indicates the extent of a dimension.

assembly drawing: A type of drawing that shows as closely as possible the way individual parts or components are placed together to produce a finished piece of equipment or result.

asymmetrical load: A load in which one-half of the load is not a mirror image of the other half.

atmospheric pressure: The force exerted by the weight of the atmosphere on Earth's surface.

atom: The smallest building block of matter that cannot be divided into smaller units without changing its basic character.

authorized individual: A knowledgeable individual to whom the authority and responsibility to perform a specific task has been given.

autoignition point: The temperature at which oil ignites by itself.

autoignition temperature: The minimum temperature at which there is sufficient energy for a material or gas to ignite spontaneously, without a spark, flame, or other source.

axial float: The axial movement of a shaft due to bearing and bearing housing clearances.

axial load: A load in which the applied force is parallel to the axis of rotation.

axial piston pump: A piston pump that consists of pistons in a rotating piston block parallel to the drive shaft.

B

babbitt metals: Alloys of soft metals such as copper, tin, and lead and a hard material such as antimony.

backlash: The play between mating gear teeth.

balanced vane pump: A vane pump that has two sets of internal ports and produces a pumping action in chambers on both sides of the rotor and shaft.

ball bearing: An antifriction bearing that permits free motion between a moving part and a fixed part by means of balls confined between inner and outer rings.

barcode: A series of vertical lines and spaces of varying widths that is used to represent data.

base: The side upon which a triangle stands.

base diameter: The diameter from which the involute portion of a tooth profile is generated.

baseline inspection: An inspection intended to establish a reference point using equipment operating under normal conditions and in good working order.

base plate: A rigid steel support for firmly coupling and aligning two or more rotating devices.

bases: The ends of a prism.

bearing: A machine part that supports another part, such as a shaft, which rotates or slides in or on it.

becket: An attachment point, usually on a block, for the dead end of a hoisting rope.

bed section: The lower section of an extension ladder.

bellows: A device that draws air in through a flapper valve when expanded and expels the air through a nozzle when contracted.

belt and sheave groove gauge: A gauge that has a male form to determine the size of a pulley and a female form to determine the size of a belt.

belt deflection method: A belt tension method in which the tension is adjusted by measuring the deflection of the belt.

belt pitch length: The total length of the timing belt measured at the belt pitch line.

belt pitch line: A line located on the same plane as the belt tension member.

bench vise: A vise that can be attached to a bench top.

bending strength: A material's resistance to bending or deflection.

bend ratio: The ratio between the diameter of a rope bend (D), such as over a pulley, and the nominal diameter of the rope (d). Also known as D/d ratio.

bent-axis piston pump: A piston pump in which the pistons and cylinders are at an angle to the drive shaft and thrust plate.

bevel gear: A gear that connects shafts at an angle in the same plane.

bevel gear drive: A pair of gears that mesh at an angle, usually 90°.

bight: A loose or slack part of a rope between two fixed ends.

binary system: A system that has two values, such as one and zero.

bird caging: A damage condition of wire rope where the strands separate and open forming a shape similar to a bird cage.

blackwall hitch: A hitch made for securing a rigging rope to a hoisting hook.

bladder gas charged accumulator: An accumulator consisting of a seamless steel shell, a rubber bladder (bag) with a gas valve, and a poppet valve.

block: An assembly of one or more sheaves in a frame.

block and tackle: A combination of ropes and sheaves (pulleys) used to improve lifting efficiency.

blocking in: A method of quickly marking the structure of an object on a sketch by breaking the subject into common shapes and lines.

blockout: The process of placing a solid object in the path of a power source to prevent accidental energy flow.

bolt: A large, partially threaded fastener with no internal drives recess that is intended to be used with a nut.

bolt bound: The prevention of the horizontal movement of a machine due to the contacting of the machine anchor bolts to the sides of the machine anchor holes.

bottom tap: A tap designed to cut threads at the bottom portion of a hole.

boundary lubrication: The condition of lubrication in which the friction between two surfaces in motion is determined by the properties of the surfaces and the properties of the lubricant other than viscosity.

Bourdon tube: The pressure-sensing element inside a mechanical pressure gauge that consists of a circular stainless steel or bronze tube that is flattened to make it flexible.

bowline knot: A knot that forms a loop that is absolutely secure.

brazing: A joining process that joins parts by heating a filler metal to temperatures greater than 840°F but less than the melting point of the base metal.

breakaway torque: The initial energy required to get a nonmoving load to turn.

break line: A line used to indicate internal features or to avoid showing continuous features of long or large objects.

breather cap: A device that allows atmospheric pressure to push the fluid up to the pump.

bridge girder: The principal horizontal beam that supports a hoist trolley and is supported by end trucks.

brush: The sliding contact that rides against the commutator segments and is used to connect the armature to the external circuit.

butt spur: A notched, pointed, or spiked end of a ladder which helps prevent the ladder butt from slipping.

C

cab: A compartment or platform attached to a crane in which an operator may ride.

cabling: A rope's attempt to rotate and untwist its strand lays while under stress.

cage: A barrier or enclosure mounted on the siderails of a fixed ladder or fastened to a structure.

calibration: The verification of graduations and incremental values of a precision measuring instrument for accuracy and adjustments.

calibration standard: A finely ground, precisely sized object that is used as a basis of dimensional comparison.

caliper: A hand tool with one fixed and one adjustable jaw.

cam ring: A metal ring that provides an area for fluid flow and a surface against which vanes ride.

cantilever: A projecting beam or member supported at only one end.

capacitor: A device that stores electrical energy in an electrostatic field.

capacitor start-and-run motor: A 1φ motor that has capacitors in both the starting and running windings.

capacitor-start motor: A 1φ motor that has a capacitor in the starting winding.

capacity: The ability to hold or contain something.

cape chisel: A chisel with a thin, tapered face and a narrow cutting edge.

capillary action: The action by which the surface of a liquid is elevated on a material due to its relative molecular attraction.

carrier: The track of a ladder safety system consisting of a flexible cable or rigid rail secured to a ladder or structure.

carrier frequency: The frequency of the short voltage pulses of varying length that simulate a lower fundamental frequency.

cat's-paw hitch: A hitch used as a light-duty, quickly formed eye for a hoisting hook.

caustic solution: A liquid that causes corrosion.

cavitation: The process in which microscopic gas bubbles expand in a vacuum and suddenly implode when entering a pressurized area.

centerline: A line that locates the center of an object.

center of gravity: The balancing point of a load.

centerpoint: The point a circle or arc is drawn around.

center punch: A steel hand tool with one end formed to a conical point of approximately 90°.

centralized system: A lubrication system that contains permanently installed plumbing, distribution valves, reservoir, and pump to provide lubrication.

centrifugal force: The outward force produced by a rotating object.

centrifugal switch: A switch that opens when a rotor reaches a certain speed and reconnects when the rotor falls below that speed.

chain: A series of connected metal links.

challenging: The process of pressing or selecting the start switch of a machine to determine if the machine starts when it is not supposed to start.

check valve: A valve that allows flow in only one direction.

chemisorption: A chemical adsorption process in which weak chemical bonds are formed between liquid or gas molecules and solid surfaces.

chisel: A hand tool with a cutting edge on one end that is used to shape, dress, or work wood, metal, or stone.

choker hook: A sliding hook used in a choker sling and hooked to a sling eye.

chord: A line from circumference to circumference not through the centerpoint.

circle: 1. A plane figure generated about a centerpoint. **2.** A plane figure formed by a cutting plane perpendicular to the axis of a cone.

circuit breaker: A manually operated switch with a mechanism that automatically opens the circuit if an overload condition occurs.

circular pitch: 1. In belts, the distance from the center of one tooth to the center of the next tooth, measured along the pitch line. **2.** In gears, the distance from a point on a gear tooth to the corresponding point on the next gear tooth, measured along the pitch circle.

circular saw: A handheld or table-mounted power saw with teeth around the circumference of a circular blade that is rotated at high speed on a central axis or shaft.

circumference: The boundary of a circle.

clean-out cover: A device used to access a reservoir when it requires cleaning from the buildup of solid materials.

clearance: The radial distance between the top of a tooth and the bottom of the mating tooth space when fully mated.

cleat: A narrow wood piece, nailed across another board or boards, to provide support or to prevent movement.

closed circuit: An electrical circuit with a complete path that allows current to flow.

clove hitch: A quick hitch used to secure a rope temporarily to an object.

coalescing filter: A device that removes submicron solids and vapors of oil or water by uniting very small droplets into larger droplets.

code: A regulation or minimum requirement.

coefficient of friction: The measure of the frictional force between two surfaces in contact.

coil: A winding of insulated conductors arranged to produce a magnetic field.

commutator: A ring made of insulated segments that keep the armature windings in the correct polarity to interact with the main fields.

competent person: A person capable of recognizing and evaluating employee exposure to hazardous substances or to other unsafe conditions and of specifying the necessary protection and precautions to be taken to ensure the safety of all employees.

complementary angles: Two angles formed by three lines in which the sum of the two angles equals 90°.

compound gear train: Two or more sets of gears where two gears are keyed and rotate on one common shaft.

computerized maintenance management system (CMMS): A software package that organizes preventive maintenance information and automatically generates reports, work orders, and other data for implementing and improving future maintenance activities.

concentric circles: Two or more circles with different diameters but the same centerpoint.

condensation: The change in state from a gas to a liquid.

conductor: A material that has very little electrical resistance and permits electrons to move through it easily.

cone: A solid generated by a straight line moving in contact with a curve and passing through the vertex.

confined space: A space large enough for an individual to physically enter and perform assigned work but has limited or restricted means for entry and exit and is not designed for continuous occupancy.

conic section: A curve produced by a plane intersecting a right circular cone.

connecting link: A three-part chain attachment used to assemble and connect the master link to a chain.

connecting rod: The rod that connects the crankshaft to a piston.

Conrad bearing: A single row ball bearing without loading slots that has deeper than normal races.

contact: A conducting part of a relay that acts as a switch to connect or disconnect a circuit or component.

contact adhesive: A flexible synthetic adhesive that is applied separately to two surfaces, which are then brought into contact.

contactor: A heavy-duty relay for controlling high-current loads.

contact tachometer: A test tool that measures the rotational or linear speed of an object through direct contact with the object.

contact temperature probe: A test tool that measures temperature at a single point through direct contact with the area being measured.

contaminant: A substance that causes harm or damage to that with which it comes in contact.

continuity: The condition of a circuit that is closed (allows current flow).

continuity tester: A device that indicates if a circuit is open or closed.

contour: An identifying outline that separates all or part of an object from the background.

control circuit: An electrical circuit consisting of switches and other control devices that is designed to operate a load in a power circuit in a particular way.

controlled flow: The fluid flow after a flow control device has reduced the flow rate of the fluid flow.

conventional current flow: A convention that shows current as flowing from positive to negative.

conventional protractor: A tool designed to measure printed angles.

core protrusion: A damage condition of wire rope where compressive forces from within the rope force the strands apart.

corrosion: The action or process of eating or wearing away gradually by chemical action.

corrosive wear: Wear resulting from metal being attacked by acid.

coupling: A device that connects the ends of rotating shafts.

coupling face: The flat surface of a coupling half, facing the flat surface of the connecting coupling half.

coupling rim: The outside diameter surface of a coupling.

coupling unbalance: An unequal radial weight distribution where the mass and coupling geometric lines do not coincide.

cow hitch: A hitch used to secure a tag line to a load.

crankshaft: A shaft that has one or more eccentric surfaces that produce a reciprocating motion when the shaft is rotated.

crescent seal: A crescent-moon-shaped seal between the gears and between the inlet and outlet sides of an internal gear pump.

cross cut: A cut that is made against the direction of the wood grain and is made with full, even strokes at about a 45° angle.

crossover: One wrap winding on top of the preceding wrap.

crowning: A reverse strand splice that is used when an enlarged rope end is desired or not objectionable.

cup seal: A lip seal whose lip forms the shape of a cup.

current (I): The flow of electrons through an electrical circuit.

curved line: A line that continually changes direction.

curvilinear belt: A timing belt containing circular-shaped teeth.

cutaway diagram: A diagram that shows the internal details of components.

cutting-plane line: A line that indicates the path through which an object will be cut so that its internal features can be seen.

cyanoacrylate adhesive: An acrylic polymer resin adhesive that bonds very quickly.

cycle: One complete positive and one complete negative alternation of a wave form.

cylinder: A solid generated by a straight line moving in contact with a curve and remaining parallel to the axis and its previous position.

cylindrical roller bearing: A roller bearing having cylinder shaped rollers.

D

DC voltage: Voltage that flows in one direction only.

D/d ratio: *See* bend ratio.

deceleration distance: The additional vertical distance a falling technician travels, excluding lifeline elongation and free-fall distance, before stopping, from the point at which the deceleration device begins to operate.

decibel (dB): A unit of measure used to express the relative intensity of sound.

decimal: A number expressed in base 10.

decimal fraction: A fraction with a denominator of 10, 100, 1000, etc.

decision: A judgment or conclusion reached or given.

deformation: The undesirable bending of a hook due to an applied force.

demulsification: The act of separating water and oil quickly.

depth micrometer: A precision measuring instrument that measures component depths and heights.

desiccant dryer: A device that removes water vapor by adsorption.

detail drawing: A type of drawing that shows as much information about a device or component as possible.

dew point: The temperature to which air must be cooled in order for the moisture in the air to begin condensing.

dial caliper: A caliper with a dial indicating gauge.

dial indicator: A device that measures the deviation from a true circular path.

diameter: The distance from circumference to circumference through the centerpoint.

diametral pitch: The ratio of the number of teeth in a gear to the diameter of the gear's pitch circle.

diamond-point chisel: A chisel with a V-shaped blade that is less than 180°.

diaphragm gas charged accumulator: An accumulator with a flexible diaphragm separating the gas and fluid.

die: A tool used to cut external threads on round rods.

digital caliper: A caliper that displays measurements with a digital electronic indicating gauge.

digital micrometer: A micrometer with a digital electronic indicating gauge.

digital multimeter (DMM): A meter that can measure two or more electrical quantities and display the measured quantities as numerical values.

dimension: A numerical value that gives the size, form, or location of objects on a drawing.

dimensioning: A method of adding dimensions to a drawing to indicate the geometrical characteristics of an object.

dimension line: A line that is used with a written dimension to indicate size or location.

direct-acting valve: A valve that is activated or directly moved by fluid pressure from the primary port.

direct current (DC): Current that flows in only one direction.

directional control valve: A valve whose primary function is to direct or prevent flow through selected passages.

direct proportion: A statement of equality between two ratios in which the first of four terms divided by the second equals the third divided by the fourth.

direct tolerancing: The practice of specifying a dimension's permissible range directly within the dimensioning lines.

dispersed solid: A solid that is finely ground in order to be spread.

displacement: 1. The volume of fluid moved during each revolution of a pump's shaft. **2.** The measurement of the distance (amplitude) an object is vibrating.

displacement transducer: A mechanical sensor whose gap to voltage output is proportional to the distance between it and the measured object (usually a shaft).

dodecahedron: A regular solid of twelve pentagons.

double-acting cylinder: A cylinder that requires fluid flow for extending and retracting.

double hitch knot: A knot with two half hitch knots.

double-pole scaffold: A wood scaffold with both sides resting on the floor or ground and is not structurally anchored to a building or other structure.

double trapezoidal belt: A timing belt containing two trapezoidal-shaped sets of teeth.

double V-belt: A belt designed to transmit power from the top and bottom of the belt.

dowel effect: A condition that exists when the bolt hole of a machine is so large that the bolt head forces the washer into the hole opening on an angle.

drain plug: A threaded device that is installed at the lowest point of a reservoir to allow for the removal of hydraulic fluid or draining of accumulated moisture.

drawing: An assembly of lines, dimensions, and notes used to convey general or specific information as required by the application and use.

drawing scale: A system of drawing representation in which drawing elements are proportional to actual elements.

drip system: A gravity-flow lubrication system that provides drop-by-drop lubrication from a manifold or manually filled cup through a needle valve.

drive gear: Any gear that turns or drives another gear.

driven gear: Any gear that is driven by another gear.

dropping point of grease: The temperature at which the oil in grease separates from the thickener and runs out, leaving just the thickener.

drum wrap: The rope length required to make one complete turn around the drum of a hoist or crane.

dry air: Air free of water vapor or oil droplets.

dryer: A device that dries air through cooling and condensing.

dynamic head: The head of fluid in motion.

dynamic head pressure: The pressure and velocity produced by a fluid in motion.

dynamic lift: The lift of fluid in motion.

dynamic range: The ratio between the smallest and largest signals that can be analyzed simultaneously.

dynamic seal: A seal used between moving parts that prevents leakage or contamination.

dynamic signal analyzer: An analyzer that uses digital signal processing and an FFT to display a dynamic vibration signal as a series of frequency components.

dynamic total column: Dynamic head plus dynamic lift.

E

earmuff: A device worn over the ears to reduce the level of noise reaching the eardrum.

earplug: A compressible device inserted into the ear canal to reduce the level of noise reaching the eardrum.

eccentric: Out-of-round or that which deviates from a circular path.

eccentric circles: Two or more circles with different diameters and different centerpoints.

eccentric surface: A surface that has a different center than the center of a crankshaft.

eddy current: An electric current that is generated and dissipated in a conductive material in the presence of an electromagnetic field.

efficiency: The measure of a component's or system's useful output energy compared to its input energy.

electrical circuit: The interconnection of conductors and electrical components through which current is designed to flow.

electrical pitting: The cratering and burn damage caused by electric arc discharge between mating metal components.

electrical power system: A system that produces, transmits, distributes, and delivers electrical power in order to operate electrical loads designed for connection to the system.

electrical shock: A shock that results any time a body becomes part of an electrical circuit.

electric arc: A discharge of electric current across an air gap.

electromagnet: A magnet created when electricity passes through a wire.

electromagnetic induction: The process by which voltage is induced in a wire by a magnetic field when lines of force cut across the wire.

electron: A particle with a negative charge.

electron current flow: A convention that shows current as flowing from negative to positive.

electron flow: The traveling of a displaced valence electron from one atom to another.

electronic reverse dial method: An alignment method that uses the reverse dial as a base method with the dial indicators replaced with electromechanical sensing devices.

element: A logic device that is capable of making a 0 or 1 output decision based on its input.

ellipse: A plane figure formed by a cutting plane oblique to the axis of a cone but at a greater angle with the axis than with the elements of the cone.

emergency maintenance: Maintenance procedures that involve reacting to operating equipment breakdowns.

emulsification: The act of mixing oil and water.

end play: The total amount of axial movement of a shaft.

end truck: A roller assembly consisting of a frame, wheels, and bearings generally installed or removed as complete units.

energy: A measure of the ability to do work.

energy-isolating device: A device that prevents the transmission or release of energy.

engineer's scale: A triangular scale used to draw large areas, such as property lines, on a building lot.

epoxy adhesive: A synthetic polymer resin adhesive that is chemically cured from two mixed liquids.

equal rotor unbalance: The unbalance of weighted force across one side of a rotor or armature.

equation: A means of showing that two numbers or two groups of numbers are equal to the same amount.

equilateral triangle: A triangle that has three equal angles and three equal sides.

equilibrium: The condition when all forces and torques are balanced by equal and opposite forces and torques.

exhaust flow: The fluid flow from an actuator, back through a valve, to a reservoir.

explosion: A sudden reaction involving rapid physical or chemical decay accompanied by an increase in temperature, pressure, or both.

explosive range: The difference between the lower explosive limit and the upper explosive limit of combustible gases.

extension ladder: An adjustable height ladder with a fixed bed section and sliding, lockable fly sections.

extension line: A line that extends from the surface features of an object; used to terminate dimension lines.

external gear pump: A gear pump that consists of two externally toothed gears that form a seal within the pump housing.

eyebolt: A bolt with a looped head.

eye loop: A rope splice containing a thimble.

F

face width: The length of gear teeth in an axial plane.

false Brinell damage: Bearing damage caused by forces passing from one ring to the other through the balls or rollers.

fastening: A process that affixes one part to another.

Fast Fourier Transform (FFT): A calculation method for converting a time waveform into a series of frequency vs. amplitude components.

Fast Fourier Transform (FFT) analyzer: A microprocessor capable of displaying the FFT of an input signal.

fatigue crack: A crack in a gear that occurs due to bending, mechanical stress, thermal stress, or material flaws.

fatigue fracture: A breaking or tearing of gear teeth.

fatigue life: The maximum useful life of a bearing.

fatigue wear: Gear wear created by repeated stresses below the tensile strength of a material.

feeler gauge: A steel leaf at a specific thickness. Also known as a thickness gauge.

ferrography: A wear particle analysis utilizing diagnostic and preventive (predictive) techniques to evaluate the on-line condition of interacting lubricated or fluid powered parts or components.

ferromagnetic material: A material, such as soft iron or steel, that is easily magnetized.

ferrous metal: A metal containing iron.

ferrule: A metal sleeve used for joining one piece of tube to another.

fiberglass ladder: A ladder constructed of fiberglass.

file: An abrasive tool with single or double rows of fine teeth cut into the surface.

filler wire: Wire rope that uses fine wires to fill the gaps between the major wires.

filter: 1. A device containing a porous substance through which a fluid can pass but particulate matter cannot. **2.** A device that limits vibration signals so only a single frequency or group of frequencies can pass.

fire point: The temperature at which oil ignites when touched with a flame.

first aid: Help for a victim immediately after an injury and before professional medical help arrives.

fixed bore pulley: A machine-bored one-piece pulley.

fixed ladder: A ladder that is permanently attached to a structure.

flared fitting: A fitting that is connected to a tubing whose end is spread outward.

flareless (compression) fitting: A fitting that seals and grips by manual adjustable deformation.

flash point: The temperature at which oil gives off enough gas vapor to ignite briefly when touched with a flame.

flat cold chisel: A chisel with a tempered cutting edge to maintain durability.

flat file: A file that is used to file flat surfaces as well as for other operations that require a fast-cutting file.

flexible belt drive: A system in which a resilient flexible belt is used to drive one or more shafts.

flexible coupling: A coupling with a resilient center, such as rubber or oil, that flexes under temporary torque or misalignment due to thermal expansion.

flow control valve: A valve whose primary function is to regulate the rate of fluid flow.

flowmeter: A test tool that measures the flow of a fluid within a system.

flow rate: The volume of fluid flow per minute.

fluid: A substance that tends to flow or conform to the outline of its container (such as a liquid or a gas).

fluid flow: The movement of fluid caused by a difference in pressure between two points.

fluting: The elongated and rounded grooves or tracks left by the etching of each roller on the rings of an improperly grounded roller bearing during welding.

fly section: The upper sections of an extension ladder.

foaming: Excessive air in hydraulic fluid.

foot pad: A metal swivel attachment with rubber or rubber-like tread that helps prevent a ladder butt from slipping.

force: An interaction that tends to change the state of rest or motion of an object.

form: The shape and structure of all or part of an object and includes a sense of mass and volume.

formula: A mathematical equation that contains a fact, rule, or principle.

foundation: An underlying base or support.

foundry hook: A hook with a wide, deep throat that fits the handles of molds or castings.

four-way directional control valve: A valve that has four main ports that change fluid flow from one port to another.

fracture: A small crack in metal caused by the stress or fatigue of repeated pulling or bending forces.

free air: Air at atmospheric pressure and ambient temperature.

free-fall distance: The vertical distance between the fall-arrest attachment point on the body harness before the fall and the attachment point when the personal fall-arrest system applies force to arrest the fall.

frequency: The number of cycles per minute (cpm), cycles per second (cps), or multiples of rotational speed (orders).

frequency domain: The amplitude versus frequency spectrum observed on an FFT analyzer.

frequency spectrum: A representation of the frequency and content of a dynamic signal.

fretting corrosion: The rusty appearance that results when two metals in contact are vibrated, rubbing loose minute metal particles that become oxidized.

friction disc: A device that transmits power through contact between two discs or plates.

frustum: The remaining portion of a pyramid or cone with a cutting plane passed parallel to the base.

fulcrum: A support on which a lever turns or pivots.

fundamental frequency: The desired voltage frequency simulated by the varying ON/OFF pulses at a higher carrier frequency.

fuse: An overcurrent protection device with a fusible link that melts and opens the circuit on an overcurrent condition.

fuse puller: A device made of a nonconductive material such as nylon that is used to grasp and remove cartridge fuses.

G

galling: A bonding, shearing, and tearing away of material from two contacting, sliding metals. Also known as adhesive wear.

gantry crane: A crane with structural beam supports for lifting equipment.

gas: A fluid that has neither independent shape nor volume and tends to expand indefinitely.

gas-charged accumulator: An accumulator that uses compressed gas over hydraulic fluid to store energy.

gasket: A seal used between machined parts or around pipe joints to prevent the escape of fluids.

gas laws: The relationships between the volume, pressure, and temperature of a gas.

gas lubricant: A lubricant that uses pressurized air to separate two surfaces.

gas metal arc welding (GMAW): An arc welding process that uses a continuous wire electrode.

gas tungsten arc welding (GTAW): An arc welding process in which a shielding gas protects the arc between a tungsten electrode and the weld area.

gate valve: A two-position valve that has an internal gate that slides over the opening through which fluid flows.

gauge pressure: Pressure above atmospheric pressure that is used to express pressures inside a closed system.

gear: A toothed machine element used to transmit motion between rotating shafts.

gear pump: A positive-displacement pump containing intermeshing gears that force the fluid from the pump.

gear train: A combination of two or more gears in mesh used to transmit motion between two rotating shafts.

general note: A note that applies to the entire print on which the note appears.

generator: A machine that converts mechanical energy into electrical energy.

globe valve: An infinite-position valve that has a disk that is raised or lowered over a port through which fluid flows.

grab hook: A hook used to adjust or shorten a sling leg through the use of two chains.

graphic diagram: A drawing that uses simple line shapes (symbols) with interconnecting lines to represent the function of each component in a circuit.

grease: A semisolid lubricant created by combining low-viscosity oils with thickeners, such as soap or other finely dispersed solids.

grease cup: A receptacle used to apply grease to bearings.

grease dropping point: The maximum temperature a grease withstands before it softens enough to flow through a laboratory testing orifice.

grease gun: A small hand-operated device that pumps grease under pressure into bearings.

great circle: The circle formed by passing a cutting plane through the center of a sphere.

ground fault: An unintended current path between a ground and an ungrounded (hot) conductor.

ground fault circuit interrupter (GFCI): An electrical device that protects personnel by detecting potentially hazardous ground faults and quickly disconnecting power from the circuit.

grounding: The connection of all exposed non-current-carrying metal parts to the earth.

guardrail: A rail secured to uprights and erected along the exposed sides and ends of a platform.

guyline: A rope, chain, rod, or wire attached to equipment as a brace or guide.

H

hacksaw: A metal-cutting hand tool with an adjustable steel frame for holding various lengths and types of blades.

half hitch knot: A binding knot where the working end is laid over the standing part and stuck through the turn from the opposite side.

halyard: A rope used for hoisting or lowering objects.

hammer: A striking or splitting tool with a hardened head fastened perpendicular to a handle.

hand chain: A continuous chain grasped by an operator to operate a pocket wheel.

hand-chain drop: The distance between the lower portion of a hand chain to the upper limit of the hoist hook travel.

hand-chain hoist: A manually operated chain hoist used for moving a load.

handsaw: A woodcutting hand tool consisting of a straight, toothed blade attached to a handle.

hasp: A multiple-lockout/tagout device.

hazardous location: A location where flammable liquids, gases, vapors, or combustible dusts exist in sufficient quantities to pose a risk of an explosion or fire.

head: The difference in the level of a liquid (fluid) between two points.

head-on view: The view when looking directly at an object from the same height as the object.

head pressure: The pressure created by fluid stacked on top of itself.

headroom: The distance from the cup of the top hook to the cup of a hoist hook when the hoist hook is at its upper limit of travel.

heat energy: The ability to do work using the heat stored or built up in a fluid.

heat exchanger: A device that transfers heat through a conducting wall from one fluid to another.

heating element: A conductor with an intentionally high resistance for producing heat when connected to an electrical power supply.

helical gear: A gear with teeth that are cut at an angle to its axis of rotation.

helical screw compressor: A compressor that contains meshing screw like helical rotors that compress air as they turn.

helix: A spiral or screw shape form.

herringbone gear: A double helical gear that contains a right- and left-hand helix.

hertz (Hz): A measurement of a frequency equal to one cycle per second.

hexahedron: A regular solid of six squares.

hidden line: A line that represents the shape of an object that cannot be seen.

hitch: The interlacing of rope to temporarily secure it without knotting the rope.

hoist: A mechanical device used to provide the lifting force on lead lines.

hoist chain: The chain that raises a load.

hoisting apparatus chain: A precisely measured chain calibrated to function in pocket type wheels used in manual or powered chain hoists.

hoisting hook: A steel alloy hook used for overhead lifting and connected directly to the piece being lifted.

hoist trolley: The unit carrying the hoisting mechanism that travels on a bridge girder.

holding contact: An auxiliary contact used to maintain current flow to a relay coil.

hook: A curved or bent implement for holding, pulling, or connecting rigging to loads or lifting equipment.

hook drift: The slippage of a hook caused by insufficient braking.

horizontal line: A line that is parallel to the horizon.

horsepower: A unit of power equal to 746 W or 33,000 lb ft per minute (550 lb ft per second).

hose: A flexible tube for carrying fluids under pressure.

humidity: The amount of moisture in the air.

hunting tooth: A tooth added to mesh with every tooth on a mating gear to produce even tooth wear.

hydraulic actuator: A device that converts hydraulic energy into mechanical energy.

hydraulic circuit: A closed path through which hydraulic fluid flows or may flow.

hydraulic cylinder: A device that converts hydraulic energy into straight-line (linear) mechanical energy.

hydraulic diagram: The layout, plan, or sketch of a hydraulic circuit that is designed to explain, demonstrate, or clarify the relationship or functions between hydraulic components.

hydraulic motor: A device that converts hydraulic energy into rotary mechanical energy.

hydraulics: The branch of science that deals with the practical application of water or other liquids at rest or in motion.

hydraulic scissor lift: A mobile hydraulically-operated platform controlled by remote switches attached at the platform.

hydrocarbon: Any substance that is composed mostly of hydrogen and carbon.

hydrodynamics: The study of the forces exerted on a solid body by the motion or pressure of a fluid.

hydrostatics: The study of liquids at rest and the forces exerted on them or by them.

hyperbola: A plane figure formed by a cutting plane that has a smaller angle with the axis than with the elements of a cone.

hypoid gear: A spiral bevel gear with curved, non-symmetrical teeth that are used to connect shafts at right angles.

hypotenuse: The side of a right triangle opposite the right angle.

I

icosahedron: A regular solid of twenty triangles.

idler gear: A gear that transfers motion and direction in a gear train but does not change speeds.

imbalance: A lack of balance.

impact flaring method: A basic flaring method in which a flaring tool is inserted into the tubing end and hammered into the tubing until the tubing end is spread (flared) as required.

implosion: An inward bursting.

inching: Slow movement in small degrees.

inclined line: A line that is slanted.

induced soft foot: Soft foot that is created by external forces such as coupling misalignment, piping strain, tight jack screws, or improper structural bracing.

induction: The process of causing electrons to align or uniformly join to create a magnetic or electrical force.

induction motor: A motor that rotates due to the interaction between the magnetic fields of the stator and rotor.

industrial crane: A crane with structural beam supports for lifting equipment.

inert gas: A gas, such as argon or helium, that does not readily combine with other elements.

infrared thermography: The science of using electronic optical devices to detect and measure radiation and correlating the radiation level to surface temperature.

infrared thermometer: A handheld device that detects infrared emissions to measure temperature.

inside micrometer: A micrometer used to measure linear dimensions between two inside points or parallel surfaces.

instructional drawing: A type of drawing that is used to indicate how to do work using the simplest and/or safest method.

instrumentation: The area of industry that deals with the measurement, evaluation, and control of process variables.

insulator: A material that has a very high electrical resistance and resists the flow of electrons.

intake filter: A filter that removes solids from free air at a compressor inlet port.

intake flow: The fluid flow from the reservoir, through the filters, to the pump.

intensifier (booster): A device that converts low pressure fluid power into high pressure fluid power.

intercooler: A heat exchanger between the discharge of one compression stage and the inlet of the next compression stage.

intercooling: The process of removing a portion of the heat of compression as the air is fed from one compression stage to another.

interference fit: A fit in which the internal member is larger than the external member so that there is always an actual interference of metal.

internal gear pump: A gear pump that consists of a small external drive gear mounted inside a large internal spur gear (ring gear).

intrinsic safety (IS): An electrical circuit design technique that provides explosion protection by eliminating arcing and heat that could ignite explosive atmospheres.

inventory control: The organization and management of commonly used consumables, vendors and suppliers, and purchasing records in a PM system.

inverse proportion: A proportion in which an increase in one quantity results in a proportional decrease in the other related quantity.

inverse ratio: The reciprocal of a given ratio.

inverter: A device that changes DC voltage into AC voltage of any frequency.

involute form: A tooth form that is curled or curved.

irregular polygon: A polygon with unequal sides and unequal angles.

irregular polyhedron: Any of a variety of solids with faces that are irregular polygons (unequal sides).

isosceles triangle: A triangle that contains two equal angles and two equal sides.

J

jack screw: A screw inserted through a block that is attached to a machine base plate allowing for ease in machine movement.

jib crane: A crane that is mounted on a single structural leg.

journal: The part of a shaft, such as an axle or spindle, that moves in a sleeve bearing.

K

key: A small removable piece of steel of standard shape and dimensions that is placed in a keyseat between a shaft and hub to provide a means for transmitting power.

keyseat: A groove along the axis of a shaft or hub.

kinetic energy: The energy of motion.

kinking: A sharp permanent bending.

knife file: A file with a blade cross-section tapering from a square to a pointed edge.

knot: The interlacing of rope to form a permanent connection.

knotting: Fastening a part of a rope to another part of the same rope by interlacing it and drawing it tight.

L

ladder: A structure consisting of two siderails joined at intervals by steps or rungs for climbing up and down.

ladder duty rating: The weight (in lb) a ladder is designed to support under normal use.

ladder jack: A ladder accessory that supports a plank to be used for scaffolding.

ladder safety system: An assembly of components whose function is to arrest the fall of a worker.

lang-lay rope: A rope in which the yarn or wires and strands are laid in the same direction.

lanyard: A flexible line of rope, wire rope, or strap that generally has a connector at each end for connecting a body harness to a deceleration device, lifeline, or anchorage point.

laser rim and face alignment method: An alignment method in which laser devices are placed opposite each other to measure alignment.

laser tachometer: A test tool that uses a laser light to measure the rotational speed of an object.

lattice-boom crane: A crane with a boom constructed from a gridwork of steel reinforcing members.

law of charges: States that opposite charges attract each other and like charges repel each other.

lay: A complete helical wrap of the strands of a rope.

leader line: A line that connects a written description such as a dimension, note, or specification with a specific feature of a drawn object.

lead line: The part of a rope to which force is applied to hold or move a load.

left lang-lay rope: A rope in which the yarn or wires are laid to the left and the strands are laid to the left.

left regular-lay rope: A rope in which the strands are laid to the left and yarn or wires are laid to the right.

lever-operated hoist: A lifting device that is operated manually by the movement of a lever.

lichen: A fungus normally seen as a growth on tree trunks or rocks.

lift: 1. In hoisting, the distance between the hoist's upper and lower limits of travel. **2.** In pumping, the height at which atmospheric pressure forces a fluid above the elevation of its supply source.

lifting: Hoisting equipment or machinery by mechanical means.

lifting lug: A thick metal loop (eyebolt) welded or screwed to a machine to allow balanced lifting.

limit dimensioning: A direct tolerancing method that includes only the maximum and minimum values of a dimension on a drawing.

limited approach boundary: The distance from an exposed energized conductor or circuit part at which a person can get an electric shock and the closest distance an unqualified person can approach.

limit switch: A device that cuts off the power automatically at or near the upper limit of hoist travel.

limit valve: A mechanically actuated 3-way valve that is used to either monitor motion or measure position of an object.

line: The boundary of a surface.

linear amplitude spectra: Amplitude signals displayed in equal increments.

line diagram: A diagram that uses lines and graphic symbols to show the logic and operation of an electrical circuit.

lip seal: A seal that is made of a resilient material that has a sealing edge formed into a lip.

liquid: A fluid that can flow readily and assume the shape of its container.

liquid lubricant: A lubricant that uses a liquid, such as oil, to separate two surfaces.

litmus paper: A color-changing, acid-sensitive paper that is impregnated with lichens.

loading slot: A groove or notch on the inside wall of each bearing ring to allow insertion of balls.

load-sensing control: A control that controls variable-displacement pump output according to the amount of pressure required at load use.

lobe: The screw helix of a rotor.

location dimension: A dimension that uses the components of an object to locate an angle or feature on an object.

location drawing: A type of drawing used to position switches, buttons, terminal connections, and other features found on a device or component.

lockout: The process of removing a source of power and installing a lock that prevents the power from being turned ON.

lockout device: A lightweight enclosure that allows the lockout of a standard control device.

logarithmic amplitude spectra: Amplitude signals displayed in powers of ten.

logarithmic scale: An amplitude or frequency displayed in powers of ten.

logic: The determination of an output based on the state(s) of a system's input(s).

loop: The folding or doubling of a line, leaving an opening through which another line may pass.

loop eye: A length of webbing folded back and spliced to a sling body, forming an opening.

lower explosive limit (LEL): The lowest concentration (air-fuel mixture) at which a gas can ignite.

lubricant: A substance placed between two solid surfaces to reduce their friction.

lubrication: The process of maintaining a fluid film between solid surfaces to prevent their physical contact.

lubricator: A device that injects atomized oil into the air sent to pneumatic components.

M

machine: An assembly of devices that transfers the force, motion, or energy input from one device to a force, motion, or energy output at another.

machine screw: A fully threaded fastener with an internal drive recess that is intended to be engaged into an internally threaded material.

machine vise: A vise made with high-grade body castings and ground-steel jaw plates.

machinist's steel protractor: A tool used to measure or mark angle measurements on rigid workpieces.

magnet: A device that attracts iron.

magnetic flux lines: The invisible lines of force that make up a magnetic field.

magnetism: A force that interacts with other magnets and ferromagnetic materials.

main header: The main air supply line that runs between a receiver and the circuits in a pneumatic system.

maintenance: The planning and action to minimize and prevent equipment breakdowns and lost production time.

manifold: A device that contains passageways that enable one input signal to be divided into several output signals.

master link: A chain attachment with a ring considerably larger than that of the chain to allow for the insertion of a hook.

maximum intended load: The total of all loads, including the working load, the weight of the scaffold, and any other loads that may be anticipated.

mechanical: Pertaining to or concerned with machinery or tools.

mechanical advantage: The ratio of the output force to the input force of a device.

mechanical drive: A system by which power is transmitted from one point to another.

mechanical puller: A tool used to remove fitted machine parts.

mercury barometer: An instrument that measures atmospheric pressure using a column of mercury.

mesh: 1. In rope, the size of the openings between the rope or twine of a net. **2.** In filters, the number of horizontal and vertical threads per square inch.

metal fatigue: The fracturing of worked metal due to normal operating conditions or overload situations.

metal ladder: A ladder constructed of metal.

metering: Regulating the amount or rate of fluid flow.

micrometer: A hand tool used to make high-accuracy measurements, often to the closest ten-thousandth of an inch (0.0001″).

micron (μm): A unit of length equal to one millionth of a meter (0.000039″).

midrail: A rail secured to uprights approximately midway between a guardrail and a platform.

mill file: An all-purpose, single-cut file especially adapted for finish filing.

misalignment: The condition where the axes of two machine shafts are not aligned within tolerances.

miter gear: A gear used at right angles to transmit horsepower between two intersecting shafts at a 1:1 ratio.

modified curvilinear belt: A timing belt containing modified circular-shaped teeth.

moisture separator: A device that separates a large percentage of water from cooled air through a series of plates or baffles.

molded notch belt: A belt that has notches molded into its cross-section along the full length of the belt.

motor starter: A heavy-duty relay that includes motor overload protection.

multimeter: A portable test tool that is capable of measuring two or more electrical quantities.

multiple point suspension scaffold: A suspension scaffold supported by four or more ropes.

multistage compressor: A compressor that uses two or three cylinders, each with a progressively smaller diameter, to produce progressively higher pressures.

N

nail: A small metal, rod-shaped spike with a broadened circular head that is driven typically into wood with a hammer.

needle bearing: A friction roller type bearing with long rollers of small diameter.

needle valve: An infinite-position valve that has a narrow tapered stem (needle) positioned in line with a tapered hole or orifice.

negative charge: An electrical charge produced when there are more electrons than protons in an atom.

neutral conductor: A wire that carries current from one side of a load to ground.

neutron: A neutral particle with a mass approximately equal to that of a proton.

nip: A pressure and friction point created when a rope crosses over itself after a turn around an object.

nominal value: A designated or theoretical value that may vary from the actual value.

noncontact temperature probe: A test tool that measures temperature using convection or radiation.

nonparallel misalignment: Misalignment where two pulleys or shafts are not parallel.

nonpositive displacement pump: A pump that is not sealed between its inlet and outlet.

nonpositive seal: A seal that allows a minute amount of fluid through to provide lubrication between surfaces.

note: A sentence that provides drawing information that does not fit within the space of the drawing.

NOT logic element: A logic element that provides an output that is the opposite of the input.

nucleus: The heavy, dense center of an atom.

nut: A small block of metal with an internally threaded hole that is threaded onto the end of a matching bolt or machine screw to assemble a stack of parts.

O

object line: A line that indicates the visible shape of an object.

oblique cylinder: A cylinder with the axis not perpendicular to the base.

oblique prism: A prism with lateral faces not perpendicular to the bases.

obtuse angle: An angle exceeding 90° but less than 180°.

obtuse triangle: A scalene triangle with one angle greater than 90°.

octahedron: A regular solid of eight triangles.

offset misalignment: 1. In flexible belt drives, a condition where two shafts are parallel but the pulleys are not on the same axis. **2.** In motor couplings, a condition where two shafts are parallel but are not on the same axis.

ohm: The resistance of a conductor in which an electrical pressure of 1 V causes an electrical current of 1 A to flow.

Ohm's law: The relationship between voltage (E), current (I), and resistance (R) in a circuit.

oil analysis: A predictive maintenance technique that detects and analyzes the presence of acids, dirt, fuel, and wear particles in lubricating oil to predict equipment failure.

oil carry-over: The released lubricating oil from the walls of a compressor cylinder and piston.

oil whirl: The buildup and resistance of a lubricant in a rolling contact bearing that is rotating at excessive speeds.

open circuit: An electrical circuit that has a gap or opening that does not allow current flow.

operational pitch point: The tangent point of two pitch circles at which gears operate.

opposing forces rotor unbalance: The unbalance of weighted forces on opposing ends and sides of a rotor or armature.

order: A multiple of a running speed (rpm) frequency.

orifice: A precisely sized hole through which fluid flows.

O-ring: A molded synthetic rubber seal having a round cross section.

OR logic element: A logic element that provides a logic level 1 if one or more inputs are at logic level 1.

orthographic projection: A type of drawing where all faces (front, top, and side) of an object are projected onto flat planes that generally are at 90° (right) angles to one another.

oscillator: A device that generates a radio frequency (RF) field that, when sent to a probe tip, creates eddy currents.

outside micrometer: A micrometer used for measuring outside diameters and thicknesses of parts.

overhead crane: A crane that is mounted between overhead runways.

overhung load: A force exerted radially on a shaft that may cause bending of the shaft or early bearing and belt failure.

overload relay: A time-delay device that senses motor current temperatures and disconnects the motor from the power supply if the current is excessive for a certain length of time.

oxidation: The combining of oxygen with oil, which breaks down the basic oil composition.

oxyacetylene welding (OAW): An oxyfuel welding process that uses oxygen mixed with acetylene.

oxyfuel welding (OFW): A welding process that produces heat from the combustion of a mixture of oxygen and a fuel gas.

P

packing: A bulk deformable material or one or more mating deformable elements reshaped by manually adjustable compression.

parabola: A plane figure formed by a cutting plane oblique to the axis and parallel to the elements of a cone.

parallel connection: A circuit connection that has two or more components connected so there is more than one path for current to flow.

parallelepiped: A prism with bases that are parallelograms.

parallel lines: Two or more lines that remain the same distance apart.

parallelogram: A four-sided plane figure with opposite sides parallel and equal.

parallel soft foot: A condition that exists when one or two machine feet are higher than the others and parallel to the base plate.

part: A rope length between the lower (hook) block and the upper block or drum.

particulate: A fine solid particle that remains individually dispersed in a gas.

pawl: A mechanism used to prevent a ratchet wheel from turning backwards.

pawl lock: A pivoting hook mechanism attached to the fly sections of an extension ladder.

peak: The absolute value from a zero point (neutral) to the maximum travel on a waveform.

peak-to-peak: The absolute value from the maximum positive travel to the maximum negative travel on a waveform.

peak-to-peak displacement: The distance from the upper limit to the lower limit of a vibration.

pendant: A pushbutton or lever control suspended from a crane or hoisting apparatus.

penetrating oil: An industrial lubricant used to clean and loosen frozen parts.

permanent chip magnet: A magnetic device placed in a reservoir to attract and hold ferrous metal particles that have contaminated the system but have not been recovered by system filters or strainers.

permit-required confined space: A confined space that has specific health and safety hazards capable of causing death or serious physical harm.

perpendicular line: A line that makes a 90° angle with another line.

personal protective equipment (PPE): Clothing and/or equipment worn by a technician to reduce the possibility of injury in the work area.

petroleum fluid: A fluid consisting of hydrocarbons.

phantom line: A line used to show a part's alternate positions or a repeated detail.

phase: The position of a vibrating part at a given moment with reference to another vibrating part at a fixed reference point.

photo tachometer: A test tool that uses light beams to measure the rotational speed of an object.

pictorial diagram: A diagram that uses drawings or pictures to show the relationship of each component in a circuit.

pictorial drawing: A three-dimensional drawing that resembles a picture.

picture plane: A two-dimensional space on paper where the flattened image of a three-dimensional object is depicted.

piezoelectric: The production of electricity by applying pressure to a crystal.

pilot line: A passage used to carry fluid to control a valve.

pilot-operated valve: A valve that is actuated by fluid in the line that is otherwise sent back to the reservoir.

pilot operation: Controlling the function of a valve using system pressure or pressure supplied by an external (pilot) source.

pin: A cylindrical fastener that is placed into a hole to secure the relative positions of two or more parts.

pinion: The smaller gear of a pair of gears, especially when engaging rack teeth.

pin punch: A punch used for removing pins from parallel holes.

pipe: A hollow cylinder of metal or other material of substantial wall thickness.

pipe vise: A vise that has a hinge at one end and a hook at the opposite end.

piston compressor: A compressor in which air is compressed by reciprocating pistons.

piston cushioning device: A device within a cylinder that provides a gradual deceleration of the piston as it nears the end of its stroke.

piston gas charged accumulator: An accumulator with a floating piston acting as a barrier between the gas and fluid.

piston pump: A hydraulic pump in which fluid flow is produced by reciprocating pistons.

pitch circle: The circle that contains the operational pitch point.

pitch diameter: The diameter of a pitch circle.

pitting: Localized corrosion that has the appearance of cavities (pits).

plain bearing: A bearing in which the shaft turns and is lubricated by a sleeve.

plane figure: A flat figure with no depth.

plank: A board 2″ to 4″ thick and at least 8″ wide.

plant survey: A complete inventory and condition assessment of a facility's equipment and structure.

platform: A landing surface which provides access/egress or rest from a fixed ladder.

pliers: A hand tool with opposing jaws for gripping and/or cutting.

plug tap: A tap used after a taper tap to start a true and straight thread.

plumb: An exact verticality (determined by a plumb bob and line) with the Earth's surface.

plus and minus tolerancing: A direct tolerancing method that provides an ideal dimension and includes the allowable deviations in the positive and negative directions.

ply: A layer of a formed material.

pneumatic circuit: A combination of air-operated components that are connected to perform work.

pneumatic hoist: A power-operated hoist operated by a geared reduction air motor.

pneumatic logic element: A miniature air valve used as a switching device to provide decision making signals in a pneumatic circuit.

pneumatics: The branch of science that deals with the transmission of energy using a gas.

pneumatic system: A system that transmits and controls energy through the use of a pressurized gas within an enclosed circuit.

pocket wheel: A pulley-like wheel with chain link pockets that is connected to a hoist mechanism.

polarity: The positive (+) or negative (–) state of an object.

pole scaffold: A wood scaffold with one or two sides firmly resting on the floor or ground.

polygon: A many-sided plane figure.

polyhedron: Any of a variety of solids bound by plane surfaces (faces).

polymer: A molecule made up of a chain of repeating units that are chemically bonded together.

polymer adhesive: A synthetic bonding substance that undergoes a chemical or physical reaction.

polytetrafluoroethylene (PTFE): A low-friction plastic material used as sealing in low-pressure hydraulic circuits. Also known as Teflon®.

polyurethane: A hard, chemical-resistant plastic used as a sealing material in hydraulic circuits.

polyurethane adhesive: A durable synthetic resin polymer adhesive available in multiple variations, each optimized for different applications.

portable band saw: A portable band saw is a handheld power saw that has a flexible metal saw blade that forms a continuous loop around two parallel pulleys.

port plate: A device that contains ports that connect the pump's internal inlet and discharge areas to the pump housing inlet and outlet ports.

position: The specific location of a spool within a valve which determines the direction of fluid flow through the valve.

positive charge: An electrical charge produced when there are fewer electrons than protons in an atom.

positive displacement: The movement of a fixed volume of fluid with each revolution of a pump shaft.

positive displacement compressor: A compressor that compresses a fixed quantity of air with each cycle.

positive displacement pump: A pump that delivers a definite volume of fluid for each cycle of the pump at any resistance encountered.

positive seal: A seal that does not allow the slightest amount of fluid to pass.

potential energy: Stored energy a body has due to its position, chemical state, or condition.

power: The rate of doing work or using energy.

power circuit: An electrical circuit that connects a load to main power lines.

power distribution: The process of delivering electrical power to where it is needed.

power drill: A power-driven rotary tool used with a bit with cutting edges for boring holes in materials such as wood, metal, or plastic.

power formula: A formula that shows the relationship between power (P), voltage (E), and current (I) in an electrical circuit.

power-operated hoist: A hoist operated by pneumatic or electric power and uses either chain or wire rope as the lifting component.

precharge pressure: The pressure of the compressed gas in an accumulator prior to the admission of hydraulic fluid.

precision: The level of accuracy or mechanical exactness.

precision measurement: A method of using measuring instruments to acquire accurate measurements.

predictive maintenance (PDM): The monitoring of wear conditions and equipment operating characteristics for comparison against a predetermined tolerance to predict potential malfunctions or failures.

preformed rope: Wire rope in which the strands are permanently formed into a helical shape during fabrication.

preloading: An initial pressure placed on a bearing when axial load forces are expected to be great enough to overcome preload force, thereby resulting in proper clearances.

pressure: The measure of force per unit area.

pressure-compensated flow control valve: A needle valve that makes allowances for pressure changes before or after an orifice through the use of a spring and spool.

pressure-compensated vane pump: A vane pump equipped with a spring on the low displacement side of the cam ring.

pressure compensator: A displacement control that alters displacement in response to pressure changes in a system.

pressure drop: The pressure differential between upstream and downstream fluid flow caused by resistance.

pressure energy: The ability to do work through the pressure of a fluid.

pressure filter: A very fine filter placed after a pump for protection of circuit components.

pressure gauge: A device that measures the intensity of a force applied to a fluid.

pressure-reducing valve: A valve that limits the maximum pressure at its outlet, regardless of the inlet pressure.

pressure regulator: A valve that restricts and/or blocks downstream air flow.

pressure-relief valve: A valve that sets a maximum operating pressure level for a circuit to protect the circuit from overpressure.

pressure switch: A device that senses a high or low pressure condition and relays an electrical signal to turn the compressor motor ON or OFF.

preventive maintenance (PM): A system that consists of maintenance procedures that involve scheduled inspections of, adjustments to, and repairs to equipment to verify that equipment is in proper working order.

preventive maintenance (PM) system: A system used to record and organize maintenance information, which is then used to make the decisions required to maintain the equipment in a facility.

prick punch: A sharp, pointed steel shaft struck with a hammer to mark centerpoints or punch holes in light-gauge metal.

primary port: A source or inlet port.

primary winding: The power input winding of a transformer and is connected to the incoming power supply.

prime mover: A device that supplies rotating mechanical energy to a fluid power circuit.

print: A reproduction of original drawings created by an architect or engineer.

print abbreviation: A letter or group of letters that represents a term or phrase.

print convention: An agreed-upon method of displaying information on prints.

print schedule: A chart used to conserve space and display information on a print in a concise and organized format.

prism: A solid with two bases that are parallel and identical polygons.

process variable: Any characteristic that changes its value during any operation within the process.

prohibited approach boundary: The distance from an exposed energized conductor or circuit part inside which any work performed is considered the same as making contact with the energized conductor or circuit part.

proportion: An expression of equality between two ratios.

protective helmet: A hard hat that is used in the workplace to prevent injury from the impact of falling and flying objects, and from electrical shock.

proton: A particle with a positive electrical charge of one unit.

protractor: A tool used for measuring and laying out angles.

pseudocavitation: Artificial cavitation caused by air being allowed into a pump suction line.

pump: A mechanical device that causes fluid to flow.

punch: A hand tool with a pointed or blunt tip for marking or making holes or driving objects when struck by a hammer.

pyramid: A solid with a base that is a polygon and sides that are triangles.

Pythagorean theorem: States that the square of the hypotenuse of a right triangle is equal to the sum of the squares of the other two sides.

Q

quadrant: One fourth of a circle containing 90°.

quadrilateral: A four-sided polygon with four interior angles.

Quad-ring®: A molded synthetic rubber seal having a basically square cross-sectional shape.

R

race: The track on which the balls of a bearing move.

rack gear: A gear with teeth spaced along a straight line.

racking: The ability to be forced out of shape or form.

rack teeth: Gear teeth used to produce linear motion.

radial bearing: A rolling contact bearing in which the load is transmitted perpendicular to the axis of shaft rotation.

radial load: A load in which the applied force is perpendicular to the axis of rotation.

radial piston pump: A piston pump that consists of a cylinder barrel, pistons with shoes, a ring, and a valve block located perpendicular to the pump shaft.

radius: The distance from the centerpoint to the circumference.

ratchet: A mechanism that consists of a toothed wheel and a spring-loaded pawl.

ratio: The relationship between two quantities or terms.

ray: A straight line or lines intersecting a point (vertex) of an angle.

reach: The distance between the cup of a top hook and the cup of a hoist hook when the hoist hook is at its lower limit of travel.

realistic goal: A goal that can be measured in terms of achievement and performance and for which sufficient means of implementation are possessed.

reciprocate: To move forward and backward alternately.

reciprocating compressor: A device that compresses gas by means of a piston that moves back and forth in a cylinder.

reciprocating saw: A multipurpose cutting tool in which the blade reciprocates (quickly moves back and forth) to create the cutting action.

rectangle: A quadrilateral with opposite sides equal and four 90° angles.

rectangular parallelepiped: A prism with bases and faces that are all rectangles.

rectifier: A device that converts AC voltage to DC voltage by allowing the voltage and current to move in only one direction.

reel: A wooden assembly on which wire rope is wound for shipping and storage.

reeving: Passing a rope through a hole or opening or around a series of pulleys.

reflex angle: An angle that exceeds 180° but is less than 360°.

refrigerant dryer: A device designed to lower the temperature of compressed air to 35°F.

regular-lay rope: A rope in which the yarn or wires in the strands are laid in the opposite direction to the lay of the strands.

regular polygon: A polygon with equal sides and equal angles.

regular polyhedron: *See* regular solid.

regular pyramid: A pyramid that has base that is a regular polygon and a vertex that is perpendicular to the center of the base.

regular solid: Any of a variety of solids with faces that are regular polygons (equal sides). Also known as a regular polyhedron.

relative humidity: The percentage of moisture contained in air compared to the maximum amount of moisture (saturation) it is capable of holding.

relay: An electrical switch that is actuated by a separate circuit.

reservoir: A container for storing fluid in a hydraulic circuit.

resilience: The capability of a material to regain its original shape after being bent, stretched, or compressed.

resistance (R): The opposition to current flow.

resonance: The magnification of vibrations and their noise by 20% or more.

respirator: A device that protects the wearer from inhaling airborne contaminants.

restricted approach boundary: The distance from an exposed energized conductor or circuit part where an increased risk of electric shock exists due to the close proximity of the person to the energized conductor or circuit part.

restrictive check valve: A check valve with a specific size hole drilled through its center.

return-line filter: A filter positioned in a circuit just before the reservoir.

reverse dial method: An alignment method that uses two dial indicators to take readings off of opposing sides of coupling rims, giving two sets of shaft runout readings.

reversible protractor: A finely graduated tool for measuring angles on workpieces with small tolerances.

revision block: A block that identifies the changes that have been marked on the drawing since its initial approval.

rhomboid: A quadrilateral with opposite sides equal and no 90° angles.

rhombus: A quadrilateral with all sides equal and no 90° angles.

rigging: Securing equipment or machinery in preparation for lifting by means of rope, chain, or webbing.

right angle: An angle formed by two perpendicular lines and measures exactly 90°.

right circular cone: A cone with the axis at a 90° angle to the circular base.

right cylinder: A cylinder with the axis perpendicular to the base.

right lang-lay rope: A rope in which the yarn or wires are laid to the right and the strands are laid to the right.

right parallelepiped: A prism with all edges perpendicular to the bases.

right prism: A prism with lateral faces perpendicular to the bases.

right regular-lay rope: A rope in which the strands are laid to the right and the yarn or wires are laid to the left.

right triangle: A triangle that contains one 90° angle.

rim-and-face alignment method: An alignment method in which the offset and angular gap of two shafts is determined using two dial indicators that measure the rim and face of a coupling.

rip cut: A cut that is made with the direction of the wood grain and is made with full, even strokes at about a 60° angle.

rivet: A permanent mechanical fastener with a cylindrical shaft, a preformed head on one end, and a head on the other that is formed in place.

roller bearing: An anti-friction bearing that has parallel or tapered steel rollers confined between inner and outer rings.

rolling: The deforming of metal on the active portion of gear teeth caused by high contact stresses.

rolling-contact (anti-friction) bearing: A bearing composed of rolling elements between an outer and inner ring.

root mean square (rms): The square root of the sum of a set of squared instantaneous values.

rope grab: A device that clamps securely to a rope.

rope lay: The length of rope in which a strand makes a complete helical wrap around the core.

rotor: The rotating part of an AC motor.

rounding off: The process of increasing or decreasing a number to the nearest acceptable number.

round-nose chisel: A chisel that has a round nose and is used for roughing out the concave surfaces of corners.

round sling: A sling consisting of one or more continuous polyester fiber yarns wound together to make a core.

routine maintenance: Maintenance procedures that involve servicing operating equipment on a scheduled basis.

rule: A measuring tool marked with even increment lines used for measuring length.

running torque: The energy that a motor develops to keep a load turning.

running winding: A coil that continues to operate a 1ϕ motor after it has started.

runout: A radial variation from a true circle.

runway: The rail and beam on which a crane operates.

rust: A form of oxidation in which metal oxides are chemically combined with water to form a reddish brown scale on metal.

S

safety factor: Ratio of a component's ultimate strength to its maximum allowable safe working load limit.

safety glasses: An eye protection device with special impact-resistant glass or plastic lenses, a reinforced frame, and side shields.

safety net: A net made of rope or webbing for catching and protecting a falling technician.

safety relief valve: A device that prevents excessive pressure from building up by venting air to the atmosphere.

safety sleeve: A moving element with a locking mechanism that is connected between a carrier and a worker's harness.

saturated air: Air that holds as much moisture as it is capable of holding.

Saybolt viscometer: A test instrument used to measure fluid viscosity.

scaffold: A temporary or movable platform and structure for workers to stand on when working at a height above the floor.

scaffold hitch: A hitch used to hold or support planks or beams.

scalene triangle: A triangle that has no equal angles or equal sides.

screw drive: A standardized internal shape recessed in a screw head that allows the screw to be rotated with a matching tool.

screwdriver: A hand tool with a tip designed to fit into a screw head for fastening operations.

screw extractor: A tool used to remove studs, bolts, or screws broken below or near the surface of a workpiece.

screw thread: A ridge in the form of a spiral on the internal or external surface of a cylinder or cone.

scuffing: The severe adhesion that causes the transfer of metal from one tooth surface to another due to welding and tearing.

seal: A device in contact between two components that contains pressure and prevents leakage.

Seale wire: Wire rope that uses different size wire in different layers.

secant: A straight line touching the circumference at two points.

secondary port: An external passage that allows fluid flow to other components.

secondary winding: The output or load winding of a transformer and is connected to the load.

sectional drawing: A type of drawing that indicates the internal features of an object.

sectional metal-framed scaffold: A metal scaffold consisting of preformed tubes and components.

section line: A line that identifies the materials cut by a cutting-plane line in a section view.

sector: A pie-shaped piece of a circle.

segment: The portion of a circle set off by a chord.

seizing: The wrapping placed around all strands of a rope near the area where the rope is cut.

seizing bar: A round bar ½″ to ⅝″ in diameter and about 18″ long used to seize rope.

self-tapping screw: A threaded fastener with a tapered point and screw threads that bite into a compressible or deformable material is it is driven.

selvedge: A knitted or woven edge of a webbing formed to prevent unraveling.

semicircle: One half of a circle containing 180°.

semisolid lubricant: A lubricant that combines low-viscosity oils with thickeners, such as soap or other finely dispersed solids.

sequence: The order of a series of operations or movements.

sequence valve: A pressure-operated valve that diverts flow to a secondary actuator while holding pressure on the primary actuator at a predetermined minimum value after the primary actuator completes its travel.

series connection: A circuit connection that has two or more components connected so there is only one path for current to flow.

series/parallel connection: A circuit connection with a combination of series- and parallel-connected components.

service life: The length of service received from a bearing.

shackle: A U-shaped metal link with the ends drilled to receive a pin or bolt.

shaft: *See* well.

shape: The extent of all or part of an object that is often contained within a contour.

shear strength: The ability of a material to withstand shear stress.

shear stress: Stress in which a material is subjected to parallel, opposing, and offset forces.

sheet note: A note that applies to a specific item in the drawing on which the note appears.

shell: An orbiting layer of electrons in an atom.

shielded metal arc welding (SMAW): An arc welding process in which the arc is shielded from impurities by the gases emitted from the decomposition of a consumable electrode covering.

shim stock: Steel material manufactured in various thicknesses, ranging from 0.0005″ to 0.125″.

sight gauge: A device used to visually inspect the hydraulic fluid level in a reservoir.

signal: In pneumatic logic, a condition that initiates a start or stop of fluid flow by opening or closing a valve.

single-acting cylinder: A cylinder in which fluid pressure moves the piston in only one direction.

single ladder: A ladder of fixed length having only one section.

single-pole scaffold: A wood scaffold with one side resting on the floor or ground and the other side structurally anchored to the building.

single-stage compressor: A compressor that uses one piston to compress air in a single stroke before it is discharged.

size dimension: A dimension that gives the overall size of an angle or feature.

sketch: A two-dimensional visual representation of an object.

slant height: The distance from the base to the vertex parallel to a side.

sling: A line consisting of a strap, chain, or rope used to lift, lower, or carry a load.

sling apex: The uppermost point where sling legs meet.

slip clutch: A spring-loaded, friction-held fiber disc that is adjusted to slip at 125% to 150% of the hoist rated load.

slip knot: A knot that slips along the rope from which it is made.

slippage: The internal leaking of fluid from a pump's outlet to a pump's inlet.

small circle: The circle formed by passing a cutting plane through a sphere but not through the center.

socket: A rope attachment through which a rope end is terminated.

soft foot: A condition that occurs when one or more machine feet do not make complete contact with its base.

soldering: A joining process that joins parts by heating a filler metal to temperatures up to 840° F but less than the melting point of the base metal.

solenoid: A device that converts electrical energy into a linear mechanical force.

solid lubricant: A material such as graphite, molybdenum disulfide, or polytetrafluoroethylene (PTFE) that shears easily between sliding surfaces.

solid punch: A hand tool with a blunt, circular end, which is used to punch small holes in light-gauge metal.

sorting hook: A hook with a tapered throat and a point designed to fit into holes.

spacer: Steel material used for filling spaces ¼″ or greater.

spalling: The flaking away of metal pieces due to metal fatigue.

specification: Additional information that is included with a set of prints.

specific goal: A goal that incorporates an action plan that outlines how the goal is to be achieved and a performance measurement that provides goal evaluation.

spectrometer: A device that vaporizes elements in the oil sample into and reads the wavelength of light given off.

spectrum: A representative combination of the amplitude (total movement) and frequency (time span) of a waveform.

speltered socket: A socket assembled by separating the wire rope ends after inserting the rope through the socket collar.

sphere: A solid generated by a circle revolving about one of its axes.

spindle: A precision-ground, moveable surface of a micrometer.

splice: The joining of two rope ends to form a permanent connection.

split-phase motor: A 1ϕ AC motor that includes a starting winding and a running winding in the stator.

springing soft foot: A condition that occurs when a dial indicator at the shaft shows soft foot, but feeler gauges show no gaps.

spring-loaded accumulator: An accumulator that applies force to a fluid by means of a spring.

spring-loaded punch: A punch that is equipped with a spring.

spur gear: A gear that has straight teeth that are parallel to the shaft axes.

square: A quadrilateral with all sides equal and four 90° angles.

staging: The process of dividing the total pressure among two cylinders by feeding the outlet from the first large (low pressure) cylinder into the inlet of a second small (high pressure) cylinder.

standard: A guideline adopted by regulating authorities.

standing end: The end of the rope that is normally fixed to a permanent apparatus or drum or that is rolled into a coil.

standing part: The portion of the rope that is not active in the knot-making process.

standoff: A ladder accessory that holds a single or an extension ladder a fixed distance from a wall.

starting torque: The energy required to start a load turning after it has been broken away from a standstill.

starting winding: A coil that is energized temporarily at startup to create the torque required to start a 1ϕ motor rotating.

static energy: The ability of a fluid to do work using the height and weight of the fluid above some reference point.

static head: The height of a fluid above a given point in a column at rest.

static head pressure: A pressure created by the weight of a fluid.

static head pressure: A pressure created by the weight of a fluid.

static lift: The height to which atmospheric pressure causes a column of fluid to rise above the supply to restore equilibrium.

static load: A load that remains steady.

static seal: A seal used as a gasket to seal nonmoving parts.

static total column: Static head plus static lift.

stator: A slotted circular iron frame that is the stationary part of an AC motor.

steel alloy: A metal formulated from the combining of iron with carbon and other elements.

stepladder: A folding ladder that stands independently of support.

stethoscope: A test tool used for detecting and locating abnormal noises within machines or equipment.

straight angle: An angle formed by two lines that intersect to form a straight line and measures exactly 180°.

straightedge alignment method: A method of coupling alignment in which an item with an edge that is straight and smooth, such as a steel rule, feeler gauge, or taper gauge, is used to align couplings.

straight line: The shortest distance between two points.

strainer: A fine metal screen that blocks contaminant particles.

strand: Several pieces of yarn helically laid about an axis.

strobe tachometer: A test tool that measures the rotational speed of an object by use of a flashing (strobe) light.

stroboscope: A test tool that can capture a motionless image of a moving object for ease of inspection by use of a flashing light or strobe light.

submersion system: A lubrication system in which the bearings are submerged below oil for lubrication.

suction strainer: A coarse filter attached to a pump inlet.

supplementary angles: Two angles formed by three lines in which the sum of the two angles equals 180°.

suspension scaffold: A scaffold supported by overhead wire ropes.

swaged socket: A compressed socket assembled to the wire rope under high pressure.

swash plate: An angled plate in contact with the piston heads that causes the pistons in the cylinders of a pump to extend and retract.

switch: A device that starts or stops the flow of electrical energy in a circuit.

symbol: 1. A graphic element that indicates a particular device. **2.** A conventional representation of a quantity or unit.

symmetrical load: A load in which one half of the load is a mirror image of the other half.

synthetic fluid: A lubricant, often with a petroleum base, that has improved heat and chemical resistance compared to straight petroleum products.

synthetic yarn: Yarn made of twisted, manufactured fibers such as nylon or polyester.

system operating pressure: The pressure of a fluid after the pump until the flow is reduced, metered, or returned to the reservoir.

T

tachometer: A test tool that measures the speed of a moving object.

tackle: The combination of ropes and block assemblies arranged to gain mechanical advantage for lifting.

tag line: A rope, handled by an individual, to control rotational movement of a load.

tagout: The process of placing a tag on a locked-out power source that indicates that the power may not be restored until the tag is removed.

tangent: A straight line touching the curve of the circumference at only one point.

tap: A tool used to cut internal threads in a predrilled hole.

tapered bore bearing: A bearing that uniformly increases or decreases from one face to the opposite face.

tapered bore pulley: A two piece pulley that consists of a tapered pulley bolted to a tapered hub (bushing).

tapered roller bearing: A roller bearing having tapered rollers.

taper gauge: A flat, tapered strip of metal with graduations in thousandths of an inch or millimeters marked along its length.

taper tap: A tap with a long, gradual taper that allows the tap to start easily.

tape rule: A measuring tool consisting of a long, continuous strip of fabric, plastic, or steel that is marked with evenly spaced increment lines.

tapping: The forming of internal threads in a material.

telescopic-boom crane: A crane with an extendable boom composed of nested sections.

tempering: The process in which metal is brought to a temperature below its critical temperature and allowed to cool slowly.

tensile strength: The maximum load in tension (pulling apart) that a material can withstand before breaking or fracturing.

tension member: The load carrying element of a belt which prevents stretching.

tetrahedron: A regular solid of four triangles.

thermal expansion: The dimensional change of a substance due to a change in temperature.

thermal imager: A device that detects heat levels in the infrared-wavelength spectrum without making direct contact with the target.

thermal signature: A false-color picture of the infrared energy emitted from an object.

thermocouple: A device that produces electricity by heating two different metals that are joined together.

thickness gauge: *See* feeler gauge.

thimble: A curved piece of metal around which the rope is fitted to form a loop.

threaded cup follower: A tapered bearing gap adjusting device that is used to adjust shaft endplay by controlling the amount of clearance between the bearings.

threaded fastener: A device that joins parts together with a screw thread.

three-square file (three-cornered file): A file that has angles of 60° and is used for filing internal angles, clearing out corners, etc.

three-way directional control valve: A valve that has three main ports that allow or stop fluid flow or exhaust.

throttling: Permitting the passing of a regulated flow.

thrust damage: Bearing damage due to axial force.

timber hitch: A binding knot and hitch combination used to wrap and drag lengthy material.

time domain: An amplitude as a function of time.

timing (synchronous) belt: A belt designed for positive transmission and synchronization between the drive shaft and the driven shaft.

title block: An area on a working drawing or print used to provide information about the drawing or print.

toeboard: A barrier to guard against the falling of tools or other objects.

tolerance: The permissible deviation from a given value or dimension.

tooth form: The shape or geometric form of a tooth in a gear when seen as its side profile.

top support: The area of a ladder that makes contact with a structure.

torque: The twisting (rotational) force of a shaft.

total column: The fluid head plus lift.

total energy: A measure of a fluid's ability to do work.

transducer: A device that converts a physical quantity into another quantity, such as an electrical signal or a graphic display.

transformer: An electric device that uses electromagnetism to change AC voltage from one level to another.

trapezium: A quadrilateral with no sides parallel.

trapezoid: A quadrilateral with two sides parallel.

trapezoidal belt: A timing belt containing trapezoidal-shaped teeth.

trending: A graphic display used for interpretation of machine characteristics.

trending inspection: An inspection performed after a baseline inspection to provide images for comparison.

triangle: A three-sided polygon with three interior angles.

truth table: A table that lists the output condition of a logic element or combination of logic elements for every possible input condition.

tube: A thin-walled, seamless or seamed, hollow cylinder.

tuck set: The wedging of each strand of a rope into and between the other rope strands.

two-point suspension scaffold: A suspension scaffold supported by two overhead wire ropes.

two-way directional control valve: A valve that has two main ports that allow or stop the flow of fluid.

tying off: Securely connecting a harness directly or indirectly to an overhead anchor point.

U

ultrasonic analysis: A nondestructive test method in which mechanical vibrational waves are used to test the integrity of a solid material.

unbalanced-vane pump: A vane pump that has one set of internal ports and produces a pumping action in the chambers on one side of the rotor and shaft.

union: A fitting used to connect or disconnect two tubes that cannot be turned.

unlay: The untwisting of the strands in a rope.

unloading valve: A device that senses a high pressure condition and removes the compression energy.

unthread rotation: Counterclockwise rotation of an eyebolt having right handed threads, or the clockwise rotation of an eyebolt having left handed threads.

upper explosive limit (UEL): The highest concentration (air-fuel mixture) at which a gas can ignite.

U-ring seal: A lip seal shaped like the letter U.

V

vacuum: A pressure lower than atmospheric pressure.

valence electron: an electron located in a valence shell.

valence shell: The outermost shell of an atom.

valve: A device that controls the pressure, direction, or rate of fluid flow.

valve actuator: A device that changes the position of a valve spool.

vane air motor: An air motor that contains a rotor with vanes that are rotated by compressed air.

vane compressor: A positive displacement compressor that has multiple vanes located in an offset rotor.

vane pump: A pump that contains vanes in an offset rotor.

variable-displacement piston pump: A piston pump in which the angle of the swash plate can be varied.

variable-displacement pump: A pump in which the displacement per cycle can be varied.

variable-frequency drive: A motor controller that is used to change the speed of AC motors by changing the frequency of the source voltage.

variable-speed belt drive: A mechanism that transmits motion from one shaft to another and allows the speed of the shafts to be varied.

V-belt: An endless power transmission belt with a trapezoidal cross section.

V-belt pulley: A pulley with a V-shaped groove.

vector: A quantity that has a magnitude and direction.

velocity: The distance a fluid travels in a specified time.

velocity transducer: An electromechanical device that is constructed of a coil of wire supported by light springs.

vernier scale: A short auxiliary scale placed along the main scale of a measuring instrument to provide accurate fractional readings of the smallest division on the main scale.

vertex: 1. The point of intersection of the sides of an angle. **2.** The common point of the triangular sides that form a pyramid.

vertical line: A line that is perpendicular to the horizon.

vibration: A continuous periodic change in displacement with respect to a fixed reference.

vibration acceleration: The increasing of vibration movement speed.

vibration amplitude: The extent of vibration movement measured from a starting point to an extreme point.

vibration analyzer: A meter that pinpoints a specific machine problem by identifying its unique vibration or noise characteristics.

vibration cycle: The complete movement from beginning to end of a vibration.

vibration signature: A set of vibration readings resulting from tolerances and play within a new machine.

vibration velocity: The rate of change of displacement of a vibrating object.

viscosity: The measure of a fluid's resistance to flow.

viscosity index: A scale used to show the magnitude of viscosity changes in lubrication oils with changes in temperature.

vise: A portable or stationary clamping device used to firmly hold work in place.

visual adjustment method: A belt tension method in which the tension is adjusted by observing the slight sag at the slack side of the belt.

voltage (*E*): The amount of electrical pressure in a circuit.

voltage tester: A device that indicates approximate voltage level and type (AC or DC) by the movement of a pointer on a scale.

volume: The size of a three-dimensional space or object as measured in cubic units.

volumetric efficiency: The percentage of actual pump output compared to the pump output if there were no slippage.

V-ring seal: A lip seal shaped like the letter V.

W

wagoner's hitch knot: A knot that creates a load-securing loop from the standing part of the rope.

Warrington wire: Wire rope constructed of strands consisting of more than one size wire staggered in layers.

washer: A small metallic disc with a hole in its center used under the head of a bolt or screw, and/or under a nut, to spread the load (tightening force) over a larger area.

waveform: A graphic presentation of an amplitude as a function of time.

way: A flow path through a valve.

wear pad: A leather or webbed pad used to protect a web sling from damage.

wear particle analysis: The study of wear particles present in lubricating oil.

webbing: A fabric of high-tenacity synthetic yarns woven into flat narrow straps.

web sling body: The part of the sling that is between the loop eyes or end fittings (if any).

web sling length: The distance between the extreme points of a web sling, including any fittings.

wedge socket: A socket with the rope looped within the socket body and secured by a wedging action.

weight-loaded accumulator: An accumulator that applies force to a fluid by means of heavy weights.

welding: A joining process that fuses materials by heating them to melting temperature.

well: A walled enclosure around a fixed ladder. Also known as a shaft.

whipping: Tightly binding the end of a rope with twine before it is cut.

wick system: A lubrication system that uses capillary action to convey oil to a bearing surface.

wiper: A seal designed to prevent foreign abrasive or corrosive material from entering a cylinder.

wire stripper/crimper/cutter: A tool used for the removal of insulation from small-diameter wire.

wiring diagram: A diagram that shows the electrical connections of a circuit.

work: The energy used when a force is exerted over a distance.

working depth: The depth of engagement of two gears.

working end: The end of the working part of a rope.

working height: The distance from the ground to the top support.

working load limit (WLL): The maximum weight that a rigging component may be subjected to.

working part: The portion of the rope where the knot is formed.

work order: Informational documentation created for communication between a technician and maintenance personnel or other relevant parties.

worm: A shank having at least one complete tooth around the pitch surface.

worm gear: A set of gears consisting of a worm (drive gear) and a wheel (driven gear) that are used extensively as a speed reducer.

worm gear drive: A pair of gears consisting of a spiral-threaded worm (worm gear) and a worm wheel (driven gear) that are used extensively as a speed reducer.

wrench: A hand tool with jaws at one or both ends; designed to turn bolts, nuts, or pipes.

Y

yarn: A continuous strand of two or more fibers twisted together.

Index

Page numbers in italic refer to figures.

Y

Z